W9-BJJ-497

Introduction to Electricity, Electronics, and Electromagnetics

Introduction to Electricity, Electronics, and Electromagnetics

Fifth Edition

Robert L. Boylestad
Louis Nashelsky

Upper Saddle River, New Jersey
Columbus, Ohio

Library of Congress Cataloging-in-Publication Data

Boylestad, Robert L.
Introduction to electricity, electronics, and electromagnetics / Robert Boylestad, Louis Nashelsky. -- 5th ed.
p. cm.
Rev. ed. of: Electronics. 4th ed., c1996.
Includes index.
ISBN 0-13-010573-2
1. Electric engineering. 2. Electronics. I. Nashelsky, Louis. II. Boylestad, Robert L. Electronics. III. Title.

TK146.B795 2002
621.3--dc21 00-053766

Vice President and Editor in Chief: Stephen Helba
Assistant Vice President and Publisher: Charles E. Stewart, Jr.
Production Editor: Alexandrina Benedicto Wolf
Production Coordination: Lisa Garboski, bookworks
Design Coordinator: Robin C. Chukes
Cover Designer: Tom Mack
Production Manager: Matthew Ottenweller

Cover painting by Sigmund Arseth, artist and teacher, Valdres, Norway

This book was set in Times Roman by The Clarinda Company. It was printed and bound by Courier/Kendallville. The cover was printed by Phoenix Color Corp.

Prentice-Hall International (UK) Limited, *London*
Prentice-Hall of Australia Pty. Limited, *Sydney*
Prentice-Hall Canada, Inc., *Toronto*
Prentice-Hall Hispanoamericana, S.A., *Mexico*
Prentice-Hall of India Private Limited, *New Delhi*
Prentice-Hall of Japan, Inc., *Tokyo*
Prentice-Hall Singapore Pte. Ltd.
Editora Prentice-Hall do Brasil, Ltda., *Rio de Janeiro*

10 9 8 7 6 5 4 3
ISBN 0-13-010573-2

Dedicated to the Memory of Our Parents:

Astrid and Anders Boylestad
Nellie and Samuel Nashelsky

Preface

This fifth edition of *Introduction to Electricity, Electronics, and Electromagnetics* provides a broad range of subject matter in the electrical and electronic technology areas. In particular, the areas of electronic devices, circuits, and applications have been updated with the inclusion of the software packages PSpice® and Electronics Workbench™ (EWB). Numerous changes in pedagogy were made in response to suggestions by current users and reviewers. The number of chapters was increased to permit the introduction of practical examples, summaries, and greater detail in areas of particular importance. Increasing the number of chapters also served to reduce the overwhelming amount of material appearing in some of them. Particular examples were introduced to provide an immediate appreciation for the principles and devices introduced in each chapter. The examples center on electronic devices encountered daily in the home and industrial environment. Each chapter now has a summary section to review the salient points of each chapter and the important equations introduced. Finally, the problems were reorganized according to their level of difficulty.

The level of presentation is suitable for students who require a background in electrical and electronic components, circuits, and applications with less emphasis on derivations and advanced mathematical techniques. Approximations are applied as often as possible, with final equations and conclusions presented in their simplest form. The authors limited the presentation to the most important equations and concepts in each area rather than burden students with material that has almost no impact on their current needs. There is sufficient material for two semesters, although the text is most frequently employed in a one-semester course that covers topics of particular interest.

Because the scope of coverage is so broad, the authors would appreciate any input on material that should be added or deleted. It was obviously an immense challenge to provide a survey of the entire field in one publication.

Acknowledgments

The authors wish to thank the following reviewers for reviewing the manuscript: Wayne Eisman, DeKalb Technical Institute; Dave Krispinsky, Rhode Island Technical Institute; Hossein Mousavine Zhad, Western Michigan University; and Alan Niemi, Lake Superior University.

We would also like to thank our Prentice Hall publisher Charles Stewart, and production editor Alex Wolf; Lisa Garboski, bookworks project coordinator; and our copyeditor Barbara Liguori, for their help with this edition.

Robert Boylestad
Louis Nashelsky

Brief Contents

Contents

4

Ac Networks 131

5

Ac Network Theorems, Polyphase Systems, and Resonance 197

Electromagnetism 255

7

Generators and Motors 293

Two-Terminal Electronic Devices 339

Introduction to Electricity, Electronics, and Electromagnetics

1 Introduction

The electrical and electronics industry continues to mature, to the point where it affects every part of our daily lives. The computer, with its e-mail and Internet access; the many devices used to communicate (pagers, cellular phones); the fax machine; answering machines; stereo systems; and television (older analog and newer digital) are all clear evidence of the important role of electronics in today's society. The engineers of the 1950s would be in awe of the amount of circuitry that can be contained in a single integrated circuit (IC), and today's engineers will probably be astounded by the technological progress that will take place by the year 2010.

Electronic systems are becoming an increasingly important component of virtually every instrument, machine, or device we use on a daily basis. For example, automotive functions previously performed by mechanical components are being replaced by equivalent electronic units. Manufacturers are even looking into using 48-V car batteries (to add to or replace the 12-V battery presently used) to provide for the increased electronic demands of cars. Power drills, lawnmowers, and chain saws all contain a large number of electronic parts for increased efficiency and power. The watch has gone from a purely mechanical unit to one that can be completely electronic. There are even instruments for playing music that use electronic components in place of the usual mechanical devices (vibrating reeds, wires, or surfaces).

The growth of the electronics industry has been due partly to the low-cost, very small, relatively complete systems referred to as *very-large-scale integrated* (VLSI) circuits. It is impressive to consider that within the typical small calculator there is essentially one small IC package, occupying a small part of the unit, containing the hundreds of thousands of interconnected components that constitute the functional electronic unit. In computer memory and microprocessors this number increases to millions of components in a single IC.

The single-chip IC is the heart of many electronic devices. Additional circuits contribute to increasing IC complexity and make possible multi-functional devices. Watches now have alarm functions, stopwatch functions, timer functions, multiple time zone displays, data storage, capacity and the ability to generate telephone tones to automatically dial a number. Electronic watches can even receive signals to allow display of selected information—stock market activity, important news items, and the like.

Electronic circuits using *complementary symmetry metal-oxide semiconductor* (CMOS) circuitry require very little battery power, permitting units to operate for many years on one small battery. With the increasing popularity of portable devices the ability to operate on a small fixed or small rechargeable battery becomes more important.

Electronic products have been developed for engineering applications as well as for special commercial products. The availability of small, low-cost electronic components now makes it practical to include electronic equipment in many new areas, including measuring instruments, and in the fields of mechanical, chemical, civil, and biomedical engineering, to name a few. In each of these areas, electronics circuitry—both linear and digital ICs—has been added to provide more automatic operation, digital display, and more central control, and to accommodate additional features not previously available. For example, digital thermometers now provide a faster, more easily visible display in a small, compact portable instrument. Similarly, instrumentation, oscilloscopes, and measurement equipment now include digital display and control circuitry to provide automatic scale selection and digital readout.

Probably the greatest advances in the VLSI circuit area have been made in microprocessor and digital memory units. The complexity of microprocessors has increased from handling 8 bits at a time to handling 16, 32, and presently 64 bits at a time. Memory capacity has also increased, so that the IC that held 1 megabyte (MB) now holds 256 MB. This is a tremendous increase in packaging density. For example, the increased storage capacity has enabled small units to provide voice output.

To better understand what is presently available in a range of areas and to develop a framework in which to understand future growth in the electronics area, this book covers a range of subject matter—from the initial chapters on fundamentals to material on the basic electronics areas, concluding with a discussion of a variety of applications. Be aware that the content of the text provides a surface treatment at best. Entire books have been devoted to the material in almost every chapter of this text. For each subject area the salient equations, concepts, and derivations have been included to establish a foundation that will permit investigation of the chapters to follow. It is therefore necessary that you examine Chapters 2 through 5 in some detail before looking at any of the advanced material. Chapters 2 and 3 provide a lengthy treatment of dc systems that may prompt some students to ask why so much time must be spent on a nonvarying system when it seems the world runs on ac and digital systems. Be aware, however, that dc networks permit an introduction to concepts that will also appear in the ac analysis but without the complexity of the required mathematical calculations. In addition, equations that appear in the dc analysis will appear in one form or another in most of the chapters to follow. It is an important, interesting fact about electrical and electronic systems that in most cases an equation learned is not replaced by another as the learning process continues—an equation learned is one that can be applied for a variety of systems.

The subject of magnetics is one that is often ignored or simply treated in a section or two in most electrical engineering texts. Consider, however, how important an understanding of magnetic effects can be in the design of any system such as a motor, generator, computer magnetic storage unit, transformer, automobile ignition, telephone speaker, or microphone, to name just a few. To spend a chapter ensuring a clear understanding of the factors that affect magnetic and electromagnetic effects seems quite warranted.

The depth to which the chapter on magnetics is covered is a matter of choice, but the following two chapters on electronics devices are essential to a clear understanding of the majority of the chapters to follow. Although a number of electronic devices to be introduced in Chapters 8 and 9 do not see application in the text, they have been included because they play an important role in some electronic systems and should be familiar to some degree. The extent to which each element is investigated is designed to provide a working knowledge of a variety of important electronic devices without becoming too involved in complex mathematical derivations and exercises. Approximations are employed as extensively as possible to establish an understanding of how a device works and how it can be used successfully without getting lost in unnecessary details and concerns.

Operational amplifiers (op-amps), covered in Chapter 10, provide the most powerful electronic circuit used in linear circuits. A single IC may contain one or many op-amps, providing merely amplifier operation or more involved signal processing operations. Chapter 11 covers multistage amplifiers and power amplifiers used in a variety of electronic units. Op-amp units can be used to provide signal amplification, signal filtering, and signal processing. Chapter 12 introduces a number of communications topics—amplitude modulation (am) and frequency modulation (fm).

Chapter 13 considers a number of topics involving digital circuits. Various types of binary data, logic circuits, and their use in transfer, storage registers, and various type counters are also introduced.

Control systems are important devices used in aircraft, cars, and manufacturing equipment, among others. The basics of this area are covered in Chapter 14. Finally, Chapter 15 discusses the basics of power supplies, converting an ac power signal into a dc voltage supply, including voltage regulation circuitry.

This fifth edition of the text includes an extensive coverage of computer methods using the most popular software packages: PSpice, Electronics Workbench (EWB), and Mathcad. These programs have proved to be of great use in analyzing and simulating the operation of both linear and digital circuits. An understanding of their capabilities and limitations will permit an appreciation of any system you may encounter or design. Although discussion of the software packages must be limited because of the enormous amount of material that must be covered in the text, there is sufficient detail to write some of the input files that permit application of the software packages. If additional information is required, the OrCad and MultiSim Corporations are anxious to provide support to the educational community.

2 Dc Networks

2.1 INTRODUCTION

This introduction to the elements of electricity and electronics will begin in the time-honored tradition with *direct-current* (dc) circuits. Systems of this type have fixed levels of electrical quantities that do not vary with time. For example, the typical 12-volt (V) car battery ideally provides 12 V independent of how long it is connected to the system or how much demand is placed on its energy resources. The electrical energy available at any home outlet is called an *alternating-current* (ac) voltage, which varies in a definite manner with time. This type of electrical supply will be introduced in Chapter 3. Since the quantities of interest in a dc network are time-independent, it is considerably easier to introduce and understand the basic laws of electrical systems. However, be assured that the similarities are so strong between the application of a theorem to a dc network as compared with an ac network that the analysis of ac systems will be made considerably easier by the knowledge gained from first examining dc networks.

Because a broad range of subject matter must be covered in a minimum number of pages, there will be little time for derivations and lengthy discussions of any particular area. Further investigation must be left to the reader, who should identify a good reference source at the earliest opportunity.

In the typical approach to this subject matter, instruments are usually considered in a separate chapter or text. In this book they are introduced the moment a measurable quantity is examined. In this way, the authors feel that the student will be best prepared for the earliest possible practical application of the concepts discussed. There will, however, be very little time to examine the internal construction of each instrument. The scales and proper use of the dials will be the common extent of the coverage of each instrument.

2.2 CURRENT

The first electrical quantity of major importance to be introduced is the *flow* variable: *current*. The *rate of flow* of charge through a conductor is a measure of the current in the conductor. The moving charges are the relatively *free* electrons found in conductors such as copper, aluminum, and gold. The term *free* simply reveals that the electrons are loosely bound to their parent atom and can be encouraged to move in a particular direction through the application of an external source of energy such as a dc battery. For the system in Fig. 2.1 the electrons are attracted to the applied positive terminal of the battery, while the negative terminal is the source of the moving *free* charge. In Fig. 2.1 the greater the quantity of charge that flows through the imaginary surface per unit time (in the same direction), the greater the current. In equation form,

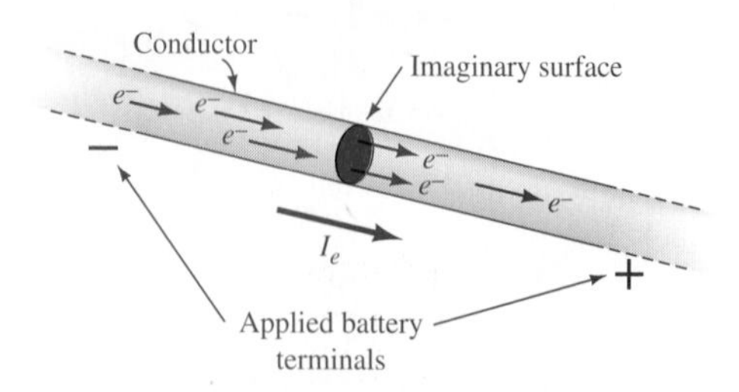

FIG. 2.1
Charge flow in a conductor.

$$I = \frac{Q}{t} \tag{2.1}$$

where I = current in amperes (A)
Q = charge in coulombs (C)
t = time in seconds (s)

An electron has an electronic charge of 1.6×10^{-19} coulomb, and conversely, one coulomb is the charge associated with 6.242×10^{18} electrons.

Note in Eq. (2.1) that current is measured in amperes, in honor of the French scientist Ampère. The equation states in words that if 6.24×10^{18} electrons (or 1 C of charge) pass through the imaginary surface of the wire in Fig. 2.1 in 1 s (all in one particular direction), the current is 1 A. Note also that time is measured in seconds, not minutes or hours. Substituting any time element other than seconds will invalidate the use of amperes and coulombs for the determined current or charge, respectively.

EXAMPLE 2.1 Determine the current in amperes through the wire in Fig. 2.1 if 18.726×10^{18} electrons pass through the cross-sectional area in one direction in 0.02 min.

Solution:

$$\text{Number of coulombs} = \frac{18.726 \times \cancel{10^{18}}\ \cancel{\text{electrons}}}{6.242 \times \cancel{10^{18}}\ \cancel{\text{electrons}}\text{/coulomb}} = 3\ \text{C}$$

$$\text{Per Eq. (2.1), } t\ (\text{s}) = 0.02\ \cancel{\text{min}}\left(\frac{60\ \text{s}}{1\ \cancel{\text{min}}}\right) = 1.2\ \text{s}$$

$$I = \frac{Q}{t} = \frac{3\ \text{C}}{1.2\ \text{s}} = \mathbf{2.5\ A}$$

EXAMPLE 2.2 How long will it take 120 C of charge to pass through a conductor if the current is 2 A?

Solution: Per Eq. (2.1),

$$t = \frac{Q}{I} = \frac{120\text{ C}}{2\text{ A}} = \mathbf{60\ s}$$

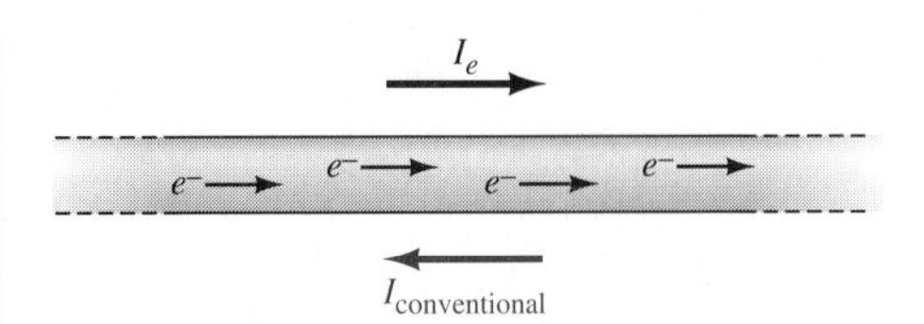

FIG. 2.2
Conventional versus electron flow.

Before continuing, we must point out that there are two fields of thought regarding the direction to be assigned to the current in a network. The vast majority of educational and industrial institutions employ the *conventional* current approach, which is defined by Fig. 2.2 to be opposite that of *electron* flow. We now know that in an electrical conductor it is the negatively charged electrons that actually move through the conductor. However, at the time the basic laws were being developed, it was believed that the charge flow was due to positive carriers. Conventional flow is used in this text to establish a foundation of understanding that will permit a dialogue with professionals in the industry that does not suffer from a mixture of definitions.

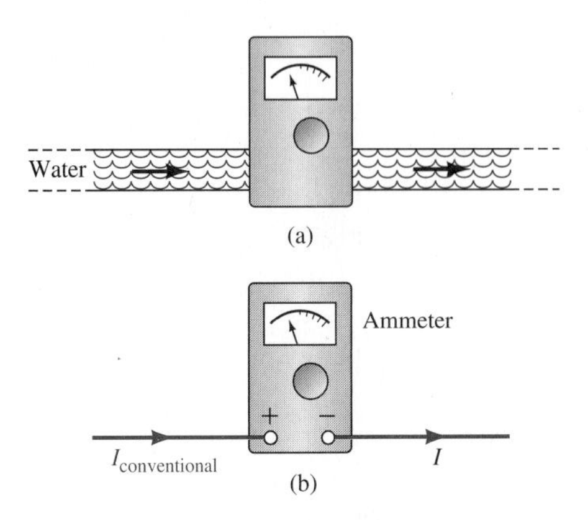

FIG. 2.3
Flow measurements: (a) water; (b) currents.

An analogue often used to develop a clearer understanding of the concept of current is the flow of water through a pipe, such as shown in Fig. 2.3. For the flow of water to be *measured* the pipe must be separated and a meter inserted as shown. The same is true for the measurement of current, as also indicated in the figure. In other words, the path of charge flow (current) must first be broken and the meter inserted between the two created (exposed) leads of the circuit. For obvious reasons, the current-measuring instrument is called an *ammeter.* Note in Fig. 2.3 that the meter is connected in a manner that causes the current to enter the positive terminal of the meter and leave the negative terminal. If hooked up in this manner, both the analog and digital meters will display a positive number. If connected in the reverse manner, the pointer of the analog meter will pin below zero, and the digital meter will have a negative sign in front of the numerical display. The significance of the positive and negative signs will become apparent in the sections to follow. The *face* of a typical ammeter is shown in Fig. 2.4. The scale appearing in Fig. 2.4 is called an *analog* scale. The accuracy of the reading is totally dependent on the ability of the user to interpret the position of the pointer on the continuous analog scale. In recent years the digital display shown in Fig. 2.5 has become increasingly popular. Although it may initially appear that a display consisting of a series of whole numbers and a properly placed decimal point is the more accurate and easier to use, it is very important that anyone even remotely associated with the engineering profession become qualified in reading both types of scales. There are relative advantages associated with each that will become apparent as the range of your applications of the field expands.

FIG. 2.4
Dc milliampere analog scale.

FIG. 2.5
Dc milliampere digital display.

The levels of current typically encountered extend from very low levels to thousands of amperes. The average home has 100-, 150-, or

200-A service. The service rating indicates the maximum current that can be drawn by that home from the power line. Considering that a single air conditioner may draw 15 A (15% of a 100-A service) makes the choice of installing a larger service in a new home an important consideration. At the other end of the scale of magnitudes is the field of electronics, where thousandths and even millionths of an ampere are encountered. In an effort to eliminate the need to carry along the string of zeros associated with very small or very large numbers, the scientific notation appearing in Table 2.1 was defined. A simple count from the decimal point to the *right* of the number 1 results in the proper power of 10. Proceeding from the left to the right results in a negative exponent, and from right to left in a positive exponent, as demonstrated by Example 2.3.

TABLE 2.1
Scientific notation

1,000,000,000,000 = 10^{12}	= terra = T
1,000,000,000 = 10^{9}	= giga = G
1,000,000 = 10^{6}	= mega = M
1000 = 10^{3}	= kilo = k
$\frac{1}{1000}$ = 0.001 = 10^{-3}	= milli = m
$\frac{1}{1,000,000}$ = 0.000001 = 10^{-6}	= micro = μ
0.000000001 = 10^{-9}	= nano = n
0.000000000001 = 10^{-12}	= pico = p

EXAMPLE 2.3 Determine the power of 10 for the following numbers.

Solution:

a. $10,000 = 10,000.0 = 10^4$. (4 places)
b. $0.00001 = 10^{-5}$. (5 places)
c. $0.004 = 4(0.001) = 4 \times 10^{-3}$.
d. $520,000 = 5.2(100,000) = 5.2 \times 10^5$.

EXAMPLE 2.4 Using the prefixes in Table 2.1, write the following quantities in the most convenient form. Take note that all the powers appearing in Table 2.1 are multiples of 3, usually requiring that the closest power of 10 be chosen as most appropriate.

Solution:

a. $0.0006\text{ A} = 0.6 \times 10^{-3}\text{ A}$ = 0.6 milliampere = **0.6 mA**
b. $4200\text{ V} = 4.2 \times 10^{3}\text{ V}$ = 4.2 kilovolts = **4.2 kV**

c. 1,200,000 V = 1.2×10^6 V = 1.2 megavolts = **1.2 MV**
d. 0.00004 A = 40×10^{-6} A = 40 microamperes = **40 μA**

The magnitude of the current, as one might expect, affects the size of wire to be employed for a particular application. The larger the current, the larger the diameter of the wire that must be employed. Obviously, therefore, the current levels encountered in the thin telephone (No. 22) or electric bell wire are very small, while those found in the heavier wire used with heavy machinery and power distribution are much higher (Nos. 1, 0, 00). Table 2.2 lists the maximum current ratings of some commercially available wire. Both solid and stranded types are available for many of the diameters listed. The stranded is made of many strands, rather than being solid, to permit increased flexibility for a wide variety of applications. The No. 12 wire is usually the type used in homes for outlets and such. Note that it can handle a maximum of 25 A. Since a

TABLE 2.2
American wire gage (AWG) sizes

AWG No.	Area (CM)	Ω/1000 ft at 20°C	Maximum Allowable Current for RHW Insulation (A)[a]	AWG No.	Area (CM)	Ω/1000 ft at 20°C
0000	211,600	0.0490	**360**	19	1288.1	8.051
000	167,810	0.0618	**310**	20	1021.5	10.15
00	133,080	0.0780	**265**	21	810.10	12.80
0	105,530	0.0983	**230**	22	642.40	16.14
1	83,694	0.1240	**195**	23	509.45	20.36
2	66,373	0.1563	**170**	24	404.01	35.67
3	52,634	0.1970	**145**	25	320.40	32.37
4	41,742	0.2485	**125**	26	254.10	40.81
5	33,102	0.3133	—	27	201.50	51.47
6	26,250	0.3951	**95**	28	159.79	64.90
7	20,816	0.4982	—	29	126.72	81.83
8	16,509	0.6282	**65**	30	100.50	103.2
9	13,094	0.7921	—	31	79.70	130.1
10	10,381	0.9989	**40**	32	63.21	164.1
11	8,234.0	1.260	—	33	50.13	206.9
12	6,529.9	1.588	**25**	34	39.75	260.9
13	5,178.4	2.003	—	35	31.52	329.0
14	4,106.8	2.525	**20**	36	25.00	414.8
15	3,256.7	3.184		37	19.83	523.1
16	2,582.9	4.016		38	15.72	659.6
17	2,048.2	5.064		39	12.47	831.8
18	1,624.3	6.385		40	9.89	1049.0

[a]From the 1965 National Electrical Code, ® published by the National Fire Protection Association.

home can have a 200-A service, it is obvious that more than one path has to be provided to carry the necessary current to all the outlets of the home.

In Table 2.2 you will note the presence of the symbol CM, which is an indication of the cross-sectional area of the wire. It is a shorthand notation for a defined unit of measure called the *circular mil.* By definition, the area of a circular wire having a diameter of 1 mil is 1 circular mil. One mil is simply one-thousandth of an inch (in.).

$$1 \text{ mil} = 0.001 \text{ in.} \tag{2.2}$$

Through a short derivation it can be shown that the area in circular mils of a wire can be determined from

$$A_{CM} = (d_{mils})^2 \tag{2.3}$$

where d_{mils} = diameter in mils.

EXAMPLE 2.5 Determine the area in CM of a $\frac{1}{16}$-in.-diameter wire.

Solution: Before converting inches to mils, you should first put the diameter in decimal form by simply performing the indicated division.

$$\tfrac{1}{16} \text{ in.} = 0.0625 \text{ in.}$$

Then move the decimal point three places to the right for the conversion to mils.

$$0.0625 \text{ in.} = 62.5 \text{ mils}$$

The area can then be determined by

$$A_{CM} = (d_{mils})^2 = (62.5 \text{ mils})^2 = \mathbf{3906.25\ CM}$$

which corresponds very closely to a **No. 14 wire.**

EXAMPLE 2.6 Determine the diameter in inches of the No. 12 wire typically used in house wiring.

Solution: From Table 2.2,

$$\text{No. 12 wire} \longrightarrow 6529.9 \text{ CM}$$

and

$$d_{mils} = \sqrt{6529.9} = 80.81 \text{ mils}$$

The diameter in inches is obtained by moving the decimal point three places to the left and

$$80.81 \text{ mils} = \mathbf{0.08081 \text{ in.}}$$

which is approximately $\frac{3}{32}$ in.

The Ω/1000-ft rating appearing in Table 2.2 will be examined in a section to follow in this chapter. The wire gage numbers were chosen so that a drop in 3 gage numbers corresponds to doubling the area of the wire. For a drop in 10 gage numbers there is a corresponding 10-fold increase in area.

Since each conductor has a maximum current rating, a circuit element had to be developed to limit the current to a safe level. The two most common protective devices are the fuse and the circuit breaker. The *fuse* contains a metal link of softer metal that melts when the current reaches a particular level. The resulting break in the current path then reduces the current level to 0 A and protects the individuals, appliances, and so on, working near or connected to the system. Once the reason for the overload is removed, service can be reestablished by inserting a new fuse such as shown in Fig. 2.6. Today, the *circuit breaker* (such as shown in Fig. 2.7) is the popular choice for such protection except in those cases where economic and space considerations prohibit the change. The fuse is still extensively used in instrumentation, automobiles, and high-power circuits. The advantage of the circuit breaker is that it can simply be reset if tripped because of an overload. The long-range result is an economically positive one, and the need to maintain a supply of fuses is eliminated. Although a household No. 12 wire can safely handle a maximum current of 25 A, a safety factor is normally introduced by using circuit breakers that trip at 15 or 20 A. Most circuit breakers have the maximum rating clearly visible on the face of the breaker. Within a home there are a number of outlets connected to each circuit breaker. Obviously, therefore, the more fixtures, appliances, and so on, attached to the outlets, the higher the current through the breaker. When a breaker trips because too many units are connected to the circuit, it is very dangerous to simply replace the circuit breaker (or fuse) with one of a higher current rating such as 30 A. Dangerous side effects such as fire and smoke could result from overheating of the conductors. The simple twisting of an extension cord is immediate evidence of the fact that the capacity of the cord is being exceeded.

The current National Electrical Code requires that outlets in the bathroom and other sensitive areas be of the *ground fault current interrupt* (GFCI) variety. They are designed to trip more quickly than the standard circuit breaker. The units in Fig. 2.8 trip in 500 ms ($\frac{1}{2}$ s). It has been determined that 6 mA is the maximum level that most individuals can be exposed to for a short period of time and not be seriously injured. A current higher than 11 mA can cause involuntary muscle contractions that could prevent a person from letting go of the conductor, and thus possibly enter a state of shock. Higher currents lasting more than a second can

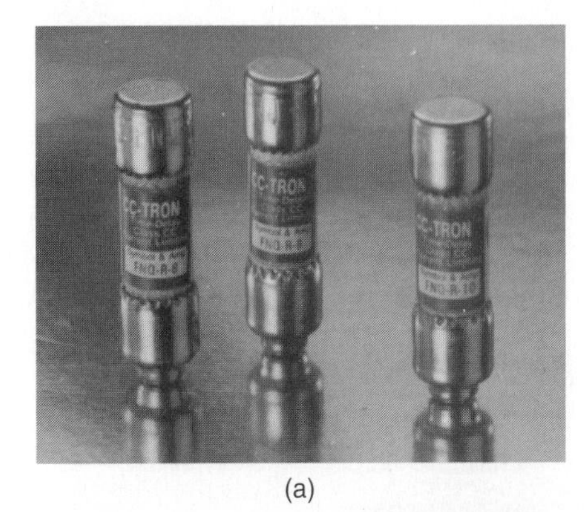

(a)

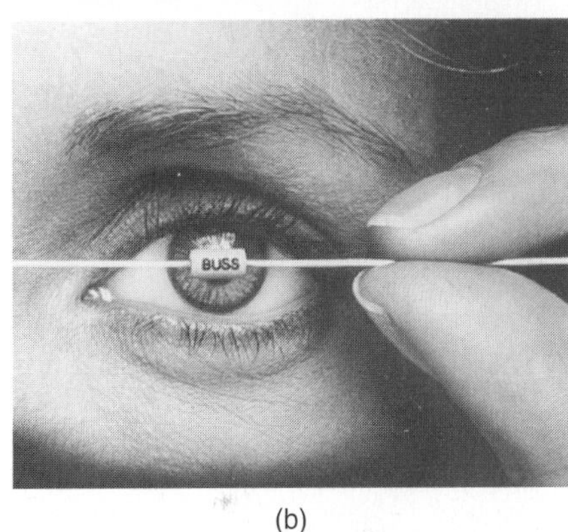

(b)

FIG. 2.6
Fuses: (a) CC-TRON® (0–10 A); (b) subminiature solid matrix.

FIG. 2.7
Circuit breaker. (Courtesy of Potter and Bromfield.)

FIG. 2.8
Ground fault current interrupter (GFCI) 125 V ac, 60 Hz, 15-A outlet. (Courtesy of Leviton, Inc.)

cause the heart to go into fibrillation and possibly cause death in a few minutes. The GFCI is able to react as quickly as it does by sensing the difference between the input and output currents to the outlet. They should be the same if everything is working properly. An errant path such as through an individual establishes a difference in the two current levels and causes the breaker to trip and disconnect the power source.

2.3 VOLTAGE

The fundamental concept of *voltage* is one that usually requires an increased measure of concentration and effort to clearly understand. For many, the term has been encountered only when referring to a 12-V car battery or reading the 120-V requirement on the label of an appliance. Unlike current, which is a *flow* variable and fairly easy to comprehend, voltage is an *across* variable that requires two points to be defined.

In Fig. 2.9 copper wire is used to make the connections between a light bulb and a battery to create the simplest of electric circuits. The battery, at the expense of chemical energy, places a net positive charge on one terminal and a net negative charge on the other. The instant the final connection is made, the free electrons (of negative charge) will drift toward the positive terminal, and the resulting positive ions in the copper wire will simply oscillate in a mean fixed position. The negative terminal is a "supply" of electrons to be drawn from when the electrons of the copper wire *drift* toward the positive terminal. The chemical activity of

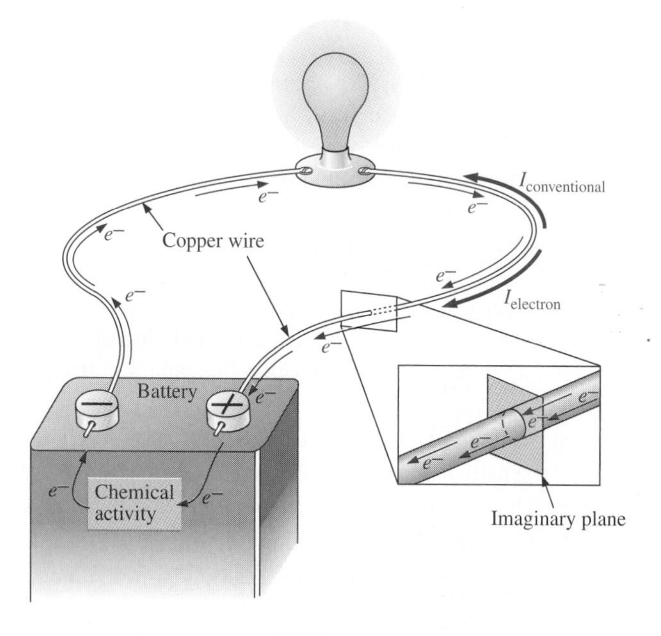

FIG. 2.9
Basic electric circuit.

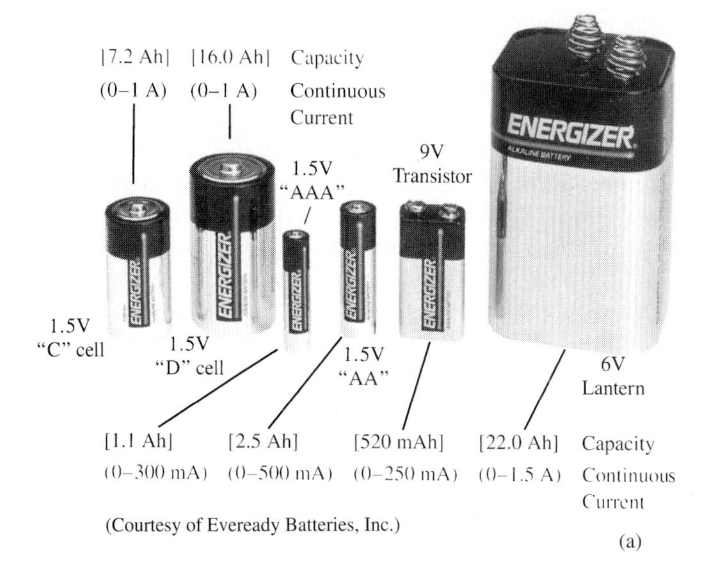

(a)

Lithiode™ lithium-iodine cell
2.8 V, 870 mAh
Long-life power sources with printed circuit board mounting capability
(Courtesy of Catalyst Research Corp.)

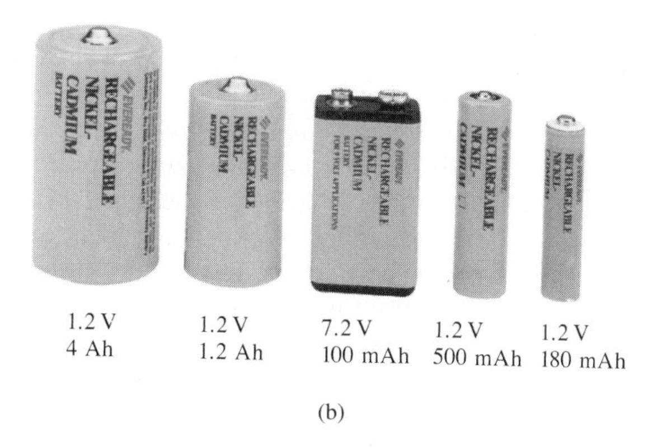

(b)

FIG. 2.11
Batteries: (a) primary; (b) secondary. (Courtesy of Eveready Batteries, Inc.) (Courtesy of Catalyst Research Corp.)

currently limited primarily to space vehicle applications where high-energy-density batteries are required that are rugged and reliable and can withstand a high number of charge/discharge cycles over a relatively long period of time. The nickel–metal hydride cell is actually a hybrid of the nickel–cadmium and nickel–hydrogen cells combining the positive characteristics of each to create a product with a high power level in a small package that has a long cycle life. Although relatively expensive, this hybrid is a valid option for such applications as portable computers.

It is important to recognize that when an appliance or system calls for a nicad battery, a primary cell should not be used. The appliance or system may have an internal charging network that would be dysfunctional with a primary cell. In addition, be aware that all nicad batteries are about 1.2 V per cell, while the most common primary cells are typically 1.5 V per cell. There is some ambiguity with regard to how often a secondary cell should be recharged. For the vast majority of situations the battery can be used until there is some indication that the energy level is low, such as a dimming light from a flashlight, less power from a drill, or a blinking light if one is provided with the equipment. Keep in mind that secondary cells do have some "memory" in that if recharged continuously after being used for a short period of time, they may begin to believe they are short-term units and actually fail to hold the charge for the rated period of time. In any event, always try to avoid a "hard" discharge, which results when every bit of energy is drained from a cell. Too many hard discharge cycles will reduce the cycle life of the battery. Finally, be aware that the charging mechanism for lead–acid batteries is quite different from that for nickel–cadmium cells. The lead–acid battery is charged by a constant voltage source, permitting the current to vary as determined by the state of the battery. The nickel–cadmium battery is charged by a constant current source with the terminal voltage staying pretty steady through the entire charging cycle. The capacity of the battery increases almost linearly throughout most of the charging cycle. You may find that nicad batteries are relatively warm when charging. The lower the capacity level of the battery when charging, the higher the temperature of the cell. As the battery approaches rated capacity the temperature of the cell approaches room temperature.

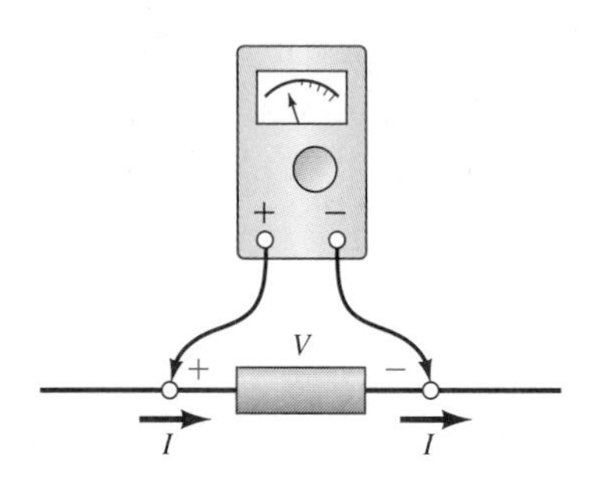

FIG. 2.12
Proper voltmeter connections.

The basic difference between current (a flow variable) and voltage (an across variable) affects the measurement of each also. The basic voltmeter is very similar to the ammeter in its basic appearance, but the measurement techniques are quite different. As shown in Fig. 2.12, the voltmeter does not break the circuit but is placed *across* the element for which the potential difference is to be determined. Like the ammeter, it is designed to affect the network as little as possible when inserted for measurement purposes. A great deal more will be said about loading effects of meters, their tolerance, and their proper use in the last section of this chapter when all the dc instruments have been introduced. For the general protection of any meter used to measure unknown voltage levels it is best to start with the highest scale, to obtain some idea of the voltage to be measured, and then work down to the best reading possible.

Like current levels, voltages can also vary from the microvolt to the megavolt range. The scientific notation introduced earlier is therefore frequently applied to voltage levels also. In radio and TV receivers, very low voltage (microvolt and millivolt) levels are encountered, while at generating stations kilovolt and megavolt readings are encountered. Typically, the bare power lines in a residential area carry 22,000 V (ac), while the covered lines into the home from the transformer on the pole carry 220 V (ac). Transformers are covered in detail in a later chapter.

A photograph of a typical laboratory dc supply appears in Fig. 2.13. The supply voltage can be taken between + and −, or + and ground, or − and ground. On most supplies the output between + and − is said to be *floating,* since it is not connected to a common ground or potential level of the network. The term *ground* simply refers to a zero or earth potential level. The chassis or outer shell of most electrical equipment, whether it be a supply or an instrument, is grounded through the power cable to the terminal board. The third prong (usually round) on any electrical equipment or appliance is the ground connection. Because of this,

FIG. 2.13
Laboratory dc supply. (Courtesy of Hewlett-Packard.)

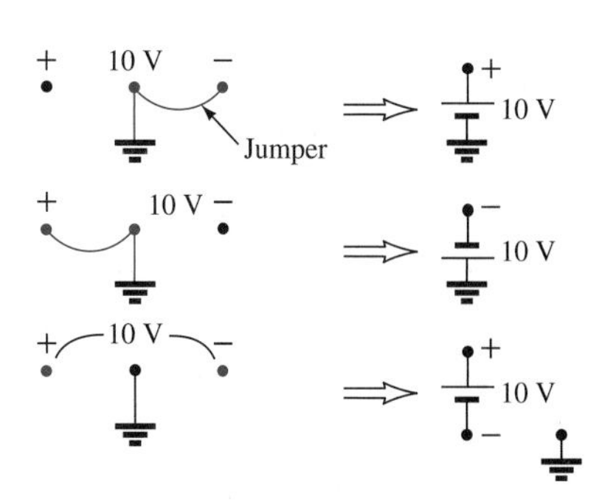

FIG. 2.14
Available voltage levels on a dc laboratory supply.

all connections in the network that are connected directly to the chassis are at ground potential. This is all done primarily for safety reasons. To prevent a high voltage from possibly finding ground through the technician, the alternative paths indicated above are made available. If the + and ground terminals are used and the dial set at 10 V, then the + terminal will be 10 V positive with respect to (w.r.t.) the ground (0-V) terminal. However, if the − and ground terminals are used, the − terminal will be 10 V negative w.r.t. the ground terminal. Figure 2.14 defines the possibilities for a three-terminal supply such as shown in Fig. 2.13. This text, and most available in the field, use the symbol E for all sources of voltage, and V for all potential rises or drops in a network. The use of each will become quite apparent in the sections to follow.

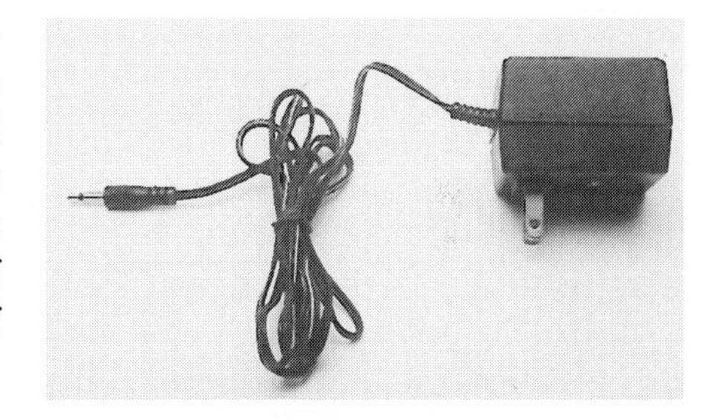

FIG. 2.15
Answering machine/phone 9-V dc supply.

2.4 APPLICATION: DC SUPPLY FOR ANSWERING MACHINES/PHONES

A wide variety of systems in the home and office receive their dc operating voltage from an ac/dc conversion system plugged right into a 120-V ac outlet. Laptop computers, answering machines/phones, radios, clocks, cellular phones, CD players, and so on, all receive their dc power from a packaged system such as appears in Fig. 2.15. The conversion from ac to dc occurs within the unit, which is plugged directly into the outlet. The dc voltage is available at the end of the long wire that is designed to be plugged into the operating unit.

In Fig. 2.16 you can see the transformer used to cut the voltage down to appropriate levels (the largest component of the system). Note that two diodes (Chapter 8) establish a dc level, and a capacitive filter (Chapter 12) is added to smooth out the dc as shown. The system can be relatively small because the operating current levels are quite small, permitting the use of thin wires to construct the transformer and limit its size. The lower currents also reduce the concerns about heating effects, permitting a small housing structure. The unit of Fig. 2.16, rated at 9 V at 200 mA, is commonly used to provide power to answering machines/phones. Further smoothing of the dc voltage is accomplished by a regulator built into the receiving unit. The regulator is normally a small IC chip placed in the receiving unit to separate the heat that it generates

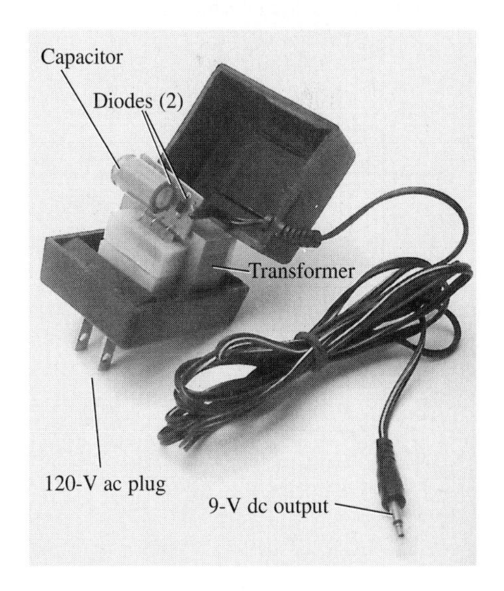

FIG. 2.16
Internal construction of the 9-V dc supply of Fig. 2.15

from the heat generated by the transformer, thereby reducing the net heat at the outlet close to the wall. In addition, its placement in the receiving unit reduces the possibility of picking up noise and oscillations along the long wire from the conversion unit to the operating unit, and ensures that the full rated voltage is available at the unit itself, not a lesser value due to losses along the line.

2.5 RESISTANCE AND OHM'S LAW

The two fundamental quantities of voltage and current are related by a third quantity of equal importance: *resistance.* In any electrical system the pressure is the applied voltage, and the result (or effect) is the flow of charge or current. The resulting level of current is controlled by the resistance of the system. The greater the resistance, the less the current, and vice versa. This effect is immediately obvious when we examine the most fundamental law of electric circuits: *Ohm's law.*

$$I = \frac{E}{R} \tag{2.5}$$

where I = amperes (A)
E = volts (V)
R = ohms (Ω)

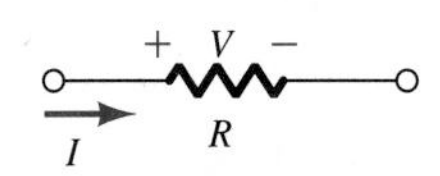

FIG. 2.17
Symbol and defined polarities for a resistor.

Note that the symbol for resistance is the capital letter R, measured in ohms, and that the Greek letter omega (Ω) is used as the symbol for its unit of measurement. The graphic symbol for the resistor appears in Fig. 2.17 with the defined polarities for the indicated current direction (conventional flow). In words, it states that the high side (+) of a potential drop across a resistor is defined by that terminal through which the current enters the resistor.

For single-source dc networks such as those in Fig. 2.18, conventional current always leaves the positive terminal of the supply. For the system in Fig. 2.18 the voltage drop across the resistor is equal to the applied voltage and

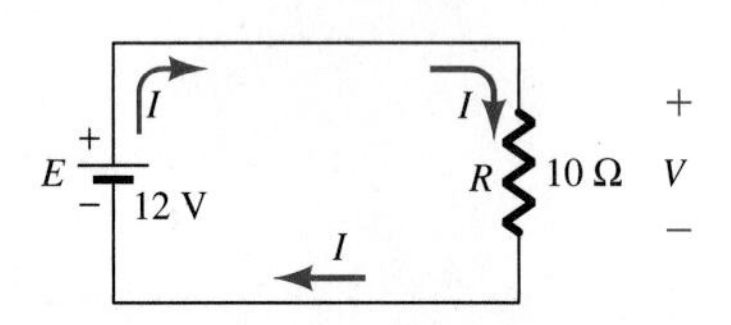

FIG. 2.18
Simplest possible dc circuit.

$$I = \frac{E}{R} = \frac{V}{R} = \frac{12\text{ V}}{10\ \Omega} = 1.2\text{ A}$$

Two rather simple algebraic manipulations result in the following equations for the voltage and resistance:

$$E = IR \quad \text{(volts)} \tag{2.6}$$

$$R = \frac{E}{I} \quad \text{(ohms)} \tag{2.7}$$

EXAMPLE 2.9 Determine the voltage drop across a 2.2-kΩ resistor if the current is 8 mA.

Solution:

$$V = IR = (8 \times 10^{-3}\ \text{A})(2.2 \times 10^{+3}\ \Omega) = (8)(2.2) \times (10^{-3})(10^{-3})\ \text{V}$$
$$= 17.6 \times 100\ \text{V} = \mathbf{17.6\ V}$$

EXAMPLE 2.10 Determine the current drawn by a toaster having an internal resistance of 22 Ω if the applied voltage is 120 V.

Solution:

$$I = \frac{E}{R} = \frac{120\ \text{V}}{22\ \Omega} = \mathbf{5.45\ A}$$

EXAMPLE 2.11 Determine the internal resistance of an alarm clock that draws 20 mA at 120 V.

Solution:

$$R = \frac{V}{I} = \frac{120\ \text{V}}{20 \times 10^{-3}\ \text{A}} = 6 \times 10^{3}\ \Omega = \mathbf{6\ k\Omega}$$

The resistance of a material is principally determined by four quantities: material, length, area, and temperature. The first three are related by the following equation at $T = 20°\text{C}$ (room temperature):

$$R = \rho \frac{l}{A} \quad \text{(ohms)} \tag{2.8}$$

where R = resistance in ohms
ρ = resistivity of the material in circular mil–ohms per foot
l = length of the sample in feet
A = area in circular mils

Note that the area is measured in circular mils, which was discussed in Section 2.2. The resistivity ρ (Greek lowercase letter rho) is a constant determined by the material used. A few are listed in Table 2.3.

In words, the equation reveals that an *increase* in length or a *decrease* in area increases the resistance. The impact of the Ω/1000-ft column in Table 2.2 is now more apparent. It provides the resistance per 1000-ft length of the conductor, which can be multiplied by the length in feet to determine the total resistance. Referring to the table, you will note the very low values encountered, although for very long lengths of any one of the wires the resistance can become a consideration. Its magnitude can usually be totally ignored for applications such as house wiring and general appliances. Note also in Table 2.2 that the resistance increases substantially with decrease in area as one progresses down the table.

TABLE 2.3
Resistivity

Material	ρ
Silver	9.9
Copper	**10.37**
Gold	14.7
Aluminum	17.0
Tungsten	33.0
Nickel	47.0
Iron	74.0
Nichrome	600.0
Carbon	21,000.0

EXAMPLE 2.12 Determine the resistance of 100 yd of copper wire having a $\frac{1}{8}$-in. diameter.

Solution:

$$l = 100 \text{ yd} = 300 \text{ ft}$$

$$\text{Diameter} = \tfrac{1}{8} \text{ in.} = 0.125 \text{ in.} = 125 \text{ mils}$$

$$A_{CM} = (d_{mils})^2 = (125 \text{ mils})^2 = 15{,}625$$

$$R = \rho\frac{l}{A} = \frac{(10.37 \text{ CM-}\Omega/\text{ft})(300 \text{ ft})}{15{,}625 \text{ CM}} = \frac{3111}{15{,}625} = \mathbf{0.199\ \Omega}$$

EXAMPLE 2.13 Determine the resistance of 1 mi of AWG 00 wire.

Solution: From Table 2.2,

$$\frac{\Omega}{1000 \text{ ft}} = 0.0780$$

$$1 \text{ mi} = 5280 \text{ ft}$$

and

$$\frac{5280 \not{\text{ft}}}{1000 \not{\text{ft}}} \times 0.0780\ \Omega = \mathbf{0.412\ \Omega}$$

EXAMPLE 2.14 A large coil of wire is lying on a factory floor. The length of the wire can be determined with an ohmmeter and a ruler using Eq. (2.8), avoiding the need to actually measure the length with the tape. An ohmmeter placed across the two ends of the wire reads 0.1 Ω, and the ruler reveals that the diameter of the wire is about $\frac{1}{8}$ in. Determine the length.

Solution:

$$R = \rho\frac{l}{A} \longrightarrow l = \frac{RA}{\rho} = \frac{(0.1)A}{10.37}$$

$$\tfrac{1}{8} \text{ in.} = 0.125 \text{ in.} = 125 \text{ mils}$$

$$A_{CM} = (125 \text{ mils})^2 = 15{,}625 \text{ CM}$$

and

$$l = \frac{RA}{\rho} = \frac{(0.1\ \Omega)(15{,}625 \text{ CM})}{10.37 \text{ CM-}\Omega/\text{ft}} = \mathbf{150.68\ ft}$$

For most conductors, as the temperature increases the increased activity of the atoms of the wire makes it more difficult for the charge carriers to pass through, and the resistance increases. The resistance-versus-temperature curve for a copper wire is shown in Fig. 2.19. Note that zero resistance is not achieved until absolute zero (−273°C) is reached. However, a straight-line approximation to the curve intersects

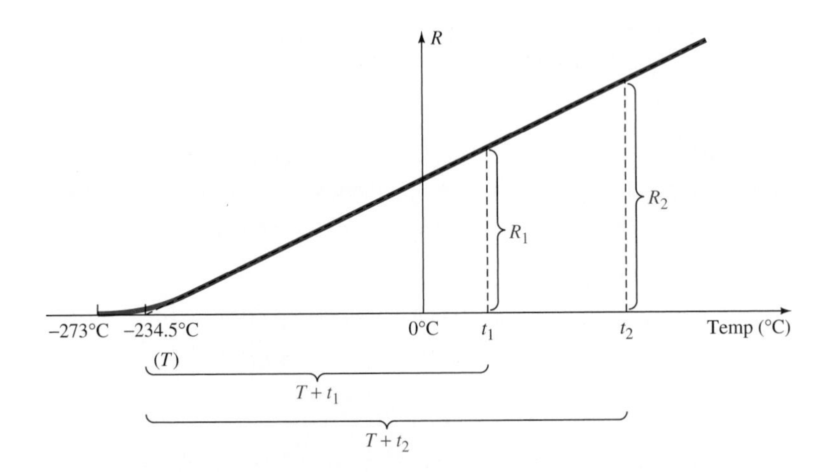

FIG. 2.19
Change in resistance of copper as a function of temperature.

at −234.5°C, the *inferred absolute temperature* of copper. The inferred absolute temperature for various conductors is provided in Table 2.4. Through similar triangles an equation can be developed from the straight-line approximation in Fig. 2.19 that will permit determining the resistance of a conductor at one temperature if its value is known at some other temperature. That is,

$$\frac{T + t_1}{R_1} = \frac{T + t_2}{R_2} \qquad \textbf{(2.9)}$$

where T = inferred absolute temperature of material (°C) (without a negative sign)
R_1 = resistance at temperature t_1
R_2 = resistance at temperature t_2

If the resistance R_1 is known at a temperature t_1, the resistance R_2 can be determined for a temperature t_2. The minus sign for the inferred absolute temperature is not included when substituting the proper value for the material of interest. Occasionally, the following equation is employed rather than Eq. (2.9) to find the resistance at another temperature:

$$R_2 = R_1[1 + \alpha_1(t_2 - t_1)] \qquad \textbf{(2.10)}$$

It utilizes a constant α_1, called the *temperature coefficient of resistance,* which has the symbol alpha (α_1) and is an indication of the rate of change of resistance for that material with change in temperature. In other words, the higher its value, the greater the change in resistance per unit change in temperature. A few values of this coefficient are provided in Table 2.5 for different materials. Note that α_1 includes the

TABLE 2.4
Inferred absolute temperature (T)

Material	Temperature (°C)
Silver	−243
Copper	**−234.5**
Gold	−274
Aluminum	−236
Tungsten	−204
Nickel	−147
Iron	−162
Nichrome	−2250

TABLE 2.5
Temperature coefficient of resistance (α_1)

Material	α_1
Silver	0.0038
Copper	**0.00393**
Gold	0.0034
Aluminum	0.00391
Tungsten	0.005
Nickel	0.006
Iron	0.0055
Nichrome	0.00044
Carbon	−0.0005

effect of the inferred absolute temperature T as determined by the material of interest.

EXAMPLE 2.15 The resistance of a copper conductor is 0.3 Ω at room temperature (20°C). Determine the resistance of the conductor at the boiling point of water (100°C).

Solution: Per Eq. (2.9),

$$\frac{T + t_1}{R_1} = \frac{T + t_2}{R_2} \rightarrow \frac{234.5° + 20°}{0.3} = \frac{234.5° + 100°}{R_2}$$

$$R_2 = \frac{(334.5°)(0.3\ \Omega)}{254.5°}$$

$$= \mathbf{0.394\ \Omega}$$

Per Eq. (2.10),

$$\begin{aligned} R_2 &= R_1[1 + \alpha_1(t_2 - t_1)] \\ &= (0.3\ \Omega)[1 + 0.00393(100° - 20°)] \\ &= (0.3\ \Omega)(1 + 0.3144) \\ &= \mathbf{0.394\ \Omega} \end{aligned}$$

EXAMPLE 2.16 If the resistance of a copper conductor is 0.6 Ω at $t = 0°C$ (freezing), at what temperature will it be 1 Ω?

Solution: Per Eq. (2.9),

$$\frac{T + t_1}{R_1} = \frac{T + t_2}{R_2}$$

Solving for t_2 (the temperature at which the resistance will be $R_2 = 1\ \Omega$) gives

$$T + t_2 = \frac{R_2}{R_1}(T + t_1)$$

$$t_2 = \frac{R_2}{R_1}(T + t_1) - T$$

Substitution of values yields

$$\begin{aligned} t_2 &= \frac{1\ \Omega}{0.6\ \Omega}(234.5° + 0°) - 234.5° \\ &= 390.83° - 234.5° \\ &= \mathbf{156.33°} \end{aligned}$$

There are numerous applications where resistive values are added to a network to perform a very specific, necessary function. For this purpose, a variety of resistors designed to suit the application have been designed with a *tolerance* (on special order) as low as 0.01%. The lower the tolerance, the more care that must go into the manufacture of the resistor. A tolerance of 0.1% (0.001) on a 50-Ω resistor indicates that the actual value of that resistor will not be off by more than $(0.001)(50) = 0.05\ \Omega$, or the manufacturer guarantees that it will fall within the range 49.95 to 50.05 Ω. Tolerances on the order of 5, 10, and 20% are more common. A few of the various types of fixed-type resistors are shown in Fig. 2.20. The resistance of the carbon resistors in Fig. 2.20(a) is controlled by an internal carbon element whose length, area, and composition can be chosen to provide the desired resistance level. All the resistors in Fig. 2.20(a) are actual size. Note the increase in size associated with an increase in wattage (power) rating, since all the electrical energy delivered to a resistor is dissipated in the form of heat. The concepts of electrical power and energy will be

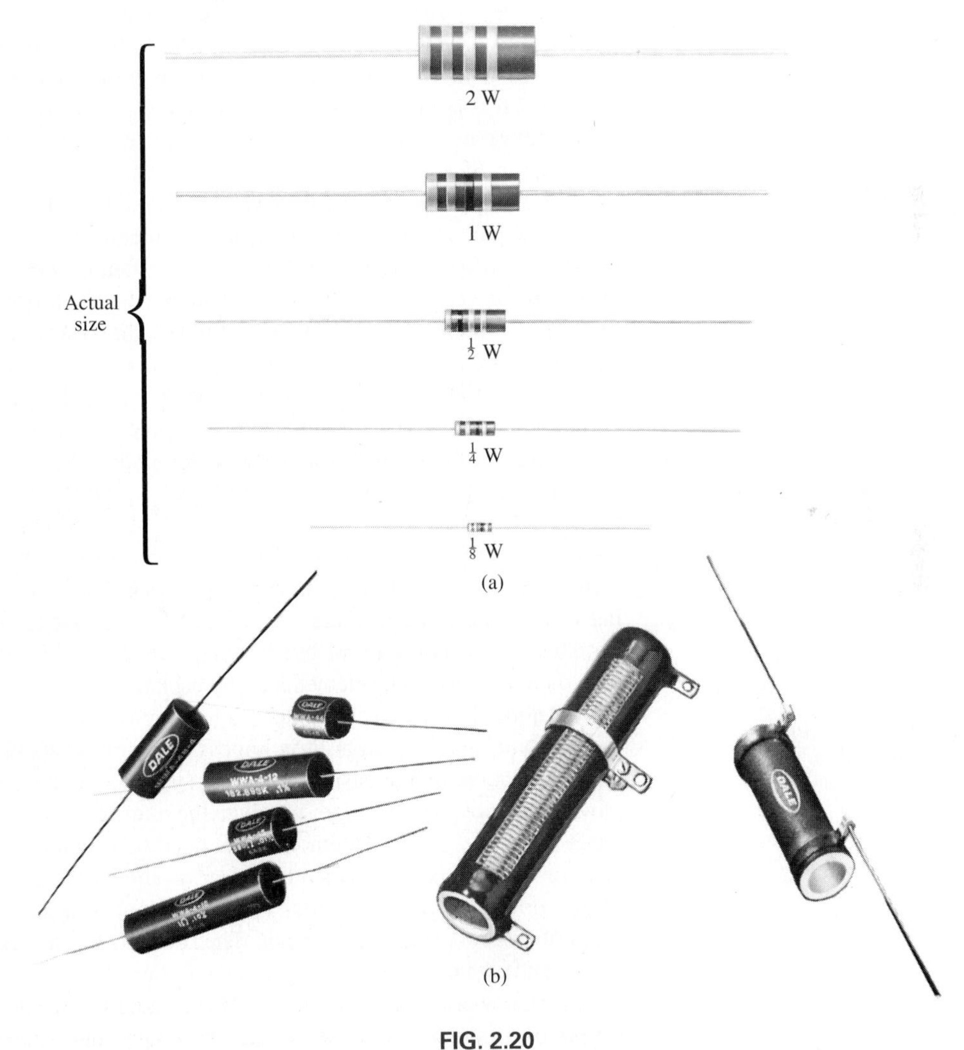

FIG. 2.20
Fixed resistors: (a) carbon; (b) wire-wound. (Courtesy of Allen Bradley Company, Ohmite Manufacturing Company.)

discussed in some detail in the next section. Metal heat conductors called *heat sinks* are often employed in conjunction with electrical and electronic elements to assist in drawing the heat away from the element and preventing unnecessary damage. A few will be shown in a later chapter dealing with electronic devices.

The wire-wound resistors in Fig. 2.20(b) employ a high-resistance wire of a specific thickness and length to establish the desired resistance. The spacing between the wire wrappings and the surface area used inside and outside the wire-wound resistors affects the power dissipation rating of the element.

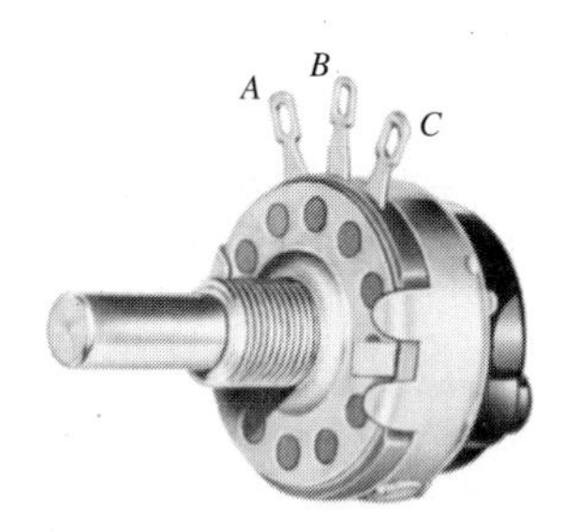

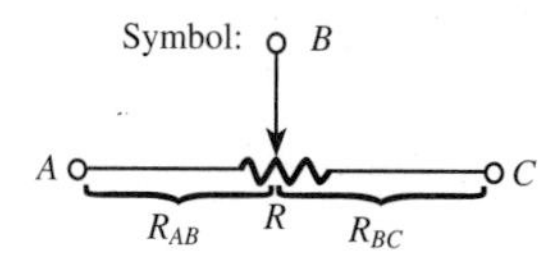

FIG. 2.21
Molded composition potentiometer. (Courtesy of Allen Bradley Company.)

A three-terminal device called a *potentiometer* can be used as a voltage- or *potential*-controlling device (whence came its name) or as a variable resistor or *rheostat* if only two of its terminals are employed. Since the device has three terminals, as shown in Fig. 2.21, it is important that the terminals be used for the variation in resistance desired. The symbol for the device appearing in Fig. 2.21 indicates quite clearly that between the two outside terminals the resistance is always the total value R even if the shaft is turned. Between the center terminal (or wiper arm) and either outside terminal the resistance varies between a minimum value of 0 Ω when the contacts touch and a maximum of R when the wiper arm reaches the other outside terminal. The sum of the resistance levels R_{AB} and R_{BC} always equals the total resistance R. Quite obviously, as R_{AB} increases in size, the resistor R_{BC} decreases by an equal amount. The manner in which the two terminals chosen are hooked up determines whether a right-hand rotation of the shaft will increase or decrease the resistance. When all three terminals are employed in an electrical system, its purpose is to control potential levels in the network as determined by the resistance between the respective terminals of the potentiometer. In Fig. 2.22 the voltage V_1 increases as the resistive component R_1 increases. When R_1 equals the total resistance R, the voltage V_1 equals the supply voltage E.

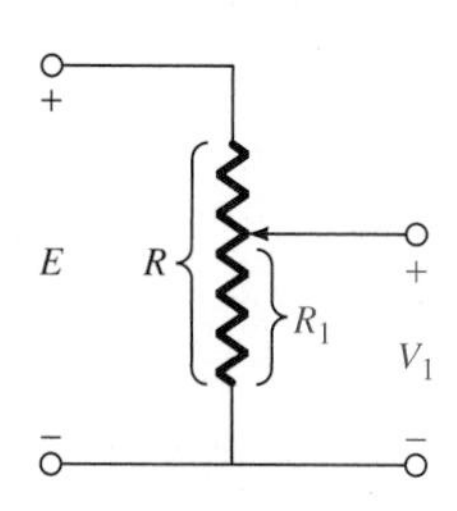

FIG. 2.22
Potentiometer as used to control the voltage level V_1.

Resistors of very small size (but not necessarily small resistance value) can be found in integrated circuits, which have become increasingly common in recent years. The resistor in this case is not a discrete element as discussed above but is manufactured within a single chip with the other electronic elements of the package. A great deal more will be said about ICs in a later chapter.

For some small resistors, it is impossible or impractical to print the numerical value on the casing. Rather, a system of color coding is employed whereby certain colors are given the numerical value indicated in Table 2.6. For the most common of the fixed resistors, the carbon composition resistor, the color bands appear as shown in Fig. 2.23. Fortunately, however, the same order is also applied to a number of other types of resistors using color bands. The first and second bands (closest to one end) determine the first and second digits, while the third determines the power of 10 to be associated with the first two digits or, more simply put, the number of zeros to follow the first two digits. The fourth band is the tolerance, which, as the table indicates, will not appear if the tolerance is ±20%.

TABLE 2.6
Color coding

0	Black	7	Violet	
1	Brown	8	Gray	
2	Red	9	White	
3	Orange	0.1	Gold	
4	Yellow	0.01	Silver	
5	Green	5%	Gold	Tolerance
6	Blue	10%	Silver	Tolerance

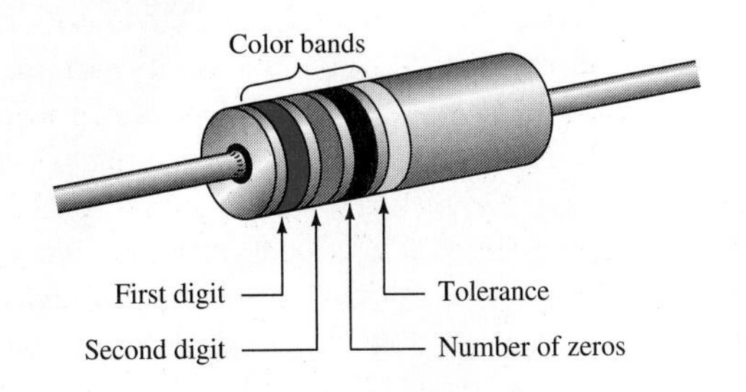

FIG. 2.23
Molded composition resistor.

EXAMPLE 2.17 Determine the manufacturer's guaranteed range of values for the carbon resistor in Fig. 2.24.

Solution:

$$\text{Blue} = 6, \quad \text{Gray} = 8, \quad \text{Black} = 0, \quad \text{Gold} = \pm 5\%$$

$$\therefore \quad 68 \times 10^0 \pm 5\% = 68 \pm 3.4 = 64.6\ \Omega \longrightarrow \mathbf{71.4\ \Omega}$$

If the measured value of the resistor was 70 Ω instead of the 68 Ω stated on the label, it would still meet the manufacturer's standards.

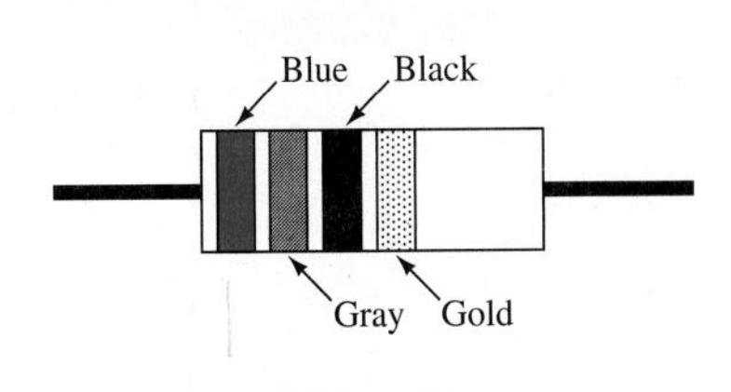

FIG. 2.24

EXAMPLE 2.18 Determine the color coding for a 100-kΩ resistor with a 10% tolerance factor.

Solution:

$$100\ \text{k}\Omega = 100{,}000 \qquad 10\%$$

$100{,}000 = 10 \times 10^4$

1st band = **brown (1)**
2nd band = **black (0)**
3rd band = **yellow (4)**
4th band = **silver**

The resistance of a resistor, network, or system can be measured using an instrument called an *ohmmeter*. The most frequently employed is the wide-range ohmmeter appearing on the typical volt–ohm milliammeter (VOM) or digital multimeter (DMM) such as shown in Fig. 2.25. The VOM and DMM can measure wide ranges of voltage, resistance, and current levels. Their versatility makes them quite popular with the laboratory

(a)

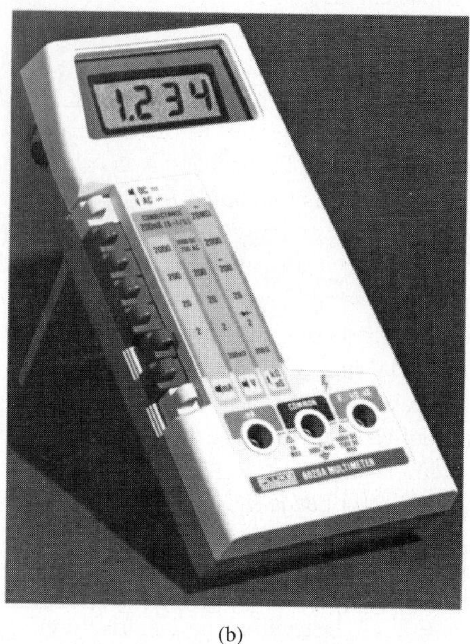

(b)

FIG. 2.25
(a) VOM (volt–ohm milliammeter); (b) DMM (digital-display meter).

technician. Note that the ohmmeter scale on the VOM is nonlinear. The scale setting therefore should be set to obtain a reading somewhere in the low- or midscale region for the best accuracy. For the $R \times 10$ setting, each reading on the scale is multiplied by a factor of 10. Similarly, the $R \times 100$ and $R \times 1000$ scales have 100- and 1000-unit multipliers. The digital display does not require the careful reading of a nonlinear analog scale, but the proper multiplying factor must still be applied. The ohmmeter is an excellent instrument for determining which terminals provide the desired resistance variation on a three-terminal potentiometer. Simply hook up the meter to the terminals of interest and turn the shaft, noting the effect on the scale indication. It must be pointed out, however, that the ohmmeter, unlike the ammeter and the voltmeter, requires an internal battery. If the instrument is left on the resistance scale and the leads should touch, the battery will be drained quite rapidly. Therefore, the VOM and DMM should be set on a high-voltage scale when not being used to prevent this possibility. It is also very important that *an ohmmeter never be connected to a live network.* The reading will be erroneous, since it is calibrated to the internal battery, and chances are that the instrument itself will be damaged. When measuring the resistance of a resistor, there is no need to be concerned about the polarity. The red lead of the meter can be connected to either end of the resistor with the black lead connected to the remaining terminal. Be aware, however, that one end of a resistor must always be disconnected from a circuit when measuring its resistance or the meter reading may include the effect of the other resistors in the circuit.

The ohmmeter can also be used for checking continuity in a network by searching for a 0-Ω (or least-resistance) indication, which indicates that the two leads of the instrument are directly connected by the leg of the network. In addition, an ohmmeter can be used to determine which lead is which when more than one wire appears in a terminal box. By connecting one lead of the ohmmeter to the wire of interest outside the box, you can find the same wire inside by touching each lead in the box and finding the 0-Ω indication. A last application of the ohmmeter is searching for a short (the high-voltage lead is touching ground) in house wiring if the circuit breaker keeps *popping* or the system was just installed and you would like to be sure the *hot* lead is not touching ground somewhere in the system. Be sure the power is off, and then simply take one ohmmeter lead and attach it to one of the wires coming out of the electrical outlet and connect the other to the metal box. Only one of the outlet wires (the ground wire) should result in a 0-Ω reading when connected in this manner. If not, the circuit and that section of wiring should be rechecked very carefully. A *short* condition is simply one in which an *unwanted* low-resistance path has been established between a voltage level and ground. From Ohm's law it is quite clear that a very low resistance (≅0 Ω) will result in a very high current that can cause dangerous conditions such as fire and smoke.

There are instruments such as shown in Fig. 2.26 that are designed specifically to measure very large resistances such as those in the

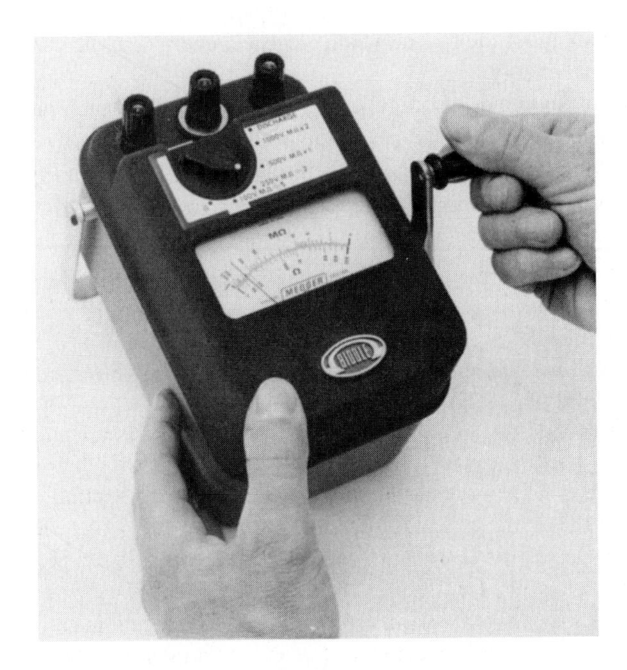

FIG. 2.26
Megger® tester. (Courtesy of James G. Biddle Company.)

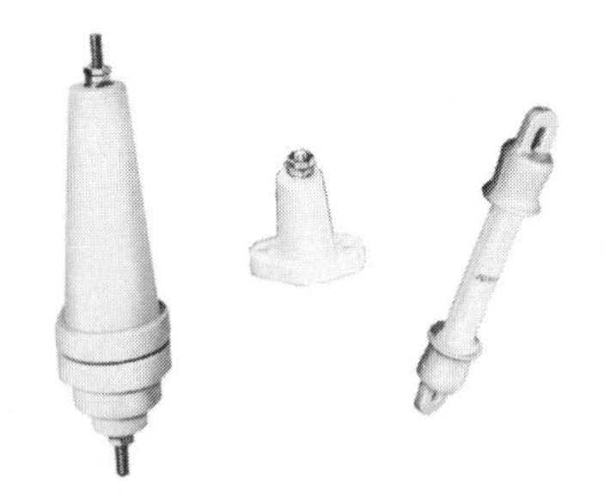

FIG. 2.27
Insulators. (Courtesy of E. F. Johnson Company.)

megohm range. One application of such an instrument is as an insulation tester. An *insulator* is any material with a very high resistance characteristic. In the rubber or plastic form it is used to insulate the wires running throughout homes and industrial plants. In the ceramic form (Fig. 2.27) it can be used to ensure that uncovered high-voltage lines do not make contact with their supporting structures. Insulating materials play a major role in a variety of ways in the electrical and electronics industries. A few will be introduced in the chapters and sections to follow.

Materials that have resistance levels somewhere between those typically associated with insulators and conductors are called *semiconductors*. The transistors, diodes, and integrated circuits that we have all been exposed to through the advertising media are constructed of these materials. More will be said about these devices in the chapters to follow. For the moment it is important to know that semiconductor materials have *negative temperature coefficients*. That is, the resistance of commonly employed materials like silicon and germanium decreases with an increase in temperature. The temperature of the element can change if the current through the device increases or decreases or if the temperature of the surrounding medium changes.

2.6 POWER, ENERGY, EFFICIENCY

For any electrical, mechanical, or input–output system *power* is a measure of the *rate* of energy conversion for that system. A motor converts electrical energy to mechanical energy at a rate determined by the applied load and the horsepower of the motor. The larger the horsepower rating, the higher the rate at which electrical energy can be converted to mechanical energy.

For any source of electrical energy such as a battery or generator, the power delivered is determined by the product of its terminal voltage and the current drain for the supply. Therefore, for the battery in Fig. 2.28(a) the power delivered is determined by

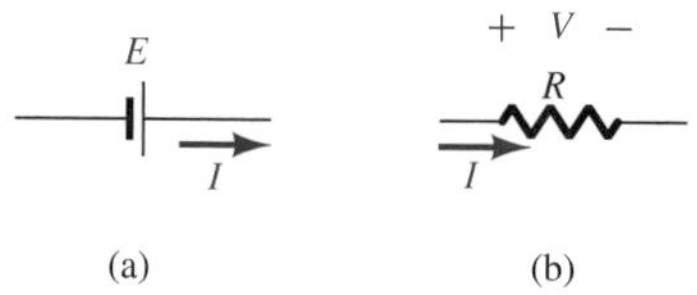

FIG. 2.28
Power: (a) supplied by a battery; (b) absorbed by a resistor.

$$\boxed{P = EI} \quad \text{(watts)} \qquad \textbf{(2.11)}$$

The unit of measurement for power is the *watt* (W), which is equivalent to a rate of energy conversion of 1 J/s.

For the resistive element in Fig. 2.28(b) the power *dissipated* by the resistive element is given by

$$\boxed{P = VI = I^2R = \frac{V^2}{R}} \quad \text{(watts)} \qquad \textbf{(2.12)}$$

where one equation is derived from the other with a simple application of Ohm's law.

Every electrical appliance in the home, from the electric shaver to the motor in the refrigerator, has a wattage rating. A few such appliances are listed in Table 2.7 with their average wattage rating.

TABLE 2.7
Average wattage ratings of some household appliances

Appliance	Wattage Rating	Appliance	Wattage Rating
Air conditioner	2000	Heating equipment	
Cassette recorder/player	5	Furnace fan	320
Clock	2	Oil-burner motor	230
Clothes dryer		Iron, dry or steam	1000
Electric, conventional	5000	Projector	1000
Electric, high-speed	8000	Radio	30
Gas	400	Range	
Clothes washer	400	Free-standing	Up to 16,000
Coffee maker	Up to 1000	Wall ovens	Up to 8000
Dishwasher	1500	Refrigerator	300
Fan		Shaver	10
Portable	160	Stereo	75
Window	200	Toaster	1200
Heater	1650	TV (color)	160

Courtesy of Con Edison.

EXAMPLE 2.19 Determine the current drawn by a 180-W television set when connected to a 120-V outlet.

Solution:

$$P = VI \longrightarrow I = \frac{P}{V} = \frac{180\text{ W}}{120\text{ V}} = \mathbf{1.5\text{ A}}$$

EXAMPLE 2.20 Determine the resistance of a 1200-W toaster that draws 10 A.

Solution:

$$P = I^2R \longrightarrow R = \frac{P}{I_2} = \frac{1200\text{ W}}{(10\text{ A})^2} = \frac{1200\ \Omega}{100} = \mathbf{12\ \Omega}$$

Power is measured by a device called (for obvious reasons) a *wattmeter*. It has two terminals for the voltage-sensing portion and two terminals for the current reading. For most wattmeters, the current terminals are heavier in appearance than those employed for the voltage sensing. In any case, the potential terminals are connected *across* the load to which the power is to be determined, while the current terminals are connected in *series* (like the ammeter) with the

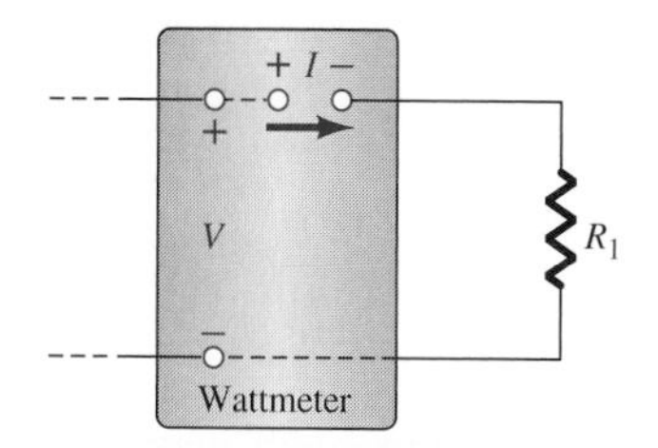

FIG. 2.29
Measuring the power to the resistor R_1.

load. The wattmeter in Fig. 2.29 is connected to measure the total power delivered to the resistor R_1. Note the polarities of the terminals of the meter with respect to the direction of assumed current and the applied potential.

The electrical bill received each month by the homeowner indicates the electrical energy used and *not* the power made available for use. It is important that the difference between power and energy be clearly understood. Consider a 10-horsepower (hp) motor, for example; unless it is used over a period of *time, no* energy conversion is effected by the machine. Energy and power are related by the following equation, which introduces the important time element ingredient.

$$W = Pt \quad \text{(joules or watt seconds)} \tag{2.13}$$

where W = energy in joules
P = power in watts (W)
t = time in seconds (s)

In other words, the longer we use an energy-converting device of a particular power rating, the greater will be the total energy converted. As indicated above, the unit of measure is the watt-second (Ws). However, this quantity is usually too small to serve most practical energy consumption measurements, and so the watt-hour and kilowatt-hour (usually written kilowatthour and abbreviated kWh) is normally employed. The meter on the side of every residential and industrial building is called a kilowatthour meter and has the appearance shown in Fig. 2.30(a). The four dials combine to indicate the total kilowatthours of energy used over a monthly period. One possible combination appears in Fig. 2.30(b).

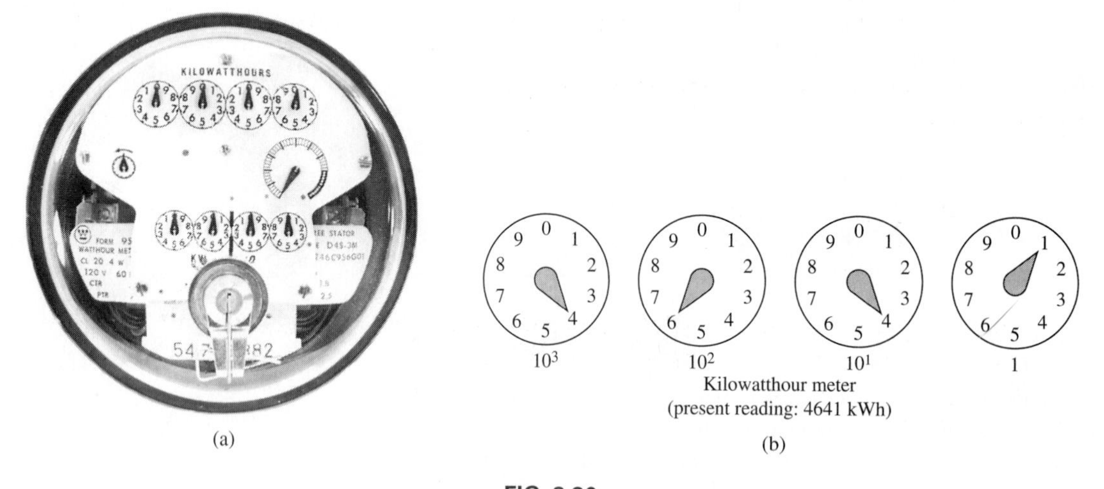

FIG. 2.30
Kilowatthour meter. (Courtesy of Westinghouse Electric Corporation.)

The number of kilowatthours is determined from the equation

$$\text{Kilowatthours} = \frac{Pt}{1000} \qquad \text{(kWh)} \tag{2.14}$$

where P = watts
t = hours

The following example should reveal how the monthly bill is determined from the kilowatthour reading. However, keep in mind that this example will use only an average cost per kilowatthour rather than using the sliding scale of cost per increase in demand. Most utilities are quite willing to forward a rate breakdown to any of their customers.

EXAMPLE 2.21 Determine the cost of using the following appliances for the time indicated if the average cost is 9 cents/kWh.

a. 1200-W iron for 2 h.
b. 160-W color TV for 3 h and 30 min.
c. Six 60-W bulbs for 7 h.

Solution:

$$\text{Kilowatthours} = \frac{(1200\text{ W})(2\text{ h}) + (160\text{ W})(3.5\text{ h}) + (6)(60\text{ W})(7\text{ h})}{1000}$$

$$= \frac{2400\text{ Wh} + 560\text{ Wh} + 2520\text{ wH}}{1000} = 5.48\text{ kWh}$$

$$(5.48\text{ kWh})\left(\frac{9\text{ cents}}{\text{kWh}}\right) \cong \mathbf{49.32\text{ cents}}$$

For any input–output electrical system, the overall efficiency of operation is a characteristic of primary importance. It is an indication of how much of the energy supplied is being used to perform the desired task. For a motor, for example, the more the output mechanical power for the same electrical power input, the higher the efficiency. In equation form, as a percentage,

$$\eta = \frac{P_o}{P_i} \times 100\% \tag{2.15}$$

Conservation of energy requires that energy be conserved and that the output energy never be greater than the supplied or input energy. The maximum efficiency is therefore 100% when $P_o = P_i$. For any system the following equation is always applicable:

$$P_i = P_o + P_l \tag{2.16}$$

where P_l is the power lost or stored in the system. For some applications the following conversion must be applied.

$$1 \text{ hp} = 746 \text{ W}$$

EXAMPLE 2.22 Determine the efficiency of operation and power lost in a 5-hp dc motor that draws 18 A at 230 V.

Solution:

$$\eta = \frac{P_o}{P_i} \times 100\% = \frac{5(746 \text{ W})}{(230 \text{ V})(18 \text{ A})} \times 100\% = \frac{3730 \text{ W}}{4140 \text{ W}} \times 100\% = \mathbf{90\%}$$

$$P_l = P_i - P_o = 4140 \text{ W} - 3730 \text{ W} = \mathbf{410 \text{ W}}$$

2.7 APPLICATION: MICROWAVE OVEN

It is probably safe to say that most modern homeowners have a microwave oven such as appears in Fig. 2.31(a)—even those of us who went through the phase of worrying about whether it was safe and whether it was a proper way to prepare food. Now we use the oven so often during the day that we wonder how we even did without it before. For most users, its operating efficiency is not the biggest concern, probably because its impact on the monthly bill is not that easy to define with so many appliances in the home. However, it might be of some interest to examine the unit in more detail and apply some of the theory presented in this chapter.

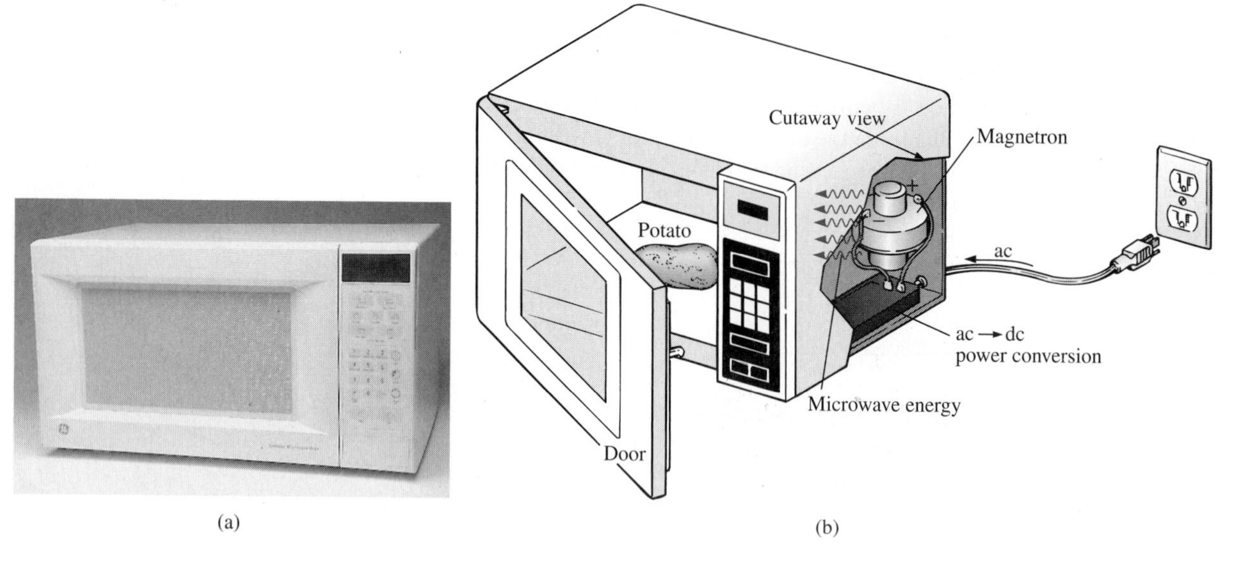

FIG. 2.31
Microwave oven: (a) photo; (b) basic construction.

First, some general comments. Most microwaves are rated at 500 W to 1200 W at a frequency of 2.45 GHz (almost 2.5 billion cycles per second compared with the 60 cycles per second for the ac voltage at the typical home outlet—details in Chapter 4). Water molecules in the food are vibrated at such a high frequency that the friction with neighboring molecules causes the heating effect. Since it is the high frequency of vibration that heats the food, there is no need for the material to be a conductor of electricity. However, any metal placed in the microwave can act as an antenna (especially if it has any point or sharp edges) that will attract the microwave energy and reach very high temperatures. In fact, a browning skillet made for microwave ovens has some metal embedded in the bottom and sides to attract the microwave energy and raise the temperature at the surface between the food and skillet to give the food a brown color and a crisp texture. Even if the metal does not act as an antenna, it is a good conductor of heat and can get quite hot as it draws heat from the food. Any container with low moisture content can be used to heat foods in a microwave oven. Because of this requirement, manufacturers have developed a whole line of microwave cookware that is very low in moisture content. Theoretically, glass and plastic contain very little moisture, but even so, when heated in the oven for a minute or so, they do get warm. The heating could be due to the moisture in the air that clings to the surface of each or perhaps to the lead used in good crystal. In any case, microwave ovens should be used only to prepare food. They were not designed to be dryers or evaporators. The instructions with every microwave oven specify that it should not be turned on when empty. Even though the oven may be empty, microwave energy will be generated and will try to find a channel for absorption. If the oven is empty, the energy might be attracted to the oven itself and could damage it. To demonstrate that a dry empty glass or plastic container will not attract a significant amount of microwave energy, place two glasses in an oven, one containing water and the other empty. After 1 minute you will find the glass with the water quite warm due to the heating effect of the hot water, while the other will be close to its original temperature. In other words, the water created a heat sink for the majority of the microwave energy, leaving the empty glass as a less attractive path for heat conduction. Dry paper towels and plastic wrap can be used in the oven to cover dishes, since they initially have low water molecule content, and paper and plastic are not good conductors of heat. However, it would be very unsafe to place a paper towel in an oven alone because, as stated above, the microwave energy will look for an absorbing medium and could set the paper on fire.

Conventional ovens cook food from the outside in. The same is true for microwave ovens, but they have the additional advantage that microwaves are able to penetrate the outside few centimeters of the food, reducing the cooking time substantially. The cooking time with a microwave oven is related to the amount of food in the oven. Two cups of water will take longer to heat than one cup, although it is not a linear relationship, so it won't take twice as long—perhaps 75% to 90% longer. Eventually, if you place enough food in the microwave oven and thus increase the cooking time, you will reach a crossover at which point it would be just as wise to use a conventional oven and get the food texture you might prefer.

The basic construction of a microwave oven is depicted in Fig. 2.31(b). It uses a 120-V ac supply that is then converted through a high-voltage transformer to one having peak values approaching 5000 V (at substantial current levels)—sufficient warning to leave microwave repair to the local service location. Through a rectifying process to be described in Chapter 8, a high dc voltage of a few thousand volts is generated that appears across a magnetron. The magnetron, through its very special design (currently the same design as invented by the British in WW II for their high-power radar units), generates the required 2.45-GHz signal for the oven. Note that the magnetron has a specific power level of operation that cannot be controlled—once it's on, it's on at a set power level. You may then wonder how the cooking temperature and duration can be controlled. This is accomplished through a controlling network that determines the amount of off and on time during the input cycle of the 120-V supply. Higher temperature are achieved by setting a high ratio of on to off time, while low temperatures are set by the reverse action.

One unfortunate characteristic of the magnetron is that in the conversion process it generates a great deal of heat that does not go toward the heating of the food and thus must be absorbed by heat sinks or dispersed by a cooling fan. Typical conversion efficiencies are between 55% and 75%. Considering other losses inherent in any operating system, it is reasonable to assume that most microwave ovens are between 50% to 60% efficient. However, the conventional oven with its continually operating exhaust fan and heating of the oven, cookware, surrounding air, and so on, also has significant losses, even if it is less sensitive to the amount of food to be cooked. All in all, the convenience factor is probably the one that weighs most heavily in this discussion. The question that remains is how our time can be figured into the efficiency equation.

For specific numbers, let us consider the energy associated with baking a 5-oz potato in a 1200-W microwave oven for 5 min if the conversion efficiency is an average value of 55%. First, it is important to realize that when a unit is rated as 1200 W, that is the rated power drawn from the line during the cooking process. If the oven is plugged into a 120-V outlet, the current drawn is

$$I = \frac{P}{V} = \frac{1200\text{ W}}{120\text{ V}} = 10\text{ A}$$

which is a significant level of current. Next, we can determine the amount of power dedicated solely to the cooking process by using the efficiency level. That is,

$$P_o = \eta P_i = (0.55)(1200\text{ W}) = 600\text{ W}$$

The energy transferred to the potato over a period of 5 min can then be determined from

$$W = Pt = (660\text{ W})(5\text{ min})(60\text{ s/1 min}) = 198\text{ kJ}$$

which is about half of the energy (nutritional value) derived from eating the potato. The number of kilowatthours drawn by the unit is determined from

$$W = \frac{Pt}{1000} = \frac{(1200\text{ W})(5/60\text{ h})}{1000} = \mathbf{0.1\ kWh}$$

At a rate of 10¢kWh we find that we can cook the potato for 1 penny—relatively speaking, pretty cheap. A typical 1550-W toaster oven would take an hour to heat the same potato, resulting in 1.55 kWH and a cost of 15.5 cents—a significant increase in cost.

2.8 SERIES DC NETWORKS

The analysis of electrical systems requires that the basic topographical definitions be clearly understood. In most configurations two adjoining elements are either in series or parallel. In this section we examine series elements and reserve parallel elements for the next section.

Two elements are in series if they have only one terminal in common that is not connected to a third current-carrying component.

In Fig. 2.32 the resistors R_1 and R_2 are in series, since they are connected only at terminal b. The other end of each resistor is connected to other elements. In addition, there are no other elements connected to terminal b. For the same reasons R_2 and R_3 are in series, R_3 and E are in series, and E and R_1 are in series. Since all the elements of the system are in series, the circuit in Fig. 2.32 is called a *series circuit*.

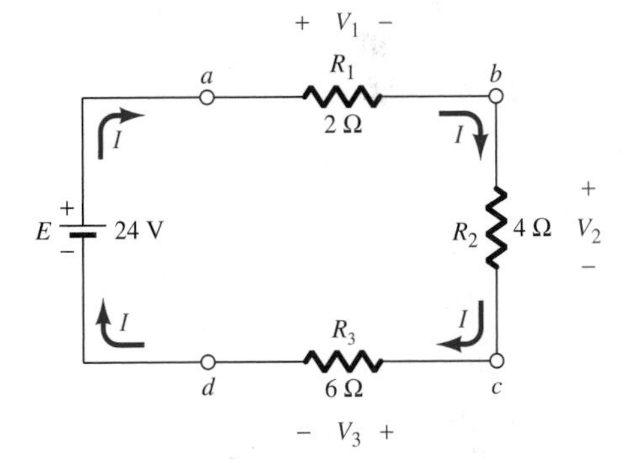

FIG. 2.32

Total Resistance

For resistors in series, the total resistance is the sum of the individual resistances. That is,

$$R_T = R_1 + R_2 + R_3 + \cdots + R_N \qquad \text{(ohms)} \qquad \mathbf{(2.17)}$$

For the circuit in Fig. 2.32,

$$R_T = 2\ \Omega + 4\ \Omega + 6\ \Omega = \mathbf{12\ \Omega}$$

Current

The current through a series circuit is the same for each element.

In Fig. 2.32, therefore, the current through E_1 or R_1 or through either of the remaining resistors is the same. Its magnitude is determined by using Ohm's law in the form

$$I = \frac{E}{R_T} \qquad \text{(amperes)} \qquad \mathbf{(2.18)}$$

For the circuit in Fig. 2.32,

$$I = \frac{24\text{ V}}{12\ \Omega} = \mathbf{2\ A}$$

Voltages

The voltage across each resistor can now be determined using Ohm's law:

$$V_1 = IR_1 = (2\text{ A})(2\ \Omega) = \mathbf{4\ V}$$
$$V_2 = IR_2 = (2\text{ A})(4\ \Omega) = \mathbf{8\ V}$$
$$V_3 = IR_3 = (2\text{ A})(6\ \Omega) = \mathbf{12\ V}$$

Note the polarity of V_1, V_2, and V_3 in Fig. 2.32 as determined by the resulting current direction.

Power

The power delivered by the source is

$$P_T = EI = (24\text{ V})(2\text{ A}) = \mathbf{48\ W}$$

and the power delivered to each element is

$$P_T = I^2R_1 = (2\text{ A})^2(2\ \Omega) = \mathbf{8\ W}$$
$$P_2 = I^2R_2 = (2\text{ A})^2(4\ \Omega) = \mathbf{16\ W}$$
$$P_3 = I^2R_3 = (2\text{ A})^2(6\ \Omega) = \mathbf{24\ W}$$

Note that $P_T = P_1 + P_2 + P_3$.

Kirchhoff's Voltage Law

Kirchhoff's voltage law states:

The algebraic sum of the voltage rises and drops around a closed path must equal zero.

When proceeding around a closed path, a change in potential from $-$ to $+$ for any element is considered positive, and a change from $+$ to $-$ is considered negative. Do not be concerned about whether the rise or drop in potential is across a load or source—it is the change in polarity that defines whether the positive or negative sign is applied. In addition, be sure that you return to the starting point before setting the resulting series of terms equal to zero.

In Fig. 2.32 if we leave point d in a clockwise direction, we obtain a rise $(+)$ in potential due to the battery since we progress from the negative $(-)$ to the positive $(+)$ terminal. For each resistive load, however, we proceed from $+$ to $-$ and a negative sign is applied to V_1, V_2, and V_3. The result is

$$+E - V_1 - V_2 - V_3 = 0$$

or

$$E = V_1 + V_2 + V_3$$

which clearly reveals that the applied voltage of a series circuit equals the sum of the voltage drops across the series elements.

Voltage-Divider Rule

It is interesting to note in Fig. 2.32 that the voltage across a series resistive element is a direct function of its magnitude compared with the other series elements. For example, since $R_3 = 3R_1$, the voltage V_3 is three times V_1, and so on.

The *voltage-divider rule* allows us to calculate the voltage across one or a combination of series resistors without first having to solve for the current. Its basic format is

$$V_x = \frac{R_x E}{R_T} \quad \text{(volts)} \tag{2.19}$$

where V_x is the voltage across a single resistor R_x or a combination of series resistors having a total resistance R_x. E is applied voltage across the series circuit, and R_T is the total resistance of the series circuit.

For Fig. 2.32,

$$V_1 = \frac{R_1(E)}{R_1 + R_2 + R_3} = \frac{(2\ \Omega)(24\ \text{V})}{12\ \Omega} = \mathbf{4\ V}$$

$$V_3 = \frac{R_3(E)}{R_T} = \frac{(6\ \Omega)(24\ \text{V})}{12\ \Omega} = \mathbf{12\ V}$$

Voltage Sources in Series

For voltage sources in series, those batteries pressuring current in one direction can be added together and the total determined by the algebraic sum of the resultant of each polarity. Consider the example in Fig. 2.33.

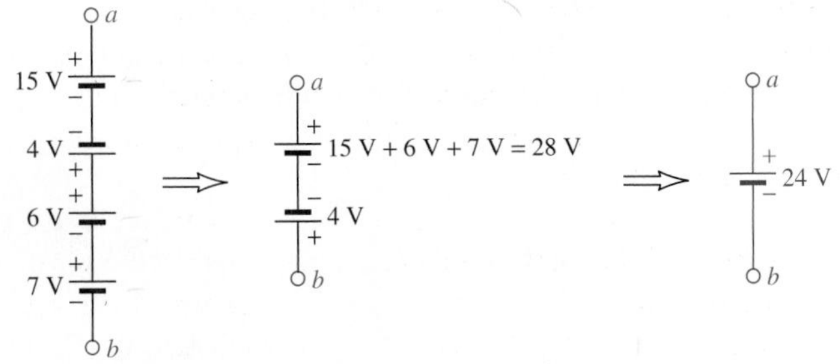

FIG. 2.33
Algebraic summation of series voltage sources.

Equal Series Resistors

For N *equal* resistors in series the total resistance is determined by

$$R_T = NR \tag{2.20}$$

where R is the value of one of the equal series resistors.

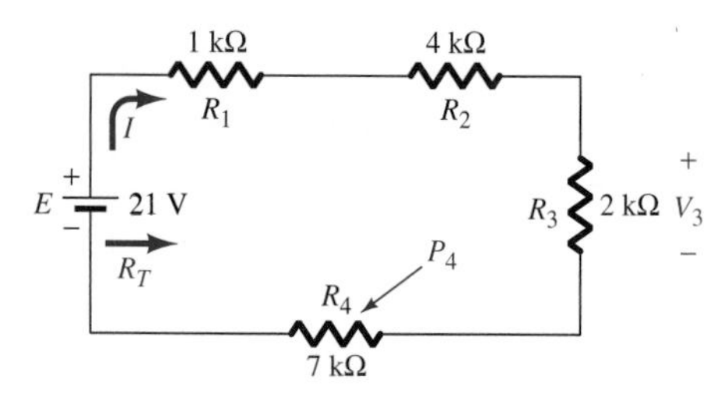

FIG. 2.34

EXAMPLE 2.23 Determine the following for the series circuit in Fig. 2.34.

a. R_T.
b. I.
c. V_3.
d. P_4.
e. V_3 using the voltage-divider rule.

Solution:

a. $R_T = R_1 + R_2 + R_3 + R_4$
$= 1\text{ k}\Omega + 4\text{ k}\Omega + 2\text{ k}\Omega + 7\text{ k}\Omega$
$= \mathbf{14\text{ k}\Omega}$

b. $I = \dfrac{E}{R_T} = \dfrac{21\text{ V}}{14\text{ k}\Omega} = \mathbf{1.5\text{ mA}}$

c. $V_3 = IR_3 = (1.5\text{ mA})(2\text{ k}\Omega) = \mathbf{3\text{ V}}$

d. $P_4 = I^2R_4 = (1.5\text{ mA})^2(7\text{ k}\Omega) = (1.5 \times 10^{-3})^2 7 \times 10^3$
$= (2.25 \times 10^{-6})(7 \times 10^3)$
$= 15.75 \times 10^{-3}\text{ W} = \mathbf{15.75\text{ mW}}$

e. $V_3 = \dfrac{R_3E}{R_T} = \dfrac{(2\text{ k}\Omega)(21\text{ V})}{14\text{ k}\Omega} = \mathbf{3\text{ V},}$ as above

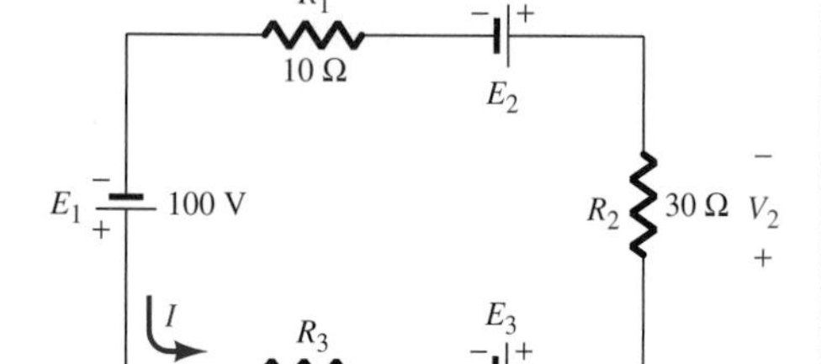

FIG. 2.35

EXAMPLE 2.24 Determine I and V_2 for the network in Fig. 2.35.

Solution: E_2 is pressuring current in the clockwise direction, while E_1 and E_3 are trying to establish I with the direction shown in Fig. 2.35. The result is a net voltage

$$E_T = E_1 + E_3 - E_2$$
$$= 100\text{ V} + 20\text{ V} - 50\text{ V}$$
$$= 70\text{ V}$$

in the counterclockwise direction. The resulting current

$$I = \frac{E_T}{R_T} = \frac{E_T}{R_1 + R_2 + R_3} = \frac{70\text{ V}}{80\ \Omega}$$
$$= \mathbf{0.875\text{ A}}$$

with

$$V_2 = IR_2 = (0.875\text{ A})(30\ \Omega) = \mathbf{26.25\text{ V}}$$

EXAMPLE 2.25 Find V_1 and V_2 of Fig. 2.36 using Kirchhoff's voltage law.

Solution: Note that there is no indication that a source or load appears within each container. For the loop *dacd*,

$$+22\text{ V} - V_1 = 0$$

and

$$V_1 = \mathbf{22\ V}$$

For the loop *cabc*,

$$V_1 - V_2 - 5\text{V} = 0$$

or

$$V_2 = V_1 - 5\text{ V}$$
$$= 22\text{ V} - 5\text{ V} = \mathbf{17\ V}$$

In addition, note that the circuit is not a simple series circuit as introduced in this section, revealing that the law can be applied to any configuration.

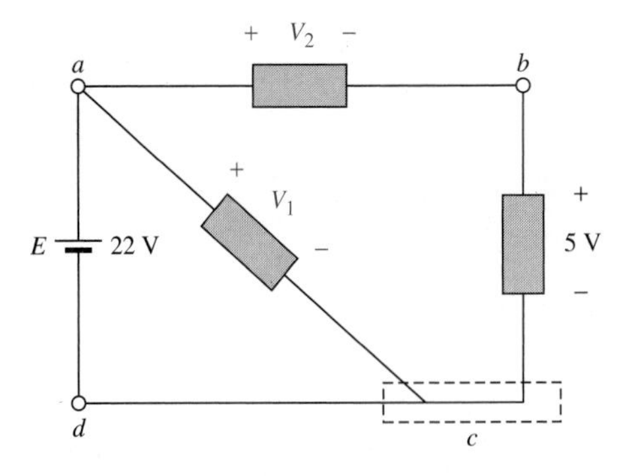

FIG. 2.36

2.9 APPLICATION: FLASHLIGHT

Although the flashlight employs one of the simplest of electrical circuits, a few fundamentals about its operation do carry over to more sophisticated systems. First, and quite obviously, it is a dc system with a lifetime totally dependent on the state of the batteries and bulb. Unless the flashlight is the rechargeable type, each time you use it, you take some of the life out of it. For many hours the brightness will not diminish noticeably. Then, however, as it reaches the end of its ampere-hour capacity, the light will become dimmer at an increasingly rapid rate (almost exponentially). For the standard two-battery flashlight appearing in Fig. 2.37(a) with its electrical schematic in Fig. 2.37(b), each 1.5-V battery has an ampere-hour rating of about 16 Ah. The single-contact miniature flange-base bulb is rated at 2.5 V and 300 mA with good brightness and a lifetime of about 30 hours. Thirty hours may not seem like a long lifetime, but you have to consider how long you usually use a flashlight on each occasion. If we assume a 300-mA drain from the battery for the bulb when in use, the lifetime of the battery is about 53 hours. Comparing the 53-hour lifetime of the battery with the 30-hour life expectancy of the bulb suggests that we normally have to replace bulbs more frequently than batteries.

However, most of us have experienced the opposite effect: We can change batteries two or three times before we need to replace the bulb. This is simply one example of the act that one cannot be guided solely by the specifications of each component of an electrical design. The operating conditions, terminal characteristics, and details about the actual

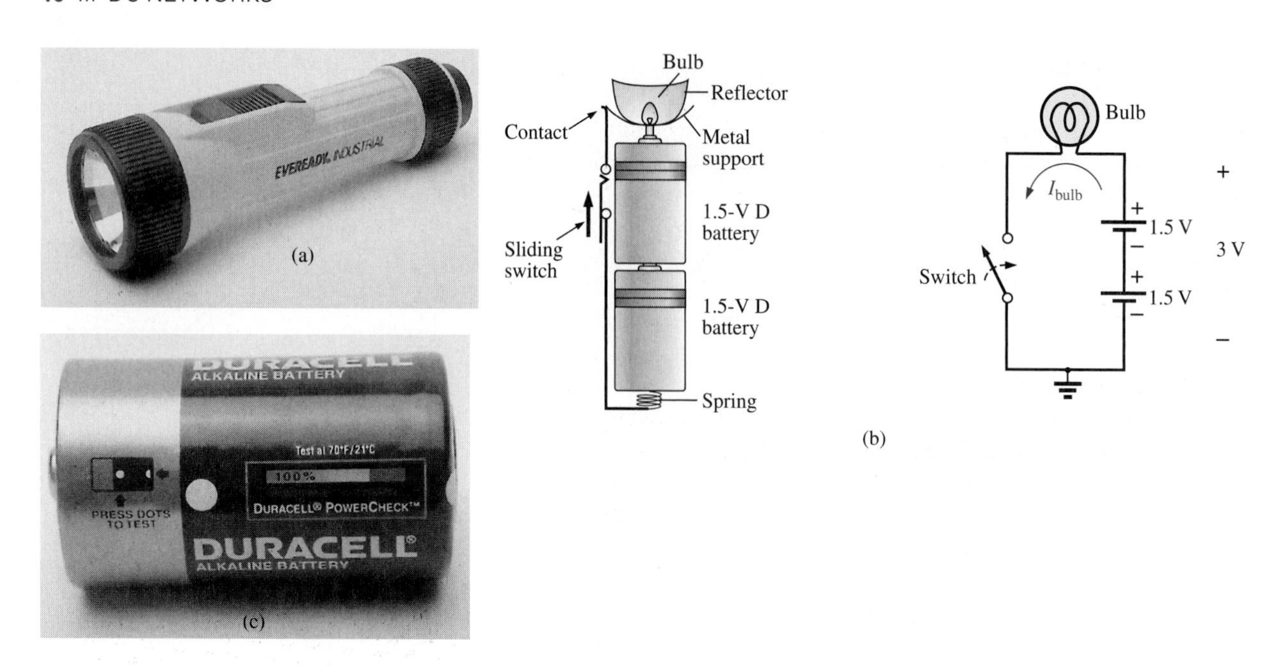

FIG. 2.37
(a) Eveready® D cell flashlight; (b) electrical schematic of flashlight of part (a); (c) Duracell® Powercheck™ D cell battery.

response of the system for short and long periods of time must be taken into account. As mentioned earlier, the battery loses some of its power each time it is used. Although the terminal voltage may not change much at first, its ability to provide the same level of current will drop with each usage. Further, batteries will slowly discharge due to "leakage currents" even if the switch is not on. The air surrounding the battery is "not clean" in the sense that moisture and other elements in the air can provide a conduction path for leakage currents through the air, through the surface of the battery itself, or through other nearby surfaces, and the battery will eventually discharge. How often have we left a flashlight with new batteries in a car for a long period of time only to find the light very dim or the batteries dead when we need the flashlight the most? An additional problem is acid leaks that appear as brown stains or corrosion on the casing of the battery. These leaks will also affect the life of the battery. Further, when the flashlight is turned on, there is an initial surge in current that will drain the battery more than continuous use for a period of time. In other words, continually turning the flashlight on and off will have a very detrimental effect on its life. We must also realize that the 30-hour rating of the bulb is for continuous use, that is 300 mA flowing through the bulb for a continuous 30 hours. Certainly, the filament in the bulb and the bulb itself will get hotter with time, and this heat has a detrimental effect on the filament wire. When the flashlight is turned on and off, it gives the bulb a chance to cool down and regain its normal characteristics, thereby avoiding any real damage. Therefore, with normal use we can expect the bulb to last longer than the 30 hours specified for continuous use.

Even though the bulb is rated for 2.5-V operation, it would appear that the two batteries would result in an applied voltage of 3 V, which suggests poor operating conditions. However, a bulb rated at 2.5 V can easily handle 2.5 V to 3 V. In addition, as was pointed out in this chapter, the terminal voltage will drop with the current demand and usage. Under normal operating conditions, a 1.5-V battery is considered to be in good condition if the loaded terminal voltage is 1.3 V to 1.5 V. When it drops to between 1 V and 1.1 V, it is weak, and when it drops to between 0.8 V and 0.9 V, it has lost its effectiveness. The levels can be related directly to the test band now appearing on Duracell® batteries, such as on the one shown in Fig. 2.37(c). In the test band on this battery, the upper voltage area (green on the actual battery) is near 1.5 V (labeled 100%); the lighter area to the right, is from about 1.3 V down to 1 V; and the replace area (red) on the far right, is below 1V.

Be aware that the total supplied voltage of 3 V will be obtained only if the batteries are connected as shown in Fig. 2.37(b). Accidentally placing the two positive terminals together will result in a total voltage of 0 V, and the bulb will not light at all. *For the vast majority of systems with more than one battery, the positive terminal of one battery will always be connected to the negative terminal of another. For all low-voltage batteries, the end with the nipple is the positive terminal, and the end with the flat end is the negative terminal. In addition, the flat or negative end of a battery is always connected to the battery casing with the helical coil to keep the batteries in place. The positive end of the battery is always connected to a flat spring connection or the element to be operated.* If you look carefully at the bulb, you will find that the nipple connected to the positive end of the battery is insulated from the jacket around the base of the bulb. The jacket is the second terminal of the battery used to complete the circuit through the on/off switch.

If a flashlight fails to operate properly, the first thin to check is the state of the batteries. It is best to replace both batteries at once. A system with one good battery and one nearing the end of its life will result in pressure on the good battery to supply the current demand, and, in fact, the bad battery will actually be a drain on the good battery. Next, check the condition of the bulb by checking the filament to see whether it has opened at some point because a long-term, continuous current level occurred or because the flashlight was dropped. If the battery and bulb seem to be in good shape, the next area of concern is the contacts between the positive terminal and the bulb and the switch. Cleaning both with emery cloth will often eliminate this problem.

2.10 PARALLEL DC NETWORKS

Each law or rule provided in the previous section for series circuits has its counterpart in parallel networks. Once the analysis of each (series and parallel) type is clearly understood, an initial reading of the most complicated system can be undertaken. It is therefore quite important that you carefully examine Sections 2.8 and 2.10 and complete a majority of the exercises at the end of the chapter.

Two elements are in parallel if they have two terminals in common.

In Fig. 2.38, R_1 and R_2 are in parallel since both ends are connected. The same is true for R_2 and R_3, R_1 and R_2, E and R_1, and so on. In fact, all the elements in Fig. 2.38 are in parallel. Figure 2.38 is therefore referred to as a *parallel network.*

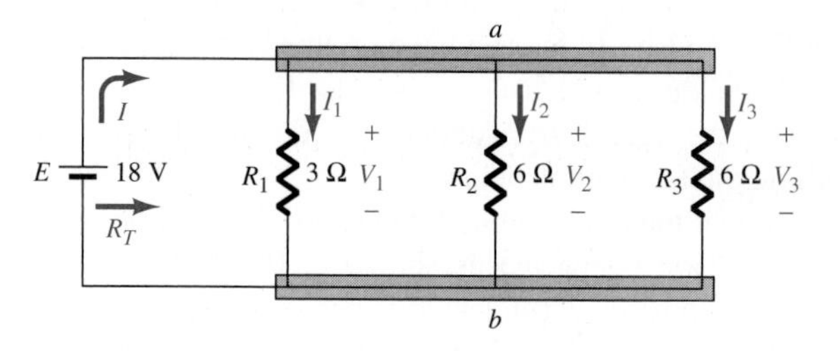

FIG. 2.38

Total Resistance

For parallel resistors the total resistance is determined by

$$\frac{1}{R_T} = \frac{1}{R_1} + \frac{1}{R_2} + \frac{1}{R_3} + \cdots + \frac{1}{R_N} \tag{2.21}$$

Note that the equation is not for R_T but for $1/R_T$, requiring that the inverse of the sum be determined. The inverse of resistance is referred to as *conductance,* since it provides a measure of the conductivity of the element. The less the resistance, the more the conductivity, and vice versa. The conductance is symbolized by G and is measured in siemens (S). In equation form,

$$G = \frac{1}{R} \quad \text{(siemens)} \tag{2.22}$$

Substituting in Eq. (2.21), we have

$$G_T = G_1 + G_2 + G_3 + \cdots + G_N \quad \text{(siemens)} \tag{2.23}$$

For the network in Fig. 2.38,

$$\frac{1}{R_T} = \frac{1}{3\,\Omega} + \frac{1}{6\,\Omega} + \frac{1}{6\,\Omega} = 0.333\text{ S} + 0.166\text{ S} + 0.166\text{ S}$$
$$= 0.666\text{ S}$$

and

$$R_T = \frac{1}{0.666\text{ S}} = \mathbf{1.5\ \Omega}$$

It is particularly important to realize that

The total resistance of parallel resistors is always less than the value of the smallest resistor.

In the preceding example, note that $R_T = 1.5\ \Omega$ is less than $R_1 = 3\ \Omega$.

Voltage

In a series circuit the current is the same through each element. For parallel networks,

The voltage across parallel elements is always the same.

For Fig. 2.38 this statement means that

$$\boxed{V_1 = V_2 = V_3 = E} \qquad \textbf{(2.24)}$$

which for Fig. 2.38 results in

$$V_1 = V_2 = V_3 = \mathbf{18\ V}$$

Currents

The source current is determined from

$$\boxed{I = \frac{E}{R_T}} \qquad \textbf{(2.25)}$$

which has the same format as applied to series circuits. For Fig. 2.38,

$$I = \frac{E}{R_T} = \frac{18\text{ V}}{1.5\ \Omega} = \mathbf{12\ A}$$

Since $V_1 = V_2 = V_3 = E$ volts,

$$I_1 = \frac{V_1}{R_1} = \frac{E}{R_1} = \frac{18\text{ V}}{3\ \Omega} = \mathbf{6\ A}$$

$$I_2 = \frac{V_2}{R_2} = \frac{E}{R_2} = \frac{18\text{ V}}{6\ \Omega} = \mathbf{3\ A}$$

$$I_3 = \frac{V_3}{R_3} = \frac{E}{R_3} = \frac{18\text{ V}}{6\ \Omega} = \mathbf{3\ A}$$

Note here that the current through equal parallel elements is the same and that current seeks the path of least resistance, as demonstrated by the fact that $I_1 > I_2 = I_3$.

Power

The power delivered by the source is

$$P_T = EI = (18\text{ V})(12\text{ A}) = \mathbf{216\ W}$$

and the power delivered to each element is

$$P_1 = \frac{V_1^2}{R_1} = \frac{E_2}{R_1} = \frac{(18\text{ V})^2}{3\Omega} = \mathbf{108\ W}$$

$$P_2 = \frac{V_2^2}{R_2} = \frac{E^2}{R_2} = \frac{(18\text{ V})^2}{6\Omega} = \mathbf{54\ W}$$

$$P_3 = \frac{V_3^2}{R_3} = \frac{E^2}{R_3} = \frac{(18\text{ V})^2}{6\ \Omega} = \mathbf{54\ W}$$

Note that $P_T = P_1 + P_2 + P_3$.

Kirchhoff's Current Law

Kirchhoff's current law states that

The sum of the currents entering a junction must equal the sum of the currents leaving.

$$\Sigma I_{\text{entering}} = \Sigma I_{\text{leaving}} \qquad \textbf{(2.26)}$$

For Fig. 2.38, note that the current I enters terminal a and the currents I_1, I_2, and I_3 leave the same junction. Therefore,

$$I = I_1 + I_2 + I_3$$
$$12\text{ A} = 6\text{ A} + 3\text{ A} + 3\text{ A}$$
$$12\text{ A} = 12\text{ A} \qquad \text{(checks)}$$

Special Case: Two Parallel Resistors

For the special case (and a frequently encountered one) of two parallel resistors as shown in Fig. 2.39, Eq. (2.21) can be employed to derive the following expression for the total resistance:

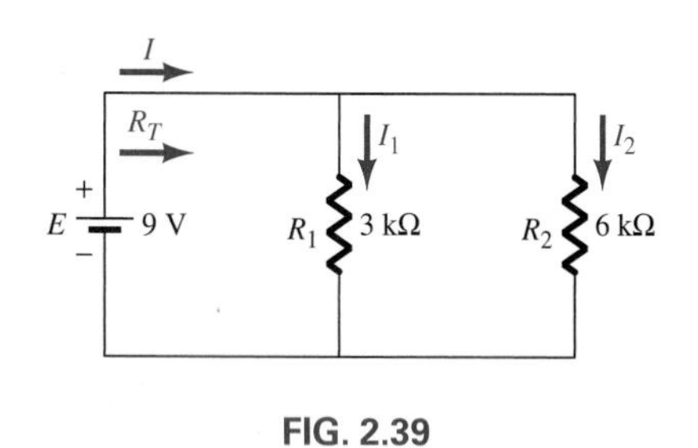

FIG. 2.39

$$R_T = \frac{R_1R_2}{R_1 + R_2} \qquad \textbf{(2.27)}$$

revealing that the total resistance of two parallel resistors is simply the product of the two resistors divided by their sum. For Fig. 2.39,

$$R_T = \frac{(3\text{ k}\Omega)(6\text{ k}\Omega)}{3\text{ k}\Omega + 6\text{ k}\Omega} = \mathbf{2\ k\Omega}$$

Current-Divider Rule

For the two parallel resistors in Fig. 2.39, the currents I_1 and I_2 can be determined from the source current I using the current-divider rule, which has the format

$$I_1 = \frac{R_2 I}{R_1 + R_2} \tag{2.28a}$$

and

$$I_2 = \frac{R_1 I}{R_1 + R_2} \tag{2.28b}$$

In words, Eq. (2.28) states that the current through one of two parallel branches is the product of the other resistor and the total entering current divided by the sum of the two parallel resistors. For Fig. 2.39, where

$$I = \frac{E}{R_T} = \frac{9\text{ V}}{2\text{ k}\Omega} = \mathbf{4.5\text{ mA}}$$

$$I_1 = \frac{R_2 I}{R_1 + R_2} = \frac{(6\text{ k}\Omega)(4.5\text{ mA})}{3\text{ k}\Omega + 6\text{ k}\Omega} = \mathbf{3\text{ mA}}$$

$$I_2 = \frac{R_1 I}{R_1 + R_{2+}} = \frac{(3\text{ k}\Omega)(4.5\text{ mA})}{9\text{ k}\Omega} = \mathbf{1.5\text{ mA}}$$

as verified by

$$I_1 = \frac{V_1}{R_1} = \frac{E}{R_1} = \frac{9\text{ V}}{3\text{ k}\Omega} = \mathbf{3\text{ mA}}$$

$$I_2 = \frac{V_2}{R_2} = \frac{E}{R_2} = \frac{9\text{ V}}{6\text{ k}\Omega} = \mathbf{1.5\text{ mA}}$$

Equal Parallel Resistors

For N equal parallel resistors the total resistance is determined by

$$R_T = \frac{R}{N} \tag{2.29}$$

where R is the magnitude of one of the equal parallel resistors.

EXAMPLE 2.26 Determine the following for the parallel network in Fig. 2.40.

a. R_T.
b. I.
c. I_2.
d. P_3.

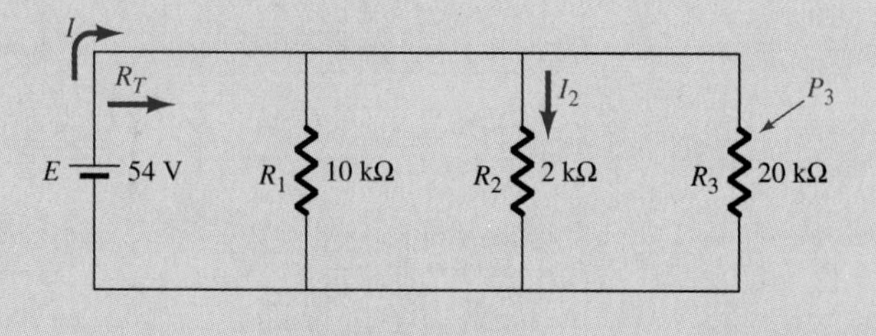

FIG. 2.40

Solution:

a. $$\frac{1}{R_T} = \frac{1}{R_1} + \frac{1}{R_2} + \frac{1}{R_3} = \frac{1}{10\text{ k}\Omega} + \frac{1}{2\text{ k}\Omega} + \frac{1}{20\text{ k}\Omega}$$
$$= 0.1 \times 10^{-3}\text{ S} + 0.5 \times 10^{-3}\text{ S} + 0.05 \times 10^{-3}\text{ S}$$
$$= 0.65 \times 10^{-3}\text{ S}$$

and

$$R_T = \frac{1}{0.65 \times 10^{-3}\text{ S}} = \frac{1}{0.65\text{ S}} \times 10^3 = \mathbf{1.538\text{ k}\Omega}$$

b. $$I = \frac{E}{R_T} = \frac{54\text{ V}}{1.538\text{ k}\Omega} = \mathbf{35.11\text{ mA}}$$

c. $$I_2 = \frac{V_2}{R_2} = \frac{E}{R_2} = \frac{54\text{ V}}{2\text{ k}\Omega} = \mathbf{27\text{ mA}}$$

d. $$P_3 = \frac{V_3^2}{R_3} = \frac{E^2}{R_3} = \frac{(54\text{ V})^2}{20\text{ k}\Omega} = \mathbf{145.8\text{ mW}}$$

The total resistance for part (a) can also be found by considering two parallel elements at a time:

$$R' = \frac{R_1 R_2}{R_1 + R_2} = \frac{(10\text{ k}\Omega)(2\text{ k}\Omega)}{10\text{ k}\Omega + 2\text{ k}\Omega} = 1.667\text{ k}\Omega$$

and

$$R_T = R' \| R_3 = \frac{R' R_3}{R' + R_3}$$
$$= \frac{(1.667\text{ k}\Omega)(20\text{ k}\Omega)}{1.667\text{ k}\Omega + 20\text{ k}\Omega}$$
$$= \mathbf{1.538\text{ k}\Omega}\text{, as obtained above}$$

In addition, note that the total resistance is less than the smallest resistance of 2 kΩ, as stated earlier.

EXAMPLE 2.27
a. Determine the current I_1 in Fig. 2.41 using the current-divider rule.
b. Determine I_2 using Kirchhoff's current law.

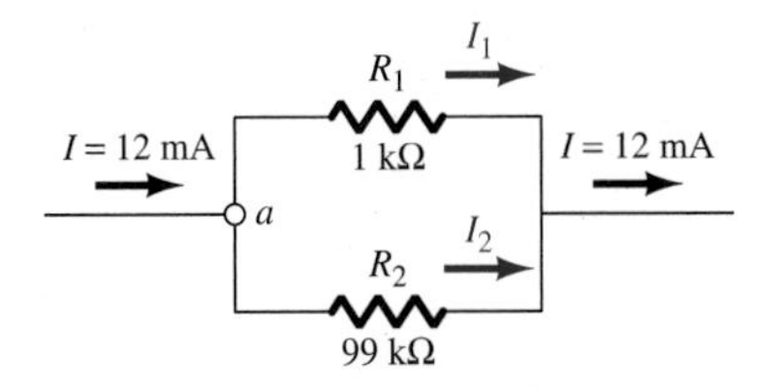

FIG. 2.41

Solution:

a. $$I_1 = \frac{R_2 I}{R_1 + R_2} = \frac{(99\,\text{k}\Omega)(12\,\text{mA})}{1\,\text{k}\Omega + 99\,\text{k}\Omega} = \frac{(99\,\text{k}\Omega)(12\,\text{mA})}{100\,\text{k}\Omega} = \mathbf{11.88\ mA}$$

b. At terminal a,

$$\Sigma I_{\text{entering}} = \Sigma I_{\text{leaving}}$$

and

$$I = I_1 + I_2$$

or

$$I_2 = I - I_1 = 12\text{ mA} - 11.88\text{ mA} = \mathbf{0.12\ mA}$$

Note in this example how the vast majority of the current takes the path of least resistance. In fact, the current divides as the ratio of resistance levels. That is, 99 kΩ : 1 kΩ = 99 : 1 and 11.88 mA : 0.12 mA = 99 : 1 also.

EXAMPLE 2.28 Using Kirchhoff's current law, determine the currents I_3 and I_6 for the system of Fig. 2.42.

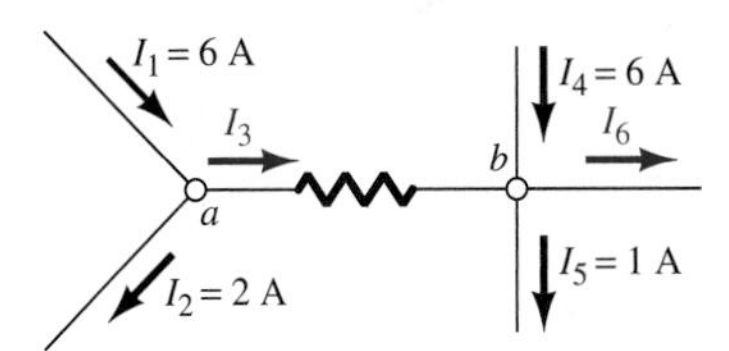

FIG. 2.42

Solution: The law must first be applied at terminal a, since there is only *one* unknown current entering or leaving this terminal. For terminal b, both I_3 and I_6 are unknown quantities. At terminal a,

$$\Sigma I_{\text{entering}} = \Sigma I_{\text{leaving}}$$
$$I_1 = I_2 + I_3$$
$$6\text{ A} = 2\text{ A} + I_3$$
$$I_3 = 6\text{ A} - 2\text{ A} = \mathbf{4\ A}$$

At terminal b,

$$\Sigma I_{\text{entering}} = \Sigma I_{\text{leaving}}$$
$$I_3 + I_4 = I_5 + I_6$$
$$4\text{ A} + 6\text{ A} = 1\text{ A} + I_6$$
$$I_6 = 10\text{ A} - 1\text{ A} = \mathbf{9\ A}$$

For home and industrial applications, all outlets, lighting, machinery, and so on, are connected in parallel. In the home every outlet has a terminal voltage of 120 V. Although the voltage is what we refer to as an ac voltage, and not dc as being considered here, it permits an examination of the effects of parallel elements. As additional appliances are connected to the same circuit the current drawn through the circuit breaker increases as determined by Kirchhoff's current law, although each appliance still receives the necessary 120-V terminal voltage for proper operation. One obvious advantage of the parallel connection of loads is that if one should fail to operate, the others will still operate properly, since the terminal voltage is still available. However, in a series connection of elements, if one should fail, the remaining appliances will cease to operate, since the current path has been broken. Consider also that with a series configuration of lighting fixtures, the more connected in the series, the greater the terminal resistance, the less the current, and the dimmer the lights. The latter condition does not occur for parallel elements, since each fixture continues to receive 120 V and to have the same brightness until the attached loads require a total current above that of the circuit breaker. Consider trying to determine which bulb in a series connection of bulbs (such as sometimes encountered with holiday tree lights) is bad if they are all out. This situation presents a definite problem not encountered with the parallel connection, since only the bad bulb fails to light.

2.11 APPLICATIONS: CAR SYSTEM AND PARALLEL COMPUTER BUS CONNECTIONS

Car System

First, it must be understood that the entire electrical system of a car is run as a *dc system*. Although the generator produces a varying ac signal, rectification converts it to one having an average dc level for charging the battery. In particular, note the filter capacitor (Chapter 3) in the alternator branch of Fig. 2.43 to smooth out the rectified ac waveform and to provide an improved dc supply. The charged battery must therefore provide the required direct current for the entire electrical system of the car.

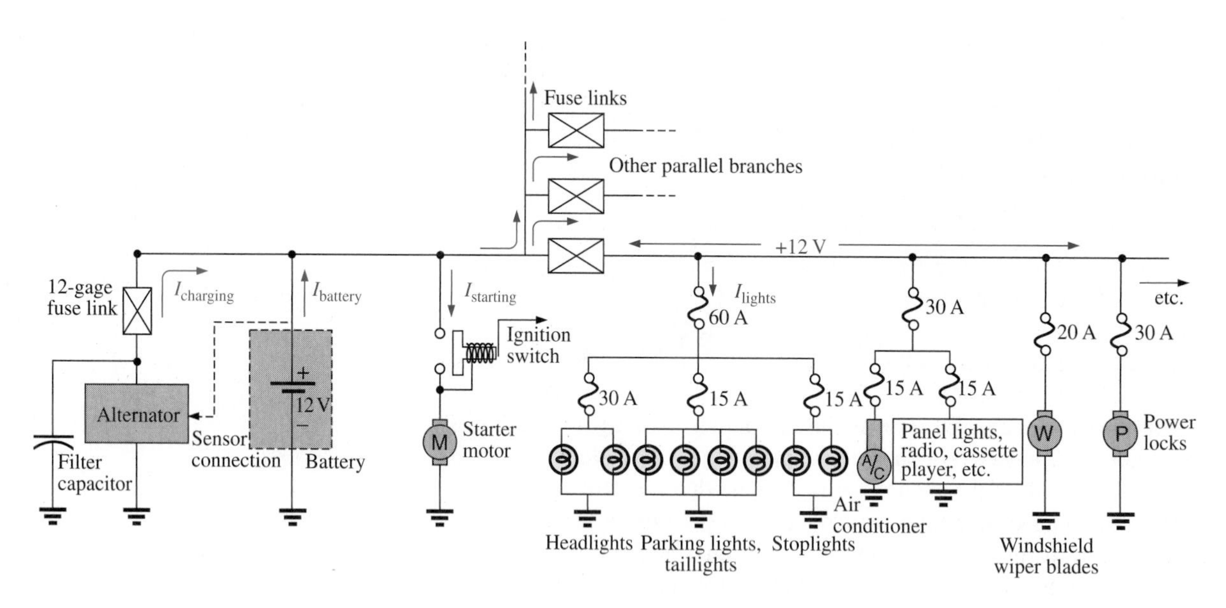

FIG. 2.43
Expanded view of an automobile's electrical system.

Thus, the power demand on the battery at any instant is the product of the terminal voltage and the current drain of the total load of every operating system of the car. This certainly places an enormous burden on the battery and its internal chemical reaction and warrants all the battery care we can provide.

Since the electrical system of a car is essentially a parallel system, the total current drain on the battery is the sum of the currents to all the parallel branches of the car connected directly to the battery. In Fig. 2.43 a few branches of the wiring diagram of a car have been sketched to provide some background information on basic wiring, current levels, and fuse configurations. Every automobile has fuse links and fuses, and some also have circuit breakers, to protect the various components of the car and to ensure that a dangerous fire situation does not develop. Expect for a few branches that may have series elements, the operating voltage for most components of a car is the terminal voltage of the battery, which we will designate as 12 V even though it typically varies between 12 V and the charging level of 14.6 V. In other words, each component is connected to the battery at one end and to the ground or chassis of the car at the other end.

Referring to Fig. 2.43, we find that the alternator or charging branch of the system is connected directly across the battery to provide the charging current as indicated. Once the car is started, the rotor of the alternator turns, generating an ac varying voltage that then passes through a rectifier network and filter to provide the dc charging voltage for the battery. Charging occurs only when the sensor, connected directly to the battery, signals that the terminal voltage of the battery is too low. Just to the right of the battery the starter branch was included to demonstrate that there is no fusing action between the battery and starter when the ignition switch is activated. The lack of fusing action is provided because enormous starting currents (hundreds of amperes) will flow through the starter to start a car that may not have been used for days and/or that may have been sitting in a cold climate—and high friction occurs between components until the oil starts flowing. The starting level can vary so much that it would be difficult to find the right fuse level, and frequent high currents might damage the fuse link and cause a failure at expected levels of current. When the ignition switch is activated, the starting relay completes the circuit between the battery and starter, and normally the car will start. If a car should fail to start, the first point of attack should be to check the connections at the battery, starting relay, and starter to be sure that they are not providing an unexpected open circuit due to vibration, corrosion, or moisture.

Once the car has started, the starting relay opens and the battery can turn its attention to the operating components of the car. Although the diagram of Fig. 2.43 does not display the switching mechanism, the entire electrical network of the car, except for the important external lights, is usually disengaged so that the full strength of the battery can be dedicated to the starting process. The lights are included for situations in which turning the lights off, even for short periods of time, could create a dangerous situation. If the car is in a safe environment, it is best to leave the lights off at starting to save the battery an additional 30 A of drain. If the lights are on at starting, a dimming of the lights can be expected due to the starter drain, which may exceed 500 A. Today, batteries are typically rated in cranking (starting) current rather than ampere-hours. Batteries rated with a cold cranking ampere rating between 700 A and 1000 A are typical today.

Separating the alternator from the battery and the battery from the numerous networks of the car are fuse links such as shown in Fig. 2.44(a). They are actually wires of specific gage designed to open at fairly high current levels of 100 A or more. They are included to protect against those situations in which there is an unexpected current drawn from the many circuits to which it is connected. That heavy drain can, of course, be from a short circuit in one of the branches, but in such cases the fuse in that branch will probably release. The fuse link is an additional protection for the line if the total current drawn by the parallel-connected branches begins to exceed safe levels. The fuses following the fuse link have the appearance shown in Fig. 2.44(b), where a gap between the legs of the fuse indicates a blown fuse. As shown in Fig. 2.43, the 60-A fuse (often called a *power distribution fuse*) for the lights is a second-tier fuse sensitive to the total drain from the three light circuits. Finally, the third fuse level is for the individual units of a car such as the lights, air conditioner, and power locks. In each case, the fuse rating exceeds the normal load (current level) of the operating component, but the level of each fuse does give some indication of the demand to be expected under normal operating conditions. For instance, the headlights will typically draw more than 10 A, the taillights more than 5 A, the air conditioner about 10 A (when the clutch engages), and the power windows 10 A to 20 A depending on how many are operated at once.

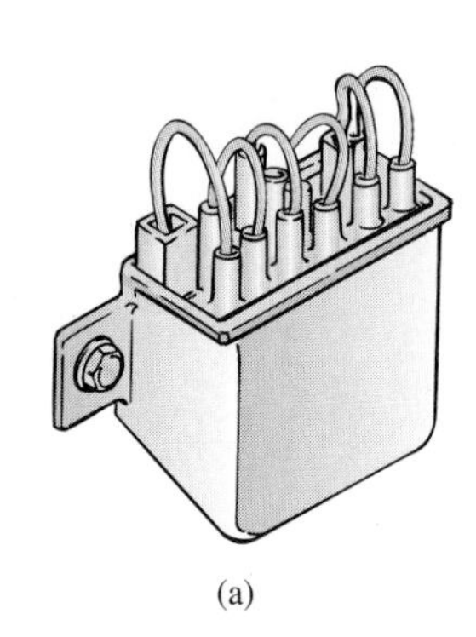

(a)

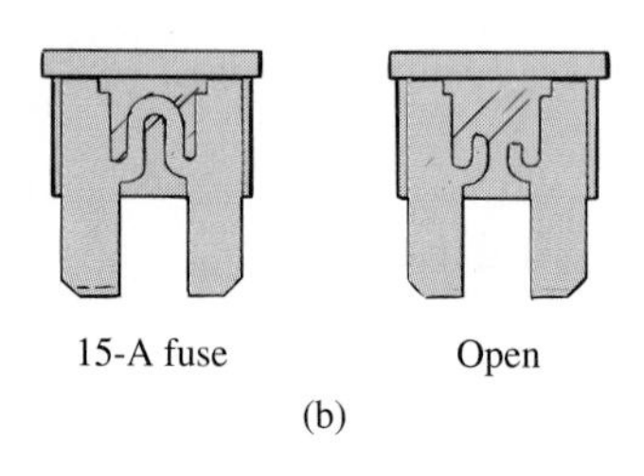

(b)

FIG. 2.44
Car fuses: (a) fuse link; (b) plug-in.

Some details for only one section of the total car network are provided in Fig. 2.43. In the same figure, additional parallel paths with their respective fuses have been provided to further reveal the parallel arrangement of all the circuits.

In all vehicles made in the United States and in some vehicles made in European countries, the return path to the battery through the ground connection is actually through the chassis of the car. That is, there is only one wire to each electrical load, with the other end simply grounded to the chassis. The return to the battery (chassis to negative terminal) is therefore a heavy-gage wire matching that connected to the positive terminal. In some European cars constructed of a mixture of materials such as metal, plastic, and rubber, the return path through the metallic chassis is lost, and two wires must be connected to each electrical load of the car.

Parallel Computer Bus Connections

The internal construction (hardware) of large mainframe computers and full-size desk models is set up to accept a variety of adapter cards such as shown in Fig. 2.45. The primary board (usually the largest), commonly called the *motherboard,* contains most of the functions required for full computer operation. Adapter cards are normally added to expand the memory, set up a network, add peripheral equipment, and so on. For

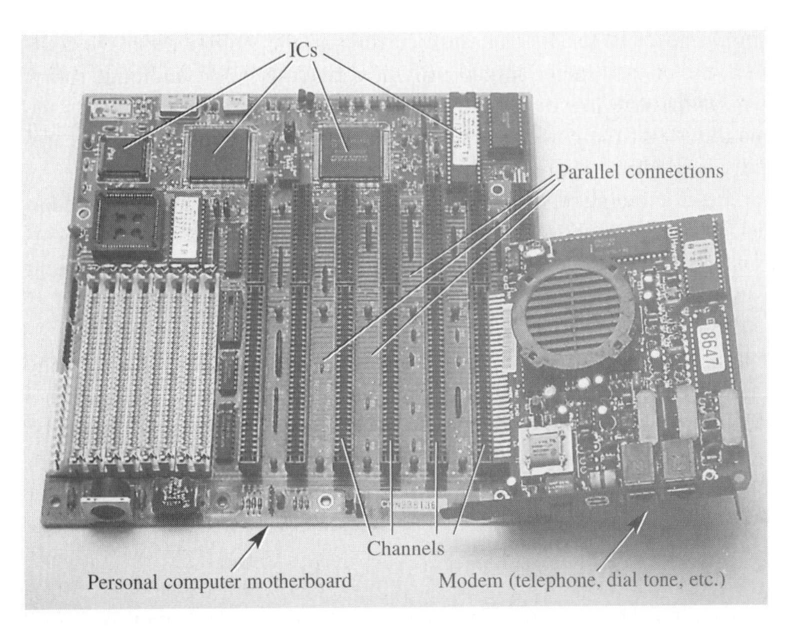

FIG. 2.45
Inserting a modem into the motherboard of a computer.

instance, if you decide to add a modem to your computer, you can simply insert the modem card into the proper channel as shown in Fig. 2.45. The bus connectors are connected in parallel with common connections to the power supply, address and data buses, control signals, ground, and so on. For instance, if the bottom connection of each bus connector is a ground connection, that ground connection carries through each bus connector and is immediately connected to any adapter card installed. Each card has a slot connector that will fit directly into the bus connector without the need for any soldering or construction. The pins for the adapter card are then designed to provide a path between the motherboard and its components to support the desired function. Since all the bus connectors are connected in parallel, an adapter card can be placed in any channel and have the same effect. In some cases, however, specific bus connectors are chosen because of other connections that must be made to the external framework or internal network.

Most small laptop computers today have all the options already installed, thereby bypassing the need for bus connectors. Additional memory and other upgrades are added as direct inserts into the motherboard.

2.12 METER CONSIDERATIONS

In the pertinent section of this chapter the ammeter, voltmeter, and ohmmeter were introduced. We must now examine those considerations that determine whether the proper meter is being used for the quantity to be measured.

For ammeters, it is usually sufficient to choose a meter that can handle the expected current level. Certainly, a milliammeter or microammeter should never be used to measure currents in the ampere range. In addition, the chosen meter should provide a reading in the midrange rather than in the very low or very high region of the scale. If in doubt as to the magnitude of the current, simply start with the highest scale and carefully work down to prevent damage through the *pinning* of the meter. An ammeter's internal resistance is usually so small that it can be ignored in comparison with those of the other series elements. *Remember:* The ammeter is *always* connected in series with the branch in which the current is to be determined.

The voltmeter presents a totally different situation, however, since its internal resistance can affect the network to which it is applied. On the face of every analog voltmeter is an ohm-per-volt rating. To determine the internal resistance of the meter the ohm-per-volt rating can simply be multiplied by the largest possible voltage reading of a particular scale. For example, if the 50-V scale of a meter with an ohm-per-volt rating of 1000 Ω is to be used, the internal resistance is simply

$$50(1000\ \Omega) = 50\ \text{k}\Omega$$

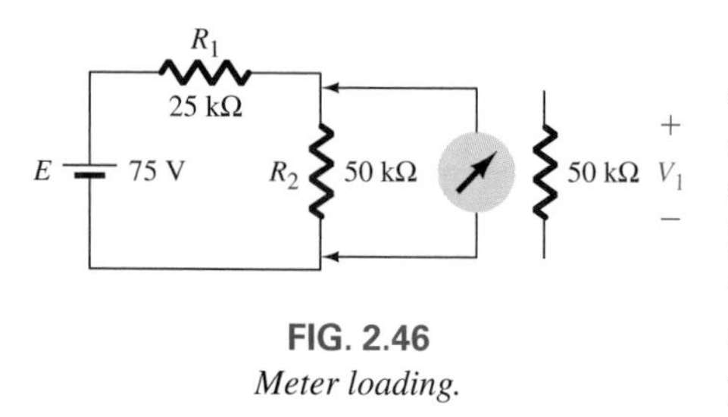

FIG. 2.46
Meter loading.

No matter what reading is obtained on that scale (whether it be 2 or 48 V) the internal resistance will be 50 kΩ. It is determined solely by the ohm-per-volt rating and the maximum reading of the scale. Why this concern? Consider the network in Fig. 2.46 in which the voltage V_1 is to be measured using the meter just described. The internal resistance of the voltmeter combines with the 50 kΩ of the network, resulting in an equivalent resistance of 25 kΩ. (Recall that voltmeters are always placed in parallel with the element across which the voltage is to be determined.)

Therefore,

$$V_1 = \frac{(25\ \text{k}\Omega)(75\ \text{V})}{25\ \text{k}\Omega\ + 25\ \text{k}\Omega} = 37.5\ \text{V}$$

rather than

$$V_1 = \frac{(50\ \text{k}\Omega)(75\ \text{V})}{50\ \text{k}\Omega + 25\ \text{k}\Omega} = 50\ \text{V}$$

as it should be.

In cases such as this it is absolutely necessary to ensure that the internal resistance of the voltmeter is *much greater* than the resistance across which the voltage is to be determined. A voltmeter with an ohm-per-volt rating of 50,000 would result in an internal resistance of (50) (50 kΩ) = 2500 kΩ, which when placed in parallel with the 50 kΩ would result in 50 kΩ ∥ 2500 kΩ ≅ 50 kΩ, and the reading would be 50 V, as it should be. The ohm-per-volt rating of 1000 is fine for networks with resistive values in the neighborhood of 25 Ω and 50 Ω rather than 25 kΩ and 50 kΩ.

Ohmmeters are never used to measure resistance in an energized network. The meter may be damaged, or at the very least the reading will be meaningless. In addition, if the resistance of a single element is to be determined, it is best to remove that element from the network to prevent the other elements from affecting the reading.

There are numerous multifunctional instruments available today. One of the most frequently employed is the volt–ohm milliammeter (in

Fig. 2.25), which typically has an ohm-per-volt rating of 20 kΩ to 50 kΩ. Before the ohmmeter section is used, the pointer on the scale should be set with the 0-Ω adjust knob. Touching the two probes together simulates 0 Ω. The 0-Ω knob can then be turned to set this indication. Remember that the ohmmeter section employs an internal battery for its measurements. If the meter should be set aside in this setting, it is possible for the probes to touch and discharge the internal battery. Generally, the meter should be left in the high-voltage range when not in use. If the meter fails to properly respond to the 0-Ω scale, the battery will probably have to be replaced, which is usually not a difficult task.

A digital-display multimeter with a current, voltage, and resistance measurement capability appears in Fig. 2.25. It has the important advantage that the internal resistance is set at 11 MΩ for each range of the voltmeter section.

On most meters, the numbers appearing opposite the dial setting are the same numbers appearing at the high end of the scale to be used, although a particular scale may be used for more than one dial setting. Consider a meter with a 5-, 50-, and 500-V dial setting. In each case, the analog scale to be used may have a maximum indication of 50 V. For example, 40 V is read as 40 V if the 50-V dial setting is used, as 4 V if the 5-V dial setting is used, and as 400 V if the 500-V setting is used. In addition, some numbers on the face of the meter are applicable to the scale above and below even though they indicate different quantities. Finally, no matter how many divisions appear between two numbers on any scale, they *equally* divide the difference between the numbers. For example, if five divisions appear between 40 and 50 V, then each division represents 2 V, while if the scale is so tight that only one division appears, then each indicates a difference of 5 V. The proper use of any of these instruments can be learned only through continual use and exposure.

2.13 CALCULATORS

In previous editions of this text, the calculator was not discussed in detail. Instead, students were left with the general exercise of choosing an appropriate calculator and learning to use it properly on their own. However, it is now clear that there must be some discussion about the use of the calculator to eliminate some of the impossible results obtained (and often strongly defended by the user—because the calculator says so) through a correct understanding of the process by which a calculator performs the various tasks. Time and space do not permit a detailed explanation of all the possible operations, but it is assumed that the following discussion will enlighten the user to the fact that it is important to understand the manner in which a calculator proceeds with a calculation and not to expect the unit to accept data in any form and always to generate the correct answer.

When choosing a calculator (scientific for our use), be absolutely sure that it has the ability to operate on complex numbers (polar and rectangular), which will be described in detail in Chapter 4. For now, simply look up the terms in the index of the operator's manual, and be sure that

FIG. 2.47
Texas Instruments TI-86 calculator. (Courtesy of Texas Instruments, Inc.)

the terms appear and that the basic operations with them are discussed. Next, be aware that some calculators perform the operations with a minimum number of steps, while others can require a lengthy or complex series of steps. Speak to your instructor if unsure about your purchase. For this text, the TI-86 of Fig. 2.47 was chosen because of its treatment of complex numbers.

Initial Settings

Format and accuracy are the first two settings that must be made on any scientific calculator. For most calculators the choices of formats are *Normal, Scientific,* and *Engineering.* For the TI-86 calculator, pressing the 2nd function (yellow) key followed by the MODE key will provide a list of options for the initial setting of the calculator. For calculators without a MODE choice, consult the operator's manual for the manner in which the format and accuracy level are set.

Examples of each are shown below:

Normal: 1/3 = 0.33
Scientific: 1/3 = 3.33E–1
Engineering: 1/3 = 333.33E–3

Note that the Normal format simply places the decimal point in the most logical location. The Scientific format ensures that the number preceding the decimal point is a single digit followed by the required power of ten. The Engineering format will always ensure that the power of ten is a multiple of 3 (whether it be positive, negative, or zero).

In the preceding examples the accuracy was hundredths place. To set this accuracy for the TI-86 calculator, return to the MODE selection and choose 2 to represent two-place accuracy or hundredths place.

In some instances, such as when dealing with finances, you may want to use the Normal mode to obtain a display similar to dollars and cents. In the majority of chapters you will probably want to stay with the Engineering mode, since it corresponds directly with those powers that have been assigned abbreviations and names. Then again, there will be occasions when Scientific mode is appropriate. In any event, you should review the mode selection sequence so that you can easily make the changes when needed.

Order of Operations

Although being able to set the format and accuracy is important, these features are not the source of the impossible results that often arise because of improper use of the calculator. Improper results occur primarily because users fail to realize that no matter how simple or complex an equation, the calculator will perform the required operations in a specific order.

For instance, the operation

$$\frac{8}{3+1}$$

is often entered as

$$\boxed{8}\boxed{\div}\boxed{3}\boxed{+}\boxed{1} = \frac{8}{3} + 1 = 2.67 + 1 = 3.67$$

which is totally incorrect (2 is the answer).

You must be aware that the calculator ***will not*** perform the addition first and then the division. In fact, addition and subtraction are the last operations to be performed in any equation. It is therefore very important that you carefully study and thoroughly understand the next few paragraphs in order to use a calculator properly.

1. The first operations to be performed by a calculator can be set using ***parentheses ( ).*** It does not matter which operations are within the parentheses. The parentheses simply dictate that this part of the equation is to be determined first. There is no limit to the number of parentheses in each equation—all operations within parentheses will be performed first. For instance, for the preceding example, if parentheses are added as shown next, the addition will be performed first and the correct answer obtained:

$$\frac{8}{(3+1)} = \boxed{8}\boxed{\div}\boxed{(}\boxed{3}\boxed{+}\boxed{1}\boxed{)} = \frac{8}{4} = \mathbf{2}$$

2. Next, ***powers and roots*** are performed, such as x^2, $\sqrt{x}$, and so on.
3. ***Negation*** (applying a negative sign to a quantity) and ***single-key operations*** such as sin, $\tan^{-1}$, and so on, are performed next.
4. ***Multiplication and division*** are then performed.
5. ***Addition and subtraction*** are performed last.

It may take a few moments and some repetition to remember the order, but at least you are now aware that there is an order to the operations and are aware that ignoring them can result in meaningless results.

EXAMPLES

a. Determine

$$\sqrt{\frac{9}{3}}$$

Solution: The following calculator operations will result in an incorrect answer of 1 because the square-root operation will be performed before the division.

$$\boxed{\sqrt{\ }}\boxed{9}\boxed{\div}\boxed{3} = \frac{\sqrt{9}}{3} = \frac{3}{3} = 1$$

However, recognizing that we must first divide 9 by 3, we can use parentheses as follows to define this operation as the first to be performed, and we will obtain the correct answer.

$$\boxed{\sqrt{\ }}\boxed{(}\boxed{9}\boxed{\div}\boxed{3}\boxed{)} = \sqrt{\left(\frac{9}{3}\right)} = \sqrt{3} = \mathbf{1.67}$$

b. Find

$$\frac{3+9}{4}$$

Solution: If the problem is entered as it appears, the incorrect answer of 5.25 will result.

$$\boxed{3}\boxed{+}\boxed{9}\boxed{\div}\boxed{4} = 3 + \frac{9}{4} = 5.25$$

Using brackets to ensure that the addition takes place before the division will result in the correct answer:

$$\boxed{(}\boxed{3}\boxed{+}\boxed{9}\boxed{)}\boxed{\div}\boxed{4} = \frac{(3+9)}{4} = \frac{12}{4} = \mathbf{3}$$

c. Determine

$$\frac{1}{4} + \frac{1}{6} + \frac{2}{3}$$

Solution: Since the division will occur first, the correct result will be obtained by simply performing the operations as indicated. That is,

$$\boxed{1}\boxed{\div}\boxed{4}\boxed{+}\boxed{1}\boxed{\div}\boxed{6}\boxed{+}\boxed{2}\boxed{\div}\boxed{3} = \frac{1}{4} + \frac{1}{6} + \frac{2}{3} = \mathbf{1.08}$$

2.14 COMPUTER ANALYSIS

In recent years, the computer has established itself as a tool of the electrical engineering profession that is not only useful, adaptive, and powerful but also an absolute necessity if recent advances in technology are to flourish and continue at their current pace. No longer can individuals associated with the field ignore the computer's potential and try to downplay its impact with comments about the possible negative effects of our dependency on this powerful instrument. There must be an acceptance of its role in the development arena with an appropriate level of awareness of its limitations. Virtually all technological and engineering programs now strive to introduce computer systems at the earliest opportunity to quickly alleviate any fears about the instrument and to develop a *friendly* line of communication. In fact, the very accreditation of an educational program may depend on the depth to which computer methods are incorporated in the curriculum.

The two fundamental components of any computer system are the hardware and the software. The *hardware* includes all the physical items such as the computer itself (and its internal components), the printer (for a *hard copy*), and a *modem* (for other lines of communication). The *software* is the numerous *programs* that control the operation of the computer and permit generation of the desired results in a variety of formats. The software can be generated by the user using one of the many *languages* developed for specific applications or purchased as a *software package.* Each will now be described in some detail.

Languages

A *language* is a set of symbols, letters, words, or statements that the user can enter in the computer in a set order and format. A computer is designed to understand the language and perform the defined operations in an order established by a series of commands called a *program.* The program tells the computer what to do on a sequential line-by-line basis in the same order one would normally approach the task using a longhand approach. The computer can respond only to the commands entered by the user, requiring that the programmer fully understand the sequence of operations and calculations needed to obtain a particular solution. In other words, the computer can only respond to the user's input—it does not have some mysterious path toward generating solutions unless told how to obtain those solutions. A lengthy analysis can result in a program having hundreds or thousands of lines. Once written, the program has to be carefully checked to be sure the results are valid for the expected range of input variables. Writing a program designed to perform a moderate or difficult task can therefore be a long, tedious process, but keep in mind that once the program is tested and the results verified, it can be stored in memory for future use. The user can then be assured that any future results obtained will have the required degree of accuracy even though they are obtained with a minimum expenditure of time and energy. Some of the popular languages applied in the electrical and electronics field include C++, Pascal, BASIC, and FORTRAN. Each has its own set of commands and statements to communicate with the computer, but each can be used to perform the same type of analysis.

Software Packages

The use of software packages eliminates the need to know a language and all its commands, statements, and structure to perform a desired task. All that is required is a knowledge of how to input the system under investigation and how to extract the results. The packaged program solves for specific unknowns as dictated by the user, eliminating any need to be aware of the details of the program within the software package. Herein lies one of the concerns of the authors with packaged programs—that a student has the ability to obtain a result for a particular analysis without knowing or understanding the steps leading to that solution. It is imperative that you realize that the computer be used as a *tool* to assist the user—it must not be allowed to control the scope and potential of the user! Therefore, as we progress through the chapters of the text, be sure you clearly understand the concepts before turning to the computer for support and efficiency.

Each software package has a menu, which defines the range of application of the package. Once entered into the computer, the system is preprogrammed to perform all the functions appearing on the menu. The user simply has to provide the parameters of the network and signal the computer that all the required data have been provided and that a solution is desired. The package then generates and prints the results for the desired unknowns. Be aware, however, that if a particular type of analysis is requested that is not on the menu, the software package cannot provide the desired results. The package is limited solely to those maneuvers developed by the team of programmers that developed the software package. In such situations you must turn to a software package that can perform the desired tasks or write a program using one of the languages listed above. All commercial software packages are ROM (read-only memory) based, and so the content cannot be accessed or changed.

In general, if a software package is available to perform a particular analysis, then you should use it rather than investing the time to develop your own routine. Most popular software packages are the result of many hours of effort by teams of programmers with years of experience who can afford to be totally dedicated to the project for long periods of time. However, if the results are not in the desired format or the software package does not provide all the desired results, then you should put your innovative talents to use to develop a software package. As noted above, any program you write that passes the tests of accuracy and reliability can be considered a software package of your authorship for extended future use (and can be modified as needed).

The software package (in CD-ROM format) employed in this text is the student version of PSpice 8.0 taken from the commercial version called OrCAD PSpice (Simulation Program with Integrated Circuit Emphasis). Appendix A provides a condensed list of information regarding the various software packages used in this text. The OrCAD office can be reached by contacting OrCAD DIRECT at 1-800-671-9505, visiting *www.orcad.com*, or writing to *info@orcad.com*.

In many ways the coverage of PSpice in this text is sufficient to act as the reference for the software package when applied to dc and ac networks. With the proper introduction to the equipment and support available when necessary, most areas of the text can be successfully examined and the problems completed using PSpice. Other software packages included in the text description are Electronics Workbench and MathCad, both of which employ the Windows format.

Computer simulation and methods are an important, integral part of the text and should not be treated as superfluous material of the lowest priority. Once a basic concept is understood, take the time to investigate computer methods, and start to develop a familiarity with the terminology and basic format of a program or use of a software package. It will be time well spent in preparation for the instruction you will eventually receive on computer systems.

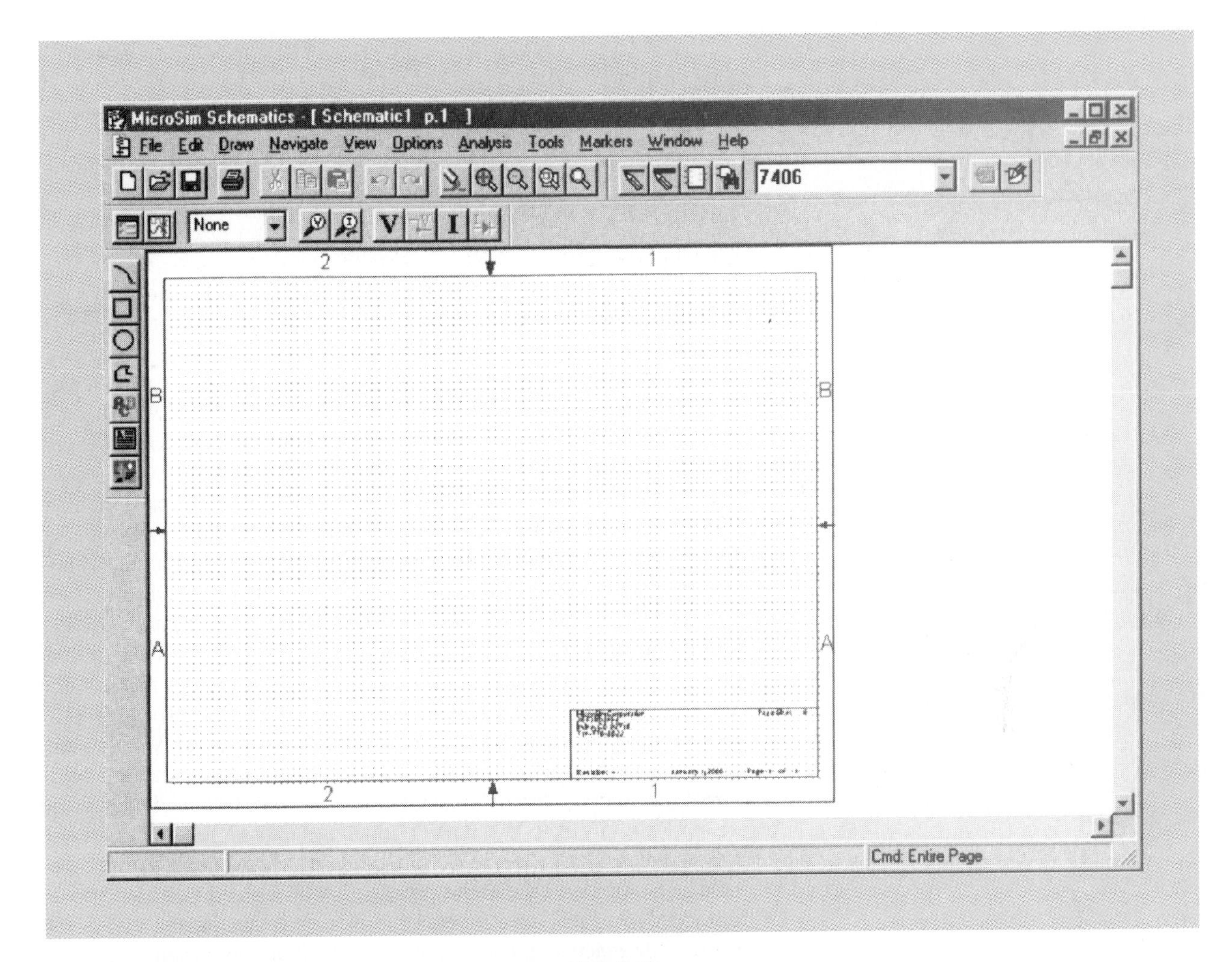

FIG. 2.48
PSpice schematics screen.

PSpice (Windows)

When using PSpice (Windows) the first step is to draw the network on the screen followed by an analysis that will provide the results required. When the Schematics icon is first chosen a full screen appears, as shown in Fig. 2.48. The menu bars provide a variety of choices for the operations that need to be performed.

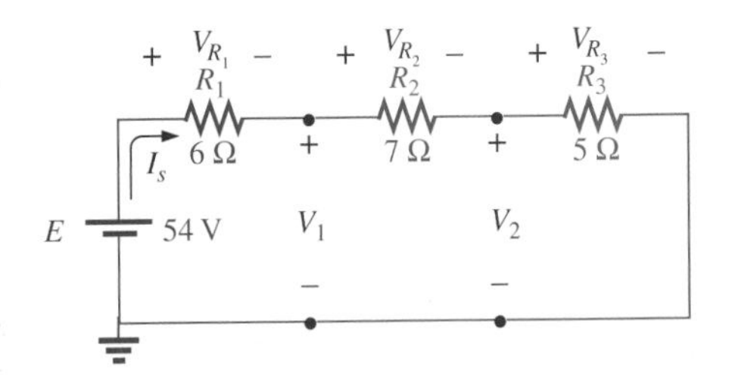

FIG. 2.49
Series dc network to be analyzed using PSpice (Windows).

For the series circuit of Fig. 2.49 the first step would be to place the elements on the screen relatively close to their final position. The sequence **Draw–Get New Part–Libraries–SOURCE.slb–OK–VDC–Place & Close** will result in a dc battery symbol displayed on the screen as shown in Fig. 2.50. This sequence will initially appear to be quite long to perform a simple operation, but be aware that each choice is simply a response to a prompt from the computer for more

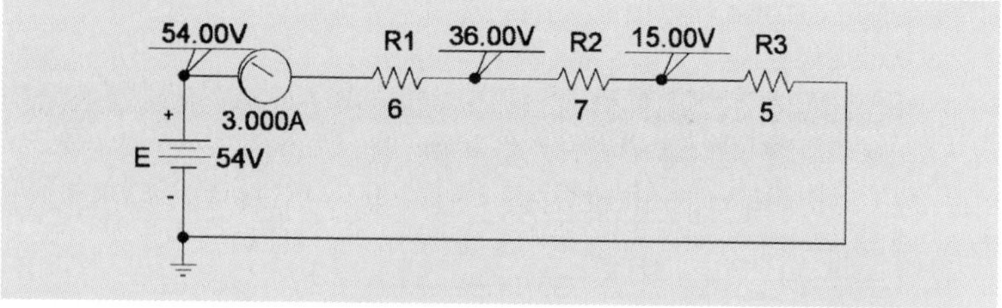

FIG. 2.50
PSpice (Windows) results for the investigation of the series dc network of Fig. 2.49

information. With practice, the process moves along quite rapidly, and shortcuts soon present themselves. A similar sequence is used to add each element on the screen. Next, the label and value of each element must be set and the elements placed in their exact final location. The final step is to connect all the components and insert the ammeter and voltage marker points. The ammeter is introduced as shown in Fig. 2.50, and the voltage marker points (horizontal bar and pointer) are placed as shown in the same figure. The analysis to be performed is then chosen, and the results appear as shown in Fig. 2.50, next to each marker. The current is 3 A, the voltage from the left of the ammeter to ground is 54 V, the voltage from a point between resistors R_1 and R_2 is 36 V to ground, and the voltage from a point between resistors R_2 and R_3 to ground is 15 V. The voltage across R_1 is then $54\text{ V} - 36\text{ V} = 18\text{ V}$, and the voltage across R_2 is $36\text{ V} - 15\text{ V} = 21\text{ V}$. The voltage across R_3 is 15 V, since one end of the resistor is at ground potential. Priorities do not permit a detailed description of the entire process, but be assured that after proper introduction of the software package in a computer laboratory session and a little practice, the network of Fig. 2.49 can be entered and analyzed in a few minutes.

Entering the parallel network of Fig. 2.51 resulted in the display of current levels for each branch next to each meter. The total current of 2.204 A is clearly the sum of the branch currents of the network, as dictated by Kirchhoff's current law.

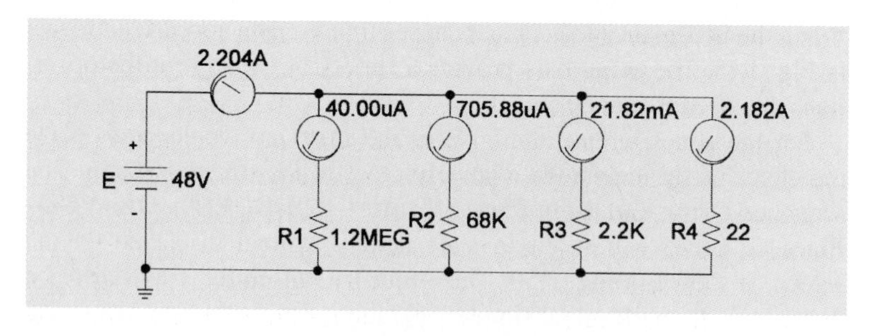

FIG. 2.51
Applying PSpice (Windows) to a parallel network with a wide range of parallel resistor values.

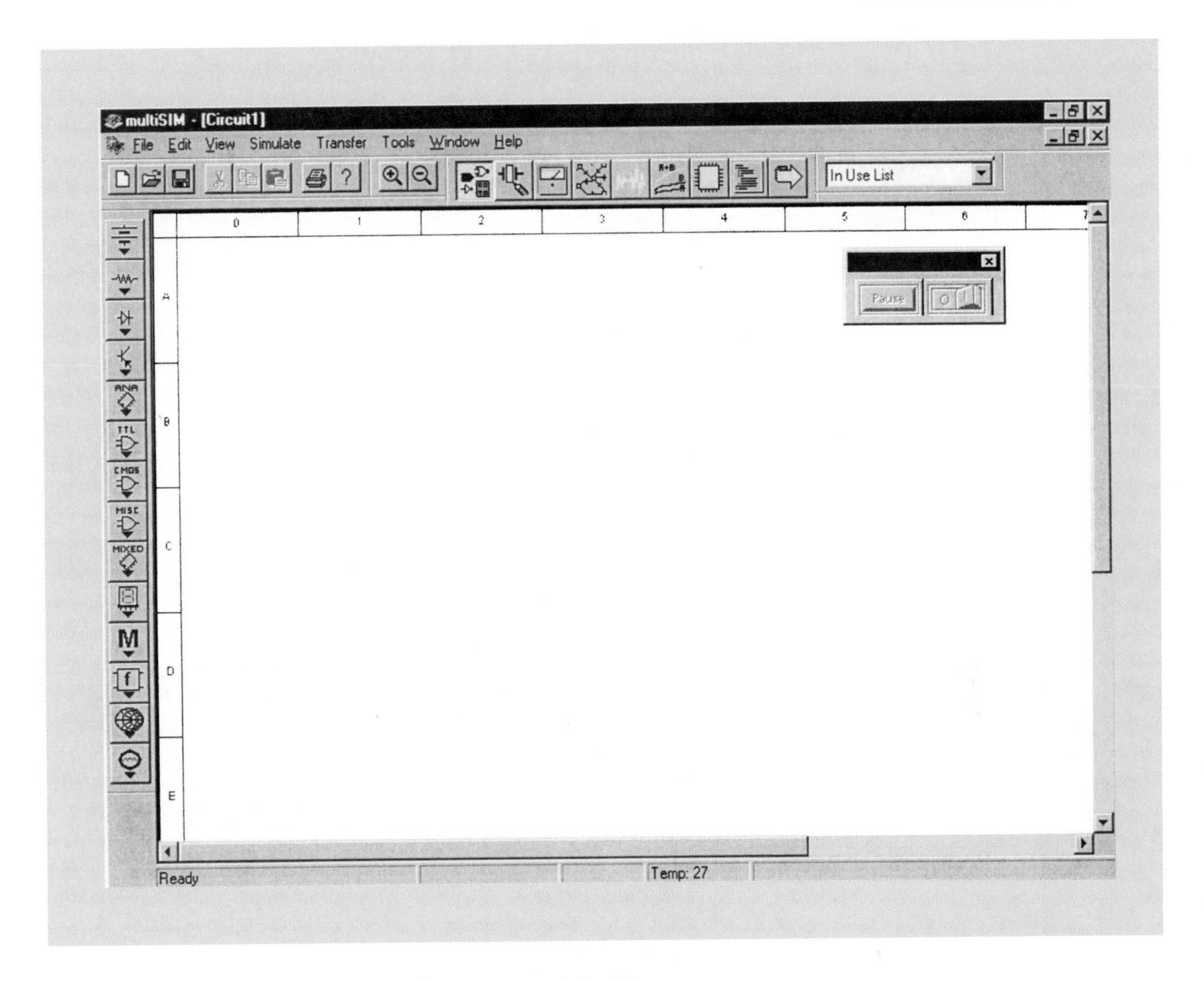

FIG. 2.52
Electronics Workbench schematics screen

Electronics Workbench

The general screen of Fig. 2.52 obtained for Electronics Workbench (EWB) is similar in many respects to that obtained using PSpice. Choosing any one of the items on the menu bar above the screen will produce a subset of other choices. Most of the controls for entering the networks are displayed as icons surrounding the top and left of the screen.

Entering the series network of Fig. 2.50 analyzed using PSpice results in the display of Fig. 2.53. Note that meters are displayed in the circuit, and the settings and results appear on the meters at the bottom of the screen. In each case, note that the meters are connected from the point of interest to ground, and the ammeter is connected in series with the other elements of the circuit. The results, of course, are the same as obtained using PSpice.

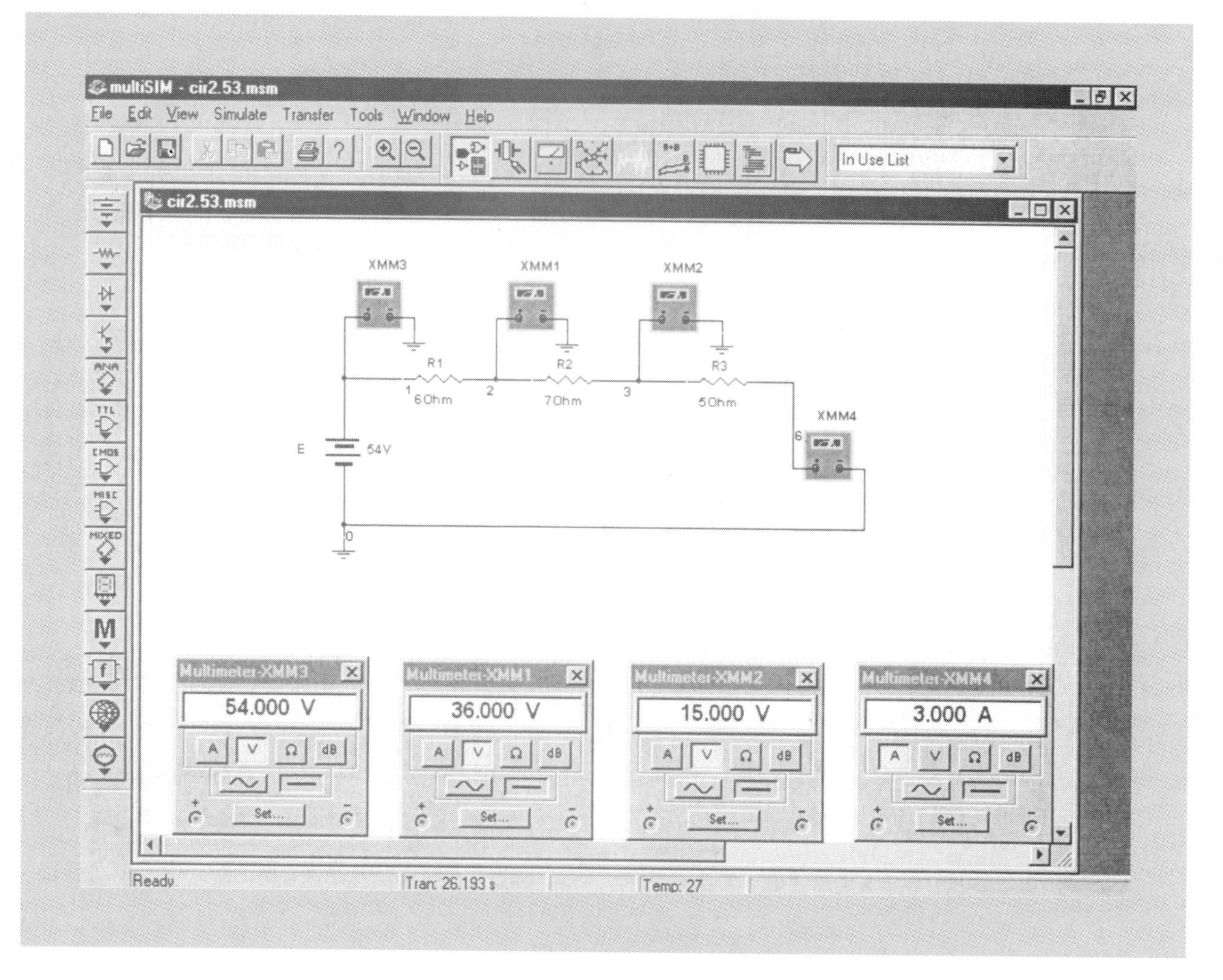

FIG. 2.53
Applying Electronics Workbench to the circuit of Fig. 2.50.

For the parallel network of Fig. 2.51 analyzed using PSpice the screen will appear as shown in Fig. 2.54. Again, we find that the results are the same, with the smallest current passing through the resistor of the largest value—current always seeks the path of least resistance! For the networks of both Figs. 2.53 and 2.54 note that all the meters at the bottom of the screens have the horizontal bar selected to represent a dc reading. The **Set** option on each meter permits changing the internal resistance of the meter to match a real-world concern.

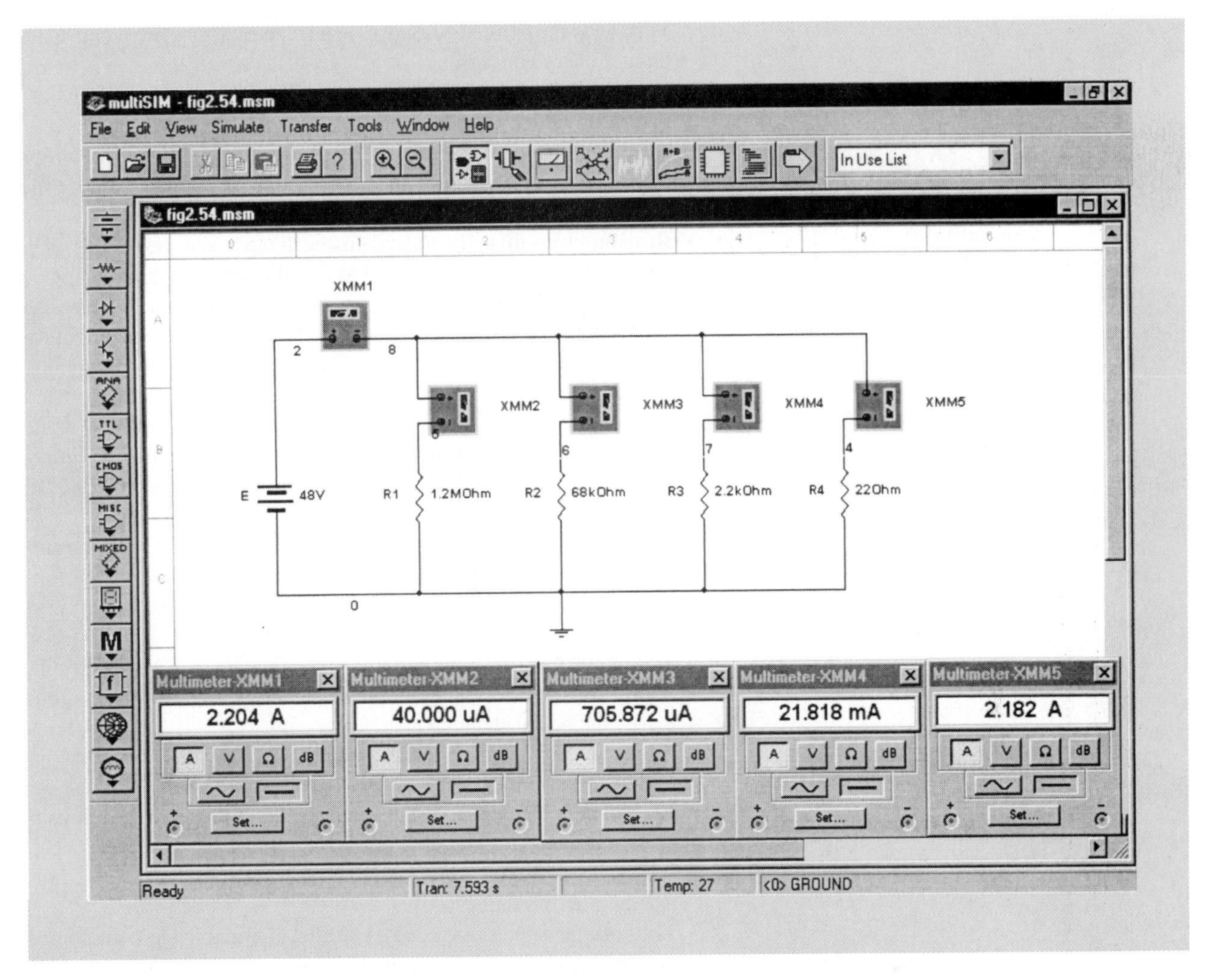

FIG. 2.54
Applying Electronics Workbench to the network of Fig. 2.51.

2.15 SUMMARY

- **Current** has a **direction defined by conventional flow.** It is the opposite of electron flow.
- **Ammeters** are always placed in **series** with the branch through which the current is to be determined, usually requiring that the branch be broken and the meter inserted.
- An **analog scale** is one in which a **pointer** indicates the measured level (requiring interpretation), whereas a **digital display** gives **numerical values.**
- **Prefixes** for powers of ten are assigned only to those powers of ten **divisible by three.**
- The **smaller** the AWG number of a wire the **larger** its diameter.

- **Voltage** is the applied **pressure** on an electrical system to **cause** the flow of charge or current.
- **Primary batteries** are **not rechargeable,** whereas **secondary batteries** can be recharged.
- **Resistance** is the **opposition** to the flow of charge, or current. For the same applied voltage, the larger the resistance, the smaller the resulting current.
- **Resistance** is **directly** related to the **length** and **resistivity** of a conductor and **inversely** related to the **diameter** in mils squared.
- For all common conductors, an **increase in temperature** causes an **increase in resistance.** The **lower** the temperature coefficient of resistance, the **less** sensitive the conductor to changes in temperature.
- The **color bands** of a resistor are always read starting with the band closest to the end. The first two bands represent **numbers,** while the third gives the **number of zeros** that follow the first two digits (or acts as a multiplying factor for resistance levels below 10 ohms).
- **Power (*P*)** is a **"rate"** of doing work, while energy (*W*) is the **number of joules converted** during the working process.
- The **higher** the efficiency the **more** output energy (or power) produced for the same applied energy (or power).
- The **current** is the **same** through each element of a **series circuit,** while the applied voltage equals the sum of the voltage drops across the series elements.
- The **total resistance** of a **series** circuit is simply the **sum** of the individual resistances.
- **Kirchhoff's voltage law** states that the **algebraic sum** of the voltage rises and drops around a closed path **must be zero.**
- For **series resistors,** the **larger** the resistance, the **larger** the voltage drop across the resistor.
- The **voltage** is the **same across parallel elements,** while the source current is the sum of the currents through the branches of the network.
- The **total resistance** of a **parallel** network is always **less** than the value of **smallest** resistance.
- The **more** resistance added in parallel, no matter the value, the **less** the total resistance.
- **Kirchhoff's current law** states that the **sum** of the currents **entering a node** must equal the **sum of the currents leaving.**
- **Current** always takes the **path of least resistance;** therefore, for two parallel elements the current through the smaller resistance will always be more than that through the larger resistance.
- For an **ammeter,** the **smaller** its internal resistance, the **less** it will affect the network to which it is applied. For a **voltmeter,** the **larger** its internal resistance, the less it will affect the network.
- A **careful** reading of the manual provided with a calculator will eliminate a wide variety of errors often encountered in its use.
- **Computers cannot think for themselves;** they can only react to the user's input. Therefore, a software program must be properly understood and applied if results are to have any meaning.

Equations:

$$I = \frac{Q}{t} \quad \text{(A)}$$

$$V = \frac{W}{Q} \quad \text{(V)}$$

$$I = \frac{E}{R} \quad \text{(A)}$$

$$R = \rho \frac{\ell}{\mathrm{A}} \quad (\Omega)$$

$$\frac{T + t_1}{R_1} = \frac{T + t_2}{R_2}$$

$$P = EI = VI = I^2R = \frac{V^2}{R}$$

$$W = Pt \quad \text{(J, W/s)}$$

$$\text{kwh} = \frac{Pt}{1000}$$

$$1 \text{ hp} = 746 \text{ W}$$

Series Circuits:

$$R_T = R_1 + R_2 + R_3 + \cdots + R_N;\ R_T = NR$$

$$E = V_1 + V_2 + V_3 + \cdots + V_N$$

$$I = \frac{E}{R_T}$$

$$P_T = EI = I^2R_1 + I^2R_2 + \cdots + I^2R_N$$

$$V_x = \frac{R_xE}{R_T}$$

Parallel Circuits:

$$\frac{1}{R_T} = \frac{1}{R_1} + \frac{1}{R_2} + \frac{1}{R_3} + \cdots + \frac{1}{R_N};\ R_T = \frac{R}{N}$$

$$R_T = \frac{R_1R_2}{R_1 + R_2}$$

$$I_T = \frac{E}{R_T}$$

$$I_T = I_1 + I_2 + I_3 + \cdots + I_N$$

$$I_1 = \frac{R_2I_T}{R_1 + R_2},\ I_2 = \frac{R_1I_T}{R_1 + R_2}$$

PROBLEMS

SECTION 2.2 Current

1. If 24×10^{16} electrons pass through a conductor in $\frac{1}{2}$ min, determine:
   - **a.** Charge in coulombs.
   - **b.** Current.

2. How long will it take 1600 mC to pass through a copper conductor if the current is 0.5 A?

3. How much charge has passed through a conductor if the current is 16 μA for 10 s?

4. For a current of 1 mA, how many electrons will pass a particular point in the circuit in 1 s? Write the number out in full decimal form (all the zeros). Is it a significant number for such a small current level?

5. Write the following quantities in the most convenient form using the prefixes in Table 2.1.
   a. 0.050 A
   b. 0.0004 V
   c. $3 + 10^4$ V
   d. 1200 V
   e. 0.0000007 A
   f. 32,000,000 V

6. What is the resistance of 1000 ft of No. 12 house wire?

7. What is the area in circular mils of wires having the following diameters?
   a. $\frac{1}{32}$ in.
   b. 0.01 ft
   c. 0.1 cm

8. What is the diameter in inches of wires having the following areas in circular mils?
   a. 10,000 CM
   b. 625 CM
   c. 50,000 CM

SECTION 2.3 Voltage

9. Determine the energy expended (in joules) to bring a charge of 40 mC through a potential difference of 120 V.

10. What is the potential difference between two points in an electric circuit if 200 mJ of energy is required to bring a charge of 40 μC from one point to the other?

11. How much energy is required to move 18×10^{18} electrons through a potential difference of 12 V?

12. How much energy is expended to maintain a current of 10 mA between two points in an electric circuit for 5 s if the potential difference between the two points is 20 mV?

SECTION 2.5 Resistance and Ohm's Law

13. Determine the internal resistance of a battery-operated clock if a current of 1.8 mA results from an applied voltage of 1.5 V.

14. Determine the current through a soldering iron if 120 V is applied. The iron has a resistance of 18 Ω.

15. Find the voltage drop across a 2.2-MΩ resistor with a current of 30 μA passing through it. What resistance would be required to limit the current to 1.5 A if the applied voltage is 64 V?

16. Determine the resistance of 50 ft of $\frac{1}{16}$-in.-diameter copper wire.

17. Calculate the resistance of 600 ft of No. 14 wire using Table 2.2.

18. Determine the diameter (in inches) of a copper inductor having a length of 200 ft and a resistance of 0.2 Ω.

19. What is the resistance of 1 mi of No. 12 house wire? How does it compare with the resistance of 1 kΩ connected to the end of the conductor?

20. If the resistance of a copper conductor is 2 Ω at room temperature ($T = 20°C$), what is its resistance at 100°C (the boiling point of water)?

21. At what temperature will the resistance of a No. 8 copper wire double if its resistance at $T = 20°C$ is 1 Ω?

22. If the resistance of a copper conductor 400 ft long is 10 Ω at room temp. ($T = 20°C$), what is its resistance at −20°C.

23. **a.** Determine the resistance of a molded composition resistor with the following color bands: first: red; second: red; third: brown; fourth: gold.
    b. Indicate its expected range of values.

24. Determine the color bands of a 100-kΩ resistor with a tolerance of 5%.

25. Determine the color bands of a 3.952 resistor with a tolerance of 10%.

SECTION 2.6 Power, Energy, Efficiency

26. Determine the power delivered by a 12-V battery at a current drain of 240 mA.

27. Calculate the power dissipated by a 2.2-kΩ resistor having a current of 4 mA passing through it.

28. A 280-W television set is connected to a 120-V outlet. Determine the current drawn by the set.

29. Determine the total energy dissipated by a 1400-W toaster used for 3.5 min.

30. Calculate the cost of using the following appliances for the indicated time period if the unit cost is 9 cents/kWh.
    a. Six 60-W bulbs for 6 h.
    b. 8-W clock for 30 days (1 month).
    c. 160-W television set for 4 h 30 min.
    d. 5000-W clothes dryer for 45 min.

31. Determine the cost of using an 8-W night light for 1 year (365 days) if the cost is 9 cents/kWh.

32. How long can we use a welding unit for $1.00 if the unit draws 14 A at 220 V and the cost is 9 cents/kWh?

33. Determine the applied power of a system that has an output power of 320 W and an efficiency of 86%.

34. **a.** A 2.2-hp motor has an input power demand of 2400 W. Determine its efficiency.
    b. If the applied voltage is 120 V, find the input current.
    c. What is the power lost in the energy transfer (in watts)?

35. The total efficiency of systems in cascade (one following the other) is the product of the individual efficiencies. With this in mind:
   a. What is the total efficiency of three systems in cascade with efficiencies of 90% each?
   b. What is the total efficiency of three systems in cascade if two have an efficiency of 90% and one has an efficiency of 20%?
   c. What conclusions can you draw from the results for parts (a) and (b)?

SECTION 2.8 Series Dc Networks

36. Find the total resistance of the networks in Fig. 2.55.

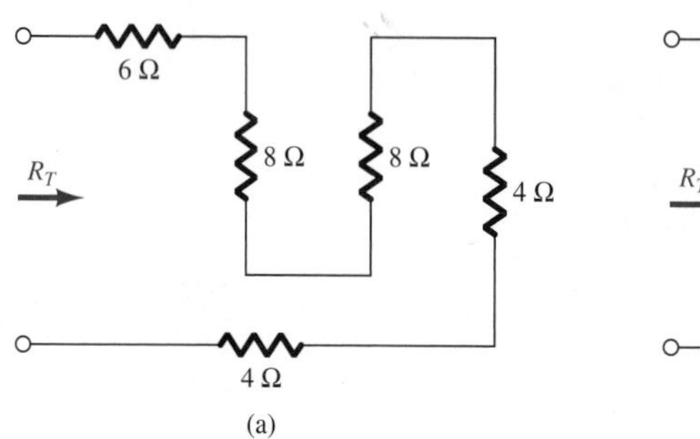

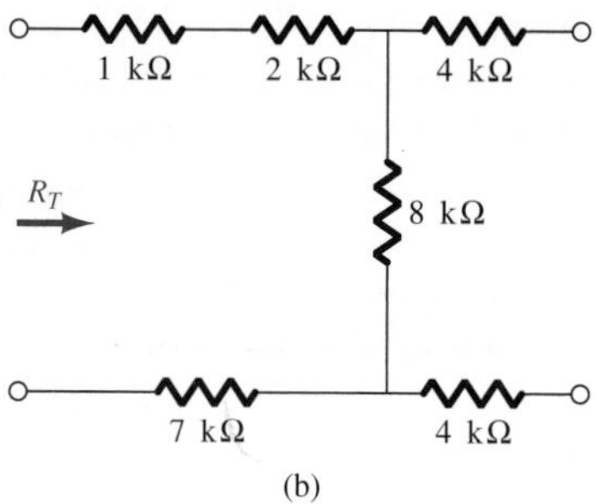

FIG. 2.55

37. Determine R_1 for the networks in Fig. 2.56.

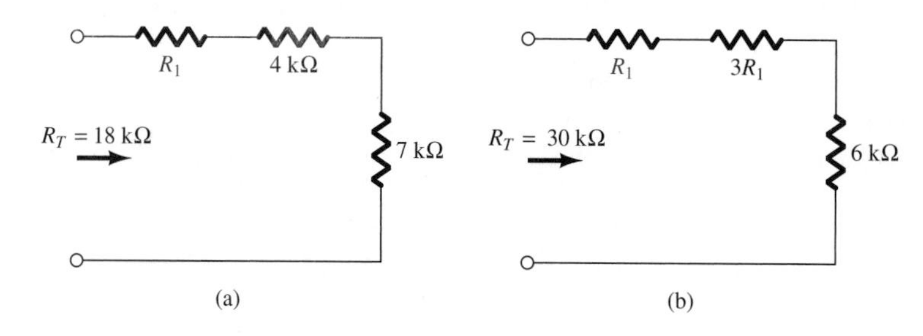

FIG. 2.56

38. For the circuit in Fig. 2.57, determine:
   a. R_T.
   b. I.
   c. V_3.
   d. P_2.

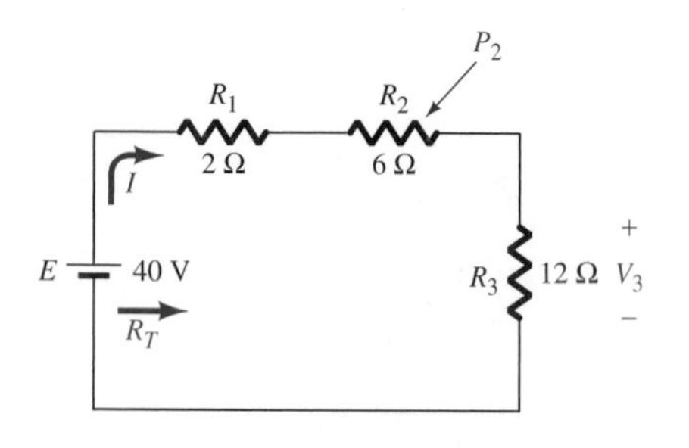

FIG. 2.57

39. Determine the unknown quantities for the networks in Fig. 2.58.

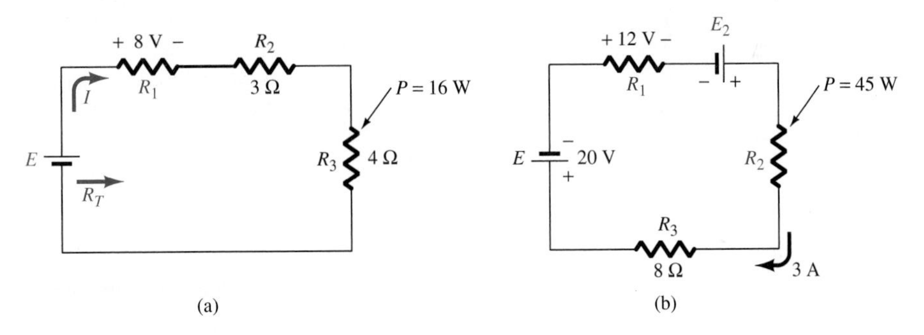

FIG. 2.58

40. Determine the unknown voltages for the circuits in Fig. 2.59 using Kirchhoff's voltage law.

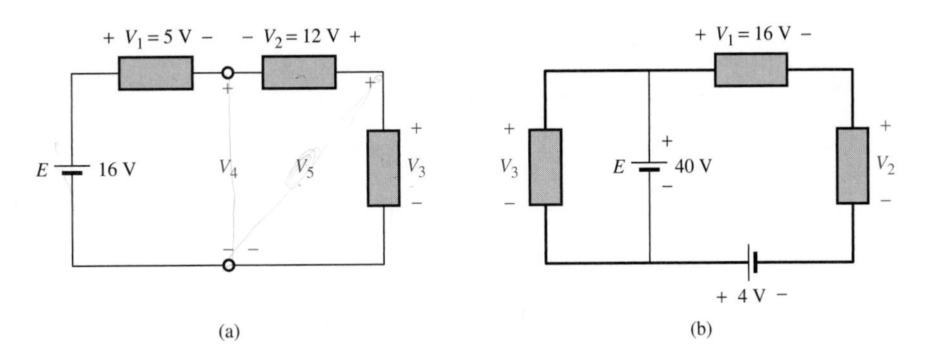

FIG. 2.59

41. Determine the voltages V_3 and V_4 using the voltage-divider rule for the network in Fig. 2.60.

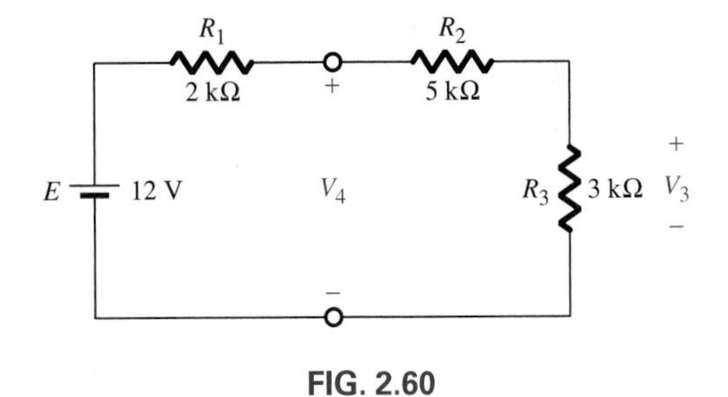

FIG. 2.60

42. Determine R_2 for the given voltage level for the network in Fig. 2.61 using the voltage-divider rule.

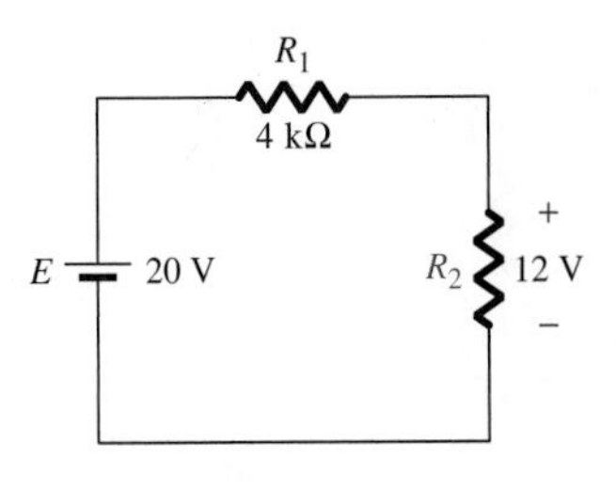

FIG. 2.61

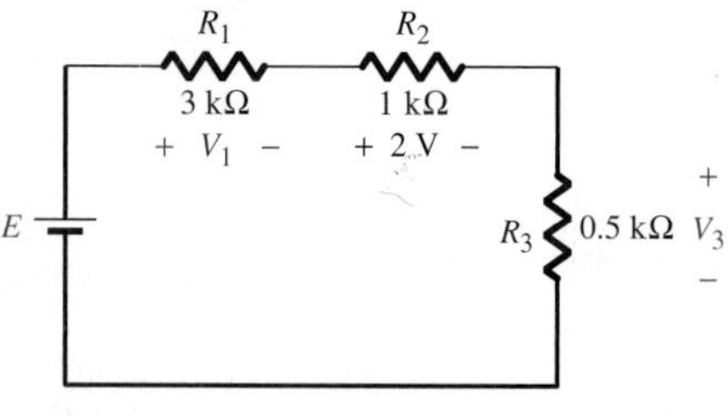

FIG. 2.62

43. Determine the voltage levels V_1 and V_3 and E for the circuit in Fig. 2.62.

SECTION 2.10 Parallel Dc Networks

44. Determine the total resistance of the networks in Fig. 2.63.

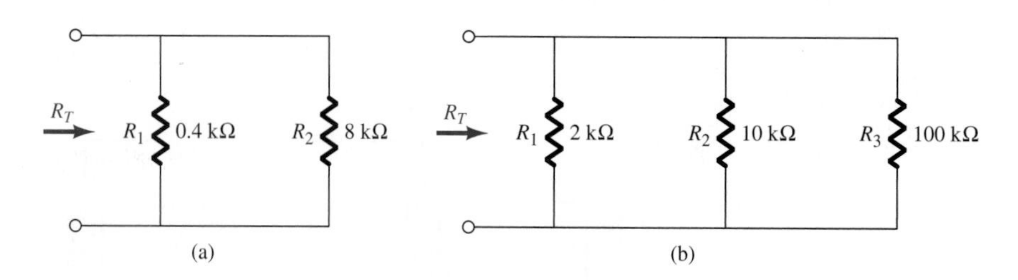

FIG. 2.63

45. Determine R_1 for the networks in Fig. 2.64.

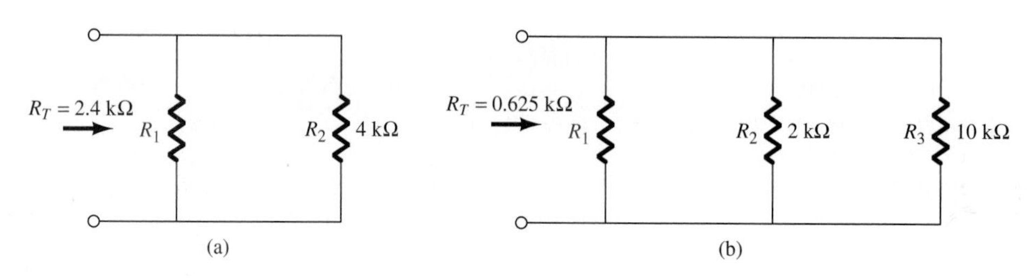

FIG. 2.64

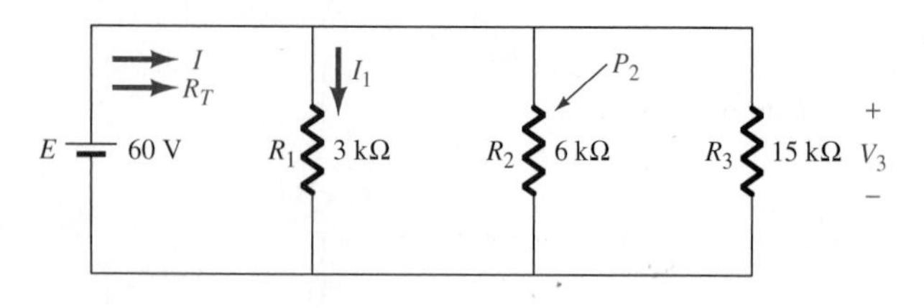

FIG. 2.65

46. For the network in Fig. 2.65 determine:

a. R_T.
b. I.
c. I_1.
d. V_3.
e. P_2.

47. Determine the unknown quantities for the networks in Fig. 2.66.

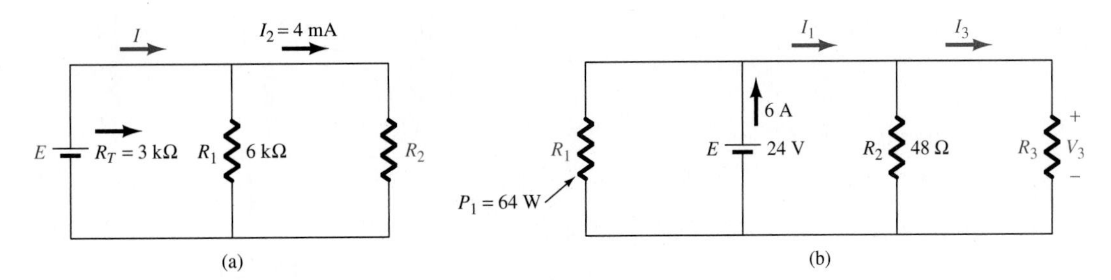

FIG. 2.66

48. Determine the unknown currents for the circuits in Fig. 2.67 using Kirchhoff's current law.

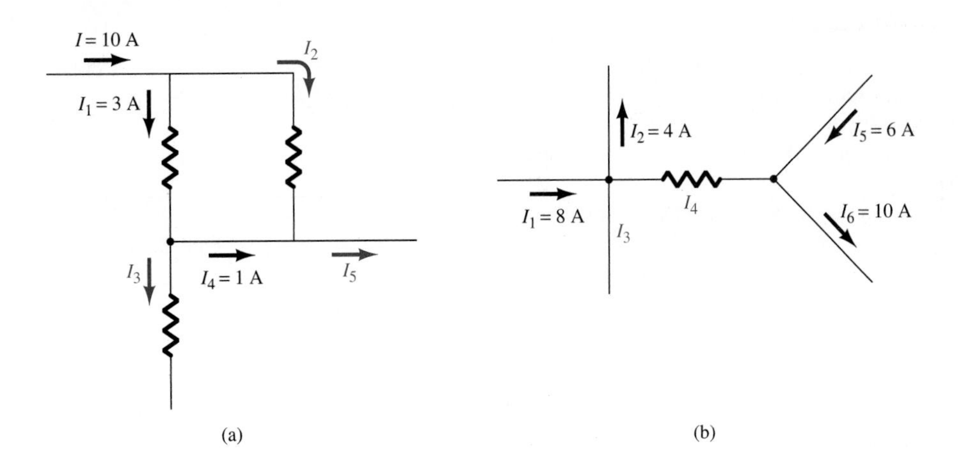

FIG. 2.67

49. Determine the currents I_1 and I_2 in Fig. 2.68 using the current-divider rule. What is the ratio R_1/R_2? How does it compare with the ratio I_1/I_2?

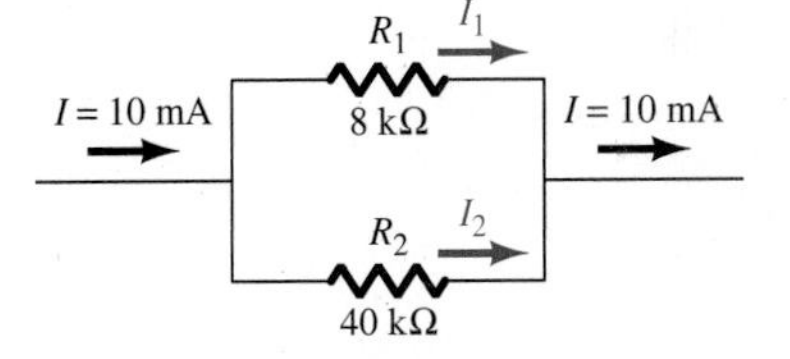

FIG. 2.68

50. Determine R_1 for the network in Fig. 2.69 using the current-divider rule.

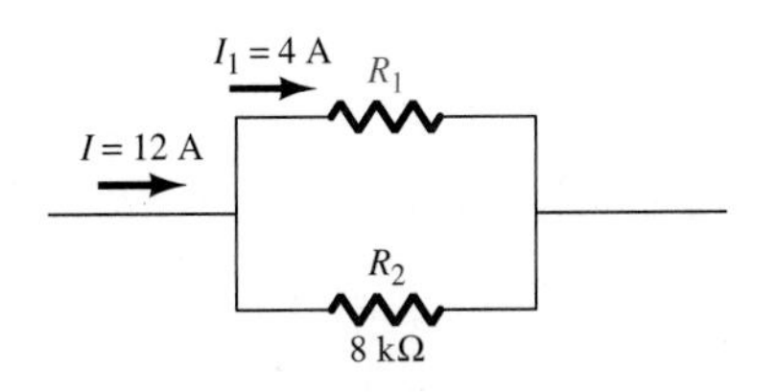

FIG. 2.69

51. Find I_1, I_3, and I for the network in Fig. 2.70.

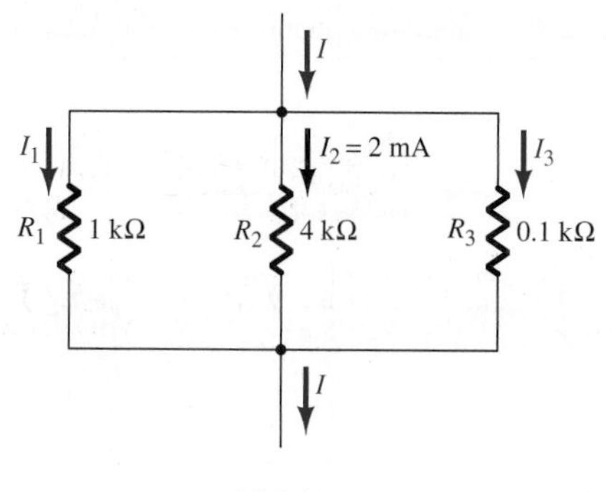

FIG. 2.70

SECTION 2.12 Meter Considerations

52. a. Sketch the location and connections of ammeters and voltmeters used to measure the currents I_1 and I_3 and voltages V_1 and V_3 in Fig. 2.71.

b. Using a voltmeter with an ohm-per-volt rating of 1000, determine the indication of the meter when it is placed across the 4-kΩ resistor if the 50-V scale is used.

c. Repeat part (b) for a meter employing an ohm-per-volt rating of 20,000.

d. Repeat part (b) for a DMM with an internal resistance of 11 MΩ.

e. Show the connection for a wattmeter reading the power delivered to R_3 and R_4.

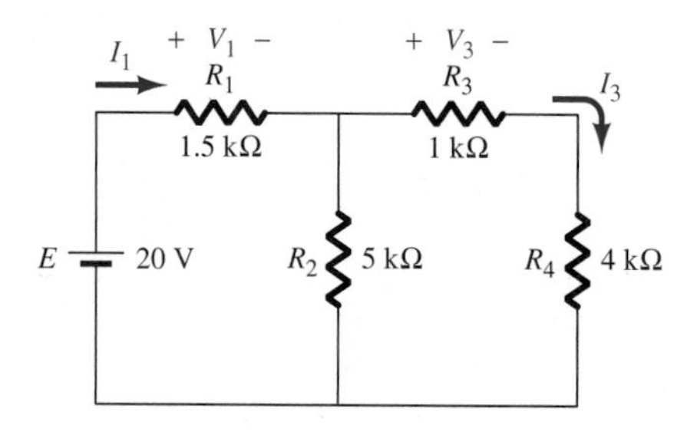

FIG. 2.71

3 Series–Parallel Dc Networks, Theorems, and Storage Elements

3.1 INTRODUCTION

One of the unique advantages of this field of study is that once an important concept is clearly understood, it will not be replaced at some later date with an updated or more complex version. Rather, there is a building process in which fundamental relationships such as Ohm's law often help define the path of investigation for more complex systems. This process will start to become obvious in the more advanced material of this chapter. Due to the enormous amount of material that has to be covered in this text the number of examples had to be limited to those that best explain the methods and theorems to be introduced. However, be aware that these are fundamental areas of study that are discussed in numerous texts available for further study. Keep in mind, also, that since space is limited, everything that appears is important and should be understood or the building process will suffer.

3.2 SERIES–PARALLEL NETWORKS

A series–parallel network is just as the name implies a combination of series and parallel elements in the same network. There is no set pattern for this type of network nor any limit on how many of each type of element can appear. The possibilities are infinite, so exposure, practice, and experience will play a key role in developing your ability to solve series–parallel dc networks. Fortunately, a firm understanding of the basic principles associated with series and parallel circuits is sufficient background to begin an investigation of any series–parallel dc network.

The following are a few steps that can be helpful in getting started on the first few exercises, although the impact and value of each will become obvious only with experience.

1. Take a moment to study the problem "in total" and make a brief mental sketch of the overall approach you plan to use. The result may be time- and energy-saving shortcuts.

2. Next examine each region of the network independently before trying them together in series–parallel combinations. This will usually simplify the network and possibly reveal a direct approach toward obtaining one or more desired unknowns. This step also eliminates many of the errors that might result due to the lack of a systematic approach.
3. Redraw the network as often as possible with the reduced branches and undisturbed unknown quantities to maintain clarity and provide the reduced networks for the trip back to unknown quantities from the source.
4. When you have a solution, check that it is reasonable by considering the magnitudes of the energy source and the elements in the network. If it does not seem reasonable, either solve the circuit using another approach or check over your work very carefully.

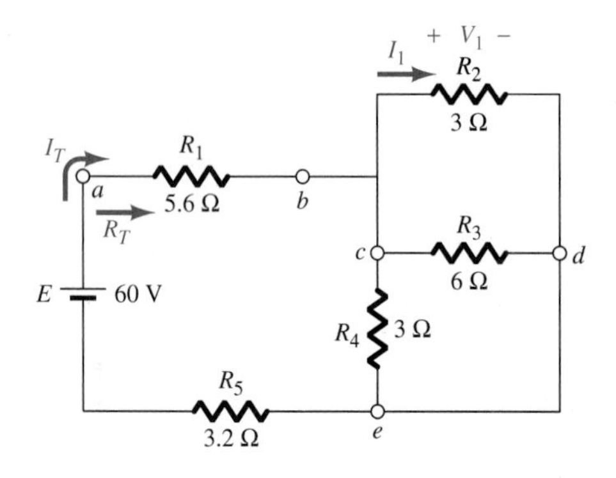

FIG. 3.1

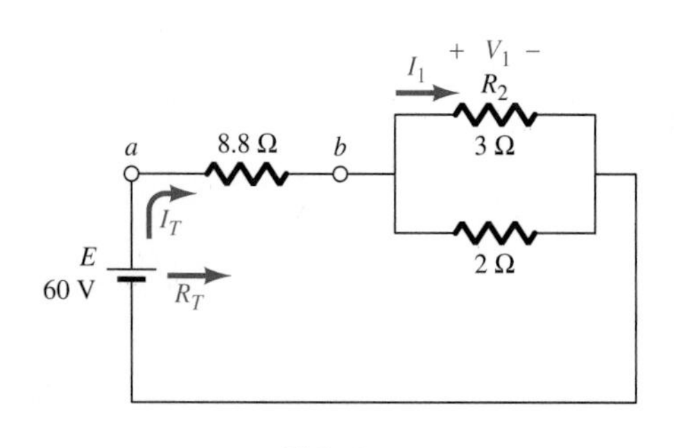

FIG. 3.2

For example, consider the network in Fig. 3.1. The battery E and the resistors R_1 and R_5 form a series path, since there is only one terminal in common between any of the two elements. The resistors R_2 and R_3 are in parallel because they have points c and d in common. Examine the network a little more carefully and you will find that R_3 and R_4 are also in parallel, since they have both c and d (or e, the same point) in common. The network can be redrawn as shown in Fig. 3.2 without losing the unknown quantities R_T, I_T, I_1, and V_1. Note that the series resistors were combined ($5.6\ \Omega + 3.2\ \Omega = 8.8\ \Omega$) along with the parallel resistors R_3 and R_4 ($6\ \Omega \parallel 3\ \Omega = 2\ \Omega$). It is now more obvious that the parallel combination of the 3- and 2-Ω resistors is in series with the 8.8-Ω resistor. For R_T we must first combine the parallel elements:

$$3\ \Omega \| 2\ \Omega = \frac{3\ \Omega(2\ \Omega)}{3\ \Omega + 2\ \Omega} = \frac{6\ \Omega}{5 \Omega} = 1.2\ \Omega$$

and the series elements,

$$R_T = 8.8\ \Omega + 1.2\ \Omega = \mathbf{10\ \Omega}$$

The source current is

$$I_T = \frac{E}{R_T} = \frac{60\ V}{10\ \Omega} = \mathbf{6\ A}$$

The current I_t through the 8.8-Ω resistor (or either of its former series components) is therefore 6 A. The current I_1 can be determined using the current-divider rule:

$$I_1 = \frac{(2\ \Omega)(I_T)}{2\ \Omega + 3\ \Omega} = \frac{2(6\ \mathrm{A})}{5} = \frac{12\ \mathrm{A}}{5} = \mathbf{2.4\ A}$$

The power dissipated in the form of heat by the 3-Ω resistor R_2 is determined by

$$P = I^2R = (2.4\ \mathrm{A})^2 \times 3\ \Omega = \mathbf{17.28\ W}$$

and the power delivered by the source voltage by

$$P_s = EI_T = (60\ \mathrm{V})(6\ \mathrm{A}) = \mathbf{360\ W}$$

The sum of the powers delivered to each resistive element equals that provided at the source. That is,

$$P_s = P_{R_1} + P_{R_2} + P_{R_3} + P_{R_4} + P_{R_5}$$

In the network just analyzed, the final configuration after the small parcels were considered is a series circuit. In Example 3.1 the final configuration is a parallel network.

EXAMPLE 3.1 For the series–parallel network in Fig. 3.3:

a. Calculate R_T.
b. Find I_T.
c. Determine I_1 and I_2.
d. Calculate V_1.
e. Calculate the power delivered to the 8-kΩ resistor.

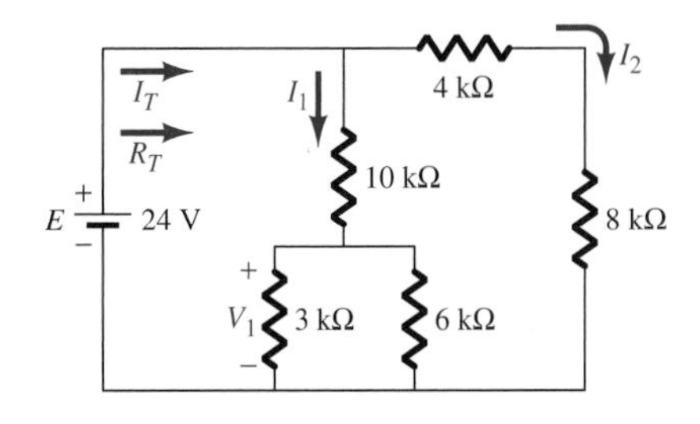

FIG. 3.3

Solution:

a. Combining the parallel 3-kΩ and 6-kΩ resistors, we have

$$3\text{ k}\Omega \| 6\text{ k}\Omega = \frac{(3\text{ k}\Omega)(6\text{ k}\Omega)}{3\text{ k}\Omega + 6\text{ k}\Omega} = 2\text{ k}\Omega$$

The equivalent 2-kΩ resistor is then in series with the 10-kΩ resistor and

$$R'_T = 10\text{ k}\Omega + 2\text{ k}\Omega = 12\text{ k}\Omega$$

The 4-kΩ and 8-kΩ resistors are in series, resulting in

$$R''_T = 4\text{ k}\Omega + 8\text{ k}\Omega = 12\text{ k}\Omega$$

The network can then be redrawn as shown in Fig. 3.4. Then

$$R_T = R'_T \| R''_T = \frac{12\text{ k}\Omega}{2} = \mathbf{6\text{ k}\Omega}$$

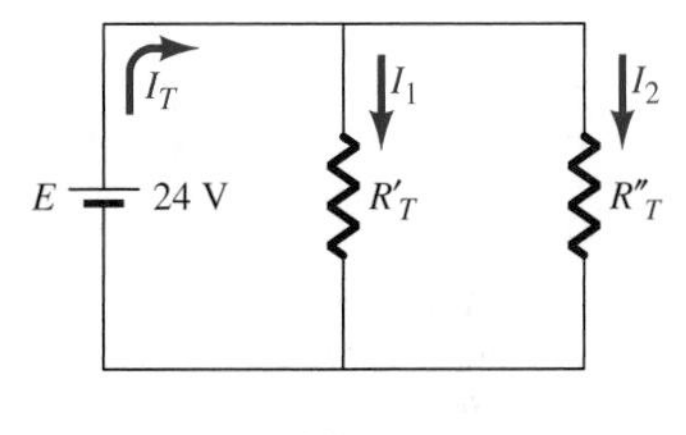

FIG. 3.4

b. $I_T = \dfrac{E}{R_T} = \dfrac{24\text{ V}}{6\text{ k}\Omega} = 4 \times 10^{-3}\text{ A} = \mathbf{4\text{ mA}}$

c. $I_1 = \dfrac{E}{R'_T} = \dfrac{24\text{ V}}{12\text{ k}\Omega} = 2 \times 10^{-3}\text{ A} = \mathbf{2\text{ mA}}$

and $I_2 = \dfrac{E}{R''_T} = \dfrac{24\text{ V}}{12\text{ k}\Omega} = 2 \times 10^{-3}\text{ A} = \mathbf{2\text{ mA}}$

Recall that the current divides equally between equal parallel resistances.

d. $V_1 = I_1(3\text{ k}\Omega \| 6\text{ k}\Omega) = I_1(2\text{ k}\Omega) = (2 \times 10^{-3}\text{ A})(2 \times 10^3\ \Omega) = \mathbf{4\text{ V}}$

e. $P_{8\text{k}\Omega} = I_2^2(8\text{ k}\Omega) = (2 \times 10^{-3}\text{ A})^2(8 \times 10^3\ \Omega) = (4 \times 10^{-6})(8 \times 10^3)$
$= 32 \times 10^{-3} = \mathbf{32\text{ mW}}$

3.3 CURRENT SOURCES

Our analysis thus far has been limited to networks energized by fixed dc voltage supplies such as provided by a battery or laboratory supply. A second source of electrical energy of equal importance, called a *current source,* will now be introduced. The past few sections have demonstrated that a dc voltage source supplies a fixed voltage to a network and that its current is determined by a resistive load to which it is applied. The *current source supplies a fixed current to a network, and its terminal voltage is determined by the network to which it is applied.*

Although not considered thus far, every source of voltage or current has some internal resistance. The voltage source has a series internal resistor such as shown in Fig. 3.5(a), and the current source a parallel resistance such as shown in Fig. 3.5(b). Ideally, the resistor R_s should be zero (like a short circuit) and the resistor R_p infinite ohms (like an open

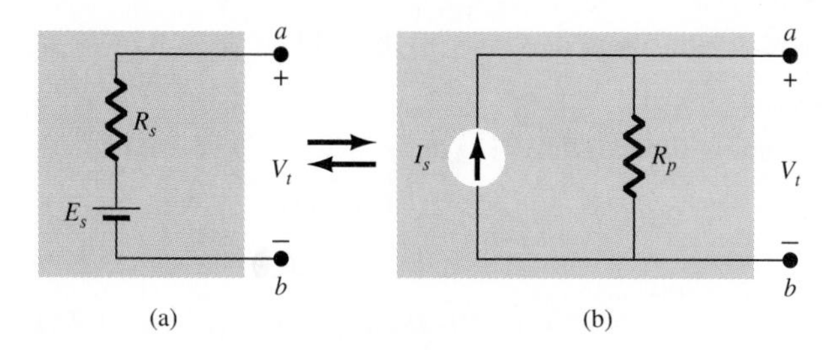

FIG. 3.5
Sources: (a) voltage; (b) current.

circuit). The representative network in Fig. 3.6 demonstrates the validity of the above. Due to the internal resistance of the source the voltage across the load is not 40 V but, as determined by the voltage-divider rule,

$$V_{R_L} = \frac{(10\text{ k}\Omega)(40\text{ V})}{10\text{ k}\Omega + 1\text{ k}\Omega} = 36.6\text{ V}$$

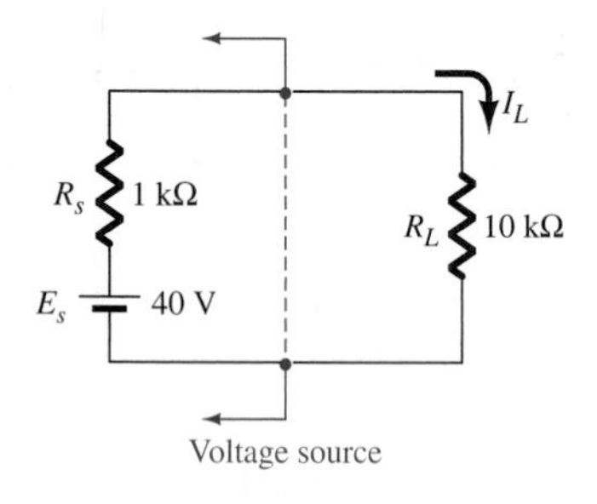

FIG. 3.6

For $R_s = 500\ \Omega$, the terminal voltage more closely approaches the ideal value of 40 V. That is,

$$V_{R_L} = \frac{(10\text{ k}\Omega)(40\text{ V})}{10\text{ k}\Omega + 0.5\text{ k}\Omega} = 38.095\text{ V}$$

For $R_s = 0\ \Omega$ (the ideal situation),

$$V_{R_L} = 40\text{ V}$$

Similarly, for the current source in Fig. 3.7, the larger the resistance R_p, the more ideal the source. That is, for $R_p = 1\text{ k}\Omega$ and using the current-divider rule, we have

$$I_L = \frac{(1\text{ k}\Omega)(5\text{ A})}{1\text{ k}\Omega + 1\text{ k}\Omega} = \frac{5\text{ A}}{2} = 2.5\text{ A}$$

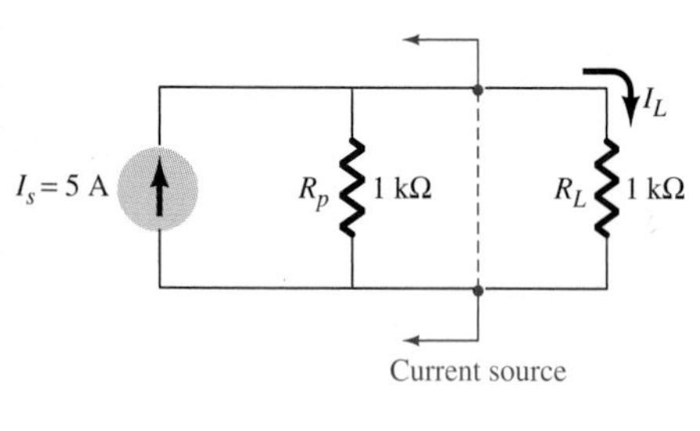

FIG. 3.7

and for $R_p = 100\text{ k}\Omega$,

$$I_L = \frac{(100\text{ k}\Omega)(5\text{ A})}{100\text{ k}\Omega + 1\text{ k}\Omega} = 4.95\text{ A}$$

and for $R_p = \infty\ \Omega$ (the ideal-case open circuit),

$$I_L = 5\text{ A}$$

The preceding discussion clearly reveals that the smaller the internal resistance R_s of the voltage source and the larger the internal resistance R_p of the current source, the more ideal the source.

The next few examples will demonstrate that a current source sets the current in the branch in which it is located and that the voltage across a current source is a direct function of the applied network.

EXAMPLE 3.2 For the network in Fig. 3.8:

a. Determine the current I_L.
b. Calculate the voltages V_L and V_s.

Solution:

a. $I_L = I_s = \mathbf{4\ mA}$
b. $V_L = I_L R_L = (4 \times 10^{-3}\text{ A})(2 \times 10^3\ \Omega) = \mathbf{8\ V}$
$V_s = V_L = \mathbf{8\ V}$

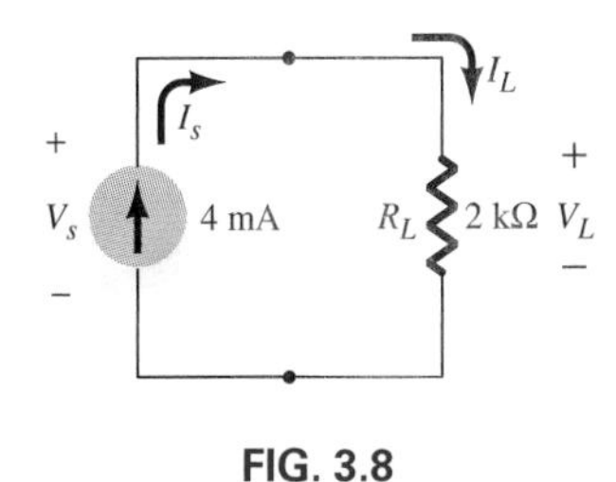

FIG. 3.8

EXAMPLE 3.3 For the network in Fig. 3.9, determine:

a. V_1.
b. I_2.
c. I_4.

Solution:

a. R_3 and R_4 are in parallel with a total resistance

$$R' = \frac{8\ \Omega}{2} = 4\ \Omega$$

R_2 is in series with R', and $R'' = R_2 + R' = 4\ \Omega + 4\ \Omega = 8\ \Omega$, resulting in the configuration in Fig. 3.10. By the current-divider rule,

$$I_1 = \frac{R''I}{R_1 + R''} = \frac{(8\ \Omega)(2\text{ A})}{6\ \Omega + 8\ \Omega} = \frac{16\text{ A}}{14} = 1.143\text{ A}$$

$$V_1 = I_1R_1 = (1.143\text{ A})(6\ \Omega) = \mathbf{6.858\ V}$$

b. Using Kirchhoff's current law, we obtain

$$I = I_1 + I_2$$
$$2\text{ A} = 1.143\text{ A} + I_2$$
$$I_2 = 2\text{ A} - 1.143\text{ A} = \mathbf{0.857\ A}$$

c. $I_4 = \dfrac{I_2}{2}$ (equal parallel resistors)

$$= \frac{0.857\text{ A}}{2} = \mathbf{0.429\ A}$$

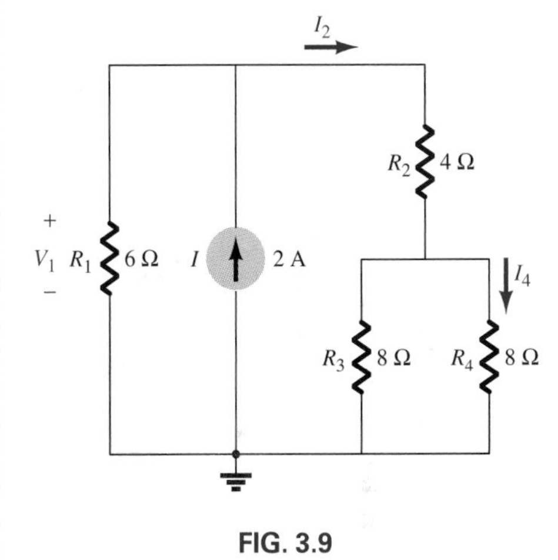

FIG. 3.9

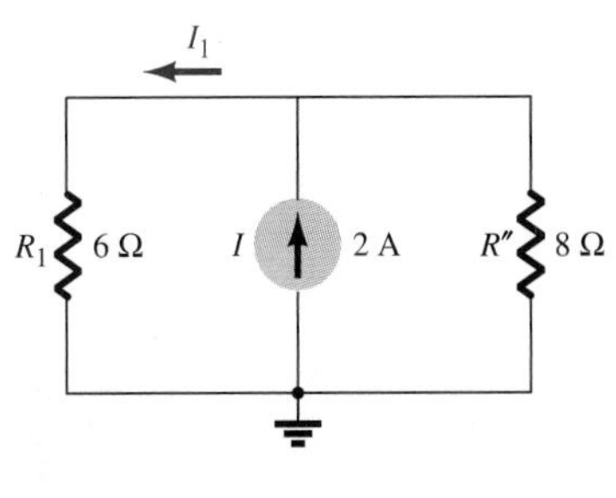

FIG. 3.10

Understandably, since the voltage source is the more common type of supply, it will take a measure of exposure before you will feel as comfortable with the current source. In a number of sections to follow, when we consider electronic devices and circuits, the current source will appear again, and quite frequently. This additional exposure will serve to clarify the basic characteristics of this source of energy. A commercially available current source is shown in Fig. 3.11. For a specified load variation, the current through the load can be set, and its terminal voltage is determined by the load applied.

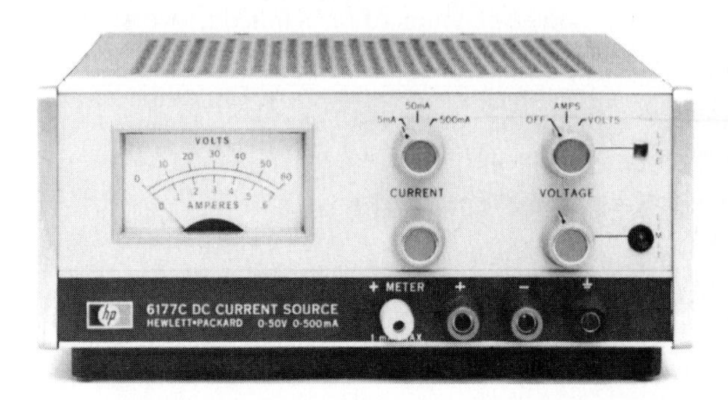

FIG. 3.11
Dc current source. (Courtesy of Hewlett-Packard.)

In circuit analysis, it is often advantageous to convert a voltage source to a current source, or vice versa. The conversion equations appear in Fig. 3.12 with their respective sources.

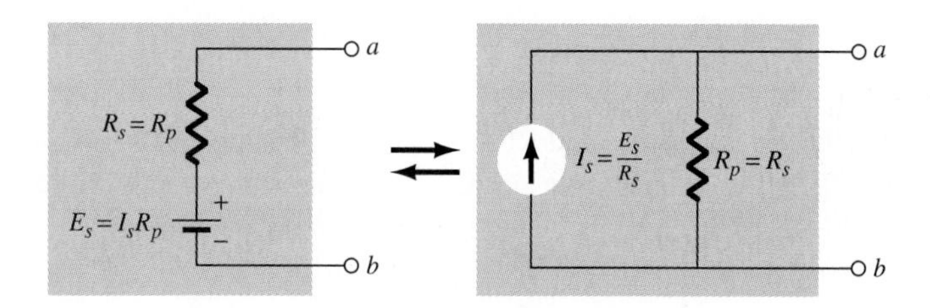

FIG. 3.12
Source conversion.

Note that the resistance is the same for each and that the current or voltage is determined by the other using Ohm's law. For the sources in Fig. 3.12, if a conversion was made, a load connected to the terminals would not be able to tell which source was present.

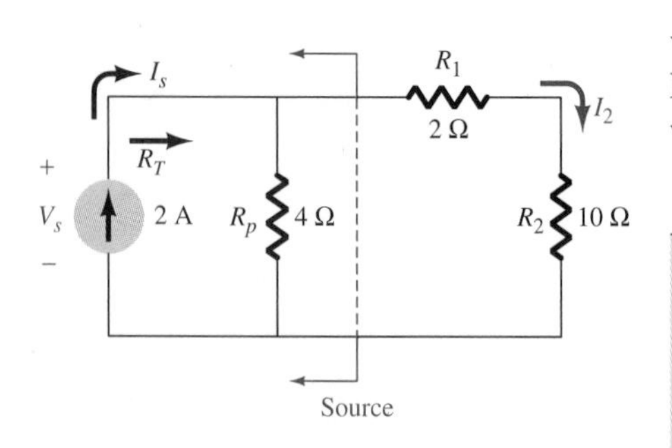

FIG. 3.13

EXAMPLE 3.4 For the network in Fig. 3.13:

a. Determine the voltage V_s.
b. Find the current I_2.
c. Convert the current source to a voltage source and calculate the current I_2. Compare with the results for part (b).

Solution:

a. $R'_T = R_1 + R_2 = 2\ \Omega + 10\ \Omega = 12\ \Omega$

$$R_T = R_p \parallel R'_T = 4\ \Omega \parallel 12\ \Omega = \frac{(4\ \Omega)(12\ \Omega)}{4\ \Omega + 12\Omega} = 3\ \Omega$$

$V_s = I_s R_T = (2\ \text{A})(3\ \Omega) = \mathbf{6\ V}$

b. Using the current-divider rule,

$$I_2 = \frac{R_P(I_s)}{R_P + R'_T} = \frac{4\ \Omega(2\ \text{A})}{4\ \Omega + 12\ \Omega} = \mathbf{0.5\ A}$$

c. Note Fig. 3.14.

$E_s = I_s R_P = (2\ \text{A})(4\ \Omega) = 8\ \text{V}$

$$I_2 = \frac{E_s}{R_s + R_1 + R_2} = \frac{8\ \text{V}}{4\ \Omega + 2\ \Omega + 10\ \Omega} = \mathbf{0.5A}$$

which agrees with part (b).

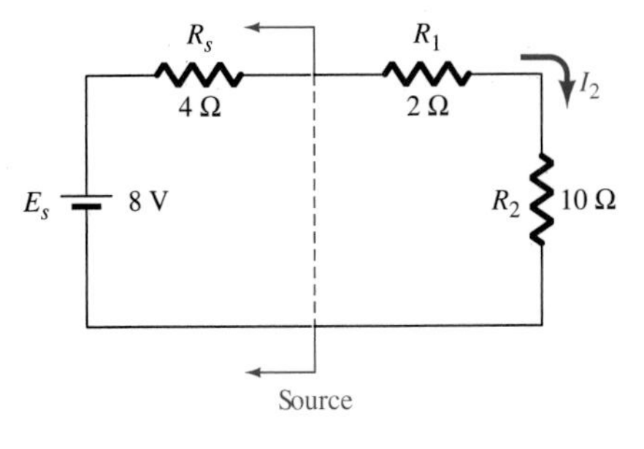

FIG. 3.14

Parallel Current Sources

For current sources in parallel, the net current is the algebraic sum of the current sources as determined by their direction. For the parallel sources in Fig. 3.15(a), the net current being supplied to the top terminal of the parallel elements is $I_1 + I_2 = 8$ A, while the current source I_3 is drawing 10 A from the same connection. The net result is a current source of

$$I_3 - (I_1 + I_2) = 10\ \text{A} - 8\ \text{A} = 2\ \text{A}$$

supplying current in the direction of I_3, as shown in Fig. 3.15(b).

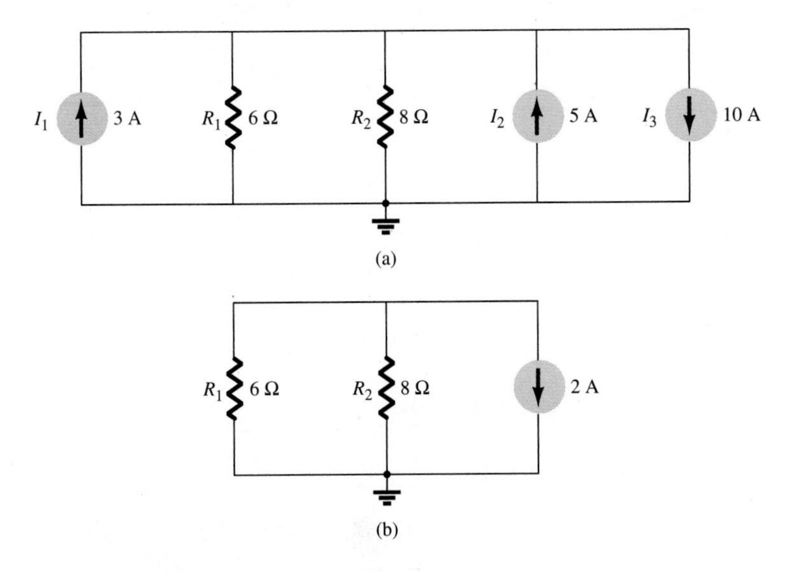

FIG. 3.15

EXAMPLE 3.5 For the network in Fig. 3.16:

a. Convert the voltage source to a current source.
b. Combine the current source in part (a) with the existing current source.
c. Find the voltage V.

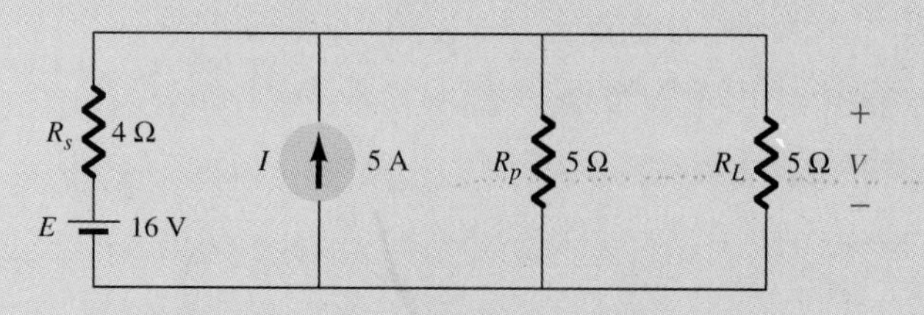

FIG. 3.16

Solution:

a. $I_s = \dfrac{E}{R_s} = \dfrac{16\text{ V}}{4\ \Omega} = \mathbf{4\ A}, \qquad R_p = R_s = \mathbf{4\ \Omega}$

b. The network is redrawn in Fig. 3.17. Combining current sources gives

$$I_T = I_s + I = 4\text{ A} + 5\text{ A} = \mathbf{9\ A}$$

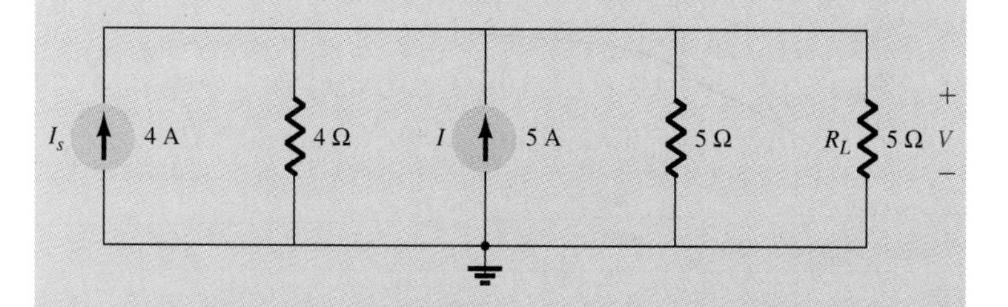

FIG. 3.17

c. Combining parallel resistive elements yields

$$R' = 5\ \Omega \parallel 5\ \Omega = \frac{5\ \Omega}{2} - 2.5\ \Omega$$

$$R_T = 4\ \Omega \parallel R' = 4\ \Omega \parallel 2.5\ \Omega = \frac{(4\ \Omega)(2.5\ \Omega)}{4\ \Omega + 2.5\ \Omega}$$

$$= \frac{10\ \Omega}{6.5} = 1.538\ \Omega$$

resulting in the network in Fig. 3.18 and

$$V = I_T R_T = (9\text{ A})(1.538\ \Omega)$$
$$= \mathbf{13.84\ V}$$

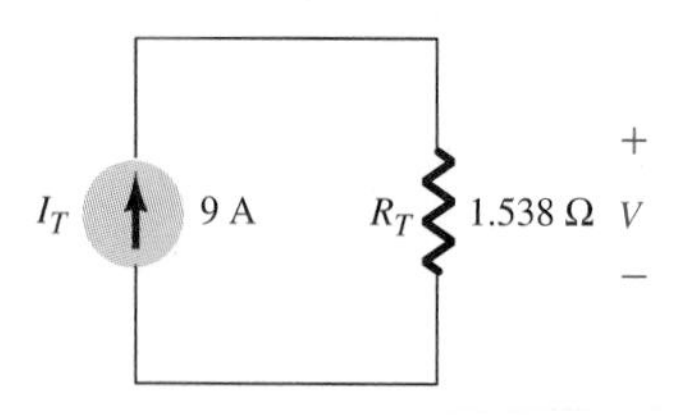

FIG. 3.18

3.4 VOLTAGE AND CURRENT REGULATION

For laboratory supplies, whether they be voltage or current, a regulation factor is provided that reveals at a glance the sensitivity of the supply to the applied load. For the ideal voltage supply the voltage set by the controls should maintain that value no matter what load current is supplied by the source. This condition is reflected by the ideal characteristic appearing in Fig. 3.19(b). However, let us define, as shown in Fig. 3.19(a), the voltage available at the supply terminals with *no load* attached as V_{NL}. Note that since a path for current is nonexistent, there is no voltage drop across R_s, and $V_{NL} = E_s$. When a load is applied, however, and adjusted so that the load current increases to its rated value, the internal voltage drop across R_s increases and the terminal voltage of the supply (or across the load) decreases. At rated conditions (full-load conditions) the voltage is denoted as V_{FL}. The graph in Fig. 3.19(b) clearly indicates that the closer the value of V_{NL} is to V_{FL}, the closer the characteristics of the supply to the ideal characteristics. The smaller the internal resistance of the supply, the closer to ideal the characteristics and the better the supply. The ideal can be obtained only with $R_s = 0\ \Omega$. A measure of the slope of this line is given on data sheets as the *voltage regulation* (VR), as defined by the equation

$$\text{voltage regulation} = \frac{V_{NL} - V_{FL}}{V_{FL}} \times 100\% \tag{3.1}$$

As indicated above, the smaller the difference between V_{NL} and V_{FL}, the better the supply, as evidenced by a very small numerator in Eq. (3.1) and a smaller voltage regulation. Supplies with voltage regulations of 0.1% or less are quite common. The current regulation for current sources is determined by an equation having the same format as Eq. (3.1).

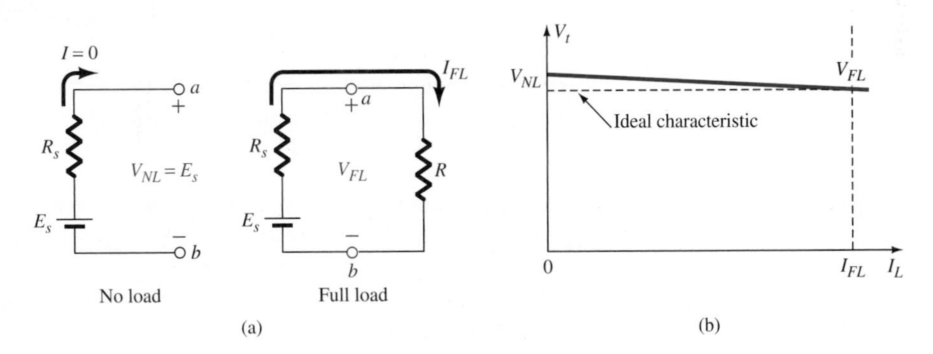

FIG. 3.19

3.5 MULTISOURCE NETWORKS

Networks with two or more voltage or current sources that are not in series or parallel cannot be analyzed using the methods described thus far.

Rather, an approach such as *superposition* or *branch-current analysis* must be used. Although the methods will be demonstrated with only two independent sources, each method can be applied to a network with any number of sources, whether they be current or voltage.

Superposition

The simplest analytical method to describe and usually the easiest to apply employs the *superposition theorem,* which states:

The current through any element in a dc network is the algebraic sum of the currents through that element due to each source independently.

The theorem can be restated, without change, for the voltage across an element in a dc network. To consider the effects of each source, the remaining sources must be properly removed. This requires that each voltage source be set to zero, reflecting a short-circuit condition, and that each current source be replaced by an open-circuit condition to indicate zero supply current. Any internal resistances associated with either type of source must remain when the effects of the source are removed.

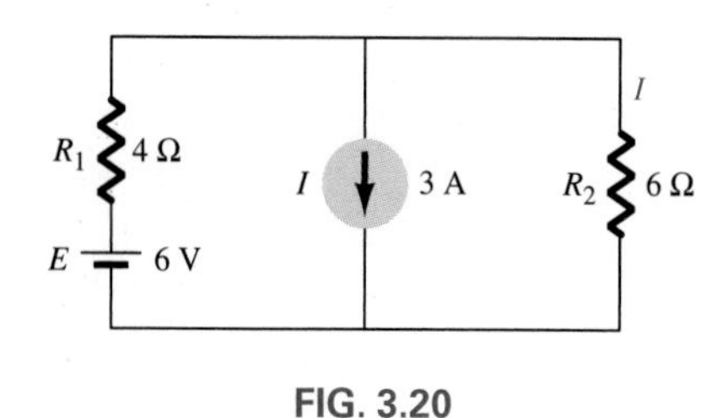

FIG. 3.20

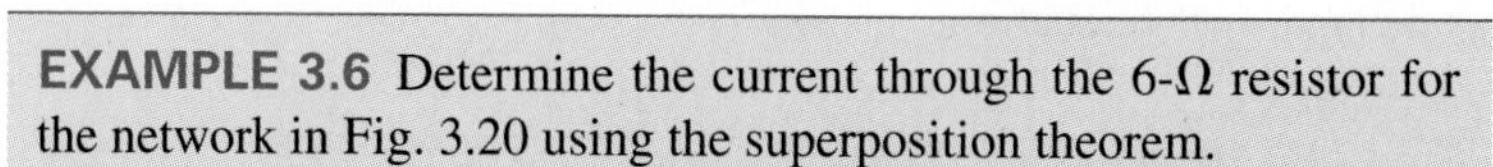

EXAMPLE 3.6 Determine the current through the 6-Ω resistor for the network in Fig. 3.20 using the superposition theorem.

Solution: We first consider the effects of the voltage source and remove the current source, as indicated in Fig. 3.21. The current

$$I' = \frac{E}{R_1 + R_2} = \frac{6\text{ V}}{4\ \Omega + 6\ \Omega} = 0.6\text{ A}$$

The effect of the current source can be examined by removing the voltage source, as shown in Fig. 3.22. Applying the current-divider rule gives

$$I'' = \frac{R_1(I)}{R_1 + R_2} = \frac{(4\ \Omega)(3\text{ A})}{4\ \Omega + 6\ \Omega} = \frac{12\text{ V}}{10\ \Omega} = 1.2\text{ A}$$

Note that I'' is opposite I' in direction, and so $I_{6\Omega} = I'' - I' = 1.2\text{ A} - 0.6\text{ A} = \mathbf{0.6\ A}$ in the direction of the larger, I''.

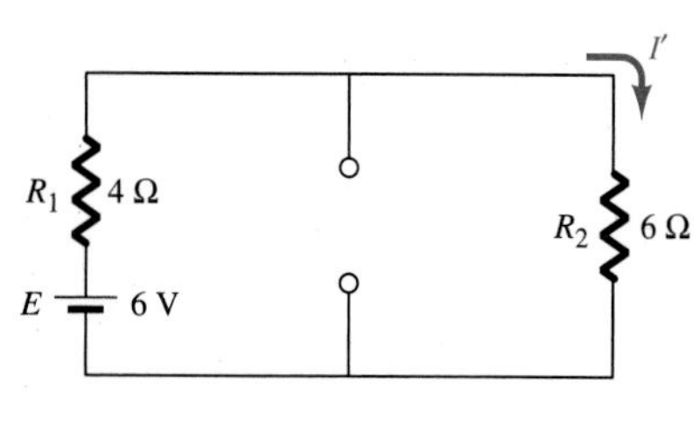

FIG. 3.21

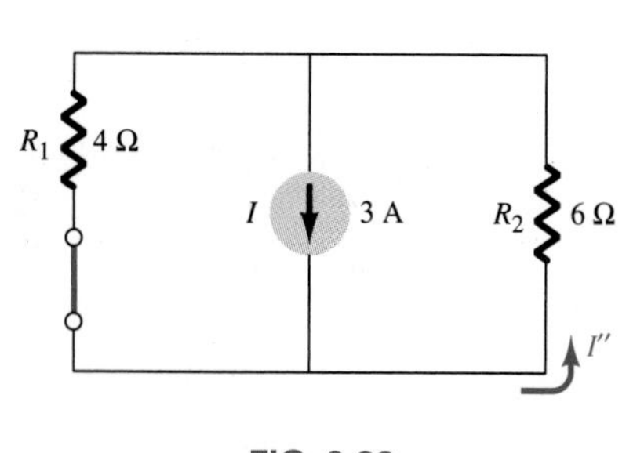

FIG. 3.22

The superposition theorem cannot be applied to power effects, since the power and the current are related by a nonlinear (I^2R) relationship. In other words, the net power cannot be determined by simply finding the sum of the powers delivered by each source. It must be determined from the final solution for the voltage or current.

Branch-Current Analysis

The second method to be applied is called *branch-current analysis.* It requires that all current sources first be converted to voltage sources. Each branch of the network is then assigned a labeled branch current, and Kirchhoff's voltage law is applied around each independent closed loop of the network. Kirchhoff's current law is then applied to provide the remaining necessary information. The term *independent* simply requires that each equation contain information not already present in the other equations. The criteria are usually satisfied by simply applying the law to each *window* of the network. In the following example there are two windows, and therefore two applications of Kirchhoff's voltage law are required. The final result is the currents of each branch, hence the name *branch-current analysis.*

EXAMPLE 3.7 Determine the current through each branch of the network in Fig. 3.23.

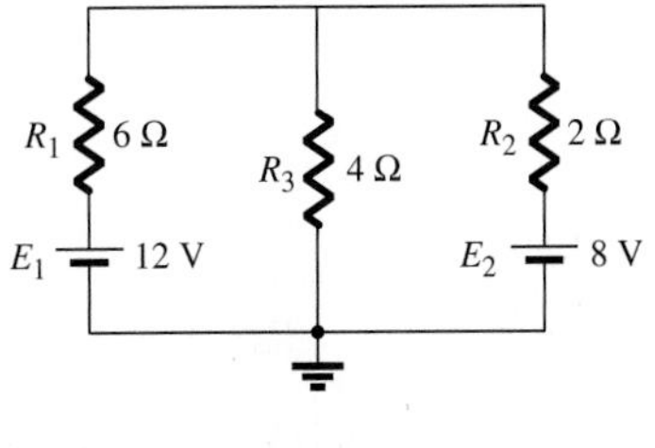

FIG. 3.23

Solution: In Fig. 3.24, the branch currents have been assigned, and the polarities across the resistors have been included for the *assumed* direction of branch current. If the solution indicates that the assumed direction was incorrect, a minus sign will appear.

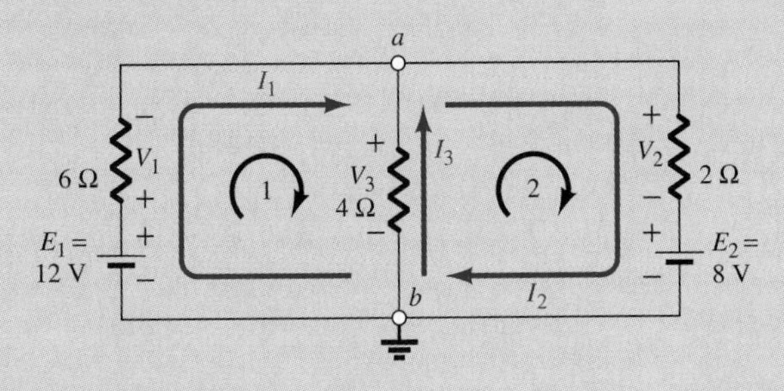

FIG. 3.24

Applying Kirchhoff's voltage law in the clockwise direction around loop 1, we have

$$E_1 - V_1 - V_3 = 0$$

and substituting, we obtain

$$+12 - 6I_1 - 4I_3 = 0$$

Note that the voltage source polarity is unaffected by the chosen branch-current direction. In addition, consider that the voltage across one resistor has a negative sign, while the other has a positive sign, evidence of the fact that the sign is determined solely by the assumed branch-current direction and not the fact that it is a resistive element.

For loop 2 (clockwise),

$$+V_3 - V_2 - E_2 = 0$$

and substituting, we obtain

$$4I_3 - 2I_2 - 8 = 0$$

Rearranging the two equations yields

$$6I_1 + 4I_3 = 12$$
$$4I_3 - 2I_2 = 8$$

We now have two equations and three unknowns. A third equation is required that is obtained by application of Kirchhoff's current law to either terminal *a* or *b* of the network in Fig. 3.24.

At node (the term for a junction of two or more branches) *a,* an application of Kirchhoff's current law results in

$$\underbrace{I_1}_{\text{Entering}} = \underbrace{I_2 + I_3}_{\text{Leaving}}$$

The result is three equations and three unknowns:

$$\begin{aligned} 6I_1 \qquad\quad + 4I_3 &= 12 \\ - 2I_2 + 4I_3 &= 8 \\ I_1 - I_2 + I_3 &= 0 \end{aligned}$$

The solutions can then be found using determinants according to the following format for I_1:

$$I_1 = \frac{\begin{vmatrix} 12 & 0 & 4 \\ 8 & -2 & 4 \\ 0 & -1 & -1 \end{vmatrix}}{\begin{vmatrix} 6 & 0 & 4 \\ 0 & -2 & 4 \\ 1 & -1 & -1 \end{vmatrix}}$$

The procedure for finding the value of I_1 from the preceding matrix pattern is covered in a number of basic mathematics courses.

Use of the computer software package MathCad 2.5 to perform the calculation results in the following display of the input data and solution:

$$n := \begin{bmatrix} 12 & 0 & 4 \\ 8 & -2 & 4 \\ 0 & -1 & -1 \end{bmatrix} \qquad d := \begin{bmatrix} 6 & 0 & 4 \\ 0 & -2 & 4 \\ 1 & -1 & -1 \end{bmatrix} \qquad Il := \frac{|n|}{|d|} \qquad Il = 0.909$$

On the TI-86 calculator the input data and solution will appear as follows:

det[[12, 0, 4][8, −2, 4][0, −1, −1]]/det[[6, 0, 4][0, −2, 4][1, −1, −1]] (ENTER)	0.909

Note that the *det* (determinant) is obtained from a Math listing under a MATRX menu, and each determinant must be determined individually. One set of brackets must encompass all the parameters of the *det,* and each internal set of brackets must include all the numbers of a row separated by commas. Obviously, the time taken to learn how to enter the parameters is minimal when you consider the savings in time and the accuracy obtained.

Instead of using third-order determinants, we could reduce the three equations to two by solving for one of the variables in one equation and substituting into the other two equations. For instance, solving for I_1 in Kirchhoff's current law equation results in

$$I_3 = I_1 - I_2$$

Substituting for I_3 in the top equation results in

$$\begin{aligned} 6I_1 + 4(I_1 - I_2) &= 12 \\ 6I_1 + 4I_1 - 4I_2 &= 12 \\ 10I_1 - 4I_2 &= 12 \end{aligned}$$

Substituting for I_3 in the bottom equation results in

$$\begin{aligned} 4(I_1 - I_2) - 2I_2 &= 8 \\ 4I_1 - 4I_2 - 2I_2 &= 8 \end{aligned}$$

or

$$6I_2 - 4I_1 = -8$$

The two resulting equations are

$$\left.\begin{aligned} 10I_1 - 4I_2 &= 12 \\ 6I_2 - 4I_1 &= -8 \end{aligned}\right. \qquad \textbf{(3.2)}$$

We now have two equations and two unknowns. The solution for I_1 or I_2 can now be obtained by using determinants or simply by solving for one of the variables and substituting into the other equation.

Using determinants:

First, the equations must be set up with the same variable in each column and the number of minus signs minimized simply to reduce the possibility of numerical errors

$$\begin{aligned} 10I_1 - 4I_2 &= 12 \\ 4I_1 + 6I_2 &= \ \ 8 \end{aligned}$$

Then,

$$I_1 = \frac{\begin{vmatrix} 12 & -4 \\ 8 & -6 \end{vmatrix}}{\begin{vmatrix} 10 & -4 \\ 4 & -6 \end{vmatrix}} = \frac{(12)(-6) - (-4)(8)}{(10)(-6) - (-4)(4)} = \frac{-72 + 32}{-60 + 16}$$

$$= \frac{-40}{-44} = \mathbf{0.909\ A} \quad \text{(as in the CALC 3.1 display)}$$

or, using the TI-86 calculator, as follows:

det[[12, −4][8, −6]]/det[[10, −4][4, −6]] (ENTER)	0.909

Continuing:

$$I_2 = \frac{\begin{vmatrix} 10 & 12 \\ 4 & 8 \end{vmatrix}}{-44} = \frac{(10)(8) - (12)(4)}{-44} = \frac{80 - 48}{-44}$$

$$= \frac{32}{-44} = \mathbf{-0.727\ A}$$

and finally,

$$I_3 = I_1 - I_2 = \mathbf{1.636\ A}$$

Using the substitution approach:

$$I_1 = \frac{12 + 4I_2}{10} = 1.2 + 0.4I_2$$

$$6I_2 - 4(1.2 + 0.4I_2) = -8$$

$$6I_2 - 4.8 - 1.6I_2 = -8$$

$$4.4I_2 = -8 + 4.8$$

$$I_2 = -\frac{3.2}{4.4} = \mathbf{-0.727\ A}$$

Solving for I_1, we obtain

$$\begin{aligned} I_1 &= 1.2 + 0.4I_2 \\ &= 1.2 + 0.4(-0.7273) \\ &= 1.2 - 0.2909 \\ &= \mathbf{0.909\ A} \end{aligned}$$

and the current I_3 is

$$\begin{aligned} I_3 &= I_1 - I_2 \\ &= 0.9091\ \text{A} - (-0.7273\ \text{A}) \\ &= \mathbf{1.636\ A} \end{aligned}$$

The negative sign associated with the current I_2 simply reveals that the *actual* current direction is the opposite of that *assumed* in Fig. 3.24. The magnitude, however, is correct.

The voltage across each resistor can then be determined using Ohm's law:

$$V_1 = I_1R_1 = (0.909\ \text{A})(6\ \Omega) = \mathbf{5.454\ V}$$

$$V_2 = I_2R_2 = (-0.727\ \text{A})(2\ \Omega) = \mathbf{-1.454\ V}$$

$$V_3 = I_3R_3 = (1.636\ \text{A})(4\ \Omega) = \mathbf{6.544\ V}$$

The minus sign for V_2 simply reveals that the *actual* polarity of V_2 is the opposite of that *assumed* in Fig. 3.24.

As a check let us apply Kirchhoff's voltage law around loop 1:

$$E - V_1 - V_3 = 0$$

or

$$E = V_1 + V_3$$

$$12\text{ V} = 5.454\text{ V} + 6.544\text{ V} = 11.998\text{ V} \cong 12\text{ V}$$

$$12\text{ V} = 12\text{ V} \qquad \text{(checks)}$$

Mesh Analysis

Mesh analysis, or *loop analysis* as it is often called, is an approach that provides all the currents of a network using a procedure that can be quite mechanical in nature. In actuality, mesh analysis is an extension of the branch-current method, as will be demonstrated in the example to follow.

In Fig. 3.25 the network in Fig. 3.24 has been redrawn with a mesh (loop) current assigned to each window of the network. A *window* is simply a framed-in area of the network (there are two in Fig. 3.25). The procedure is not limited to a window approach, but it is a useful method at the introductory level to ensure that all the resulting equations are independent. Note, in particular, that each mesh current loops around the entire frame. In actuality the current through E_1 and R_1 is I_1, I_2 through E_2 and R_2, and $I_1 - I_2$ for R_3 if we assume $I_1 > I_2$. By using these currents, we can apply Kirchhoff's voltage law to each window as follows.

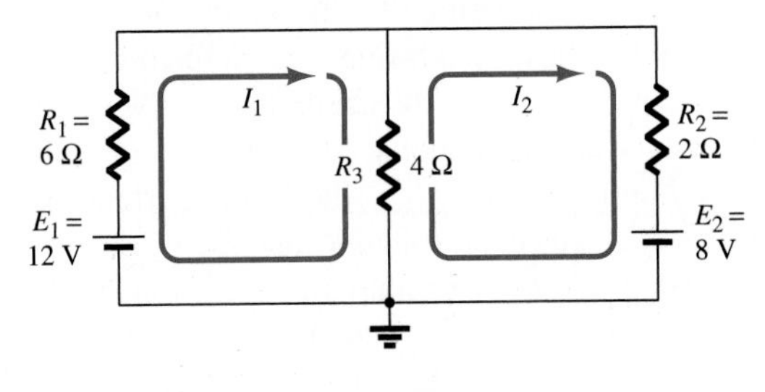

FIG. 3.25

For loop 1,

$$E_1 - I_1R_1 - (I_1 - I_2)R_3 = 0$$

resulting in

$$I_1(R_1 + R_3) - I_2R_3 = E_1$$

For loop 2,

$$(I_1 - I_2)R_3 - I_2R_2 - E_2 = 0$$

resulting in

$$I_2(R_2 + R_3) - I_1R_3 = -E_2$$

The following two equations with two unknowns result:

$$\begin{aligned} I_1(R_1 + R_3) - I_2R_3 &= E_1 \\ I_2(R_2 + R_3) - I_1R_3 &= -E_2 \end{aligned} \tag{3.3}$$

Substitution of numerical values gives

$$\begin{aligned} 10I_1 - 4I_2 &= 12 \\ 6I_2 - 4I_1 &= -8 \end{aligned}$$

which are exactly the same as Eq. (3.2) obtained using the branch-current method. In essence, therefore, by letting the mesh currents define the current through R_3, we have eliminated the need to apply Kirchhoff's current law.

There is another important benefit to be derived from Eq. (3.3) that will allow us to write the mesh equations by inspection in the future. In the first equation, for instance, the mesh current I_1 is multiplied by the sum of the resistors it "touches" in its window. The mutual resistor (the resistor with both mesh currents through it) is multiplied by the other mesh current and subtracted from the first element of the equation. The equation is then equal to the voltage source supporting the assigned direction of I_1. For the second equation, note that I_2 is multiplied by only R_2 and R_3, since it does not touch R_1. The mutual term is again subtracted using the other mesh current. The negative sign is associated with the 8-V source, revealing that the source does not support the direction of I_2. The preceding few sentences essentially define how the mesh equations for any dc network can be written by inspection. It is important at this stage to point out that these conclusions and defined procedure are valid only if all the mesh currents are chosen in the clockwise direction.

Once the results are obtained, it is important to fully understand how they should be interpreted. I_1 is the actual current through R_1 and E_1, and I_2 is the actual current through R_2 and E_2. The current through R_3 is determined by $I_1 - I_2$, assuming the direction of I_1. A negative result for I_3 simply means that the actual current is that of I_2 (the opposite of I_1).

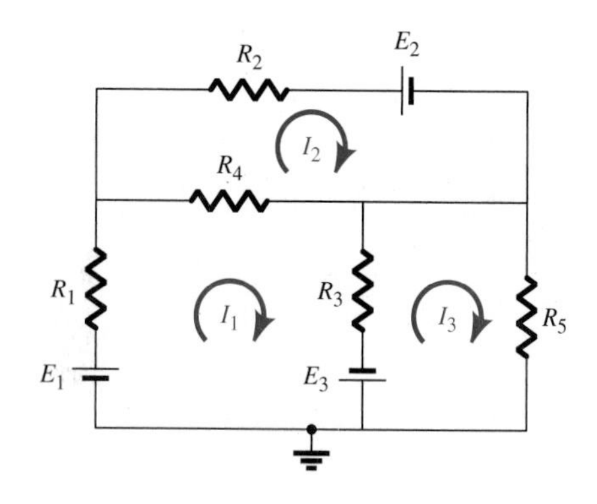

FIG. 3.26

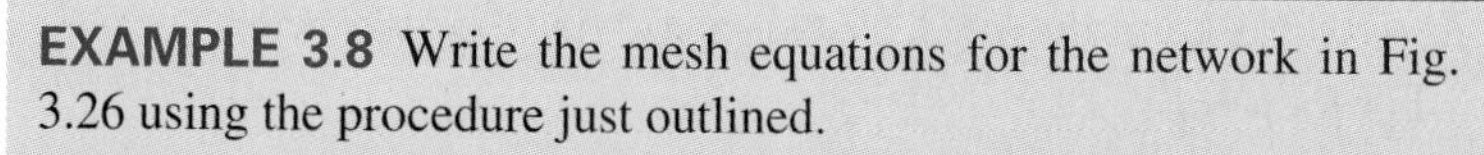

EXAMPLE 3.8 Write the mesh equations for the network in Fig. 3.26 using the procedure just outlined.

Solution:

$$\begin{aligned} I_1(R_1 + R_4 + R_3) - I_2R_4 - I_3R_3 &= E_1 + E_3 \\ I_2(R_2 + R_4) - I_1R_4 &= -E_2 \\ I_3(R_3 + R_5) - I_1R_3 &= -E_3 \end{aligned}$$

Nodal Analysis

Nodal analysis is an approach to the analysis of multiloop networks that can determine all the voltages of a network. The currents can then be determined using Ohm's law. The term *nodal* is derived from the fact that each voltage is defined between a *node* and some reference point, a node simply being a connection point in the network for two or more elements.

Although the method can be applied to networks with both voltage and current sources, the format approach (parallel to the method established for mesh analysis) requires that only current sources be present. As a general first step, therefore, all voltage sources should be converted to current sources before the method is applied.

Consider, for example, the network in Fig. 3.27 with three nodes (three junction points holding the network together). The bottom node at ground potential (0 V) is defined as the reference node, and the other two nodes are assigned the labels V_1 and V_2. Nodal analysis will now determine the voltages V_1 and V_2 with respect to the reference level.

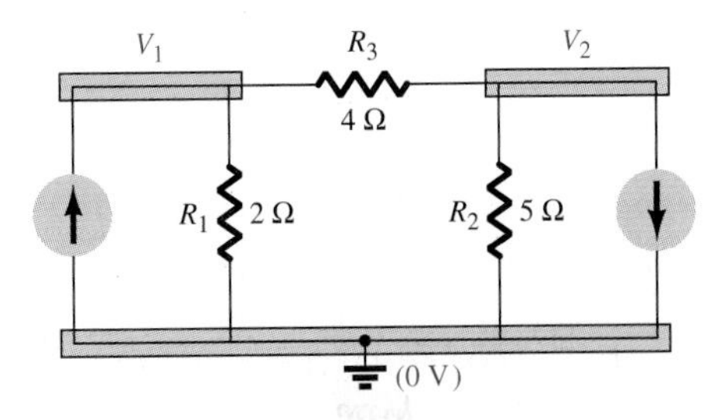

FIG. 3.27

Applying Kirchhoff's current law at node V_1, we have

$$\Sigma I_i = \Sigma I_o$$

$$I_1 = \frac{V_{R_1}}{R_1} + \frac{V_{R_3}}{R_3} = \frac{V_1}{R_1} + \frac{V_1 - V_2}{R_3}$$

and

$$V_1\left(\frac{1}{R_1} + \frac{1}{R_3}\right) - V_2\left(\frac{1}{R_3}\right) = I_1$$

Applying Kirchhoff's current law at node V_2, we have

$$\Sigma I_i = \Sigma I_o$$

$$I_{R_3} = I_{R_2} + I_2$$

$$\frac{V_1 - V_2}{R_3} = \frac{V_2}{R_2} + I_2$$

and

$$V_2\left(\frac{1}{R_2} + \frac{1}{R_3}\right) - V_1\left(\frac{1}{R_3}\right) = -I_2$$

The result is the following two equations with two unknowns:

$$V_1\left(\frac{1}{R_1} + \frac{1}{R_3}\right) - V_2\left(\frac{1}{R_3}\right) = I_1$$
$$V_2\left(\frac{1}{R_2} + \frac{1}{R_3}\right) - V_1\left(\frac{1}{R_3}\right) = -I_2 \qquad \textbf{(3.4)}$$

Substituting numerical values, we obtain

$$V_1(\tfrac{1}{2} + \tfrac{1}{4}) - V_2(\tfrac{1}{4}) = 6$$
$$V_2(\tfrac{1}{5} + \tfrac{1}{4}) - V_1(\tfrac{1}{4}) = -1$$

$$0.75V_1 - 0.25V_2 = 6$$
$$0.45V_2 - 0.25V_1 = -1$$

Rearranging and applying determinants, we have

$$0.75V_1 - 0.25V_2 = 6$$
$$0.25V_1 - 0.45V_2 = 1$$

$$V_1 = \frac{\begin{vmatrix} 6 & -0.25 \\ 1 & -0.45 \end{vmatrix}}{\begin{vmatrix} 0.75 & -0.25 \\ 0.25 & -0.45 \end{vmatrix}} = \frac{(6)(-0.45) - (-0.25)(1)}{(0.75)(-0.45) - (-0.25)(0.25)}$$

$$= \frac{-2.7 + 0.25}{-0.3375 + 0.0625}$$

$$= \frac{-2.45}{-0.275} = \mathbf{8.909\ V}$$

$$V_2 = \frac{\begin{vmatrix} 0.75 & 6 \\ 0.25 & 1 \end{vmatrix}}{-0.275} = \frac{(0.75)(1) - (6)(0.25)}{-0.275} = \frac{0.75 - 1.5}{-0.275}$$

$$= \frac{-0.75}{-0.275} = \mathbf{2.727\ V}$$

Since V_1 is 8.909 V above the reference level of 0 V, the voltage across the current source I_1 and resistor R_1 is 8.909 V. Similarly, the voltage across R_2 and I_2 is 2.727 V. The voltage across R_3 is $V_1 - V_2 = 6.182$ V.

The currents can then be determined using Ohm's law:

$$I_{R_1} = \frac{V_{R_1}}{R_1} = \frac{V_1}{R_1} = \frac{8.909\ \text{V}}{2\ \Omega} = \mathbf{4.455\ A}$$

$$I_{R_2} = \frac{V_{R_2}}{R_2} = \frac{V_2}{R_2} = \frac{2.727\ \text{V}}{5\ \Omega} = \mathbf{0.545\ A}$$

$$I_{R_3} = \frac{V_{R_3}}{R_3} = \frac{V_1 - V_2}{R_3} = \frac{6.182\ \text{V}}{4\ \Omega} = \mathbf{1.546\ A}$$

As a check, let us apply Kirchhoff's current law at node V_1:

$$I_1 = I_{R_1} + I_{R_3}$$
$$6\ \text{A} = 4.455\ \text{A} + 1.546\ \text{A}$$
$$6\ \text{A} = 6\ \text{A} \qquad \text{(checks)}$$

A format approach to writing the nodal equations by inspection can be developed by noting that in Eq. (3.4) each nodal voltage is multiplied by the sum of the conductances G touching that node (recall that $G = 1/R$). The second term in each equation is called the *mutual term* because R_3

forms a bridge between V_1 and V_2. If V_1 and V_2 are assumed positive with respect to ground, the mutual term is always subtracted from the node of interest. The right side of each equation is dependent on the current sources connected to the node. If supplying current to the node, a positive sign is assigned, while if drawing current from the node, a negative sign is applied.

EXAMPLE 3.9 Write the nodal equations for the network in Fig. 3.28 using the format approach just described.

Solution:

$$V_1\left(\frac{1}{R_1} + \frac{1}{R_2}\right) - V_2\left(\frac{1}{R_2}\right) = -I_1$$

$$V_2\left(\frac{1}{R_2} + \frac{1}{R_3} + \frac{1}{R_4}\right) - V_1\left(\frac{1}{R_1}\right) - V_3\left(\frac{1}{R_4}\right) = 0$$

$$V_3\left(\frac{1}{R_4} + \frac{1}{R_5}\right) - V_2\left(\frac{1}{R_4}\right) = I_2$$

Note that only R_1 and R_2 touch V_1, V_1 has only one mutual term with V_2 (through R_2), and I_1 is drawing current from node 1. V_2 has mutual terms with V_1 and V_3 but is not connected to either current source.

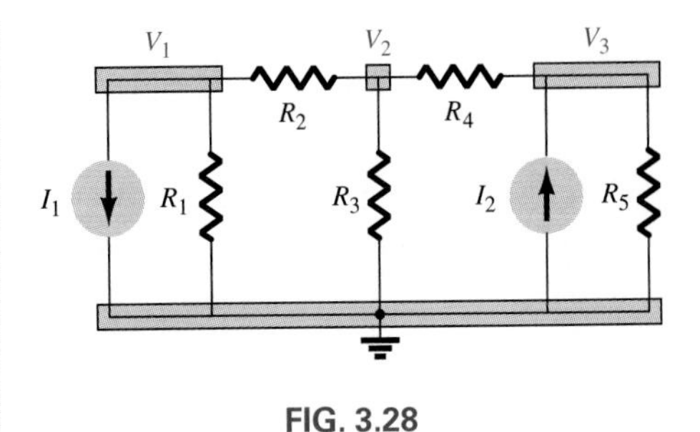

FIG. 3.28

3.6 COMPUTER ANALYSIS

We shall now determine a mesh current and all the nodal voltages for the complex dc network of Fig. 3.29 using both PSpice and Electronics Workbench. This example provides an opportunity to compare two of the most popular software programs for analyzing electrical and electronic networks.

For PSpice the components of the network are placed on the screen as shown in Fig. 3.30 using the procedure described in Chapter 2. When the analysis is requested the magnitude of the mesh current I_3 will appear next to the ammeter (called **IPROBE**), as shown in the figure. The voltage from each node to ground will also appear on the horizontal bar of the nodal voltage indicators (called **VIEWPOINT**), as shown in the same figure. Since each voltage is with respect to ground, the voltage across R_2 is 3.699 V, and across R_4, 2.8 V, since one end of each is connected to ground potential (0 V). The voltage across R_5 is 15 V $-$ 2.8 V $=$ 12.2 V; across R_1, 15 V $-$ 3.699 V $=$ 11.301 V; and across R_3, 3.699 V $-$ 2.8 V $=$ 0.899 V. The mesh current through R_1 can be calculated using $I_{R_1} = V_{R_1}/8\ \Omega = 11.301\ \text{V}/8\ \Omega = 1.413$ A, and through R_3, $I_{R_3} = V_{R_3}/5\ \Omega = 0.899\ \text{V}/5\ \Omega = 0.18$ A. The current through R_2 is then simply $I_{R_1} - I_{R_3} = 1.233$ A, and so on.

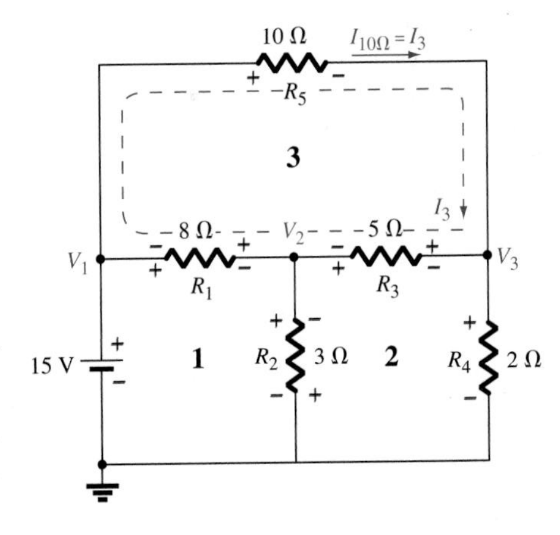

FIG. 3.29

The screen obtained using Electronics Workbench will be quite different, with meters placed as shown in Fig. 3.31 to indicate the level of

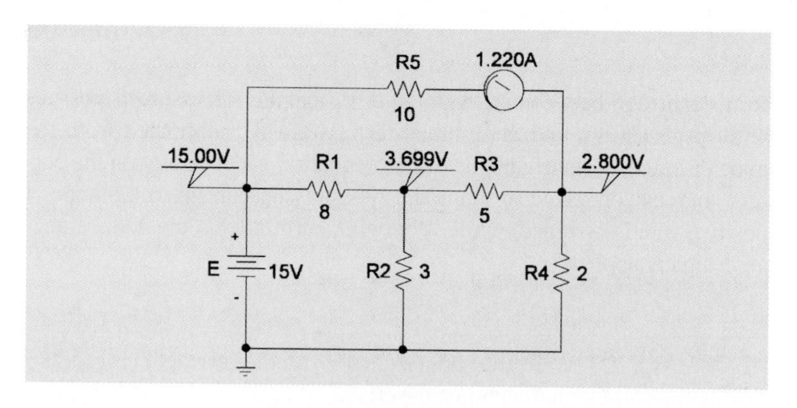

FIG. 3.30
Determining the mesh (loop) current $I_{10\Omega}$ and the nodal voltages for the network of Fig. 3.29 using PSpice (Windows).

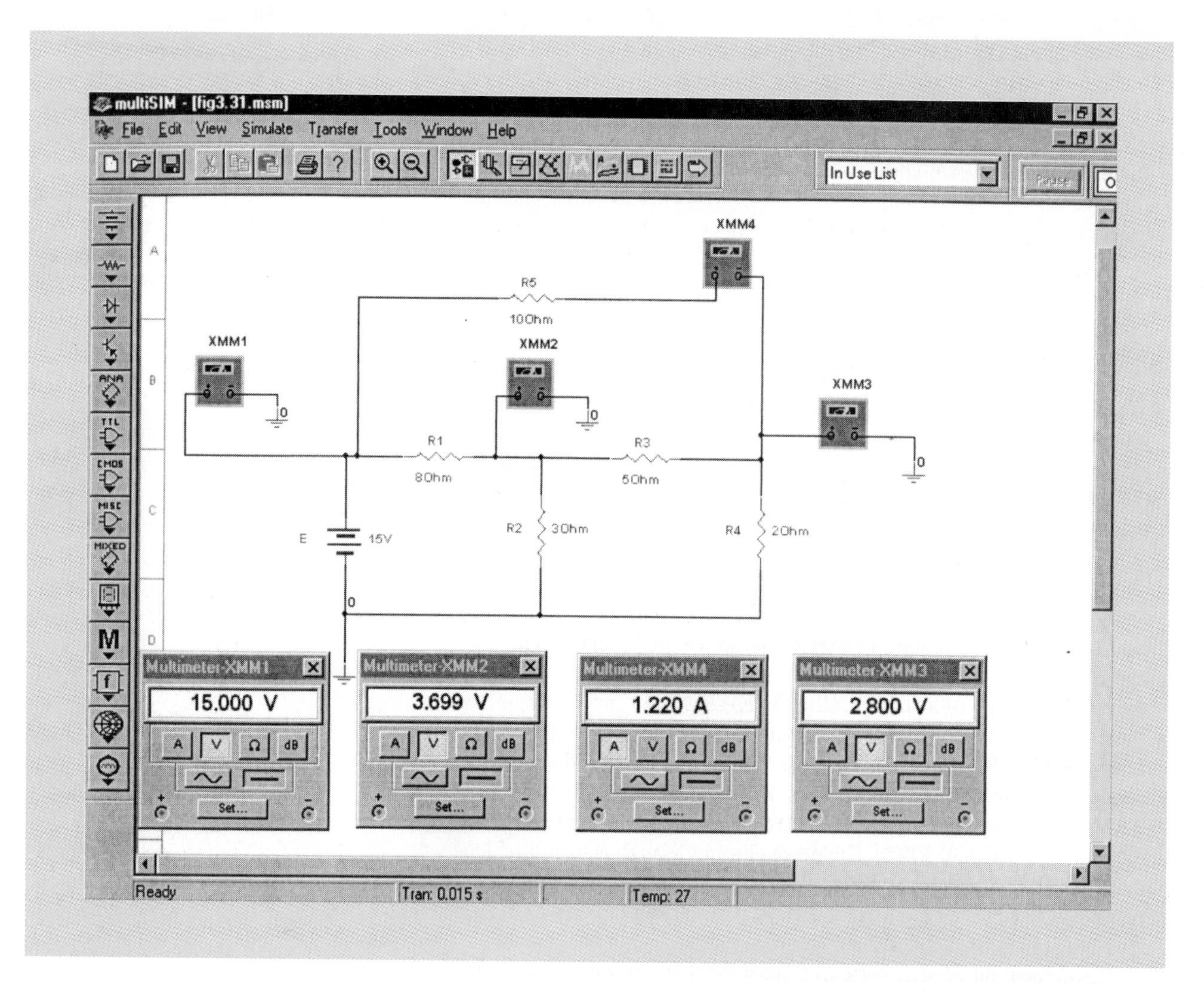

FIG. 3.31

each quantity. The network has to be drawn on the screen using a procedure that is similar in many ways to that used for PSpice. The results, as indicated, are the same for each approach.

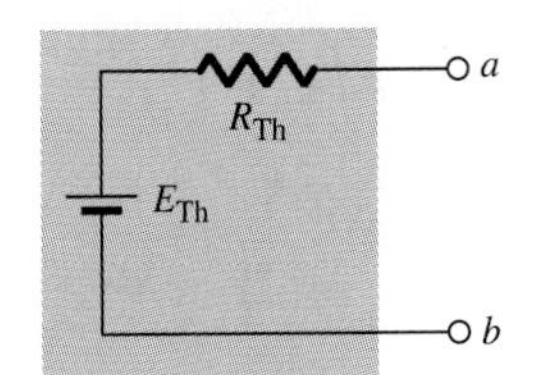

FIG. 3.32
Thévenin equivalent circuit.

3.7 NETWORK THEOREMS

There are a number of theorems that are extremely useful in the analysis and design of electrical systems. Although the analysis here is for dc networks, the theorems can also be applied to ac networks (Chapter 5) with very little change in their mode of application. The theorems to be included are *Thévenin's theorem* and the *maximum power theorem.*

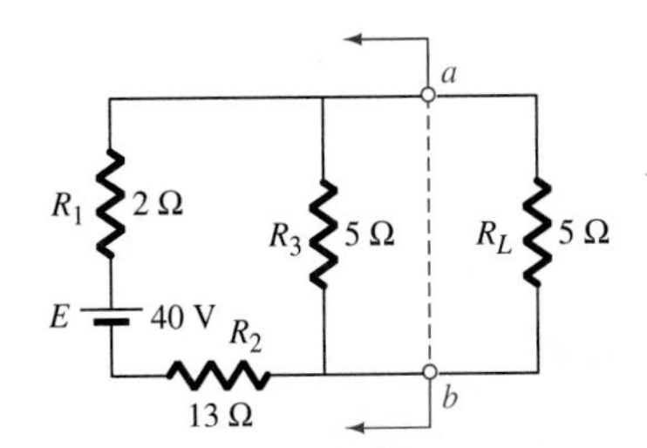

FIG. 3.33

Thévenin's Theorem

Thévenin's theorem permits the reduction of a two-terminal dc network with any number of resistors and sources to one having only one source and one internal resistor in the series configuration in Fig. 3.32.

The Thévenin resistance R_{Th} is the dc resistance between the output terminals of the network to be reduced, with all sources (current and voltage) set to zero. The Thévenin voltage E_{Th} is the open-circuit voltage between the output terminals, with all sources present as in the original network. Consider, as an example, the network in Fig. 3.33 in which the network to the left of points *a* and *b* is to be replaced by a Thévenin equivalent circuit. We shall then see that the current though R_L for various values of R_L can be determined very quickly from the reduced Thévenin equivalent circuit rather than analyzing the entire network again for each R_L.

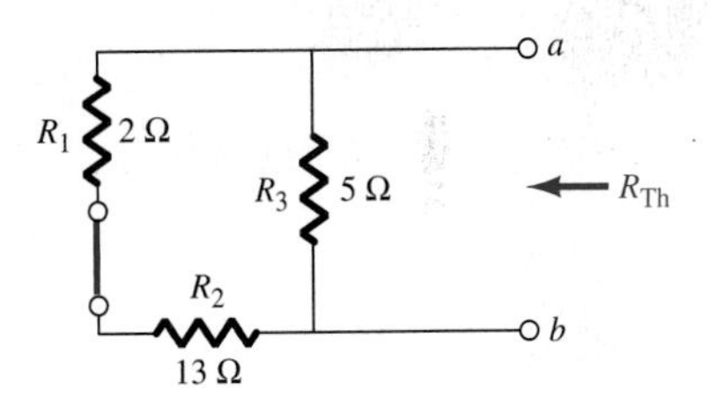

FIG. 3.34
Determining R_{Th}.

To determine R_{Th}, we set all voltage sources to zero by replacing each with a short-circuit equivalent, which results in the network in Fig. 3.34.

$$R_{Th} = R_3 \parallel (R_1 + R_2) = \frac{(5\ \Omega)(15\ \Omega)}{5\ \Omega + 15\ \Omega} = \mathbf{3.75\ \Omega}$$

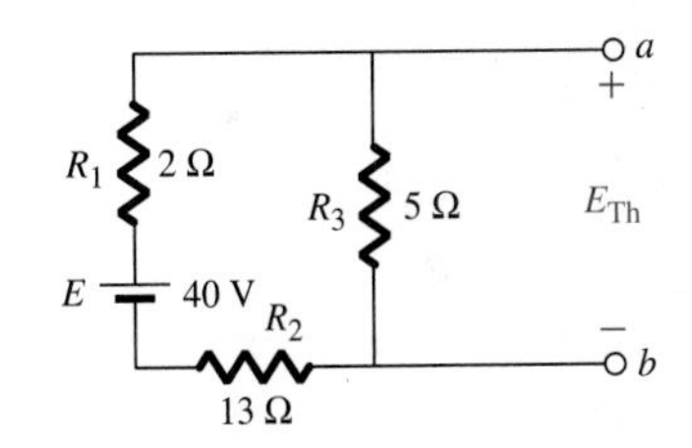

FIG. 3.35
Determining E_{Th}.

To determine E_{Th}, we replace the sources and determine the open-circuit voltage as shown in Fig. 3.35.

$$E_{Th} = V_{R_3} = \frac{R_3E}{R_T} = \frac{(5\ \Omega)(40\ \text{V})}{20\ \Omega}$$

$$= \frac{200\ \text{V}}{20} = \mathbf{10\ V}$$

The Thévenin circuit appears in Fig. 3.36 with the load resistor R_L replaced between terminals *a* and *b*.

Insofar as the load resistor R_L is concerned, the network to the left of *a* and *b* is the same as before the reduction. The current through R_L is

$$I_L = \frac{E_{Th}}{R_{Th} + R_L} = \frac{10\ \text{V}}{3.75\ \Omega + 5\ \Omega} = \mathbf{1.143\ A}$$

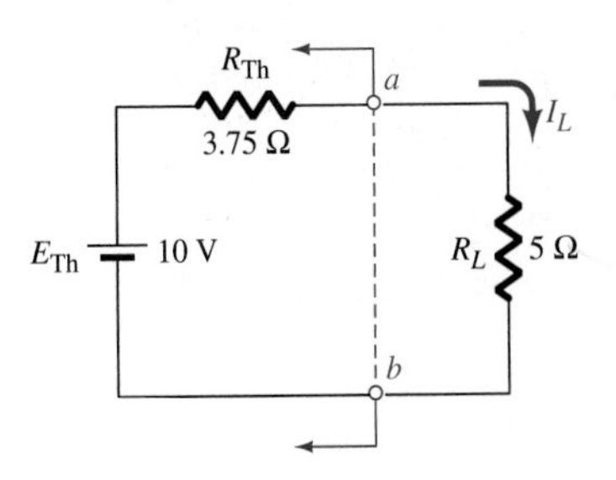

FIG. 3.36

An application of series–parallel techniques to the original network would result in the same solution for I_L. However, if R_L should now be changed to 30 Ω (for example), the entire series–parallel network would have to be reexamined if it were not for our Thévenin equivalent, which permits the following Ohm's law application for the new load current:

$$I_L = \frac{E_{Th}}{R_{Th} + R_L} = \frac{10\ \text{V}}{3.75\ \Omega + 30\ \Omega} = \mathbf{0.296\ A}$$

Consider also the savings in components if the elements of the original network could be replaced by the two required for the Thévenin equivalent. Additional examples involving the use of this theorem will appear throughout the book.

Maximum Power Transfer Theorem

The *maximum power transfer theorem* is used to ensure that a load receives maximum power from the supply. In words, it states that a load receives maximum power when its terminal resistance is equal to the Thévenin resistance seen by the load. In the preceding example, R_L should be 3.75 Ω in order to receive maximum power. For greater or lesser values the power delivered to the load decreases. In general, for maximum power to a load,

$$\boxed{R_L = R_{Th}} \tag{3.5}$$

In Fig. 3.37, where $R_L = R_{Th}$, we find that

$$I_L = \frac{E_{Th}}{R_L + R_{Th}} = \frac{E_{Th}}{2R_{Th}}$$

and

$$P_L = I_L^2 R = \left(\frac{E_{Th}}{2R_{Th}}\right)^2 R_{Th}$$

or

$$\boxed{P_{L\max} = \frac{E_{Th}^2}{4R_{Th}}} \tag{3.6}$$

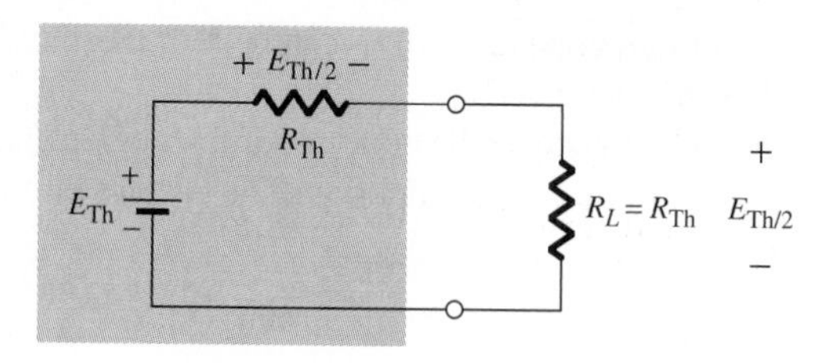

FIG. 3.37
Maximum power transfer conditions.

For the preceding example, the maximum power that can be delivered to R_L is

$$P_{L\max} = \frac{E_{\text{Th}}^2}{4R_{\text{Th}}} = \frac{(10\text{ V})^2}{4(3.75\ \Omega)} = \mathbf{6.667\ W}$$

Always keep in mind that the power that does not reach a load connected directly to a supply is lost in the internal resistance of the source and serves no useful purpose (reducing the overall efficiency of the system). The accepted level of efficiency must be carefully weighed, particularly if large amounts of power are involved.

3.8 Δ–Y CONVERSIONS

Occasionally, a network configuration is encountered where the elements do not appear to be in series or parallel. One such configuration is shown in Fig. 3.38. Note, in particular, that R_1 and R_3 are not in parallel, since they do not have two points in common, and yet they are not in series, since a third element is connected to the common point between the two resistors. In fact, in the entire network no two elements are in series or parallel. A solution can be obtained by converting the delta (Δ) configuration in the upper part of the figure to a wye (Y) configuration, as shown in Fig. 3.39. This description accounts for the section title, Δ–Y conversions. It is necessary only to have the following converting equations, which refer to the resistors as labeled in Fig. 3.39.

Δ→Y *conversion*

$$R_6 = \frac{R_1R_3}{R_1 + R_3 + R_5} \qquad R_7 = \frac{R_1R_5}{R_1 + R_3 + R_5} \qquad R_8 = \frac{R_3R_5}{R_1 + R_3 + R_5} \tag{3.7}$$

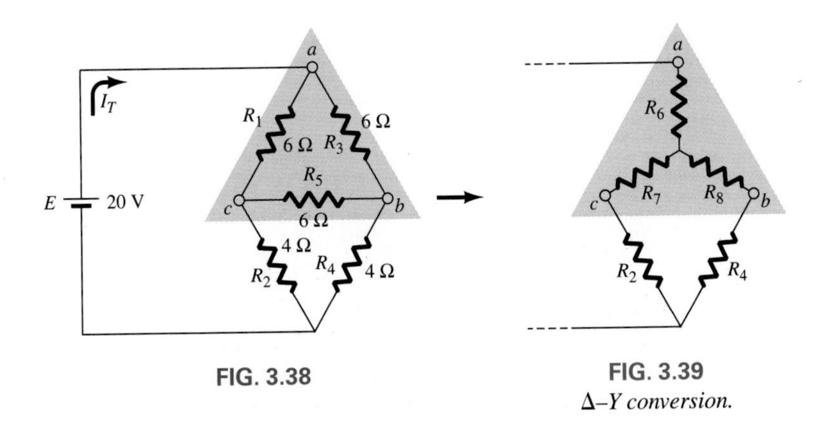

FIG. 3.38

FIG. 3.39
Δ–Y conversion.

It is sometimes necessary to convert from a Y to a Δ configuration. The necessary converting equations follow.

Y → Δ conversion

$$R_1 = \frac{R_6R_8 + R_7R_8 + R_6R_7}{R_8} \qquad R_3 = \frac{R_6R_8 + R_7R_8 + R_6R_7}{R_7}$$
$$R_5 = \frac{R_6R_8 + R_7R_8 + R_6R_7}{R_6} \qquad \textbf{(3.8)}$$

The relative positions of the resistors with regard to terminals *a*, *b*, and *c* must be carefully noted to use the conversion equations properly. Once the resistor values of the Y configuration are known, the delta (R_1, R_3, and R_5) configuration is completely removed and *replaced* between the same three terminals with the Y (R_6, R_7, and R_8) configuration. You will then find that R_4 and R_8 are in series and as a unit in parallel with the series combination of R_2 and R_7. R_T and I_T can then be obtained using series–parallel techniques, discussed earlier. A solution could have been obtained using a method of analysis such as branch-current or nodal analysis, but either of these techniques would have resulted in a set of at least three simultaneous equations, which would prove very time-consuming to those without a background in determinants.

If the configuration is such that $R_1 = R_3 = R_5$, or, in words, that each element of the delta has the same value, then the equations reduce to the simple form

$$R_Y = \frac{R_\Delta}{3} \quad \textit{or} \quad R_\Delta = 3R_Y \qquad \textbf{(3.9)}$$

For the network in Fig. 3.38, where this condition is satisfied,

$$R_Y = \frac{R_\Delta}{3} = \frac{6\ \Omega}{3} = 2\ \Omega$$

and the configuration in Fig. 3.40 results, where $R'_T = R_4 + R_8 = R_2 + R_7 = 6\ \Omega$, and $R_T = R_6 + (R'_T/2) = 2\ \Omega + \frac{6}{2}\ \Omega = 5\ \Omega$, with $I_T = E/R_T = 20\text{V}/5\Omega = 4\text{ A}$.

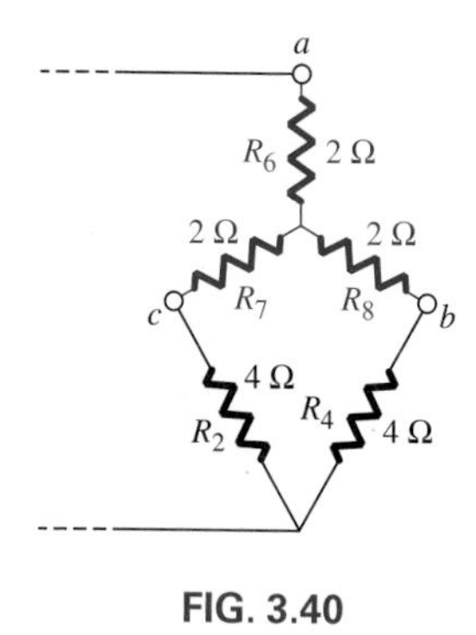

FIG. 3.40

The unique configuration in Fig. 3.38 is referred to as a *bridge* network. A very interesting and useful condition results when the following relationship is satisfied:

$$\frac{R_1}{R_2} = \frac{R_3}{R_4} \qquad \textbf{(3.10)}$$

When the ratio condition in Eq. (3.10) is satisfied, the current through (and consequently the voltage across) the connecting arm R_5 is zero. The resistor R_5 can then be replaced with either an open circuit ($i = 0$) or a short circuit ($v = 0$), and the analysis leading to R_T or I_T is unaffected. The bridge network is said to be *balanced* when Eq. (3.10) is satisfied. Note in Fig. 3.38 that the ratios result in

$$\frac{R_1}{R_2} = \frac{6\,\Omega}{4\,\Omega} = \frac{R_3}{R_4}$$

and if we substitute an open circuit for R_5, we have

$$R_T = (6\,\Omega + 4\,\Omega) \parallel (6\,\Omega + 4\,\Omega) = 10\,\Omega \parallel 10\,\Omega = 5\,\Omega$$

obtained earlier, or using the short-circuit equivalent of R_5, we have

$$R_T = 6\,\Omega \parallel 6\,\Omega + 4\,\Omega \parallel 4\,\Omega = 3\,\Omega + 2\,\Omega = 5\,\Omega$$

as above.

This condition can be put to good use in a resistance-measuring device called a *Wheatstone bridge*. A galvanometer (a sensitive ammeter) is inserted in the network in place of R_5, and the unknown resistance substituted for R_4 (or any of the other three), as shown in Fig. 3.41 with a photograph of a typical commercial bridge. The resistors R_1, R_2, and R_3 are then adjusted until the current through the movement is zero. The equation can then be applied, and the unknown resistor determined from $R_4 = R_{\text{unknown}} = R_2R_3/R_1$. As indicated in the figure, the actual instrument is not as simple as indicated by the network in Fig. 3.41, but the principle of operation is exactly the same. The bridge configuration will see further application in this chapter and in those to follow.

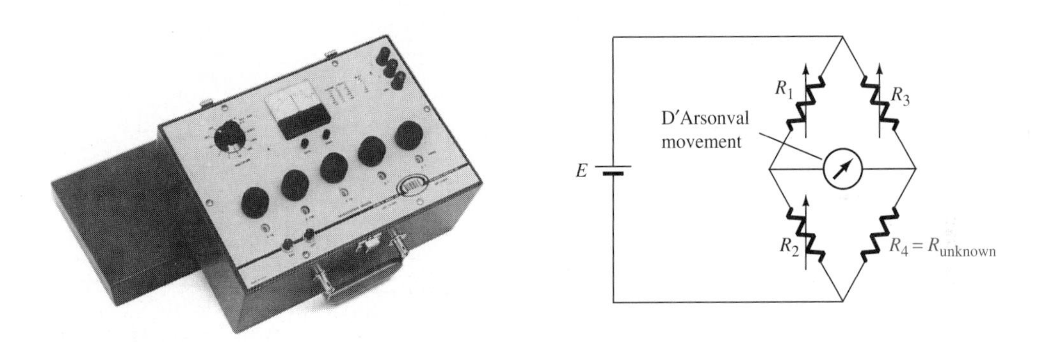

FIG. 3.41
Wheatstone bridge. (Courtesy of James G. Biddle Company.)

3.9 APPLICATIONS: SMOKE DETECTOR AND LOGIC PROBE

Wheatstone Bridge Smoke Detector

The Wheatstone bridge is a popular network configuration whenever detection of small changes in a quantity is required. In Fig. 3.42(a), the dc bridge configuration is employing a photoelectric device to detect the presence of smoke and to sound the alarm. A photograph of an actual photoelectric smoke detector appears in Fig. 3.42(b), and the internal

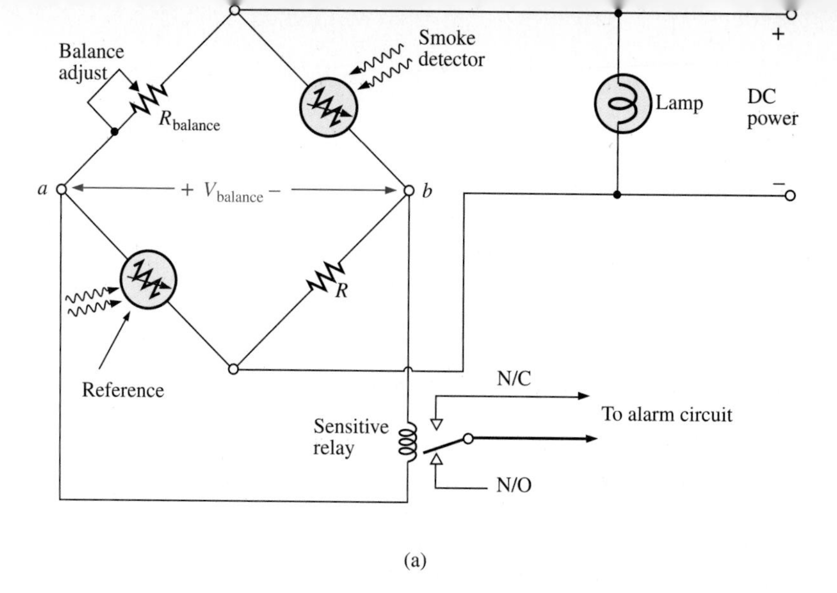

(a)

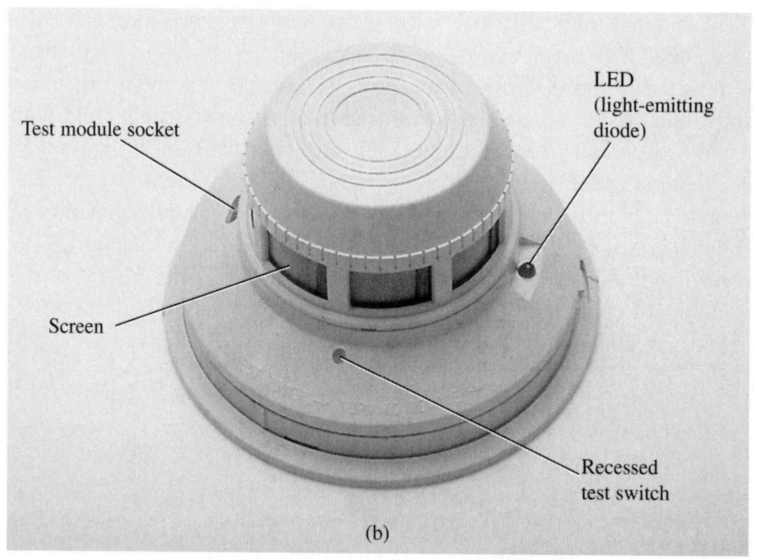

(b)

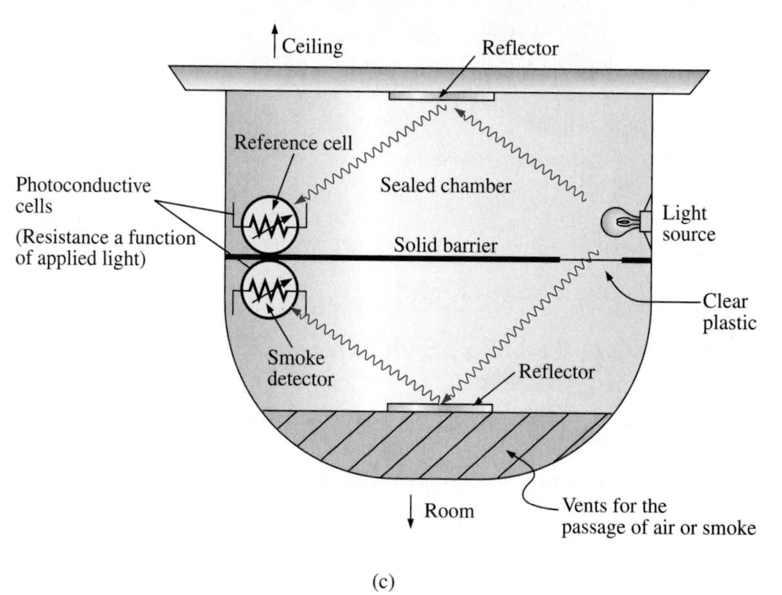

(c)

FIG. 3.42

Wheatstone bridge smoke detector: (a) dc bridge configuration; (b) outside appearance; (c) internal construction.

construction of the unit is shown in Fig. 3.42(c). First note that air vents are provided to permit the smoke to enter the chamber below the clear plastic. The clear plastic prevents the smoke from entering the upper chamber but permits the light from the bulb in the upper chamber to bounce off the lower reflector to the semiconductor light sensor (a cadmium photocell) at the left side of the chamber. The clear plastic separation ensures that the light hitting the light sensor in the upper chamber is not affected by the entering smoke. It establishes a reference level to compare against the chamber with the entering smoke. If no smoke is present, the difference in response between the sensor cells is registered as the normal situation. Of course, if both cells were exactly identical, and if the clear plastic did not cut down on the light, both sensors would establish the same reference level, and their difference would be zero. However, this is seldom the case, so a reference difference is recognized as the sign that smoke is not present. However, once smoke is present, there is a sharp difference in the sensor reaction from the norm, and the alarm sounds.

In Fig. 3.42(a), we find that the two sensors are located on opposite arms of the bridge. With no smoke present the balance-adjust rheostat is used to ensure that the voltage V between points a and b is 0 V, and the resulting current through the primary of the sensitive relay is 0 A. Taking a look at the relay, we find that the absence of a voltage from a to b leaves the relay coil unenergized and the switch in the N/O position (recall that the position of a relay switch is always drawn in the unenergized state). An unbalanced situation results in a voltage across the coil and activation of the relay, and the switch moves to the N/C position to complete the alarm circuit and activate the alarm. Relays with two contacts and one movable arm are called *single-pole–double-throw* (SPDT) relays. The dc power is required to set up the balance situation, energize the parallel bulb so we know that the system is on, and provide the voltage from a to b if an unbalanced situation should develop.

One may ask why only one sensor isn't used, since its resistance would be sensitive to the presence of smoke. The answer lies in the fact that the smoke detector might generate a false readout if the supply voltage or output light intensity of the bulb should vary. Smoke detectors of the type just described must be used in gas stations, kitchens, dentist offices, and the like, where the range of gas fumes present may set off an ionizing type smoke detector.

Schematic with Nodal Voltages

When an investigator is presented with a system that is down or not operating properly, one of the first options is to check the system's specified voltages on the schematic. These specified voltage levels are actually the nodal voltages determined in this chapter. *Nodal voltage* is simply a special term for a voltage measured from that point to ground. The technician will attach the negative or lower-potential lead to the ground of the network (often the chassis) and then place the positive or higher-potential lead on the specified points of the network to check the nodal voltages. If they match, it is a good sign that that section of the system is operating properly. If one or more fail to match the given values, the problem area can usually be identified. Be aware that a reading of −15.87 V is significantly different from an expected reading of +16 V if the leads have been properly attached. Although the actual numbers seem close, the difference is actually more than 30 V. You must expect some deviation from the given value as shown, but always be very sensitive to the resulting sign of the reading.

The schematic of Fig. 3.43(a) includes the nodal voltages for a logic probe used to measure the input and output states of integrated circuit logic chips. In other words, the probe determines whether the measured voltage is one of two states: high or low (often referred to as "on" or "off" or 1 or 0). If the LOGIC IN terminal of the probe is placed on a chip at a location where the voltage is between 0 and 1.2 V, the voltage is considered a low level, and the green LED will light. (LEDs are light-emitting semiconductor diodes that emit light when current is passed through them.) If the measured voltage is between 1.8 V and 5 V, the reading is considered high, and the red LED will light. Any voltage between 1.2 V and 1.8 V is considered a "floating level" and is an indication that the system being measured is not operating correctly. Note that the reference levels just mentioned are established by the voltage divider network to the right of the schematic. The op-amps employed are of such high input impedance that their loading on the voltage divider network can be ignored and the voltage divider network considered a network unto itself. Even though three 5.5-V dc supply voltages are indicated on the diagram, be aware that all three points are connected to the same supply. The other voltages provided (the nodal voltages) are the voltage levels that should be present from that point to ground if the system is working properly.

The op-amps are used to sense the difference between the reference at points 3 and 6 and the voltage picked up in LOGIC IN. Any difference will result in an output that will light either the green or the red LED. Be aware, because of the direct connection, that the voltage at point 3 is the same as shown by the nodal voltage to the left, or 1.8 V. Likewise, the voltage at point 6 is 1.2 V for comparison with the voltages at points 5 and 2, which reflect the measured voltage. If the input voltage happened to be 1.0 V, the difference between the voltages at points 5 and 6 would be 0.2 V, which ideally would appear at point 7. This low potential at point 7 would cause a current to flow from the much higher 5.5-V dc supply through the green LED, causing it to light and indicating a low condition. By the way, LEDs, like diodes, permit current through them only in the direction of the arrow in the symbol. Also note that the voltage at point 6 must be higher than that at point 5 for the output to turn on the LED. The same is true for point 2 over point 3, which reveals why the red LED does not light when the 1.0-V level is measured.

Oftentimes it is impractical to draw the full network as shown in Fig. 3.43(b) because there are space limitations or because the same voltage divider network is used to supply other parts of the system. In such cases you must recognize that points having the same shape are connected, and the number in the figure reveals how many connections are made to that point.

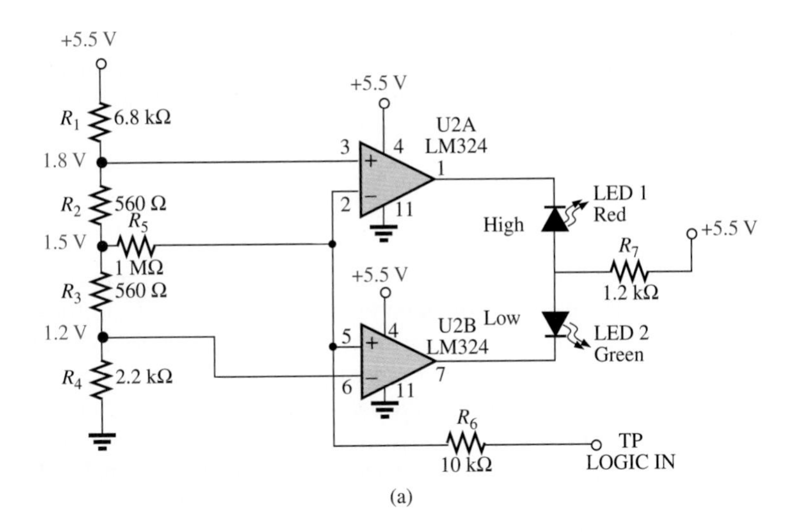

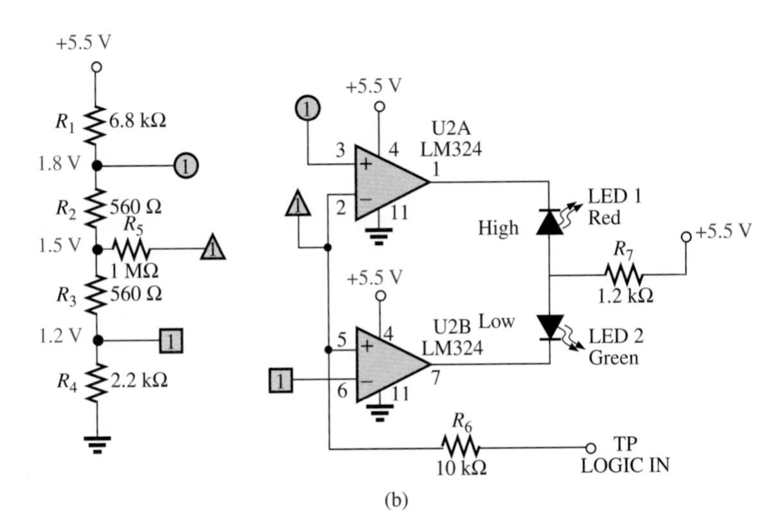

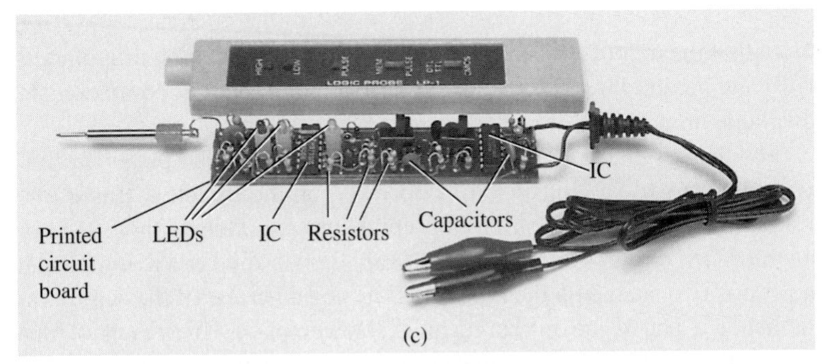

FIG. 3.43

Logic probe: (a) schematic with nodal voltages; (b) network with global connections; (c) photograph of commercially available unit.

A photograph of the outside and inside of a commercially available logic probe is provided in Fig. 3.43(c). Note the increased complexity of system because of the variety of functions that the probe can perform.

3.10 CAPACITORS

Thus far the resistor has been the only passive device examined in detail. Two other elements of importance in the design of electrical systems are the *capacitor* and the *inductor*. Both are quite different from the resistor in the sense that the resistor dissipates all the energy delivered to it in the form of heat, while the capacitor and the inductor ideally store the electrical energy delivered to them. That energy can then be returned to the electrical system when called for by the design. In addition, while the voltage and current of a resistor are related by the constant R (Ohm's law), the voltage and current of a capacitor or inductor are related by either a differential or an integral equation. Although this latter revelation suggests an increased measure of mathematical difficulty associated with these elements, there are methods to be introduced that minimize the complexity. This section discusses the fundamental construction, characteristics, and response of capacitive elements, while the following section is devoted to inductive elements.

The capacitor in its simplest form is two conducting surfaces separated by a dielectric (a type of insulator) as shown in Fig. 3.44. The label *capacitor* comes from the *capacity* of the element to store a charge on its plates. The larger its capacitance (C), the more charge (Q) deposited on its plates for the same voltage (V) across the plates. In equation form, capacitance is defined as

$$C = \frac{Q}{V} \qquad \text{(farads)} \qquad \textbf{(3.11)}$$

Note that the unit of measurement is the *farad* (F), although microfarads (μF) and picofarads (pF) are more common magnitudes of commercially available units.

For the system in Fig. 3.44 we shall assume that the plates are initially uncharged and the switch is open. When the switch is closed, the relatively free electrons on the top conducting surface of the capacitor are attracted to the positive side of the applied dc source, leaving behind a net positive charge on the top plate. The negative side of the supply establishes a net negative charge on the lower plate. The result of this charge flow from source to capacitor is a current i_C that initially jumps to a value limited by the resistor R and then decreases toward zero. In other words, the charging rate of the plates is initially very heavy and then decreases quite rapidly to zero. The result of the deposited charge is an *electric field*, shown between the plates of the capacitor in Fig. 3.44, that extends from the positive to the negative charges.

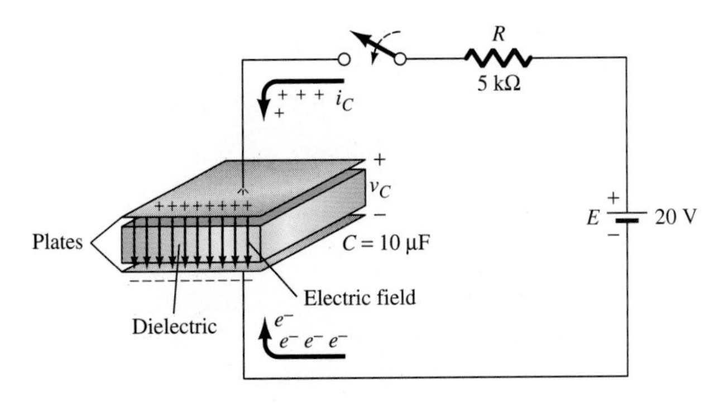

FIG. 3.44
Charging capacitor.

Charge cannot flow from one plate to the other because of the insulator between the plates. Since the voltage is directly related to the charge on the plates through Eq. (3.11) and the charging of the plates cannot occur instantaneously,

the voltage v_C of a capacitor cannot change instantaneously.

It takes a period of time determined by the circuit elements, as described in the next paragraph.

A plot of the current i_C versus time results in the exponential curve in Fig. 3.45. Note that the current jumps to a peak value of E/R = 20 V/5 kΩ = 4 mA and then decreases to zero exponentially. For most practical applications the time required to decay to zero is approximately 5τ, where τ (tau), called the *time constant* of the circuit, is determined by

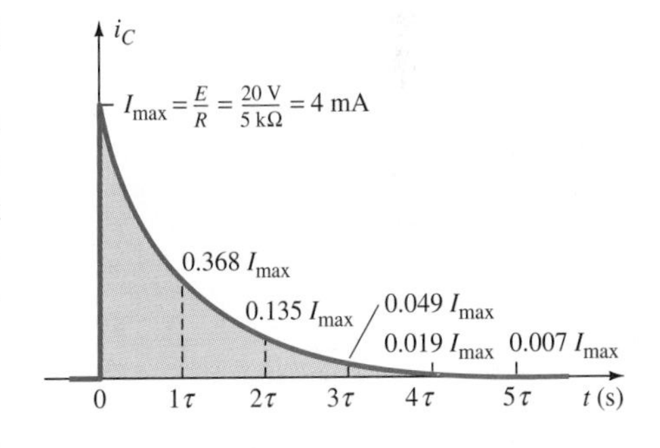

FIG. 3.45
Charging current i_C in Fig. 3.44.

$$\tau = RC \quad \text{(seconds)} \tag{3.12}$$

For the network in Fig. 3.44,

$$\tau = RC = (5\text{ k}\Omega)(10\ \mu\text{F}) = (5 \times 10^3\Omega)(10 \times 10^{-6}\text{F})$$
$$= 50 \times 10^{-3}\text{s} = 50\text{ ms}$$

and the discharge time is

$$5\tau = 5(50\text{ ms}) = 250\text{ ms} = 0.25\text{ s} = \tfrac{1}{4}\text{ s}$$

In equation form, the current i_C is

$$i_C = \frac{E}{R}e^{-t/\tau} \tag{3.13}$$

Since the exponential function in the form e^{-x}, or $1 - e^{-x}$, appears quite frequently in the analysis of electrical systems, a plot of the

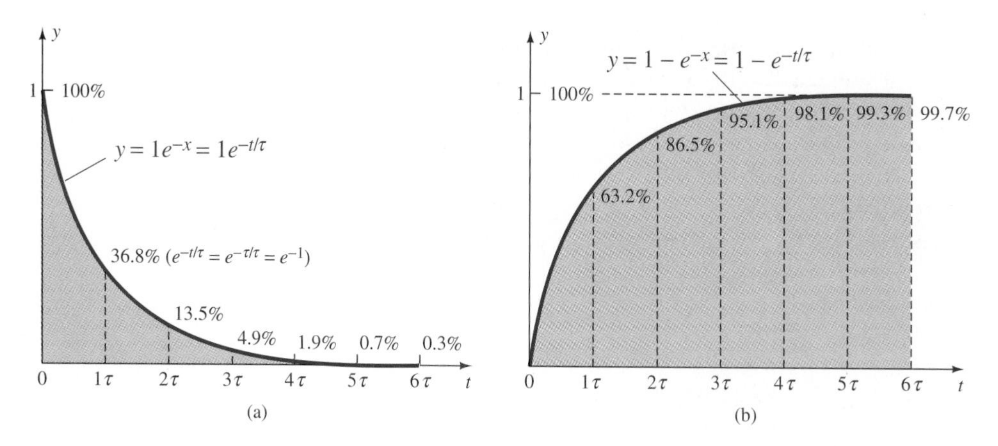

FIG. 3.46
Functions: (a) $e^{-t/\tau}$; (b) $1 - e^{-t/\tau}$.

function and its value at each time constant appears in Fig. 3.46. Note that after one time constant, the curve, and therefore i_C, drops to 36.8% of its peak value, while at five time constants it is down to 0.7% of its peak value (less than 1%).

The voltage v_C builds up to a maximum value of $E = 20$ V at a rate directly related to the rate in which charge is deposited on the plates. Note from the plot of v_C in Fig. 3.47 that it reaches 63.2% of its final value in only one time constant, and 99.3% in five time constants. The curve in Fig. 3.47 can be described by the equation

$$v_c = E(1 - e^{-t/\tau}) \tag{3.14}$$

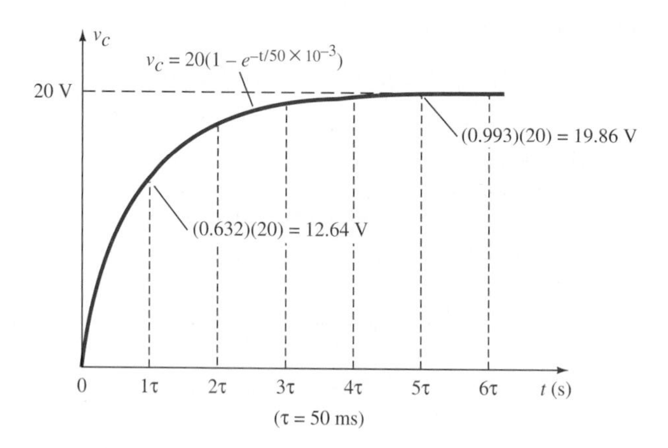

FIG. 3.47
Increasing voltage v_C in Fig. 3.44.

Note the presence of the same exponential function that dictates that v_C will reach its final value of E volts at the same instant that i_C decays to zero. Consider the importance of the following statement, which will be used to full advantage in the chapters to follow: when $v_C = E = 20$ V and $i_C = 0$ A (after approximately five time constants), the capacitor has the characteristics of an *open circuit*. In electronic systems this characteristic can be used to *block* or isolate direct current from other parts of the network.

Since $v_R = i_R R = i_C R$, and R is a constant, the voltage v_R has the same shape as the i_C curve and a peak value of 20 V. The general behavior of the capacitor in a simple dc switching situation is thus described.

The factors that affect the capacitance of a capacitor appear in the equation

$$C = \epsilon_0 \epsilon_r \frac{A}{d} \qquad \text{(farads, F)} \qquad \textbf{(3.15)}$$

where ϵ_0 = permittivity of air, 8.85×10^{-12} F/m
ϵ_r = relative permittivity
A = area in square meters
d = distance between plates in meters

The equation clearly reveals that the larger the area of the plates or the smaller the distance between the plates, the larger the capacitance. The factor ϵ, called the *permittivity* of the material, is a measure of how well the dielectric between the plates "permits" the establishment of field lines between the plates. Insertion of the insulators between the plates increases the capacitance above that with simply air between the plates by a factor of ϵ_r, the *relative permittivity* given by $\epsilon_r = \epsilon/\epsilon_0$. Table 3.1 lists a few dielectrics with their ϵ_r values.

TABLE 3.1
Relative permittivity

Dielectric	ϵ_r
Vacuum	1.0
Air	1.0006
Teflon	2.0
Paper, paraffined	2.5
Rubber	3.0
Transformer oil	4.0
Mica	5.0
Porcelain	6.0
Bakelite	7.0
Glass	7.5
Water	80.0

A few of the more common types of capacitors include the air, ceramic, mica, and electrolytic units in Fig. 3.48. The symbol for the fixed and variable varieties is also provided in the same figure. The basic construction of every commercial capacitor leans toward making the capacitance its largest value while limiting its size. For the ceramic type, two conducting surfaces are rolled together (with a ceramic dielectric between them) to make the area factor a maximum. For the mica capacitor, maximum area is obtained by stacking the plates (with the mica dielectric between them) and connecting the proper plates together. For the air capacitor the common area between the stacked plates is made adjustable (for various values of capacitance) by a control knob that can move one set of plates.

If possible, the capacitance value is stamped right on the capacitor. However, some very small mica and ceramic capacitors may require the use of a color-coding system. The numerical values associated with each color are exactly the same as applied to the color coding of resistors.

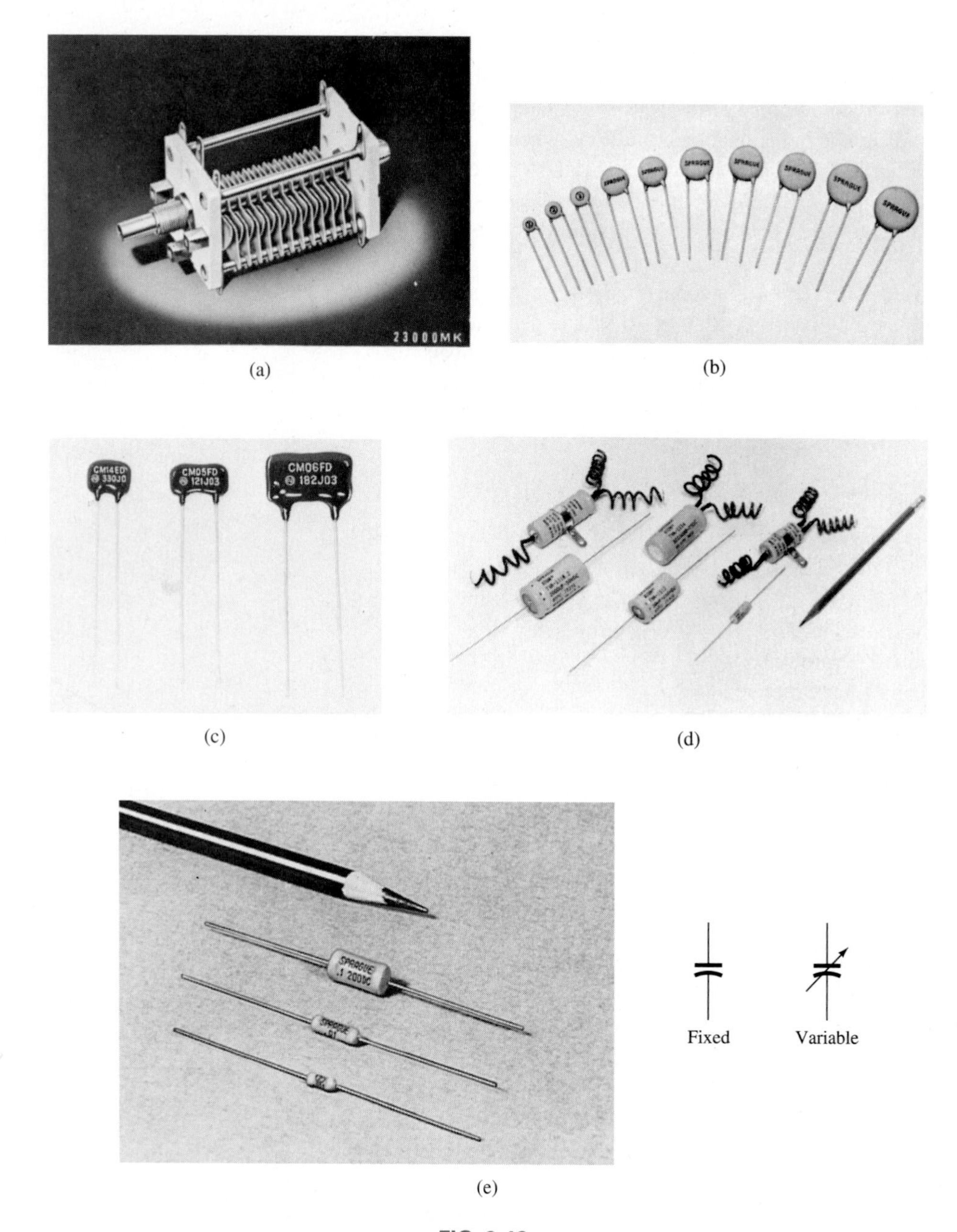

FIG. 3.48
Commercially available capacitors with their graphic symbols: (a) air; (b) ceramic; (c) mica; (d) electrolytic; (e) symbols.

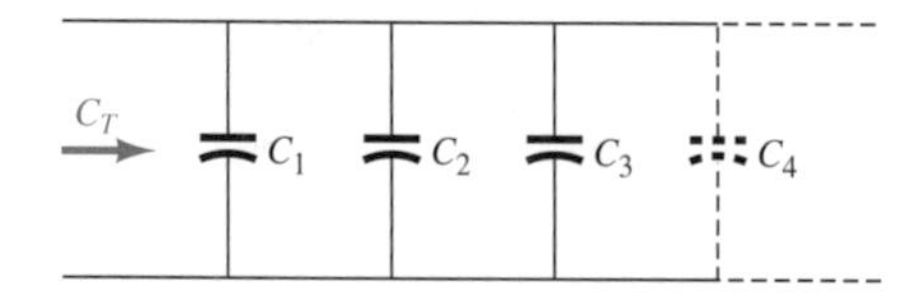

FIG. 3.49
Parallel capacitors.

Capacitors connected in parallel are treated as resistors in series when it comes to determining the total capacitance. That is, for the parallel capacitors in Fig. 3.49, the total capacitance is determined from

$$C_T = C_1 + C_2 + C_3 + \cdot \cdot \cdot + C_v \qquad \textbf{(3.16)}$$

For capacitors in series, the equation relates directly to that obtained for parallel resistors. That is, for the capacitors in Fig. 3.50, the total capacitance is determined from

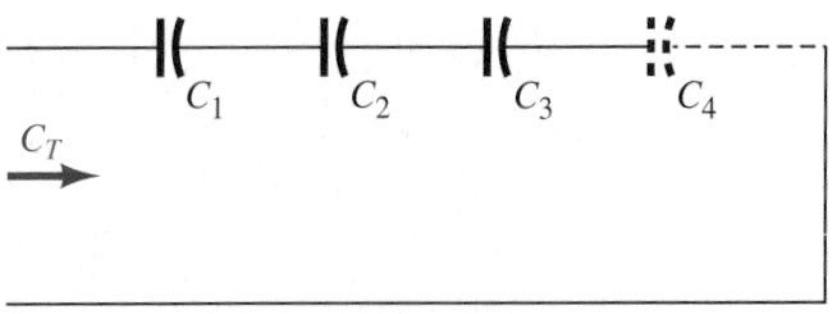

$$\frac{1}{C_T} = \frac{1}{C_1} + \frac{1}{C_2} + \frac{1}{C_3} + \cdots + \frac{1}{C_N} \tag{3.17}$$

FIG. 3.50
Series capacitors.

For two series capacitors (as derived for two parallel resistors),

$$C_T = \frac{C_1 C_2}{C_1 + C_2} \tag{3.18}$$

The preceding equations can understandably be quite useful if a particular capacitance value is not available and must be obtained through a combination of available values.

Although an insulator (the dielectric) is placed between the plates, there is a very small flow of charge between the plates called *leakage current*. If a capacitor is fully charged and then removed from the network and set aside, it will be able to hold onto its charge for only a limited amount of time because of the resulting leakage current through the dielectric that will deplete the plates of their charge. The type of capacitor determines the discharge rate. Mica and ceramic capacitors have very small leakage currents, while electrolytics can discharge in a very small period of time.

In addition to the capacitance obtained through commercial design, there is a measure of capacitance between any two conducting or charged surfaces in an electronic system. This usually undesirable capacitance is called *stray capacitance,* and if it is not carefully considered, it can have a pronounced effect on the performance of some electronic systems. In both the transistor and tube the stray capacitance between the elements of each can severely limit the range of operation of each in an electronic system unless properly considered.

For networks in which the values of R and C to be used in determining τ are not evident, an application of Thévenin's theorem can often be quite helpful, as demonstrated in Example 3.10.

EXAMPLE 3.10 Determine the mathematical expression for v_C if the switch is closed at $t = 0$ in Fig. 3.51.

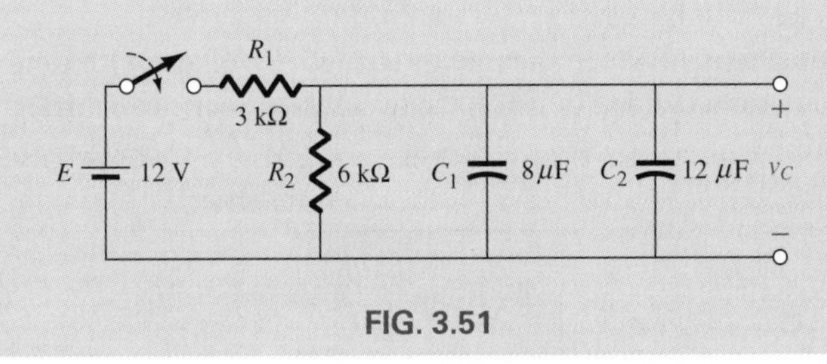

FIG. 3.51

Solution: The total capacitance of the parallel capacitors is given by

$$C = C_1 + C_2 = 8\ \mu\text{F} + 12\ \mu\text{F} = 20\ \mu\text{F}$$

The value of R to be used is not evident. Applying Thévenin's theorem to the network of E_1, R_1, and R_2, we find that $R_{Th} = R_1 \parallel R_2 = 3\ \text{k}\Omega \parallel 6\ \text{k}\Omega = 2\ \text{k}\Omega$ (remember that the source E is set to zero when R_{Th} is determined).

$$E_{Th} = \frac{R_2(E)}{R_2 + R_1} = \frac{(6\ \text{k}\Omega)(12\ \text{V})}{6\ \text{k}\Omega + 3\ \text{k}\Omega} = \frac{2}{3}(12\ \text{V}) = 8\ \text{V}$$

and

$$\tau = R_{Th}C = (2\ \text{k}\Omega)(20 \times 10^{-6}\ \text{F}) = 40 \times 10^{-3}\ \text{s} = 40\ \text{ms}$$

with

$$v_C = E(1 - e^{-t/\tau}) = E_{Th}(1 - e^{-t/\tau})$$

or

$$\mathbf{v_C = 8(1 - e^{-t/40\times10^{-3}})}$$

A plot of v_C appears in Fig. 3.52.

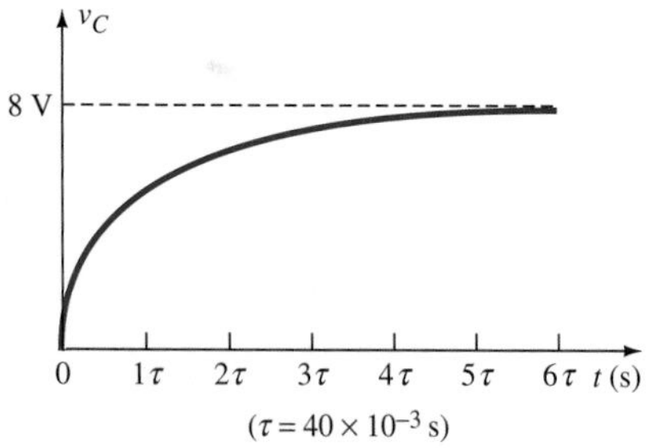

FIG. 3.52
Voltage v_C in Fig. 3.51.

One very interesting characteristic of the capacitor is that the current i_C is not related directly to the magnitude of the voltage across the capacitor but to its *rate of change* of voltage across the capacitor. That is, the faster the rate of change of voltage across the capacitor, the greater the resulting i_C. An applied dc voltage of 100 V would (after 5τ of switching the network on) result in effectively zero current, since the voltage does not change with time. In equation form, i_C and v_C are related by the derivative

$$i_C = C\frac{dv_C}{dt} \qquad \textbf{(3.19)}$$

where dv_C/dt is a measure of the instantaneous rate of change of v_C with time. The equation clearly indicates that if v_C fails to change, $dv_C/dt = 0$ and hence i_C will be zero.

For situations in which the rate of change of voltage with time has a fixed value for a defined interval, the average current can be determined from

$$i_{C_{av}} = C\frac{\Delta v_C}{\Delta t} \qquad \textbf{(3.20)}$$

Example 3.11 will demonstrate the use of Eq. (3.20) and clarify the necessity for a change in voltage with time in order to establish a current i_C.

EXAMPLE 3.11 Determine $i_{C_{av}}$ for a 20-μF capacitor having the voltage in Fig. 3.53 applied across its plates.

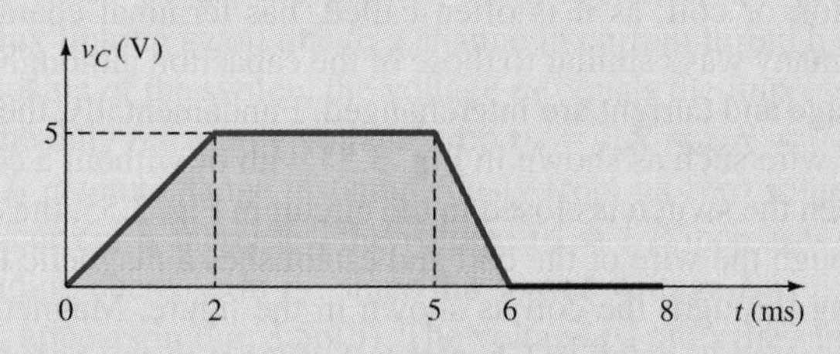

FIG. 3.53

Solution:

$$0 \rightarrow 2 \text{ ms: } i_C = C\frac{\Delta v_C}{\Delta t} = (20 \times 10^{-6}\text{ F})\left(\frac{5\text{ V}}{2 \times 10^{-3}\text{ s}}\right) = 50\text{ mA}$$

$$2 \rightarrow 5 \text{ ms: } i_C = (20 \times 10^{-6}\text{ F})\left(\frac{0\text{ V}}{3 \times 10^{-3}\text{ s}}\right) = 0\text{ A}$$

Note that the failure of the voltage to change results in $i_{C_{av}} = 0$ A.

$$5 \rightarrow 6 \text{ ms: } i_C = -(20 \times 10^{-6}\text{ F})\left(\frac{5\text{ V}}{1 \times 10^{-3}\text{ s}}\right) = -100\text{ mA}$$

Note the negative sign used to indicate decaying levels of applied voltage and a reversal in current direction.

$$6 \rightarrow 8 \text{ ms: } \Delta v_C = 0 \quad \text{and} \quad i_C = 0\text{ A}$$

A plot of $i_{C_{av}}$ versus time appears in Fig. 3.54.

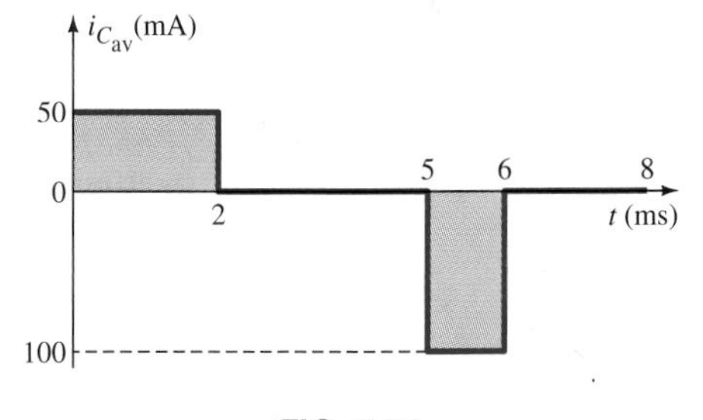

FIG. 3.54

The previous sections revealed that the resistor dissipates the energy delivered to it in the form of heat. The ideal capacitor, however (if we ignore the leakage current), *does not dissipate* the energy delivered to it but simply stores it in the form of an electric field that through design can be returned to the system as electrical energy. In other words, the capacitor, unlike the resistor, is an energy-storing element and not a dissipative one like the resistor.

The energy stored is given by

$$\boxed{W_C = \tfrac{1}{2}CV^2} \qquad \text{(joules)} \qquad \textbf{(3.21)}$$

where V is the steady-state voltage across the capacitor. For the capacitor C_1 in Fig. 3.51,

$$W_C = \tfrac{1}{2}C_1V^2 = \tfrac{1}{2}(8 \times 10^{-6}\text{ F})(8\text{ V})^2 = \mathbf{256\ \mu J}$$

where μ_0 is the permeability of air ($4\pi \times 10^{-7}$ Wb/Am for the SI system) and μ_r is the *relative permeability* of the material as compared with air. The greater the magnetic properties of the material, the higher the relative permeability factor. Some materials such as steel and iron have permeabilities hundreds and even thousands of times that of air. That is, $\mu_r \geq 100$. For coils without a core, $\mu_r = 1$.

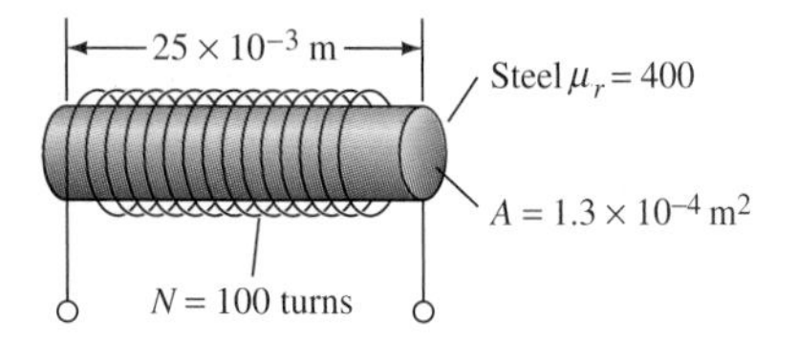

FIG. 3.58

EXAMPLE 3.12 Determine the inductance L of the coil in Fig. 3.58.

Solution:

$$L = \frac{N^2 \mu A}{l} = \frac{N^2 \mu_r \mu_0 A}{l}$$

$$= \frac{(100)^2(400)(4\pi \times 10^{-7})(1.3 \times 10^{-4})}{25 \times 10^{-3}} = 0.0261 \text{ H}$$

$$= \mathbf{26.1 \text{ mH}}$$

The circuit symbol for the inductor (fixed and variable) appears in Fig. 3.59 with photographs of a number of types of commercially available inductors. In most cases, the inductance can be stamped right on the element, although in some cases a color-coding system very similar to that described for capacitance is employed. One technique of obtaining a

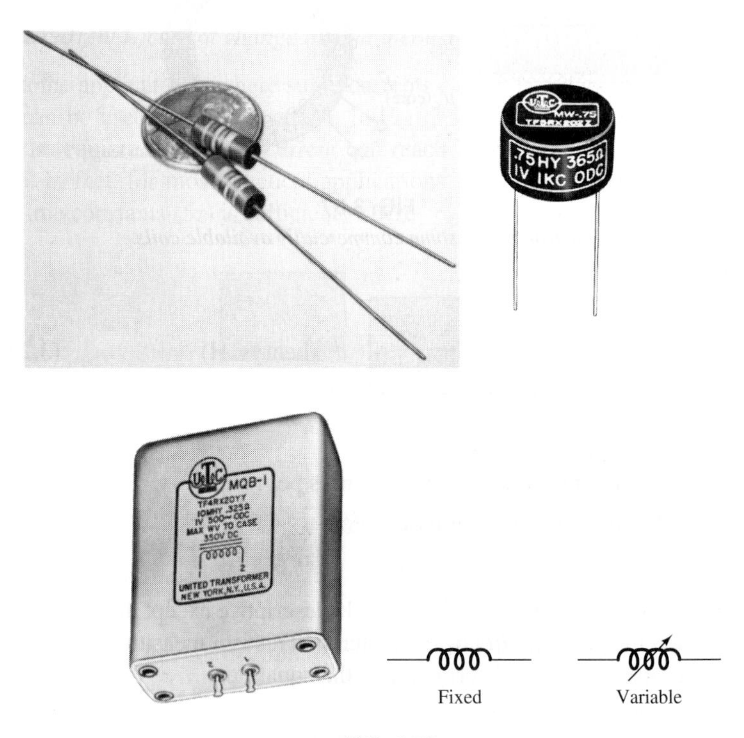

FIG. 3.59
Commercially available inductors with their graphic symbols.

variable inductance is to have a movable core within the coil. The more the core is inserted in the coil, the more the flux linking the coil and the higher the inductance level.

Inductors in *series* (Fig. 3.60) have a total inductance value determined in the same manner as for series resistors. That is,

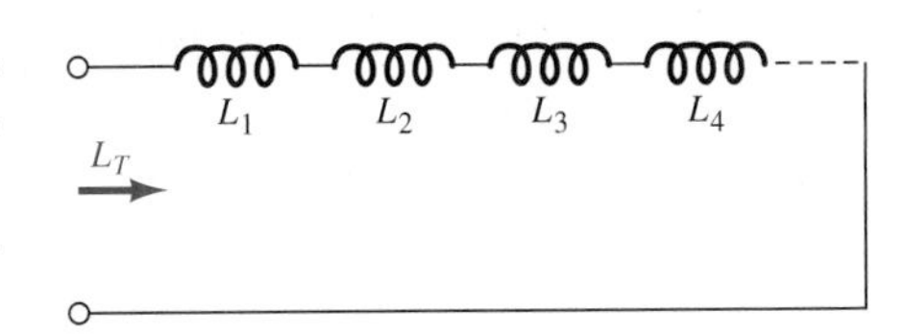

FIG. 3.60
Series inductors.

$$L_T = L_1 + L_2 + L_3 + \cdots + L_N \qquad \textbf{(3.26)}$$

For parallel inductors, the total inductance is found in the same manner as for parallel resistors. That is, for the coils in Fig. 3.61 the total inductance is given by

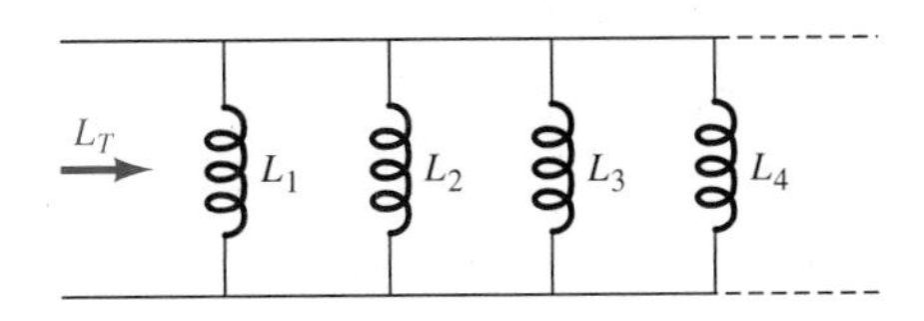

FIG. 3.61
Parallel inductors.

$$\frac{1}{L_T} = \frac{1}{L_1} + \frac{1}{L_2} + \frac{1}{L_3} + \cdots + \frac{1}{L_N} \qquad \textbf{(3.27)}$$

while for two parallel inductors,

$$L_T = \frac{L_1 L_2}{L_1 + L_2} \qquad \textbf{(3.28)}$$

EXAMPLE 3.13 Determine the total inductance of the series–parallel connection of elements in Fig. 3.62.

Solution: For the parallel inductors L_2 and L_3,

$$L'_T = \frac{L_2 L_3}{L_2 + L_3} = \frac{(40\text{ mH})(20\text{ mH})}{40\text{ mH} + 20\text{ mH}}$$
$$= \frac{800}{60}\text{ mH} = 13.3\text{ mH}$$

and

$$L_T = L_1 + L'_T = 10\text{ mH} + 13.3\text{ mH} = \mathbf{23.3\ mH}$$

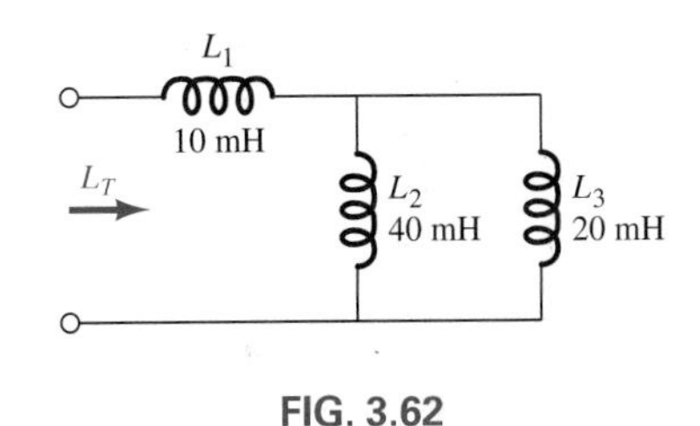

FIG. 3.62

Like the capacitor, the inductor is not an *ideal* device. In addition to its inductance there is a measure of dc resistance due to the many turns of fine wire. In addition, stray capacitance exists between the parallel conducting turns of wire of the coil. The addition of these two elements would result in the equivalent circuit in Fig. 3.63. However, for many practical applications the inductor can still be considered ideal.

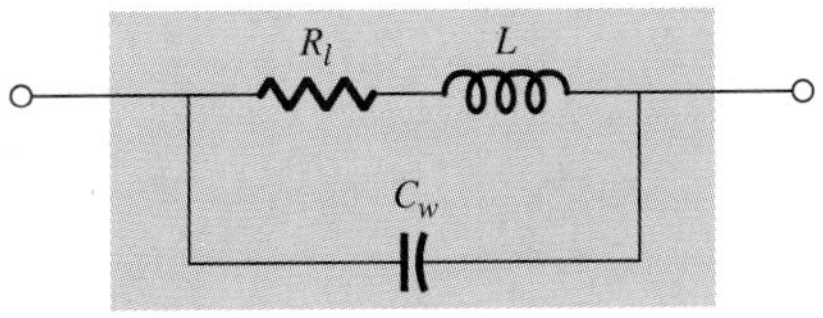

FIG. 3.63
Total inductor equivalent circuit.

For networks in which it is difficult to determine the proper value of L and R for the time constant, an application of Thévenin's theorem may be the solution, as demonstrated by the following example.

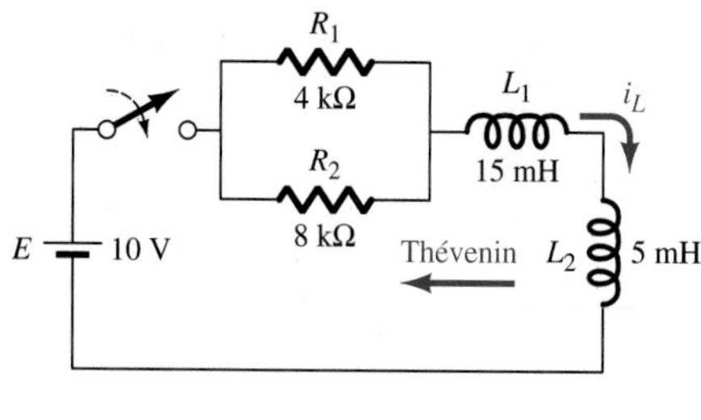

FIG. 3.64

EXAMPLE 3.14 Sketch the curve for i_L in Fig. 3.64 if the switch is closed at $t = 0$.

Solution: The total inductance is

$$L_T = L_1 + L_2 = 15 \text{ mH} + 5 \text{ mH} = 20 \text{ mH}$$

Application of Thévenin's theorem gives

$$R_{\text{Th}} = R_1 \parallel R_2 = \frac{(4 \text{ k}\Omega)(8 \text{ k}\Omega)}{4 \text{ k}\Omega + 8 \text{ k}\Omega} = \frac{32 \text{ k}\Omega}{12} = 2.66 \text{ k}\Omega$$

$$E_{\text{Th}} = E = 10 \text{ V}$$

Therefore,

$$\tau = \frac{L}{R} = \frac{20 \times 10^{-3} \text{ H}}{2.66 \times 10^3 \ \Omega} = 7.5 \ \mu\text{s}$$

and

$$i_L = \frac{E}{R}(1 - e^{-t/\tau}) = \frac{10 \text{ V}}{2.66 \times 10^3 \ \Omega}(1 - e^{-t/7.5\times10^{-6}})$$

$$= \mathbf{3.76 \times 10^{-3}(1 - e^{-t/7.5\times10^{-6}})}$$

The plot of i_L appears in Fig. 3.65.

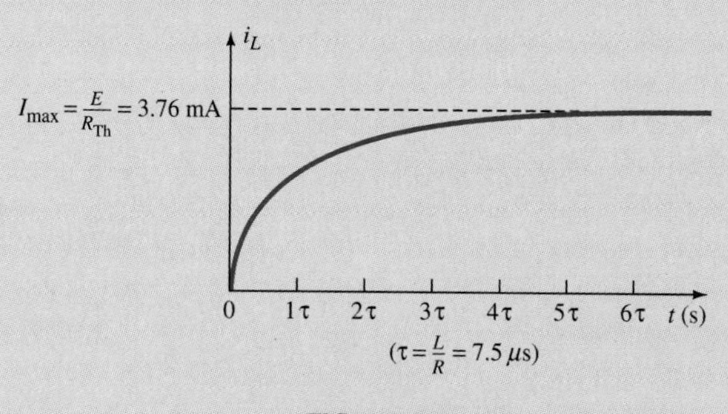

FIG. 3.65
Current i_L for the network in Fig. 3.64.

For the inductor, the instantaneous voltage and current are related by

$$v_L = L\frac{di_L}{dt} \tag{3.29}$$

Note again the appearance of the derivative. It clearly indicates that the magnitude of the voltage v_L is directly proportional to the rate of change of current through the coil and not simply the magnitude of the current through the coil. In other words, for an applied dc voltage the current fails to change after steady-state conditions are reached, and the voltage v_L decays to zero.

For situations in which the rate of change of current with time has a fixed value for a defined interval, the average induced voltage can be determined from

$$v_{L_{av}} = L\frac{\Delta i_L}{\Delta t} \qquad \textbf{(3.30)}$$

Note the similarities between this equation and Eq. (3.20) for the capacitor. In fact, the application of Eq. (3.30) matches the technique introduced in Example 3.11.

In an effort to emphasize the effect of a coil and a capacitor in a dc network after the changing (transient) behavior has passed, consider the network in Fig. 3.66. As indicated in the adjoining figure, the capacitor

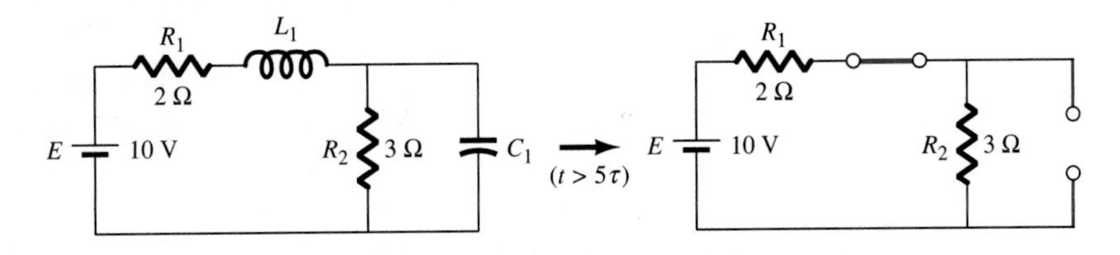

FIG. 3.66
Steady-state behavior of a dc network ($t>5\tau$).

has been replaced by an open circuit and the inductor by a short circuit. The resulting currents and voltages are

$$I_1 = \frac{E}{R_1 + R_2} = \frac{10\text{ V}}{2\ \Omega + 3\ \Omega} = \frac{10\text{ V}}{5\ \Omega} = 2\text{ A} = I_L$$

$$V_C = V_{R_2} = \frac{R_2(E)}{R_2 + R_1} = \frac{(3\ \Omega)(10\text{ V})}{3\ \Omega + 2\ \Omega} = \frac{30}{5}\text{ V} = 6\text{ V}$$

We shall find when we analyze electronic networks that the capacitor can play an important role in isolating dc and ac signals in the same system by acting (on an approximate basis) as an open circuit to direct current and a short circuit for alternating current. The energy stored by an inductor is given by

$$W_L = \tfrac{1}{2}LI^2 \quad \text{(joules)} \qquad \textbf{(3.31)}$$

where I is the steady-state current of the inductor.

3.12 APPLICATION: LINE CONDITIONER (SURGE PROTECTOR)

In recent years we have all become familiar with the line conditioner as a safety measure for our computers, TVs, CD players, and other sensitive instrumentation. In addition to protecting equipment from unexpected surges in voltage and current, most quality units will also filter out (remove) electromagnetic interference (EMI) and radio-frequency interference (RFI). EMI encompasses any unwanted disturbances down the power line established by any combination of electromagnetic effects such as those generated by motors on the line, power equipment in the area emitting signals picked up by the power line acting as an antenna, and so on. RFI includes all signals in the air in the audio range and beyond that may also be picked up by power lines inside or outside the house.

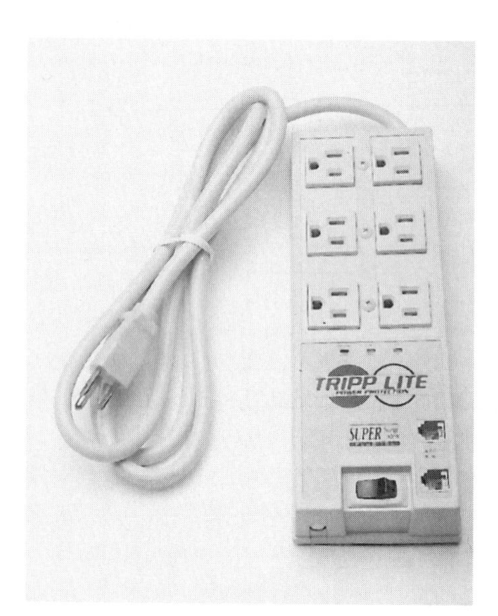

FIG. 3.67
Line conditioner: general appearance.

The unit of Fig. 3.67 has all the design features expected in a good line conditioner. The figure reveals that it can handle the power drawn by six outlets and that it is set up for fax/modem protection. Also note that it has both LED displays that reveal whether there is fault on the line or whether the line is OK and an external circuit breaker to reset the system. In addition, when the surge protector is on, a red light is visible at the power switch.

The schematic of Fig. 3.68 does not include all the details of the design, but it does include the major components that appear in most good line conditioners. First, note in the photograph of Fig. 3.69 that the outlets are all connected in parallel, with a ground bar used to establish a ground connection for each outlet. The circuit board had to be flipped over to show the components, so it will take some adjustment to relate the position of the elements on the board to the casing. The *feed line* or *hot lead wire* (black in the actual unit) is connected directly from the line to the circuit breaker. The other end of the circuit breaker is connected to the other side of the circuit board. All the large discs that you see are 20-μF capacitors (not all have been included in Fig 3.68 for clarity). There are quite a few capacitors to handle all the possibilities. For instance, there are capacitors from line to return (black wire to white wire in actual unit), from line to ground (black to green), and from return to ground (white to ground). Each has two functions. The first and most obvious function is to prevent any spikes in voltage that may come down the line because of external effects such as lightning from reaching the equipment plugged into the unit. Recall from this chapter that the voltage across capacitors cannot change instantaneously and in fact will act to squelch any rapid change in voltage across its terminals. The capacitor, therefore, will prevent the line to neutral voltage from changing too quickly, and any spike that tries to come down the line will have to find another point in the feed circuit to fall across. In this way the appliances to the surge protector are well protected.

The second function requires some knowledge of the reaction of capacitors to different frequencies and will be discussed in more detail in later chapters. For the moment, let it suffice to say that the capacitor will have a different impedance to different frequencies, thereby preventing undesired frequencies, such as those associated with EMI and RFI disturbances, from affecting the operation of units connected to the line conditioner. The rectangular-shaped capacitor of 1 μF near the center of the board is connected directly across the line to take the brunt of a

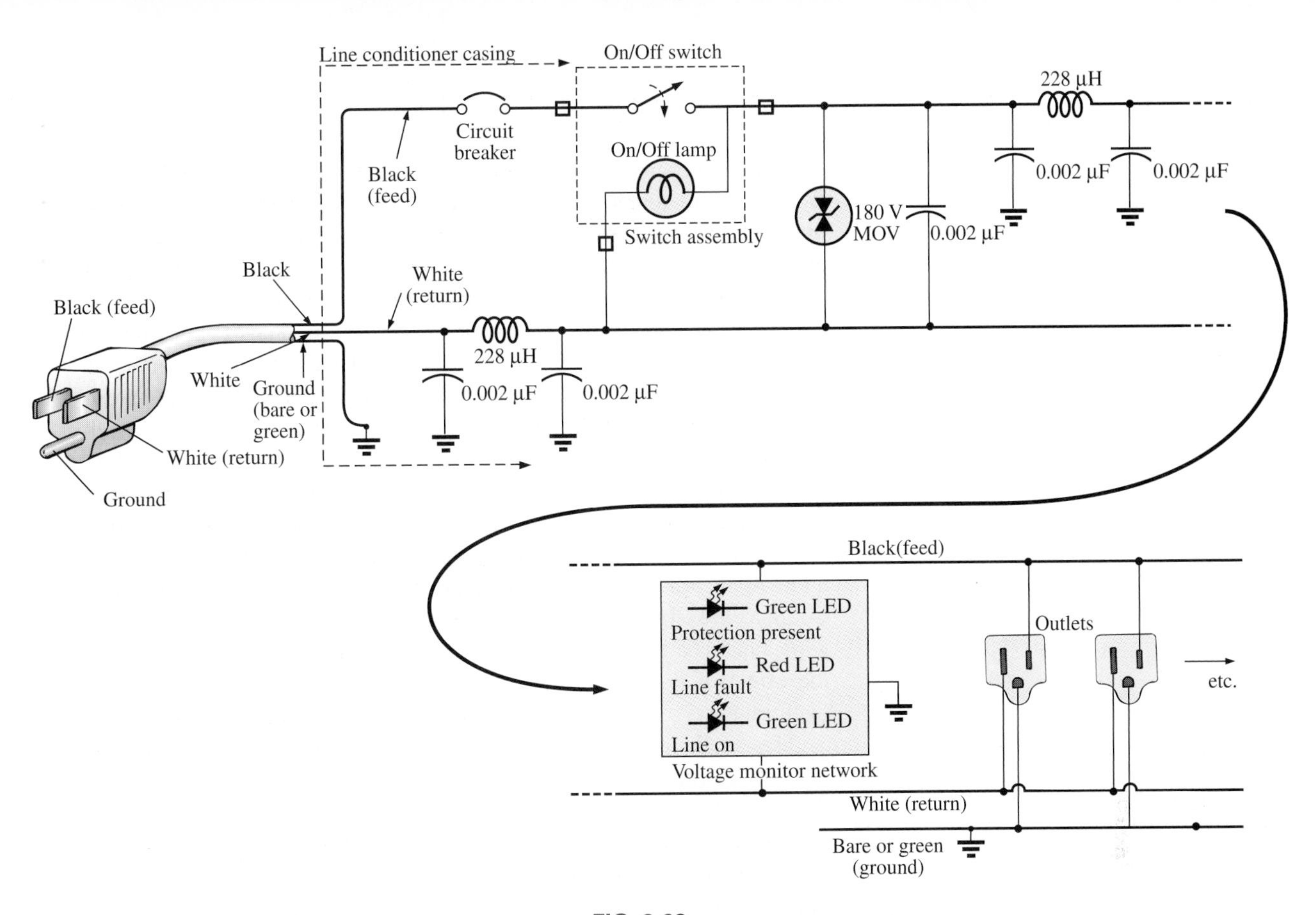

FIG. 3.68
Line conditioner: electrical schematic.

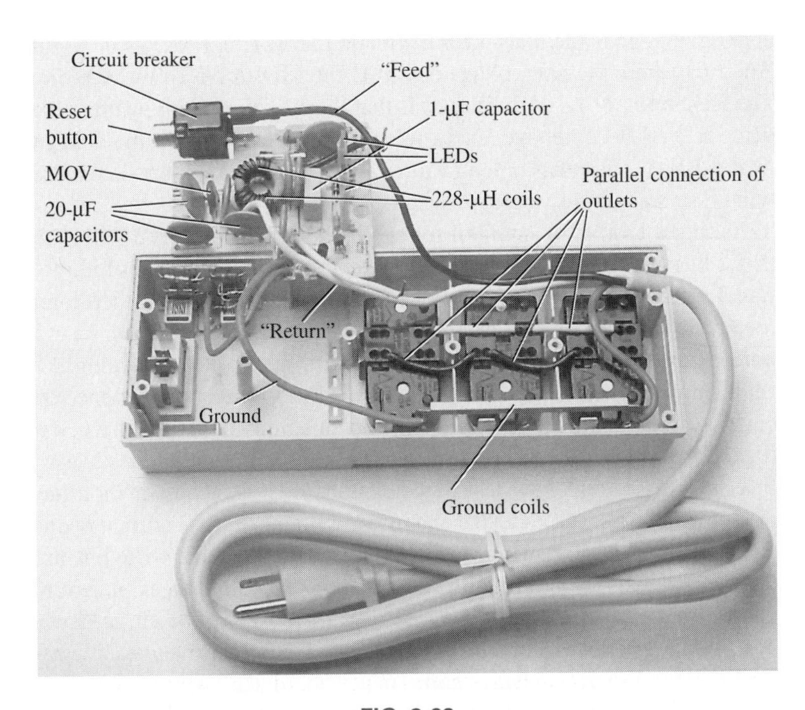

FIG. 3.69
Line conditioner: internal construction.

strong voltage spike down the line. Its larger size is clear evidence that it is designed to absorb a fairly high energy level that may be established by a large voltage—significant current over a period of time that might exceed a few milliseconds.

The large, toroidal-shaped, ferromagnetic structure in the center of the circuit board of Fig. 3.69 has two coils of 228 μH that appear in the line and neutral of Fig. 3.68. Their purpose, like that of the capacitors, is twofold: to block spikes in current from coming down the line and to block unwanted EMI and RFI frequencies from getting to the connected systems. In the next chapter you will find that coils act as "chokes" to quick changes in current; that is, the current through a coil cannot change instantaneously. For increasing frequencies, such as those associated with EMI and RFI disturbances, the reactance of a coil increases and will absorb the undesired signal rather than let it pass down the line. Using a choke in both the line and the neutral makes the conditioner network balanced to ground. In total, capacitors in a line conditioner have the effect of *bypassing* the disturbances, whereas inductors *block* the disturbance.

The smaller disc (blue in the actual unit) between two capacitors and near the circuit breaker is an MOV (metal-oxide varistor), which is the heart of most line conditioners. It is an electronic device whose terminal characteristics change with the voltage applied across its terminals. For the normal range of voltages down the line, its terminal resistance is sufficiently large to be considered an open circuit, and its presence can be ignored. However, if the voltage is too large, its terminal characteristics change from a very large resistance to a very small resistance that can essentially be considered a short circuit. This variation in resistance with applied voltage is the reason for the name *varistor*. For MOVs in North America where the line voltage is 120 V, the MOVs are 180 V or more. The reason for the 60-V difference is that the 120-V rating is an effective value related to dc voltage levels, whereas the waveform for the voltage at any 120-V outlet has a peak value of about 170 V. A great deal more will be said about this topic in Chapter 4.

Taking a look at the symbol for an MOV in Fig. 3.68, you will note that it has an arrow in each direction, revealing that the MOV is bidirectional and will block voltages with either polarity. In general, therefore, for normal operating conditions, the presence of the MOV can be ignored, but if a large spike should appear down the line, exceeding the MOV rating, it will act as a short across the line to protect the connected circuitry. It is a significant improvement over simply putting a fuse in the line because it is voltage sensitive, can react much more quickly than a fuse, and displays its low-resistance characteristics for only a short period of time. After the spike has passed, it will return to its normal open-circuit characteristic. If you're wondering where the spike will go if the load is protected by a short circuit, remember that all sources of disturbance, such as lightning, generators, inductive motors (such as in air conditioners, dishwashers, and power saws), have their own "source resistance," and there is always some resistance down the line to absorb the disturbance.

Most line conditioners, as part of their advertising, like to mention their energy absorption level. The rating of the unit of Fig. 3.67 is 1200 J, which is actually higher than most. Remembering that $W = Pt = EIt$, we now realize that if a 5000-V spike came down the line, we would be left with the product $It = W/E = 1200\text{ J}/5000\text{ V} = 240$ mAs. If we assume a linear relationship between all quantities, the rated energy level reveals that a current of 100 A could be sustained for $t = 240\text{ mAs}/100\text{ A} = 2.4$ ms, a current of 1000 A for 240 μs, and a current of 10,000 A for 24 μs. Obviously, the higher the power product of E and I, the smaller the time element.

The technical specifications of the unit of Fig. 3.67 include an instantaneous response time of 0 ns, with a phone line protection of 5 ns. The unit is rated to dissipate surges up to 6000 V and current spikes up to 96,000 A. It has a very high noise suppression ratio (80 dB; see Chapter 11) at frequencies from 50 kHz to 1000 MHz, and (a credit to the company) it has a lifetime warranty.

3.13 COMPUTER ANALYSIS: PSpice (Windows)

In all the examples that have appeared in this text so far the transient effect was initiated by the closing of a switch at $t = 0$ s, as shown in Fig. 3.70(a). When applying PSpice, the instantaneous change in level is produced by applying a pulse waveform, as shown in Fig. 3.70(b) with a duration longer than the peirod of interest for the network. In PSpice the parameters of the pulse are defined as shown in Fig. 3.71.

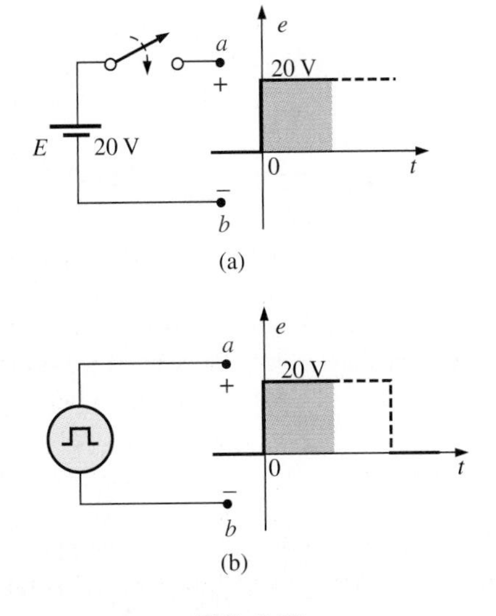

FIG. 3.70
Establishing a switching dc voltage level: (a) series dc voltage–switch combination; (b) PSpice (Windows) pulse option.

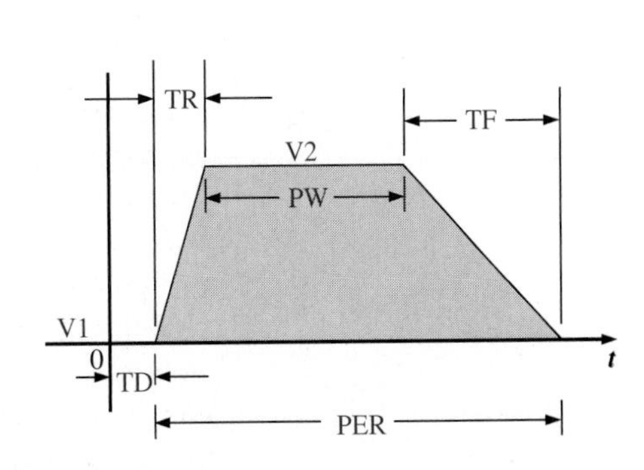

FIG. 3.71
The defining parameters of PSpice (Windows) ***VPULSE.***

For the network of Fig. 3.72 the time constant $\tau = (5\text{ k}\Omega)(8\ \mu\text{F}) = 40$ ms, and $5\tau = 200$ ms $= 0.2$ s, so a pulse width (PW) of 1 s was chosen to ensure that the transient phase had passed before the pulse ended. The other parameters of the pulse are defined in the PSpice screen of Fig. 3.73, with the resulting transient response appearing in Fig. 3.74.

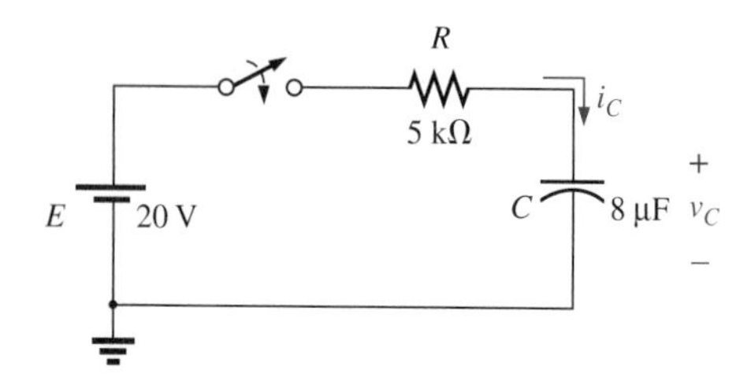

FIG. 3.72
Circuit to be analyzed using PSpice (Windows).

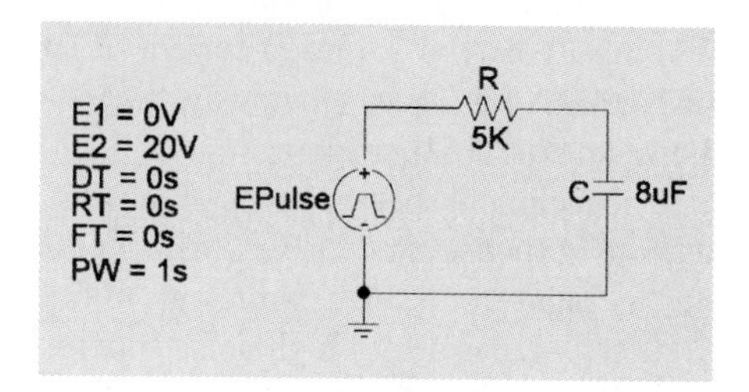

FIG. 3.73
PSpice (Windows) equivalent of the circuit of Fig. 3.72.

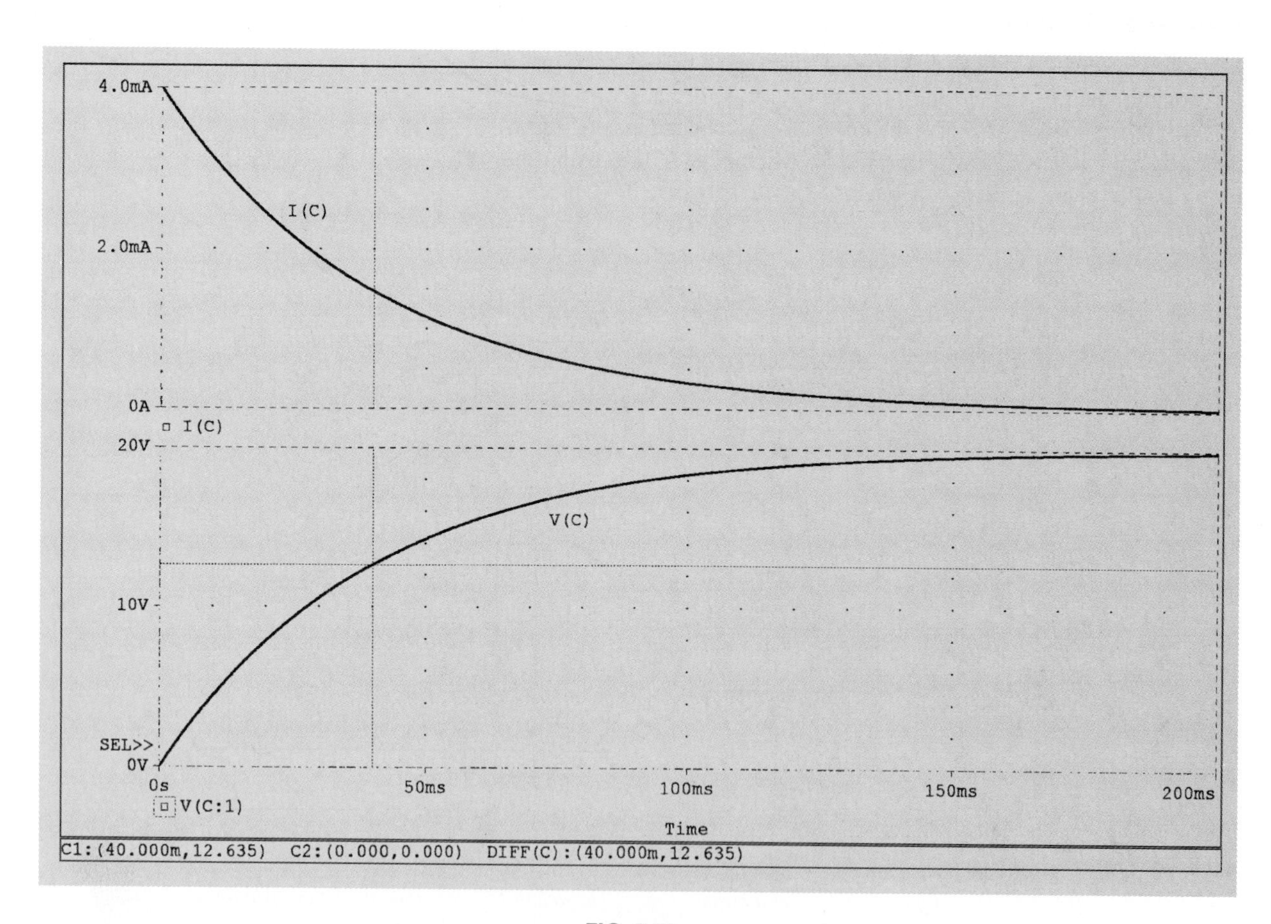

FIG. 3.74
v_C and i_C for the circuit of Fig. 3.72.

For the transient response of Fig. 3.74 a number of parameters have to be set up under the analysis and transient dialog boxes of the software program. Although not discussed in detail in this text, be assured that the selection process is fairly obvious. In particular, note in Fig. 3.74 that the peak value of the waveform is $I = E/R = 20\text{ V}/5\text{ k}\Omega = 4\text{ mA}$, and the final value of v_C is 20 V. In addition, note that the transient effect has essentially ended after 200 ms = 0.2 s.

Electronics Workbench

The resulting schematic for Electronics Workbench appears in Fig. 3.75 with the required signal and oscilloscope for viewing the response. Note that the levels of the applied pulse are 0 V and 20 V, as used for the PSpice response, but the pulse duration is now defined by the 1-Hz frequency, which results in a period of 1 s. The voltage across the capacitor will appear on channel A, and the voltage across the resistor Rs will appear across channel B. The voltage across the 1-Ω resistor will actually have the same shape and magnitude as the current due to the use of a 1-Ω resistor. Simply read the vertical scale on channel B as mA rather than mV (on the right side of Fig. 3.76) to obtain the transient waveform for the current of the circuit. Note in Fig. 3.76 that the voltage v_C has the same shape as obtained using PSpice as it decays from 20 V to 0 V in about 200 ms. An added feature of this type of setup is the response of the circuit when the pulse drops from 20 V to 0 V. Note that the voltage

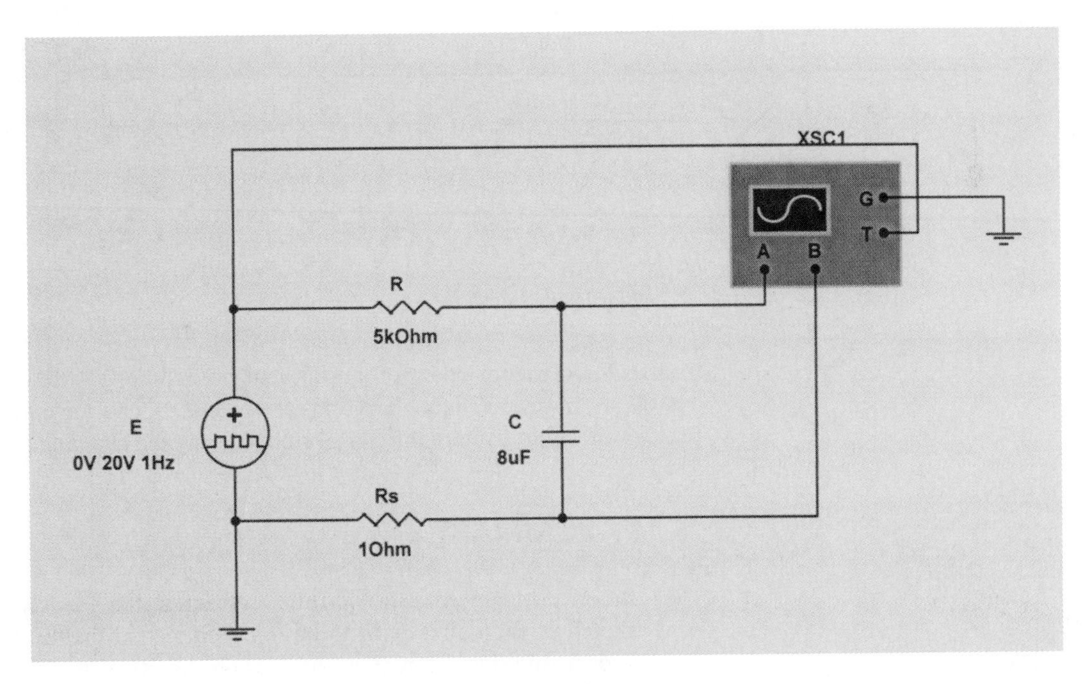

FIG. 3.75

Electronics Workbench setup for viewing i_C and v_C for the circuit of Fig. 3.72.

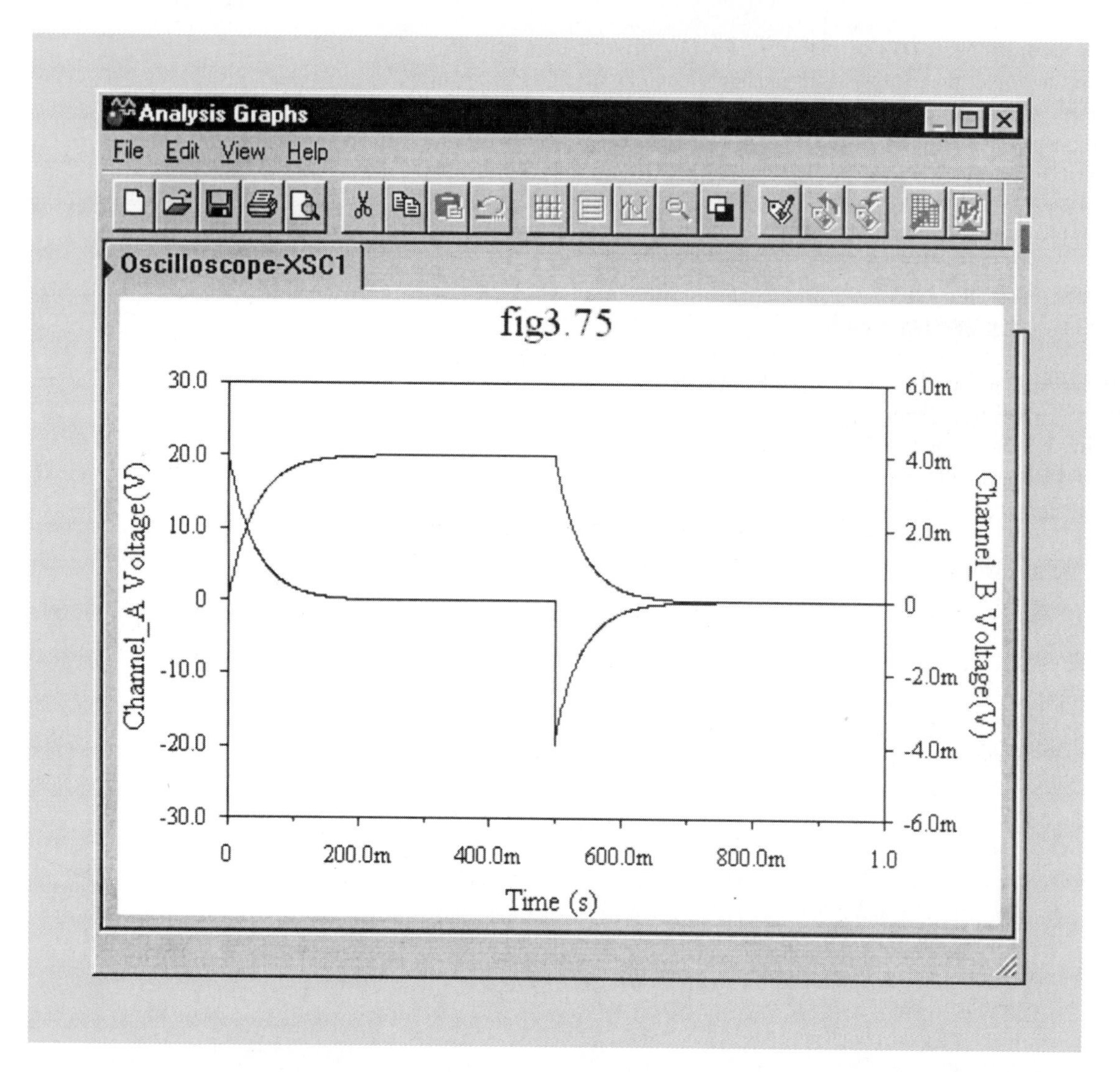

FIG. 3.76
Electronics Workbench response for the setup of Fig. 3.75. v_C is on channel A, and i_C is on channel B.

across the capacitor does not drop instantaneously to 0 V (the voltage across a capacitor cannot change instantaneously!), and the current reverses direction to return the charge stored during the charging phase.

3.14 SUMMARY

- Before analyzing a **series–parallel network** make a **brief mental sketch of the entire path to be followed** to find the desired unknown. In addition, redraw the network as often as possible to eliminate foolish errors and so as not to lose sight of the best path to follow.

- For **any series–parallel network,** the **basic laws** associated with series or parallel networks are applicable.
- **Current sources** are the **converse** of voltage sources. Whereas a voltage source has a fixed voltage and supplies a current determined by the network to which it is attached, a current source provides a fixed current to the branch in which it is located and has a terminal voltage determined by the network to which it is attached.
- The **lower** the voltage regulation the **less** the terminal voltage changes as the current through the load is increased.
- The **superposition theorem** states that the current through or voltage across any element of a multisource dc network is the **algebraic sum** of the currents or voltages for that element due to each source **independently.**
- **Branch-current analysis** requires that a **distinct current be assigned to each branch** of a network, where a branch is defined as a continuous connection of elements having the same current.
- **Branch-current analysis** requires the application of **both Kirchhoff's voltage and current laws** to develop the required number of independent equations.
- **Mesh analysis** is simply a **shorthand method** for obtaining a set of equations for a network in a more mechanical manner by applying **Kirchhoff's voltage law.** In addition, fewer equations are obtained than with branch-current analysis, thereby reducing the mathematical complexity. Be aware, however, that some branch currents will be a combination of mesh currents.
- **Nodal analysis** is a **shorthand method** for finding a set of equations relating all the nodal voltages of the network by applying **Kirchhoff's current law.** The voltage across a praticular element may require the writing of an equation relating two or more nodal voltages.
- Through the application of **Thévenin's theorem** any two-terminal linear multisource network can be reduced to one having only **one voltage source and a series resistor.**
- **Maximum power** will be transmitted to a load when its terminal resistance **equals** the Thévenin resistance of the network applied to the load.
- **Capacitors** have the ability to **store charge** on their plates—hence the term *storage capacity*.
- The **voltage** across a capacitor **cannot** change instantaneously.
- The **transient phase** of a capacitive or inductive network is primarily defined by the first **five time constants.**
- The **higher** the **permittivity** of a material the **greater** the charge that can be stored on the plates of the capacitor for the same applied voltage across the plates.
- The **current of a capacitor** is directly related to the **rate of change** of voltage across the capacitor. The greater the rate of change the more current, and if the voltage fails to change, the associated current is zero.
- **Inductors,** like capacitors, can **store energy;** however, inductors store energy in the form of a magnetic field.
- The **current of an inductor cannot** change instantaneously.

- The **voltage** across an inductor is directly related to the **rate of change** of cureent through the inductor. The greater the rate of change the larger the resulting terminal voltage, and if the current fails to change, the associated voltage is zero.
- The **inductance of a coil** is **directly** determined by the **square** of the number of turns, the **permeability** of the core material, and **area** of the coil, and **inversely** proportional to the **length.**

Equations:

$$VR\% = \frac{V_{NL} - V_{FL}}{V_{FL}} \times 100\%$$

$$R_L = R_{Th},\ P_{L_{max}} = \frac{E_{Th}^2}{4R_{Th}}$$

$$R_Y = \frac{R_\Delta}{3},\ R_\Delta = 3R_Y$$

$$\frac{R_1}{R_2} = \frac{R_2}{R_4}$$

$$C = \frac{Q}{V}$$

$$\tau = RC\ \text{(s)}$$

$$i_C = \frac{E}{R}e^{-t/\tau},\ v_C = E(1 - e^{-t/\tau})\ \text{(charging phase)}$$

$$C = \epsilon_0\epsilon_r\frac{A}{a}\ \text{(farads)},\ \epsilon_0 = 8.85 \times 10^{-12}\ \text{F/m}$$

$$C_T = C_1 + C_2 + C_3 + \cdots + C_N\ \text{(parallel capacitors)}$$

$$\frac{1}{C_T} = \frac{1}{C_1} + \frac{1}{C_2} + \frac{1}{C_3} + \cdots + \frac{1}{C_N}\ \text{(series capacitors)}$$

$$C_T = \frac{C_1 \cdot C_2}{C_1 + C_2}\ \text{(series capacitors)}$$

$$i_C = C\frac{dv_C}{dt}$$

$$W_C = \tfrac{1}{2}CV^2\ \text{(joules)}$$

$$\tau = \frac{L}{R}$$

$$i_L = \frac{E}{R}(1 - e^{-t/\tau}),\ v_L = Ee^{-t/\tau}\ \text{(setting up magnetic fields)}$$

$$L = \frac{N^2\mu A}{l}\ \text{(henries)}$$

$$L_T = L_1 + L_2 + L_3 + \cdots + L_N\ \text{(series inductors)}$$

$$\frac{1}{L_T} = \frac{1}{L_1} + \frac{1}{L_2} + \frac{1}{L_3} + \cdots + \frac{1}{L_N}\ \text{(parallel inductors)}$$

$$L_T = \frac{L_1 \cdot L_2}{L_1 + L_2}\ \text{(parallel inductors)}$$

$$v_L = L\frac{di_L}{dt}$$

$$W_L = \tfrac{1}{2}LI^2\ \text{(joules)}$$

PROBLEMS

SECTION 3.2 Series–Parallel Networks

1. For the series-parallel network in Fig. 3.77:
   a. Determine R_T.
   b. Calculate I.
   c. Find I_1, I_2, and I_3.
   d. Calculate the power to R_3.

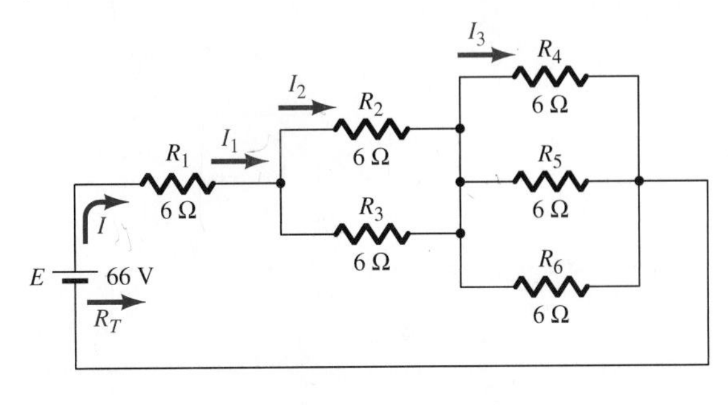

FIG. 3.77

2. For the series-parallel network in Fig. 3.78:
   a. Determine I_4.
   b. Determine I.
   c. Find P_3.

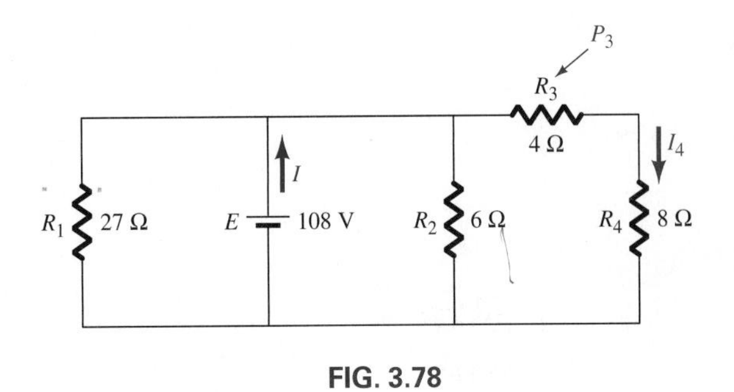

FIG. 3.78

3. For the "ladder" network in Fig. 3.79 find I_6.

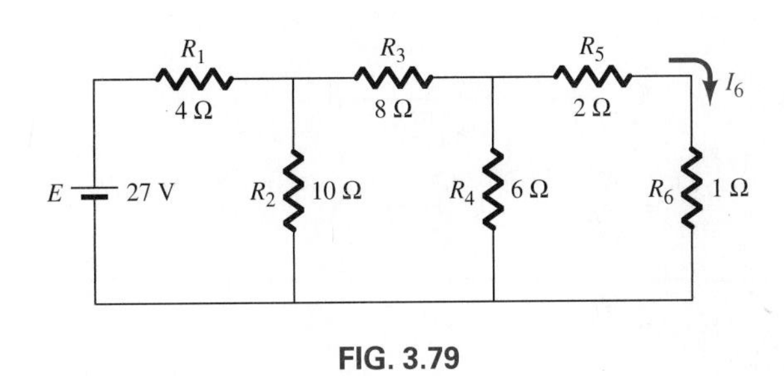

FIG. 3.79

SECTION 3.3 Current Sources

4. For the network in Fig. 3.80, find:
   a. V_s.
   b. I_1.
   c. I_4.

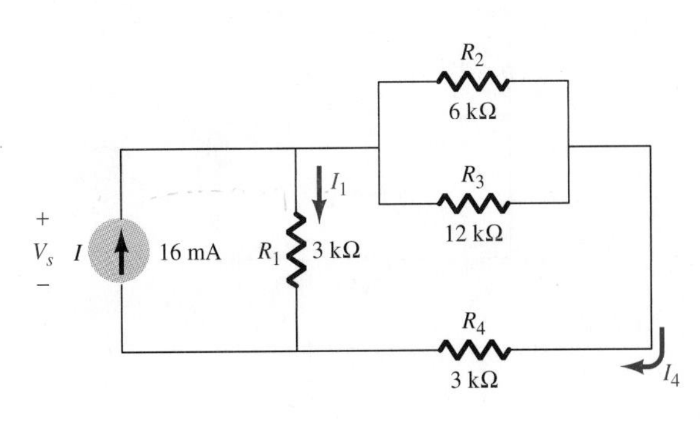

FIG. 3.80

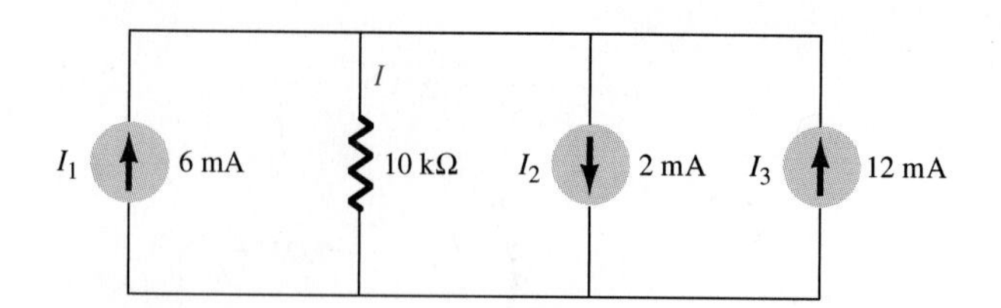

FIG. 3.81

5. Find the current I for the network in Fig. 3.81.

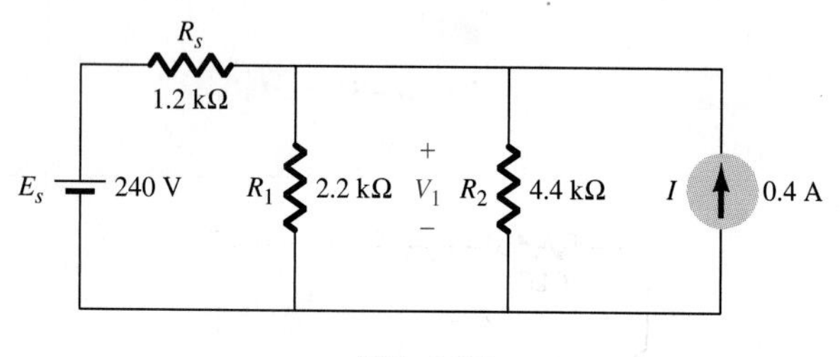

FIG. 3.82

6. **a.** Convert the voltage source in Fig. 3.82 to a current source.
   b. Find the voltage V_1.

SECTION 3.4 Voltage and Current Regulation

7. Determine the voltage regulation of a supply whose terminal voltage drops 1.8 V from open-circuit to full-load conditions if the full-load voltage is 220 V.

8. A 12-V supply has an internal resistance of 0.05 Ω. An applied load of 1 Ω draws the full-load current from the supply. What is the percent voltage regulation of the supply?

9. A current source has a current regulation of 1.2%. If the full-load voltage draws a current of 200 mA, what is the no-load current?

SECTION 3.5 Multisource Networks

10. Using the superposition theorem, determine the current I for the network in Fig. 3.83.

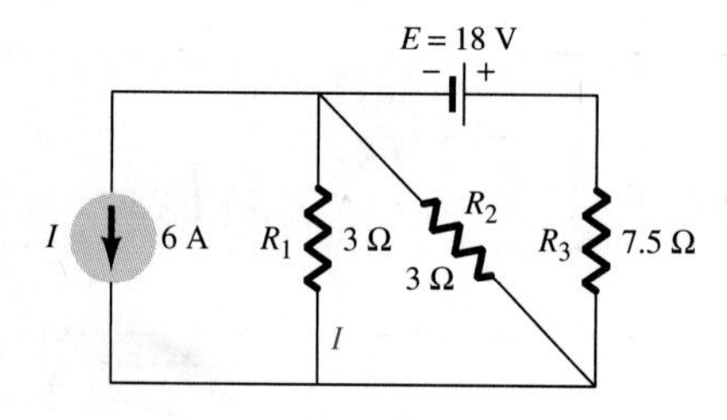

FIG. 3.83

11. Using the superposition theorem, determine the current I for the network in Fig. 3.84.

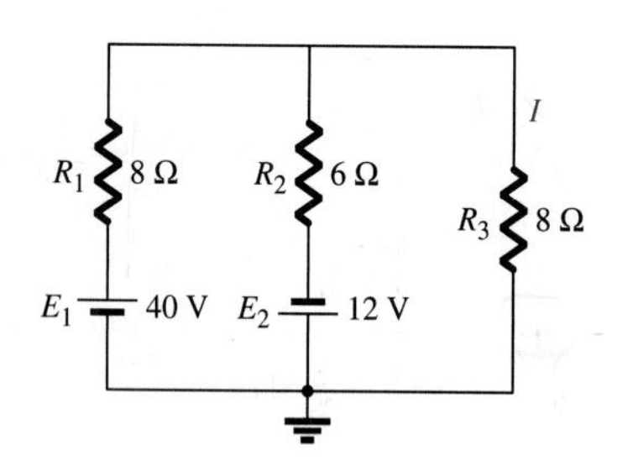

FIG. 3.84

12. Repeat Problem 11 using branch-current analysis and compare the results.

13. Determine the branch currents for the network in Fig. 3.85 using branch-current analysis.

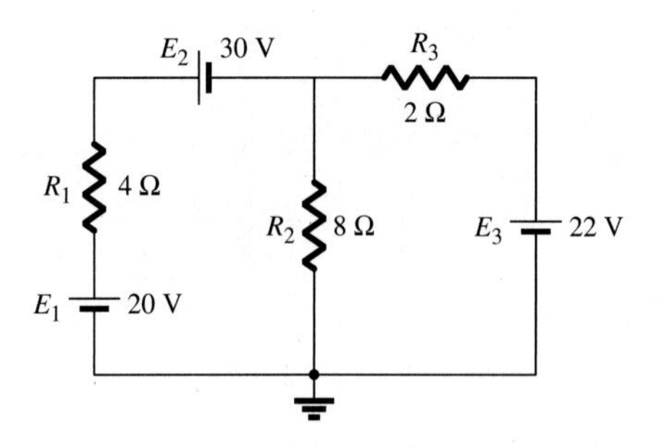

FIG. 3.85

14. Repeat Problem 11 using mesh analysis.

15. Determine the current I_1 for the network in Fig. 3.86 using mesh analysis.

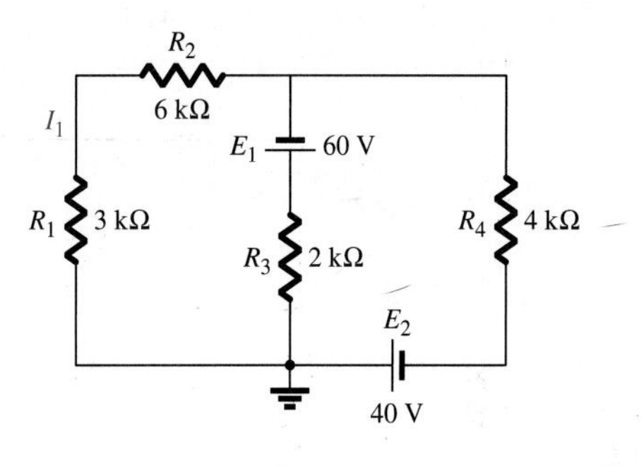

FIG. 3.86

16. Determine the current I_6 for the network in Fig. 3.79 using mesh analysis.

17. **a.** Determine the nodal voltages for the network in Fig. 3.87.

b. Using the nodal voltages, determine the current (magnitude and direction) through each resistive element.

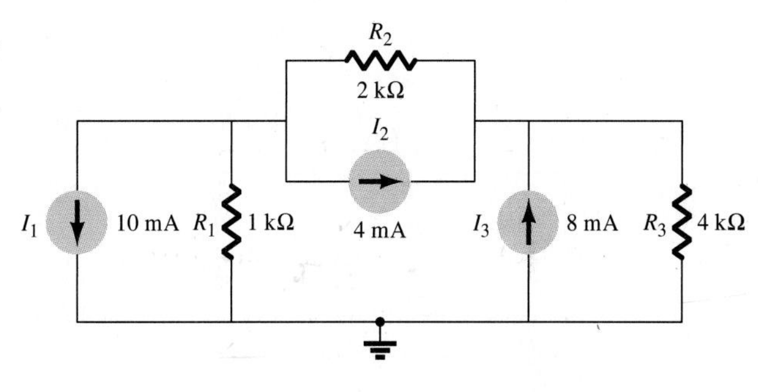

FIG. 3.87

18. **a.** Convert the voltage sources in Fig. 3.84 to current sources.

b. Determine the nodal voltage for the resulting network.

19. Determine the nodal voltages for the network in Fig. 3.88.

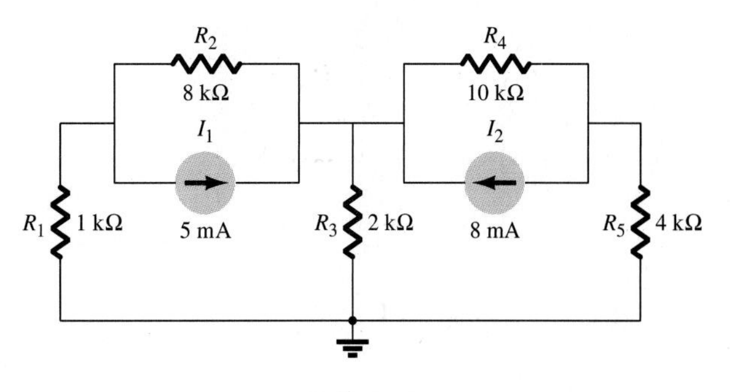

FIG. 3.88

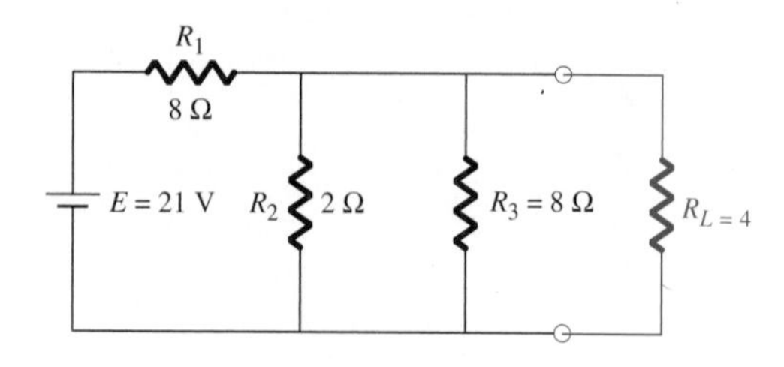

FIG. 3.89

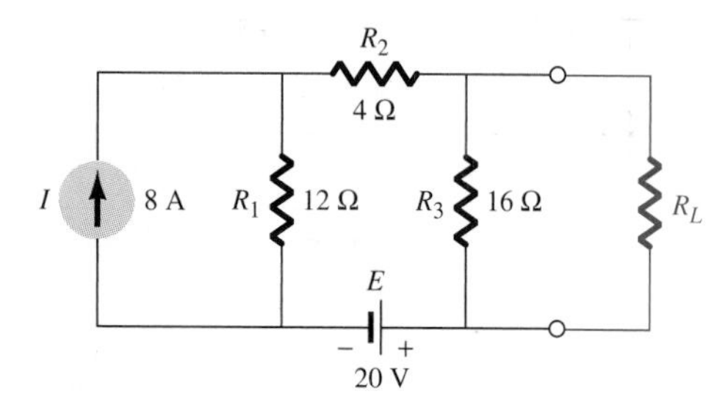

FIG. 3.90

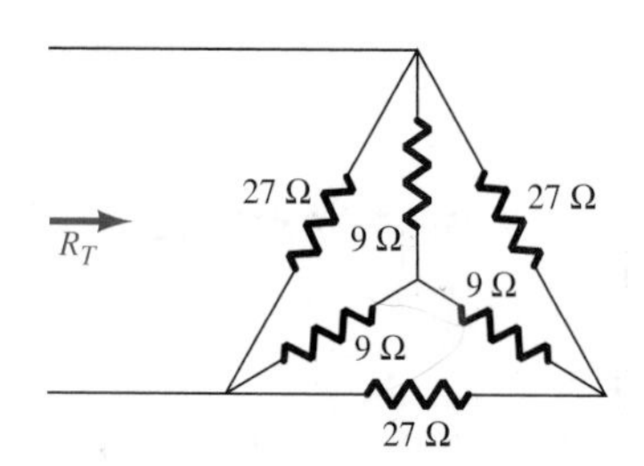

FIG. 3.91

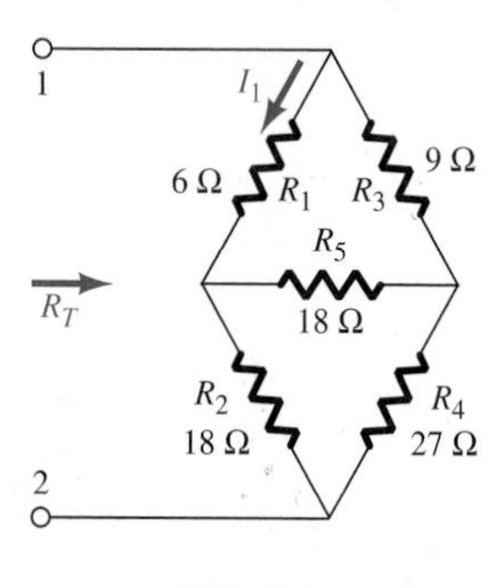

FIG. 3.92

SECTION 3.7 Network Theorems

20. **a.** Find the Thévenin network for the portion of the network in Fig. 3.89 to the left of resistor R_L.
b. Using the Thévenin network, determine I_L.
c. Use series–parallel techniques on the original network to determine I_L, and compare the results with those for part (b).

21. **a.** Determine the value of R_L in Fig. 3.89 that would result in maximum power to R_L.
b. Calculate the maximum power that could be delivered to R_L if it were changed to that value.
c. Find the power delivered to R_L if $R_L = 4\ \Omega$ and verify that it is less than the maximum value in part (b).

22. **a.** Find the Thévenin network for the network external to the resistor R_L in Fig. 3.90.
b. Find R_L for maximum power to R_L.
c. Find the maximum power to R_L.

SECTION 3.8 Δ–Y Conversions

23. Determine the total resistance of the network in Fig. 3.91.

24. Calculate the total resistance of the network in Fig. 3.92.

25. Is the bridge in Problem 24 balanced? Balanced or not, find the current I_1 with 30 V applied between points 1 and 2.

26. **a.** If $R_1 = 10\ \Omega$, $R_2 = 40\ \Omega$, $R_3 = 4\ \Omega$, $R_4 = 16\ \Omega$, and $R_5 = 18\ \Omega$ in Fig. 3.92, is the bridge balanced?

b. If 20 V is applied to points 1 and 2, determine the current through R_4 [for the values in part (a)].
c. With the 20 V applied, determine the current through R_1 [for the values of part (a)].

SECTION 3.10 Capacitors

27. If 3600 μC of charge is deposited on the plates of a capacitor having a potential drop of 120 V across the plates, determine the capacitance of the capacitor.

28. For the network in Fig. 3.93:
a. Determine the time constant of the network.
b. For the charging phase, write the mathematical expression for the current i_C and voltages v_C and v_R.
c. Sketch the waveform of each quantity in part (b).
d. What is i_C in magnitude after one time constant?
e. At what instant of time will $v_C = 50$ V?
f. Find the energy stored by the capacitor at steady state.

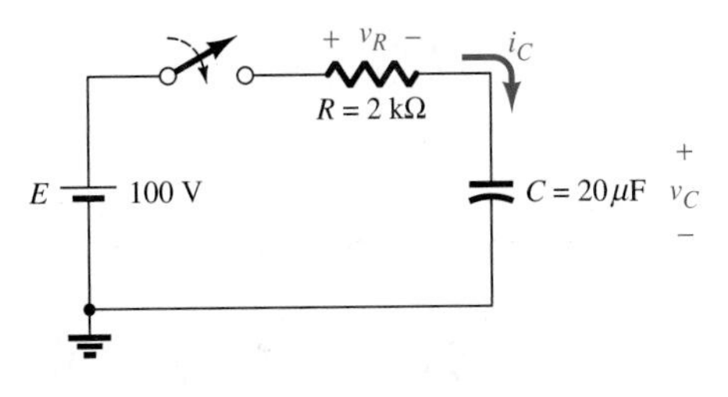

FIG. 3.93

29. Find the capacitance of a capacitor having $2 \times 10^{-4}\ m^2$ plates, a dielectric of mica, and a d (distance between plates) of 50 μm.

30. a. Find the total capacitance of four 10-μF capacitors in series.
b. Repeat part (a) for capacitors in parallel.

31. a. Find the expression for v_C in the network in Fig. 3.94 following the closing of the switch.
b. Sketch the waveform of i_C.

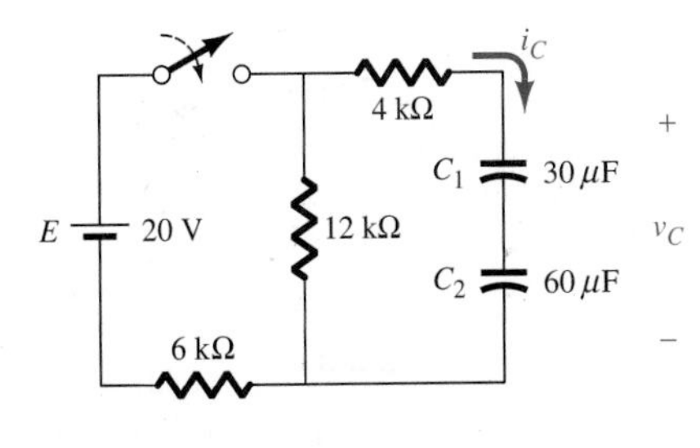

FIG. 3.94

32. Determine the waveform for the current i_C of a 2-μF capacitor for the applied voltage v_C in Fig. 3.95.

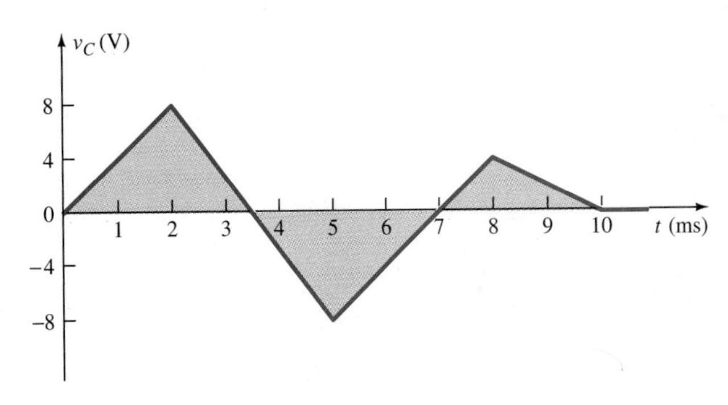

FIG. 3.95

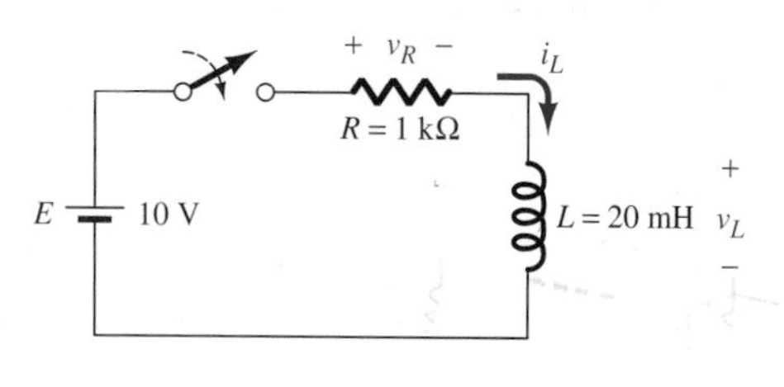

FIG. 3.96

SECTION 3.11 Inductors

33. For the network in Fig. 3.96:
 a. Determine the time constant.
 b. Determine the mathematical expressions for i_L, v_L, and v_R during the charging phase.
 c. Sketch the waveforms for part (b).
 d. What is the magnitude of v_R after one time constant?
 e. At what instant will $i_L = 2$ mA?
 f. Find the energy stored by the inductor.

34. Determine the inductance of an inductor with 250 turns, $\mu_r = 2000$, area $= 1.5 \times 10^{-4}\ \text{m}^2$, and $l = 50$ mm.

35. **a.** Determine the total inductance of two series coils with values of 10 mH and 90 mH.
 b. Repeat part (a) for the coils in parallel.

36. **a.** Determine the expression for i_L when the switch is closed in Fig. 3.97.
 b. Repeat part (a) for v_L.

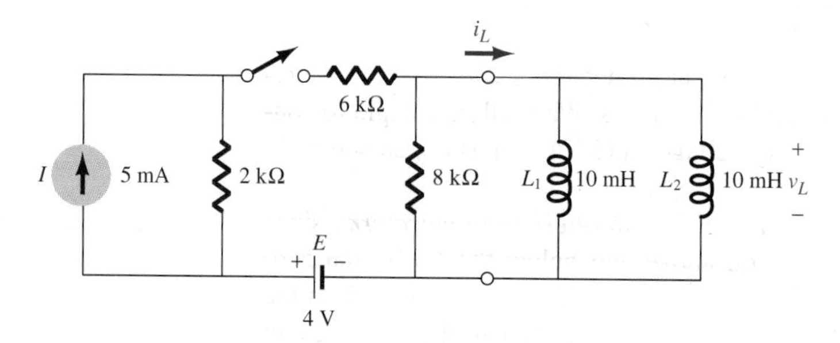

FIG. 3.97

37. Determine the waveform for the voltage v_L of a 0.5-H inductor for the current i_L in Fig. 3.98.

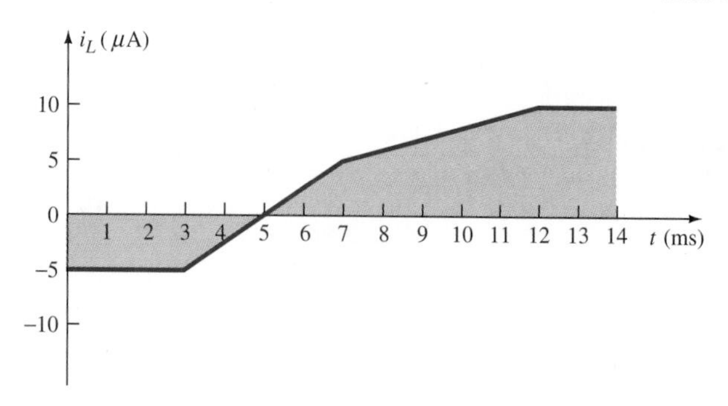

FIG. 3.98

38. **a.** Determine I_1 and V_1 for the network in Fig. 3.99 under steady-state conditions.
 b. Find V_C and I_L (steady state).
 c. Determine the energy stored by L and C.

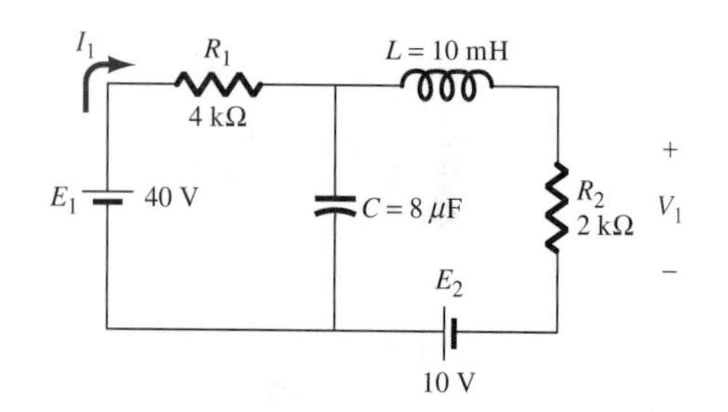

FIG. 3.99

4 Ac Networks

4.1 INTRODUCTION

Our analysis thus far has been limited to networks with fixed, nonvarying, time-independent currents and voltages. We shall now begin to consider the effects of alternating voltages and currents such, as shown in Fig. 4.1.

Each waveform in Fig. 4.1 is called an *alternating waveform,* since it alternates between the region above and below the horizontal zero axis. The first [Fig. 4.1(a)] is called a *sinusoidal ac voltage* and is the type universally available at home outlets, at industrial locations, and

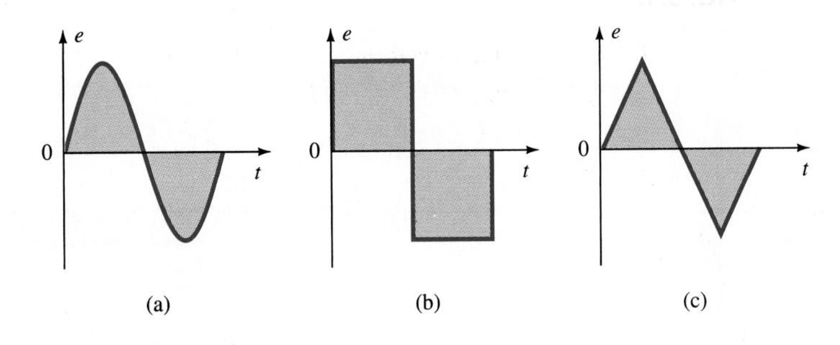

FIG. 4.1
Alternating waveforms: (a) sinusoidal; (b) square wave; (c) triangular.

so on. In this chapter we shall deal specifically with this type of waveform. The second [Fig. 4.1(b)] is called, for obvious reasons, a *square wave,* which has its application in a number of areas to be introduced in later chapters. The last of the three [Fig. 4.1(c)] and perhaps the least encountered is called a *triangular* waveform. Compare each with the dc voltage in Fig. 4.2, which is fixed in magnitude and does not vary with time (the horizontal axis).

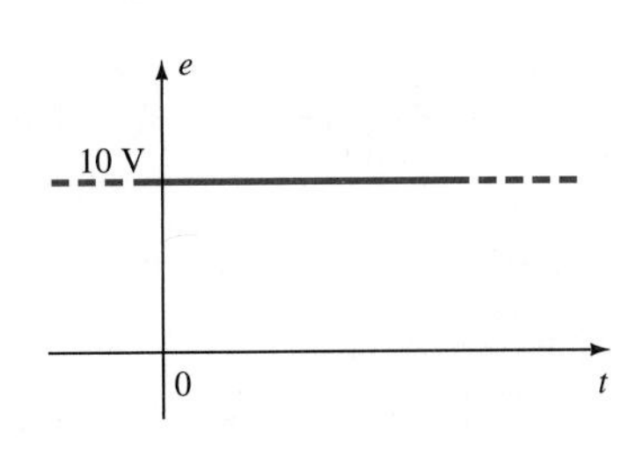

FIG. 4.2
Dc voltage.

Although a multitude of questions about how to analyze networks with an applied signal that varies with time may now begin to surface, be assured that after a careful examination of this chapter the analysis of ac networks will be on a level of difficulty only slightly greater than that encountered for dc networks. In fact, the only added difficulty will be mathematical and not in the application of the theorems already described for dc networks.

For the remainder of this chapter the phrase *ac voltage* or *current* will refer specifically to the sinusoidal alternating waveform in Fig. 4.1(a).

4.2 SINUSOIDAL (AC) WAVEFORM

Sinusoidal ac voltages are available from a variety of sources. The most common source is the typical home outlet, which provides an ac voltage that originates at a power plant; such a power plant is most commonly fueled by coal, water power, oil, gas, or nuclear fission. In each case an *ac generator* (also called an *alternator*), as in Fig. 4.3(a), is the primary component in the energy-conversion process. The power to the shaft developed by one of the energy sources listed turns a *rotor* (constructed of alternating magnetic poles) inside a set of windings housed in the *stator* (the stationary part of the dynamo) and induces a voltage across the windings of the stator. Through proper design of the generator, a sinusoidal ac voltage is developed that can be transformed to higher levels for distribution through the power lines to the consumer. In isolated locations where power lines have not been installed, portable ac generators [Fig. 4.3(b)] that run on gasoline are available. As in the larger power plants, however, an ac generator is an integral part of the design.

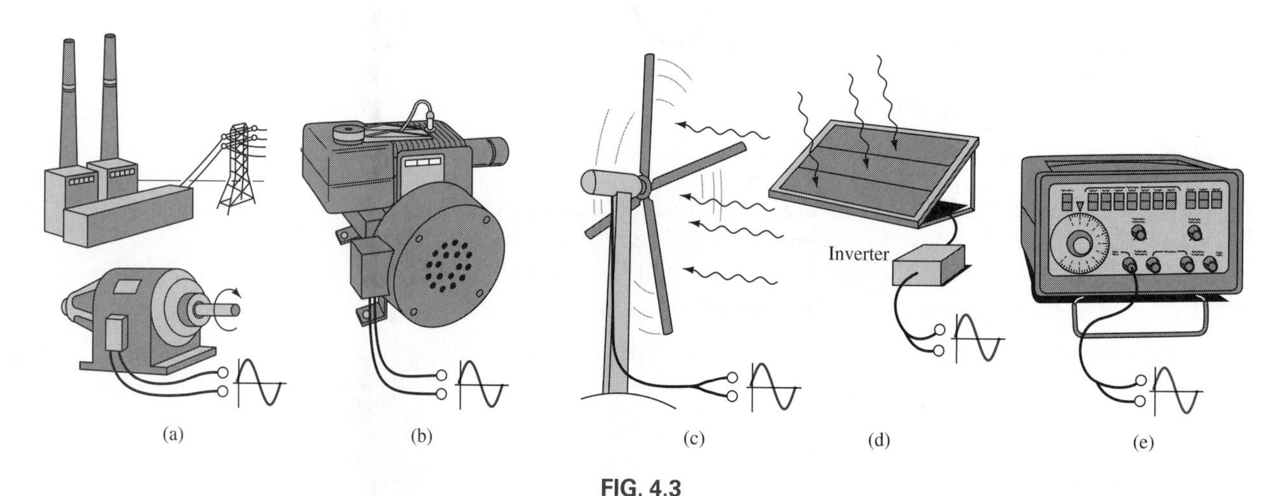

FIG. 4.3
Various sources of ac power: (a) generating plant; (b) portable ac generator; (c) wind-power station; (d) solar panel; (e) function generator.

In an effort to conserve our natural resources, wind power and solar energy are receiving increasing attention from various districts of the world that have such energy sources available in level and duration that make the conversion process practicable. The turning propellers of a windpower station [Fig. 4.3(c)] are connected directly to the shaft of an ac generator to provide the ac voltage described above. By converting light energy absorbed in the form of *photons,* solar cells [Fig. 4.3(a)] generate dc voltages. Through an electronic package called an *inverter,* the dc voltage can be converted to one of a sinusoidal nature. Boats, recreational vehicles (RVs), and so on, make frequent use of the inversion process in isolated areas.

Sinusoidal ac voltages with characteristics that can be controlled by the user are available from *function generators* such as the one in Fig. 4.3(e). By setting the various switches and controlling the position of the knobs on the face of the instrument, the user can generate sinusoidal voltages of different peak values and different repetition rates. The function generator plays an integral role in the investigation of the variety of theorems, methods of analysis, and topics to be introduced in the chapters to follow.

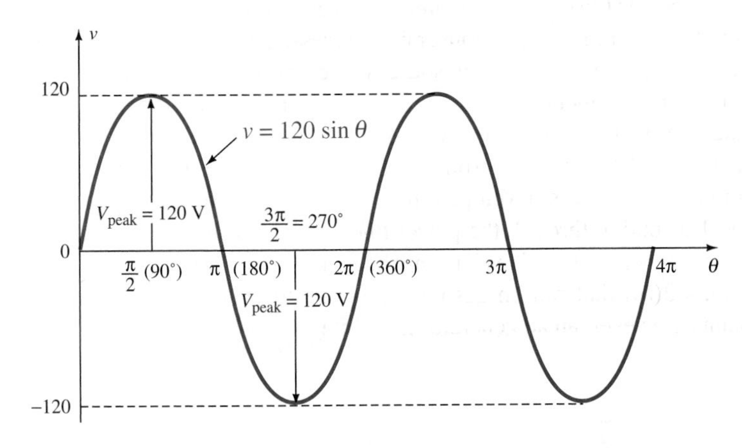

FIG. 4.4
Sinusoidal waveform.

The expanded view of a sinusoidal voltage in Fig. 4.4 reveals that the waveform repeats itself in the negative sense from π to 2π and has the same peak value of 120 V above and below the axis. In its simplest mathematical form, an equation for the sinusoidal voltage is

$$v = V_{peak} \sin \theta \tag{4.1}$$

Note that at 90° $\sin \theta = \sin 90° = 1$ and that $v = V_{peak}(1) = V_{peak} = 120$ V. The same occurs at 270°, but $\sin 270° = -\sin 90° = -1$. Certainly at 0°, 180°, and 360° $\sin \theta = 0$, and therefore $v = 0$ V. The axis in Fig. 4.4 is more frequently labeled in radians than degrees. The equivalent radian values appear for 90°, 180°, 270°, and 360°. The following equations permit a direct conversion from one to the other:

$$\text{radians} = \left(\frac{\pi}{180°}\right)(\text{degrees}) \tag{4.2}$$

or

$$\text{degrees} = \left(\frac{180°}{\pi}\right)(\text{radians}) \tag{4.3}$$

Figure 4.5 pictorially compares degree and radian measure.

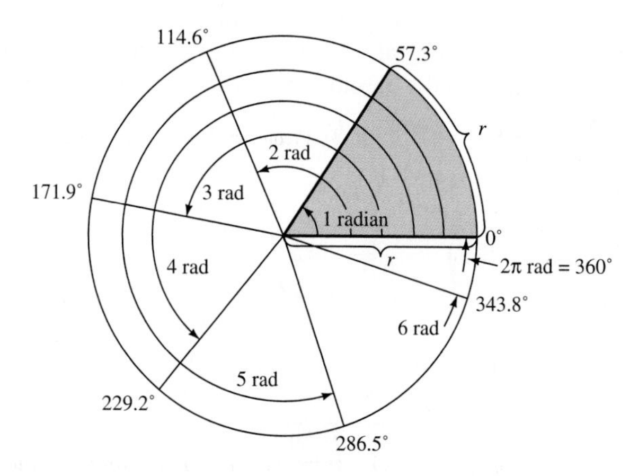

FIG. 4.5
Radian versus degree measure.

EXAMPLE 4.1

a. Convert the following from degrees to radians:

$$45°\text{: radians} = \left(\frac{\pi}{180°}\right)(45°) = \boldsymbol{\frac{\pi}{4}}$$

$$270°\text{: radians} = \left(\frac{\pi}{180°}\right)(270°) = \boldsymbol{\frac{3}{2}\pi}$$

b. Convert the following from radians to degrees:

$$\frac{\pi}{3}\text{: degrees} = \left(\frac{180°}{\pi}\right)\left(\frac{\pi}{3}\right) = \mathbf{60°}$$

$$\frac{4}{3}\pi\text{: degrees} = \left(\frac{180°}{\pi}\right)\left(\frac{4}{3}\pi\right) = \mathbf{240°}$$

By definition, the *period (T)* of a sinusoidal waveform is the time required for one complete appearance of the waveform, which is called a *cycle.* For the waveform in Fig. 4.6(a) the period is 1 s. For the adjoining waveform [Fig. 4.6(b)] the period is $\frac{1}{2}$ s. The *frequency (f)* of an alternating waveform is the number of cycles that appear in a time span of 1 s. For Fig. 4.6(a) the frequency is 1 cycle per second, and for Fig. 4.6(b) it

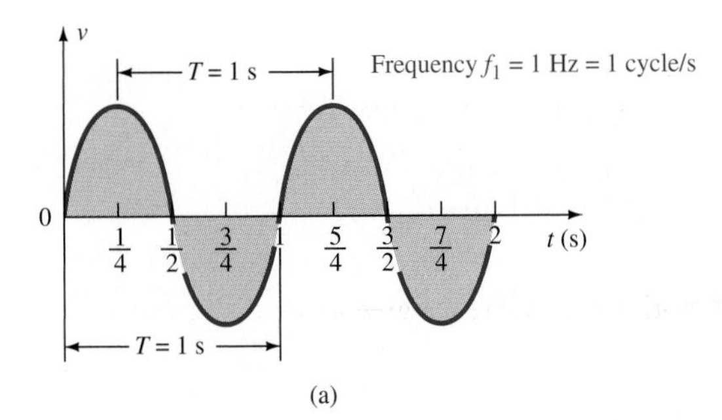

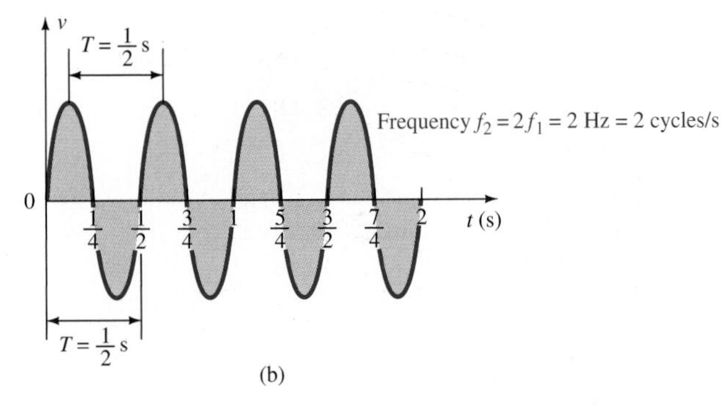

FIG. 4.6
Sinusoidal voltages of different frequencies plotted against time.

is 2 cycles per second. The unit *cycle per second* has almost universally been replaced by the *hertz (Hz)*; that is,

$$1 \text{ hertz} = 1 \text{ cycle per second} \tag{4.4}$$

Note from the examples in Fig. 4.6 that the period and frequency are inversely related. That is, as the time per cycle decreases, the frequency increases by a corresponding amount. In equation form, the two are related by

$$T = \frac{1}{f} \tag{4.5}$$

where T = period in seconds
f = frequency in hertz

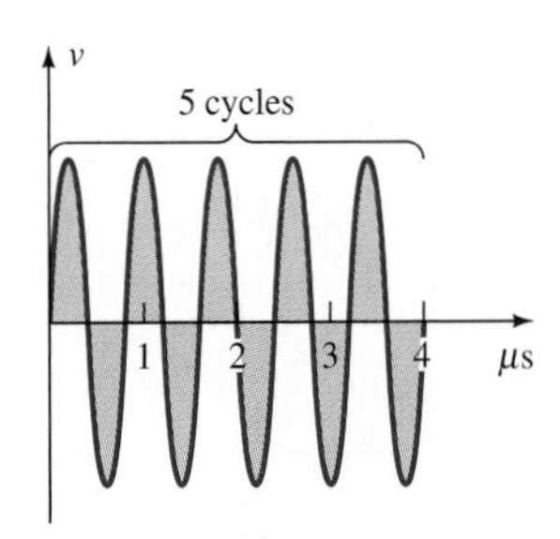

FIG. 4.7
Waveform for Example 4.2.

EXAMPLE 4.2 Determine the period and the frequency of the waveform in Fig. 4.7.

Solution: In Fig. 4.7 five cycles encompass the indicated time of 4 μs. Therefore,

$$T = \tfrac{1}{5}(4\ \mu\text{s}) = \mathbf{0.8\ \mu s}$$

and

$$f = \frac{1}{T} = \frac{1}{0.8 \times 10^{-6}\ \text{s}} = \frac{10^6}{0.8} = 1.2 \times 10^6\ \text{Hz} = \mathbf{1.2\ MHz}$$

EXAMPLE 4.3 How long will it take a sine wave with a frequency of 0.2 kHz to complete 10 cycles?

Solution:

$$T = \frac{1}{f} = \frac{1}{0.2 \times 10^3\ \text{Hz}} = \frac{1}{0.2} \times 10^{-3}$$
$$= 5\ \text{ms}$$

For 10 cycles:

$$10(5\ \text{ms}) = \mathbf{50\ ms}$$

EXAMPLE 4.4 For $v = 20 \sin \theta$, at what angle will the voltage rise to 5 V?

Solution:

$$v = 20 \sin \theta$$
$$5 = 20 \sin \theta$$
$$0.25 = \sin \theta$$

and

$$\theta = \sin^{-1} 0.25$$
$$= \mathbf{14.48°}$$

The use of log scale permits a frequency spectrum from 1 GHz to 1000 GHz to be scaled off on the same axis, as shown in Fig 4.8. A number of terms in the various spectra are probably familiar to you from everyday experiences. Note that the audio range (human ear) extends from only 15 Hz to 20 kHz, but radio signals can be transmitted between 3 kHz and 300 GHz. The uniform process of defining the intervals of the radio-frequency spectrum from VLF to EHF is quite evident from the length of the bars in the figure (although keep in mind that it is a log scale, so the frequencies encompassed within each segment are quite different). Other frequencies of particular interest (TV, CB, microwave, etc.) are also included for reference purposes. Although it is numerically easy to talk about frequencies in the megahertz and gigahertz range, keep in mind that a frequency of 100 MHz, for instance, represents a sinusoidal waveform that passes through 100,000,000 cycles in only 1 s —an incredible number when we compare it with the 60 Hz of our conventional power sources. The Pentium III chip manufactured by Intel can run at speeds up to 450 MHz. Imagine a product able to handle 450,000,000 instructions per second—an incredible achievement.

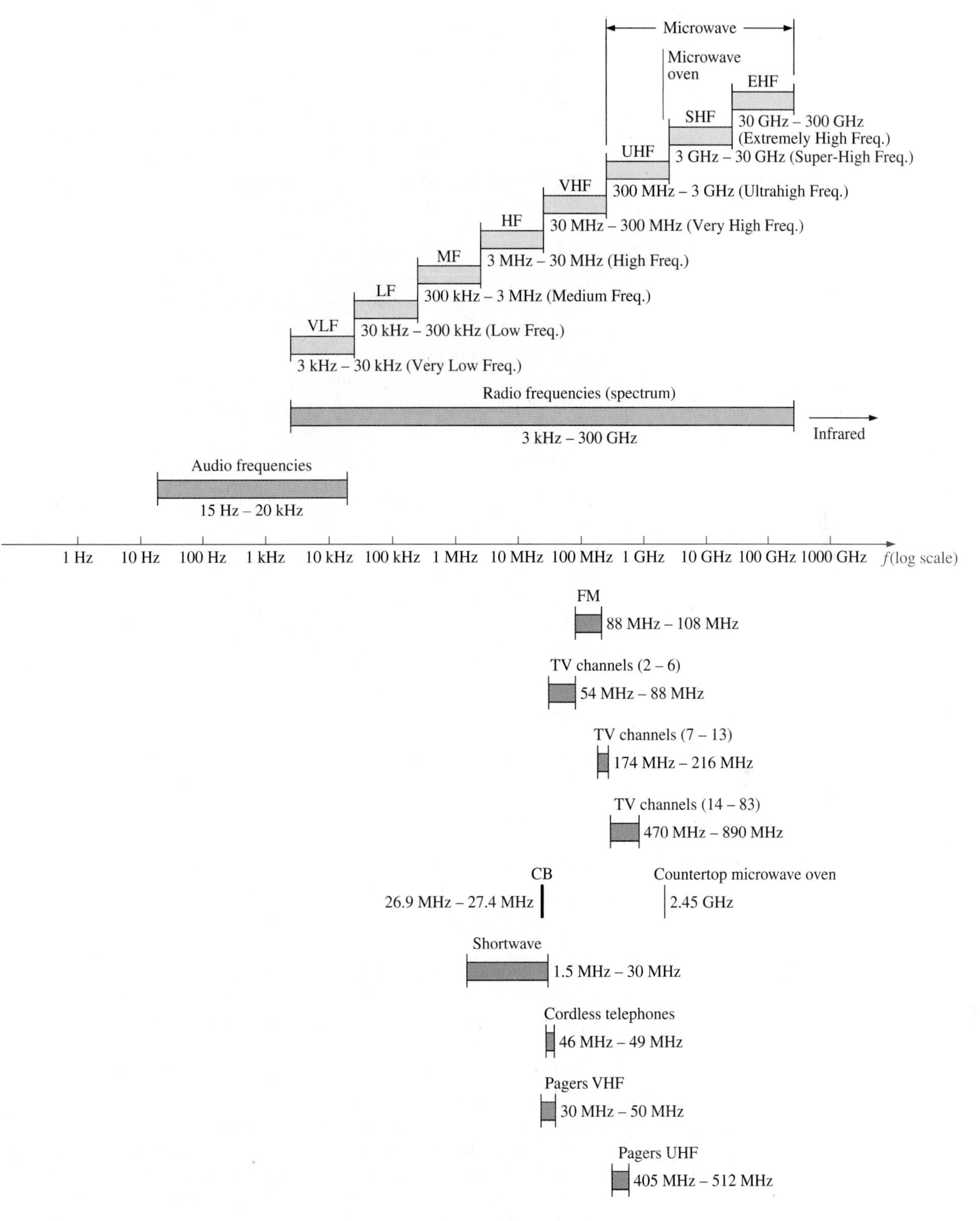

FIG. 4.8
Areas of application for specific frequency bands.

The sinusoidal waveform can be generated by following the vertical projection of a rotating vector as shown in Fig. 4.9. At $t = 0\text{s}$ ($\theta = 0°$) the vector is on the horizontal axis, and the vertical projection is zero, as shown in Fig. 4.9(a).

As the vector continues to rotate it reaches $\theta = 30°$, with the height of the tip of the vector from the horizontal axis defining the height of the sine wave in the right-hand figure. Note also the horizontal shift of 30° from the vertical axis in the right figure to define the horizontal parameter (θ or t).

At $\theta = 90°$ ($t = T/4$) the vertical projection is a maximum, while at $\theta = 180°$ it drops to zero again. As the vector continues to rotate in the counterclockwise direction it eventually generates a negative maximum and then returns to zero at $\theta = 360°$ ($t = T$).

If the vector in Fig. 4.9 rotates at an increased speed, more cycles will be generated for a fixed time interval. The angular velocity of the vector is therefore directly related to the frequency of the sinusoidal waveform generated.

In fact,

$$\text{velocity} = \frac{\text{distance}}{\text{time}}$$

and angular velocity

$$\omega = \frac{2\pi}{T} \quad \text{(rad/s)} \tag{4.6}$$

since 2π radians are traversed in one period (T) of the waveform. Substituting $f = 1/T$, we have

$$\omega = 2\pi f \quad \text{(rad/s)} \tag{4.7}$$

Since distance = velocity × time, the angle θ (in radians) traversed at an angular velocity ω for a period of time t can be determined from

$$\theta = \omega t \quad \text{(radians)} \tag{4.8}$$

Keep in mind, however, that if ω is given in radians per second and t is in seconds, θ is provided in radians and not degrees.

Equation (4.1) for the sinusoidal waveform can therefore be written as

$$v = V_p \sin \omega t \tag{4.9}$$

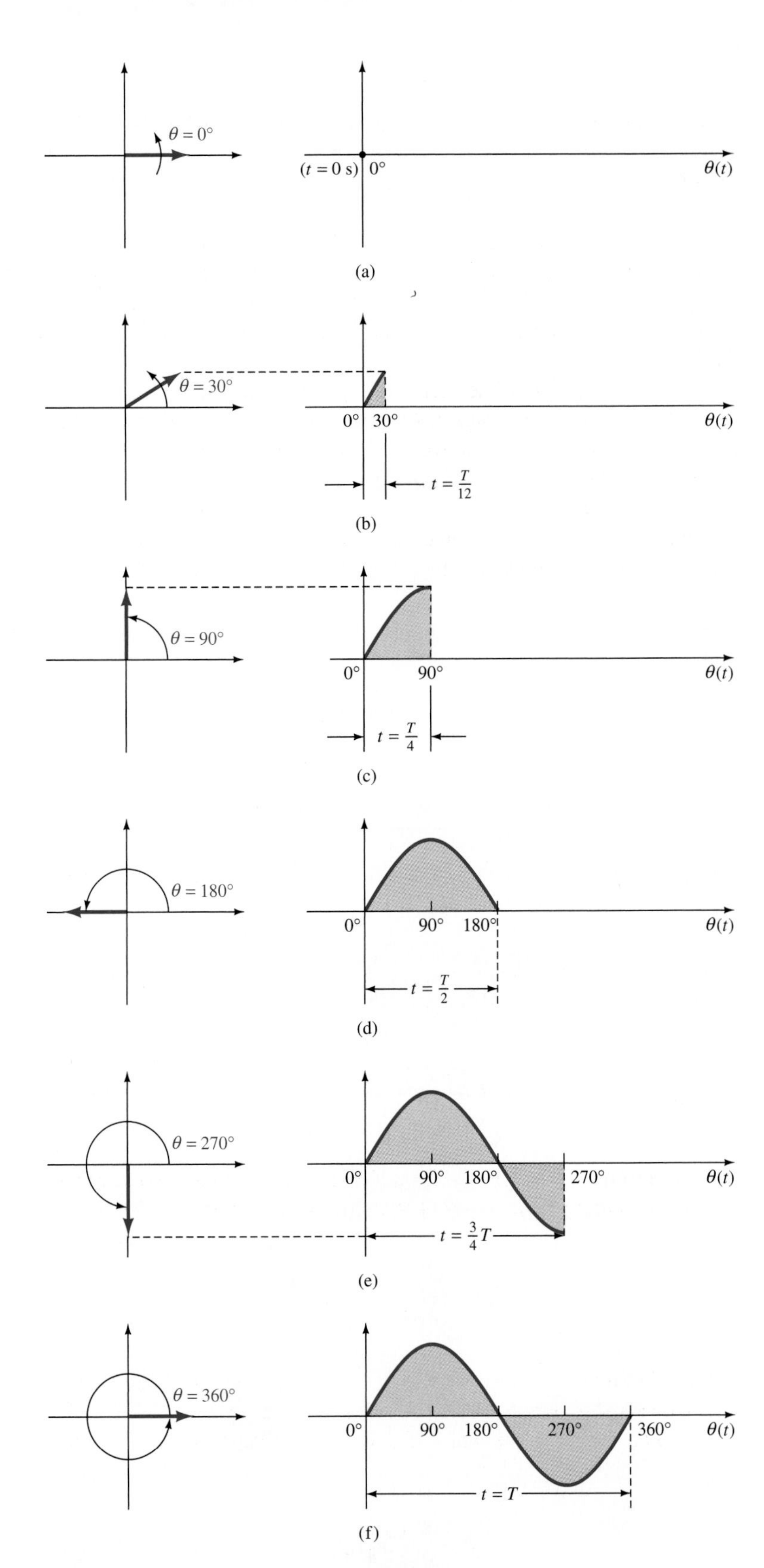
θ = 0°
(t = 0 s)
0°
θ(t)
(a)
θ = 30°
0°
30°
θ(t)
t = T/12
(b)
θ = 90°
0°
90°
θ(t)
t = T/4
(c)
θ = 180°
0°
90°
180°
θ(t)
t = T/2
(d)
θ = 270°
0°
90°
180°
270°
θ(t)
t = 3/4 T
(e)
θ = 360°
0°
90°
180°
270°
360°
θ(t)
t = T
(f)

FIG. 4.9

or

$$v = V_p \sin 2\pi ft \qquad \textbf{(4.10)}$$

For home and industrial use in North America, 60 Hz is the available frequency, while in Europe and surrounding areas 50 Hz is the chosen frequency. From the preceding we now know that the sinusoidal voltage available at a home outlet in North America completes 60 cycles in 1 s and has a period

$$T = \frac{1}{f} = \frac{1}{60\text{ Hz}} = \mathbf{16.67\ ms}$$

and an angular velocity

$$\omega = 2\pi f = (6.28)(60\text{ Hz}) \cong \mathbf{377\ rad/s}$$

It also has a peak voltage of about 170 V, and so the entire sinusoidal function can be represented by

$$v = 170 \sin 377t$$

EXAMPLE 4.5 Given $i = 20 \times 10^{-3} \sin 400\ t$, find the time t at which i rises to 10 mA.

Solution:

$$i = 20 \times 10^{-3} \sin 400t$$

$$10\text{ mA} = 20\text{ mA} \sin 400t$$

$$0.5 = \sin 400t = \sin \theta$$

and

$$\theta = \sin^{-1} 0.5 = 30°$$

$$\text{with } 30° = \omega t = 400t$$

but $\omega = 400$ is in radians per second. Therefore, the 30° must be converted to radians before t can be determined.

$$\text{radians} = \frac{30°}{180°} \times \pi = \frac{\pi}{6} \text{ radians}$$

and

$$\frac{\pi}{6} = 400t$$

or

$$t = \frac{\pi}{2400} = \mathbf{1.31\ ms}$$

Checking, we find that

$$T = \frac{2\pi}{\omega} = \frac{2\pi}{400} = \frac{\pi}{200}$$

and 30° is 30°/360° = $\frac{1}{12}$ of a full cycle, with

$$t = \frac{1}{12}(T) = \frac{1}{12}\left(\frac{\pi}{200}\right) = \frac{\pi}{2400} = \mathbf{1.31\ ms}$$

The waveform in Fig. 4.4 cuts through the horizontal axis with increasing magnitude at time t or θ equal to zero. If the waveform cuts through the axis before or ager 0°, a phase-shift term must be added to the general mathematical expression for a sine wave. For intersections before 0°, such as shown in Fig. 4.10(a), the following format is used:

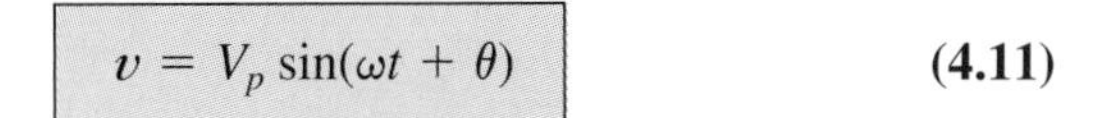

$$v = V_p \sin(\omega t + \theta) \qquad \textbf{(4.11)}$$

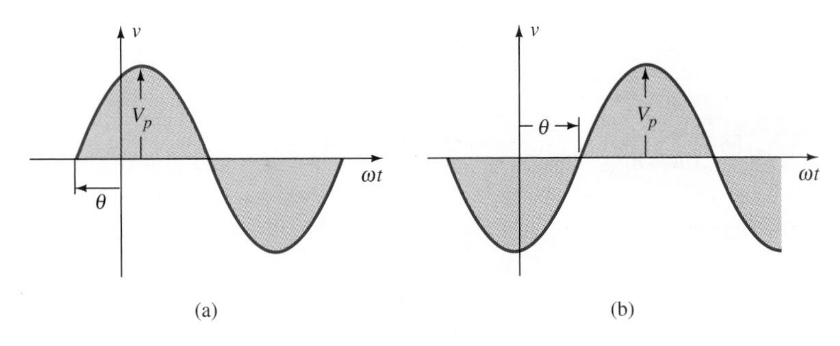

FIG. 4.10
Initial phase angles.

while for intersections to the right of the zero axis [Fig. 4.10(b)] the following format is employed:

$$v = V_p \sin(\omega t - \theta) \qquad \textbf{(4.12)}$$

For the two sinusoidal waveforms plotted on the same horizontal axis, a *phase relationship* between the two can be determined. For the waveforms in Fig. 4.11(a) the sinusoidal voltage crosses the horizontal axis 90° before the sinusoidal current. The voltage v is said to *lead* the current i by 90°, or the current i to *lag* the voltage v by 90°. Note that positive (increasing values with time) slope crossings must be compared. In Fig. 4.11(b), since positive slopes are 210° apart, v is said to lead i by 210° or i to lag v by the same angle. Note that if one positive slope intersection and one negative slope intersection were compared, an incorrect answer of 30° would be obtained. It is also correct to say that i leads v by 360° − 210° = 150°, as shown in the figure.

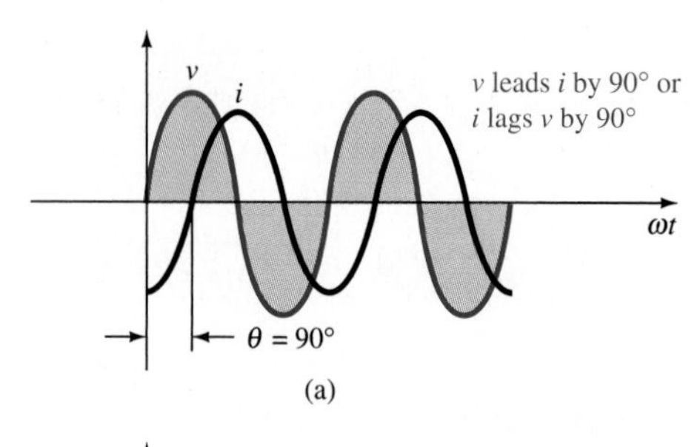

(a)

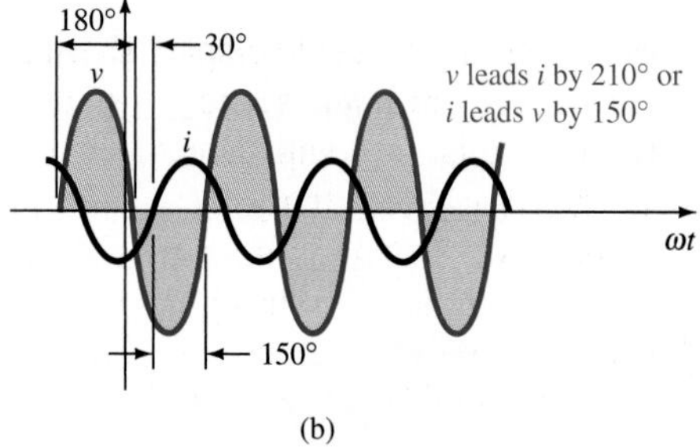

(b)

FIG. 4.11
Phase relationships.

EXAMPLE 4.6

a. Write the sinusoidal expression for each waveform appearing in Fig. 4.12.
b. What is the phase relationship between the two waveforms?

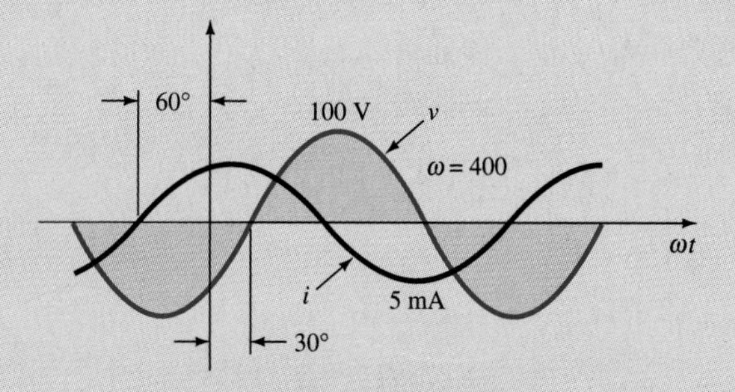

FIG. 4.12
Waveforms for Example 4.6.

Solution:

a. $i = \mathbf{5 \times 10^{-3} \sin(400t + 60°)}$
   $v = \mathbf{100 \sin(400t - 30°)}$
b. i leads v by $60° + 30° = \mathbf{90°}$

EXAMPLE 4.7 Determine the phase relationship between the following two waveforms:

$$v = 8.6 \sin (300t + 80°)$$
$$i = 0.12 \sin (300t + 10°)$$

Solution: Both waveforms cut the horizontal axis with a positive slope before $\theta = 0°$ ($t = 0$ s). Therefore, v leads i by the angle with which it leads the current i.

$$v \text{ leads } i \text{ by } (80° - 10°) = \mathbf{70°}$$

4.3 OSCILLOSCOPE

The **oscilloscope** is an instrument used to display alternating waveforms such as those described in the preceding section. A sinusoidal pattern appears on the oscilloscope of Fig. 4.13 with the indicated vertical and horizontal sensitivities. The vertical sensitivity of 0.1 V/div. defines the voltage associated with each vertical division of the display. Virtually all oscilloscope screens are cut into a crosshatch pattern of lines separated by 1 cm in the vertical and horizontal directions. The horizontal sensitivity of 50 μs/div. defines the time period associated with each horizontal division of the display.

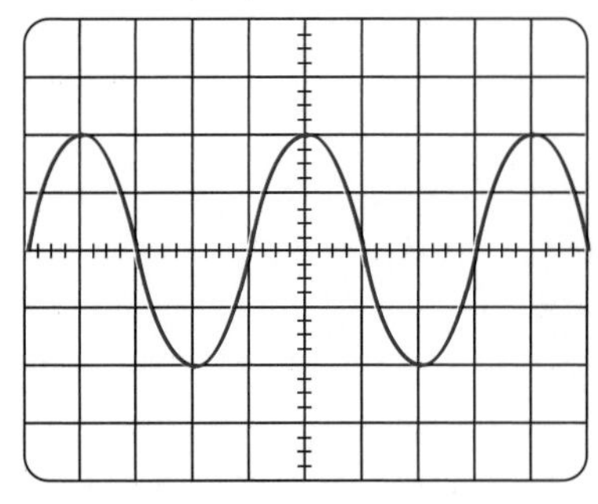

Vertical sensitivity = 0.1 V/div.
Horizontal sensitivity = 50 μs/div.

FIG. 4.13

For the pattern of Fig. 4.13 one cycle spans 4 divisions. The period is therefore

$$T = 4 \text{ div.}\left(\frac{50\ \mu\text{s}}{\text{div.}}\right) = \mathbf{200\ \mu s}$$

and the frequency is

$$f = \frac{1}{T} = \frac{1}{200 \times 10^{-6}\ \text{s}} = \mathbf{5\ kHz}$$

The vertical height above the horizontal axis encompasses 2 divisions. Therefore,

$$V_m = 2 \text{ div.}\left(\frac{0.1\ \text{V}}{\text{div.}}\right) = \mathbf{0.2\ V}$$

4.4 EFFECTIVE (RMS) VALUE

Since a sinusoidal voltage or current has the same shape above and below the axis, the question of how power can be delivered to a load may be a bothersome one because it appears that net flow to a load over one full cycle is zero. However, one must simply keep in mind that *at each instant* of the positive or negative portion of the waveform, power is being delivered and dissipated by the load. Although the current may reverse direction during the negative portion, power is still delivered and absorbed by the load at each instant of time. In other words, the power delivered at each instant of time is additive even though the current may reverse in direction.

In an effort to determine a single numerical value to associate with the time-varying sinusoidal voltage and current, a relationship was developed experimentally that correlates the amount of ac voltage that needs to be applied in order to deliver the *same* power to a load as a given dc voltage. The results indicate that if a 10-V dc source is applied to a load, the same power can be delivered to that load with a sinusoidal voltage having a peak value of 14.14 V. The waveforms of the two voltages just described appear in Fig. 4.14. In equation form, the *equivalent dc* or *effective* value of a sinusoidal voltage is equal to 0.707 times the peak value. For the preceding example,

$$0.707(V_p) = 0.707(14.14) = \mathbf{10\ V}$$

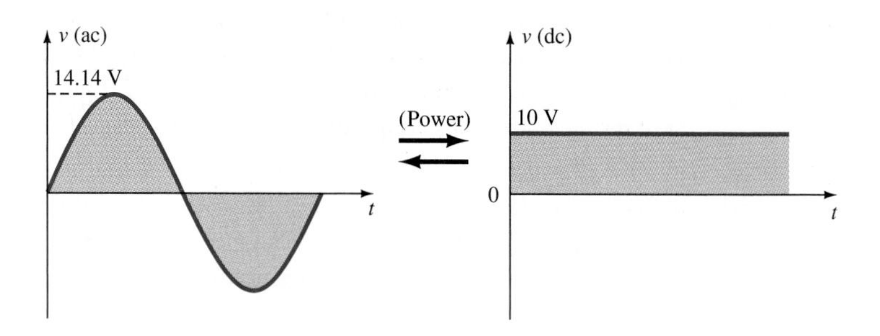

FIG. 4.14
Relating effective values.

In equation form,

$$V_{\text{dc equivalent}} = V_{\text{eff}} = 0.707(V_{\text{peak}}) = \frac{1}{\sqrt{2}}(V_{\text{peak}}) \quad \textbf{(4.13)}$$

and

$$I_{\text{dc equivalent}} = I_{\text{eff}} = 0.707(I_{\text{peak}}) = \frac{1}{\sqrt{2}}(I_{\text{peak}}) \quad \textbf{(4.14)}$$

For an equivalent dc voltage an ac voltage must satisfy the relationship

$$V_{\text{peak}} = 1.414V_{\text{eff}} = \sqrt{2}V_{\text{eff}} \quad \textbf{(4.15)}$$

or

$$I_{\text{peak}} = 1.414I_{\text{eff}} = \sqrt{2}I_{\text{eff}} \quad \textbf{(4.16)}$$

For the function introduced earlier,

$$v = 170 \sin 377t$$

$$V_{\text{eff}} = 0.707(170) = \mathbf{120\ V}$$

Note that the frequency does not play a part in determining the equivalent dc value.

Another label for the effective value is the *rms* value, which is derived from the mathematical procedure for determining the effective value of any waveform. First, the *square* of the waveform is found and then the *mean* value by finding the area under the squared curve and dividing by the period T. The square *root* of the resulting value is then obtained, which equals the effective value. The abbreviation rms is derived from the first letters of the preceding italicized words in the reverse order that they appear (root mean square).

EXAMPLE 4.8 Calculate the effective values of the waveforms in Fig. 4.12.

Solution:

$$I_{\text{eff}} = (0.707)(5 \times 10^{-3}\ \text{A}) = \mathbf{3.535\ mA}$$

$$V_{\text{eff}} = (0.707)(100\ \text{V}) = \mathbf{70.7\ V}$$

EXAMPLE 4.9 Write the sinusoidal expression for a voltage having an rms value of 40 mV, a frequency of 500 Hz, and an initial phase shift of +40°.

Solution:

$$V_p = 1.414(V_{\text{rms}}) = 1.414(40\ \text{mV}) = \mathbf{56.56\ mV}$$

$$\omega = 2\pi f = (6.283)(0.5\ \text{kHz}) = \mathbf{3.142 \times 10^3\ rad/s}$$

$$v = \mathbf{56.56 \times 10^{-3} \sin(3142}t\ \mathbf{+ 40°)}$$

The manufacturer's label on all electrical equipment, machinery, instruments, appliances, and so on, lists the effective values of the voltage or current. The home outlet is rated 120 V, 60 Hz, indicating an effective value of 120 V and a peak value of $\sqrt{2}(120) \cong 170$ V. Some appliances, such as air conditioners, are rated at 220 V. For such units a separate line must be run from the meter panel to provide the increased voltage. All ac voltmeters and ammeters indicate effective values or peak-to-peak (p-p) values unless otherwise specified.

Ac meters that employ D'Arsonval movements (such as used to measure dc quantities) must have an additional electronic circuitry to give the ac waveform an average (or dc) level so that the meter can be calibrated and a meaningful reading obtained. It should be somewhat obvious that the average value (and therefore the indication of a D'Arsonval movement) of the sinusoidal waveform over one full cycle is zero. There are other types of movements that do not require any additional circuitry to measure ac quantities properly. Two such movements are the *iron vane* and *dynamometer* types. They read the effective or dc equivalent value of any waveform including sinusoidal, dc, square, and triangular. Digital-display meters employ advanced electronic circuitry to establish a comparison between internal standards and the measured quantities. A digital multimeter capable of reading a wide range of dc and ac quantities, including frequency, appears in Fig. 4.15.

The *wattmeter,* in Fig. 4.16, can read the power dissipated by a dc or ac network. Recall that the voltage terminals are always placed in parallel with the load for which the power is to be determined, while the current terminals are connected in series with that load.

Incidentally, you will note as we progress through the book that very little is said about the internal construction of meters. Priorities require that this type of detail be left to the reader using one of the many references available in most libraries.

FIG. 4.15
Digital Multimeter (DMM). (Courtesy of John Fluke Manufacturing Company, Inc.)

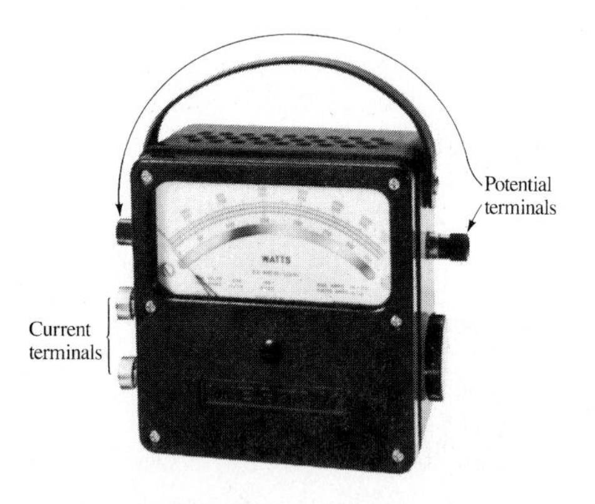

FIG. 4.16
Wattmeter. (Courtesy of Electrical Instrument Service, Inc.)

4.5 AVERAGE VALUES

The average value or dc component of a waveform will be of some importance in the more advanced analysis to follow. For the individual waveform of Fig. 4.17(a) the average value (as indicated earlier) is zero over one cycle. For the waveform of Fig. 4.17(b), the average value is 5 V. The waveform of Fig. 4.17(b) can be obtained by simply placing a dc voltage of 5 V in series with a sinusoidal waveform with a peak value of 10 V as shown in Fig. 4.17(c). For electronic systems where both dc and ac signals are present, waveforms such as Fig. 4.17(b) will appear quite frequently.

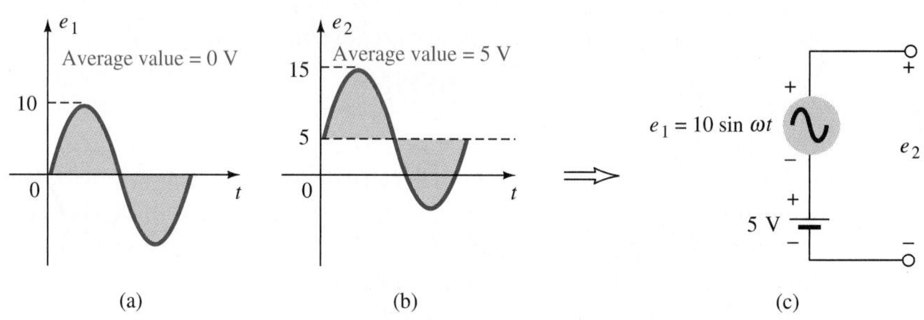

FIG. 4.17
Effect of a dc voltage on the average value.

For waveforms such as shown in Fig. 4.18(a), the average value is not so obvious, but it can be determined using the equation

$$G \text{ (average value)} = \frac{\text{area (algebraic sum)}}{T \text{ (period)}} \tag{4.17}$$

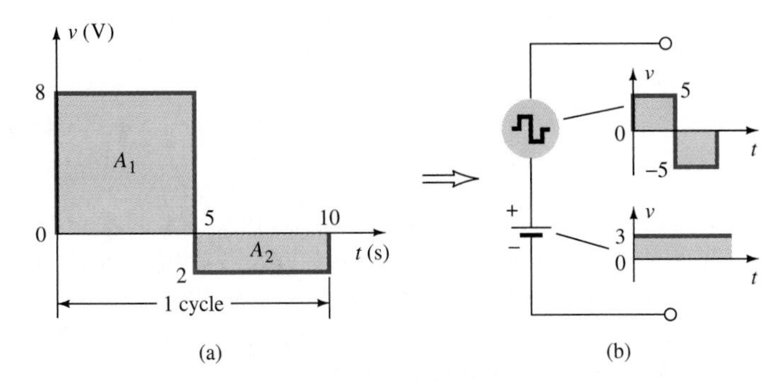

FIG. 4.18
Determining the average value of a waveform.

The algebraic sum is the sum of the areas above the axis less the sum of the areas below the axis. The time T is the time interval for which the average value is to be determined.

For the complete cycle in Fig. 4.18(a),

$$G = \frac{A_1 - A_2}{T} = \frac{(8\text{ V})(5\text{ s}) - (2\text{ V})(5\text{ s})}{10\text{ s}} = \frac{40 - 10}{10}\text{V} = \frac{30}{10}\text{V} = \mathbf{3\ V}$$

indicating that the square wave in Fig. 4.18(a) can be generated using the two components in Fig. 4.18(b).

For waveforms whose area cannot be obtained as directly as in the preceding example, integral calculus can be employed, or you can try to approximate the odd shape using figures such as rectangles and triangles for which the area can easily be obtained.

EXAMPLE 4.10 Calculate the average value of the waveform in Fig. 4.19 over one full cycle.

Solution:

$$G = \frac{A_1 + A_2 - A_3}{T}$$

$$= \frac{\frac{1}{2}(2\times10^{-3}\text{s})(10\text{mV}) + \frac{1}{2}(2\times10^{-3}\text{s})(10\text{mV}) - (4\times10^{-3}\text{s})(5\text{mV})}{10\times10^{-3}\text{ ms}}$$

$$= \frac{10\times10^{-3} + 10\times10^{-3} - 20\times10^{-3}}{10\times10^{-3}}\text{V}$$

$$= \mathbf{0\ V}$$

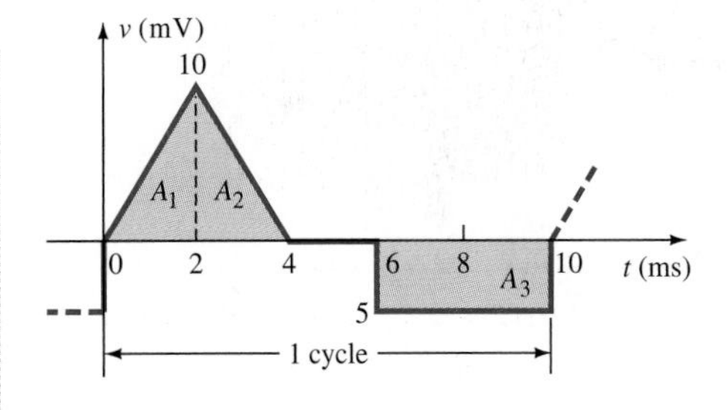

FIG. 4.19
Waveform for Example 4.10.

For the positive pulse of a sine wave the area under the wave can be determined using integral calculus. Thus, for the pulse in Fig. 4.20,

$$\text{area} = \int_0^{\pi} A_m \sin\theta\, d\theta = A_m \int_0^{\pi} \sin\theta\, d\theta$$

$$= A_m[-\cos\theta]_0^{\pi} = -A_m(\cos\pi - \cos 0°)$$

$$= -A_m[-1 - (+1)] = -A_m(-2)$$

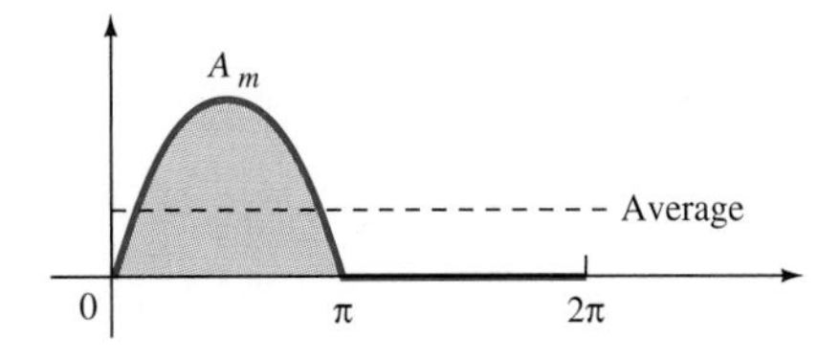

FIG. 4.20
Defining the average value of the positive region of a sinusoidal waveform over one full period.

and

$$\text{area} = 2A_m \tag{4.18}$$

and the area of a pulse of a sine wave is simply twice the peak value.

The waveform in Fig. 4.20 is referred to as a *half-wave rectified waveform* and is frequently employed in the conversion of ac to dc.

EXAMPLE 4.11 Determine the average value of the waveform in Fig. 4.20 over one full cycle ($0 \rightarrow 2\pi$).

Solution:

$$G = \frac{2A_m}{2\pi} = \mathbf{0.318A_m} \tag{4.19}$$

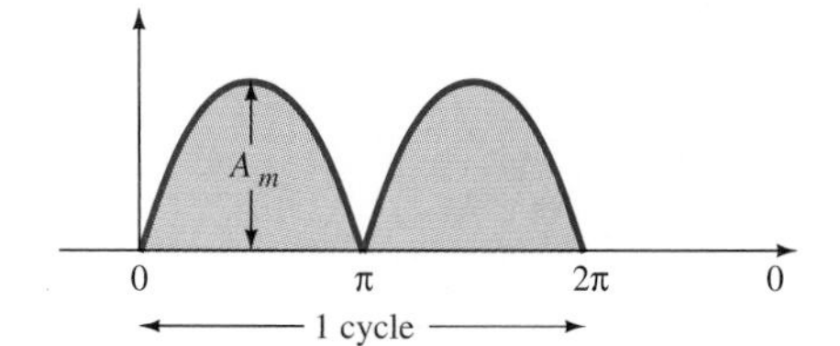

FIG. 4.21
Full-wave-rectified waveform.

EXAMPLE 4.12 Determine the average value of the *full-wave-rectified waveform* in Fig. 4.21 over one full cycle. This waveform is also frequently encountered when converting alternating current to direct current.

Solution:

$$G = \frac{2A_m + 2A_m}{2\pi} = \frac{4A_m}{2\pi} = \mathbf{0.636A_m} \tag{4.20}$$

which is twice that resulting for the half-wave-rectified waveform.

4.6 SAFETY CONCERNS (HIGH VOLTAGES AND DC VERSUS AC)

Be aware that any "live" network should be treated with a calculated level of respect. Electricity in its various forms is not to be feared but should be employed with some awareness of its potentially dangerous side effects. It is common knowledge that electricity and water do not mix (never use extension cords or plug in TVs or radios in the bathroom) because a full 120 V in a layer of water of any height (from a shallow puddle to a full bath) can be *lethal*. However, other effects of dc and ac voltages are less known. In general, as the voltage and current increase, your concern about safety should increase exponentially. For instance, under dry conditions, most human beings can survive a 120-V ac shock such as obtained when changing a light bulb, turning on a switch, and so on. Most electricians have experienced such a jolt many times in their careers. However, ask an electrician to relate how it feels to hit 220 V, and the response (if he or she has been unfortunate to have had such an experience) will be totally different. How often have you heard of a backhoe operator hitting a 220-V line and having a fatal heart attack? Remember, the operator is sitting in a metal container on a damp ground,

which provides an excellent path for the resulting current to flow from the line to ground. If only for a short period of time, with the best environment (rubber-soled shoes, etc.), in a situation that allows a quick escape, most human beings can survive a 220-V shock. However, as mentioned above, it is one a person will not quickly forget. For voltages beyond 220 V rms, the chances of survival decrease exponentially with increase in voltage. It takes only about 10 mA of steady current through the heart to put it into fibrillation. In general, therefore, always be sure that the power is disconnected when repairing electrical equipment. Don't assume that throwing a wall switch will disconnect the power. Throw the main circuit breaker and test the lines with a voltmeter before working on the system. Since voltage is a two-point phenomenon, don't be a hero and work with one line at at time—accidents happen!

You should also be aware that the reaction to dc voltages is quite different from that to ac voltages. You have probably seen in movies or comic strips that people are often unable to let go of a *hot* wire. This is an illustration of the most important difference between the two types of voltages. As mentioned previously, if you happen to touch a hot 120-V ac line, you will probably get a good sting, but *you can let go.* If it happens to be a hot 120-V dc line, you will probably not be able to let go, and you could die. Time plays an important role when this happens, because the longer you are subjected to the dc voltage, the more the resistance in the body decreases, and a fatal current can be established. The reason that we can let go of an ac line is best demonstrated by carefully examining the 120-V rms, 60-Hz voltage in Fig. 4.22. Since the voltage is oscillating, there is a period of time when the voltage is near zero or less than, say, 20 V, and is reversing in direction. Although this time interval is very short, it appears every 8.3 ms and provides a window to *let go*.

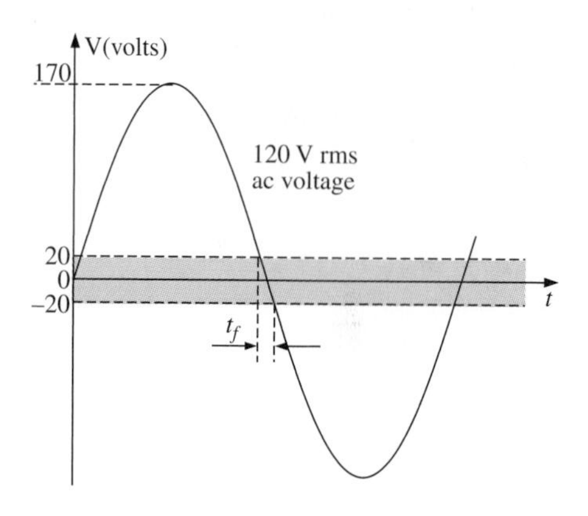

FIG 4.22
Interval of time when sinusoidal voltage is near zero volts.

Now that you are aware of the additional dangers of dc voltages, it is important to mention that under the wrong conditions, dc voltages as low as 12 V such as from a car battery can be quite dangerous. If you happen to be working on a car under wet conditions, or if you are sweating profusely for some reason or, worse yet, wearing a ring that may have moisture and body salt underneath, touching the positive terminal may initiate the process whereby the body resistance begins to drop and serious injury can take place. This is one of the reasons you seldom see a professional electrician wearing any rings or jewelry—it is just not worth the risk.

Before leaving this topic of safety concerns, you should also be aware of the dangers of high-frequency supplies. We are all aware of what 2.45 GHz at 120 V can do to a meat product in a microwave oven. The seal, therefore, should be as tight as possible. However, don't ever assume that anything is absolutely perfect in design—so don't make it a habit to view the cooking process in the microwave 6 in. from the door on a continuing basis. Find something else to do, and check the food only when the cooking process is complete. If you ever visit the Empire State Building, you will notice that you are unable to get close to the antenna on the dome due to the high-frequency signals being emitted with a great deal of power. Also note the large KEEP OUT signs near radio transmission towers for local radio stations. Standing within 10 ft of an AM transmitter operating at 540 kHz would bring on disaster. Simply holding (not

to be tried!) a fluorescent bulb near the tower could make it light up due to the excitation of the molecules inside the bulb.

In summary, therefore, treat any situation with high ac voltages or currents, high-energy dc levels, and high frequencies with added care.

4.7 THE *R, L,* AND *C* ELEMENTS

The effect of an ac sinusoidal signal on the basic *R, L,* and *C* elements will now be examined. Although certain statements of fact will have to be made without full derivation or explanation, be assured that the material will be more than enough to permit a full understanding of the more advanced material to follow.

In Fig. 4.23, a sinusoidal voltage has been applied across the resistor. As shown by the resulting current in the same figure, the peak values are related directly by Ohm's law:

$$I_{peak} = \frac{E_{peak}}{R} \tag{4.21}$$

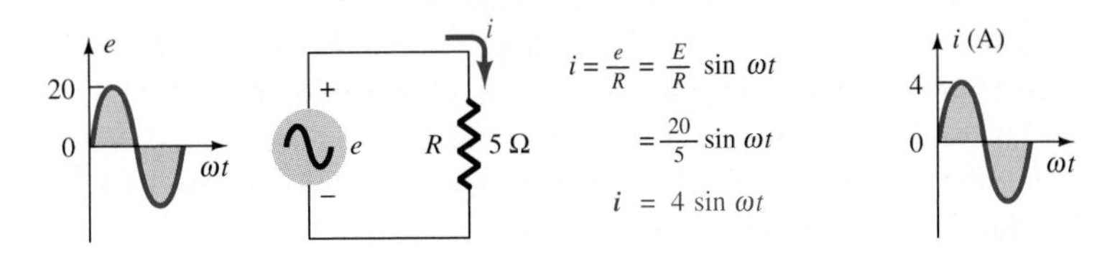

FIG. 4.23
Simple ac resistive network.

In addition, there is no phase shift introduced by the resistor, and so v_R and i_R are said to be *in phase*. Note that the frequencies of both v_R and i_R are also the same.

The power to the resistor can be determined from the *effective* values of the voltage and current in the following manner:

$$P_R = I_R^2 R = \frac{V_R^2}{R} = V_R I_R \qquad \text{(watts)} \tag{4.22}$$

For the case in Fig. 4.23,

$$P_R = V_R I_R = \left(\frac{20}{\sqrt{2}}\text{ V}\right)\left(\frac{4}{\sqrt{2}}\text{ A}\right) = \frac{80}{2}\text{ W} = \mathbf{40\ W}$$

Note the similarities between the forms of Eq. (4.22) and those used for dc networks. The *only* difference is the substitution of effective values rather than dc values.

For the *ideal* resistor the resistance does not change with frequency, as indicated by the plot in Fig. 4.24. In the real (practical) world, however, there are parasitic capacitive and inductive elements in every resis-

tor that affect its characteristics at very high and very low frequencies. For the moment, however, we will assume the resistors are all ideal with the response curve in Fig. 4.24.

The reaction of a coil or capacitor to an ac signal is quite different from that of a resistor. Both limit the magnitude of the current, but neither (ideally speaking) dissipates any of the electrical energy delivered to it. They simply store it in the form of a magnetic field for the inductor, or an electric field for the capacitor, with the ability to return it to the electrical system as required by design.

For the inductor the *reactance* to an ac signal is determined by

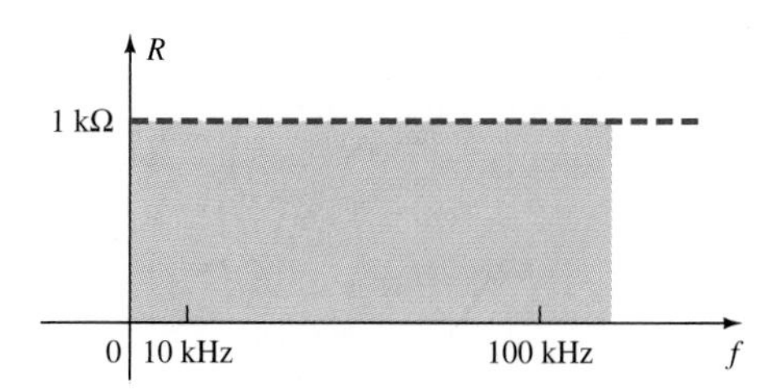

FIG. 4.24
Resistance versus frequency for an ideal resistor.

$$X_L = \omega L = 2\pi f L \quad \text{(ohms)} \tag{4.23}$$

Although reactance is similar to resistance in its ability to limit the current, always keep in mind that it is not an energy-dissipating form of opposition as is encountered for resistive elements.

Equation (4.23) reveals that the inductive reactance is directly proportional to the frequency of the applied signal. Recall that the ideal inductor had a short-circuit equivalent in dc steady-state systems. Certainly for dc, $f = 0$, and we find that that reactance of the inductor is $X_L = 2\pi fL = 2\pi(0)L = 0\ \Omega$, supporting the conclusion. At very high frequencies, the equivalent circuit takes on the characteristics of an open circuit, since the resulting reactance is so high.

A plot of X_L versus frequency is provided in Fig. 4.25. Note, as discussed above, that the plot starts from $X_L = 0$ at $f = 0$ Hz and increases linearly to very high values at high frequencies. Writing $X_L = \omega L$ as $X_L = L\omega = L2\pi f = (2\pi L)f$ puts it in the form of the basic straight-line equation $y = mx + b$, with $b = 0$ and $m = 2\pi L$ as the slope. In other words, the higher the inductance, the steeper the slope and the more quickly the reactance increases with frequency, as demonstrated in Fig. 4.25.

Ohm's law continues to be applicable to inductive elements with

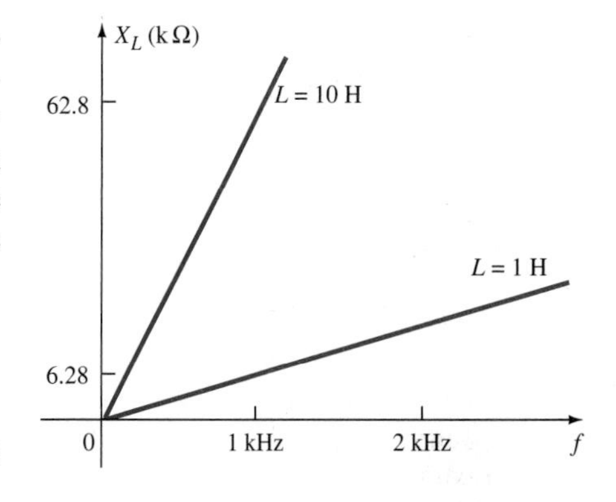

FIG. 4.25
X_L versus frequency for ideal inductors.

$$I_{\text{peak}} = \frac{V_{\text{peak}}}{X_L} \tag{4.24}$$

A sinusoidal voltage has been applied across a 0.5-H inductor in Fig. 4.26. The reactance of the inductor at the applied frequency is

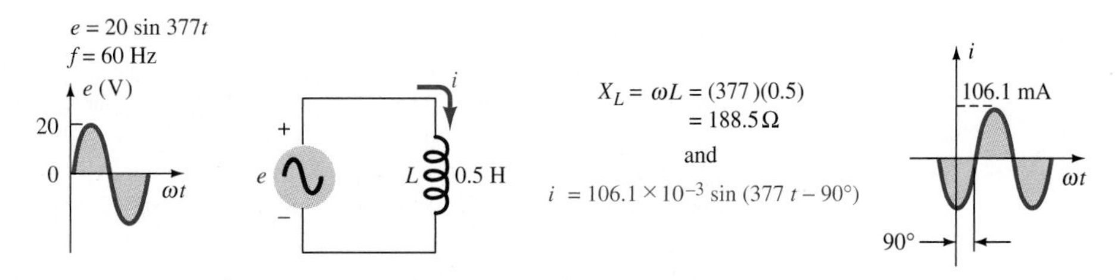

FIG. 4.26
Simple ac inductive network.

$X_L = 2\pi fL = 2\pi(60 \text{ Hz})(0.5 \text{ H}) = 188.5\ \Omega$. Ohm's law can then be applied to determine the peak value of the current. That is,

$$I_{peak} = \frac{V_{peak}}{X_L} = \frac{20 \text{ V}}{188.5\ \Omega} = 106.1 \times 10^{-3} \text{ A} = \mathbf{106.1\ mA}$$

as shown in Fig. 4.26. Note that the voltage applied across the inductor *leads* the resulting current by 90°. The inductor, therefore, introduces a 90° phase shift between the two quantities not present for the pure resistor.

For an ac system the basic power equation is

$$P = \frac{V_p I_p}{2} \cos\theta = V_{eff} I_{eff} \cos\theta \qquad \text{(watts)} \qquad \mathbf{(4.25)}$$

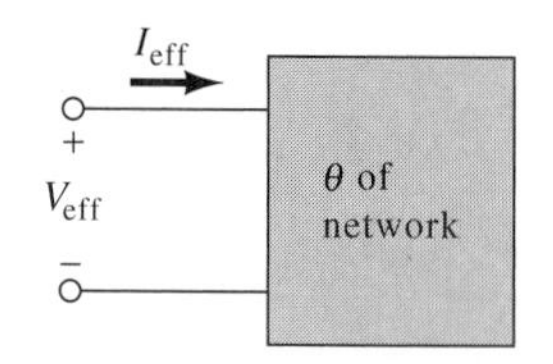

FIG. 4.27
Determining the power delivered to an ac network.

As indicated by Fig. 4.27, V_{eff} is the voltage across the elements or network to which the power is to be determined, while I_{eff} is the current drawn by the network. The angle θ is the phase angle between V and I. For a pure resistor in the container, we found that v and i were *in phase* and that $\theta = 0°$. Substitution into the equation results in $P = VI \cos\theta = VI(1) = VI$ (from this point on the symbols V and I will refer to V_{eff} and I_{eff}, respectively), as indicated earlier. For the pure inductor $\theta = 90°$ and $P = VI \cos 90° = VI(0) = 0$ W, indicating, as stated earlier, that the pure inductor does not dissipate electrical energy but simply stores it in the form of a magnetic field. Networks with both resistance and inductance result in an angle θ between 0° and 90°.

For the pure capacitor, the reactance is determined from

$$X_C = \frac{1}{\omega C} = \frac{1}{2\pi fC} \qquad \text{(ohms)} \qquad \mathbf{(4.26)}$$

indicating that as the frequency increases, the reactance of a capacitor decreases (the opposite of an inductor). In addition, for $f = 0$, corresponding to the dc condition, $X_C = \dfrac{1}{2\pi(0)C} \Rightarrow \infty\ \Omega$ = a very large value, corresponding to an open-circuit equivalent as described for dc networks. At very high frequencies the characteristics of a capacitor approach those of a short circuit.

A plot of $X_C = 1/\omega C = 1/2\pi fC$ in Fig. 4.28 reveals that the resulting curve will have a parabolic shape starting at very high levels near 0 Hz and dropping very rapidly to lower levels as the frequency increases. As noted by the plots provided, an increase in capacitance causes the curve to drop more quickly with frequency. Ohm's law is also applicable to capacitive elements:

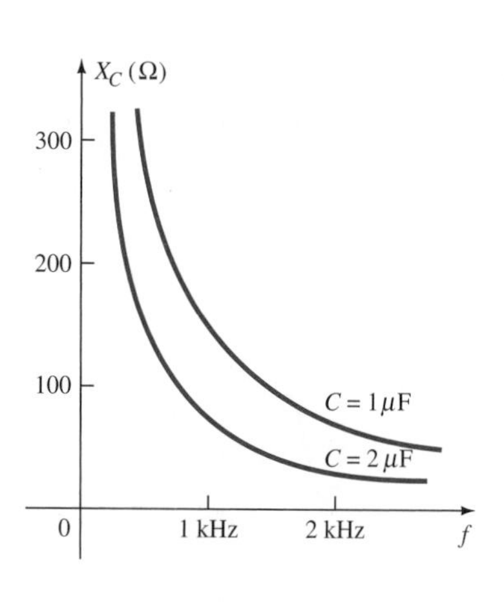

FIG. 4.28
X_C versus frequency for ideal capacitors.

$$I_{peak} = \frac{V_{peak}}{X_C} \qquad \mathbf{(4.27)}$$

A sinusoidal voltage has been applied across a 10-μF capacitor in Fig. 4.29. The reactance X_C as shown in the figure is

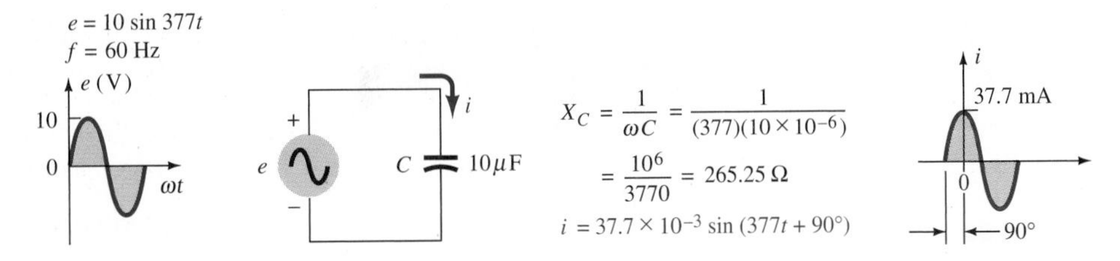

FIG. 4.29

$$X_C = \frac{1}{2\pi fC} = \frac{1}{2\pi(60\text{ Hz})(10\ \mu\text{F})} = 265.25\ \Omega$$

and the peak value of the current can be determined through a simple application of Ohm's law. That is,

$$I_p = \frac{V_p}{X_C} = \frac{10\text{ V}}{265.25\ \Omega} = 0.0377\text{ A} = \mathbf{37.7\text{ mA}}$$

as shown in Fig. 4.29. Note in this case that a 90° phase shift has been introduced but that i_C *leads* v_C by 90°, which is the reverse of that encountered for the inductor. Substituting into the general power equation, we have

$$P_C = VI\cos\theta = VI\cos 90° = VI(0) = 0\text{ W}$$

substantiating the previous comments for the capacitor also.

The factor $\cos\theta$ in the general power equation is called the *power factor* of the network and is abbreviated

$$\boxed{\text{power factor} = \cos\theta = F_p} \qquad \textbf{(4.28)}$$

Its largest value is 1, which results when the network appears to be purely resistive at its input terminals and $\theta = 0°$. The smallest value is zero, as obtained from a network whose terminal characteristics are purely reactive (capacitive or inductive). For a network composed of resistors and reactive elements the power factor can vary from 0 to 1. For more reactive loads the power factor drifts toward zero, while for more resistive loads it drifts toward 1.

The following example demonstrates the effect of a phase shift between the applied voltage and current.

EXAMPLE 4.13 The current i_L through a 10-mH inductor is 5 sin $(200t + 30°)$. Find the voltage v_L across the inductor.

Solution: The peak value of the voltage is determined by Ohm's law:

$$X_L = \omega L = (200\text{ rad/s})(10 \times 10^{-3}\text{ H}) = 2\ \Omega$$

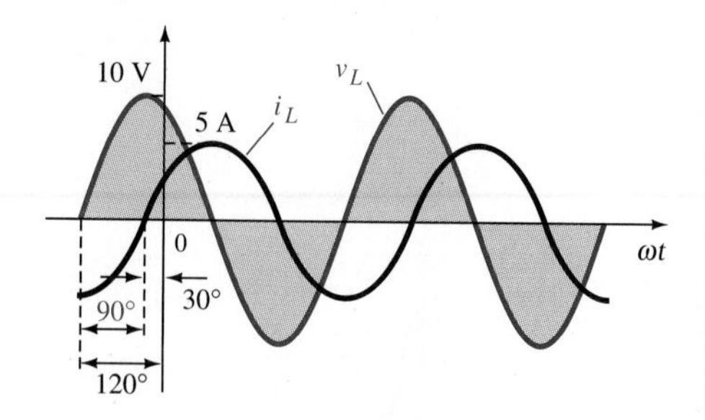

FIG. 4.30
v_L and i_L for Example 4.13.

and

$$V_p = I_p X_L = (5\text{ A})(2\ \Omega) = 10\text{ V}$$

Since v_L always leads i_L by 90°, an additional 90° must be added to the 30° already present in the sinusoidal expression for the current. Therefore,

$$\mathbf{v_L = 10 \sin(200t + 120°)}$$

The curves for v_L and i_L appear in Fig. 4.30, clearly indicating that the voltage leads the current by 90°.

EXAMPLE 4.14 The voltage across a 2-μF capacitor is 4 mV (rms) at a phase angle of −60°. If the applied frequency is 100 kHz, find the sinusoidal expression for the current i_C.

Solution:

$$V_p = 1.414V_{\text{rms}} = 1.414(4\text{ mV}) = 5.656\text{ mV}$$

$$X_C = \frac{1}{2\pi fC} = \frac{1}{(2\pi)(100 \times 10^3\text{ Hz})(2 \times 10^{-6}\text{ F})}$$

$$= 0.796\ \Omega$$

$$I_p = \frac{V_p}{X_C} = \frac{5.656\text{ mV}}{0.796\ \Omega} = 7.11\text{ mA}$$

and since i_C leads v_C by 90°, the angle associated with i_C is −60° + 90° = +30° and

$$\omega = 2\pi f = (2\pi)(100 \times 10^3\text{ Hz}) = 6.283 \times 10^5\text{ rad/s}$$

so that

$$\mathbf{i_C = 7.11 \times 10^{-3} \sin(6.283 \times 10^5 t + 30°)}$$

4.8 PHASORS AND COMPLEX NUMBERS

For single-element networks the proper phase angle can be determined with a limited amount of effort. For complex networks, however, this could become a serious undertaking if it were not for a technique to be introduced in this section. This technique incorporates the use of vectors, which can be used to represent ac voltages or currents and element reactances. Recall that a *vector* is a quantity that has both *magnitude* and *direction.* In Fig. 4.31 we have the vector representation of resistance, inductive reactance, and capacitive reactance. The angle shown with each is determined by the phase shift introduced between the voltage and current for each element. For the resistor, recall that the voltage and current are in phase; hence, there is no phase shift, and the associated angle is 0°. Since the angle is measured from the right horizontal axis, the resistance vector appears on the axis. Its length is determined by the magnitude of the resistance R. For X_L and X_C the angle included is the angle by which the voltage across the element leads the current through

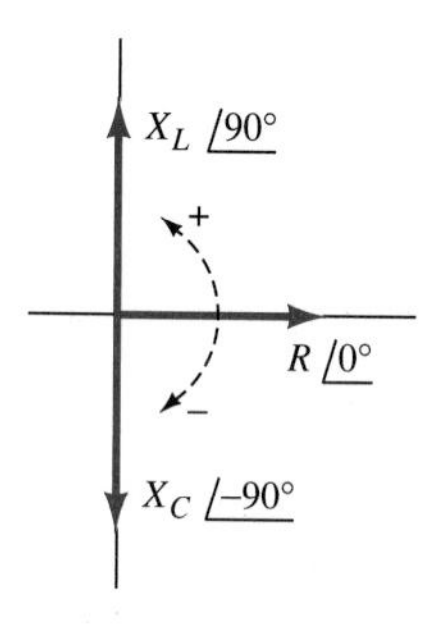

FIG. 4.31
Impedance vectors.

the component. For X_L it is $+90°$, and for X_C it is $-90°$. The magnitudes (or lengths) of the vectors are determined by the reactance of each element. Note that the angle is always measured from the same axis. Any one or a combination of the resistive or reactive elements in Fig. 4.31 is called an *impedance* and is represented by the symbol **Z**. It is a measure of the ability of the ac network to *impede* the flow of charge or current through the network. The diagram in Fig. 4.31 with one or any number of the elements is called an impedance *diagram*. Only resistances and reactances appear in an impedance diagram. Voltages and currents appear in a *phasor* diagram such as shown in Fig. 4.32 for each element. The angle associated with each quantity is the phase angle appearing in the sinusoidal time domain. The magnitude is the effective value of the sinusoidal quantity. Each magnitude and associated angle is written in boldface roman notation and is called a *phasor.* The phasor diagram for the pure resistor indicates quite clearly that v_R and i_R are in phase, since they have the same angle and therefore direction for leading and lagging situations. The counterclockwise direction defines the leading vector. For instance, for the inductor in Fig. 4.32(b), if $\mathbf{V}_L$ and $\mathbf{I}_L$ were rotating vectors in the clockwise direction as defined in Fig. 4.9, $\mathbf{V}_L$ would lead $\mathbf{I}_L$ by 90°. For the capacitor in Fig. 4.32(c), $\mathbf{I}_C$ leads $\mathbf{V}_C$ by 90°. The clockwise direction defines lagging quantities as shown in Fig. 4.32(c). Keep in mind that the vector quantities in Fig. 4.32 are "snapshots" of the rotating vectors in Fig. 4.9 as they pass through the horizontal axis at $t = 0$ s ($\theta = 0°$). In other words, for the inductor in Fig. 4.32(b), the voltage across the coil is a maximum (vertical projection is a maximum), while the current is just passing through the horizontal axis (vector on the horizontal) with a magnitude of zero and an angle of 0°.

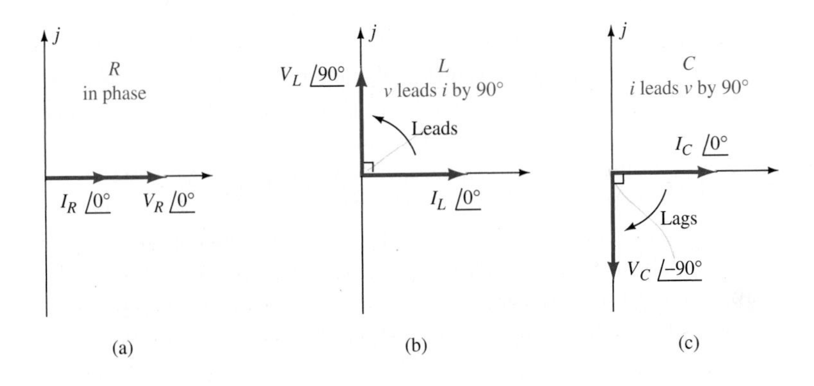

FIG. 4.32
Phasor diagrams: (a) pure resistor; (b) pure inductor; (c) pure capacitor.

If the phasor or vector notation introduced is to be of any assistance in our future analysis, we must be able to perform certain basic mathematical operations with the phasors or vectors. Keep in mind, however, that the term *phasor* is applied solely to those vectors representing a sinusoidal quantity. The vector components of an impedance diagram are not phasors.

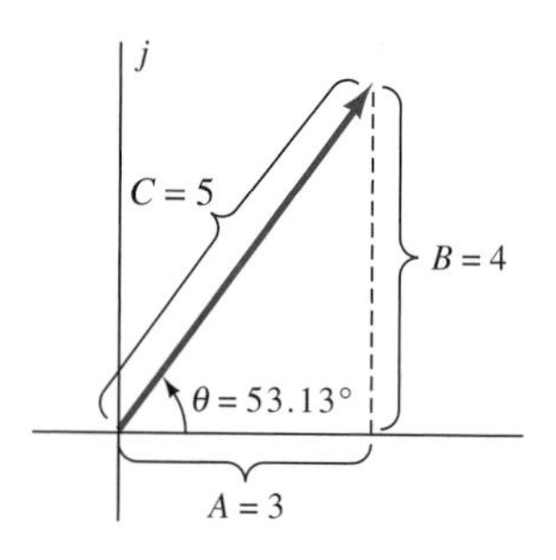

FIG. 4.33

A vector as shown in Figure 4.33 can be determined (or represented) either by providing its magnitude and angle measured from the positive (+) (right) horizontal axis or by providing the horizontal and vertical components of the vector. The former is called the *polar* form, and the latter is called the *rectangular* form. The equations necessary to convert from one form to the other are

Polar → Rectangular	*Rectangular → Polar*
$A = C\cos\theta$	$C = \sqrt{A^2 + B^2}$
$B = C\sin\theta$	$\theta = \tan^{-1}\dfrac{B}{A}$

(4.29)

The letter j is included in the rectangular form to distinguish between the real (horizontal) and imaginary (vertical) components of the rectangular form. The terms real and imaginary are related solely to mathematical definition and do not require further discussion here. To perform the mathematical operations to be described, the letter j, defined mathematically by $\sqrt{-1}$, must be examined in its various forms. That is,

$$j = \sqrt{-1}$$

so that $$j^2 = (\sqrt{-1})^2 = -1$$

and $$j^3 = j^2 j^1 = (-1)(\sqrt{-1}) = -\sqrt{-1} = -j$$

with $$j^4 = j^2 j^2 = (-1)(-1) = +1$$

etc.

Although all four operations of addition, subtraction, multiplication, and division can be performed using the rectangular form, only the operations of addition and subtraction will be described (except for special cases), since they are accomplished most directly. The multiplication and division operations will be described in the polar form (except for special cases) for the same reason. This will avoid a measure of mathematical confusion caused by considering all possibilities. If both quantities to be added appear in the polar form, they will simply have to be converted to the rectangular form before the operation can be performed.

First, let us examine a few descriptive examples of conversion.

EXAMPLE 4.15

Convert the following from polar to rectangular form.

a. $10\angle 53.13^\circ$:

$$A = 10\cos 53.13^\circ = 10(0.6) = 6$$
$$B = 10\sin 53.13^\circ = 10(0.8) = 8$$

and

$$10\angle 53.13^\circ = \mathbf{6 + j8}$$

b. $16\angle -30°$:

$$A = 16 \cos 30° = 16(0.866) = 13.86$$
$$B = 16 \sin 30° = 16(0.5) = 8$$

and

$$16\angle -30° = \mathbf{13.86 - j8}$$

Convert the following from rectangular to polar form:

a. $30 + j40$:

$$C = \sqrt{(30)^2 + (40)^2} = 50$$
$$\theta = \tan^{-1}\tfrac{40}{30} = \tan^{-1} 1.333 = 53.13°$$

and

$$30 + j40 = \mathbf{50\angle 53.13°}$$

b. $4 - j20$:

$$C = \sqrt{(4)^2 + (20)^2} = 20.4$$
$$\theta = \tan^{-1}\tfrac{20}{4} = 78.69°$$

and

$$4 - j20 = \mathbf{20.4\angle -78.69°}$$

Addition (Rectangular Form)

In rectangular form, addition is carried out simply by algebraically (paying heed to the signs of the quantities to be added) adding the real and imaginary components independently.

$$(A_1 + jB_1) + (A_2 + jB_2) = (A_1 + A_2) + j(B_1 + B_2) \quad \textbf{(4.30)}$$

EXAMPLE 4.16 Determine the applied voltage $\mathbf{E}_{in}$ for the network in Fig. 4.34.

Solution: Applying Kirchhoff's voltage law, we have

$$e_{in} = v_1 + v_2$$

In phasor form (using effective values):

$$\mathbf{V}_1 = 0.707(10 \text{ V})\angle 0° = 7.07 \text{ V} \angle 0°$$
$$\mathbf{V}_2 = 0.707(20 \text{ V})\angle 60° = 14.14 \text{ V} \angle 60°$$

Conversion to rectangular form for addition yields

$$\mathbf{V}_1 = 7.07 \text{ V} + j0$$
$$\mathbf{V}_2 = 14.14 \cos 60° + j14.14 \sin 60°$$
$$= 14.14(0.5) + j14.14(0.866)$$
$$= 7.07 \text{ V} + j12.25 \text{ V}$$

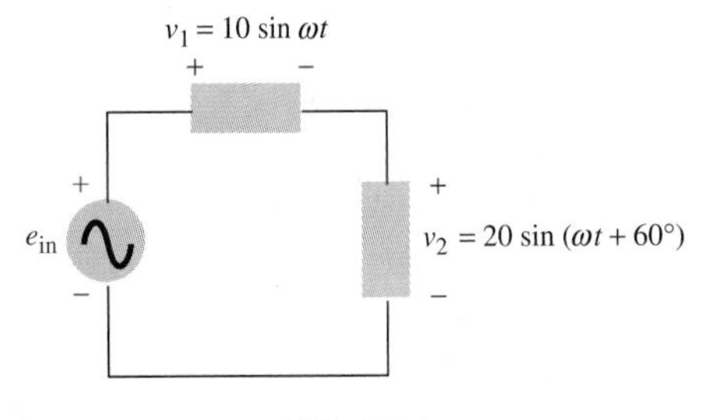

FIG. 4.34
Series ac circuit.

and

$$\mathbf{E}_{\text{in}} = \mathbf{V}_1 + \mathbf{V}_2 = (7.07 + j0) + (7.07 + j12.25)$$
$$= (7.07 + 7.07) + j(0 + 12.25)$$
$$= 14.14 \text{ V} + j12.25 \text{ V}$$

For polar form,

$$|\mathbf{E}_{\text{in}}| = \sqrt{(14.14 \text{ V})^2 + (12.25 \text{ V})^2} = 18.71 \text{ V}$$

$$\theta = \tan^{-1}\frac{12.25}{14.14} = \tan^{-1} 0.866 = 40.9^\circ$$

and

$$\mathbf{E}_{\text{in}} = \mathbf{18.71\ V\angle 40.9^\circ}$$

In the sinusoidal time domain,

$$e_{\text{in}} = \sqrt{2}(18.71)\sin(\omega t + 40.9^\circ)$$
$$= \mathbf{26.46\ sin(\omega t + 40.9^\circ)}$$

Now, you might ask at this point if vector algebra is of any use whatsoever if such a lengthy series of calculations is required to simply add two sinusoidal functions. However, what are the alternatives? Certainly, each function could be carefully graphed on the same axis in the proper phase relationship and the two waveforms added on a point-by-point basis. This would obviously prove to be a very lengthy procedure, and the accuracy would be totally dependent on the care with which the artwork was handled. Admittedly, the introduction to vector algebra and its application to electrical systems has been brief and is still somewhat vague. However, in the analysis to follow there will be continual application of this technique, and in time the confusion should vanish.

If both phasors are drawn on the same set of axes as shown in Fig. 4.35, the resultant can also be obtained by vector addition. However, this still requires the use of the phasor domain and a measure of artwork.

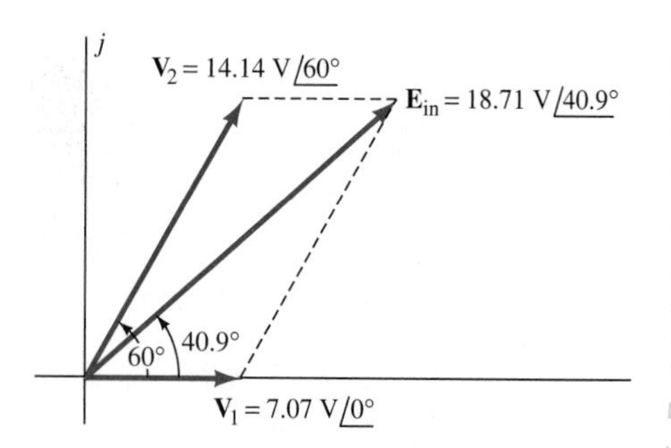

FIG. 4.35
Phasor algebra solution for $\mathbf{E}_{in}$.

If both sinusoidal functions have the same phase angle or are 180° out of phase, the summation can be determined in the polar form, since the vectors have the same or directly opposite directions. For instance, if $\mathbf{V}_1 = 10 \text{ V}\angle 30^\circ$ and $\mathbf{V}_2 = 3.6 \text{ V}\angle 30^\circ$, then

$$\mathbf{E}_{\text{in}} = \mathbf{V}_1 + \mathbf{V}_2 = 13.6 \text{ V}\angle 30^\circ$$

If $\mathbf{V}_1 = 0.6 \text{ V}\angle -20^\circ$ and $\mathbf{V}_2 = 1.8 \text{ V}\angle 160^\circ$ (180° phase shift), then

$$\mathbf{E}_{\text{in}} = \mathbf{V}_1 + \mathbf{V}_2 = 1.2 \text{ V}\angle 160^\circ$$

taking on the direction of the larger and a magnitude equal to the difference.

Subtraction (Rectangular Form)

Like addition, subtraction requires that the real part be treated separately from the imaginary component. That is, in equation form,

$$(A_1 + jB_1) - (A_2 + jB_2) = (A_1 - A_2) + j(B_1 - B_2) \quad \textbf{(4.31)}$$

EXAMPLE 4.17 Determine the current $\mathbf{I}_1$ for the network in Fig. 4.36 in rectangular form.

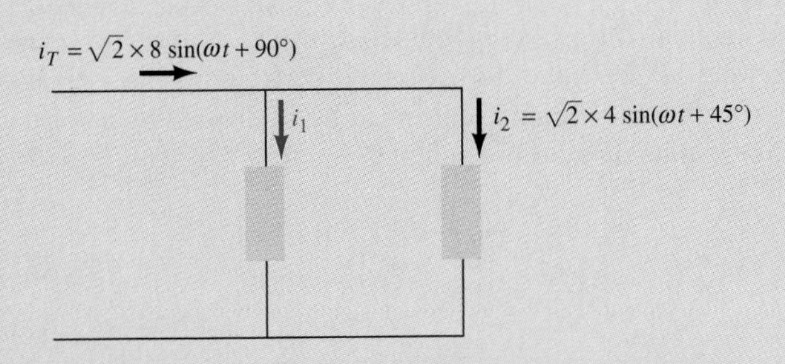

FIG. 4.36

Solution: Application of Kirchhoff's current law results in

$$\mathbf{I}_T = \mathbf{I}_1 + \mathbf{I}_2$$

and

$$\begin{aligned}
\mathbf{I}_1 &= \mathbf{I}_T - \mathbf{I}_2 \\
&= 8\text{ A}\angle 90° - 4\text{ A}\angle 45° \\
&= (0 + j8) - (4 \cos 45° + j4 \sin 45°) \\
&= (0 + j8) - (2.828 + j2.828) \\
&= (0 - 2.828) + j(8 - 2.828)
\end{aligned}$$

$$\mathbf{I}_1 = \mathbf{-2.828\text{ A} + j5.172\text{ A} = 5.895\text{ A}\angle 118.67°}$$

and

$$i_1 = \mathbf{8.34 \sin(\omega t + 118.67°)}$$

Note in Fig. 4.37 the location of the current vector. The real part is negative and therefore to the left of the vertical axis, while the imaginary part is positive and above the horizontal axis. In other words, it is in the third quadrant. In Fig. 4.37 the result is obtained through a vector addition.

Again, vectors in polar form that have the same angle or are 180° out of phase can be subtracted without first converting to the rectangular form. For example, if $\mathbf{I}_1 = 0.2\text{ A}\angle 60°$ and $\mathbf{I}_T = 0.6\text{ A}\angle 60°$, then

$$\mathbf{I}_2 = \mathbf{I}_T - \mathbf{I}_1 = 0.6\text{ A}\angle 60° - 0.2\text{ A}\angle 60° = 0.4\text{ A}\angle 60°$$

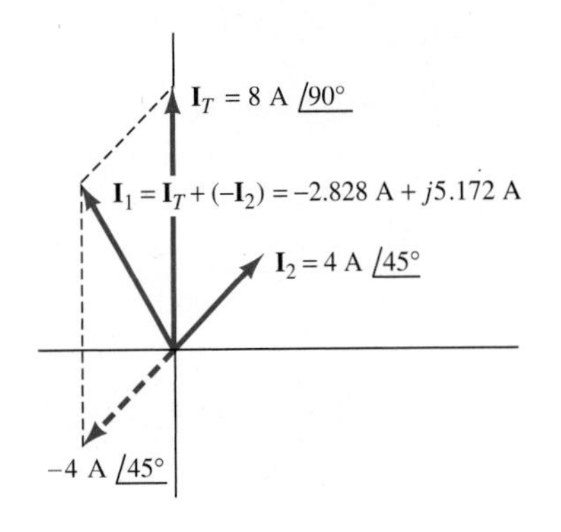

FIG. 4.37
Vector solution for Example 4.17.

Multiplication and Division (Polar Form)

The following equations are applied for the indicated operations in the polar form:

$$(C_1\angle\theta_1)(C_2\angle\theta_2) = C_1C_2\angle\theta_1 + \theta_2 \tag{4.32}$$

$$\frac{C_1\angle\theta_1}{C_2\angle\theta_2} = \frac{C_1}{C_2}\angle\theta_1 - \theta_2 \tag{4.33}$$

If rectangular forms are encountered, it will simply be necessary to first convert to the polar form before performing the operation. Of course, if a phasor is simply multiplied by or divided by a constant, the operation is quite direct. For example,

$$\frac{5 + j6}{2} = 2.5 + j3$$

and

$$6(2 - j4) = 12 - j24$$

EXAMPLE 4.18 Determine the result of the following operations.

a. $(10\angle 60^\circ)(6\angle -20^\circ) = (10)(6)\angle 60^\circ + (-20^\circ) = \mathbf{60\angle 40^\circ}$
b. $(0.4\angle -30^\circ)(600\angle 60^\circ) = (0.4)(600\angle -30^\circ + 60^\circ) = \mathbf{240\angle +30^\circ}$
c. $(40\angle 20^\circ)/(5\angle -30^\circ) = \frac{40}{5}\angle 20^\circ - (-30^\circ) = \mathbf{8\angle 50^\circ}$

EXAMPLE 4.19 Determine the result of the following mixed operations.

a. $\dfrac{(5 + j10)(6\angle 20^\circ)}{0.2 - j0.8}$
b. $(0.2\angle 30^\circ)^2(8 + j6)$

Solution:

a. $5 + j10 = 11.18\angle 63.43^\circ$
$0.2 - j0.8 = 0.825\angle -75.96^\circ$
$$\frac{(11.18\angle 63.43^\circ)(6\angle 20^\circ)}{0.825\angle -75.96^\circ} = \frac{67.08\angle 83.43^\circ}{0.825\angle -75.96^\circ} = \mathbf{81.31\angle 159.39^\circ}$$
b. $(0.2\angle 30^\circ)^2 = (0.2\angle 30^\circ)(0.2\angle 30^\circ) = 0.04\angle 60^\circ$
$8 + j6 = 10\angle 36.87^\circ$
$(0.04\angle 60^\circ)(10\angle 36.87^\circ) = \mathbf{0.4\angle 96.87^\circ}$

FIG. 4.38
TI-86 scientific calculator. (Courtesy of Texas Instruments, Inc.)

Calculators

The TI-86 calculator of Fig. 4.38 is only one of numerous calculators that can convert from one form to another and perform lengthy calculations with complex numbers in a concise, neat form. Not all the details of using a specific calculator will be included here because each has its own format and sequence of steps. However, the basic operations with the TI-86 will be included primarily to demonstrate the ease with which the conversions can be made and the format for more complex operations.

For the TI-86 calculator, you must first call up the 2nd function CPLX from the keyboard, which results in a menu at the bottom of the display including conj, real, imag, abs, and angle. If you choose the key MORE, ▶ Rec and ▶ Pol will appear as options (for the conversion process). To convert from one form to another, simply enter the current form in brackets with a comma between components for the rectangular form and an angle symbol for the polar form. Follow this form with the operation to be performed, and press the ENTER key—the result will appear on the screen in the desired format.

EXAMPLE 4.20 This example is for demonstration purposes only. It is not expected that all readers will have a TI-86 calculator. The sole purpose of the example is to demonstrate the power of today's calculators.

Using the TI-86 calculator, perform the following conversions:

a. $3 - j4$ to polar form.

b. $0.006 \angle 20.6°$ to rectangular form.

Solutions:

a. The TI-86 display for part (a) appears as follows:

```
(3, −4) ▶ Pol (ENTER)
(5.000E0∠−53.130E0)
```

b. The TI-86 display for part (b) appears as the follows:

```
(0.006∠20.6) ▶ Rec (ENTER)
(5.616E−3, 2.111E−3)
```

EXAMPLE 4.21 Using the TI-86 calculator, perform the desired operations required in part (a) of Example 4.19, and compare solutions.

Solution: You must now be aware of the hierarchy of mathematical operations, in other words, the sequence in which the calculator will perform the desired operations? In most cases, the sequence is the same as that used in longhand calculations, although you must become adept at setting up the parentheses to ensure the correct order of operations. For this example, the TI-86 display appears as follows:

```
((5, 10) * (6∠ 20))/(0.2, −0.8) ▶ POL (ENTER)
(81.349E0∠ 159.399E0)
```

which is exactly the same as the earlier solution.

Computer Methods

MathCad is not limited to the dc analysis performed in the earlier chapters but can perform the same operations on complex numbers. The format is slightly different because MathCad employs the letter *i* to represent the imaginary component, and it is placed after the magnitude. The *j* can be employed instead of the *i*, however, simply by using <Esc>, entering FORMAT, and responding as prompted.

EXAMPLE 4.22 Use MathCad to perform the operations required in part (b) of Example 4.19.

Solution: The printout for the solution appears below as follows:

$$r := 0.2 \qquad a := 30\cdot\frac{\pi}{180}$$
$$x := r\cdot\cos(a) + r\cdot\sin(a)\cdot i$$
$$x := 0.173 + 0.1i$$
$$R := x^2\cdot(8 + 6j)$$
$$|R| = 0.4 \qquad \arg(R) = 96.87\ ^\circ\text{deg}$$

The Basic Elements

Let us now apply the phasor algebra to the basic *RLC* elements and note the effect of associating an angle with the resistive and reactive elements. Recall the following from Fig. 4.31:

$$\begin{aligned} \mathbf{Z}_R &= R\angle 0^\circ \\ \mathbf{Z}_L &= X_L\angle 90^\circ \\ \mathbf{Z}_C &= X_C\angle -90^\circ \end{aligned} \tag{4.34}$$

By associating an angle with each impedance vector, it will no longer be necessary to remember the phase relationships between the currents and voltages of each element.

Consider the resistor in Fig. 4.39 for which the voltage is provided and the current is to be determined:

$$\mathbf{I}_R = \frac{\mathbf{V}}{\mathbf{Z}_R} = \frac{V\angle\theta}{R\angle 0^\circ} = \frac{V}{R}\angle\theta - 0^\circ = \frac{V}{R}\angle\theta$$

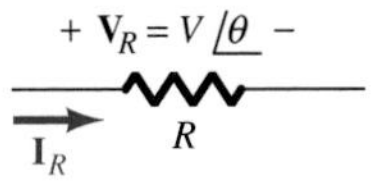

Note that **V** and **I** are in phase, since they have the same angle θ. Associating 0° with the resistance permitted the use of phasor algebra and maintained the proper phase relationships between the current and voltage.

For the inductor in Fig. 4.40,

$$\mathbf{I}_L = \frac{\mathbf{V}}{\mathbf{Z}_L} = \frac{V\angle\theta}{X_L\angle 90^\circ} = \frac{V}{X_L}\angle\theta - 90^\circ$$

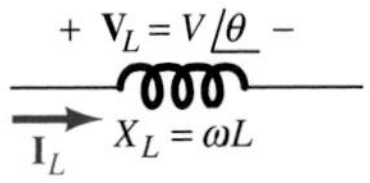

The solution indicates that **I** lags the voltage by 90° and that its magnitude (effective value) is V/X_L.

Finally, for the capacitor in Fig. 4.41,

$$\mathbf{I}_C = \frac{\mathbf{V}}{\mathbf{Z}_C} = \frac{V\angle\theta}{X_C\angle -90^\circ} = \frac{V}{X_C}\angle\theta + 90^\circ$$

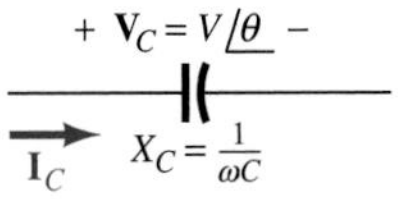

The current leads the voltage by 90°, and its effective value is determined by an application of Ohm's law.

A few examples will clarify the use of phasor notation.

$+\ \mathbf{V}_R = V\angle\theta\ -$; $\mathbf{I}_R$; R

FIG. 4.39
Pure resistor.

$+\ \mathbf{V}_L = V\angle\theta\ -$; $\mathbf{I}_L$; $X_L = \omega L$

FIG. 4.40
Pure inductor.

$+\ \mathbf{V}_C = V\angle\theta\ -$; $\mathbf{I}_C$; $X_C = \frac{1}{\omega C}$

FIG. 4.41
Pure capacitor.

EXAMPLE 4.23 Determine the current through a 20-Ω resistor if the voltage across the resistor is 40 sin (200t + 20°).

Solution: In phasor notation,

$$\mathbf{V} = (0.707)(40\text{ V})\angle 20^\circ = \mathbf{28.28\ V\angle 20^\circ}$$

By Ohm's law:

$$\mathbf{I} = \frac{\mathbf{V}}{\mathbf{Z}_R} = \frac{28.28\text{ V}\angle 20^\circ}{20\ \Omega\angle 0^\circ} = \mathbf{1.414\ A\angle 20^\circ}$$

In the time domain,

$$i = (\sqrt{2})(1.414)\sin(200t + 20^\circ)$$
$$= \mathbf{2\sin(200t + 20^\circ)}$$

EXAMPLE 4.24 Determine the voltage across a 20-mH coil if the current through the coil is 10 × 10^{-3} sin(500t + 60°).

Solution:

$$X_L = \omega L = (500\text{ rad/s})(20 \times 10^{-3}\text{ H}) = 10{,}000 \times 10^{-3}\ \Omega = 10\ \Omega$$

In phasor notation,

$$\mathbf{I} = (0.707)(10\text{ mA})\angle 60^\circ = 7.07\text{ mA}\angle 60^\circ$$

By Ohm's law:

$$\mathbf{V} = \mathbf{I}\mathbf{Z}_L = (7.07\ \text{mA}\ \underline{/60^\circ})(10\ \Omega\ \underline{/90^\circ}) = 70.7\ \text{mV}\ \underline{/150^\circ}$$

In the time domain,

$$\begin{aligned} v &= (\sqrt{2})(70.7 \times 10^{-3}) \sin(500t + 150^\circ) \\ &= 100 \times 10^{-3} \sin(500t + 150^\circ) \\ &= \mathbf{0.1 \sin(500}\boldsymbol{t}\mathbf{ + 150^\circ)} \end{aligned}$$

The waveforms appear in Fig. 4.42.

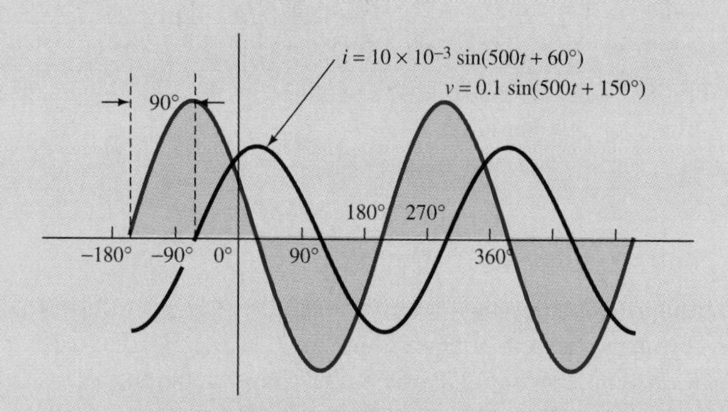

FIG. 4.42
Solution for Example 4.24.

Note how nicely the impedance notation has carried along the proper phase shift.

EXAMPLE 4.25 Determine the current through a 5-μF capacitor if the voltage across the capacitor is 20 sin 377t.

Solution:

$$X_C = \frac{1}{\omega C} = \frac{1}{(377\ \text{rad/s})(5 \times 10^{-6}\ \text{F})} = 530.5\ \Omega$$

In phasor notation:

$$\mathbf{V} = (0.707)(20\ \text{V})\ \underline{/0^\circ} = 14.14\ \text{V}\ \underline{/0^\circ}$$

By Ohm's law:

$$\mathbf{I} = \frac{\mathbf{V}}{\mathbf{Z}_C} = \frac{14.14\ \text{V}\ \underline{/0^\circ}}{530.5\ \Omega\ \underline{/-90^\circ}} = 0.0267\ \text{A}\ \underline{/90^\circ} = \mathbf{26.7\ mA\ \underline{/90^\circ}}$$

In the time domain,

$$\begin{aligned} i_C &= (\sqrt{2})(26.7 \times 10^{-3}) \sin(377t + 90^\circ) \\ &= \mathbf{37.75 \times 10^{-3} \sin(377}\boldsymbol{t}\mathbf{ + 90^\circ)} \end{aligned}$$

Many textbooks written for this area totally avoid an introduction to vector algebra. Rather, they simply state the solution for very particular situations and assume that other possibilities will not be encountered. However, it is a continually applied technique in the field, and you should be somewhat familiar with the notation and technique. For those who fully understand the preceding examples there is no limitation on the type of single-element networks that may be encountered. Consider also, How would you perform the necessary task of simply adding or subtracting two sinusoidal waveforms? For these reasons, the authors felt that a few pages of introduction to phasor algebra would be in the best interests of the reader. Incidentally, now that the technique has been introduced, the material to be considered will follow a more logical sequence, and many unanswered questions resulting from a "given solution" technique will not result.

4.9 SERIES AC NETWORKS

In this section it will be demonstrated that the analysis of series ac networks is very similar to that applied to series dc networks. It is simply necessary to use the vector notation for each quantity rather than just the magnitude, as employed for dc networks.

For a series ac network, *the current is the same through each element,* and the total impedance is the vector sum of the impedances of the series elements. That is,

$$\mathbf{Z}_T = \mathbf{Z}_1 + \mathbf{Z}_2 + \mathbf{Z}_3 + \cdots + \mathbf{Z}_N \tag{4.35}$$

Consider the series RL network in Fig. 4.43. The reactance of the inductor is

$$X_L = \omega L = (377 \text{ rad/s})(10.61 \text{ mH}) = 4\ \Omega$$

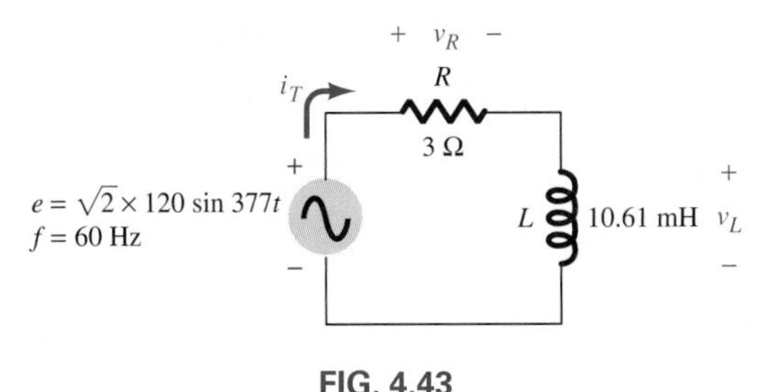

FIG. 4.43
Series RL network.

For more complex networks, it is usually quite helpful to find the solutions for a network in terms of block impedances, as shown in Fig. 4.44 before substituting magnitudes and angles. This type of approach usually results in fewer errors. There is also a more direct relationship with the analysis of dc networks. The content of $\mathbf{Z}_1$ and $\mathbf{Z}_2$ as defined by Fig. 4.43 appears in Fig. 4.44.

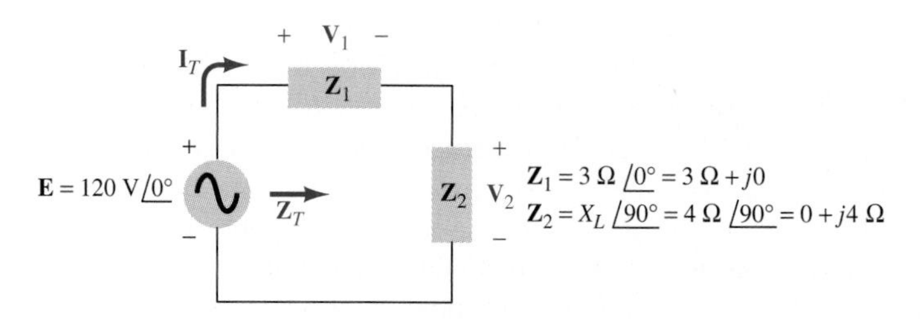

FIG. 4.44

If $\mathbf{Z}_1$ and $\mathbf{Z}_2$ in Fig. 4.44 were resistors R_1 and R_2, respectively, in a dc network, the total resistance would simply be the sum of the two: $R_T = R_1 + R_2$. Treating $\mathbf{Z}_1$ and $\mathbf{Z}_2$ as if they were pure resistors would result in

$$\mathbf{Z}_T = \mathbf{Z}_1 + \mathbf{Z}_2$$

Substitution gives

$$\begin{aligned}\mathbf{Z}_T &= (3\ \Omega + j0) + (0 + j4\ \Omega)\\ &= \mathbf{3\ \Omega + j4\ \Omega = 5\ \Omega\ \angle 53.13^\circ}\end{aligned}$$

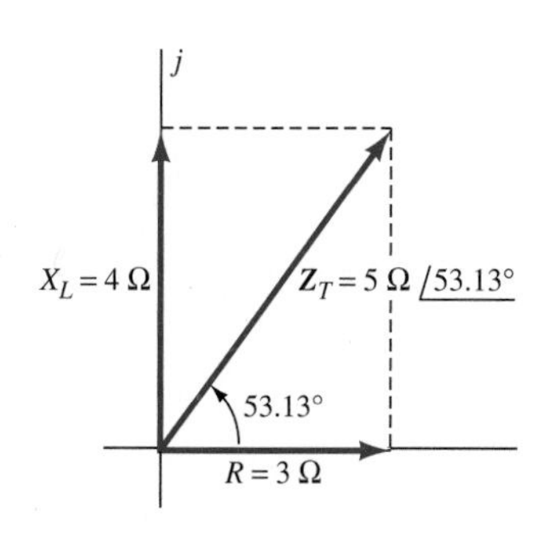

FIG. 4.45
Impedance diagram for a series RL network.

The impedance diagram in Fig. 4.45 clearly reveals that the total impedance $\mathbf{Z}_T$ can also be determined graphically through simple vector addition.

Ohm's law:

$$\mathbf{I} = \frac{\mathbf{E}}{\mathbf{Z}_T} = \frac{120\ \text{V}\angle 0^\circ}{5\ \Omega\angle 53.13^\circ} = \mathbf{24\ A\angle -53.13^\circ}$$

which in the time domain is

$$\begin{aligned}i &= \sqrt{2}(24)\sin(\omega t - 53.13^\circ)\\ &= \mathbf{33.94\sin(\omega t - 53.13^\circ)}\end{aligned}$$

For the resistor:

$$\mathbf{V}_R = \mathbf{V}_1 = \mathbf{IZ}_1 = (24\ \text{A}\angle -53.13^\circ)(3\ \Omega\angle 0^\circ) = \mathbf{72\ V\angle -53.13^\circ}$$

which in the time domain is

$$\begin{aligned}v_R &= \sqrt{2}(72)\sin(\omega t - 53.13^\circ)\\ &= \mathbf{101.81\sin(\omega t - 53.13^\circ)}\end{aligned}$$

Note that v_R and i are *in phase,* since they have the same phase angle. For the inductor:

$$\mathbf{V}_L = \mathbf{V}_2 = \mathbf{IZ}_2 = (24\ \text{A}\angle -53.13^\circ)(4\ \Omega\angle 90^\circ) = \mathbf{96\ V\angle 36.87^\circ}$$

which in the time domain is

$$v_L = \sqrt{2}(96)\sin(\omega t + 36.87°)$$
$$= \mathbf{133.74\ sin(\omega t + 36.87°)}$$

A phasor diagram of the voltages and current appears in Fig. 4.46. Note that the applied voltage **E** is the *vector* sum of $\mathbf{V}_R$ and $\mathbf{V}_L$ as determined by Kirchhoff's voltage law:

$$\mathbf{E} = \mathbf{V}_R + \mathbf{V}_L$$

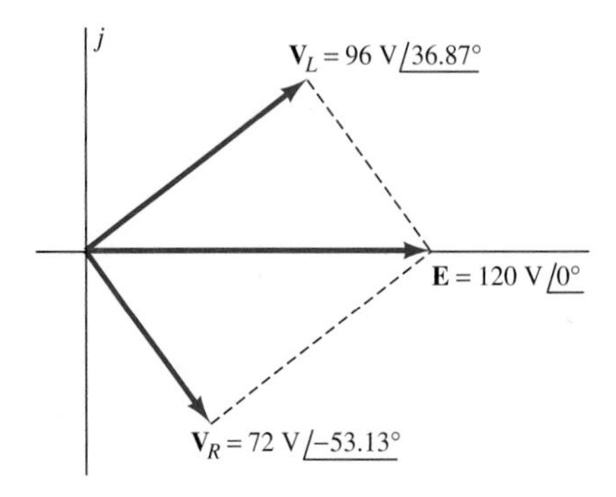

FIG. 4.46
Phasor diagram for a series RL network.

The *voltage-divider rule* is applied in exactly the same manner as described for dc networks. In this example, if we were interested in $\mathbf{V}_R$, the following would result from using the block impedances:

$$\mathbf{V}_R = \frac{\mathbf{Z}_1(\mathbf{E})}{\mathbf{Z}_1 + \mathbf{Z}_2} = \frac{(3\angle 0°)(120\angle 0°)}{3 + j4} = \frac{360\angle 0°}{5\angle 53.13°}$$
$$= \mathbf{72\ V\angle -53.13°}$$

as obtained before. Compare this calculation with an application of the rule to a dc network with two series resistors.

Before continuing with the analysis of this network, take a moment to carefully examine the waveforms of the voltages and current plotted on the same axis in Fig. 4.47. Note especially that v_R and i_R are in phase and that v_L leads i_L by 90°. Consider also that since the network is *inductive,*

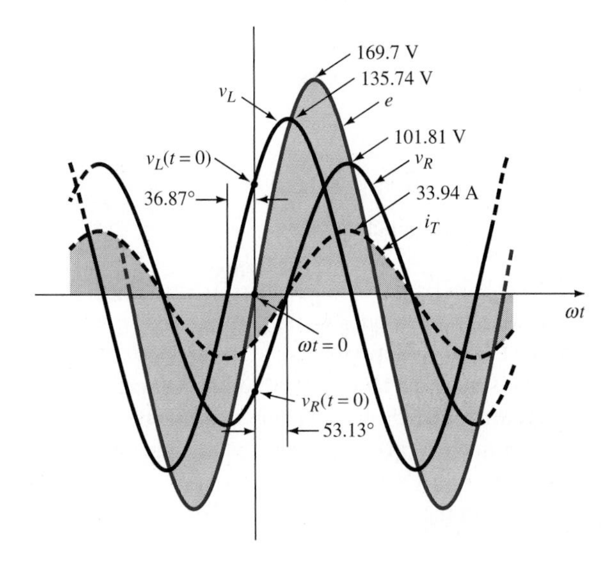

FIG. 4.47
Waveforms for the series RL network in Fig. 4.43.

the input current also lags the applied voltage by an angle of 53.13°. The more inductive the network (or the less resistive), the larger the angle by which i lags the applied voltage e. At any point on the axis, the instantaneous values of e, v_R, and v_L satisfy Kirchhoff's voltage law. For instance, at $t = 0$ or $\theta = 0°$, $e = 0$, and

$$e = v_R + v_L$$

becomes

$$0 = v_R + v_L$$

and

$$v_R = v_L$$

which is reflected at this point in Fig. 4.47.

The power to the network can be determined by using any one of the following equations:

$$P = EI\cos\theta = I^2R = V_RI_R = \frac{V_R^2}{R} \quad \text{(watts)} \qquad \textbf{(4.36)}$$

In each case, the same result will be obtained, as shown next for two of the equations. Keep in mind that power is delivered to (or dissipated by) only the resistive elements.

$$P = EI\cos\theta = (120\text{ V})(24\text{ A})\cos 53.13°$$

where θ is the phase angle between the applied voltage and current drawn by the network.

$$P = (120\text{ V})(24\text{ A})(0.6)$$
$$= \mathbf{1728\ W}$$

and

$$P = I_T^2R = (24\text{ A})^2(3\ \Omega) = \mathbf{1728\ W}$$

The power factor of the network is

$$F_P = \cos\theta = \mathbf{0.6}$$

indicating that it is far from purely resistive ($F_p = 1$) but is certainly not purely reactive ($F_p = 0$). Networks (such as this one) where the applied voltage leads the source current are said to have a *lagging* F_p to indicate the inductive characteristics. For capacitive networks the label *leading* F_p is applied. For the network just investigated a lagging power factor of 0.6 is said to exist.

For series circuits the network power factor can also be determined by the ratio

$$F_p = \frac{R}{Z_T} \qquad \textbf{(4.37)}$$

which for this example is

$$F_p = \tfrac{3}{5} = \mathbf{0.6}$$

Occasionally, in practical applications, a network is designed so that for a particular frequency range the inductive reactance is much greater than the impedance of any elements in series with it. In this example, for instance, if $X_L = 400\ \Omega$ and $R = 3\ \Omega$, the total impedance is

$$Z_T = \sqrt{R^2 + X_L^2} = \sqrt{(3\ \Omega)^2 + 400\ \Omega^2} \cong 400\ \Omega$$

and the network from a practical viewpoint is purely inductive. In addition,

$$F_p = \frac{R}{Z} \cong \frac{3}{400} = 0.0075 \cong 0$$

If time and space allocation permitted, a simple *RC* series network would now be examined. However, this exercise has been left to the problems at the end of the chapter. The only major differences will appear in the impedance diagram, where the capacitive reactance vector appears below the axis, and in the phasor diagram, where the series current leads the capacitive voltage by 90° and the applied voltage by a number of degrees determined by the network.

Let us now go a step further and examine the series *RLC* network in Fig. 4.48. Replacing each element with a block impedance will result in the configuration in Fig. 4.49. The total impedance is the vector sum, and

$$\begin{aligned}
\mathbf{Z}_T &= \mathbf{Z}_1 + \mathbf{Z}_2 + \mathbf{Z}_3 = (5\ \text{k}\Omega + j0) + (0 + j4\ \text{k}\Omega) + (0 - j16\ \text{k}\Omega) \\
&= 5\ \text{k}\Omega + j4\ \text{k}\Omega - j16\ \text{k}\Omega \\
&= 5\ \text{k}\Omega - j12\ \text{k}\Omega \\
&= \mathbf{13\ k\Omega \angle -67.38^\circ}
\end{aligned}$$

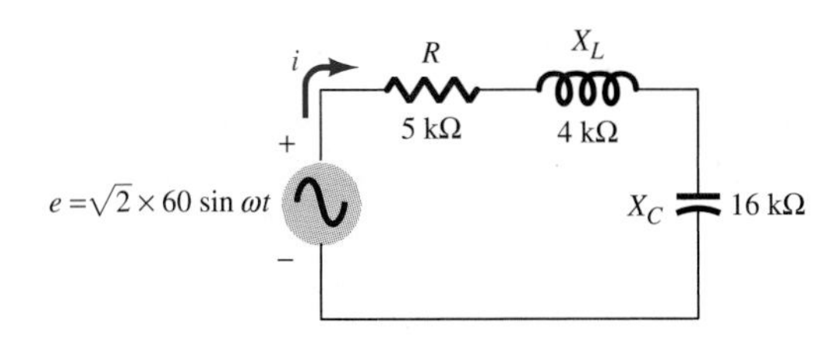

FIG. 4.48
Series RLC network.

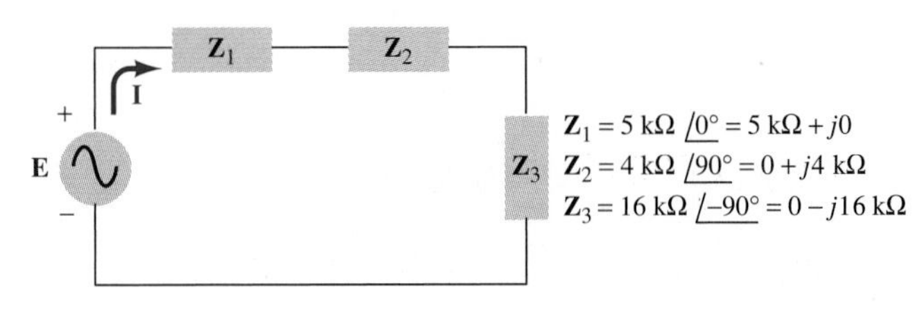

FIG. 4.49

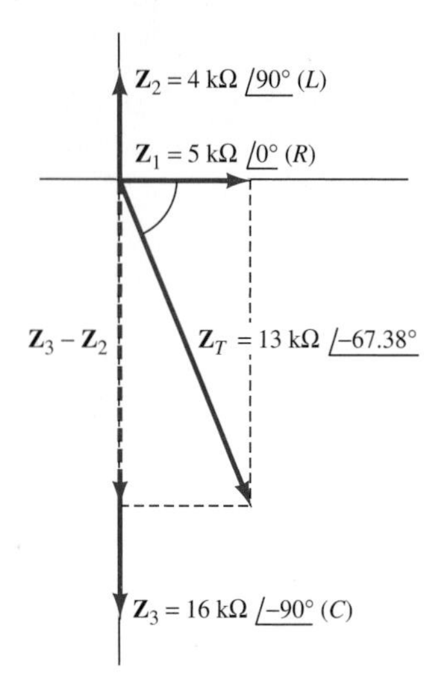

FIG. 4.50
Impedance diagram for the series RLC network in Fig. 4.48.

The impedance diagram appears in Fig. 4.50. Note that the capacitive and inductive reactances are directly opposing and that the difference is the net reactance of the network. The current I is

$$\mathbf{I} = \frac{\mathbf{E}}{\mathbf{Z}_T} = \frac{60\ \text{V}\angle 0°}{13\ \text{k}\Omega\angle -67.38°} = \mathbf{4.615\ mA\angle 67.38°}$$

which in the time domain is

$$\begin{aligned} i &= \sqrt{2}(4.615 \times 10^{-3})\sin(\omega t + 67.38°) \\ &= \mathbf{6.53 \times 10^{-3}\sin(\omega t + 67.38°)} \end{aligned}$$

The voltage across each element can then be obtained directly using Ohm's law:

$$\begin{aligned} \mathbf{V}_R &= \mathbf{IR} = (4.615\ \text{mA}\ \angle 67.38°)(5\ \text{k}\Omega\ \angle 0°) \\ &= \mathbf{23.075\ V\angle 67.38°} \\ \mathbf{V}_L &= \mathbf{IZ}_L = (4.615\ \text{mA}^{-3}\angle 67.38°)(4\ \text{k}\Omega\ \angle 90°) \\ &= \mathbf{18.46\ V\angle 157.39°} \\ \mathbf{V}_C &= \mathbf{IZ}_C = (4.615\ \text{mA}\ \angle 67.38°)(16\ \text{k}\Omega^3\angle -90°) \\ &= \mathbf{73.84\ V\angle -22.62°} \end{aligned}$$

The phasor diagram of the voltages and current appears in Fig. 4.51. Note that, like the reactances, the voltages $\mathbf{V}_L$ and $\mathbf{V}_C$ are in vector opposition and that the current **I** lags the voltage $\mathbf{V}_L$ by 90°, leads the voltage $\mathbf{V}_C$ by 90°, and is in phase with $\mathbf{V}_R$.

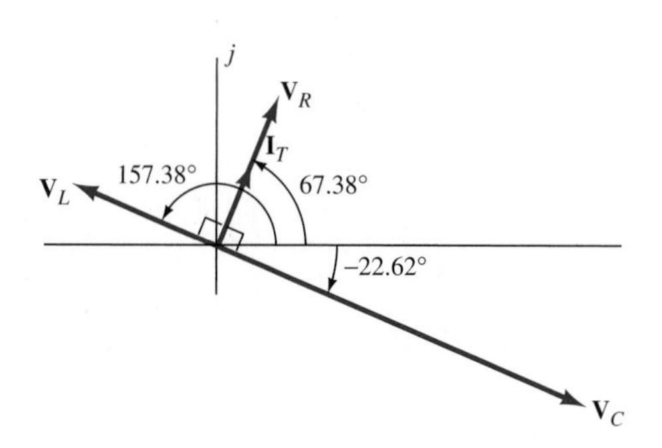

FIG. 4.51
Phasor diagram for the series RLC network in Fig. 4.48.

Any of the voltages found above could also be obtained using the voltage-divider rule, which would remove the necessity to find **I** first. For $\mathbf{V}_L$,

$$\begin{aligned} \mathbf{V}_L &= \frac{\mathbf{Z}_2(\mathbf{E})}{\mathbf{Z}_1 + \mathbf{Z}_2 + \mathbf{Z}_3} = \frac{(4\ \text{k}\Omega\angle 90°)(60\ \text{V}\angle 0°)}{13 \times 10^3\angle -67.38°} = \frac{240\ \text{V}\angle 90°}{13\angle -67.38°} \\ &= \mathbf{18.46\ V\angle 157.38°} \end{aligned}$$

In the time domain,

$$v_L = \sqrt{2}(18.46)\sin(\omega t + 157.38°)$$
$$= \mathbf{26.1\sin(\omega t + 157.38°)}$$

For the power to the network,

$$P = EI_T \cos\theta$$
$$= (60\text{ V})(4.615\text{ mA})\cos 67.38°$$
$$= \mathbf{106.5\text{ mW}}$$

or

$$P = I^2R$$
$$= (4.615\text{ mA})^2(5\text{ k}\Omega)$$
$$= \mathbf{106.5\text{ mW}}$$

For the network power factor,

$$F_p = \cos\theta = \cos(67.38°) = \mathbf{0.3846\text{ leading}}$$

or

$$F_p = \frac{R}{Z_T} = \frac{5\text{ k}\Omega}{13\text{ k}\Omega} = \mathbf{0.3846\text{ leading}}$$

The low power factor indicates that the terminal characteristics reflect a highly reactive (leading) network.

The entire analysis thus far has been for a fixed frequency. Even though the resistance does not change with frequency (ideally), the reactance of an inductor increases and that of a capacitor decreases as the frequency increases.

The impedance versus the frequency for each element of a series *RL* circuit is provided in Fig. 4.52.

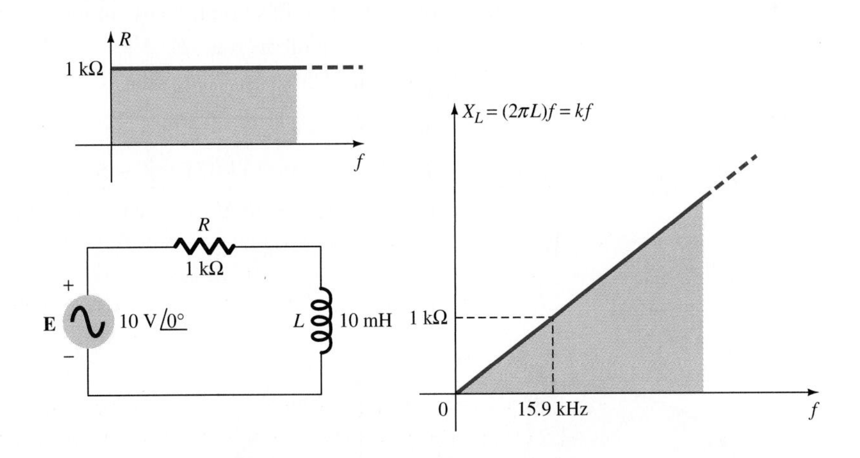

FIG. 4.52
Effect of frequency on the impedance of each element of a series RL circuit.

At $f = 0$ Hz, the impedance of the inductor is 0 Ω and the total impedance is simply that of the resistor, as shown in Fig. 4.53. At very high

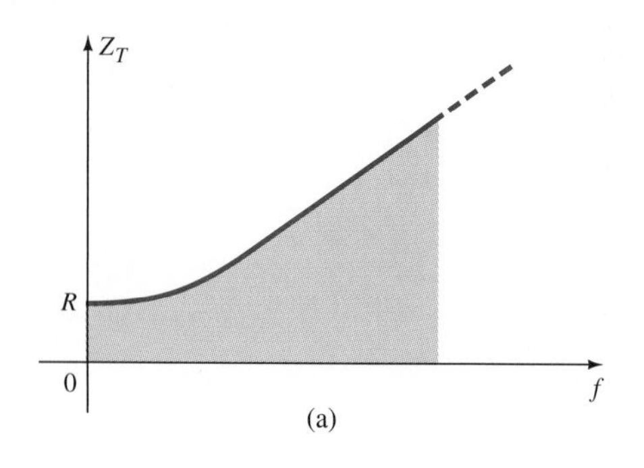

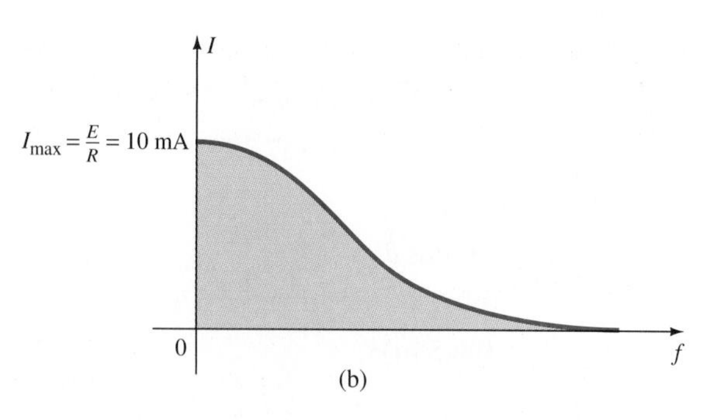

FIG. 4.53
Effect of frequency on the total impedance Z_T (Fig. 4.53(a)) and the current I (Fig. 4.53(b)) for the series RL circuit in Fig. 5.43.

frequencies $X_L = 2\pi fL \gg R$, and the total impedance defined by $Z_T = \sqrt{R^2 + X_L^2}$ is simply X_L. For the full frequency range the total impedance varies as shown in Fig. 4.53(a).

Since the magnitude of $I = E/Z_T$, the plot of I versus frequency appears as shown in Fig. 4.53(b).

At $f = 0$ Hz, $X_L = 0\ \Omega$ (a short-circuit equivalent), and the entire applied voltage appears across the resistor. However, as the frequency increases, X_L will eventually be equal to R, and the magnitude of V_L and VR will be the same. An application of the voltage-divider rule results in

$$\mathbf{V}_L = \frac{\mathbf{Z}_L\mathbf{E}}{\mathbf{Z}_R + \mathbf{Z}_L} = \frac{(1\ \text{k}\Omega\angle 90^\circ)(10\ \text{V}\angle 0^\circ)}{1\ \text{k}\Omega + j1\ \text{k}\Omega} = 7.071\ \text{V}\angle{+45^\circ}$$

with $\mathbf{V}_R = 7.071\ \text{V}\angle{-45^\circ}$ revealing that the two voltages are out of phase by 90°, and V_L and V_R do not equal $\frac{1}{2}(10\ \text{V}) = 5$ V but 7.071 V. The application of Kirchhoff's voltage law might suggest an error, since $E \neq V_L + V_R$, but remember that $\mathbf{E}$, $\mathbf{V}_L$, and $\mathbf{V}_R$ are vectors, and $\mathbf{E} = \mathbf{V}_R + \mathbf{V}_L$, which will be satisfied. That is,

$$\begin{aligned}
\mathbf{E} &= \mathbf{V}_R + \mathbf{V}_L \\
10\ \text{V}\angle 0^\circ &= 7.071\ \text{V}\angle{-45^\circ} + 7.071\ \text{V}\angle{+45^\circ} \\
&= (5\ \text{V} - j5\ \text{V}) + (5\ \text{V} + j5\ \text{V}) = 10\ \text{V} + j(-5\ \text{V} + 5\ \text{V}) \\
10\ \text{V}\angle 0^\circ &= 10\ \text{V}\angle 0^\circ \qquad \text{(checks)}
\end{aligned}$$

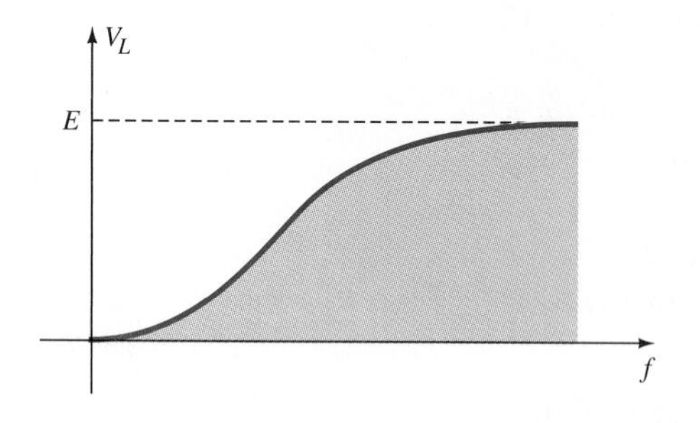

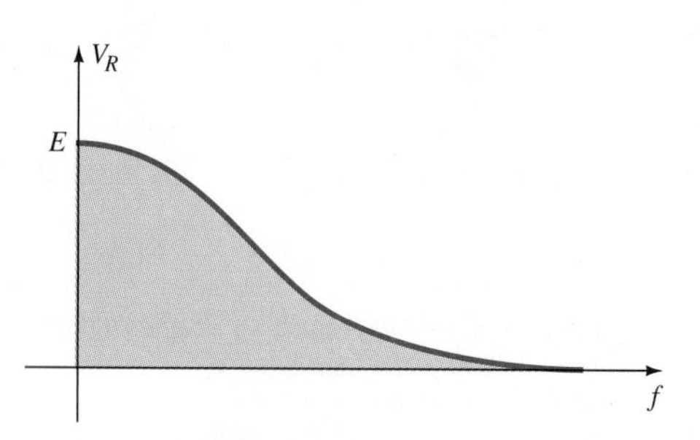

FIG. 4.54
Effect of frequency on the magnitude of $\mathbf{V}_L$ and $\mathbf{V}_R$ for the circuit in Fig. 4.52.

As the frequency continues to increase, the voltage across the inductor increases, since X_L becomes increasingly larger than R. Eventually, $V_L \cong E$ and $V_R = 0$ V. A plot of the magnitude of $\mathbf{V}_L$ and $\mathbf{V}_R$ with frequency is provided in Fig. 4.54.

4.10 PARALLEL AC NETWORKS

The analysis of parallel ac networks is also very similar to that introduced for dc networks. The reciprocal of impedance, called *admittance,* is defined by the following equation and, as indicated, is measured in *siemens.*

$$\mathbf{Y} = \frac{1}{\mathbf{Z}} \quad \text{(siemens, s)} \tag{4.38}$$

For parallel ac networks such as appears in Fig. 4.55 the total admittance is determined from

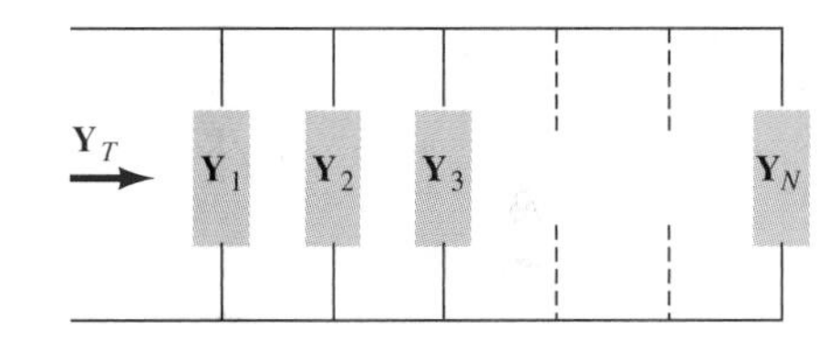

FIG. 4.55
Parallel admittances.

$$\mathbf{Y}_T = \mathbf{Y}_1 + \mathbf{Y}_2 + \mathbf{Y}_3 + \cdots + \mathbf{Y}_N \tag{4.39}$$

or

$$\frac{1}{\mathbf{Z}_T} = \frac{1}{\mathbf{Z}_1} + \frac{1}{\mathbf{Z}_2} + \frac{1}{\mathbf{Z}_3} + \cdots + \frac{1}{\mathbf{Z}_N} \tag{4.40}$$

which for two parallel impedances becomes

$$\mathbf{Z}_T = \frac{\mathbf{Z}_1\mathbf{Z}_2}{\mathbf{Z}_1 + \mathbf{Z}_2} \tag{4.41}$$

The *voltage is the same across parallel ac branches,* and the total input current can be determined using Kirchhoff's current law or by first finding the total input impedance (or admittance) and then applying Ohm's law.

The reciprocal of resistance in an ac network is called *conductance* (as for dc networks) and has the angle 0° associated with it as shown:

$$\mathbf{Y}_R = G\angle 0^\circ = \frac{1}{R\angle 0^\circ} \quad \text{(siemens, s)} \tag{4.42}$$

The reciprocal of reactance is called *susceptance* and is also measured in siemens. For each element it has the notation and angle appearing

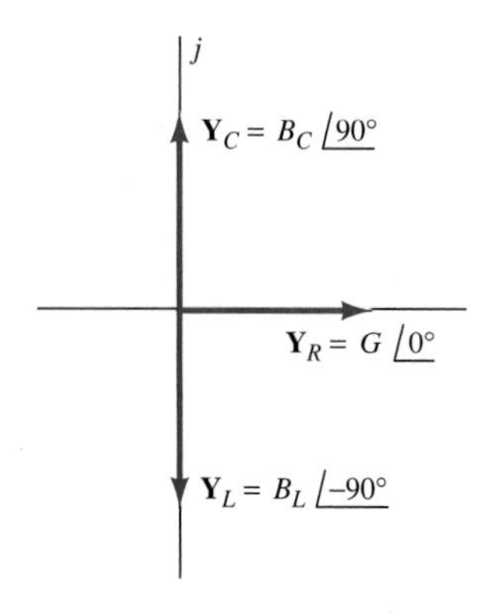

FIG. 4.56
Admittance diagram.

in Eqs. (4.43) and (4.44). The term *susceptance* comes from the word *susceptible,* since the reciprocal of the reactance is a measure of how susceptible the branch or element is to the flow of charge through it. An increase in its value parallels a rise in the branch current.

$$\mathbf{Y}_L = B_L \angle -90° = \frac{1}{X_L \angle 90°} \quad \text{(siemens, s)} \quad \textbf{(4.43)}$$

$$\mathbf{Y}_C = B_C \angle -90° = \frac{1}{X_C \angle 90°} \quad \text{(siemens, s)} \quad \textbf{(4.44)}$$

An *admittance diagram* is defined as shown in Fig. 4.56 for a parallel *RLC* network.

Consider the parallel *RL* network in Fig. 4.57. Inserting block impedances results in the configuration in Fig. 4.58. The total admittance and impedance can be determined from

$$\mathbf{Y}_R = \frac{1}{\mathbf{Z}_R} = \frac{1}{R\angle 0°} = \frac{1}{3\ \text{k}\Omega \angle 0°} = 0.333\ \text{mS} \angle 0°$$

$$\mathbf{Y}_L = \frac{1}{\mathbf{Z}_L} = \frac{1}{X_L \angle 90°} = \frac{1}{4\ \text{k}\Omega \angle 90°} = 0.250\ \text{mS} \angle -90°$$

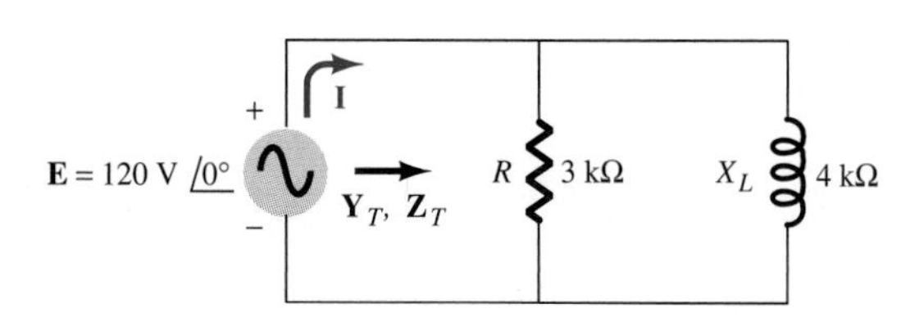

FIG. 4.57
Parallel RL network.

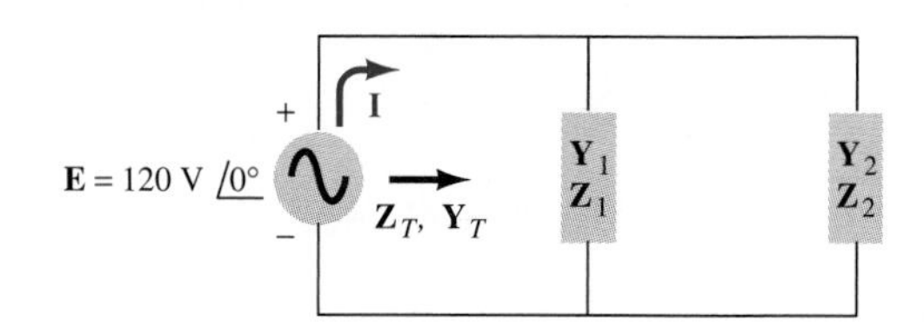

FIG. 4.58

and
$$\begin{aligned}\mathbf{Y}_T &= \mathbf{Y}_R + \mathbf{Y}_L \\ &= 0.333\ \text{mS} \angle 0° + 0.250\ \text{mS} \angle -90° \\ &= 0.333\ \text{mS} - j0.250\ \text{mS} \\ &= \mathbf{0.416\ mS \angle -36.9°}\end{aligned}$$

and
$$\mathbf{Z}_T = \frac{1}{\mathbf{Y}_T} = \frac{1}{0.416\ \text{mS} \angle -36.9°} = \mathbf{2.4\ k\Omega \angle 36.9°}$$

or
$$\mathbf{Z}_T = \frac{\mathbf{Z}_R\mathbf{Z}_L}{\mathbf{Z}_R + \mathbf{Z}_L} = \frac{(3\ \text{k}\Omega \angle 0°)(4\ \text{k}\Omega \angle 90°)}{3\ \text{k}\Omega + j4\ \text{k}\Omega} = \mathbf{2.4\ k\Omega \angle 36.9°}$$

The admittance diagram is provided in Fig. 4.59.

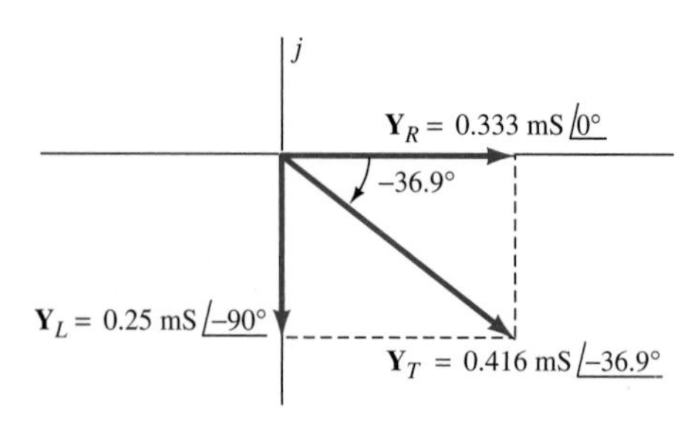

FIG. 4.59
Admittance diagram for the parallel RL network in Fig. 4.57.

Note again that the total admittance can be determined using relatively simple vector algebra. The current **I** is

$$\mathbf{I} = \frac{\mathbf{E}}{\mathbf{Z}_T} = \frac{120 \text{ V} \angle 0^\circ}{2.4 \text{ k}\Omega \angle 36.9^\circ} = \mathbf{50 \text{ mA} \angle -36.9^\circ}$$

or

$$\mathbf{I} = \frac{\mathbf{E}}{\mathbf{Z}_T} = \mathbf{E}(\mathbf{Y}_T) = (120 \text{ V} \angle 0^\circ)(0.416 \text{ mS} \angle -36.9^\circ)$$
$$= \mathbf{50 \text{ mA} \angle -36.9^\circ}$$

The current through either element can then be determined by Ohm's law:

$$\mathbf{I}_R = \mathbf{E}\mathbf{Y}_R = \frac{\mathbf{E}}{\mathbf{Z}_R} = \frac{120 \text{ V} \angle 0^\circ}{3 \text{ k}\Omega \angle 0^\circ} = \mathbf{40 \text{ mA} \angle 0^\circ}$$

and

$$\mathbf{I}_L = \mathbf{E}\mathbf{Y}_L = \frac{\mathbf{E}}{\mathbf{Z}_L} = \frac{120 \text{ V} \angle 0^\circ}{4 \text{ k}\Omega \angle 90^\circ} = \mathbf{30 \text{ mA} \angle -90^\circ}$$

A phasor diagram of the currents and voltage can then be drawn as shown in Fig. 4.60. Note that $\mathbf{I}_R$ is in phase with **E** and that $\mathbf{I}_L$ lags **E** by 90°. The diagram itself demonstrates the validity of Kirchhoff's current law as a method for determining **I** from

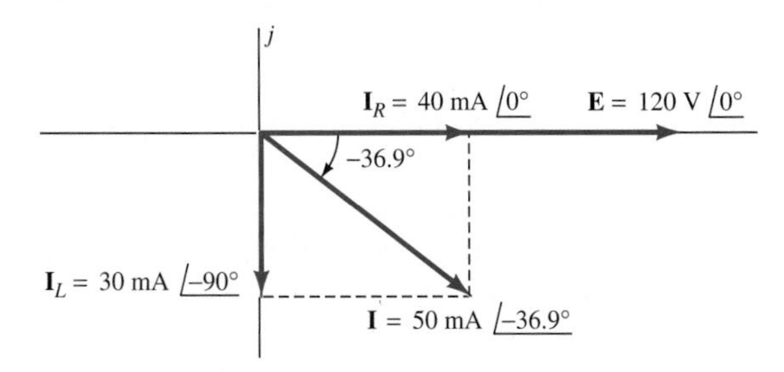

FIG. 4.60
Phasor diagram for the parallel RL network in Fig. 4.57.

$$\mathbf{I} = \mathbf{I}_R + \mathbf{I}_L$$

The power to the network can be determined using the same equation applied to series ac networks. That is,

$$P = EI \cos \theta_T = I_R^2 R = \frac{V_R^2}{R} = V_R I_R \qquad \textbf{(4.45)}$$

where all voltages and currents are effective values.

For this example,

$$P = EI\cos\theta_T = (120\text{ V})(50\text{ mA})\cos 36.9^\circ = (6)(0.7997)$$
$$= \mathbf{4.8\ W}$$

or

$$P = \frac{V_R^2}{R} = \frac{E^2}{R} = \frac{(120\text{ V})^2}{3\text{ k}\Omega} = \mathbf{4.8\ W}$$

The power factor for parallel ac networks can be determined using the equation

$$F_p = \cos\theta_T = \frac{G}{Y_T} \tag{4.46}$$

which for this example results in

$$F_p = \cos\theta_T = \cos 36.9^\circ = \mathbf{0.8\ lagging}$$

or

$$F_p = \frac{G}{Y_T} = \frac{0.333 \times 10^{-3}}{0.416 \times 10^{-3}} = \mathbf{0.8\ lagging}$$

The term *lagging* reflects the fact that the input voltage leads the input current by a number of degrees—in other words, an *inductive* network.

It should by now be somewhat obvious that the analysis of dc and ac networks is quite similar once the vector notation is introduced. This will continue to be true as we progress through the text.

A final example on parallel ac networks will be a detailed examination of the parallel *RLC* network in Fig. 4.61(a).

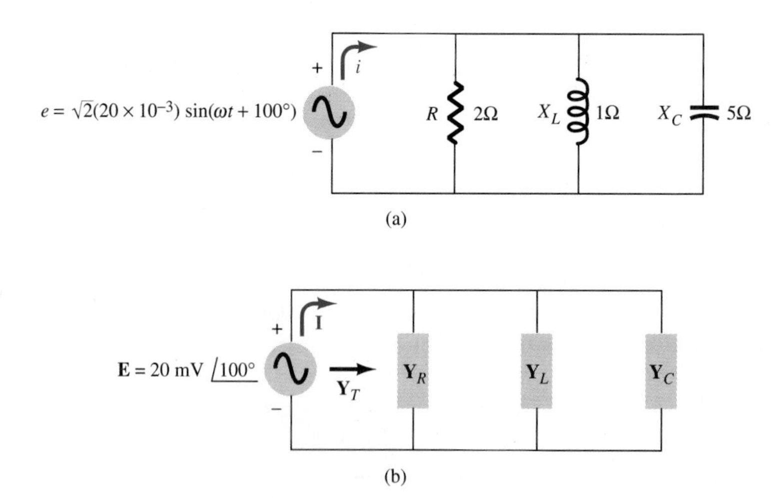

FIG. 4.61
Parallel RLC network.

Applying the block admittance labels results in the configuration in Fig. 4.61(b), where

$$\mathbf{Y}_R = \frac{1}{\mathbf{Z}_R} = \frac{1}{2\ \Omega\angle 0^\circ} = 0.5\ \text{S}\angle 0^\circ$$

$$\mathbf{Y}_L = \frac{1}{\mathbf{Z}_L} = \frac{1}{1\ \Omega\angle 90^\circ} = 1\ \text{S}\angle -90^\circ$$

$$\mathbf{Y}_C = \frac{1}{\mathbf{Z}_L} = \frac{1}{5\ \Omega\angle -90^\circ} = 0.2\ \text{S}\angle 90^\circ$$

The total admittance is

$$\begin{aligned}\mathbf{Y}_T &= \mathbf{Y}_R + \mathbf{Y}_L + \mathbf{Y}_C \\ &= (0.5\ \text{S} + j0) + (0 - j1\ \text{S}) + (0 + j0.2\ \text{S}) \\ &= 0.5\ \text{S} + j(-1\ \text{S} + 0.2\ \text{S}) \\ &= 0.5\ \text{S} - j0.8\ \text{S} \\ &= \mathbf{0.943\ S\ \angle -58^\circ}\end{aligned}$$

and

$$\mathbf{Z}_T = \frac{1}{\mathbf{Y}_T} = \frac{1}{0.943\ \text{S}\angle -58^\circ} = \mathbf{1.06\ \Omega\angle 58^\circ}$$

or combining two parallel elements at a time, we obtain

$$\mathbf{Z'}_T = \mathbf{Z}_R \| \mathbf{Z}_L = \frac{\mathbf{Z}_R\mathbf{Z}_L}{\mathbf{Z}_R + \mathbf{Z}_L}$$

and

$$\mathbf{Z}_T = \mathbf{Z'}_T \| \mathbf{Z}_C = \frac{\mathbf{Z'}_T\mathbf{Z}_C}{\mathbf{Z'}_T + \mathbf{Z}_C} = \mathbf{1.06\ \Omega\angle 58^\circ}$$

The admittance diagram appears in Fig. 4.62.

Note that the susceptances are in direct opposition. Since the inductive susceptance is the larger, the network has lagging characteristics. That is, the applied voltage **E** leads the input current **I** by a number of degrees determined by the driving point impedance of the network, $\mathbf{Z}_T$.

The input current can again be determined by Ohm's law:

$$\begin{aligned}\mathbf{I} &= \frac{\mathbf{E}}{\mathbf{Z}_T} = \mathbf{E}(\mathbf{Y}_T) = (20\ \text{mV}\angle 100^\circ)(0.943\ \text{S}\angle -58^\circ) \\ &= \mathbf{18.86\ mA\angle 42^\circ}\end{aligned}$$

which in the time domain is

$$\begin{aligned}i &= \sqrt{2}(18.86 \times 10^{-3})\sin(\omega t + 42^\circ) \\ &= \mathbf{26.67 \times 10^{-3}\sin(\omega t + 42^\circ)}\end{aligned}$$

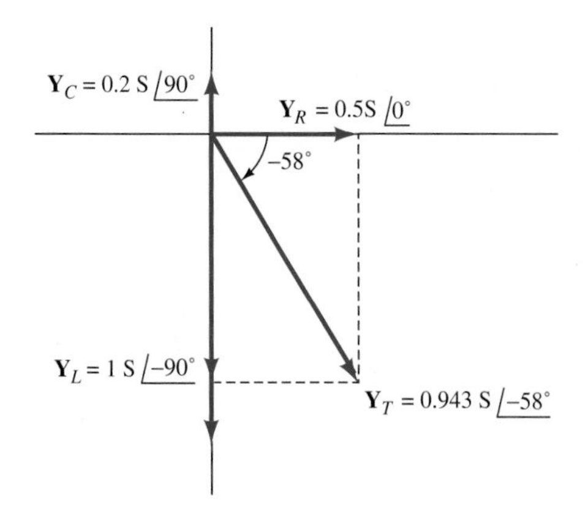

FIG. 4.62
Admittance diagram for the parallel RLC network in Fig. 4.61.

The current through each element can again be determined by Ohm's law:

$$\mathbf{I}_R = \frac{\mathbf{E}}{\mathbf{Z}_R} = \frac{20\text{ mV}\angle 100^\circ}{2\,\Omega\angle 0^\circ} = \mathbf{10\text{ mA}\angle 100^\circ}$$

$$\mathbf{I}_L = \frac{\mathbf{E}}{\mathbf{Z}_L} = \frac{20\text{ mV}\angle 100^\circ}{1\,\Omega\angle 90^\circ} = \mathbf{20\text{ mA}\angle 10^\circ}$$

$$\mathbf{I}_C = \frac{\mathbf{E}}{\mathbf{Z}_C} = \frac{20\text{ mV}\angle 100^\circ}{5\,\Omega\angle -90^\circ} = \mathbf{4\text{ mA}\angle 190^\circ}$$

The current **I** can then be determined using Kirchhoff's current law,

$$\mathbf{I} = \mathbf{I}_R + \mathbf{I}_L + \mathbf{I}_C$$

or as determined above.

The phasor diagram of the currents and voltages appears in Fig. 4.63. Note that the phase angle of **E** has simply rotated (in the counterclockwise direction) the entire phasor diagram by 100°. The current $\mathbf{I}_R$ is still in phase with $\mathbf{V}_R(=\mathbf{E})$, $\mathbf{I}_L$ lags the voltage $\mathbf{V}_L\,(=\mathbf{E})$ by 90°, and $\mathbf{I}_C$ leads the voltage $\mathbf{V}_C\,(=\mathbf{E})$ by 90°. The vector addition required to determine **I** is shown in the figure. Compare its vector solution with that obtained above.

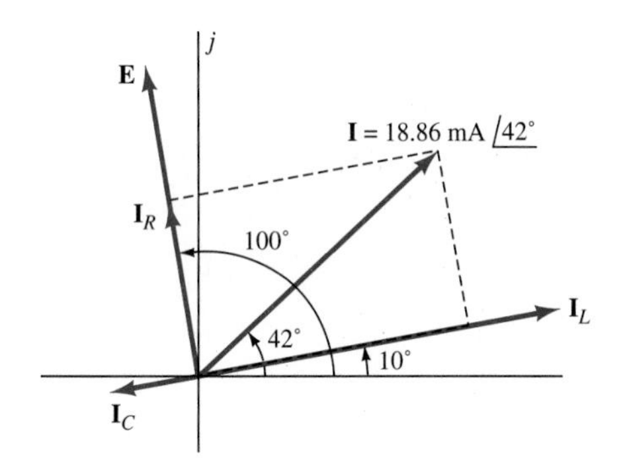

FIG. 4.63
Phasor diagram for the parallel RLC network in Fig. 4.61.

The power to the network can be determined using any one of the equations introduced earlier. That is,

$$\begin{aligned} P &= EI\cos\theta_T \\ &= (20\text{ mV})(18.86\text{ mA})\cos(100^\circ - 42^\circ) \\ &= (377.2\times 10^{-6})(\cos 58^\circ) = (377.2\times 10^{-6})(0.5299) \\ &= \mathbf{200\ \mu W} \end{aligned}$$

or

$$P = \frac{E^2}{R} = \frac{(20\text{ mV})^2}{2\,\Omega} = \frac{400\ \mu\text{W}}{2} = \mathbf{200\ \mu W}$$

The power factor of the network is

$$F_p = \cos\theta_T = \cos 58^\circ = \mathbf{0.5299\ lagging}$$

For any series or parallel network a change in frequency changes the reactance of the capacitive and inductive elements and therefore the response of the network. In fact, at another frequency, a network may have a leading rather than a lagging characteristic, indicating an increase in the terminal effect of the capacitive elements over those of the inductive components. For the range of frequencies being considered here, the effect of frequency on resistive elements can be ignored.

Consider the parallel *RC* network in Fig. 4.64, for instance. The impedance versus the frequency of each element is included in the same figure.

For parallel elements one must keep in mind that the smallest parallel impedance has the greatest impact. Therefore, at low frequencies where $X_C \gg R$ the network is primarily resistive and about 1 kΩ. At very high

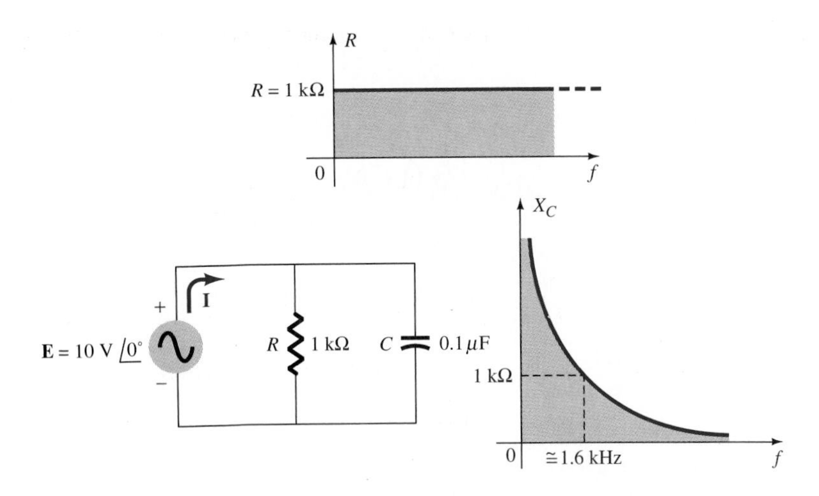

FIG. 4.64
Effect of frequency on the impedance of each element of a parallel RC network.

frequencies $R \gg X_C$, and the network is primarily capacitive with a very small total impedance because X_C becomes progressively smaller. In fact, at very high frequencies the capacitive reactance is so small that the element essentially creates a short circuit across the network. The result is very high source and capacitive currents that can damage the equipment or cause dangerous side effects.

The total impedance and source current vary as shown in Fig. 4.65 for the full frequency range.

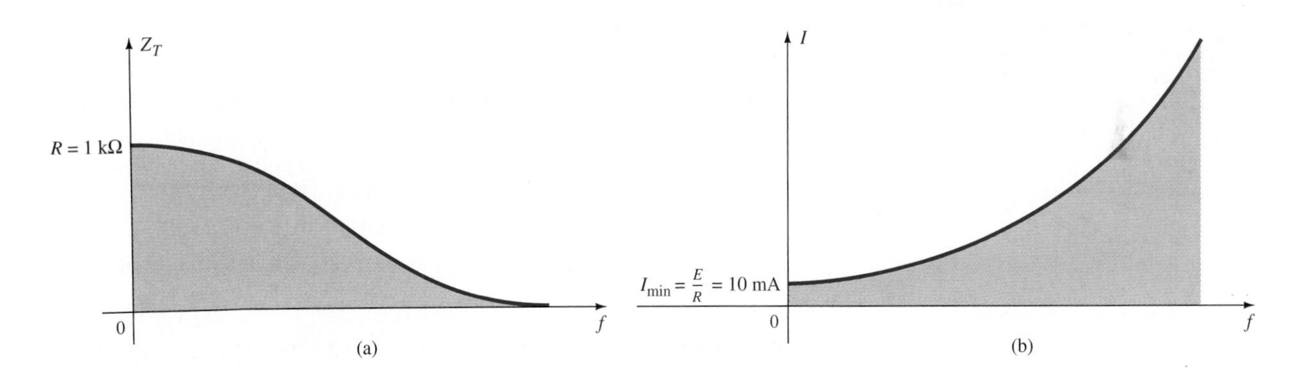

FIG. 4.65
Effect of frequency on the total impedance (Fig. 4.65(a) and source current (Fig. 4.65(b) of a parallel RC network.

The dc current source has its ac equivalent as shown in Fig. 4.66(a). The characteristics and impact are very similar to those of the dc equivalent; however, the ac current source also has an associated frequency and phase angle.

An ac current source can be converted to a voltage source using the same equations applied to dc circuits. The parallel impedance is simply

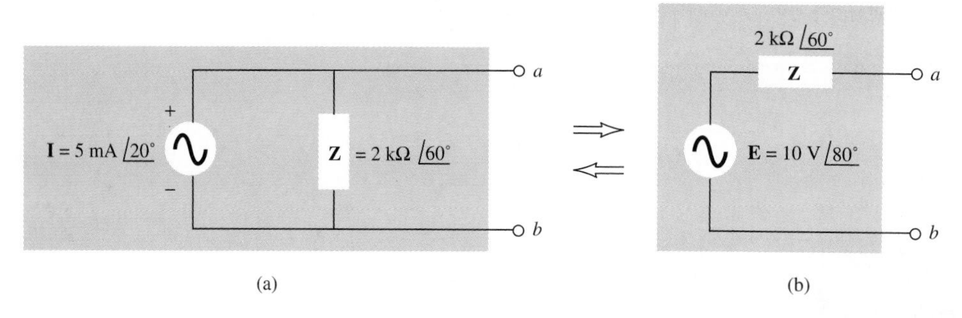

FIG. 4.66
Source conversions.

placed in series with the voltage source, as shown in Fig. 4.66(b), and the voltage source is determined by Ohm's law. That is

$$\mathbf{E} = \mathbf{IZ} = (5\ \text{mA}\angle 20^\circ)(2\ \text{k}\Omega\angle 60^\circ) = 10\ \text{V}\angle 80^\circ$$

An ac current source can be derived from a voltage source by simply using Ohm's law in the form $\mathbf{I} = \mathbf{E}/\mathbf{Z}$.

4.11 SHORT CIRCUITS AND OPEN CIRCUITS

The terms *short circuit* and *open circuit* are two network conditions that the general public has heard about more than once but probably remain a mystery to most except for the disastrous consequences often associated with each. There are times in the operation of many electronic systems that a short or an open circuit is part of the normal system operation. However, it is the unexpected damaging effect of each that concerns laypeople because they do not work with electrical or electronic systems on a daily basis.

In general, a short circuit is nothing more than a connection (desired or not) of 0 Ω between two points of a network. An open circuit is simply the lack of any connection whatsoever between two points of a system.

The effect of a short circuit can be demonstrated using the simple circuit in Fig. 4.67(a)—a toaster having an internal resistance of 12 Ω, a power cord of 0.1-Ω resistance, and a typical household outlet of 120 V.

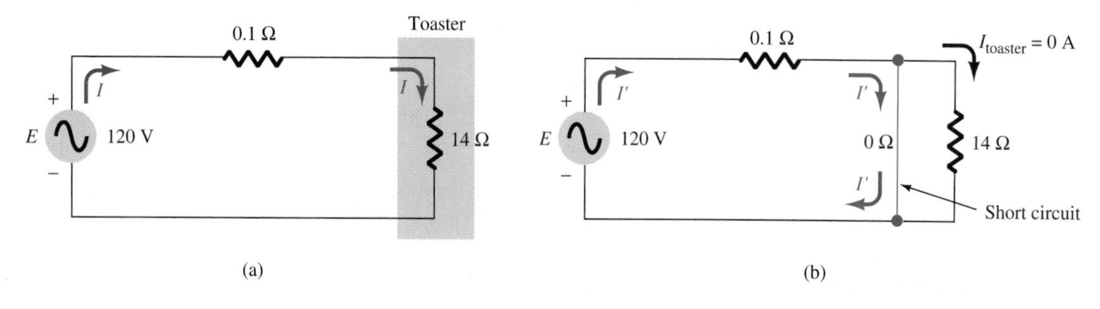

FIG. 4.67
Demonstrating the effect of a short circuit on the operation of a toaster: (a) normal operation; (b) short-circuit condition.

Under normal operating conditions the current through the network is

$$I = \frac{E}{R_T} = \frac{120\text{ V}}{0.1\ \Omega + 14\ \Omega} = \frac{120\text{ V}}{14.1\ \Omega} = 8.51\text{ A}$$

which is certainly in the safe range of a typical household 20-A breaker.

However, if the power cord becomes frayed at the point where it enters the toaster, the two wires in the cord may touch and introduce a short circuit, as shown in Fig. 4.67(b). The 14-Ω load presented by the toaster has been short-circuited, and the total resistance of the circuit drops to that of the line cord, which is simply 0.1 Ω. Recall from the discussion of parallel networks that the total resistance of two parallel elements is the product over the sum. The product of 0 Ω and 12 Ω obviously results in a total resistance for the parallel combination of 0 Ω.

The resulting source current now attempts to jump to a level limited only by the resistance of the line cord or

$$I' = \frac{E}{R_T} = \frac{120\text{ V}}{0.1\ \Omega} = 1200\text{ A}$$

The resulting increase immediately causes the 20-A breaker to open and, it is hoped, prevent any serious damage.

The effect of an undesirable open circuit can be demonstrated by the simple series–parallel network in Fig. 4.68(a) having two parallel load elements that call for a current of 5 A each. With the use of series–parallel network analysis techniques, the source current is determined to be

$$I = \frac{E}{R_T} = \frac{120\text{ V}}{10\ \Omega + 4\ \Omega \| 4\Omega} = \frac{120\text{ V}}{10\ \Omega + 2\ \Omega} = \frac{120\text{ V}}{12\ \Omega} = 10\text{ A}$$

and the current through each 4-Ω load is 5 A, since current splits equally between equal parallel loads.

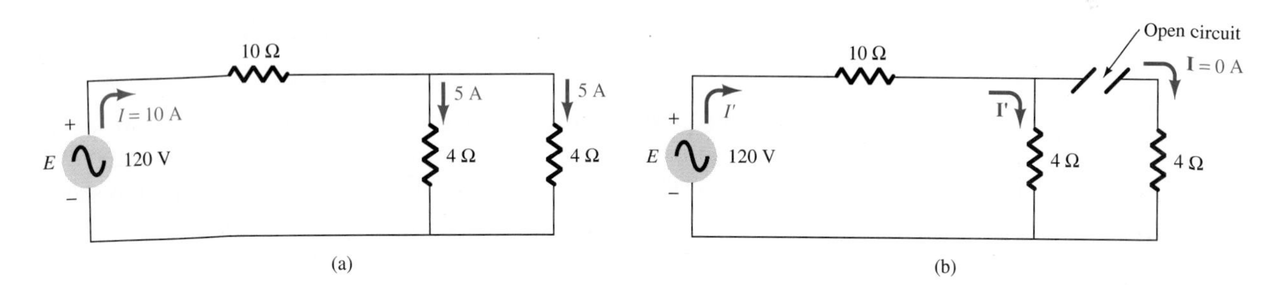

FIG. 4.68

Demonstrating the effect of an open circuit on the current to a load: (a) normal operation; (b) open-circuit operation.

If a terminal becomes disconnected from the second 4-Ω load as shown in Fig. 4.68(b), the total resistance of the network increases to 14 Ω, reducing the source current to 8.57 A. However, note that because of the open circuit all the current must pass through the remaining 4-Ω

Although the preceding example provides some exposure to how an undesirable open circuit can affect the normal behavior of a network, it is normally the open circuit associated with a lost connection to a power source that often causes the most havoc. The lack of power to any system such as a train, air-conditioning unit, or elevator will obviously cause it to shut down.

4.12 PHASE MEASUREMENTS

The phase shift between the voltages of a network or between the voltages and currents of a network can be found using a dual-trace (two signals displayed at the same time) oscilloscope. Phase-shift measurements can also be performed using a single-trace oscilloscope by properly interpreting the resulting Lissajous patterns obtained on the screen. This latter approach, however, will be left for the laboratory experience.

In Fig. 4.69, channel 1 of the dual-trace oscilloscope is hooked up to display the applied voltage e. Channel 2 is connected to display the voltage across the inductor v_L. Of particular importance is the fact that the ground of the scope is connected to the ground of the oscilloscope for both channels. In other words, there is only one common ground for the circuit and oscilloscope. The resulting waveforms may appear as shown in Fig. 4.70.

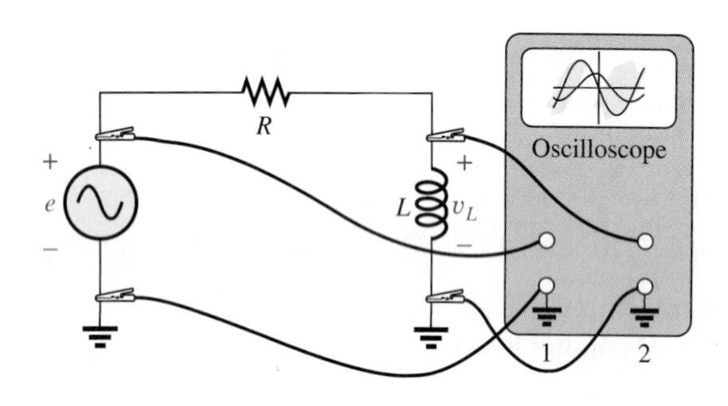

FIG. 4.69
Determining the phase relationship between e and v_L.

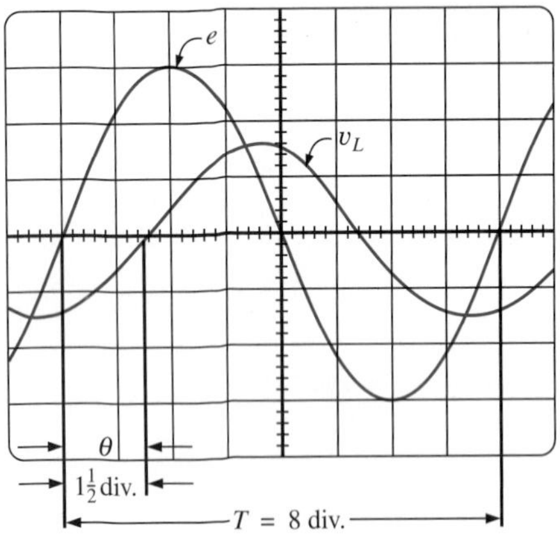

FIG. 4.70
Determining the phase angle between e and v_L.

For the chosen horizontal sensitivity, each waveform of Fig. 4.70 has a period T defined by eight horizontal divisions, and the phase angle between the two waveforms is defined by $1\frac{1}{2}$ divisions. Using the fact that each period of a sinusoidal waveform encompasses 360°, we can set up the following ratios to determine the phase angle θ:

$$\frac{8 \text{ div.}}{360^\circ} = \frac{1.5 \text{ div.}}{\theta}$$

and

$$\theta = \left(\frac{1.5}{8}\right)360^\circ = \mathbf{67.5^\circ}$$

In general,

$$\theta = \frac{(\text{div. for } \theta)}{(\text{div. for } T)} \times 360^\circ \tag{4.47}$$

If the phase relationship between e and v_R is required, the oscilloscope *must not* be hooked up as shown in Fig. 4.71. Points a and b have a common ground that will establish a 0-V drop between the two points; this drop will have the same effect as a short-circuit connection between a and b. The resulting short circuit will "short out" the inductive element, and the current will increase due to the drop in impedance for the circuit. A dangerous situation can arise if the inductive element has a high impedance and the resistor has a relatively low impedance. The current, controlled solely by the resistance R, could jump to dangerous levels and damage the equipment.

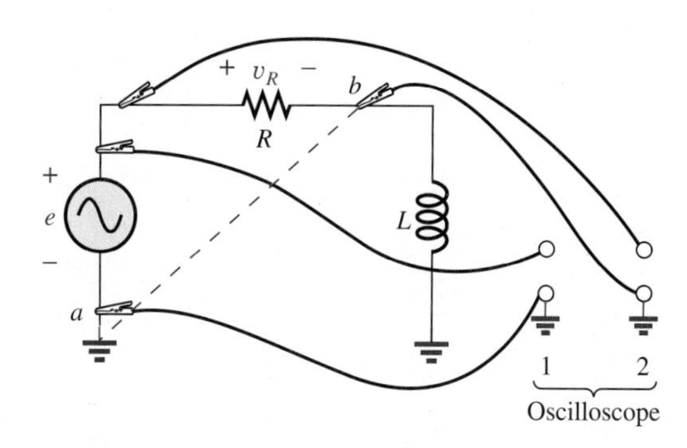

FIG. 4.71
An improper phase-measurement connection.

The phase relationship between e and v_R can be determined by simply interchanging the positions of the coil and resistor or by introducing a sensing resistor, as shown in Fig. 4.72. A sensing resistor is exactly

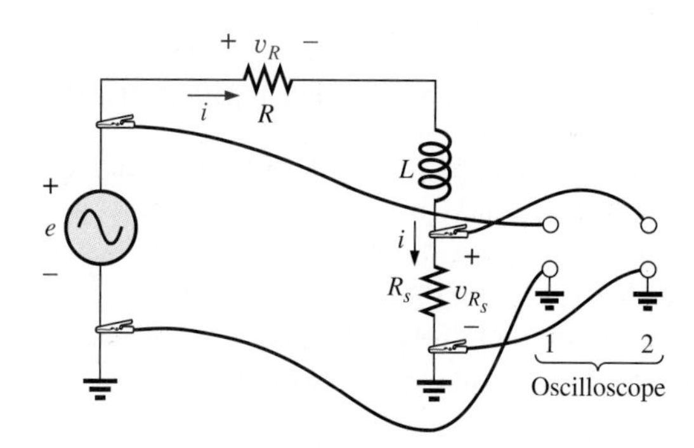

FIG. 4.72
Determining the phase relationship between e and v_R or e and i using a sensing resistor.

that: introduced to "sense" a quantity without adversely affecting the behavior of the network. In other words, the sensing resistor must be small enough compared with the other impedances of the network not to cause a significant change in the voltage and current levels or phase relationships. Note that the sensing resistor is introduced in a way that causes one end to be connected to the common ground of the network. In Fig. 4.72, channel 2 will display the voltage v_{R_s}, which is in phase with the current i. However, the current i is also in phase with the voltage v_R across the resistor R. The net result is that the voltages v_{R_s} and v_R are in phase, and the phase relationship between e and v_R can be determined from the waveforms e and v_{R_s}. Since v_{R_s} and i are in phase, this procedure will also determine the phase angle between the applied voltage e and the source current i. If the magnitude of R_s is sufficiently small compared with R or X_L, the phase measurements of Fig. 4.69 can be performed with R_s in place. That is, channel 2 can be connected to the top of the inductor and to ground, and the effect of R_s can be ignored. In this application, the sensing resistor will not reveal the magnitude of the voltage v_R but simply the phase relationship between e and v_R.

For the parallel network of Fig. 4.73, the phase relationship between two of the branch currents, i_R and i_L, can be determined using a sensing

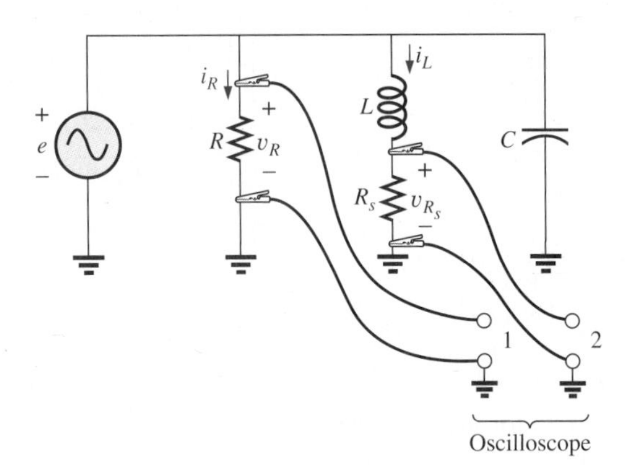

FIG. 4.73
Determining the phase relationship between i_R and i_L.

resistor, as shown in the figure. Channel 1 will display the voltage v_R, and channel 2 will display the voltage v_{R_s}. Since v_R is in phase with i_R, and v_{R_s} is in phase with the current i_L, the phase relationship between v_R and v_{R_s} will be the same as that between i_R and i_L. In this case, the magnitudes of the current levels can be determined using Ohm's law and the resistance levels R and R_s, respectively.

If the phase relationship between e and i_s of Fig. 4.73 is required, a sensing resistor can be employed, as shown in Fig. 4.74.

In general, therefore, for dual-trace measurements of phase relationships, be particularly careful of the grounding arrangement, and fully utilize the in-phase relationship between the voltage and current of a resistor.

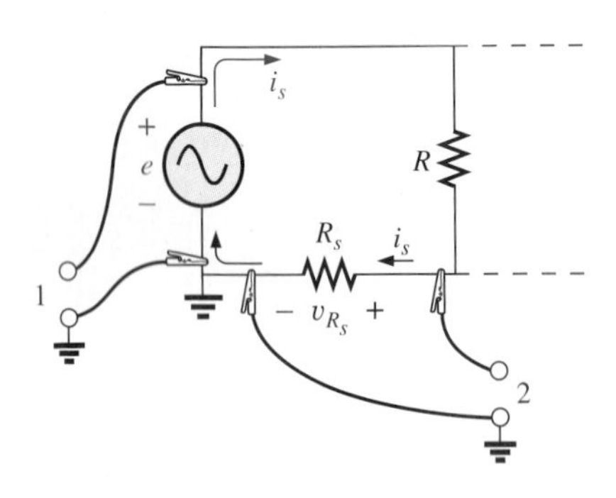

FIG. 4.74
Determining the phase relationship between e and i_s.

4.13 APPLICATION: HOUSE WIRING

An expanded view of house wiring is provided in Fig. 4.75 to permit a discussion of the entire system. The house panel has been included with the "feed" and the important grounding mechanism. In addition, a number of typical circuits found in the home have been included to provide a sense for the manner in which the total power is distributed.

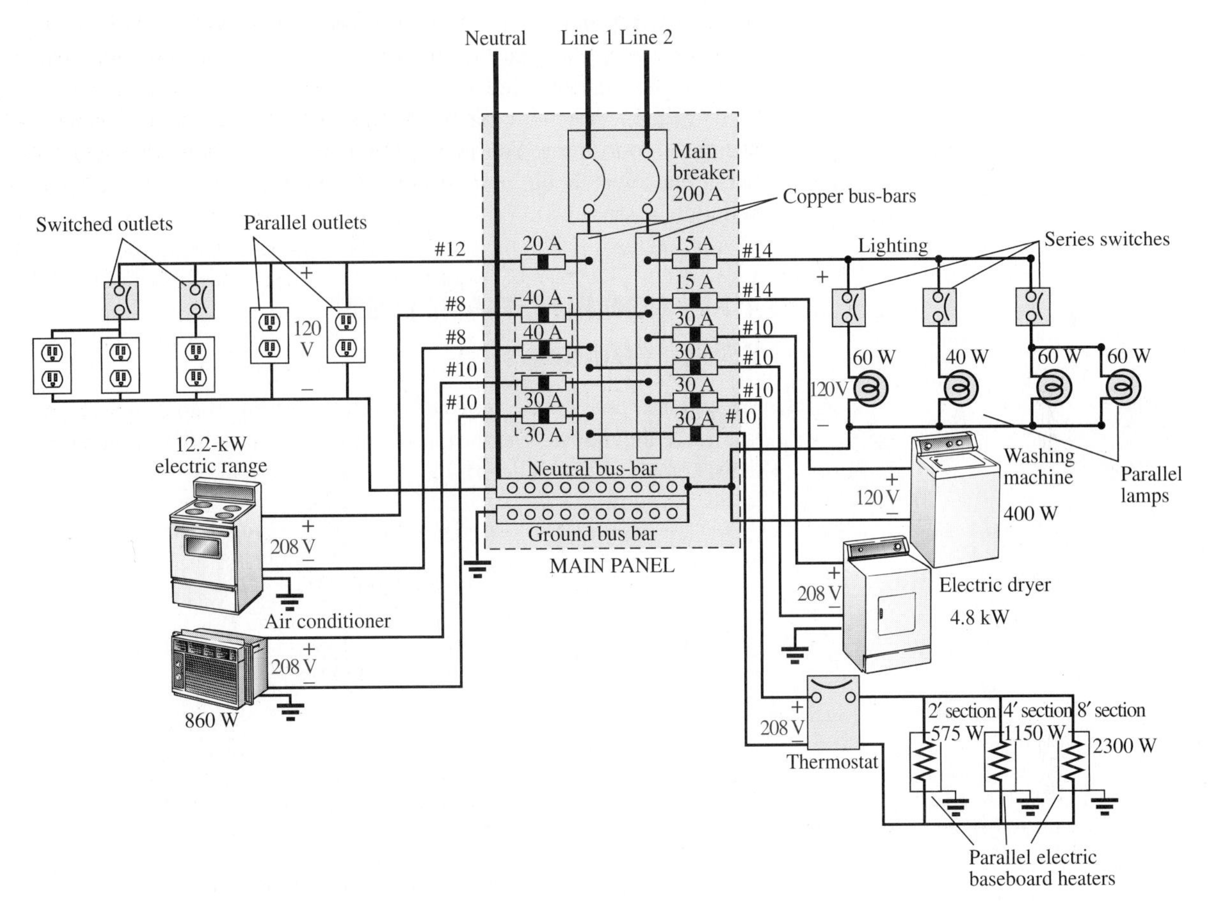

FIG. 4.75
Home wiring diagram.

First note how the copper bars in the panel are laid out to provide both 120 V and 208 V. Between any one bar and ground is the single-phase 120-V supply. However, the bars have been arranged so that 208 V can be obtained between two vertical adjacent bars using a double-gang circuit breaker. When time permits, examine your own panel (but do not remove the cover), and note the dual circuit breaker arrangement for the 208-V supply.

For appliances such as fixtures and heaters that have a metal casing, the ground wire is connected to the metal casing to provide a direct path to ground path for a "shorting" or errant current as described in Section

4.11. For outlets and such that do not have a conductive casing, the ground lead is connected to a point on the outlet that distributes to all important points of the outlet.

Note the series arrangement between the thermostat and the heater but the parallel arrangement between heaters on the same circuit. In addition, note the series connection of switches to lights in the upper right corner but the parallel connection of lights and outlets. Due to high current demand the air conditioner, heaters, and electric stove have 30-A breakers. Keep in mind that the total current does not equal the product of the two (or 60 A), since the breakers are in a line, and the same current will flow through each breaker.

In general, you now have a surface understanding of the general wiring in your home. You may not be a qualified, licensed electrician, but at least you should now be able to converse with some intelligence about the system.

4.14 COMPUTER ANALYSIS

PSpice (Windows)

We will now analyze the series *RLC* network of Fig. 4.76 using schematics. Since the inductive and capacitive reactances cannot be entered into the schematic, we first determine the associated inductive and capacitive levels as follows:

$$X_L = 2\pi f L \Rightarrow L = \frac{X_L}{2\pi f} = \frac{7\ \Omega}{2\pi(1\text{ kHz})} = 1.114\text{ mH}$$

$$X_C = \frac{1}{2\pi f C} \Rightarrow C = \frac{1}{2\pi f X_C} = \frac{1}{2\pi(1\text{ kHz})3\ \Omega} = 53.05\ \mu\text{F}$$

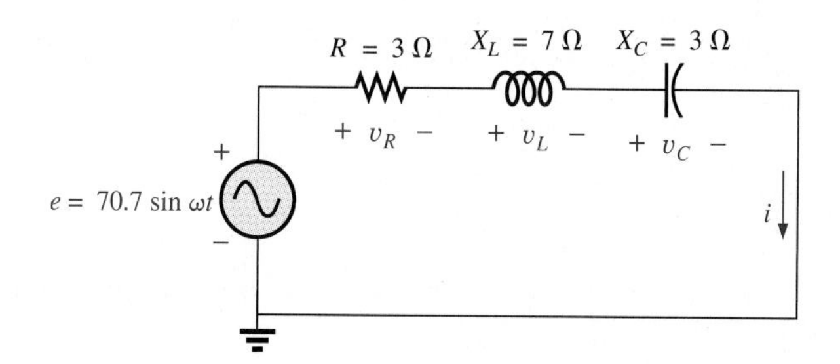

FIG. 4.76
Series RLC ac circuit.

These values were then entered into the schematic as shown in Fig. 4.77 using the same procedure described numerous times in the text. The ac source chosen was **VSIN** with the following settings made in sequential order: **DC** = 0 V, **AC** = 70.70 A, **VOFF** = 0 V, **VAMPL** = 70.70 A, **FREQ** = 1 kHz, **PHASE** = 0.

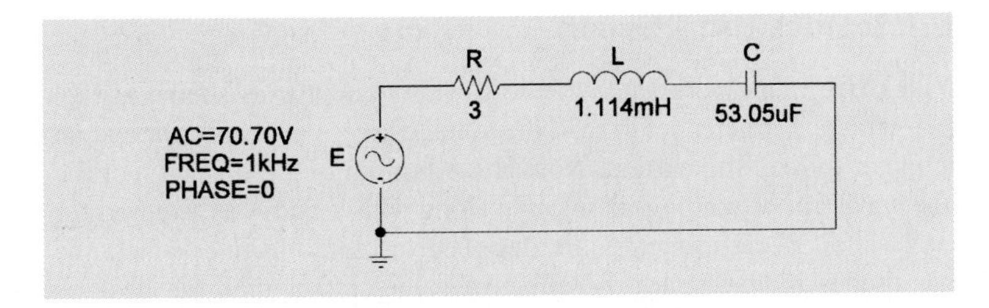

FIG. 4.77
Series RLC ac circuit to be analyzed using PSpice (Windows).

Using PSpice we can plot the waveform for each voltage and the current on the same graph, as shown in Fig. 4.78. The peak value of the current is 14.14 A; for the voltage v_R, 42.42 V; for v_L, 98.98 V; and for v_C, 42.42 V. In addition, note that v_R and i are in phase, v_L leads i by 90°, and v_C leads i by 90°. The applied voltage e leads that current i by 53.13°.

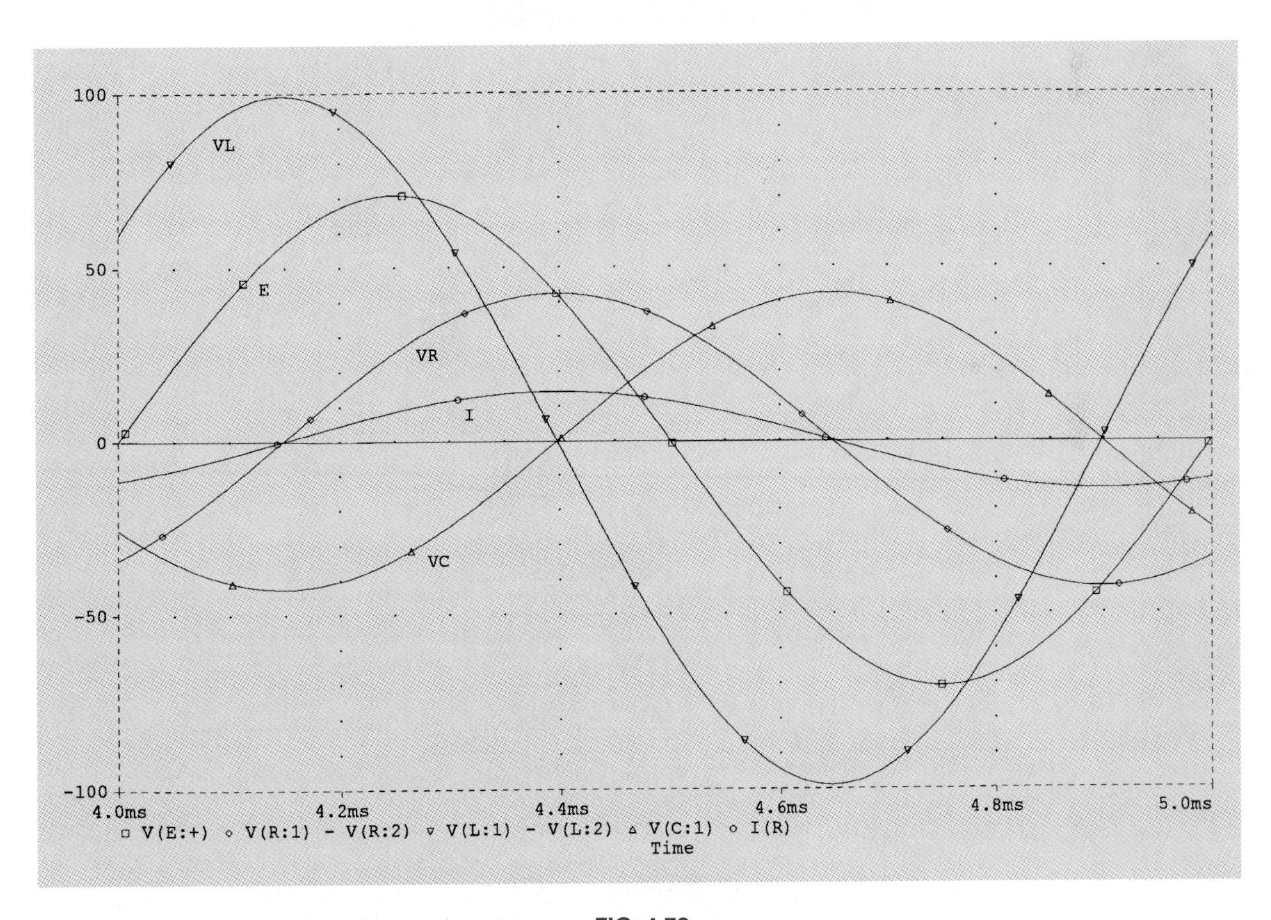

FIG. 4.78
A plot of all the voltages and current of interest for the circuit of Fig. 4.76.

Electronics Workbench

With Electronics Workbench the network will appear as shown in Fig. 4.79 with a voltmeter set to read the voltage across the capacitor and an ammeter to read the current. Note at the bottom of the display that the sine wave (for ac reading) is selected along with V and A as required for each meter. Because meters are designed to read effective values, the magnitudes indicated are the effective values rather than the peak as shown in Fig. 4.78.

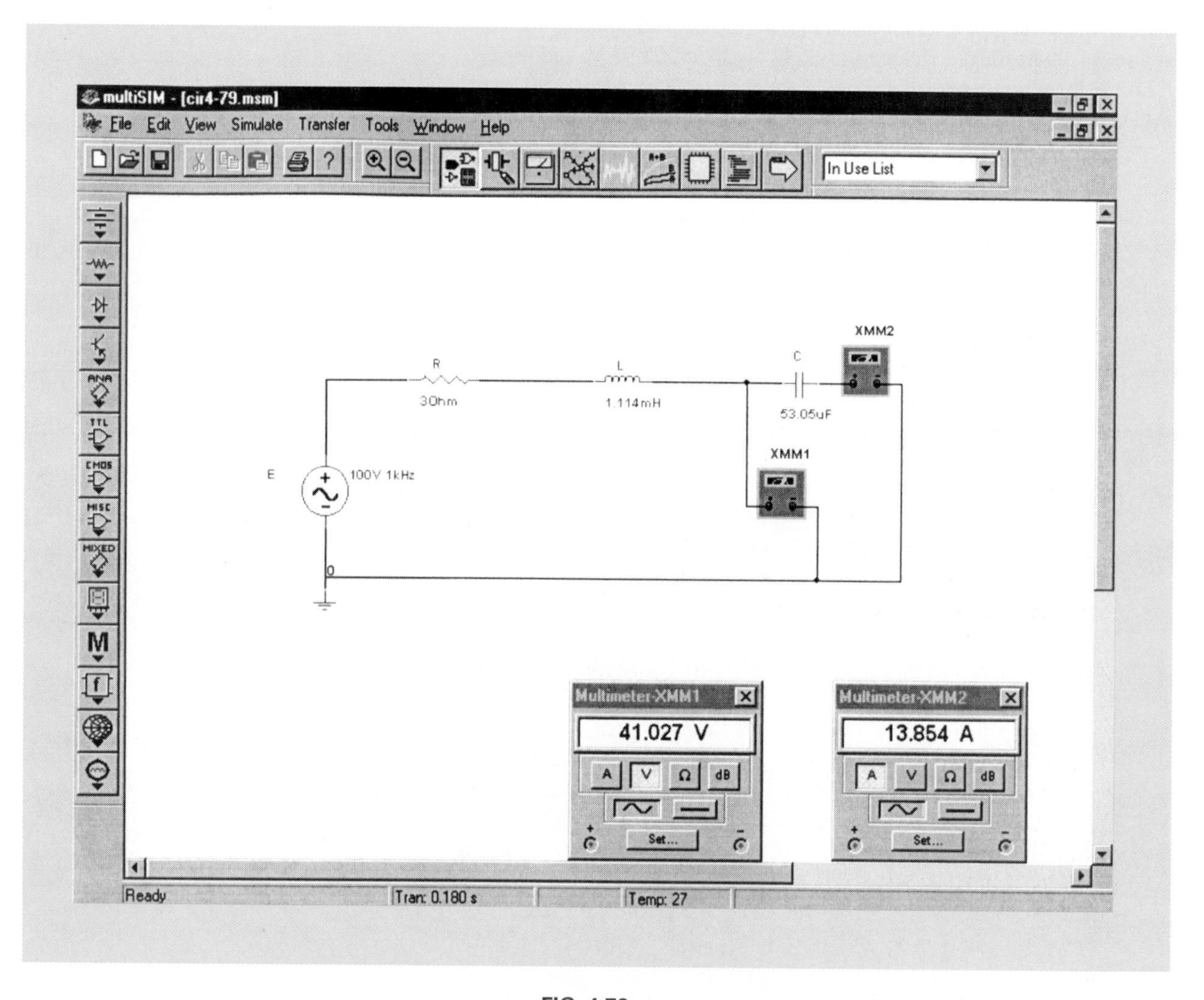

FIG. 4.79
Applying Electronics Workbench to the circuit of Fig. 4.76 to determine the current and voltage across the capacitor.

4.15 SUMMARY

- The **frequency** of a sinusoidal ac waveform is the **number of cycles** that appear in **1 s.**
- The **period** of a sinusoidal ac waveform is the time required to **complete one cycle** of the waveform measured from any point on the waveform.
- The **phase relationship** between two waveforms is determined by the **angle between corresponding slopes** of the two waveforms.
- The **vertical scale** of an oscilloscope is set in units of **voltage,** and the **horizontal axis** is scaled in units of **time.**
- The **effective value** of a sine wave is the **equivalent dc value.**
- The **average value** of a **sinusoidal waveform** over one full period is **zero.**
- The **voltage and current** of a **resistive** element are **in phase.**
- The **reactance** of an inductor is **directly** related to the applied **frequency** and the **inductance** level.
- The **voltage** across a coil **leads** the **current** through the coil by 90°.
- The **average power** delivered to a pure **inductor** or **capacitor** over one full cycle of the applied signal is always **0 W**.
- The **reactance** of a capacitor is **inversely** related to the applied **frequency** and the **capacitance** level.
- The **current** of a capacitive element **leads** the **voltage** across the capacitor by 90°.
- The **power factor** of a network is an immediate indication of the terminal impedance of a network. The closer the value to **1** the more **resistive,** and the closer to **0** the more **reactive.**
- **Phasors** are a **representation** of sinusoidal voltages or currents that permit a path of investigation of ac networks that parallels that of dc networks.
- In general, **complex numbers** should be **added** and **subtracted** in the **rectangular form** and multiplied or divided in the polar form.
- **Calculators** such as the TI-86 and **software packages** such as MathCad are extremely **useful tools** for working with complex numbers.
- **Ohm's law and the general laws, rules, and equations** associated with series and parallel dc networks **can also be applied** to ac networks.
- A **lagging** network is one in which the input **current lags** the applied **voltage,** and a **leading** network is one in which the input **current leads** the applied **voltage.**
- As the **frequency increases,** the **reactance of inductive elements increases,** and the **reactance of capacitive elements decreases.**
- A **short circuit** is a **direct connection** of very low resistance between two points in a network. If unexpected, it can have a devastating effect on the behavior of the network.
- When an **oscilloscope** is used for phase measurements all the supplies, meters, and the oscilloscope must share a **common ground.**

Equations:

$$v = V_{\text{peak}} \sin \theta = V_{\text{peak}} \sin \omega t$$

$$T = \frac{1}{f} \quad \text{(s)}$$

$$\omega = \frac{2\pi}{T} = 2\pi f \qquad \text{(rad/s)}$$

$$\text{Degrees} = \frac{180^\circ}{\pi}\text{(radians)}$$

$$V_{\text{eff}} = \frac{1}{\sqrt{2}} V_{peak} = 0.707\, V_{\text{peak}}$$

$$G = \frac{\text{Area (algebraic sum)}}{T\ \text{(period)}}$$

$$I_{\text{peak}} = \frac{V_{\text{peak}}}{R} = \frac{V_{\text{peak}}}{X_L} = \frac{V_{\text{peak}}}{X_C}$$

$$P_R = I^2 R = \frac{V^2}{R} = VI$$

$$X_L = \omega L = 2\pi f L \quad (\Omega)$$

$$X_C = \frac{1}{\omega C} = \frac{1}{2\pi f C} \quad (\Omega)$$

$$F_p = \cos \theta = \frac{R}{Z_T} = \frac{G}{Y_T}$$

$$\mathbf{Z}_R = R\angle 0^\circ,\ \mathbf{Z}_L = X_L\angle 90^\circ,\ \mathbf{Z}_C = X_C\angle -90^\circ$$

Series Circuits:

$$\mathbf{Z}_T = \mathbf{Z}_1 + \mathbf{Z}_2 + \mathbf{Z}_3 + \cdot \cdot \cdot + \mathbf{Z}_N$$

$$\mathbf{I}_T = \frac{\mathbf{E}}{\mathbf{Z}_T}$$

$$\mathbf{V}_R = \mathbf{I}\mathbf{Z}_R,\ \mathbf{V}_L = \mathbf{I}\mathbf{Z}_L,\ \mathbf{V}_C = \mathbf{I}\mathbf{Z}_C$$

$$P_T = E_T I_T \cos \theta = I_T^2 R = V_R I_R = \frac{{V_R}^2}{R}$$

$$\mathbf{E} = \mathbf{V}_1 + \mathbf{V}_2 + \mathbf{V}_3 + \cdot \cdot \cdot + \mathbf{V}_N$$

Parallel Networks:

$$\mathbf{Y}_T = \mathbf{Y}_1 + \mathbf{Y}_2 + \mathbf{Y}_3 + \cdot \cdot \cdot + \mathbf{Y}_N$$

$$\frac{1}{\mathbf{Z}_T} = \frac{1}{\mathbf{Z}_1} + \frac{1}{\mathbf{Z}_2} + \frac{1}{\mathbf{Z}_3} + \cdot \cdot \cdot + \frac{1}{\mathbf{Z}_N},\ \mathbf{Y}_T = \frac{1}{\mathbf{Z}_T}$$

$$\mathbf{Z}_T = \frac{\mathbf{Z}_1 \cdot \mathbf{Z}_2}{\mathbf{Z}_1 + \mathbf{Z}_2}$$

$$\mathbf{I}_T = \frac{\mathbf{E}}{\mathbf{Z}_T} = \mathbf{E}\mathbf{Y}_T$$

$$\mathbf{I}_R = \frac{\mathbf{E}}{\mathbf{Z}_R},\ \mathbf{I}_L = \frac{\mathbf{E}}{\mathbf{Z}_L},\ \mathbf{I}_C = \frac{\mathbf{E}}{\mathbf{Z}_C}$$

$$P_T = E_T I_T \cos \theta = \frac{E^2}{R} = E I_R = {I_R}^2 R$$

PROBLEMS

SECTION 4.2 Sinusoidal (Ac) Waveform

1. Sketch the following waveforms.

a. $v = 110 \sin \theta$ **b.** $i = -6 \times 10^{-3} \sin \theta$

2. Sketch the following waveform versus θ, radians, and time.

$$v = 0.8 \sin 200t$$

3. Sketch the following waveform versus θ, radians, and time.

$$i = 0.02 \sin(1000t + 30°)$$

4. How long will it take the following waveform to pass through 10 cycles?

$$v = 120 \sin(2 \times 10^3 t - 30°)$$

5. At what instant of time will the following waveform be 6 V? Associate $t = 0$ s with θ of $\sin \theta$ equal to 0°.

$$v = 12 \sin(200t + 30°)$$

6. a. At what angle θ (closest to 0°) will the following waveform reach 3 mA?

$$i = 8.6 \times 10^{-3} \sin 500t$$

b. At what time t will it occur?

7. What is the phase relationship between the following pairs of waveforms?

a. $v = 12 \sin(400t - 72°)$
$i = 0.4 \sin(400t - 16°)$

b. $v = 0.05 \sin(\omega t - 120°)$
$i = 5 \times 10^{-6} \sin(\omega t + 20°)$

8. Write the sinusoidal expression for a current i that has a peak value of 6 μA and leads the following voltage by 40°.

$$v = 16 \sin(1000t + 6°)$$

9. Write the sinusoidal expression for a voltage v that has a peak value of 48 mV and lags the following current i by 60°.

$$i = 4 \times 10^{-3} \sin(\omega t - 30°)$$

SECTION 4.4 Effective (rms) Value

10. Determine the effective value of each of the following.

a. $v = 64 \sin(5 \times 20^3\, t + 30°)$

b. $i = 6 \times 10^{-3} \sin(\omega t - 20°)$

c. $V_{dc} = 20$ V

11. Write the sinusoidal expression for each quantity using the information provided.

a. $I_{eff} = 36$ mA, $f = 1$ kHz, phase angle $= 60°$

b. $V_{eff} = 8$ V, $f = 60$ Hz, phase angle $= -10°$

SECTION 4.5 Average Values

12. Determine the average value of the following.

a. $v = 96 \sin(1000t + 60°)$

b. $V_{dc} = 12.2$ V

c. $i = 5 + 6 \sin \omega t$

d. $v = 0.6 + 2.8 \sin(400t - 30°)$

e. $i = 10 \sin \omega t + 4 \sin \omega t$

13. Determine the average value of the waveform in Fig. 4.80.

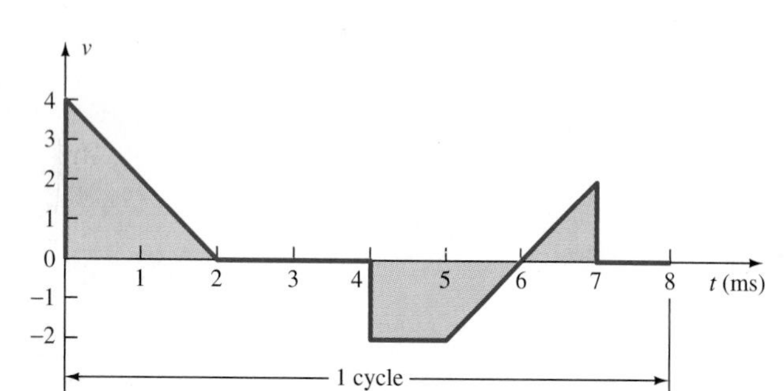

FIG. 4.80

14. Determine the average value of the waveform in Fig. 4.81.

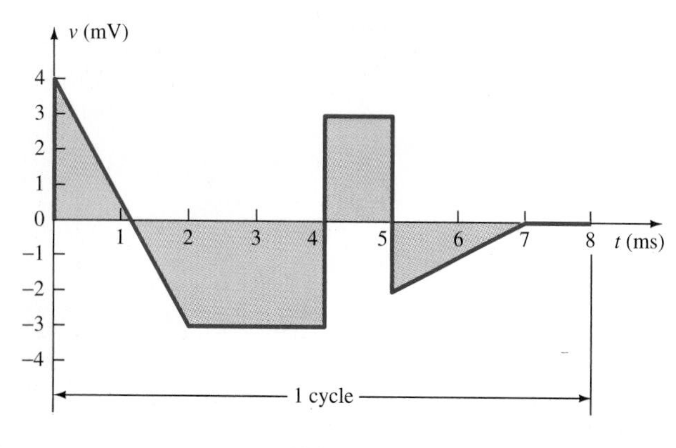

FIG. 4.81

15. Determine the average value of i^2 from $\theta = 0$ to π if $i = 6 \sin \theta$ (integral calculus required).

SECTION 4.7 The *R*, *L*, and *C* Elements

16. a. Determine the sinusoidal expression for the voltage drop across a 1.2-kΩ resistor if the current i_R is $8 \times 10^{-3} \sin 200t$.

b. Find the power delivered to the resistor.

c. What is the power factor of the load?

17. a. Find the sinusoidal expression for the current through a 2.2-kΩ resistor if the power delivered to the resistor is 3.6 W at a frequency of 1000 Hz.

b. Find the sinusoidal expression for the voltage across the resistor.

18. a. Find the sinusoidal expression for the voltage drop across a 20-mH coil if the current i_L is $4 \sin(500t + 60°)$.

b. Find the power delivered to the coil.

c. What is the power factor of the load?

19. Determine the sinusoidal expression for the current i_C of a 10-μF capacitor if the voltage across the capacitor is $v_C = 20 \times 10^{-3} \sin(2000t + 30°)$

20. a. For the following pairs determine whether the element is a resistor, inductor, or capacitor.

b. Determine the resistance, inductance, or capacitance.

1. $v = 16 \sin(200t + 80°)$
   $i = 0.04 \sin(200t - 10°)$
2. $v = 0.12 \sin(1000t + 10°)$
   $i = 6\ 3\ 10^{-3} \cos(1000t + 10°)$

21. a. For the following pairs determine the power delivered to the load.

b. Find the power factor and indicate whether it is inductive or capacitive.

1. $v = 1600 \sin(377t + 360°)$
   $i = 0.8 \sin(377t + 60°)$
2. $v = 100 \sin(10^6 t - 10°)$
   $i = 0.2 \sin(10^6 t - 40°)$

SECTION 4.8 Phasors and Complex Numbers

22. Convert the following to the other domain.

a. $40\angle 45°$

b. $0.5\angle 120°$

c. $6400\angle -90°$

d. $4 - j10$

e. $0.3 + j0.8$

f. $-0.1 + j8.6$

23. Perform the following operations. State your answer in polar form.

a. $(60\angle 40°) - (6 + j14)$

b. $(1 - j6) + (5\angle 30°)$

c. $(10\angle 60°)(4 + j12)$

d. $(100 + j300)/(20 + j60)$

e. $(6 + j6)/(40\angle 10° + 10\angle +30°)$

f. $(0.002\angle 90°)(4 \times 10^3\angle +60°)(30 - j30)$

24. Using phasor notation, determine the voltage (in the time domain) across a 2.2-kΩ resistor if the current through the resistor is $i = 20 \times 10^{-3} \sin(400t + 30°)$.

25. Using phasor notation, determine the current (in the time domain) through a 20-mH coil if the voltage across the coil is $v_L = 4 \sin(1000t + 10°)$.

26. Using phasor notation, determine the voltage (in the time domain) across a 10-μF capacitor if the current $i_C = 40 \times 10^{-3} \sin(10t + 40°)$.

27. For the system in Fig. 4.82, determine the voltage v_1 in the time domain.

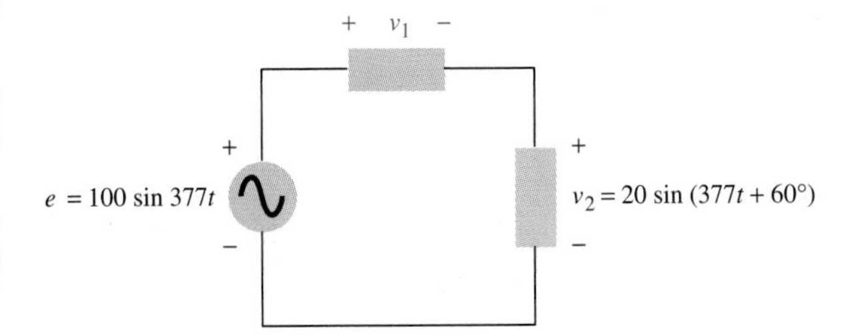

FIG. 4.82

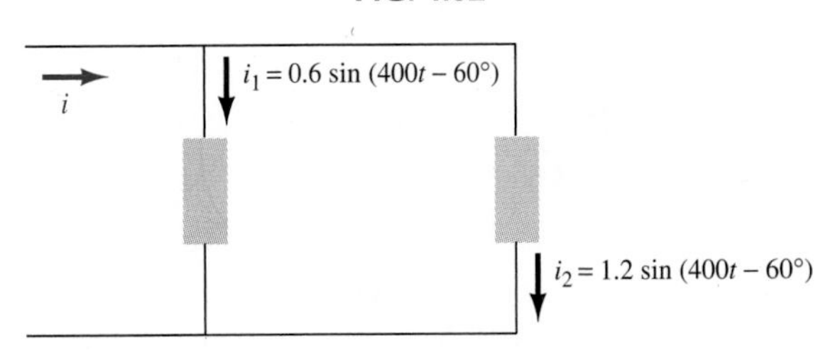

FIG. 4.83

28. For the system in Fig. 4.83, determine the current i in the time domain.

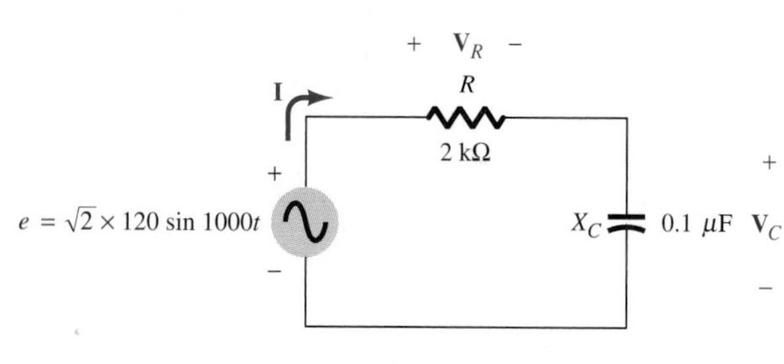

FIG. 4.84

SECTION 4.9 Series Ac Networks

29. For the series ac network in Fig. 4.84, determine:

a. The reactance of the capacitor.
b. The total impedance and the impedance diagram.
c. The current **I**.
d. The voltages $\mathbf{V}_R$ and $\mathbf{V}_C$ using Ohm's law.
e. The voltages $\mathbf{V}_R$ and $\mathbf{V}_C$ using the voltage-divider rule.
f. The power to R.
g. The power supplied by the voltage source e.
h. The phasor diagram.
i. The F_p of the network.
j. The current and voltages in the time domain.

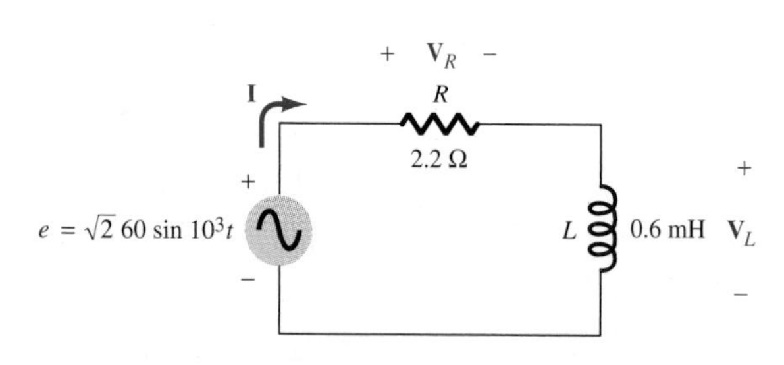

FIG. 4.85

30. Repeat Problem 29 for the network in Fig. 4.85, after making the appropriate changes in parts (a), (d) and (e).

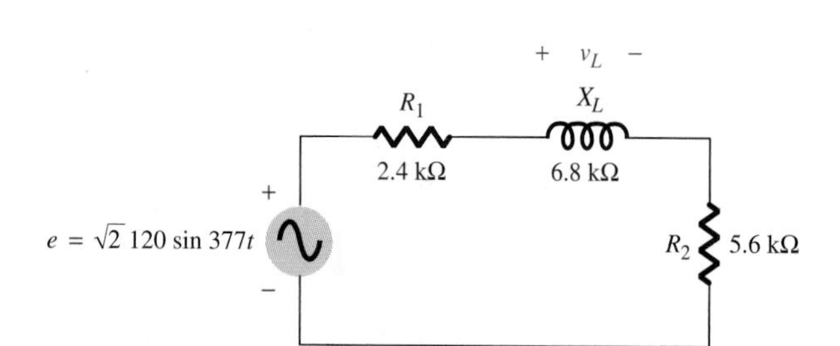

FIG. 4.86

31. Determine the voltage v_L (in the time domain) for the network in Fig. 4.86 using the voltage-divider rule.

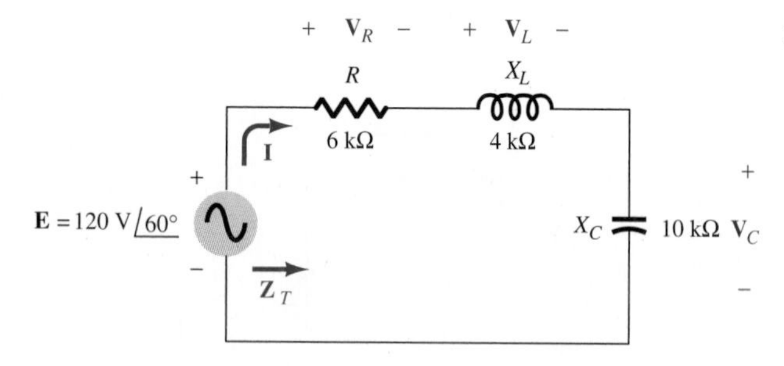

FIG. 4.87

32. For the series *RLC* network in Fig. 4.87, determine:

a. $\mathbf{Z}_T$.
b. **I**.
c. $\mathbf{V}_R$, $\mathbf{V}_L$, $\mathbf{V}_C$ using Ohm's law.
d. $\mathbf{V}_L$ using the voltage-divider rule.
e. The power to R.
f. The F_p.
g. The phasor and impedance diagrams.

33. Determine the voltage $\mathbf{V}_C$ for the network in Fig. 4.88 using the voltage-divider rule.

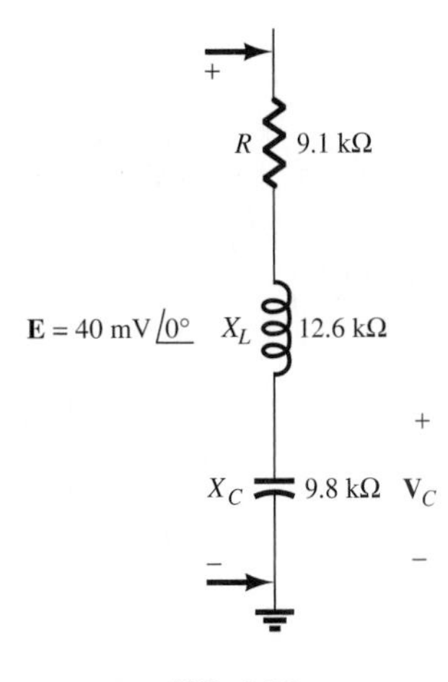

FIG. 4.88

SECTION 4.10 Parallel Ac Networks

34. For the parallel *RC* network in Fig. 4.89, determine:

a. The admittance diagram.
b. $\mathbf{Y}_T$, $\mathbf{Z}_T$.
c. **I**.
d. $\mathbf{I}_R$, $\mathbf{I}_C$ using Ohm's law.
e. The total power delivered to the network.
f. The power factor of the network.
g. The admittance diagram.

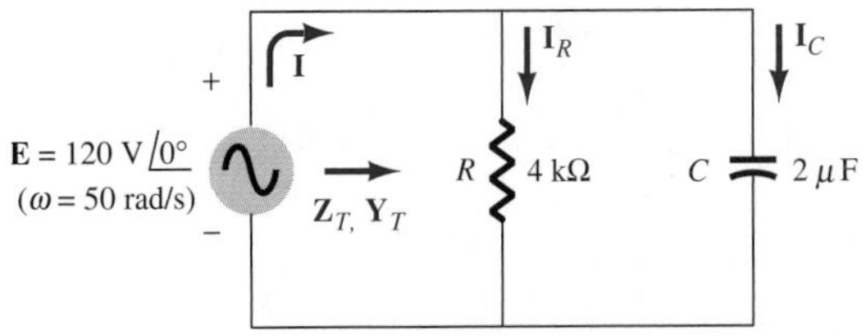

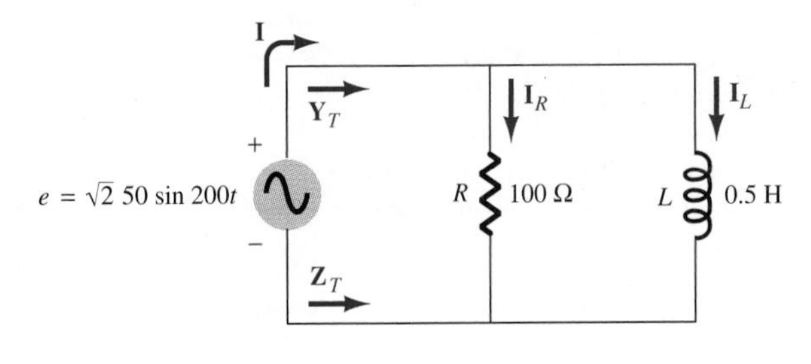

FIG. 4.89

35. Repeat Problem 34 for the network in Fig. 4.90, replacing $\mathbf{I}_C$ with $\mathbf{I}_L$ in part (d).

FIG. 4.90

36. Find the currents $\mathbf{I}_1$ and $\mathbf{I}_2$ in Fig. 4.91 using the current-divider rule. If necessary, review Section 2.10.

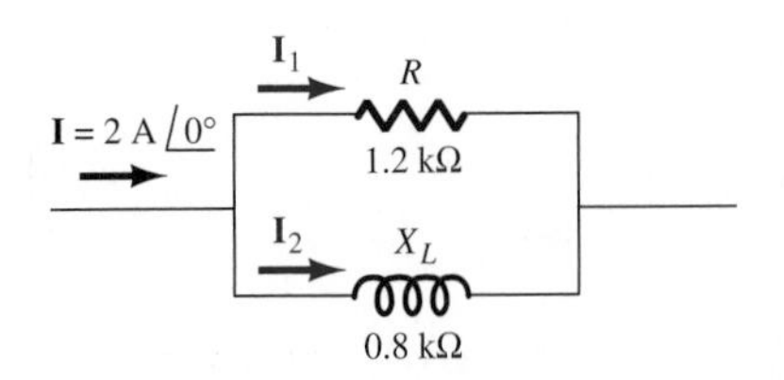

FIG. 4.91

37. For the parallel RLC network in Fig. 4.92, determine:
   a. The admittance diagram.
   b. $\mathbf{Y}_T$, $\mathbf{Z}_T$.
   c. $\mathbf{I}$, $\mathbf{I}_R$, $\mathbf{I}_L$, and $\mathbf{I}_C$.
   d. The total delivered power.
   e. The power factor of the network.
   f. The sinusoidal format of $\mathbf{I}$, $\mathbf{I}_R$, $\mathbf{I}_L$, and $\mathbf{I}_C$.
   g. The phase relationship between e and i_L.

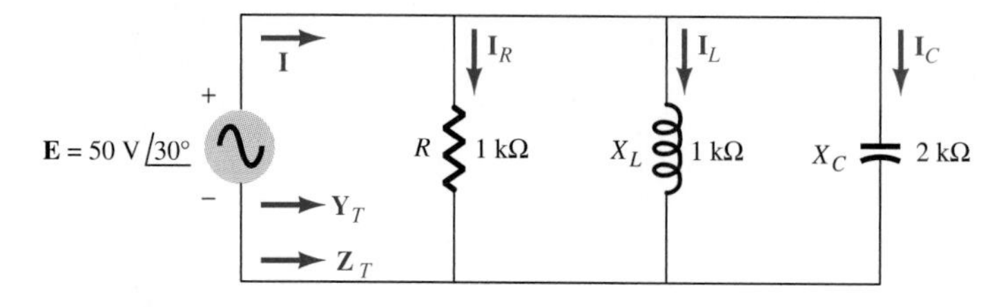

FIG. 4.92

SECTION 4.11 Short Circuits and Open Circuits

38. For the network in Fig. 4.93, determine:
   a. The short-circuit currents I_1 and I_2.
   b. The voltages V_1 and V_2.
   c. The source current I.

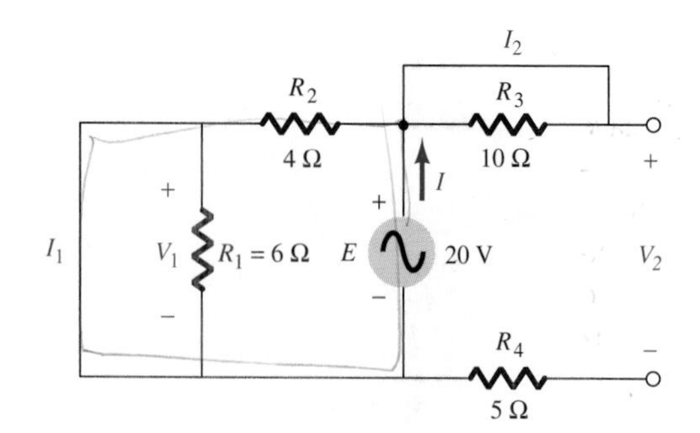

FIG. 4.93

39. Determine the current I and the voltage V for the network in Fig. 4.94.

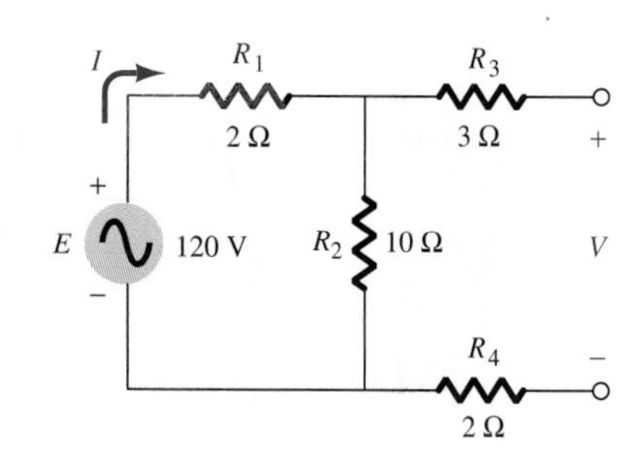

FIG. 4.94

5 Ac Network Theorems, Polyphase Systems, and Resonance

5.1 INTRODUCTION

Now that the fundamental laws of series and parallel ac networks have been described in some detail, series–parallel networks can be examined. The approach is the same as that applied to dc networks with the major difference being the need to work with complex numbers throughout. To this end, the calculator and computer methods such as MathCad can be very helpful because the resulting equations can be more complex than those encountered for series or parallel ac networks.

An investigation of the same theorems introduced for dc networks follows. Again, the only major difference is the need to work with complex numbers throughout the analysis—the application of each theorem is essentially the same.

Polyphase systems and their use in the distribution of power will then be introduced, followed by series and parallel resonant networks and their ability to define particular frequency bands as those of primary interest.

5.2 SERIES–PARALLEL AC NETWORKS

As pointed out for dc networks, there exists an infinite variety of series–parallel configurations. The approach is one that will develop primarily from exposure to a number and variety of problems. However, in the early stages one can first look for the obvious series or parallel combinations and work toward the total impedance or admittance of the system and then find the source current or voltage. It is not always necessary to work back toward the source, but in the majority of problems it is the path to follow.

We use the techniques developed for dc systems simply by changing resistances to impedance levels and maintaining an awareness of phasor relationships. We continue to use the block diagram technique for clarity and efficiency.

The first network to be analyzed is equivalent to the first two loops of the *ladder* network in Fig. 5.1(b). A reactive element appears in both

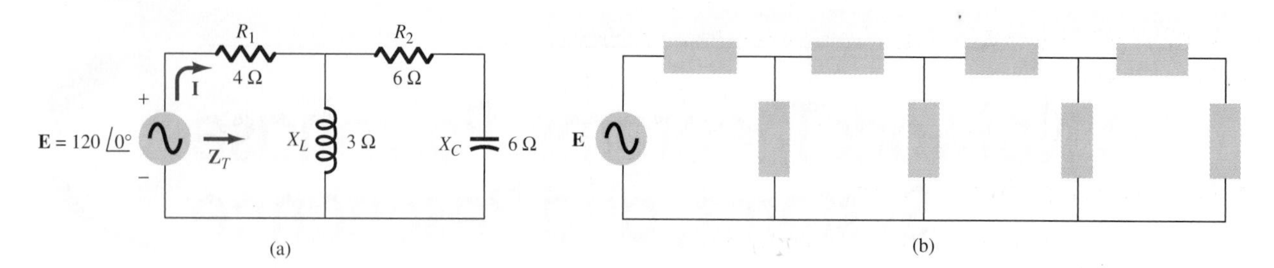

FIG. 5.1
Series–parallel ac networks: (a) two-loop configuration; (b) ladder network.

vertical lags in Fig. 5.1(a), with resistive elements on the horizontal. Working toward the source, we find that R_2 and X_C are in series and that the resultant impedance is in parallel with X_L. The parallel combination is then in series with R_1, permitting a determination of $\mathbf{Z}_T$, $\mathbf{I}$, and so on.

By employing the block impedance approach, we redraw the network in Fig. 5.2 with the following defined impedances:

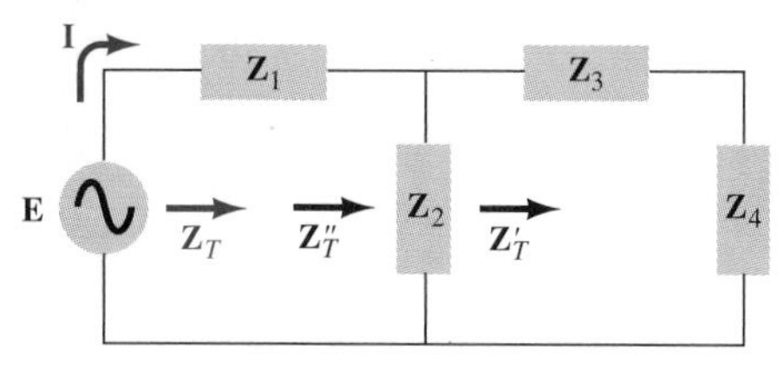

FIG. 5.2

$$\mathbf{Z}_1 = R_1\angle 0^\circ = 4\ \Omega\angle 0^\circ$$
$$\mathbf{Z}_2 = X_L\angle 90^\circ = 3\ \Omega\angle 90^\circ$$
$$\mathbf{Z}_3 = R_2\angle 0^\circ = 6\ \Omega\angle 0^\circ$$
$$\mathbf{Z}_4 = X_C\angle -90^\circ = 6\ \Omega\angle -90^\circ$$

First, we determine the series combination of $\mathbf{Z}_3$ and $\mathbf{Z}_4$:

$$\mathbf{Z}'_T = \mathbf{Z}_3 + \mathbf{Z}_4 = (6\ \Omega + j0) + (0 - j6\ \Omega) = 6\ \Omega - j6\ \Omega = 8.485\ \Omega\angle -45^\circ$$

This total impedance is in parallel with $\mathbf{Z}_2$, and

$$\mathbf{Z}''_T = \mathbf{Z}_2 \| \mathbf{Z}'_T = \frac{\mathbf{Z}_2\mathbf{Z}'_T}{\mathbf{Z}_2 + \mathbf{Z}'_T} = \frac{(3\ \Omega\angle 90^\circ)(8.485\ \Omega\angle -45^\circ)}{(0 + j3\ \Omega) + (6\ \Omega - j6\ \Omega)}$$
$$= \frac{25.455\ \Omega\angle 45^\circ}{6 - j3} = \frac{25.455\ \Omega\angle 45^\circ}{6.708\angle -26.57^\circ}$$
$$= 3.795\ \Omega\angle 71.57^\circ = 1.2\ \Omega + j3.6\ \Omega$$
$$\mathbf{Z}_T = \mathbf{Z}_1 + \mathbf{Z}''_T = (4\ \Omega + j0) + (1.2\ \Omega + j3.6\ \Omega)$$
$$= 5.2\ \Omega + j3.6\ \Omega = 6.32\ \Omega\angle 34.7^\circ$$

and

$$\mathbf{I} = \frac{\mathbf{E}}{\mathbf{Z}_T} = \frac{120\ \text{V}\angle 0^\circ}{6.32\ \Omega\angle 34.7^\circ} = 18.99\ \text{A}\angle -34.7^\circ$$

which in the time domain is

$$i = \sqrt{2}(18.99)\sin(\omega t - 34.7^\circ)$$
$$= \mathbf{26.85\sin(\omega t - 34.7^\circ)}$$

If the voltage across $\mathbf{Z}_4$ was desired, we would have to work back to that load element. That is,

$$\mathbf{V}_{Z_2} = \mathbf{I}\mathbf{Z}''_T = (18.99\ \text{A}\angle{-34.7^\circ})(3.795\ \Omega\angle{71.57^\circ}) = 72.07\ \text{V}\angle{36.87^\circ}$$

and using the voltage-divider rule, we obtain

$$\mathbf{V}_4 = \frac{\mathbf{Z}_4(\mathbf{V}_{Z_2})}{\mathbf{Z}_4 + \mathbf{Z}_3} = \frac{(6\ \Omega\angle{-90^\circ})(72.07\ \text{V}\angle{36.87^\circ})}{-j6\ \Omega + 6\ \Omega}$$
$$= \frac{432.42\ \text{V}\angle{-53.13^\circ}}{8.49\angle{-45^\circ}}$$
$$= \mathbf{50.93\ V\angle{-8.13^\circ}}$$

EXAMPLE 5.1 For the network in Fig. 5.3, determine:

a. $\mathbf{I}_C$.
b. $\mathbf{V}_s$.
c. $\mathbf{V}_L$.

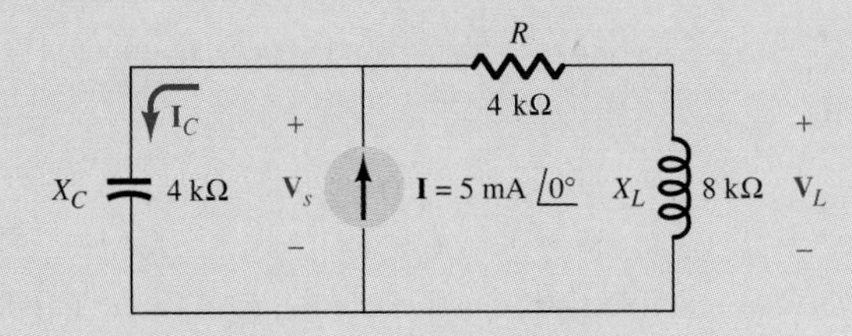

FIG. 5.3

Solution:

a. The block diagram equivalent appears in Fig. 5.4. Using the current-divider rule, we have

$$\mathbf{I}_C = \frac{\mathbf{Z}_2\mathbf{I}}{\mathbf{Z}_1 + \mathbf{Z}_2} = \frac{(8.94\ \text{k}\Omega\angle{63.43^\circ})(5\ \text{mA}\angle{0^\circ})}{-j4\ \text{k}\Omega + 4\ \text{k}\Omega + j8\ \text{k}\Omega}$$
$$= \frac{44.7\angle{63.43^\circ}}{4 \times 10^3 + j4 \times 10^3} = \frac{44.7\angle{63.43^\circ}}{5.657 \times 10^3\angle{45^\circ}} = \mathbf{7.9\ mA\angle{18.43^\circ}}$$

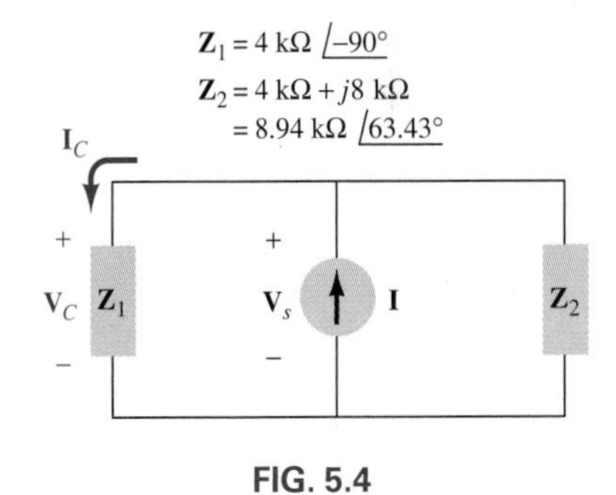

FIG. 5.4

The TI-86 calculator would give the following result (Fig. 5.5) for the current $\mathbf{I}_C$:

```
((8.94E3∠ 63.43) * (5E – 3∠ 0))/((0, –4E3) + (4E3,8E3)) ▶ POL  (ENTER)
(7.902E – 3∠ 18.430EO)
```

FIG. 5.5
Determining I_C for the network of Fig. 5.3 using the TI-86 calculator.

Using MathCad for the same operation, we would obtain the following result (Fig. 5.6) for the current $\mathbf{I}_C$:

$$Ic := \frac{(4000 + 8000j)\,(5 \cdot 10^{-3})}{-4000j + 4000 + 8000j} \qquad Ic = 7.5 \bullet 10^{-3} + 2.5 \bullet 10^{-3}\,j$$

$$|Ic| = 7.906 \bullet 10^{-3} \quad arg(\,Ic\,) = 18.435\,°$$

FIG. 5.6
Using MathCad to determine $\mathbf{I}_C$ for the network of Fig. 5.3.

b. Since $\mathbf{V}_s = \mathbf{V}_C$,

$$\mathbf{V}_s = \mathbf{I}_C\mathbf{X}_C = (7.9\text{ mA}\angle 18.43°)(4\text{ k}\Omega\angle -90°)$$
$$= \mathbf{31.6\text{ V}\angle -71.57°}$$

c. Since $\mathbf{V}_s$ is the total voltage across the series RL combination, the voltage-divider rule can be employed to determine $\mathbf{V}_L$. That is,

$$\mathbf{V}_L = \frac{\mathbf{Z}_L\mathbf{V}_s}{\mathbf{Z}_R + \mathbf{Z}_L} = \frac{(8\text{ k}\Omega\angle 90°)(31.6\text{ V}\angle -71.57°)}{4\text{ k}\Omega + j8\text{ k}\Omega}$$
$$= \frac{252.8 \times 10^3\angle 18.43°}{8.94 \times 10^3\angle 63.43°} = \mathbf{28.28\text{ V}\angle -45°}$$

5.3 MULTISOURCE AC NETWORKS

Once the various reactive and resistive elements of an ac network have been replaced by their block impedances, the superposition, branch-current mesh, and nodal methods introduced for dc networks can be applied to ac systems with a minimum level of confusion and difficulty. Recall that both methods permit the analysis of networks with two or more sources that are not in series or in parallel.

Superposition

The superposition theorem permits an analysis of the effect of each source on a particular current or voltage of the network. The resultant current or voltage is the algebraic sum of the contribution of each source.

EXAMPLE 5.2 Determine the current $\mathbf{I}_1$ for the network in Fig. 5.7.

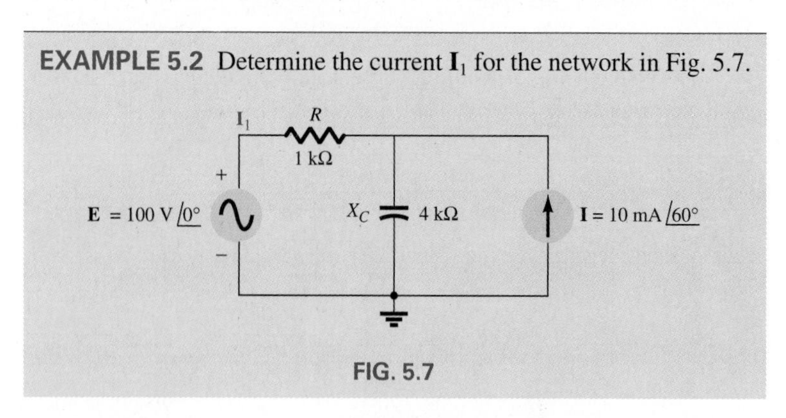

FIG. 5.7

Solution: The block impedances have been assigned in Fig. 5.8. Considering the effects of the voltage source results in the configuration in Fig. 5.9 in which the current source has been replaced by its open-circuit equivalent.

$\mathbf{Z}_1 = 1\text{ k}\Omega\angle 0°$
$\mathbf{Z}_2 = 4\text{ k}\Omega\angle -90°$

FIG. 5.8

The current $\mathbf{I}'_1$ is

$$\mathbf{I}'_1 = \frac{\mathbf{E}}{\mathbf{Z}_1 + \mathbf{Z}_2}$$
$$= \frac{100\text{ V}\angle 0°}{1\text{ k}\Omega - j4\text{ k}\Omega} = \frac{100\text{ V}\angle 0°}{4.123\text{ k}\Omega\angle -75.963°}$$
$$= 24.254\text{ mA}\angle 75.963°$$

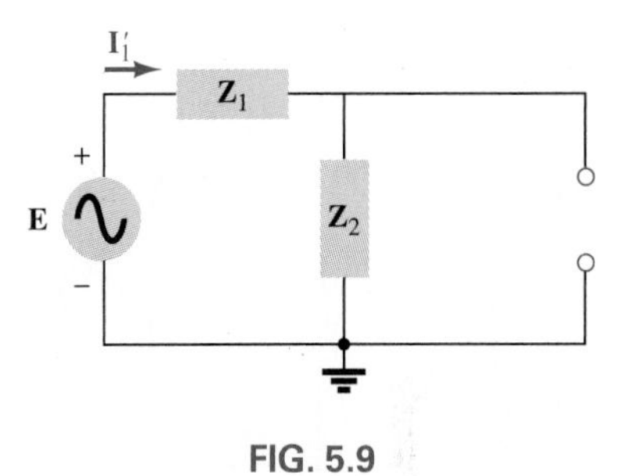

FIG. 5.9

For the current source the configuration in Fig. 5.10 results, in which the voltage source has been replaced by a short circuit. Application of the current-divider rule yields

$$\mathbf{I}''_1 = \frac{\mathbf{Z}_2(\mathbf{I})}{\mathbf{Z}_1 + \mathbf{Z}_2}$$
$$= \frac{(4\text{ k}\Omega\angle -90°)(10\text{ mA}\angle 60°)}{-j4\text{ k}\Omega + 1\text{ k}\Omega}$$
$$= \frac{40\text{ A}\angle -30°}{4.123 \times 10^3\angle -75.96°} = 9.7\text{ mA}\angle 45.96°$$

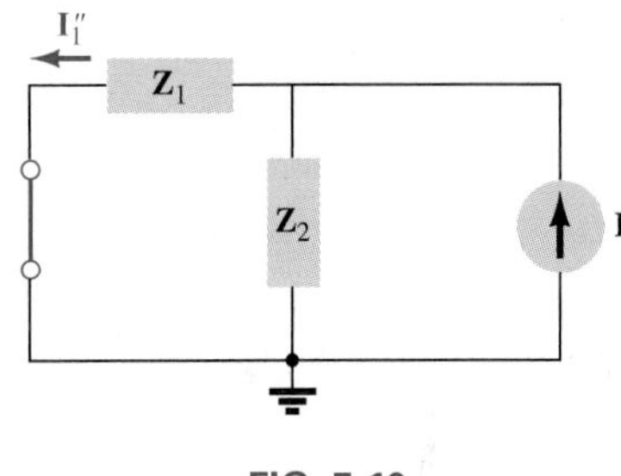

FIG. 5.10

Using $\mathbf{I}'_1$ as the desired direction the current $\mathbf{I}_1$ is defined by $\mathbf{I}_1 = \mathbf{I}'_1 - \mathbf{I}''_1$.

$$\mathbf{I}_1 = 24.254\text{ mA}\angle 75.963° - 9.7\text{ mA}\angle 45.96°$$
$$= (5.88 \times 10^{-3} + j23.53 \times 10^{-3})$$
$$- (6.743 \times 10^{-3} + j6.972 \times 10^{-3})$$
$$= -0.863\text{ mA} + j16.59\text{ mA}$$
$$= \mathbf{16.61\text{ mA}\angle 92.98°}$$

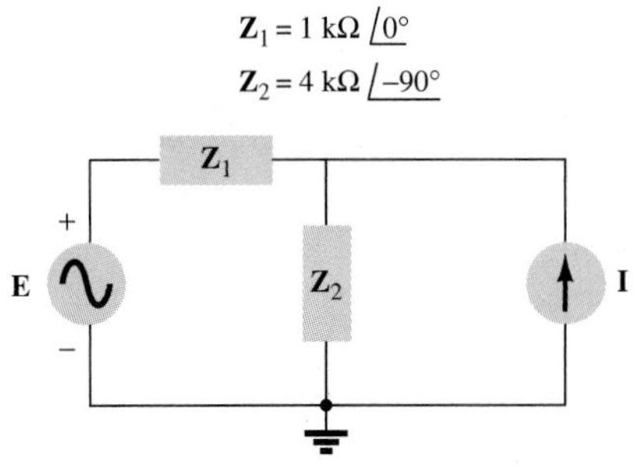

Branch-Current Analysis

Consider the network in Fig. 5.11 with two ac voltage sources that are not in series or parallel. Substitution of the block impedances results in

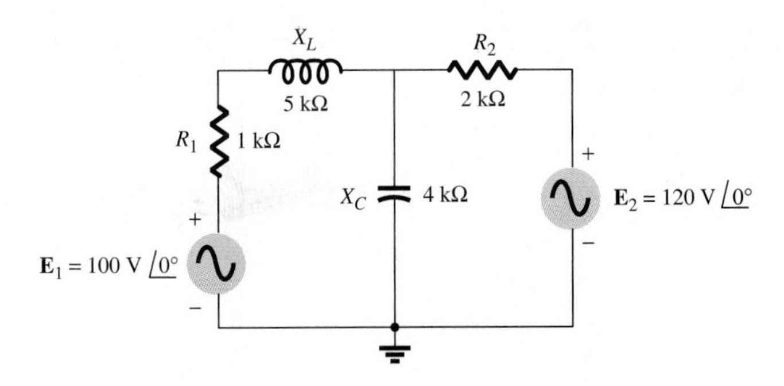

FIG. 5.11

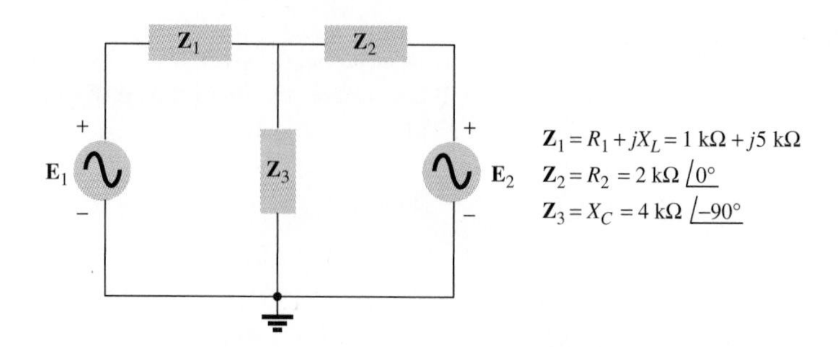

FIG. 5.12

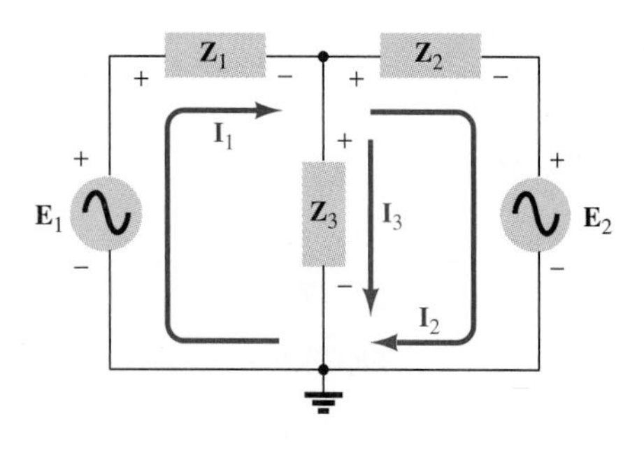

FIG. 5.13

the network in Fig. 5.12. A review of the parallel dc section may be necessary before continuing with this approach, since it is assumed that the general procedure is understood. Assigning the labeled branch currents an assumed direction and including the polarities gives us the configuration in Fig. 5.13.

Writing Kirchhoff's voltage law around each closed loop in the clockwise direction, we have

$$\mathbf{E}_1 - \mathbf{I}_1\mathbf{Z}_1 - \mathbf{I}_3\mathbf{Z}_3 = 0$$
$$\mathbf{I}_3\mathbf{Z}_3 - \mathbf{I}_2\mathbf{Z}_2 - \mathbf{E}_2 = 0$$

and Kirchhoff's current law at the top node:

$$\mathbf{I}_1 = \mathbf{I}_2 + \mathbf{I}_3$$

or

$$\mathbf{I}_3 = \mathbf{I}_1 - \mathbf{I}_2$$

Substitution for $\mathbf{I}_3$ in the first two equations gives

$$\mathbf{E}_1 - \mathbf{I}_1\mathbf{Z}_1 - (\mathbf{I}_1 - \mathbf{I}_2)\mathbf{Z}_3 = 0$$
$$(\mathbf{I}_1 - \mathbf{I}_2)\mathbf{Z}_3 - \mathbf{I}_2\mathbf{Z}_2 - \mathbf{E}_2 = 0$$

Multiplying through and rearranging terms, we have

$$(\mathbf{Z}_1 + \mathbf{Z}_3)\mathbf{I}_1 \qquad - \mathbf{Z}_3\mathbf{I}_2 = \mathbf{E}_1$$
$$\mathbf{Z}_3\mathbf{I}_1 - (\mathbf{Z}_2 + \mathbf{Z}_3)\mathbf{I}_2 = \mathbf{E}_2$$

Using determinants, we obtain

$$\frac{\begin{vmatrix} \mathbf{E}_1 & -\mathbf{Z}_3 \\ \mathbf{E}_2 & -(\mathbf{Z}_2 + \mathbf{Z}_3) \end{vmatrix}}{\begin{vmatrix} (\mathbf{Z}_1 + \mathbf{Z}_3) & -\mathbf{Z}_2 \\ \mathbf{Z}_3 & -(\mathbf{Z}_2 + \mathbf{Z}_3) \end{vmatrix}} = \frac{-\mathbf{E}_1(\mathbf{Z}_2 + \mathbf{Z}_3) + \mathbf{E}_2\mathbf{Z}_3}{-(\mathbf{Z}_1 + \mathbf{Z}_3)(\mathbf{Z}_2 + \mathbf{Z}_3) + \mathbf{Z}_2\mathbf{Z}_3}$$
$$= \frac{\mathbf{E}_1(\mathbf{Z}_2 + \mathbf{Z}_3) - \mathbf{E}_2\mathbf{Z}_3}{\mathbf{Z}_1\mathbf{Z}_2 + \mathbf{Z}_2\mathbf{Z}_3 + \mathbf{Z}_1\mathbf{Z}_3}$$

Substituting values, we have

$$\mathbf{I}_1 = \frac{(100\text{ V}\angle 0°)(2\text{ k}\Omega - j4\text{ k}\Omega) - (120\text{ V}\angle 0°)(4\text{ k}\Omega\angle -90°)}{(1\text{ k}\Omega + j5\text{ k}\Omega)(2\text{ k}\Omega\angle 0°) + (2\text{ k}\Omega\angle 0°)(4\text{ k}\Omega\angle -90°) + (1\text{ k}\Omega + j5\text{ k}\Omega)(4\text{ k}\Omega\angle -90°)}$$

$$= \frac{(100\angle 0°)(4.472 \times 10^3\angle -63.43°) - (480 \times 10^3\angle -90°)}{(5.1 \times 10^3\angle 78.69°)(2 \times 10^3\angle 0°) + 8 \times 10^6\angle -90° + (5.1 \times 10^3\angle 78.69°)(4 \times 10^3\angle -90°)}$$

$$= \frac{447.2 \times 10^3\angle -63.43° - 480 \times 10^3\angle -90°}{10.2 \times 10^6\angle 78.69° - j8 \times 10^6 + 20.4 \times 10^6\angle -11.31°}$$

$$= \frac{(200 \times 10^3 - j400 \times 10^3) - (-j480 \times 10^3)}{2 \times 10^6 + j10 \times 10^6 - j8 \times 10^6 + 20 \times 10^6 - j4 \times 10^6}$$

$$= \frac{200 \times 10^3 + j80 \times 10^3}{22 \times 10^6 - j2 \times 10^6} = \frac{215.4 \times 10^3\angle 21.8°}{22.09 \times 10^6\angle -5.19°}$$

$$= \mathbf{9.75\text{ mA}\angle 26.99°}$$

The TI-86 calculator gives the following result (Fig. 5.14) for $\mathbf{I}_1$:

```
((100∠0) * (2E3, –4E3) – (120∠0) * (4E3∠–90°))/((1E3, 5E3) * (2E3∠0)
+ (2E3∠0) * (4E3∠–90) + (1E3, 5E3) * (4E3∠–90°)) ► POL  (ENTER)

(9.751E – 3∠26.996EO)
```

FIG. 5.14
Using the TI-86 calculator to determine I_1 *for the network of Fig. 5.11.*

Using MathCad gives us the display shown in Fig. 5.15.

$$I1 := \frac{(100 + 0j)\,(2000 - 4000j) - (120 + 0j)\,(0 - 4000j)}{((1000 + 5000j)\,(2000 + 0j)\,(2000 + 0j)\,(0 - 4000j) + (1000 + 5000j)\,(0 - 4000j))}$$

$I1 = 8.689 \cdot 10^{-3} + 4.426 \cdot 10^{-3}j$ $\quad |I1| = 9.751 \cdot 10^{-3}$ $\quad \arg(I1) = 26.996\,°$

FIG. 5.15
Using MathCad to determine I_1 *for the network of Fig. 5.11.*

The current $\mathbf{I}_2$ can be determined in the same manner utilizing the fact that the denominators of $\mathbf{I}_1$ and $\mathbf{I}_2$ are the same.

There is no question that even the use of block impedances requires extensive use of complex algebra to determine the required current levels. However, keep in mind that the equations for $\mathbf{I}_1$ and $\mathbf{I}_2$ have the same format for any impedance level $\mathbf{Z}_1$, $\mathbf{Z}_2$, and $\mathbf{Z}_3$ and any voltage levels $\mathbf{E}_1$ and $\mathbf{E}_2$.

Mesh Analysis

The application of mesh analysis to ac networks is the same as for dc networks with the exception that now we are dealing with impedances and phasors. The format procedure is the same, as will be demonstrated by the following example. We simply treat each block impedance of a branch as if it is a resistor in a dc network.

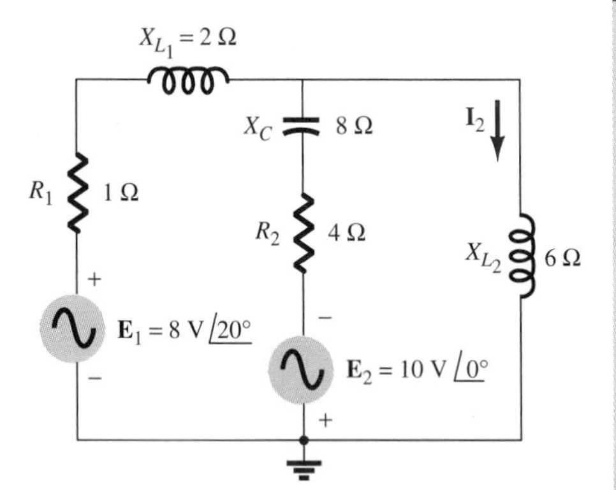

FIG. 5.16

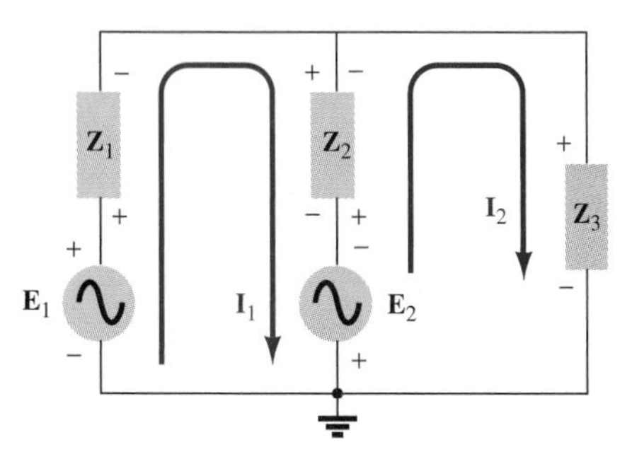

FIG. 5.17

EXAMPLE 5.3 Using the format approach, find the current **I** for the network in Fig. 5.16.

Solution: The network is redrawn in Fig. 5.17.

$$\mathbf{Z}_1 = R_1 + jX_{L_1} = 1\ \Omega + j2\ \Omega \qquad \mathbf{E}_1 = 8\ \text{V} \angle 20°$$
$$\mathbf{Z}_2 = R_2 - jX_C = 4\ \Omega - j8\ \Omega \qquad \mathbf{E}_2 = 10\ \text{V} \angle 0°$$
$$\mathbf{Z}_3 = +jX_{L_2} = +j\,6\ \Omega$$

Note the reduction in the complexity of the problem with substitution of the subscripted impedances.

Two loop currents are assigned in the clockwise direction within each window. Each mesh current is first multiplied by the sum of the impedances it "touches." The mutual terms are then subtracted, and each equation is set equal to the applied voltage sources. A source is given a positive sign if it "supports" the direction of the mesh current, and a minus sign if it opposes.

The resulting equations are

$$\mathbf{I}_1(\mathbf{Z}_1 + \mathbf{Z}_2) - \mathbf{I}_2\mathbf{Z}_2 = \mathbf{E}_1 + \mathbf{E}_2$$
$$\mathbf{I}_2(\mathbf{Z}_2 + \mathbf{Z}_3) - \mathbf{I}_1\mathbf{Z}_2 = -\mathbf{E}_2$$

which can be rewritten as

$$\begin{aligned} \mathbf{I}_1(\mathbf{Z}_1 + \mathbf{Z}_2) - \mathbf{I}_2\mathbf{Z}_2 \qquad\qquad &= \mathbf{E}_1 + \mathbf{E}_2 \\ -\mathbf{I}_1\mathbf{Z}_2 \qquad\qquad + \mathbf{I}_2(\mathbf{Z}_2 + \mathbf{Z}_3) &= -\,\mathbf{E}_2 \end{aligned}$$

Using determinants, we have

$$\mathbf{I}_2 = \frac{\begin{vmatrix} \mathbf{Z}_1 + \mathbf{Z}_2 & \mathbf{E}_1 + \mathbf{E}_2 \\ -\mathbf{Z}_2 & -\,\mathbf{E}_2 \end{vmatrix}}{\begin{vmatrix} \mathbf{Z}_1 + \mathbf{Z}_2 & -\,\mathbf{Z}_2 \\ -\mathbf{Z}_2 & \mathbf{Z}_2 + \mathbf{Z}_3 \end{vmatrix}}$$
$$= \frac{-(\mathbf{Z}_1 + \mathbf{Z}_2)\mathbf{E}_2 + \mathbf{Z}_2(\mathbf{E}_1 + \mathbf{E}_2)}{(\mathbf{Z}_1 + \mathbf{Z}_2)(\mathbf{Z}_2 + \mathbf{Z}_3) - \mathbf{Z}_2^2}$$
$$= \frac{-\mathbf{Z}_1\mathbf{E}_2 + \mathbf{Z}_2\mathbf{E}_1}{\mathbf{Z}_1\mathbf{Z}_2 + \mathbf{Z}_1\mathbf{Z}_3 + \mathbf{Z}_2\mathbf{Z}_3}$$

Substitution of numerical values yields

$$\mathbf{I}_2 = \frac{-(1\ \Omega + j2\ \Omega)(10\ \text{V}\angle 0^\circ) + (4\ \Omega - j8\ \Omega)(8\ \text{V}\angle 20^\circ)}{(1\ \Omega + j2\ \Omega)(4\ \Omega - j8\ \Omega) + (1\ \Omega + j2\ \Omega)(+j6\ \Omega) + (4\ \Omega - j8\ \Omega)(+j6\ \Omega)}$$

$$= \frac{-(10 + j20) + (4 - j8)(7.52 + j2.74)}{20 + (j6 - 12) + (j24 + 48)}$$

$$= \frac{-(10 + j20) + (52.0 - j49.20)}{56 + j30} = \frac{+42.0 - j69.20}{56 + j30} = \frac{80.95\ \text{A}\angle -58.74^\circ}{63.53\angle 28.18^\circ}$$

$$= \mathbf{1.27\ A\angle -86.92^\circ}$$

Nodal Analysis

The application of nodal analysis is also essentially the same as for dc networks except that we are now dealing with impedances and phasors rather than solely resistors and dc sources.

EXAMPLE 5.4 Using the format approach, find the voltage across the 4-Ω resistor in Fig. 5.18.

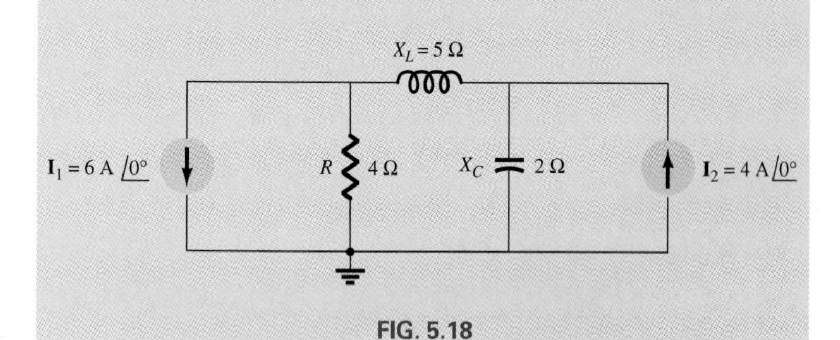

FIG. 5.18

Solution: Choosing nodes (Fig. 5.19) and writing the nodal equations, we have $\mathbf{Z}_1 = R = 4\ \Omega$, $\mathbf{Z}_2 = jX_L = j5\ \Omega$, and $\mathbf{Z}_3 = -jX_C = -j2\ \Omega$.

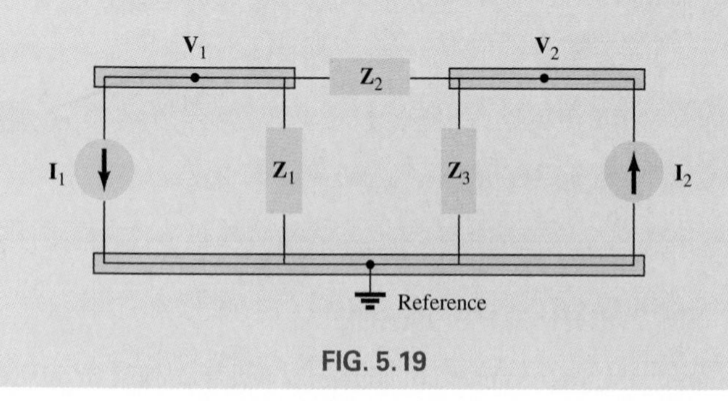

FIG. 5.19

Two nodal voltages are assigned, and the reference node is chosen. Each nodal voltage is then multiplied by the sum of the admittances connected to that node. The mutual terms are then subtracted, and each equation is set equal to the applied current sources. A source is given a positive sign if it "supplies" current to the node, and a negative sign if it "draws" current from the node.

The resulting equations are

$$\mathbf{V}_1(\mathbf{Y}_1 + \mathbf{Y}_2) - \mathbf{V}_2(\mathbf{Y}_2) = -\mathbf{I}_1$$
$$\mathbf{V}_2(\mathbf{Y}_3 + \mathbf{Y}_2) - \mathbf{V}_1(\mathbf{Y}_2) = +\mathbf{I}_2$$

or

$$\begin{aligned} \mathbf{V}_1(\mathbf{Y}_1 + \mathbf{Y}_2) - \mathbf{V}_2(\mathbf{Y}_2) \qquad &= -\mathbf{I}_1 \\ -\mathbf{V}_1(\mathbf{Y}_2) \qquad + \mathbf{V}_2(\mathbf{Y}_3 + \mathbf{Y}_2) &= +\mathbf{I}_2 \end{aligned}$$

$$\mathbf{Y}_1 = \frac{1}{\mathbf{Z}_1} \qquad \mathbf{Y}_2 = \frac{1}{\mathbf{Z}_2} \qquad \mathbf{Y}_3 = \frac{1}{\mathbf{Z}_3}$$

Using determinants, we obtain

$$\begin{aligned} \mathbf{V}_1 &= \frac{\begin{vmatrix} -\mathbf{I}_1 & -\mathbf{Y}_2 \\ +\mathbf{I}_2 & \mathbf{Y}_3 + \mathbf{Y}_2 \end{vmatrix}}{\begin{vmatrix} \mathbf{Y}_1 + \mathbf{Y}_2 & -\mathbf{Y}_2 \\ -\mathbf{Y}_2 & \mathbf{Y}_3 + \mathbf{Y}_2 \end{vmatrix}} \\ &= \frac{-(\mathbf{Y}_3 + \mathbf{Y}_2)\mathbf{I}_1 + \mathbf{I}_2\mathbf{Y}_2}{(\mathbf{Y}_1 + \mathbf{Y}_2)(\mathbf{Y}_3 + \mathbf{Y}_2) - \mathbf{Y}_2^2} \\ &= \frac{-(\mathbf{Y}_3 + \mathbf{Y}_2)\mathbf{I}_1 + \mathbf{I}_2\mathbf{Y}_2}{\mathbf{Y}_1\mathbf{Y}_3 + \mathbf{Y}_2\mathbf{Y}_3 + \mathbf{Y}_1\mathbf{Y}_2} \end{aligned}$$

Substituting numerical values, we have

$$\begin{aligned} \mathbf{V}_1 &= \frac{-[(1/-j2\ \Omega) + (1/j5\ \Omega)]6\ \text{A}\angle 0^\circ + 4\ \text{A}\angle 0^\circ(1/j5\ \Omega)}{(1/4\Omega)(1/-j2\ \Omega) + (1/j5\ \Omega)(1/-j2\ \Omega) + (1/4\Omega)(1/j5\ \Omega)} \\ &= \frac{-(+j0.5 - j0.2)6\angle 0^\circ + 4\angle 0^\circ(-j0.2)}{(1/-j8) + (1/10) + (1/j20)} \\ &= \frac{(-0.3\angle 90^\circ)(6\angle 0^\circ) + (4\angle 0^\circ)(0.2\angle -90^\circ)}{j0.125 + 0.1 - j0.05} \\ &= \frac{-1.8\angle 90^\circ + 0.8\angle -90^\circ}{0.1 + j0.075} \\ &= \frac{2.6\ \text{V}\angle -90^\circ}{0.125\angle 36.87^\circ} \\ &= \mathbf{20.80\ V\angle -126.87^\circ} \end{aligned}$$

5.4 NETWORK THEOREMS

We examine the Thévenin and maximum power theorems in this section to parallel the coverage for dc networks.

Thévenin's Theorem

Thévenin's theorem permits the reduction of a linear ac network to the series configuration of Fig. 5.20. Any network connected to this reduced network is totally unaware of the change. In addition to providing a network composed of fewer elements that has the same terminal characteristics, the theorem provides a technique for quickly determining the effect of changing the load on the load current or voltage.

The Thévenin voltage or impedance can be determined in the same manner as for dc networks. Consider the network in Fig. 5.21, where the Thévenin network is to be determined for the network external to the load R_L. The block impedances have been assigned in Fig. 5.22.

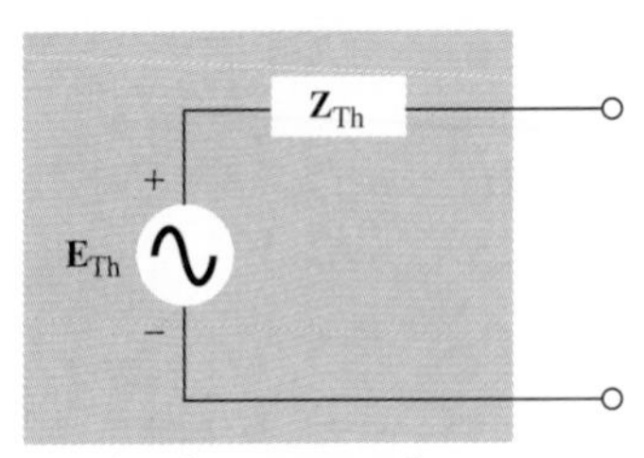

FIG. 5.20
Thévenin equivalent circuit.

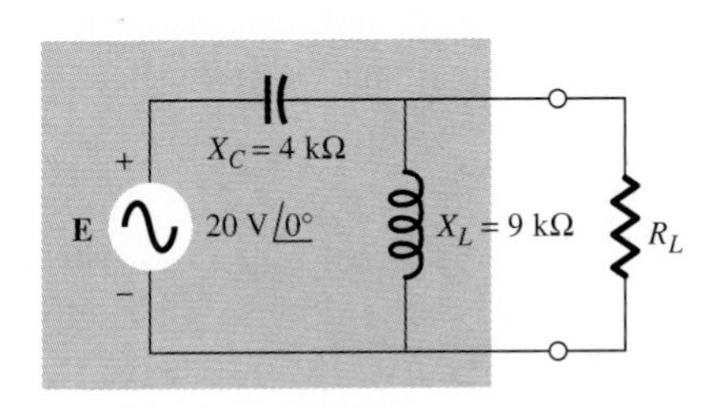

FIG. 5.21

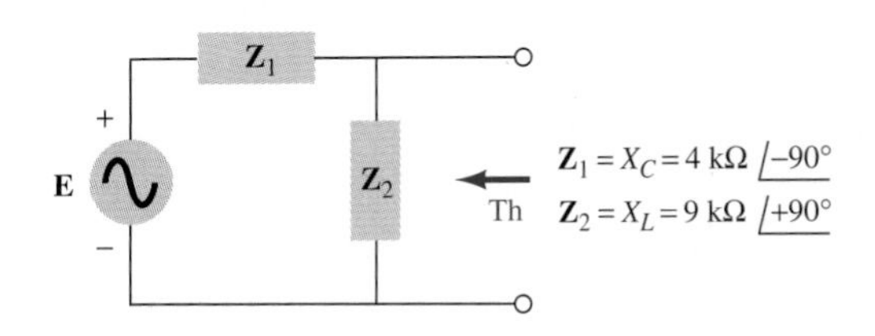

FIG. 5.22
Assigning the block impedances.

The Thévenin impedance is determined by setting **E** to zero as shown in Fig. 5.23 and finding the terminal impedance. That is,

$$\mathbf{Z}_{Th} = \mathbf{Z}_1 \| \mathbf{Z}_2 = \frac{\mathbf{Z}_1\mathbf{Z}_2}{\mathbf{Z}_1 + \mathbf{Z}_2}$$

$$= \frac{(4\text{ k}\Omega\angle -90^\circ)(9\text{ k}\Omega\angle +90^\circ)}{-j4\text{ k}\Omega + j9\text{ k}\Omega}$$

$$= \frac{36\text{ k}\Omega\angle 0^\circ}{5\angle 90^\circ} = \mathbf{7.2\ k\Omega\angle -90^\circ}$$

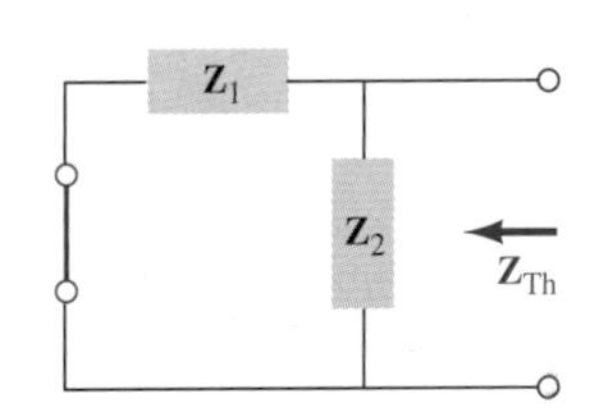

FIG. 5.23
Determining the Thévenin impedance.

The Thévenin voltage is the open-circuit voltage between the network terminals, as shown in Fig. 5.24.

$$\mathbf{E}_{Th} = \frac{\mathbf{Z}_2\mathbf{E}}{\mathbf{Z}_2 + \mathbf{Z}_1} \quad \text{(voltage-divider rule)}$$

$$= \frac{(9\text{ k}\Omega\angle +90^\circ)(20\text{ V}\angle 0^\circ)}{9 \times 10^3\angle +90^\circ + 4 \times 10^3\angle -90^\circ} = \frac{180\text{ V}\angle 90^\circ}{5\angle 90^\circ}$$

$$= \mathbf{36\ V\angle 0^\circ}$$

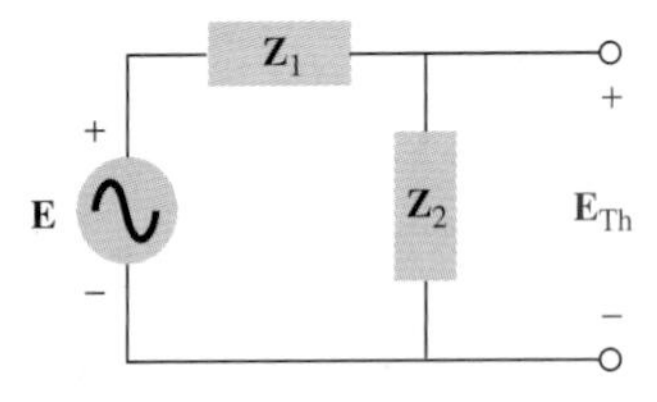

FIG. 5.24
Determining the Thévenin voltage.

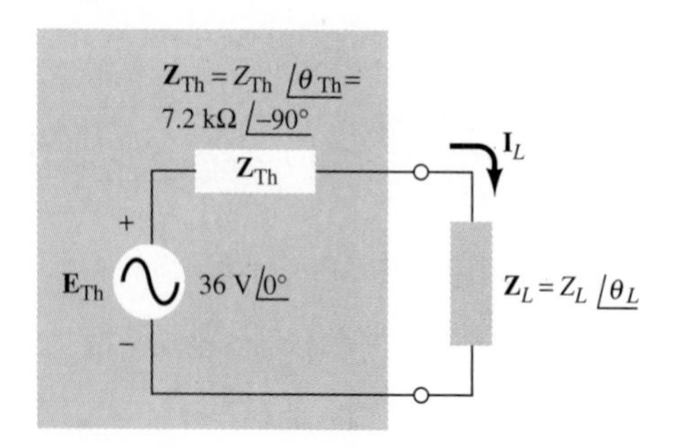

FIG. 5.25
Construction of the Thévenin network.

The Thévenin equivalent circuit appears in Fig. 5.25. The current through $\mathbf{Z}_L$ can now be determined from

$$\mathbf{I}_L = \frac{\mathbf{E}_{Th}}{\mathbf{Z}_{Th} + \mathbf{Z}_L} \tag{5.1}$$

Any change in $\mathbf{Z}_L$ would simply require changing $\mathbf{Z}_L$ in Eq. (5.1), removing any necessity to analyze the entire network again—hence an obvious advantage of the Thévenin equivalent circuit.

Maximum Power Theorem

For the network in Fig. 5.26, maximum power is transferred to $\mathbf{Z}_L$ when the magnitude of $\mathbf{Z}_L$ equals that of $\mathbf{Z}_{Th}$, and their associated angles are such that $\theta_L = -\theta_{Th}$. That is, for maximum power transfer,

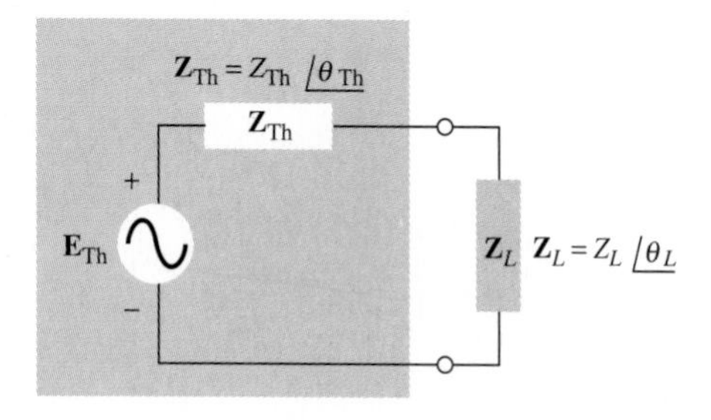

FIG. 5.26
Establishing maximum transfer criteria.

$$\begin{aligned} |\mathbf{Z}_L| &= |\mathbf{Z}_{Th}| \\ \theta_L &= -\theta_{Th} \end{aligned} \tag{5.2}$$

Since the Thévenin network can be found for any two-terminal linear network, it is necessary only to connect a load that satisfies the preceding conditions to transfer maximum power.

In rectangular form the conditions are

$$\begin{aligned} & R_L = R_{Th} \\ \text{and} \quad & X_{L_{(\text{load})}} = X_{C_{\text{Th}}} \\ \text{or} \quad & X_{C_{(\text{load})}} = X_{L_{\text{Th}}} \end{aligned} \tag{5.3}$$

In Fig. 5.27 these conditions have been satisfied for the Thévenin impedance resulting from an inductive network.

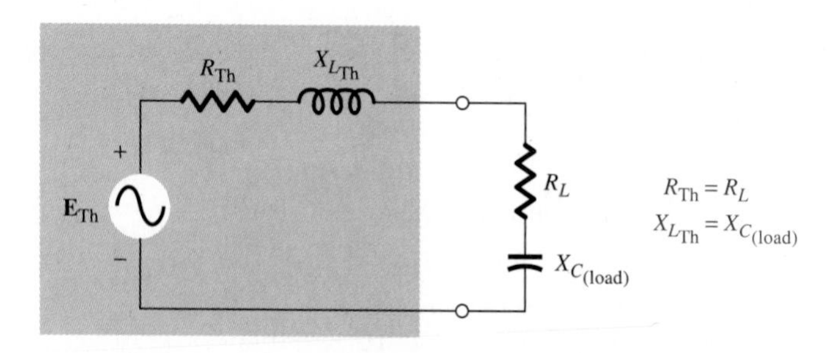

FIG. 5.27
Conditions for maximum power transfer to the load.

The resulting impedance is

$$\mathbf{Z} = (R_{\text{Th}} + jX_{L_{\text{Th}}}) + (R_L - jX_{C_{\text{(load)}}})$$
$$= (R_{\text{Th}} + R_L) + \underbrace{j(X_{L_{\text{Th}}} - X_{C_{\text{(load)}}})}_{\text{0 for maximum power conditions}}$$

$$\boxed{\mathbf{Z} = 2R_{\text{Th}}} \qquad (\text{for } R_{\text{Th}} = R_L) \tag{5.4}$$

and

$$I_L = \frac{E_{\text{Th}}}{Z} = \frac{E_{\text{Th}}}{2R_{\text{Th}}}$$

with the power to the load (the maximum value) determined by

$$P_L = I_L^2 R_L = I_L^2 R_{\text{Th}}$$
$$= \left(\frac{E_{\text{Th}}}{2R_{\text{Th}}}\right)^2 R_{\text{Th}}$$

$$\boxed{P_{L\,\text{max}} = \frac{E_{\text{Th}}^2}{4R_{\text{Th}}}} \tag{5.5}$$

EXAMPLE 5.5 For the network in Fig. 5.28 determine the load for maximum power transfer and the maximum power delivered.

Solution:

$$\mathbf{Z}_{\text{Th}} = R\,\angle 0^\circ \| X_L\,\angle 90^\circ = \frac{(500\ \Omega\,\angle 0^\circ)(2000\ \Omega\,\angle 90^\circ)}{500\ \Omega + j2000\ \Omega}$$
$$= 485.07\ \Omega\,\angle 14.94^\circ$$
$$= 470.58\ \Omega + j117.68\ \Omega$$

Therefore,

$$R_L = \mathbf{470.58\ \Omega} \text{ and } X = X_C = \mathbf{117.68\ \Omega}$$
$$P_{\text{max}} = \frac{E_{\text{Th}}^2}{4R_{\text{Th}}} = \frac{(120\text{ V})^2}{4(470.58\ \Omega)} = \mathbf{7.65\ W}$$

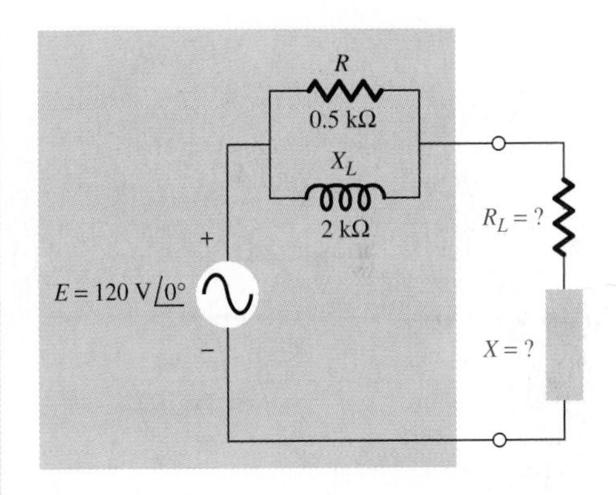

FIG. 5.28

5.5 APPLICATION: SPEAKER SYSTEMS

One of the most common applications of the maximum power transfer theorem is speaker systems. An audio amplifier (amplifier with a frequency range matching the typical range of the human ear) with an output impedance of 8 Ω is shown in Fig. 5.29(a). Every speaker has an internal resistance that can be represented as shown in Fig. 5.29(b) for a

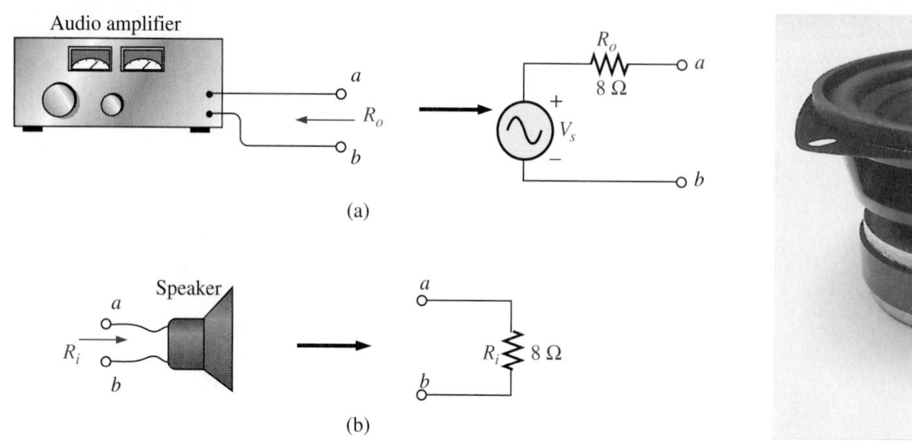

(c)

FIG. 5.29
Components of a speaker system: (a) amplifier; (b) speaker; (c) commercially available unit.

standard 8-Ω speaker. Figure 5.29(c) is a photograph of a commercially available 8-Ω woofer (for very low frequencies). The primary purpose of the following discussion is to shed some light on how the audio power can be distributed and which approach would be the most effective.

Since the maximum power theorem states that the load impedance should match the source impedance for maximum power transfer, let us first consider the case of a single 8-Ω speaker as shown in Fig. 5.30(a) with an applied amplifier voltage of 12 V. Since the applied voltage will split equally, the speaker voltage is 6 V, and the power to the speaker is a maximum value of $P = V^2/R = (6\text{ V})^2/8\ \Omega = 4.5\text{ W}$.

If we have two 8-Ω speakers that we would like to hook up, we have the choice of hooking them up in series or parallel. For the series configuration of Fig. 5.30(b), the resulting current will be $I = E/R = 12\text{ V}/24\ \Omega = 500\text{ mA}$, and the power to each speaker will be $P = I^2R = (500\text{ mA})^2(8\ \Omega) = 2\text{ W}$, which is a drop of over 50% from the maximum output level of 4.5 W. If the speakers are hooked up in parallel as shown in Fig. 5.30(c), the total resistance of the parallel combination will be 4 Ω, and the voltage across each speaker as determined by the voltage-divider rule will be 4 V. The power to each speaker will be $P = V^2/R = (4\text{ V})^2/8\ \Omega = 2\text{ W}$, which, interestingly enough, is the same power delivered to each speaker whether in series or parallel. However, the parallel arrangement is normally chosen for a variety of reasons. First, when the speakers are connected in parallel, if a wire should become disconnected from one of the speakers due simply to the vibration caused by the emitted sound, the other speakers will continue to operate—although perhaps not at maximum efficiency. In series, they would all fail to operate. A second reason relates to the general wiring procedure. When all the speakers are in parallel, all their red wires and all their black wires can be connected together from various parts of a room. If the speakers were in series, and you were presented with a bundle of red and black wires in the basement, you would first have to determine which wires went with which speakers.

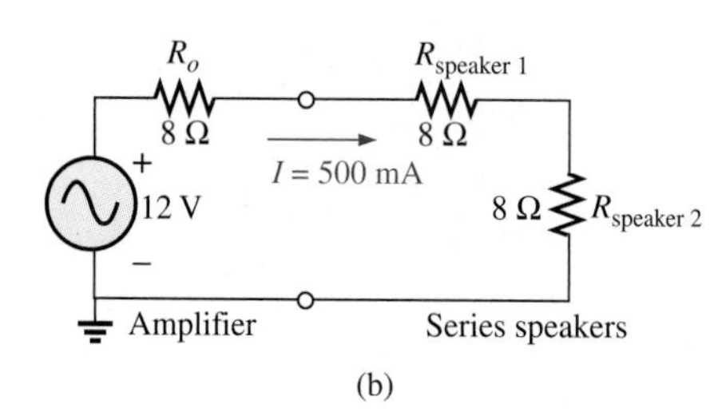

(b)

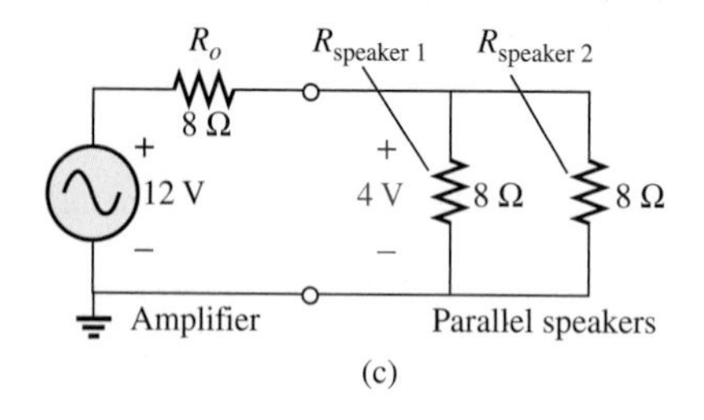

(c)

FIG. 5.30
Speaker connections: (a) single unit; (b) in series; (c) in parallel.

Speakers are also available with input impedances of 4 Ω and 16 Ω. If you know that the output impedance is 8 Ω, purchasing either two 4-Ω speakers or two 16-Ω speakers will result in maximum power to the speakers, as shown in Fig. 5.31. Connecting the 16-Ω speakers in parallel and the 4-Ω speakers in series will establish a total load impedance of 8 Ω.

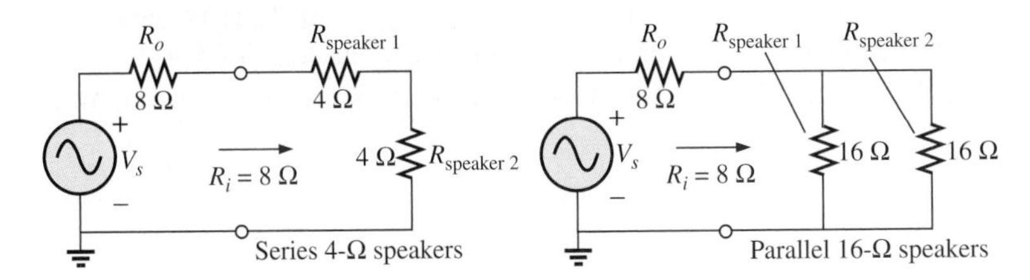

FIG. 5.31
Applying 4-Ω and 16-Ω speakers to an amplifier with an output impedance of 8 Ω.

In any case, always try to match the total resistance of the speaker load to the output resistance of the supply. Yes, a 4-Ω speaker can be placed in series with a parallel combination of 8-Ω speakers for maximum power transfer from the supply, since the total resistance will be 8 Ω. However, the power distribution will not be equal, with the 4-Ω speaker receiving 2.25 W and the 8-Ω speakers each 1.125 W for a total of 4.5 W. The 4-Ω speaker is therefore receiving twice the audio power of the 8-Ω speakers, and this difference may cause distortion or imbalance in the listening area.

All speakers have maximum and minimum levels. A 50-W speaker is rated for a maximum output power of 50 W and will provide that level on demand. However, in order to function properly, it will probably need to be operating at least at the 1- to 5-W level. A 100-W speaker typically needs between 5 W and 10 W of power to operate properly. It is also important to realize that power levels less than the rated value (such as 40 W for the 50-W speaker) will not result in an increase in distortion, but simply in a loss of volume. However, distortion will result if you exceed the rated power level. For example, if you apply 2.5 W to a 2-W speaker, you will definitely have distortion. However, applying 1.5 W will simply result in less volume. A rule of thumb regarding audio levels states that the human ear can sense changes in audio level only if the applied power is doubled (a 3-dB increase; decibels (dB) will be introduced in Chapter 11). The doubling effect is always with respect to the initial level. For instance, if the original level was 2 W, you would have to go to 4 W to notice the change. If starting at 10 W, you would have to go to 20 W to appreciate the increase in volume. An exception to this rule occurs at very low power levels or very high power levels. For instance, a change from 1 W to 1.5 W may be discernible, just as a change from 50 W to 80 W may be noticeable.

5.6 Δ-Y CONVERSIONS

Perhaps you will recall from the treatment of Δ and Y configurations in dc networks that neither configuration had series or parallel combinations of elements as their building blocks. In fact, without the conversions from one form to another it appeared as if a solution for the total impedance or driving current could not be obtained.

Consider the configuration in Fig. 5.32, for example. A close examination reveals that within the right-hand structure there are no series or parallel elements. However, a conversion from one form of Fig. 5.33 to the other would permit a reduction in the network (as described in detail for dc networks).

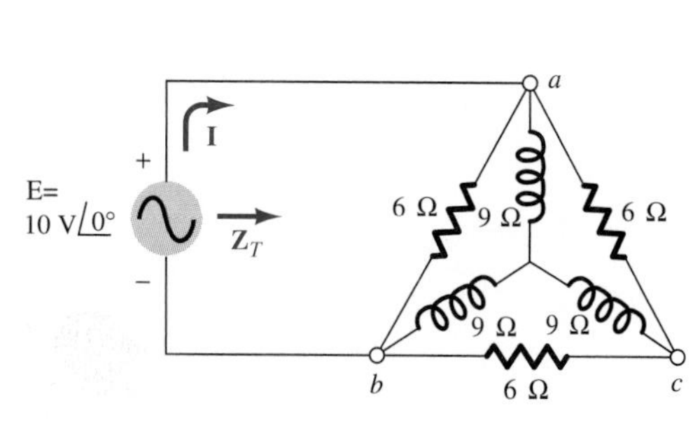

FIG. 5.32

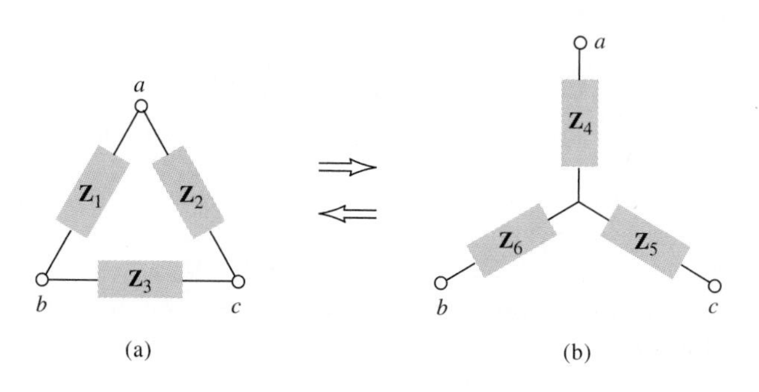

FIG. 5.33
(a) Δ configuration; (b) Y configuration.

The conversion equations for the Δ and Y configurations in Fig. 5.33 are

$$\mathbf{\Delta \rightarrow Y} \qquad\qquad \mathbf{Y \rightarrow \Delta}$$

$$\mathbf{Z}_4 = \frac{\mathbf{Z}_1\mathbf{Z}_2}{\mathbf{Z}_1 + \mathbf{Z}_2 + \mathbf{Z}_3} \qquad \mathbf{Z}_1 = \frac{\mathbf{Z}_4\mathbf{Z}_5 + \mathbf{Z}_4\mathbf{Z}_6 + \mathbf{Z}_5\mathbf{Z}_6}{\mathbf{Z}_5}$$

$$\mathbf{Z}_5 = \frac{\mathbf{Z}_2\mathbf{Z}_3}{\mathbf{Z}_1 + \mathbf{Z}_2 + \mathbf{Z}_3} \qquad \mathbf{Z}_2 = \frac{\mathbf{Z}_4\mathbf{Z}_5 + \mathbf{Z}_4\mathbf{Z}_6 + \mathbf{Z}_5\mathbf{Z}_6}{\mathbf{Z}_6} \qquad (5.6)$$

$$\mathbf{Z}_6 = \frac{\mathbf{Z}_1\mathbf{Z}_3}{\mathbf{Z}_1 + \mathbf{Z}_2 + \mathbf{Z}_3} \qquad \mathbf{Z}_3 = \frac{\mathbf{Z}_4\mathbf{Z}_5 + \mathbf{Z}_4\mathbf{Z}_6 + \mathbf{Z}_5\mathbf{Z}_6}{\mathbf{Z}_4}$$

If all three impedances of a Δ or Y are equal, the equation for conversion reduces to

$$\boxed{\mathbf{Z}_Y = \frac{\mathbf{Z}_\Delta}{3}} \quad \text{or} \quad \boxed{\mathbf{Z}_\Delta = 3\mathbf{Z}_Y} \qquad (5.7)$$

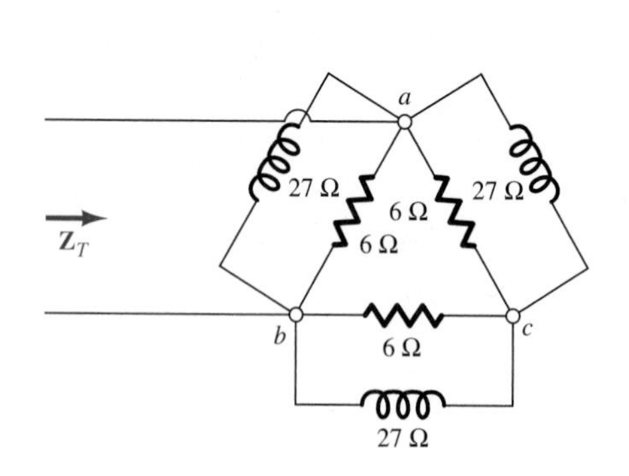

FIG. 5.34

For the example provided, the Y will be converted to a Δ between the same three terminals, resulting in the configuration in Fig. 5.34.

The parallel combination of each 6-Ω resistor and 3-Ω inductor results in the following total impedance:

$$\mathbf{Z} = \frac{(6\ \Omega\angle 0^\circ)(27\ \Omega\angle 90^\circ)}{6\ \Omega + j27\ \Omega} = \frac{162\ \Omega\angle 90^\circ}{27.66\angle 77.47^\circ} = 5.857\ \Omega\angle 12.53^\circ$$

and

$$\begin{aligned}\mathbf{Z}_T &= \mathbf{Z} \| (\mathbf{Z} + \mathbf{Z}) \\ &= \frac{\mathbf{Z}(2\mathbf{Z})}{\mathbf{Z} + 2\mathbf{Z}} = \frac{2\mathbf{Z}^2}{3\mathbf{Z}} = \frac{2}{3}\mathbf{Z} = \frac{2}{3}5.857\ \Omega\angle 12.53^\circ \\ &= 3.9\ \Omega\angle 12.53^\circ\end{aligned}$$

and

$$\begin{aligned}\mathbf{I} &= \frac{\mathbf{E}}{\mathbf{Z}_T} = \frac{10\ \text{V}\angle 0^\circ}{3.9\ \Omega\angle 12.53^\circ} \\ &= \mathbf{2.564\ A\angle -12.53^\circ}\end{aligned}$$

The Δ could also be converted to a Y. The two center points (of each Y) could then be connected, since the loads are balanced (same impedance in each branch). For unbalanced situations the Y–Δ conversion should be used, since the center points cannot be connected using a Δ–Y conversion.

The Δ–Y conversion can also be applied to the *bridge* configuration in Fig. 5.35, which was also introduced for dc networks. As in the network just examined, there are no two elements in series or parallel. Since all three elements of the upper delta are equal, we use the Δ–Y conversion to obtain the network in Fig. 5.36.

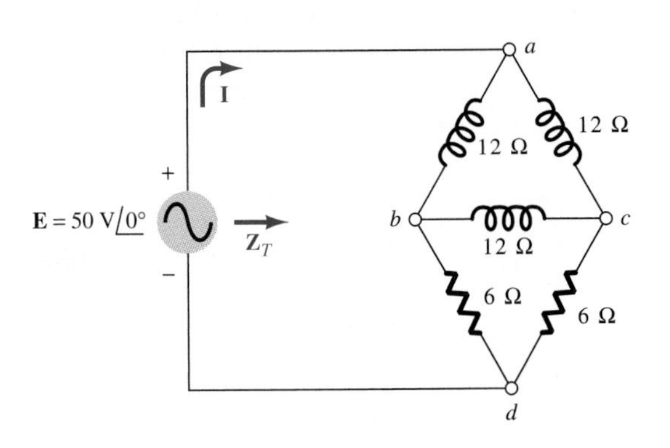

FIG. 5.35
Bridge network.

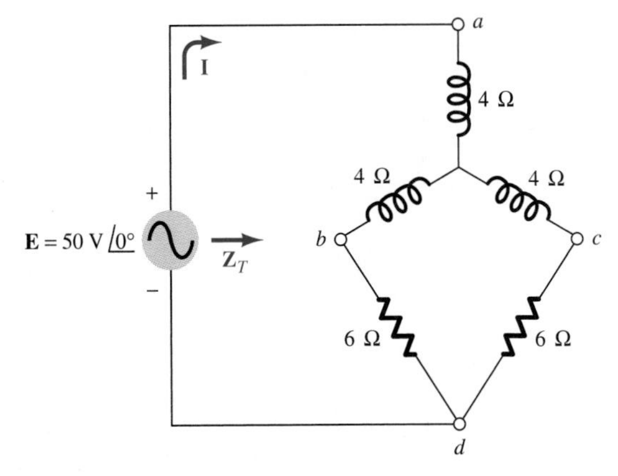

FIG. 5.36

The impedance of each branch of the new Y was obtained from Eq. (5.7):

$$\mathbf{Z}_Y = \frac{\mathbf{Z}_\Delta}{3} = \frac{12\ \Omega\angle 90^\circ}{3} = 4\ \Omega\angle 90^\circ$$

We can now use series–parallel techniques to find the total impedance and current. That is, for the series 4-Ω inductance and 6-Ω resistance, we have

$$\mathbf{Z}_1 = 6\ \Omega + j4\ \Omega$$

The parallel branches of $\mathbf{Z}_1$ result in a total impedance determined by

$$\mathbf{Z}_2 = \frac{\mathbf{Z}_1}{2} = \frac{6\ \Omega + j4\ \Omega}{2} = 3\ \Omega + j2\ \Omega$$

The total impedance is then

$$\begin{aligned}\mathbf{Z}_T &= 4\ \Omega\angle 90^\circ + \mathbf{Z}_2 \\ &= j4\ \Omega + (3\ \Omega + j2\ \Omega) \\ &= 3\ \Omega + j6\ \Omega \\ &= \mathbf{6.71\ \Omega\angle 63.43^\circ}\end{aligned}$$

with the result that

$$\mathbf{I} = \frac{\mathbf{E}}{\mathbf{Z}_T} = \frac{50\ \text{V}\angle 0^\circ}{6.71\ \Omega\angle 63.43^\circ} = \mathbf{7.45\ A\angle -63.43^\circ}$$

5.7 BRIDGE CONFIGURATION

In our analysis of dc networks we found that there was a condition set by the branch impedances of the bridge configuration that resulted in zero voltage drop across the bridge arm or zero current through that branch impedance. For ac networks, the general form of the equation is the same, except now we are dealing with impedances rather than fixed resistive values as encountered for dc networks. For the general network in Fig. 5.37, the balance condition is determined by

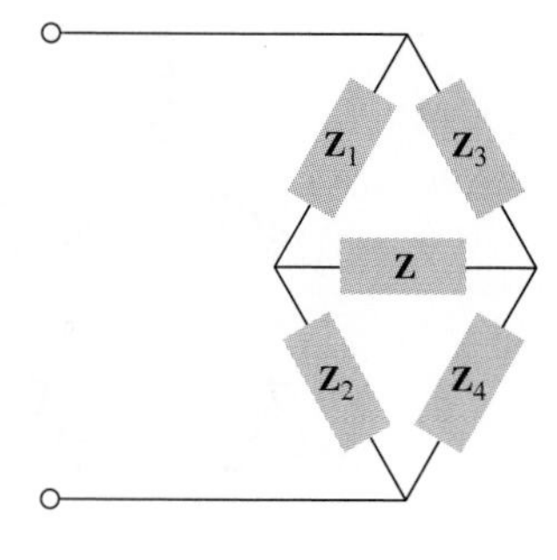

FIG. 5.37

$$\frac{\mathbf{Z}_1}{\mathbf{Z}_2} = \frac{\mathbf{Z}_3}{\mathbf{Z}_4} \qquad \textbf{(5.8)}$$

For the network in Fig. 5.35 the following ratios result:

$$\frac{12\ \Omega\angle 90^\circ}{6\ \Omega\angle 0^\circ} = \frac{12\ \Omega\angle 90^\circ}{6\ \Omega\angle 0^\circ}$$

and

$$2\ \Omega\angle 90^\circ = 2\ \Omega\angle 90^\circ \qquad \text{(checks)}$$

Since Eq. (5.8) is satisifed, there is zero voltage drop across branch *bc* or zero current through that branch impedance. The latter condition permits substitution of an open circuit for that branch, as shown in Fig. 5.38.

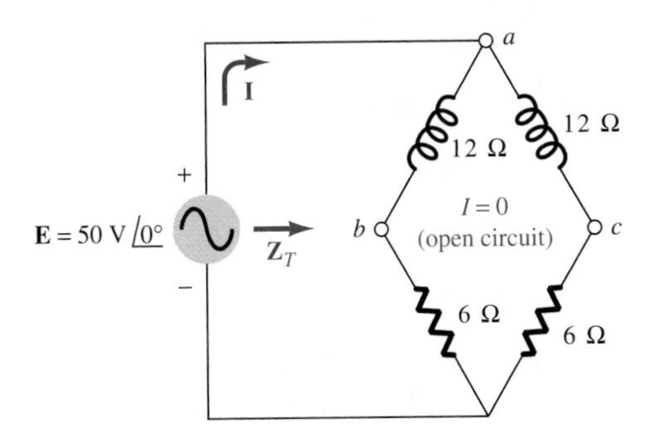

FIG. 5.38

Network in Fig. 5.35 with open circuit substituted for "balance" arm.

Now the 12-Ω inductance and 6-Ω resistor are in series, and the series combination of each results in

$$\mathbf{Z}_1 = 6\ \Omega + j12\ \Omega = 13.42\ \Omega\angle 63.43^\circ$$

and

$$\mathbf{Z}_T = \frac{\mathbf{Z}_1}{2} = \frac{13.42\ \Omega\angle 63.43^\circ}{2} = \mathbf{6.71\ \Omega\angle 63.43^\circ}$$

as obtained before. Placing a short between terminals a and b (for the condition $v = 0$) also results in the same solution, but this will be left as an exercise in the problems of this chapter.

For a situation in which the ratios of Eq. (5.8) result in

$$2\ \Omega\angle 90^\circ \neq 2\ \Omega\angle -90^\circ$$

the equality condition does not hold. Both the magnitude and the angle must be equal for the condition $v = 0$ or $i = 0$ to be applicable. In rectangular form both the real and imaginary components must be equal.

When employed in a resistance-measuring capacity for dc networks, the configuration was referred to as a *Wheatstone bridge.* You will recall that it was necessary to ensure only that the bridge current from a to b was zero before the equation could be applied to determine the unknown resistance from the other three known values. For ac networks, this bridge can be used to measure resistance, inductance, and capacitance.

Three commercially employed bridge configurations appear in Table 5.1. When the bridge arms are adjusted until $\boldsymbol{I}_{\text{gal}} = 0$, the unknown quantity denoted by the x subscripts can be determined using the equations provided.

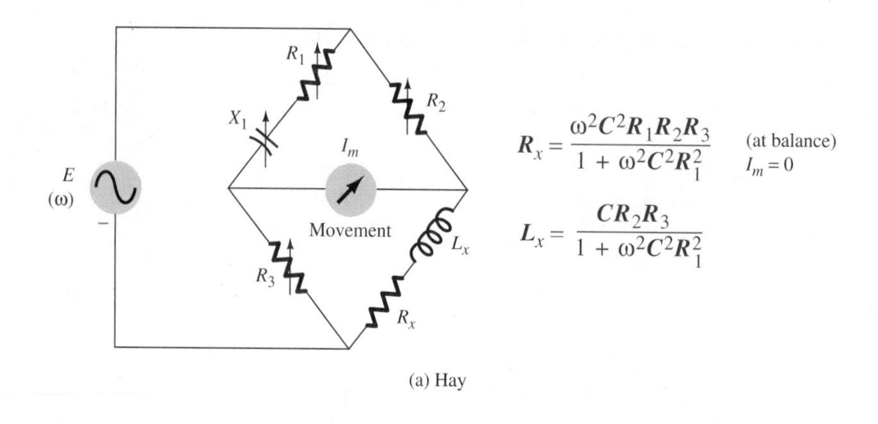

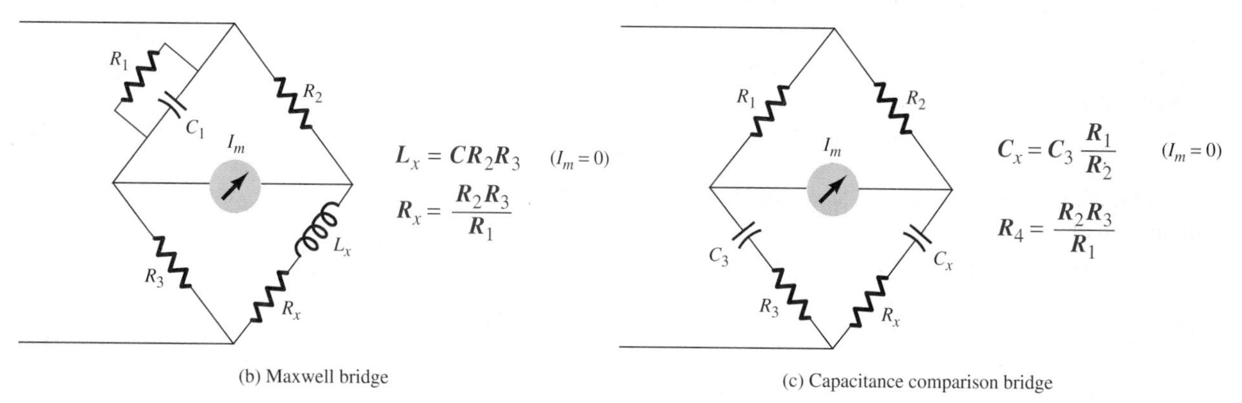

TABLE 5.1

A photograph of a commercially available *RLC* bridge appears in Fig. 5.39. The user's manual provided by the manufacturer will make the determination quite straightforward for each of the three elements.

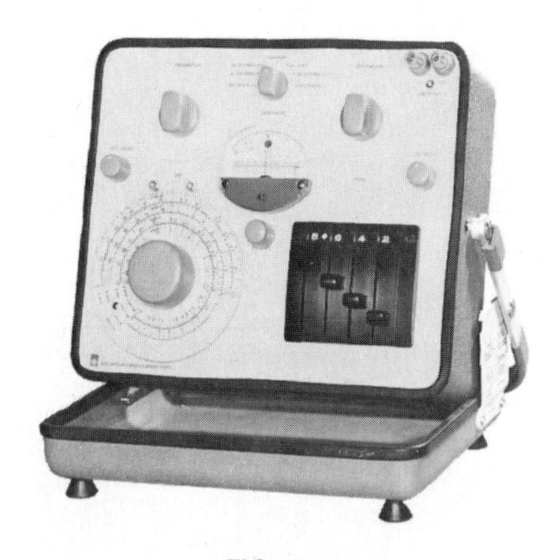

FIG. 5.39
Impedance bridge. (Courtesy of General Radio Company.)

5.8 POWER (AC)

In an ac network, the resistive elements are the only components that dissipate electrical energy. The purely reactive elements simply store the energy in a form that can be returned to the electrical system when required by the circuit design. For the total watts dissipated, therefore, it is necessary only to find the sum of the watts dissipated by the resistive elements. The power dissipated can also be determined at the driving point of the network as shown in Fig. 5.40.

The total watts dissipated is determined by

$$P = EI \cos \theta \tag{5.9}$$

The wattmeter connections are also shown in Fig. 5.40 for determining the total watts dissipated. The voltage terminals sense the voltage level, while the current terminals reflect the magnitude of I. The meter includes the effect of $\cos\theta$ to indicate the proper wattage level. A dynamometer wattmeter can read the watts dissipated in a dc or in an ac network. In fact, it can properly indicate the wattage level for any input voltage no matter how distorted it may appear.

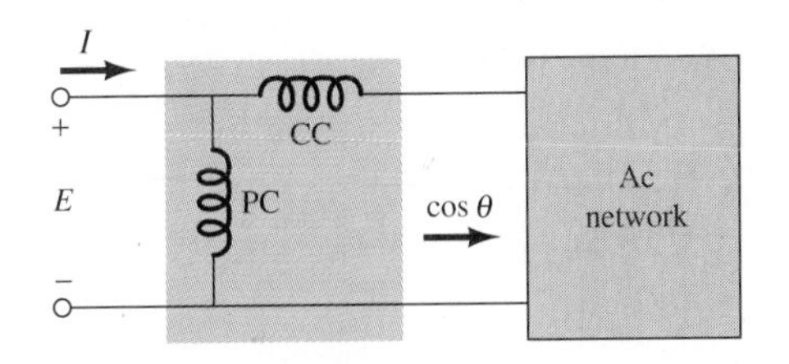

FIG. 5.40
Using a wattmeter to measure the power delivered to an ac network.

Although there is no dissipation of electrical energy by the reactive elements, a level of electrical energy is drawn from the supply and stored in the electric or magnetic fields of the reactive elements. This additional energy can be supplied only through an increase in the current drawn from the supply. Of course, the reactive elements can return the energy to the system, but at particular instants of time there is an increase in the supply current due to the reactive elements. This increase in current requires that the supply or generator be designed to handle the increased currents. For fixed voltage levels this increase in current results in an increase in the maximum instantaneous power the supply must be able to provide. An increased current or power rating results in increased costs for the generating equipment and production of the necessary electrical energy. This difference in demand from that dissipated by the resistive elements is accounted for in the power factor ($F_p = \cos\theta$) of the network. For $F_p = 1$, all the power supplied is dissipated, while for more reactive elements F_p approaches zero, and more energy is stored in the reactive (storage) elements.

The product EI, which is independent of whether dissipation or storage is the result, is called the *apparent* power of an ac system and is measured in volt-amperes (VA). That is, for the network in Fig. 5.40 the apparent power is determined by

$$S = EI \quad \text{(volt-amperes)} \tag{5.10}$$

The current I is the current that must be supplied by the power station even if a major portion is diverted to the storage elements. Larger industrial outfits pay for the apparent power demand rather than the average or real power dissipated because of the current requirement above that of the resistive elements. Any power factor less than 1 will certainly result in a wattage demand less than the volt-ampere demand.

On a trigonometric basis, the apparent power and *real, average,* or *dissipated* power are related as shown in Fig. 5.41. The third component of the triangle is similar to the real power equation except for the sine rather than the cosine term. It is called the *reactive* power and is measured in volt-ampere reactive (VAR) units. That is,

Reactive power:
$$Q = EI\sin\theta \quad \text{(volt-ampere reactive units)} \tag{5.11}$$

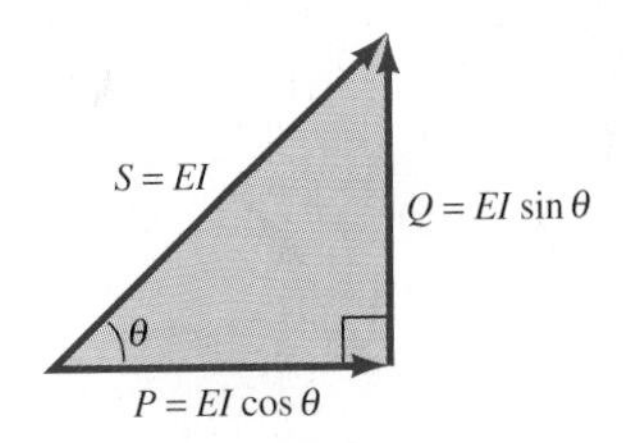

FIG. 5.41
Power triangle.

The reactive power is a measure of the input power absorbed (but not dissipated) by the reactive elements. For a fixed supply voltage the smaller this component, the smaller the source current. Peak efficiency of the system is obtained (insofar as the source is concerned) when $Q = 0$ and $P = S$.

For any network, the total reactive power is simply the difference between the capacitive and inductive components as determined by any one of the following equations:

Inductor:
$$Q_L = I_L^2 X_L = \frac{V_L^2}{X_L} = V_L I_L \quad \text{(VAR)} \tag{5.12}$$

Capacitor:
$$Q_C = I_C^2 X_C = \frac{V_C^2}{X_C} = V_C I_C \quad \text{(VAR)} \tag{5.13}$$

For a network in which the capacitive VAR equals the inductive VAR, the net reactive power is zero and the real and apparent powers are equal. Since

$$P_T = EI\cos\theta = S_T\cos\theta = S_T F_p$$

we find that

$$F_p = \frac{P_T}{S_T} \tag{5.14}$$

where P_T and S_T represent the total of each quantity of the system.

EXAMPLE 5.6 The currents and voltages of the network in Fig. 5.42 are indicated. Determine:

a. The total power dissipated.
b. The net reactive power.
c. The total apparent power.
d. The power factor of the network.

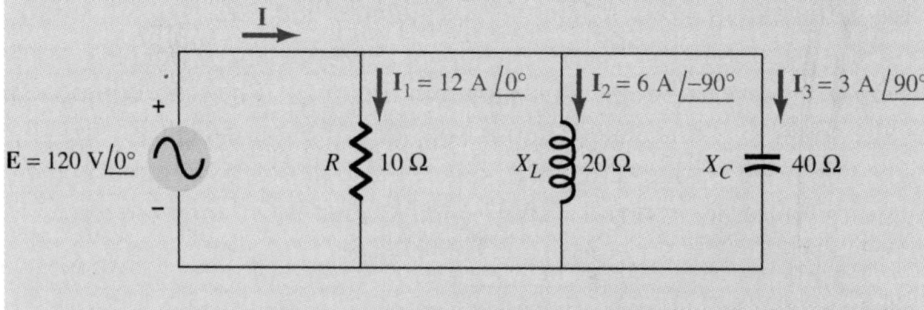

FIG. 5.42

Solution:

a. Power is dissipated only by the resistive element:

$$P_T = I^2R = (12\text{ A})^2(10\ \Omega) = (144)(10) = \mathbf{1440\ W}$$

b. $Q_C = I^2X_C = (3\text{ A})^2(40\ \Omega) = (9)(40) = \mathbf{360\ VAR\ (capacitive)}$
$Q_L = I^2X_L = (6\text{ A})^2(20\ \Omega) = (36)(20) = \mathbf{720\ VAR\ (inductive)}$
$Q_T = Q_L - Q_C = 720\text{ VAR} - 360\text{ VAR} = \mathbf{360\ VAR\ (inductive)}$

c. $S_T = \sqrt{P_T^2 + Q_T^2}$ (from a right triangle)
$= \sqrt{(1440\text{ W})^2 + (360\text{ VAR})^2}$
$\cong \mathbf{1484\ VA}$

d. $F_p = \dfrac{P_T}{S_T} = \dfrac{1440\text{ W}}{1484\text{ VA}} = \mathbf{0.97\ (inductive,\ lagging)}$

Note that in the solution to Example 5.6 no consideration was given to the type of network. Each element was treated individually to determine either the real or reactive power. There was no concern (once the currents were known) for the fact that it was a parallel network in the determination of the total watts or reactive power. In addition, note that the total apparent power was determined from the total real and reactive power. It cannot be determined by a simple algebraic addition of the real and reactive power to each element.

Power Factor Correction

For any electrical power system to work at peak efficiency, the current drawn from the supply should be reduced to a minimum, since the line voltage is typically fixed in magnitude. The result is that the apparent power rating of the system determined by the product of the source voltage and current should also be kept at a minimum. Since $S_T = EI_T = \sqrt{P_T^2 + Q_T^2}$, the smaller the net reactive component of the load for a fixed dissipative level P_T, the smaller the net apparent powers and the higher the network power factor $(=P_T/S_T)$. The terminology *power-factor correction* is applied to any system when an effort is made to ensure that the power factor is a maximum (as close to 1 as possible) by reducing the net reactive component of the loading. As noted previously, the result is a reduction in current drawn from the supply.

The typical example is to use capacitive elements to improve the power factor of a system with a lagging power factor due to inductive loads such as motors, transformers, and general lighting.

EXAMPLE 5.7 A 2.2-hp motor has a 0.8 lagging power factor and an efficiency of 76% when connected to a 208-V, 60-Hz supply. Determine the level of capacitance placed in parallel with the motor that will raise the power factor of the system to unity.

Solution: For the 2.2-hp motor,

$$P_o = 2.2\text{ hp} = (2.2\text{ hp})(746\text{ W/hp}) = 1641.2\text{ W}$$

$$\eta = \frac{P_o}{P_i} \quad \text{and} \quad P_i = \frac{P_o}{\eta} = \frac{1641.25\text{ W}}{0.76} = 2159.47\text{ W}$$

$$F_p = \cos\theta = 0.8$$
$$\theta = \cos^{-1} 0.8 = 36.87^\circ$$
$$\tan\theta = \frac{Q_L}{P_i}$$
$$Q_L = P_i \tan\theta = 2159.47(\tan 36.87^\circ) = 2159.47(0.75) = \mathbf{1619.6\ VAR}$$

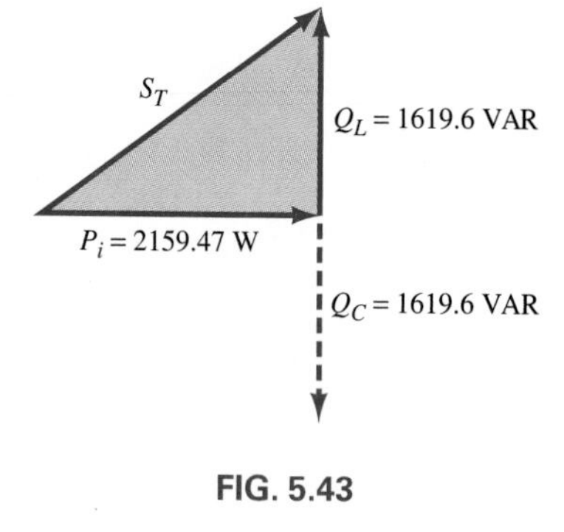

FIG. 5.43

The power triangle of the load appears in Fig. 5.43. To establish that $S_T = P_i$ and $F_p = 1$, a capacitive reactance must be introduced as shown in Fig. 5.43 such that $Q_C = Q_L = 1619.6$ VAR. Since

$$Q_C = \frac{V_C^2}{X_C} = \frac{E^2}{X_C} \qquad \text{(parallel network)}$$

and

$$X_C = \frac{E^2}{Q_C} = \frac{(208\ \text{V})^2}{1619.6\ \text{VAR}} = \frac{43{,}264\Omega}{1619.0} = 26.71\ \Omega$$

but

$$X_C = \frac{1}{2\pi fC}$$

and

$$C = \frac{1}{2\pi fX_C} = \frac{1}{2\pi(60\ \text{Hz})(26.71\ \Omega)} = \mathbf{99.31\ \mu F}$$

5.9 POLYPHASE SYSTEMS

Throughout the analysis of ac networks we have been concerned with the response of a network to a single sinusoidal signal as generated by a single-phase ac generator. There are *polyphase* generators available that can develop two, three, or more phases simultaneously. Each generated voltage has a definite phase relationship with the other phase voltages. For power applications where large voltages are generated, the three-phase generator is employed almost exclusively. Each of the three ac voltages is 120° out of phase, as shown in Fig. 5.44.

In the phasor domain,

$$\begin{aligned} \mathbf{E}_{AN} &= E_{AN}\angle 0^\circ \\ \mathbf{E}_{AN} &= E_{BN}\angle -120^\circ \\ \mathbf{E}_{CN} &= E_{CN}\angle +120^\circ \end{aligned} \qquad \textbf{(5.15)}$$

which in the time domain are

$$\begin{aligned} e_{AN} &= \sqrt{2}E_{AN}\sin(\omega t + 0^\circ) \\ e_{BN} &= \sqrt{2}E_{BN}\sin(\omega t - 120^\circ) \\ e_{CN} &= \sqrt{2}E_{CN}\sin(\omega t + 120^\circ) \end{aligned} \qquad \textbf{(5.16)}$$

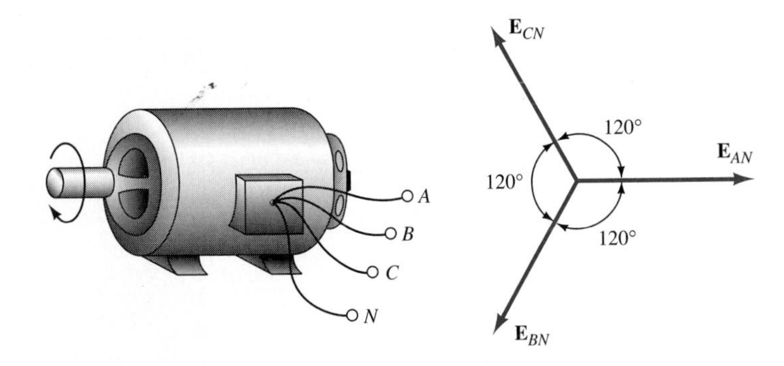

FIG. 5.44
Three-phase generator and the voltages developed.

and appear in Fig. 5.45. Note that at any instant of time three voltage levels are available and that each has the same peak value.

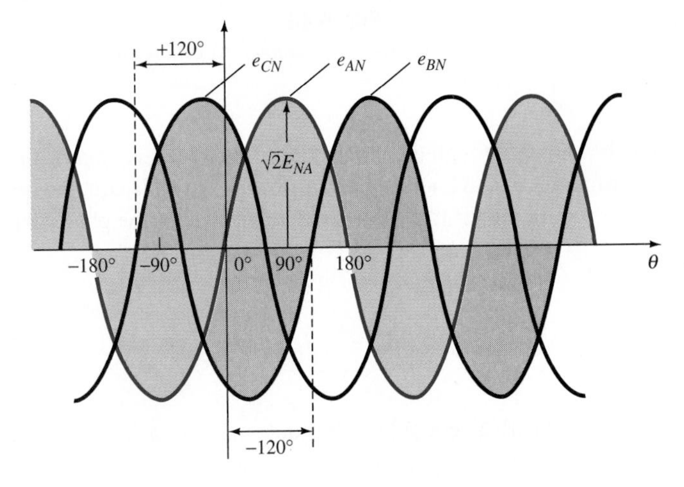

FIG. 5.45
Phase voltages of a three-phase generator.

The schematic representation of a three-phase generating system appears in Fig. 5.46. The letter N is a shorthand notation for the *neutral* or common terminal of the generated voltages. Since the induced voltages appear across the armature *coils* of the generator, the symbol for each phase employs the coil notation. The system in Fig. 5.46 is referred to as a 3-ϕ,Y-*connected, four-wire system.* The choice of the terminology Y-*connected* should be obvious from the schematic representation. There are four lines to the load, resulting in the four-wire label. From the diagram it should be obvious that the line current equals the generated phase current. That is,

$$I_L = I_{\phi_g} \tag{5.17}$$

for each phase.

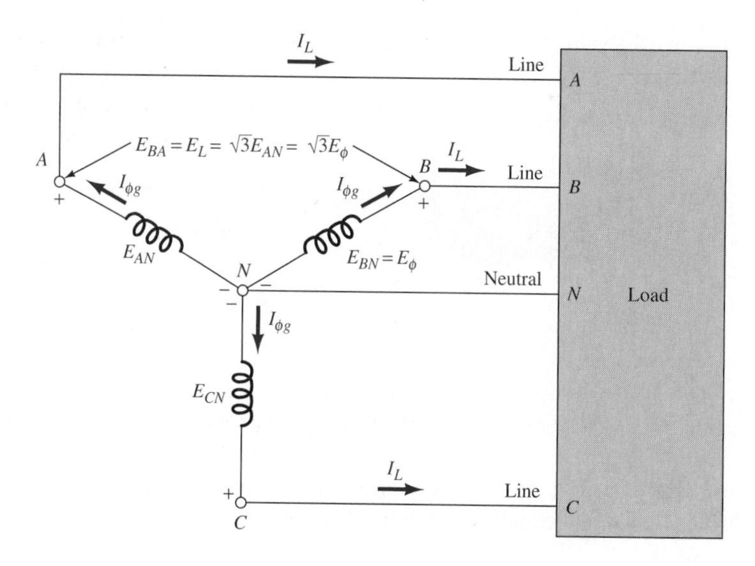

FIG. 5.46
A three-phase, four-wire, Y-connected generator.

It can be shown through the vector addition of any two phases that the voltage from one line to another (such as from A to B) is equal in magnitude to the square root of three times the magnitude of the phase voltage. That is,

$$E_L = \sqrt{3}E_\Phi \tag{5.18}$$

The line voltages are all equal in magnitude and have a 120° phase angle between any two, as shown in Fig. 5.47(b) and 5.47(c).

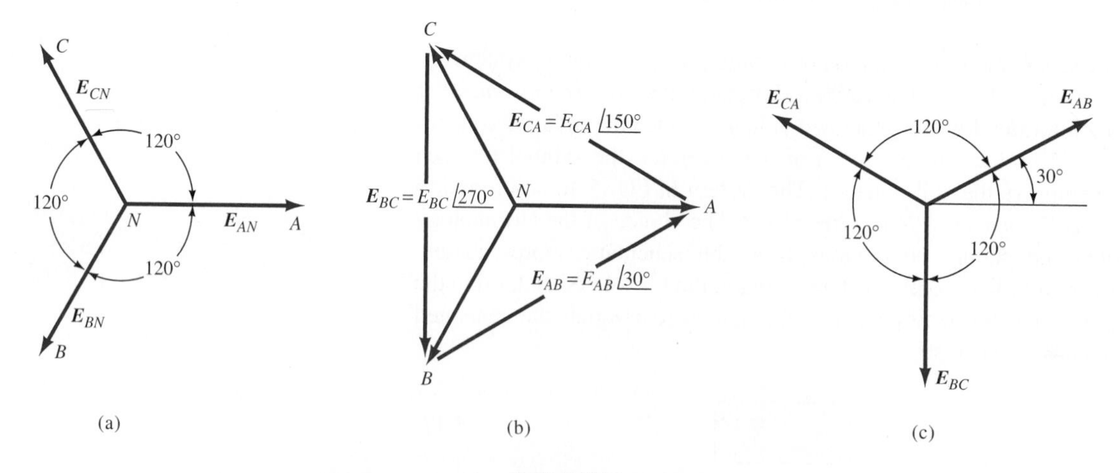

FIG. 5.47
Three-phase and line voltages.

For a three-phase supply, the *phase sequence* can be of particular importance. It is the order in which the phase voltages appear at the output terminals of the generator in Fig. 5.44. In Fig. 5.47(a) the phases appear in the order $\mathbf{E}_{AN}\mathbf{E}_{BN}\mathbf{E}_{CN}$, and the phase sequence is said to be *ABC*, as determined by the order of the first subscripts. For the only other possibility, $\mathbf{E}_{AN}\mathbf{E}_{CN}\mathbf{E}_{BN}$, the phase sequence would be *ACB*. Placing a mark on the phasor diagram [such as an *x* in Fig. 5.47(a)] and rotating the vectors counterclockwise, gives the phase sequence from the sequence of the first or second subscripts of the voltages. The operation of certain electrical equipment is quite dependent on the phase sequence. The direction of rotation of a shaft of a three-phase motor can be reversed simply by changing the phase sequence.

In a three-phase system the load can be connected in a Y or Δ configuration. The equations for the Y-connected load as shown in Fig. 5.48 are fundamentally the same as introduced for the generator. For obvious reasons, the system is referred to as Y–Y-*connected.*

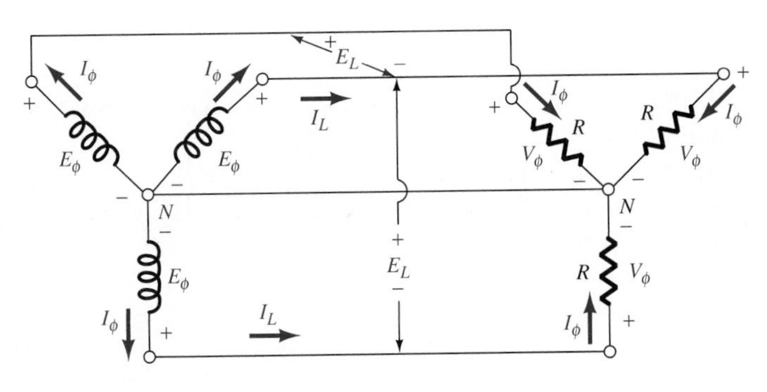

FIG. 5.48
Y–Y-connected system.

The relating equations are

$$E_L = \sqrt{3}V_\phi \qquad (5.19)$$

$$\mathbf{I}_\phi = \mathbf{I}_L \qquad (5.20)$$

and for a Y–Y system,

$$\mathbf{V}_\phi = \mathbf{E}_\phi \qquad (5.21)$$

Note that Eq. (5.19) is in magnitude form, while Eqs. (5.20) and (5.21) are in phasor form.

The load network in Fig. 5.48 is said to be *balanced,* since the same impedance appears in each branch. The result is zero neutral current ($I_N = 0$), since each phase current is equal but out of phase by 120°. The result of the following vector addition is therefore 0 A:

$$\mathbf{I}_N = \mathbf{I}_{\Phi 1} + \mathbf{I}_{\Phi 2} + \mathbf{I}_{\Phi 3} = 0$$

An attempt is always made to balance the load as much as possible to reduce current $\mathbf{I}_N$ to its minimum value so that the line loss $I_N^2 R_{\text{line}}$ has the smallest possible value.

The obvious advantages of the three-phase system are two available voltage levels (phase and line), a phase shift between them of 120°, and a zero return current (for minimum line loss) for balanced loads. For industrial and commercial use, two of the three phases are provided so that the larger line-to-line voltage of 208 V is available for heavy machinery, welders, and so on. The 120-V outlet voltage for the same plant is available from each phase. The same is true in the home, where 208 V may be necessary for an air conditioner, while only 120 V is necessary for the outlets.

EXAMPLE 5.8 For the Y–Y connected, three-phase, four-wire system in Fig. 5.49, determine:

a. The load phase voltages.
b. The magnitude of the load line voltages.
c. The phase current of each load element.
d. The vector sum of the three phase currents.
e. The line currents.

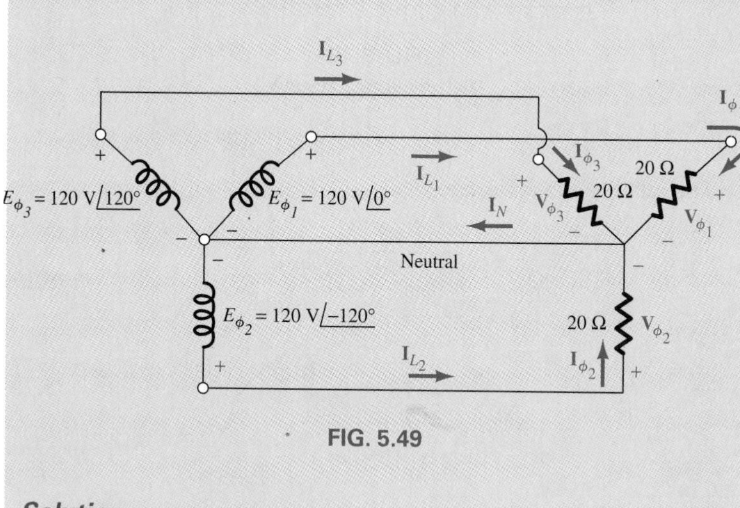

FIG. 5.49

Solution:

a. $\mathbf{V}_{\Phi 1} = \mathbf{E}_{\Phi 1} = \mathbf{120\ V\angle 0°}$
$\mathbf{V}_{\Phi 2} = \mathbf{E}_{\Phi 2} = \mathbf{120\ V\angle -120°}$
$\mathbf{V}_{\Phi 3} = \mathbf{E}_{\Phi 3} = \mathbf{120\ V\angle +120°}$

b. $V_L = \sqrt{3}V_\Phi = (1.73)(120\text{ V}) \cong \mathbf{208\ V}$

c. $\mathbf{I}_{\Phi 1} = \dfrac{\mathbf{V}_{\Phi 1}}{\mathbf{Z}_1} = \dfrac{120\text{ V}\angle 0^\circ}{20\ \Omega\angle 0^\circ} = \mathbf{6 A\angle 0^\circ}$

$\mathbf{I}_{\Phi 2} = \dfrac{\mathbf{V}_{\Phi 2}}{\mathbf{Z}_2} = \dfrac{120\text{ V}\angle -120^\circ}{20\ \Omega\angle 0^\circ} = \mathbf{6\text{ A}\angle -120^\circ}$

$\mathbf{I}_{\Phi 3} = \dfrac{\mathbf{V}_{\Phi 3}}{\mathbf{Z}_3} = \dfrac{120\text{ V}\angle +120^\circ}{20\ \Omega\angle 0^\circ} = \mathbf{6\text{ A}\angle +120^\circ}$

d. $\mathbf{I}_N = \mathbf{I}_{\Phi 1} + \mathbf{I}_{\Phi 2} + \mathbf{I}_{\Phi 3} = 6\text{ A}\angle 0^\circ + 6\text{ A}\angle -120^\circ + 6\text{ A}\angle +120^\circ$
$= (6\text{ A} + j0) + (-3\text{ A} - j5.196\text{ A}) + (-3\text{ A} + j5.196\text{ A})$
$= (6\text{ A} - 3\text{ A} - 3\text{ A}) + j(-5.196\text{ A} + 5.196\text{ A})$
$= \mathbf{0}$ (balanced load)

e. $\mathbf{I}_{L_1} = \mathbf{I}_{\Phi 1} = \mathbf{6\text{ A}\angle 0^\circ}$, $\mathbf{I}_{L_2} = \mathbf{I}_{\Phi_2} = \mathbf{6\text{ A}\angle -120^\circ}$,
$\mathbf{I}_{L_3} = \mathbf{I}_{\Phi_3} = \mathbf{6\text{ A}\angle +120^\circ}$

The three phases of a polyphase generator may be connected in a Δ configuration as shown in Fig. 5.50 for reduced line voltage but increased line current (for the same induced voltages and load). In this case it is immediately obvious that

$$\mathbf{E}_{\Phi} = \mathbf{E}_L = \mathbf{V}_{\Phi} \tag{5.22}$$

and through vector algebra it can be shown that

$$I_L = \sqrt{3} I_{\Phi} \tag{5.23}$$

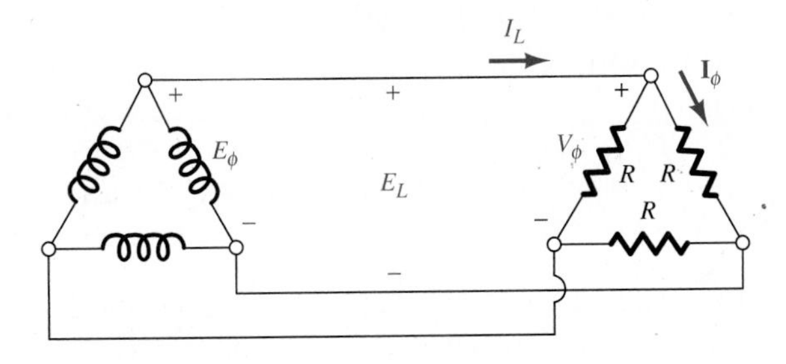

FIG. 5.50
Δ–Δ configuration.

EXAMPLE 5.9 For the Δ-Δ-connected, three-phase, three-wire system in Fig. 5.51, determine:

a. The line and phase voltages.
b. The load phase currents.
c. The magnitude of the line currents.

d. The relationship between the line current and voltage of a system as compared with a Y–Y system.
e. The power to each load element for the Δ–Δ system as compared with the Y–Y system.

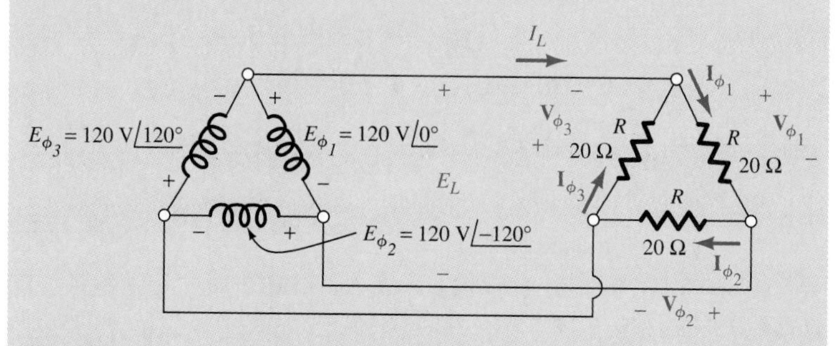

FIG. 5.51

Solution:

a. $\mathbf{V}_{\Phi_1} = \mathbf{E}_L = \mathbf{E}_{\Phi_1} = \mathbf{120\ V\angle 0°}$
$\mathbf{V}_{\Phi_2} = \mathbf{E}_L = \mathbf{E}_{\Phi_2} = \mathbf{120\ V\angle -120°}$
$\mathbf{V}_{\Phi_3} = \mathbf{E}_L = \mathbf{E}_{\Phi_3} = \mathbf{120\ V\angle +120°}$

b. $\mathbf{I}_{\Phi_1} = \dfrac{\mathbf{V}_{\Phi_1}}{\mathbf{Z}_1} = \dfrac{120\text{ V}\angle 0°}{20\ \Omega\angle 0°} = \mathbf{6\ A\angle 0°}$

$\mathbf{I}_{\Phi_2} = \dfrac{\mathbf{V}_{\Phi_2}}{\mathbf{Z}_2} = \dfrac{120\text{ V}\angle -120°}{20\ \Omega\angle 0°} = \mathbf{6\ A\angle -120°}$

$\mathbf{I}_{\Phi_3} = \dfrac{\mathbf{V}_{\Phi_3}}{\mathbf{Z}_3} = \dfrac{120\text{ V}\angle +120°}{20\ \Omega\angle 0°} = \mathbf{6\ A\angle +120°}$

c. $I_L = \sqrt{3}I_\Phi = (1.73)(6\text{ A}) = \mathbf{10.39\ A}$

d. Y–Y: $I_L = \mathbf{6\ A}$ and $E_L = \mathbf{208\ V}$
Δ–Δ: $I_L = \mathbf{10.39\ A}$ and $E_L = \mathbf{120\ V}$
Note the increased current but reduced voltage for the Δ–Δ configuration.

e. Δ–Δ: $P = I_\Phi^2 R = (6\text{ A})^2 \times 20\ \Omega = (36)(20) = \mathbf{720\ W}$
Y–Y: $P = I_\Phi^2 R = (6\text{ A})^2 \times 20\ \Omega = (36)(20) = \mathbf{720\ W}$

Two other configurations are possible, the Y–Δ and the Δ–Y. In each case, it is necessary to apply only those equations that apply to that configuration for the generator or load. For instance, for a Y–Δ system, $E_L = \sqrt{3}E_\Phi$ but $V_\Phi = E_L$, and so on.

The total real, apparent, and reactive power to a balanced load can be determined using the concepts introduced in the previous section. In general,

$$P_T = 3I_\Phi^2 R \qquad \text{(watts)} \qquad \mathbf{(5.24)}$$

$$Q_T = 3I_\Phi^2 X \qquad \text{(VAR)} \qquad \mathbf{(5.25)}$$

$$S_T = 3V_\Phi I_\Phi \quad \text{(VA)} \tag{5.26}$$

$$F_p = \frac{P_T}{S_T} \tag{5.27}$$

For a Δ- or Y-connected, balanced or unbalanced load, the total power to the load can be determined if the wattmeters are hooked up as shown in Fig. 5.52.

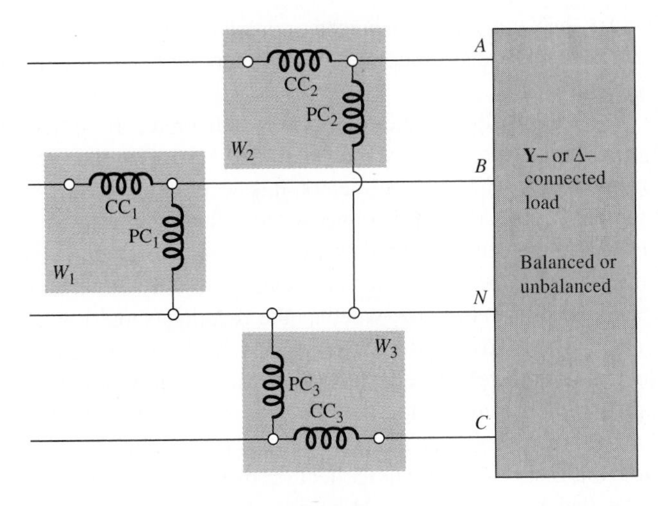

FIG. 5.52
Three-wattmeter measurement of total power to a three-phase load.

EXAMPLE 5.10 For a balanced Y–Y system in which the load of each branch is $\mathbf{Z} = 4\ \Omega + j8\ \Omega = R + jX_L$ and the line voltage is 420 V, determine:

a. P_T.
b. Q_T.
c. S_T.
d. F_p.

Solution:

a. $Z = \sqrt{R^2 + X_L^2} = \sqrt{(4\ \Omega)^2 + (8\ \Omega)^2} = 8.944\ \Omega$

with $$V_\Phi = \frac{V_L}{\sqrt{3}} = \frac{420\text{ V}}{1.732} = 242.49\text{ V}$$

and $$I_\phi = \frac{V_\Phi}{Z} = \frac{242.49\text{ V}}{8.944\ \Omega} = 27.11\text{ A}$$

with

$$P_T = 3P_\Phi = 3I_\Phi^2 R = 3(27.11\ \text{A})^2\, 4\ \Omega$$
$$= \mathbf{8.819\ kW}$$

b. $Q_T = 3I_\Phi^2 X_L = 3(27.11\ \text{A})^2\, 8\ \Omega$
$$= \mathbf{17.64\ kVAR}$$

c. $S_T = \sqrt{P_T^2 + Q_T^2} = \sqrt{(8.819\ \text{kW})^2 + (17.64\ \text{kVAR})^2}$
$$= \mathbf{19.72\ kVA}$$

d. $F_p = \dfrac{P_T}{S_T} = \dfrac{8.819\ \text{kW}}{19.72\ \text{kVA}} \cong \mathbf{0.447}$ (lagging)

For a Δ- or Y-connected, balanced or unbalanced load, the total power to the load can be determined if the wattmeters are hooked up as shown in Fig. 5.52.

For a three-phase load the total power delivered can be determined using only two wattmeters as shown in Fig. 5.53. Two other combinations are possible with two wattmeters. It is necessary to ensure only that both potential coils (PC) are connected to a line without a current coil (CC) and that the current terminals are properly connected in series with the line current to be sensed by that meter. With only two meters, however, it is necessary to know whether the readings should be added or subtracted. The following simple test makes the determination: After ensuring that both meters have an upscale reading, remove that terminal of the potential coil connected to the line without a current coil of the low-reading wattmeter and touch the line that has the current coil of the high-reading wattmeter. An upscale deflection of the low-reading wattmeter indicates that the reading should be added, while the reverse indicates that their difference is the total wattage.

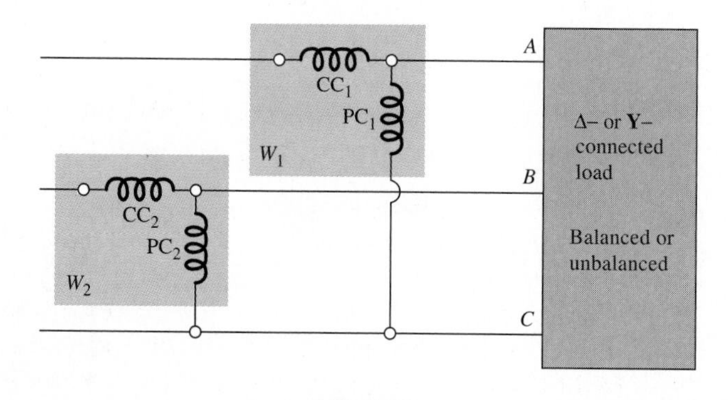

FIG. 5.53
Two-wattmeter measurement of total power to a three-phase load.

Polyphase wattmeters are available that have, in a single package, the terminals necessary for measuring the total power to a three-phase load. Further comment on three-phase systems will be provided when we consider the role of transformers in the distribution of power to homes and industries.

5.10 GROUNDING

Although usually treated too lightly in most introductory electrical or electronics texts, the impact of the ground connection and how it can provide a measure of safety to a design are very important topics. Ground potential is 0 V at every point in a network that has a ground symbol. Since all grounds are at the same potential, they can all be connected together, but for purposes of clarity most are left isolated on a large schematic. On a schematic the voltage levels provided are always with respect to ground. A system can therefore be checked quite rapidly by simply connecting the black lead of the voltmeter to the ground connection and placing the red lead at the various points where the typical operating voltage is provided. A close match normally implies that that portion of the system is operating properly.

There are various types of grounds depending on the application. An *earth ground* is one that is connected directly to the earth by a low-impedance connection. Under typical environmental conditions, local ground potentials are uniform and typically are defined as equal to zero. It is the same at every point because there are sufficient conductive agents in the soil such as water and electrolytes to ensure that any difference in voltage on the surface is equalized by a flow of charge between the two points. Every home has an earth ground usually established by a long conductive rod driven into the ground and connected to the power panel. The electrical code requires a direct connection from earth ground to the cold-water pipes of a home for safety reasons. A "hot" wire touching a cold-water pipe draws sufficient current because of the low-impedance ground connection to throw the breaker. Otherwise, someone in the bathroom could pick up the voltage when they touched the cold-water faucet and risk bodily harm. Because water is a conductive agent, any area of the home with water such as the bathroom or kitchen is of particular concern. Most electrical systems are connected to earth ground primarily for safety reasons. All the power lines in a laboratory, at industrial locations, or in the home are connected to earth ground.

A second type is referred to as a *chassis ground,* which may be *floating* or connected directly to an earth ground. A chassis ground simply stipulates that the chassis has a reference potential for all points of the network. If the chassis is not connected to earth potential (0 V), it is said to be floating and can have any other reference voltage to which the other voltages can be compared. For instance, if the chassis is sitting at 120 V, all measured voltages of the network will be referenced to this level. A reading of 32 V between a point in the network and the chassis ground will therefore actually be at 152 V with respect to earth potential. Most high-voltage systems are not left floating, however, because of loss of the safety factor. For instance, if someone should touch the chassis and be standing on a suitable ground, the full 120 V would fall across that individual.

The National Electrical Code requires that the "hot" (or feeder line) that carries current to a load be *black,* and the line (called the neutral) that carries the current back to the supply be *white.* Three-wire conductors have a ground wire that must be *green* or *bare* that ensures a common ground but is not designed to carry current. The components of a three-prong extension cord and wall outlet are shown in Fig. 5.54. Note that on both fixtures the connection to the hot lead is smaller than the return leg and that the ground connection is circular.

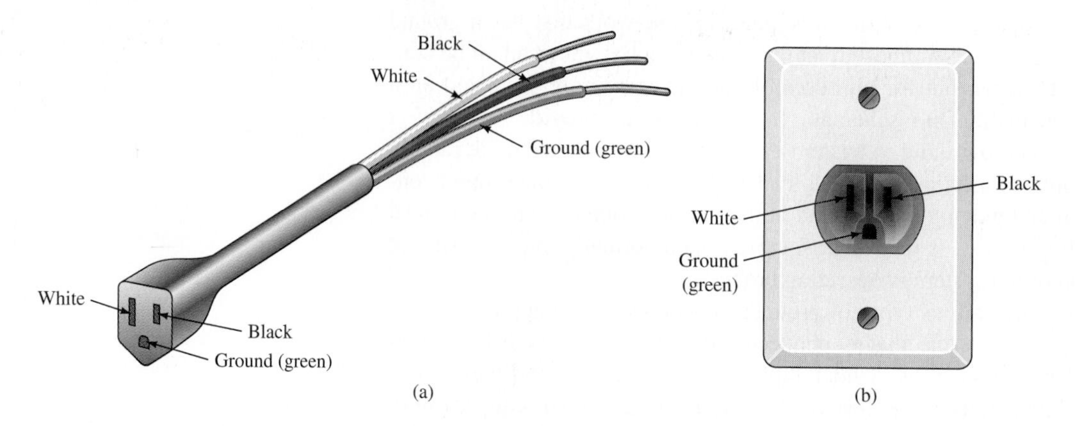

FIG. 5.54
Three-wire conductors: (a) extension cord; (b) home outlet.

The complete wiring diagram for a household outlet is shown in Fig. 5.55. Note that the current through the ground wire is zero and that both the return wire and ground wire are connected to an earth ground. The full current to the loads flows through the feeder and return lines.

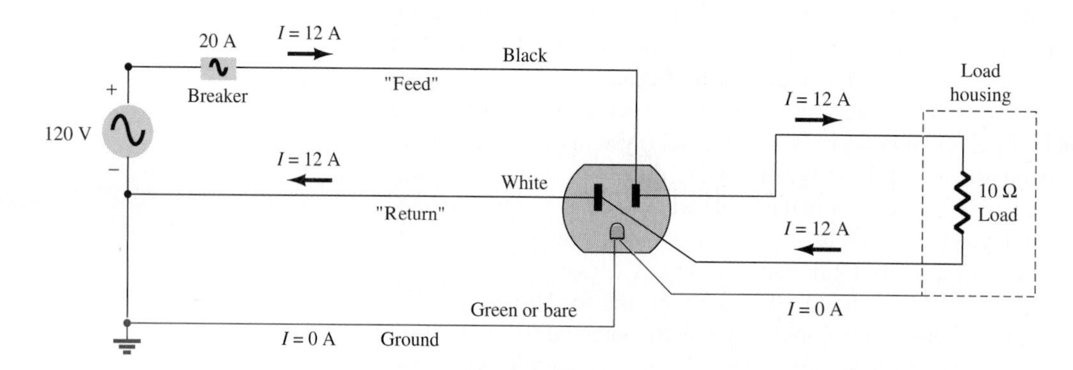

FIG. 5.55
Complete wiring diagram for a household outlet with a 10-Ω load.

The importance of the ground wire in a three-wire system can be demonstrated by the toaster in Fig. 5.56 rated 1200 W at 120 V. From the power equation $P = EI$, the current drawn under normal operating conditions is $I = P/E = 1200\ \text{W}/120\ \text{V} = 10\ \text{A}$. If a two-wire line were employed as shown in Fig. 5.56(a), the 20-A breaker would be quite comfortable with the 10-A current and the system would perform normally. However, if abuse to the feeder (or return line) caused it to become frayed and to touch the metal housing of the toaster, the situation depicted in Fig. 5.56(b) would result. The housing would become "hot,"

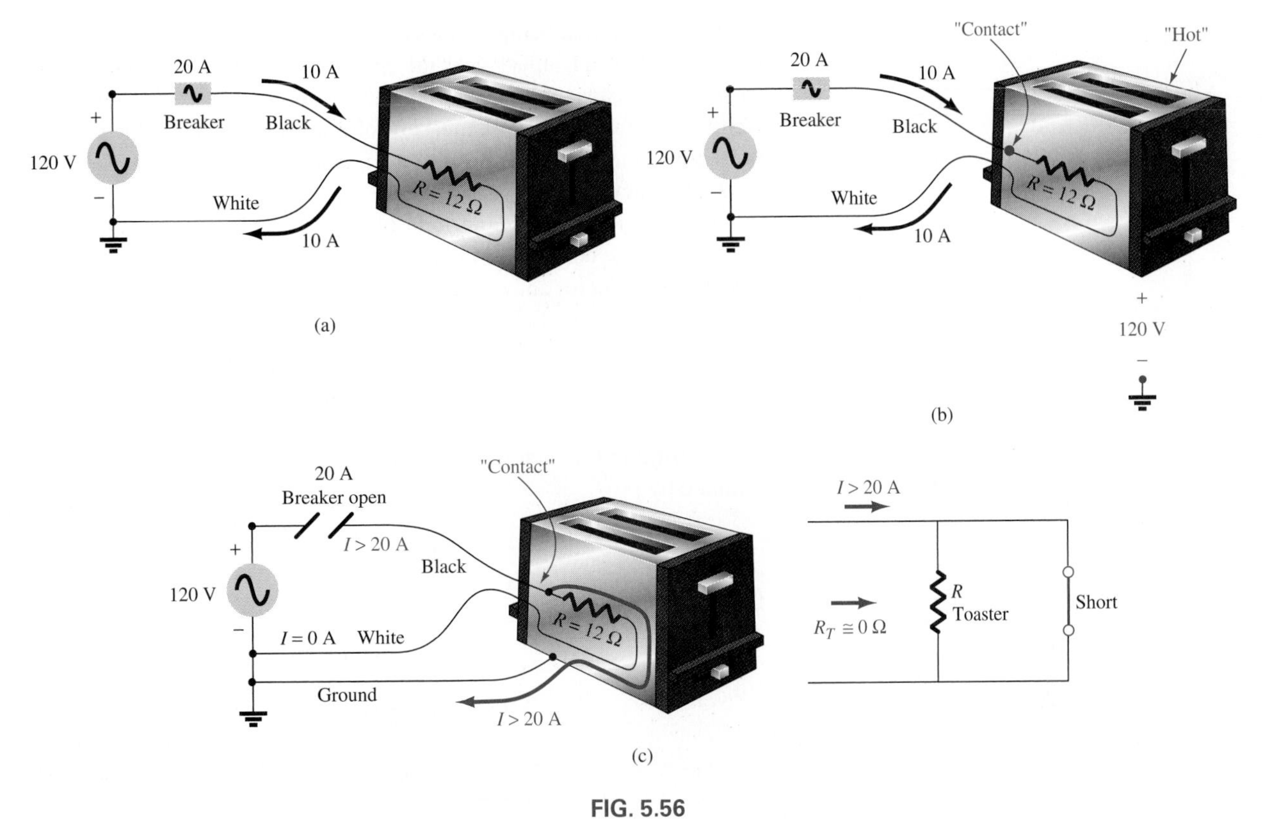

FIG. 5.56

Demonstrating the importance of a properly grounded appliance: (a) ungrounded; (b) ungrounded and undesirable contact; (c) grounded appliance with undesirable contact.

yet the breaker would not "pop" because the current would still be the rated 10 A. However, a dangerous condition would now exist because anyone touching the toaster would feel the full 120 V to ground. If the ground wire were attached to the chassis as shown in Fig. 5.56(c), a low-resistance path would be created between the short-circuit point and ground, and the current would jump to very high levels. The breaker would "pop," and the user would be warned that a problem existed.

Although this discussion does not cover all possible areas of concern with proper grounding or introduce all the nuances associated with the effect of grounds on a system's performance, it is hoped that it provides an awareness of the importance of understanding its impact. Further comment on the effects of grounding will be made in the chapters to follow as the need arises.

5.11 TUNED (RESONANT) NETWORKS

Series Resonance

For every series *RLC* network there is a particular frequency that will result in a minimum input impedance and maximum power transfer to the network. In addition, the energy released by one reactive element is exactly equal to that being absorbed by the other. In other words, during the first half of an input cycle the inductor absorbs all the energy released by the capacitor, and during the second half of the cycle the capacitor recaptures the same measure of energy from the inductor. This oscillatory condition is referred to as a state of *resonance,* and the applied frequency is called the *resonant frequency.*

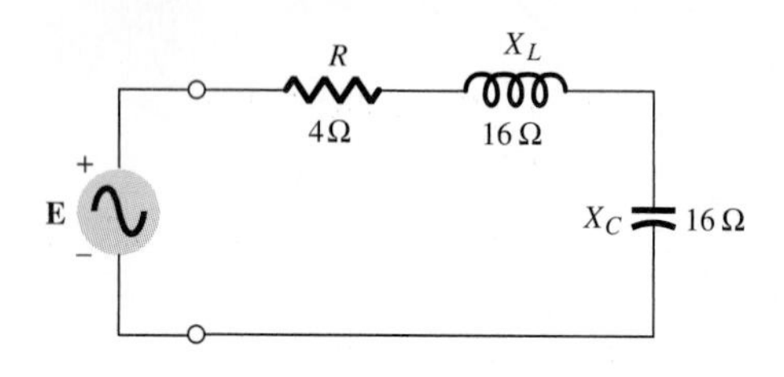

FIG. 5.57
Series resonant circuit.

The series *RLC* configuration in Fig. 5.57 is called a *series resonant circuit* when the condition of maximum power transfer is satisfied. Note that *both* an inductor and capacitor are required to establish the resonant state. The condition of resonance occurs when

$$X_L = X_C \tag{5.28}$$

If we substitute for each term, we find the resonant frequency to be determined by

$$2\pi f_s L = \frac{1}{2\pi f_s C}$$

$$f_s^2 = \frac{1}{4\pi^2 LC}$$

$$f_s = \frac{1}{2\pi\sqrt{LC}} \quad \text{(hertz)} \tag{5.29}$$

At resonance, the impedance of the network is determined by

$$\mathbf{Z}_{T_s} = R + j(X_L - X_C) = R + j0$$

$$\mathbf{Z}_{T_s} = R \quad \text{(at resonance)} \tag{5.30}$$

Since the net reactance at resonance is zero, the apparent power equals the real power, and the power factor of the network $P_T/S_T = 1$.

The current-versus-frequency curve appears in Fig. 5.58. Note that since the net impedance is a minimum value (*R*), the current is a maxi-

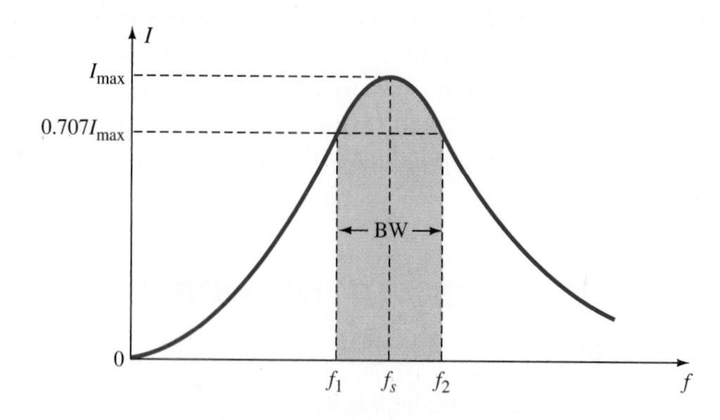

FIG. 5.58
Series resonance curve.

mum. At frequencies to the left of the resonant value the reactance of the capacitor increases above that of the inductor, and the net impedance increases, causing a decrease in I. For frequencies above the resonant value the inductive reactance is greater than that of the capacitor, causing an increase in impedance and a decrease in I toward zero.

The power delivered at resonance is determined by

$$P_{max} = I_{max}^2 R \tag{5.31}$$

where $I_{max} = E/Z_T = E/R$.

If the current drops to 0.707 of its peak value as indicated on the curve, the power delivered will drop to

$$P = (0.707I_{max})^2 R = 0.5I_{max}^2 R = 0.5P_{max}$$

or half the maximum at resonance. The frequencies f_1 and f_2 are called the *half-power, cutoff,* and *band* frequencies. The last term comes from the fact that f_1 and f_2 define a *bandwidth* (BW) as shown in Fig. 5.58. For most applications, it is assumed that the resonant frequency bisects the bandwidth, so that f_1 and f_2 are equidistant from f_s.

In equation form,

$$\text{BW} = f_2 - f_1 \qquad \text{(hertz)} \tag{5.32}$$

It can be shown through a sequence of substitutions that the bandwidth is related to the circuit elements in the following manner:

$$\text{BW} = f_2 - f_1 = \frac{R}{2\pi L} \qquad \text{(hertz)} \tag{5.33}$$

It is important to realize from the preceding that any chosen frequency in the range f_1 to f_2 results in a power transfer that is at least half the maximum level.

If we define

$$Q_s = \frac{X_L}{R} \tag{5.34}$$

where X_L is the resonant frequency value, to be the *quality factor* of the network, the following relationship results:

$$\text{BW} = \frac{f_s}{Q_s} \qquad \text{(hertz)} \tag{5.35}$$

The quality factor of a network provides information about the shape of the curve in Fig. 5.58, which is also the *selectivity* curve. The term comes from the fact that one must be selective in the choice of frequencies to ensure that one is within the bandwidth. As the Q factor increases, the peaking curve narrows and the bandwidth decreases, as indicated by Eq. (5.35).

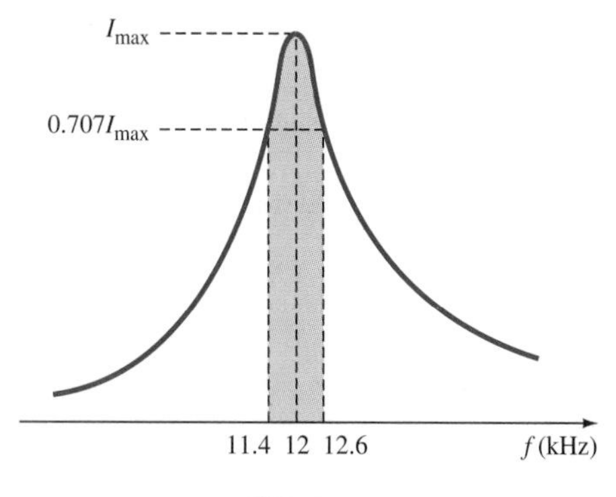

FIG. 5.59

EXAMPLE 5.11 For the resonance curve in Fig. 5.59, determine:

a. The resonant frequency.
b. The bandwidth.
c. The Q of the network.
d. The inductance of the network if the resistance of the network is 4 Ω.

Solution:

a. $f_s = 11.4\text{ kHz} + (12.6\text{ kHz} - 11.4\text{ kHz})/2 = 11.4\text{ kHz} + 0.6\text{ kHz} = \mathbf{12\text{ kHz}}$

b. $\text{BW} = 12.6\text{ kHz} - 11.4\text{ kHz} = \mathbf{1.2\text{ kHz}}$

c. $\text{BW} = \dfrac{f_s}{Q_s}$ and $Q_s = \dfrac{f_s}{\text{BW}} = \dfrac{12{,}000\text{ Hz}}{1200\text{ Hz}} = \mathbf{10}$

d. $Q_s = \dfrac{X_L}{R}$ and $X_L = QR = (10)(4\ \Omega) = \mathbf{40\ \Omega}$

$$X_L = 2\pi f_s L \text{ and } L = \frac{X_L}{2\pi f_s} = \frac{40\ \Omega}{(6.28)(12 \times 10^3\text{ Hz})} = \frac{40 \times 10^{-3}}{75.36}\text{ H} = \mathbf{0.531\text{ mH}}$$

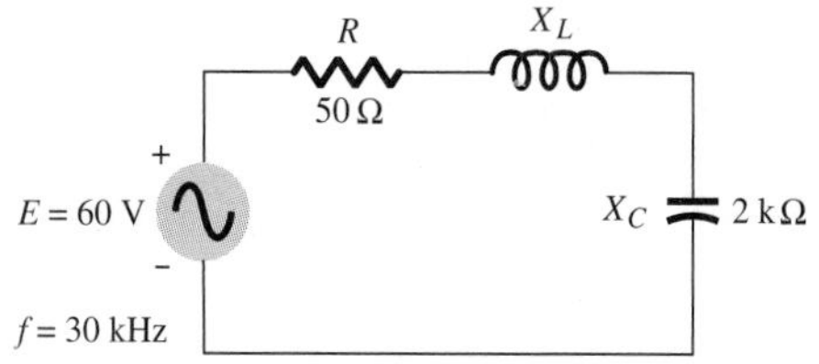

FIG. 5.60

EXAMPLE 5.12 For the series *RLC* circuit in Fig. 5.60, determine:

a. X_L for resonance.
b. The Q of the network at resonance.
c. The bandwidth.
d. The power delivered at resonance.
e. The power delivered at the half-power frequencies (HPFs).
f. The general shape of the resonance curve.

Solution:

a. $X_L = X_C = \mathbf{2\text{ k}\Omega}$

b. $Q = \dfrac{X_L}{R} = \dfrac{2000\ \Omega}{50\ \Omega} = \mathbf{40}$

c. $\text{BW} = \dfrac{f_s}{Q} = \dfrac{30{,}000\text{ Hz}}{40} = \mathbf{750\text{ Hz}}$

d. $P_{max} = I^2_{max}R = \left(\dfrac{E}{R}\right)^2 R = \dfrac{E^2}{R} = \dfrac{(60\text{ V})^2}{50\ \Omega} = \dfrac{3600}{50}\text{ W} = \mathbf{72\text{ W}}$

e. $P_{HPF} = \dfrac{P_{max}}{2} = \dfrac{72\text{ W}}{2} = \mathbf{36\text{ W}}$

f. See Fig. 5.61.

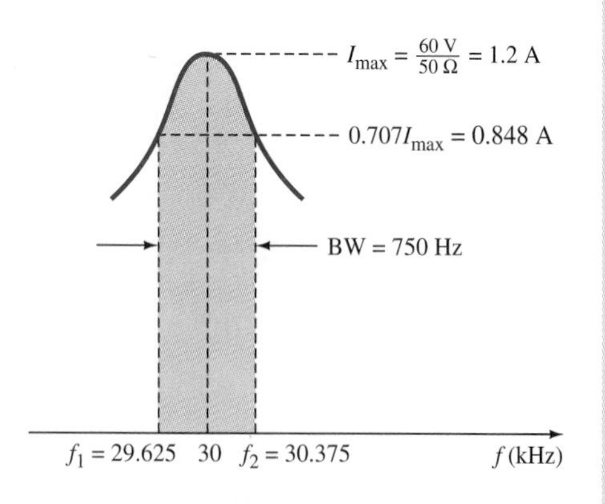

FIG. 5.61

Parallel Resonance

In addition to the series resonant circuit there is a parallel resonant network such as shown in Fig. 5.62 that will also develop a frequency curve similar to that obtained for I in Fig. 5.58. However, for this network the curve is that of the voltage V_C versus frequency. Note also that both a coil and a capacitor are required to obtain the resonant condition. The resistance R can be either the resistance of the coil or a combination of that resistance and some added resistance chosen to affect the shape of the curve in a particular manner. At the condition of resonance there is an equilateral transfer of energy between the reactive elements. This particular resonant circuit is also referred to as a *tank* circuit because of its storage of energy in the reactive elements. As shown in Fig. 5.62, the source applied to the parallel resonant circuit has a constant-current characteristic rather than the voltage source appearing for the series resonant circuit. This fact, in combination with the shape of the impedance-versus-frequency curve for this network, will result in the desired frequency curve. As shown in Fig. 5.63 the impedance of the parallel resonant circuit is a maximum at the resonant frequency and drops off to the right and left of this frequency much like the current characteristics of the series resonant circuit. The maximum impedance is given by

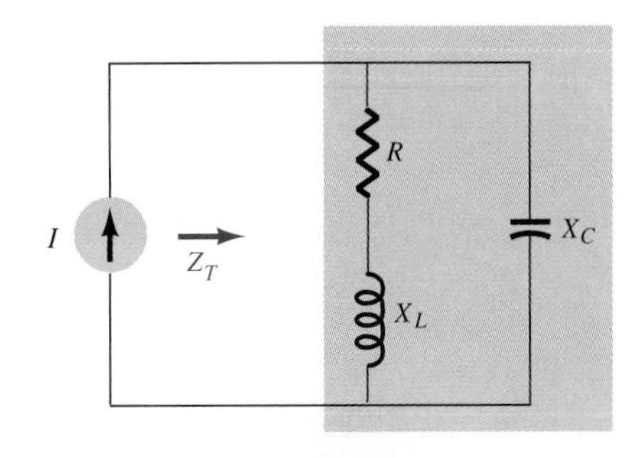

FIG. 5.62
Parallel resonant circuit.

$$Z_{T_{\max}} = Q_p^2 R \qquad \textbf{(5.36)}$$

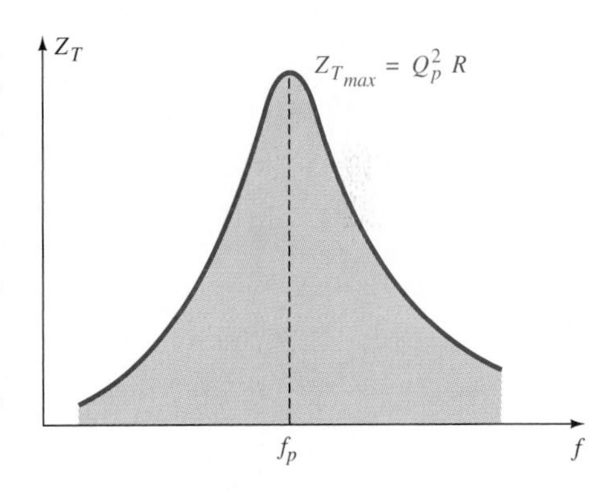

FIG. 5.63
Impedance characteristics of a parallel resonant circuit.

where Q_p is in the quality factor of the network and for most networks is determined by

$$Q_p = \frac{X_L}{R} \qquad \textbf{(5.37)}$$

In this book our interest lies in networks in which $Q_p \geq 10$. When this condition is satisfied, the related equations take a much simpler form. The majority of practical situations encountered satisfy this condition. For quality factors less than 10 there are numerous excellent references available. In the analysis to follow note the similarity between the equations for parallel resonance and those provided for series resonance. For example, parallel resonance occurs when

$$X_L = X_C \qquad (Q_p \geq 10) \qquad \textbf{(5.38)}$$

and the resonant frequency is determined by

$$f_p = \frac{1}{2\pi\sqrt{LC}} \qquad (Q_p \geq 10) \qquad \textbf{(5.39)}$$

The voltage across the capacitor is determined by

$$V_C = IZ_T$$

where I is the constant magnitude of the current source and Z_T is the frequency-dependent impedance of the tank circuit, as depicted by Fig. 5.63. Since I is a constant, the voltage V_C has the same shape as the impedance curve, as shown in Fig. 5.64.

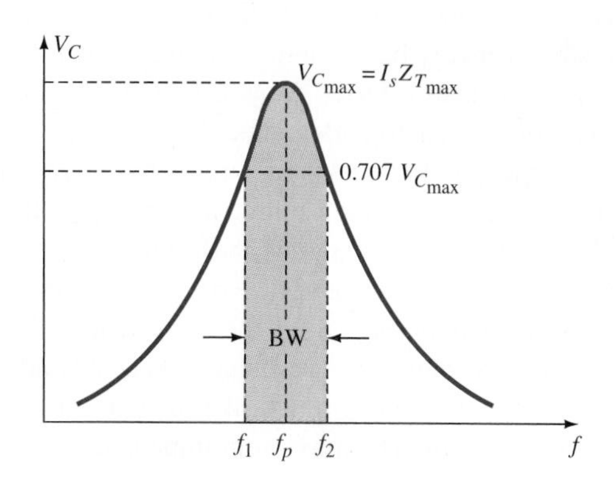

FIG. 5.64
Resonance curve for a "tank" circuit.

The bandwidth is determined by

$$\text{BW} = \frac{f_p}{Q_p} \tag{5.40}$$

and the half-power frequencies are defined as for the series resonant circuit. This network therefore allows us to choose the frequencies that will result in maximum output voltage for the next stage. The tuning dial on a radio adjusts the capacitance, and therefore the resonant frequency, to the station we are interested in listening to. All other stations at the low end of the curve receive less voltage and therefore less power and are not audible.

EXAMPLE 5.13 For the tank circuit in Fig. 5.65, determine:

a. X_C for resonance.
b. The impedance at resonance.
c. The resonant frequency if $L = 10\ \mu\text{H}$.
d. The cutoff frequencies.
e. The shape of the resonant curve.

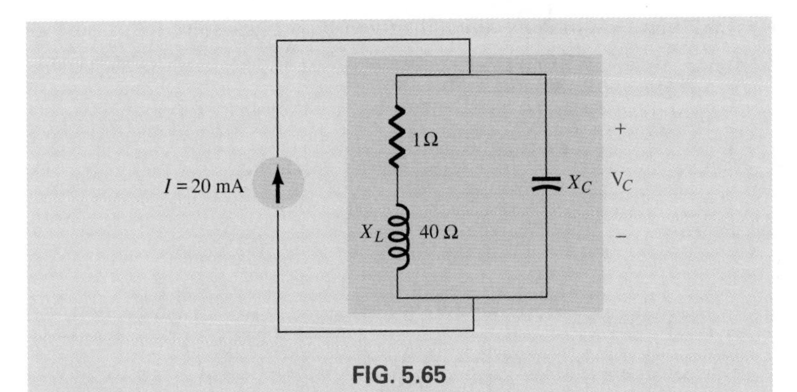

FIG. 5.65

Solution:

a. $Q_p = \dfrac{X_L}{R} = \dfrac{40\ \Omega}{1\ \Omega} = 40 > 10$; therefore, $X_C = X_L = \mathbf{40\ \Omega}$

b. $Z_{T_{max}} = Q_p^2 R = (40)^2 \times 1\ \Omega = 1600(1\ \Omega) = \mathbf{1600\ \Omega}$

c. $X_L = 2\pi f_p L \Rightarrow f_p = \dfrac{X_L}{2\pi L} = \dfrac{40\ \Omega}{(6.28)(10 \times 10^{-6}\ \text{H})} = \mathbf{636.94\ kHz}$

d. $\text{BW} = \dfrac{f_p}{Q_p} = \dfrac{636{,}940\ \text{Hz}}{40} = 15{,}923.50\ \text{Hz}$, and

$$f_1 = f_p - \frac{\text{BW}}{2} = 636{,}940\ \text{Hz} - \frac{15{,}923.50\ \text{Hz}}{2} = \mathbf{628{,}978\ Hz}$$

$$f_2 = f_p + \frac{\text{BW}}{2} = 637{,}000\ \text{Hz} + \frac{15923.50\ \text{Hz}}{2} = \mathbf{644{,}902\ Hz}$$

e. The resonant curve appears in Fig. 5.66.

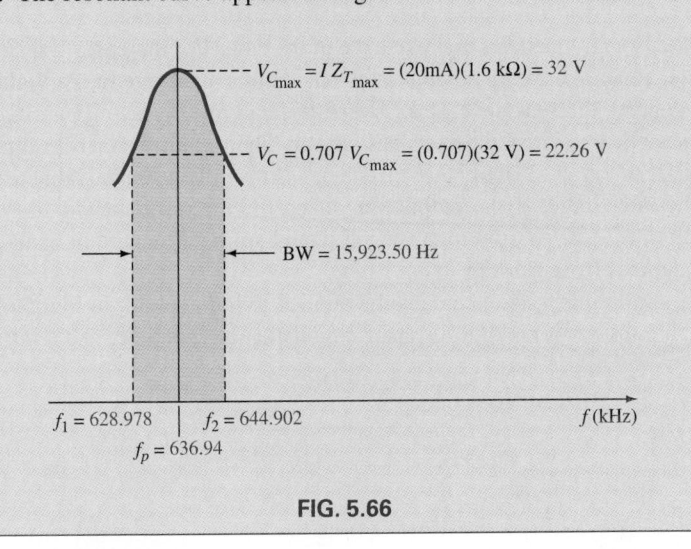

FIG. 5.66

5.12 FILTERS

Filters are simply an extension of the tuned networks introduced in the previous section. As the name implies, filters pick out a range of frequencies for passage or blockage. They *filter out* the unwanted frequencies.

The first to be described is called the *band-pass* filter, since it "passes" a particular range of frequencies. Figure 5.67 shows a series resonant

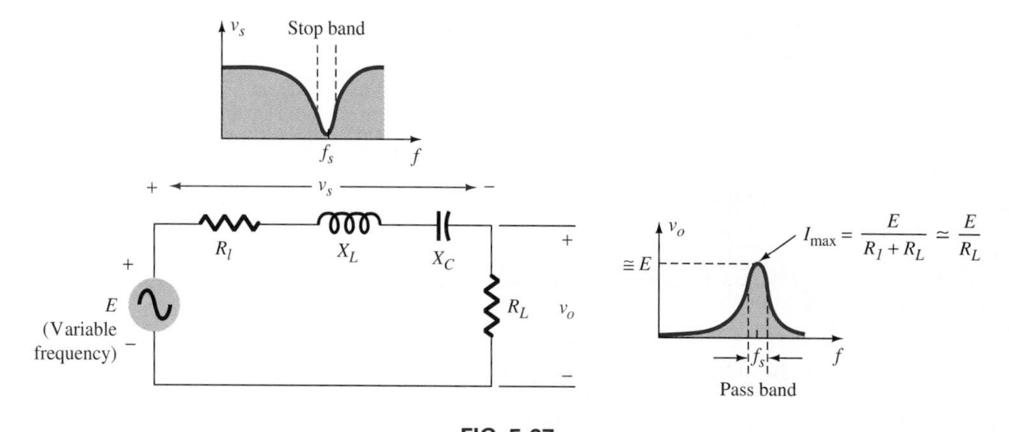

FIG. 5.67
Series resonant filter.

circuit designed to perform this function. The output is taken off the resistor R_L. The resistance R_l is the internal resistance of the inductor. Usually, the resistance R_L is considerably larger than R_l, causing a major portion of the applied voltage to appear across R_L. At resonance, the impedance of the series R_l, X_L, X_C circuit is a minimum, and

$$V_o = \frac{R_L E}{R_L + R_l} \cong E$$

At lower frequencies the reactance of the capacitor increases, causing an increasing part of E to appear across the resonant circuit. At higher frequencies the reactance of X_L overrides, and an increasing part of E again appears across the resonant circuit. The output frequency curve appears in Fig. 5.67. Only those frequencies near the resonant value pass to the next stage. If the output was taken off the resonant circuit, we would have a *band-stop* filter, which would stop only a certain band of frequencies from passing.

A similar discussion can also be applied to the network in Fig. 5.68, which employs a parallel resonant circuit for its selectivity. As a *pass-*

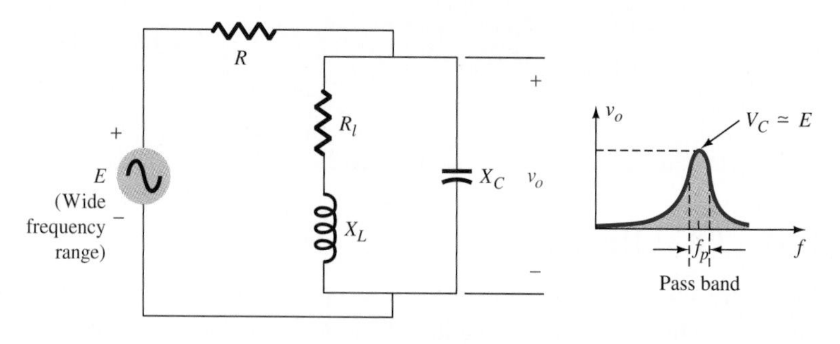

FIG. 5.68
Parallel resonant filter.

band filter, the output is taken across the tank circuit since the impedance of the network is a maximum only near resonance. This results in $V_o \cong E$, since the impedance at the resonant frequency is much greater than that of R. The frequency characteristics appear in Fig. 5.68. The band-stop characteristics could be obtained by taking the output off the resistor R.

EXAMPLE 5.14 Sketch the output frequency characteristic for the network in Fig. 5.69.

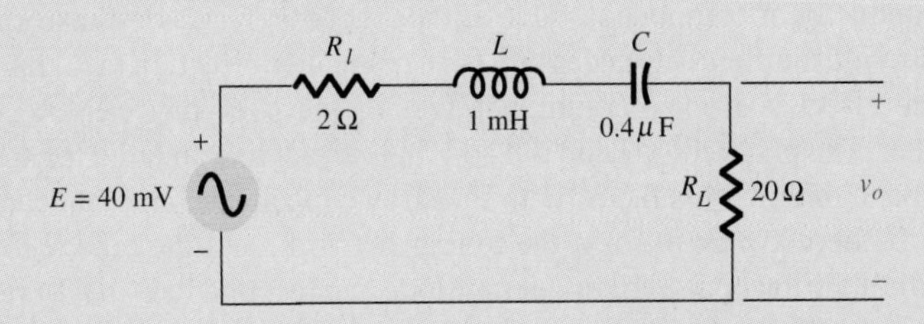

FIG. 5.69

Solution: The resonant frequency is

$$f_s = \frac{1}{2\pi\sqrt{LC}} = \frac{1}{(6.28)\sqrt{(1 \times 10^{-3}\text{ H})(0.4 \times 10^{-6}\text{ F})}}$$

$$= \frac{1}{6.28\sqrt{4 \times 10^{-10}}} = \frac{1}{6.28(2 \times 10^{-5})} = \frac{10^4}{1.256}$$

$$= \mathbf{7.96\ kHz}$$

$$\text{BW} = \frac{R}{2\pi L} = \frac{2\ \Omega + 20\ \Omega}{(6.28)(1 \times 10^{-3}\text{ H})} = \frac{22}{6.28} \times 10^3\text{ Hz} = \mathbf{3.5\ kHz}$$

$$f_1 = f_s - \frac{\text{BW}}{2} = 7.96\text{ kHz} - \frac{3.5\text{ kHz}}{2} = \mathbf{6.21\ kHz}$$

$$f_2 = f_s + \frac{\text{BW}}{2} = 7.96\text{ kHz} + \frac{3.5\text{ kHz}}{2} = \mathbf{9.71\ kHz}$$

$$V_{\max} = \frac{(20\ \Omega)(40\text{ mV})}{20\ \Omega + 2\ \Omega} = \frac{800}{22}\text{ mV} = \mathbf{36.36\ mV}$$

The pass-band output appears in Fig. 5.70.

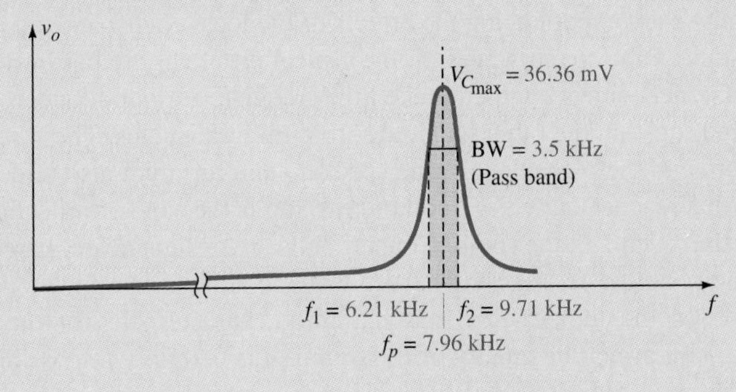

FIG. 5.70

5.13 APPLICATIONS: STRAY RESONANCE AND GRAPHIC AND PARAMETRIC EQUALIZERS

Stray Resonance

Stray resonance, like stray capacitance and inductance and unexpected resistance levels, can occur in totally unexpected situations and can severely affect the operation of a system. All that is required to produce stray resonance is, for example, a level of capacitance introduced by parallel wires or copper leads on a printed circuit board, or simply two parallel conductive surfaces with residual charge and inductance levels associated with any conductor or components such as tape recorder heads or transformers that provide the elements necessary for a resonance effect. In fact, this resonance effect is a very common effect in the everyday cassette tape recorder. The play/record head is a coil that can act like an inductor and an antenna. Combine this factor with the stray capacitance and real capacitance in the network to form the tuning network, and the tape recorder with the addition of a semiconductor diode can respond like an AM radio. As you plot the frequency response of any transformer, you will normally find a region where the response has a peaking effect (look ahead to Fig. 6.29). This peaking is due solely to the inductance of the coils of the transformer and the stray capacitance between the wires.

In general, any time you see an unexpected peaking in the frequency response of an element or a system, it is normally caused by a resonance condition. If the response has a detrimental effect on the overall operation of the system, a redesign may be in order, or a filter can be added that will block the frequencies that result in the resonance condition. Of course, when you add a filter composed of inductors and/or capacitors, you must be careful that you don't add another unexpected resonance condition. It is a problem that can be properly weighed only by constructing the system and exposing it to the full range of tests.

Graphic and Parametric Equalizers

We have all noticed at one time or another that the music we hear in a concert hall doesn't quite sound the same when we play it at home on our entertainment center. Even after we check the specifications of the speakers and amplifiers and find that both are nearly perfect (and the most expensive we can afford), the sound is still not what it should be. In general, we are experiencing the effects of the local environmental characteristics on the sound waves. Some typical problems are hard walls or floors (stone, cement), which make high frequencies sound louder. Curtains and rugs, on the other hand, absorb high frequencies. The shape of the room and the placement of the speakers and furniture also affect the sound that reaches our ears. Another criterion is the echo or reflection of sound that occurs in the room. Concert halls are designed very carefully with their vaulted ceilings and curved walls to allow a certain amount of echo. Even the temperature and humidity characteristics of the surrounding air affect the quality of the sound. It is certainly impossible, in most cases, to redesign your listening area to match a concert hall, but with the proper use of electronic systems you can develop a response that will have all the qualities that you can expect from a home entertainment center.

For a quality system you can take a number of steps: *characterization and digital delay (surround sound)* and *proper speaker and amplifier selection and placement.* Characterization is a process whereby a thorough sound absorption check of the room is performed and the frequency response determined. A *graphic equalizer* such as appears in Fig. 5.71(a) is then used to make the response "flat" for the full range of frequencies. In other words, the room is made to appear as though all the frequencies receive equal amplification in the listening area. For instance, if the room is fully carpeted with full draping curtains, there will be a lot of high-frequency absorption, requiring that the high frequencies have additional amplification to match the sound levels of the mid and low frequencies. To *characterize* the typical rectangular-shaped room, a setup such as shown in Fig. 5.71(b) may be used. The amplifier and speakers are placed in the center of one wall, with additional speakers in the corners of the room facing the reception area. A mike is then placed in the reception area about 10 ft from the amplifier and centered between the two other speakers. A *pink noise* is then sent out from a spectrum analyzer (often an integral part of the graphic equalizer) to the amplifier and speakers. Pink noise is actually a square-wave signal whose amplitude and frequency can be controlled. A square-wave signal was chosen because a Fourier breakdown of a square-wave signal results in a broad range of frequencies for the system to check. It can be shown that a square wave can be constructed of an infinite series of sine waves of different frequencies. Once the proper volume of pink noise is established, the spectrum analyzer can be used to set the response of each slide band to establish the desired flat response. The center frequencies for the slides of the graphic equalizer of Fig. 5.71(a) are provided in Fig. 5.71(c), along with the frequency response for a number of adjoining frequencies evenly spaced on a logarithmic scale. Note that each center frequency is actually the resonant frequency for that slide. The design is such that each slide can control the volume associated with that frequency, but the bandwidth and frequency response stay fairly constant. A good spectrum analyzer will have each slide set against a decibel (dB) scale. The decibel scale simply establishes a scale for the comparison of audio levels. At a normal listening level, usually a change of about 3 dB is necessary for the audio change to be detectable by the human ear. At low levels of sound, a 2-dB change may be detectable, but at loud sounds probably a 4-dB change would be necessary for the change to be noticed. These are not strict laws but simply rules of thumb commonly used by audio technicians. For the room in question, the mix of settings may be as shown in Fig. 5.71(c). Once set, the slides are not touched again. A flat response has been established for the room for the full audio range so that every sound or type of music is covered.

A *parametric equalizer* such as appears in Fig. 5.72 is similar to a *graphic equalizer,* but instead of separate controls for the individual frequency ranges, it uses three basic controls over three or four broader frequency ranges. The typical controls—the *gain, center frequency,* and *bandwidth*—are typically available for the *low-, mid-,* and *high-frequency*

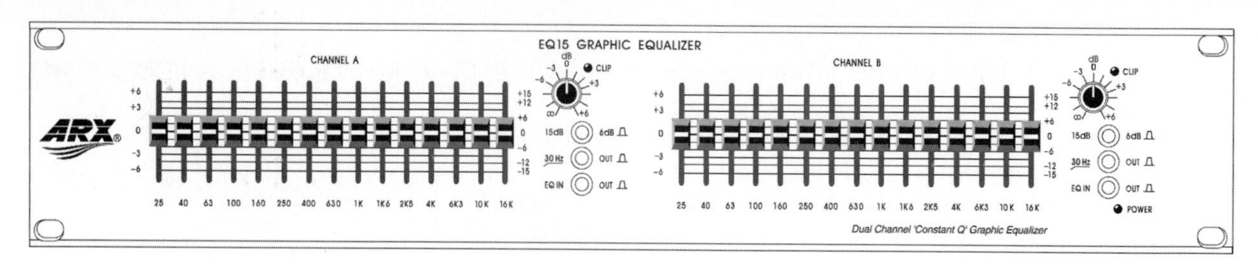
EQ15 GRAPHIC EQUALIZER
CHANNEL A
CHANNEL B
ARX
CLIP
15dB
6dB
30 Hz
OUT
EQ IN
POWER
25 40 63 100 160 250 400 630 1K 1K6 2K5 4K 6K3 10K 16K
Dual Channel 'Constant Q' Graphic Equalizer

(a)

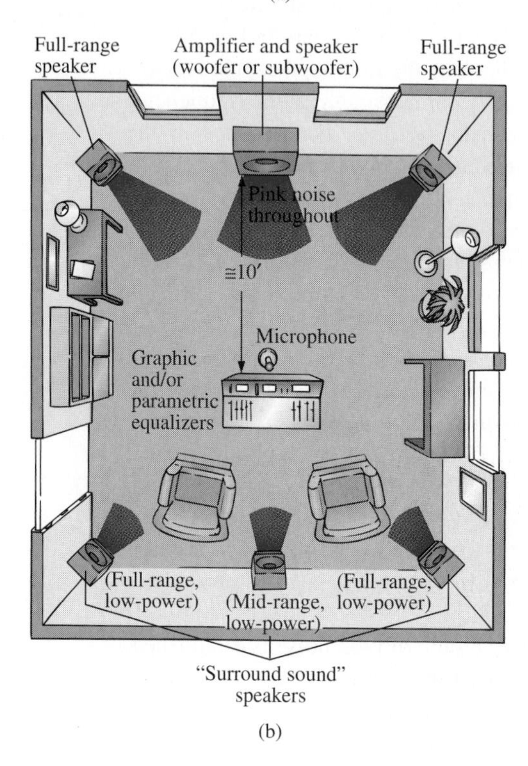
Full-range speaker
Amplifier and speaker (woofer or subwoofer)
Full-range speaker
Pink noise throughout
≅10′
Microphone
Graphic and/or parametric equalizers
(Full-range, low-power)
(Mid-range, low-power)
(Full-range, low-power)
"Surround sound" speakers

(b)

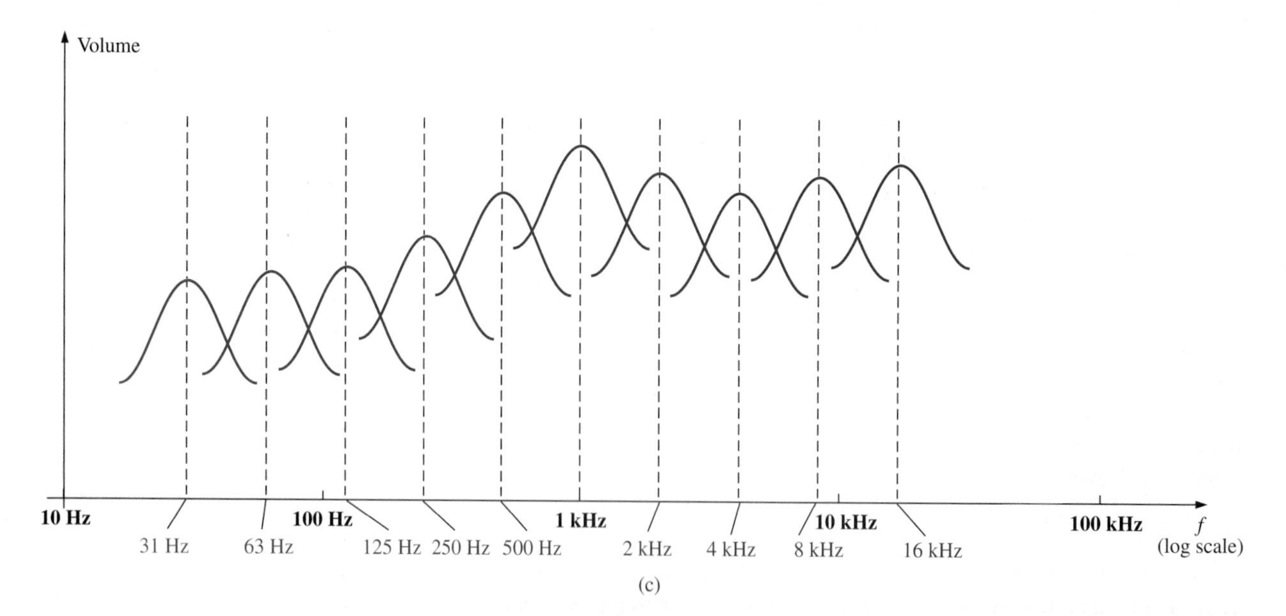
Volume
10 Hz
100 Hz
1 kHz
10 kHz
100 kHz
f (log scale)
31 Hz
63 Hz
125 Hz
250 Hz
500 Hz
2 kHz
4 kHz
8 kHz
16 kHz

(c)

FIG. 5.71

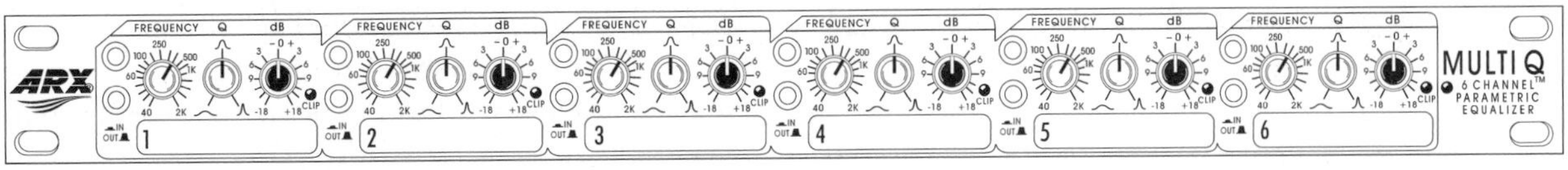

FIG. 5.72
Six-channel parametric equalizer. (Courtesy of ARX Systems.)

ranges. Each is fundamentally an independent control; that is, a change in one can be made without affecting the other two. For the parametric equalizer of Fig. 5.72, each of the six channels has a frequency control switch that, in conjunction with the $f \times 10$ switch, will give a range of center frequencies from 40 Hz through 16 kHz. It has controls for BW ("Q") from 3 octaves to $\frac{1}{20}$ octave, and ± 18 dB cut and boost. Some like to refer to the parametric equalizer as a *sophisticated tone control* and actually use them to enrich the sound after the flat response has been established by the graphic equalizer. The effect achieved with a standard tone control knob is sometimes referred to as "boring" compared with the effect established by a good parametric equalizer, primarily because the former can control only the volume and not the bandwidth or center frequency. In general, graphic equalizers establish the important *flat response,* while parametric equalizers are adjusted to provide the *type* and *quality* of sound you like to hear. You can "notch out" the frequencies that bother you and remove tape "hiss" and the "sharpness" often associated with CDs.

One characteristic of concert halls that is more difficult to fake is the fullness of sound that concert halls are able to provide. In the concert hall you have the direct sound from the instruments and the reflection of sound off the walls and the vaulted ceilings that were all carefully designed expressly for this purpose. Any reflection results in a delay in the sound waves reaching the ear, creating the fullness effect. Through digital delay, speakers can be placed to the back and side of a listener to establish the surround-sound effect. In general, the delay speakers are much lower in wattage, with 20-W speakers typically used with a 100-W system. The echo response is one reason that people often like to play their stereos louder than they should for normal hearing. By playing the stereo louder, they create more echo and reflection off the walls, bringing into play some of the fullness heard at concert halls.

It is probably safe to say that any system composed of quality components, a graphic and parametric equalizer, and surround sound will have all the components necessary to have a quality reproduction of the concert hall effect.

5.14 COMPUTER ANALYSIS

PSpice (Windows)

The beauty of the **Probe** response within PSpice can now be fully appreciated in the plots that are produced in only a few seconds for series and parallel resonant circuits. The construction of the network, followed by the setting up of the analysis and the resulting curves, will immediately reveal the resonant frequency, the bandwidth, and the quality factor. In general, PSpice produces an enormous savings in time and energy compared with longhand methods.

For the first time, the horizontal axis will be in the frequency domain rather than in the time domain, as in all prior examples. For the series resonant circuit of Fig. 5.73, the magnitude of the source was chosen to

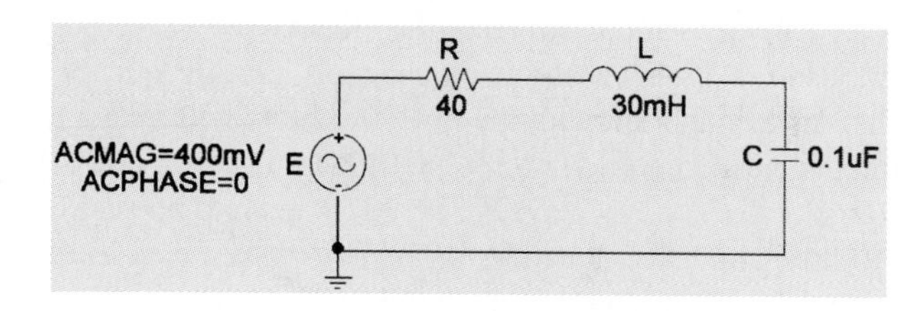

FIG. 5.73
Applying PSpice (Windows) to a series resonant circuit.

produce a maximum current of $I = 400 \text{ mV}/40\ \Omega = 10 \text{ mA}$ at resonance, and the reactive elements were chosen to establish a frequency of

$$f_s = \frac{1}{2\pi\sqrt{LC}} = \frac{1}{2\pi\sqrt{(30 \text{ mH})(0.1\ \mu\text{F})}} \cong \mathbf{2.91\ kHz}$$

The quality factor is

$$Q_l = \frac{X_L}{R_l} = \frac{546.64\ \Omega}{40\ \Omega} \cong \mathbf{13.7}$$

and the bandwidth is

$$BW = \frac{f_s}{Q_l} = \frac{2.91 \text{ kHz}}{13.7} \cong \mathbf{212\ Hz}$$

The chosen source was **VAC,** since the **AC Sweep** will take care of the frequency range of interest. The source **VAC** does not have a frequency attribute. Under **AC Sweep** the following choices were made: **Total Pts:** 1000, **Start Freq:** 1 kHz, and **End Freq:** 10 kHz. A **Simulation** then resulted in the **Probe** screen, and **Trace-Add-I(R)** produced the waveform of Fig. 5.74. As indicated, the response peaked at a frequency of 2.91 kHz with a peak current of 9.992 mA, corresponding very well with the calculated resonant frequency of 2.91 kHz and current of 10 mA. The bandwidth was defined by clicking the right mouse button and placing a cursor as close to 7.07 mA as possible. Note in the cursor dialog box at the bottom of the screen that the upper cutoff frequency is 3.01 kHz at a level of 7.068 mA. Since the Q_l of the network is quite high, if we assume that the resonant frequency is bisecting the bandwidth, the total bandwidth is $2 \times (3.0142 \text{ kHz} - 2.91 \text{ kHz}) = 208.4 \text{ Hz}$, which is very close to the calculated level of 212 Hz. The labels **Imax, fs,** and so on, were all added using the **ABC** icon.

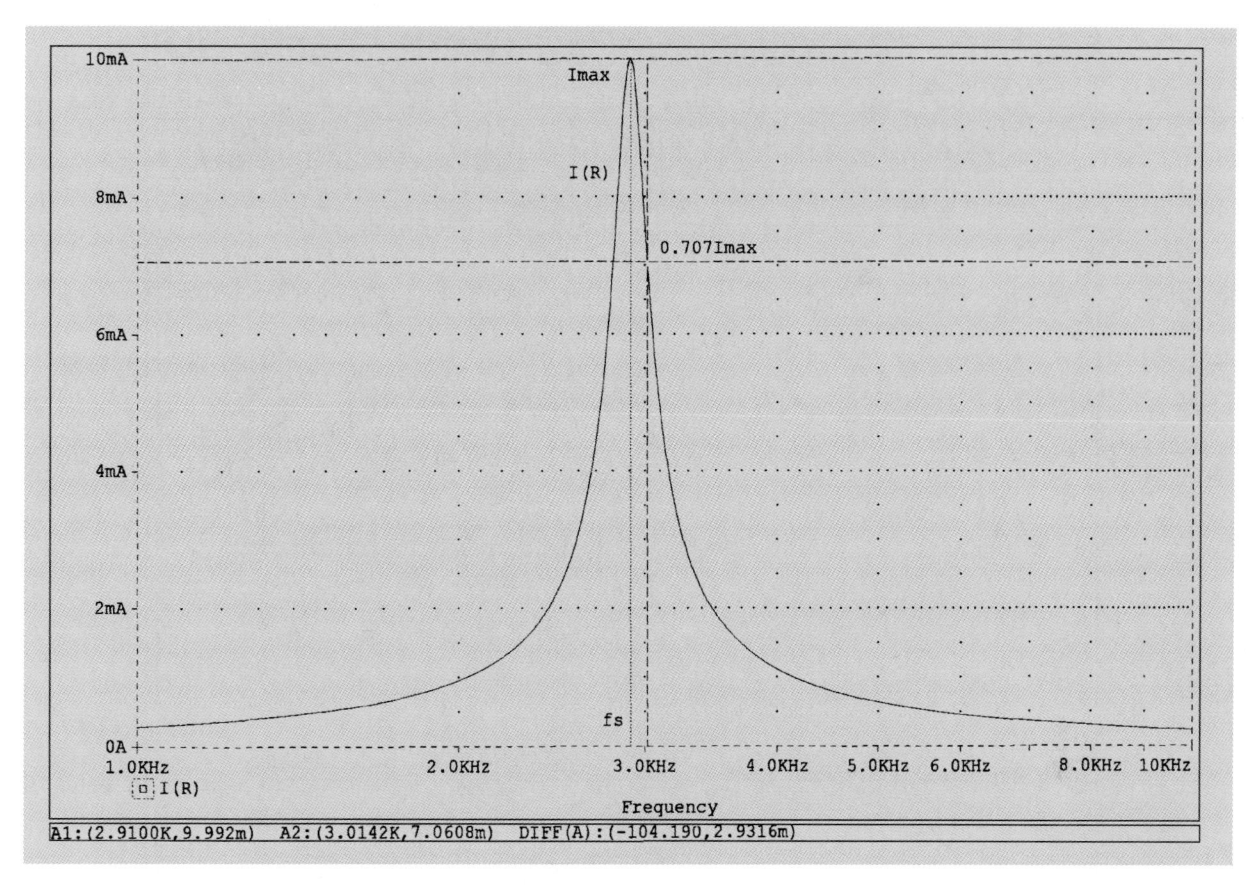

FIG. 5.74
Frequency response for the series resonant circuit of Fig. 5.73.

Electronics Workbench

Next, we employed Electronics Workbench to obtain the response for the parallel resonant network of Fig. 5.75.

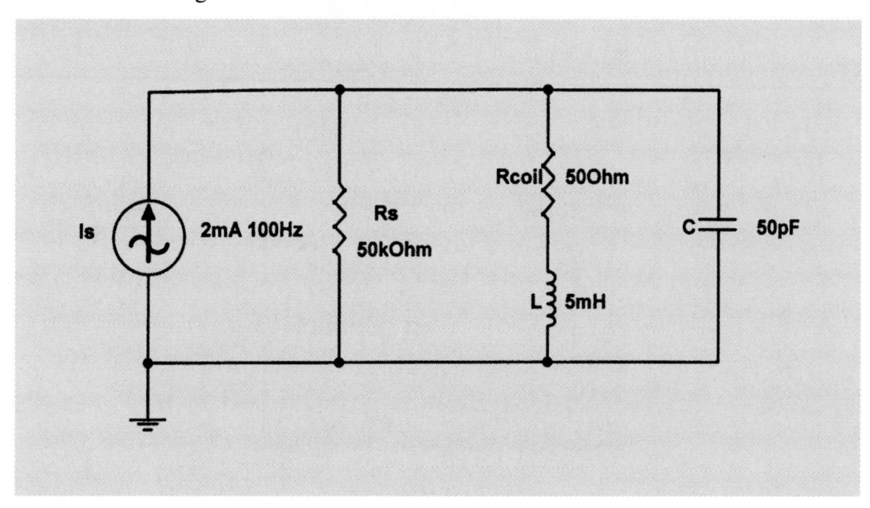

FIG. 5.75
Application of Electronics Workbench to obtain the frequency response of a parallel resonant network.

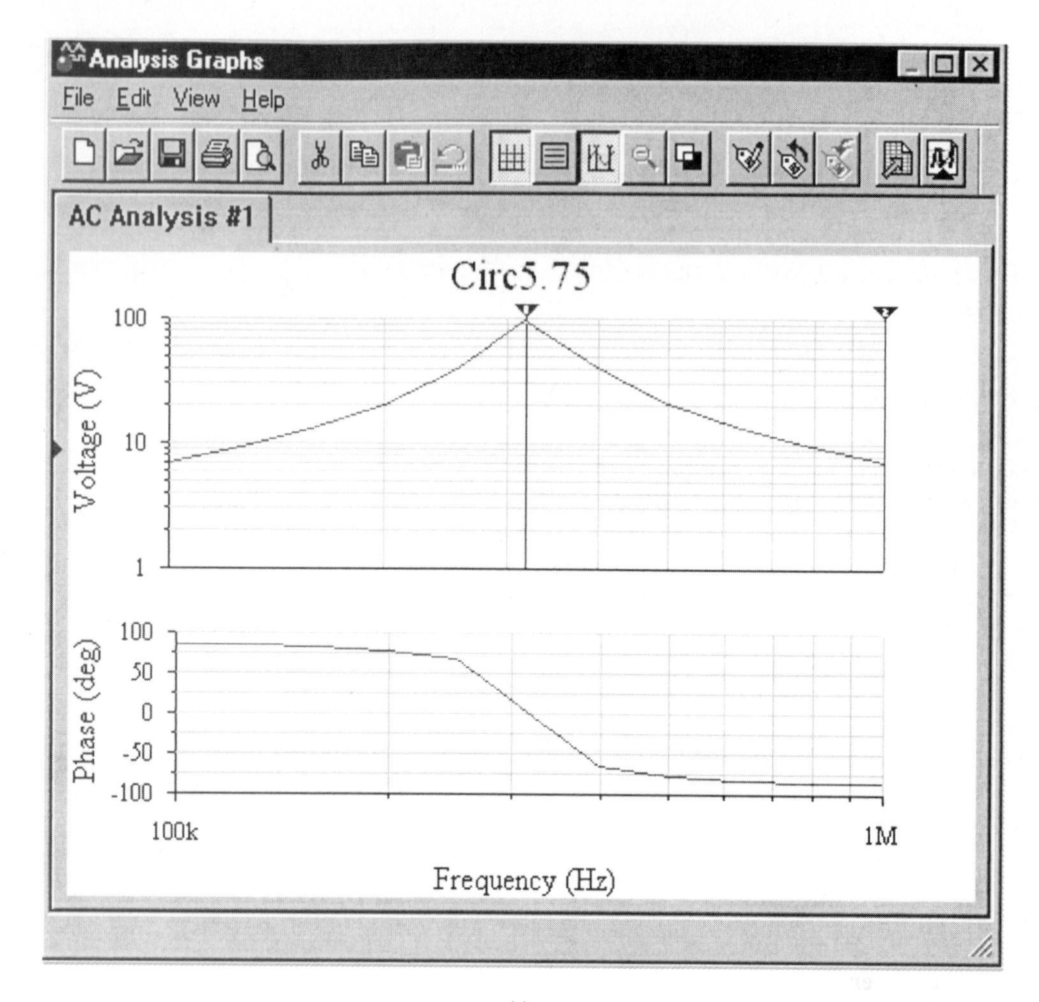

(a)

AC Analysis

	— 1
x1	318.4407k
y1	95.7931
x2	1.0000M
y2	7.0662
dx	681.5593k
dy	-88.7269
1/dx	1.4672u
1/dy	-11.2705m
min x	100.0000k
max x	1.0000M
min y	6.9467
max y	97.3312

(b)

FIG. 5.76

Electronics Workbench response for the parallel resonant network of Fig. 5.75.

The response of Fig. 5.76(a) clearly reveals that the peak response for the voltage V occurs very close to 320 kHz at a level of about 98 V. Using the cursor option, which reveals a great deal about the point of intersection of a vertical line and the graph, we find in Fig. 5.76(b) that the intersection of cursor 1 when placed at the peak is at 316.2 kHz at a level of 97.3 V. The results compare very favorably with a longhand solution of 318.31 kHz and 97.56 V.

5.15 SUMMARY

- The **block impedance format** for the analysis of series–parallel ac networks results in an analysis that is **very similar** to that employed for series–parallel dc networks.
- The **basic procedure** for applying superposition, branch-current analysis, mesh analysis, and nodal analysis is **essentially the same** as for dc networks. The only major difference is the need to work with complex numbers that have both a magnitude and angle.
- For **maximum power transfer** in the ac domain the **magnitude** of the **load** must **equal** that of the series **Thévenin impedance,** and the **angle** associated with the **load** must be the **negative** of that associated with the series **Thévenin impedance.**
- When a **Δ–Y conversion** is performed, the new configuration must be connected between the **same three terminals.**
- When the **balance equation** is applied to the bridge configuration, the **angles** associated with the impedances must **satisfy** the basic equations also.
- The **apparent power** of an ac network is the power that would be delivered if it were **not** for the effect of the **phase angle** between the terminal voltage and current.
- The **reactive power** is a measure of the power **absorbed** by the reactive elements.
- The **real** or **average power** is the power **dissipated** by the network.
- **Power-factor correction** is the act of introducing capacitive elements to an electrical system to bring the **total power factor** closer to **unity.**
- **Polyphase systems** can generate **more than one** sinusoidal voltage for each rotation of the rotor. The only difference between the voltages generated is the phase angle associated with each.
- For a **balanced three-phase system** the current through the neutral is **0 A.**
- For a series *RLC* circuit, **resonance** is achieved at that frequency that results in the **smallest input impedance** and **maximum current.**
- The **half-power, cutoff,** or **band frequencies** are defined as those frequencies at which the **current** of a series resonant circuit is **0.707 of its maximum value.**
- The **higher** the **quality factor** for a series resonant circuit the **narrower** the peak and the smaller the bandwidth.
- For a **parallel resonant network,** resonance is achieved when the **terminal impedance** and **output voltage** are a **maximum.**

- The **half-power, band,** or **cutoff frequencies** for a parallel resonant network are defined as those frequencies at which the **output voltage** drops to **0.707 of its maximum value.**
- **Band-pass filters** are filters designed to **pass** (at high power levels) a specific band of frequencies, while **band-stop filters** are filters designed to **block** (to low power levels) a specific band of frequencies.

Equations:

Thévenin's Theorem

$$|Z_L| = |Z_{\text{Th}}|$$
$$\theta_L = -\theta_{\text{Th}}$$
$$P_{L_{\max}} = \frac{{E_{\text{Th}}}^2}{4R_{\text{Th}}}$$

Δ–Y Conversions

$$\mathbf{Z}_{\text{Y}} = \frac{\mathbf{Z}_{\Delta}}{3}, \mathbf{Z}_{\Delta} = 3\mathbf{Z}_{\text{Y}}$$

Balanced Bridge

$$\frac{\mathbf{Z}_1}{\mathbf{Z}_2} = \frac{\mathbf{Z}_3}{\mathbf{Z}_4}$$

Power

$$P = EI \cos \theta \quad \text{(W)}$$
$$S = EI \quad \text{(VA)}$$
$$P = S \cos \theta \quad \text{(W)}$$
$$Q = EI \sin \theta \quad \text{(VAR)}$$
$$Q = S \sin \theta \quad \text{(VAR)}$$
$$Q_L = I_L^2 X_L = \frac{V_L^2}{X_L} = V_L I_L \quad \text{(VAR)}$$
$$Q_C = I_C^2 X_C = \frac{V_C^2}{X_C} - V_C I_C \quad \text{(VAR)}$$
$$F_P = \frac{P_T}{S_T}$$

Y–Y System

$$I_\phi = I_L$$
$$E_L = \sqrt{3}\, V_\phi$$
$$V_\phi = E_\phi$$

Δ–Δ System

$$E_\phi = E_L$$
$$I_L = \sqrt{3}\, I_\phi$$

Power (Balanced Load)

$$P_T = 3I_\phi^2 R$$
$$Q_T = 3I_\phi^2 X$$
$$S_T = 3V_\phi I_\phi$$
$$F_P = \frac{P_T}{S_T}$$

Resonance

Series

$$f_s = \frac{1}{2\pi\sqrt{LC}}$$
$$Z_{T_s} = R$$
$$P_{\max} = I_{\max}^2 R$$
$$\mathrm{BW} = f_1 - f_2 = \frac{R}{2\pi L} = \frac{f_s}{Q_s}$$
$$Q_s = \frac{X_L}{R}$$

Parallel

$$Z_{T_{\max}} = Q_P^2 R$$
$$Q_p = \frac{X_L}{R}$$
$$f_p = \frac{1}{2\pi\sqrt{LC}}$$
$$\mathrm{BW} = \frac{f_p}{Q_p}$$

PROBLEMS

SECTION 5.2 Series–Parallel Ac Networks

1. a. Determine the currents **I** and $\mathbf{I}_L$ for the network in Fig. 5.77.
   b. Find the power delivered to the resistor R_2.

2. a. Determine the voltage $\mathbf{V}_s$ for the network in Fig. 5.78.
   b. Find the power delivered to R_2.

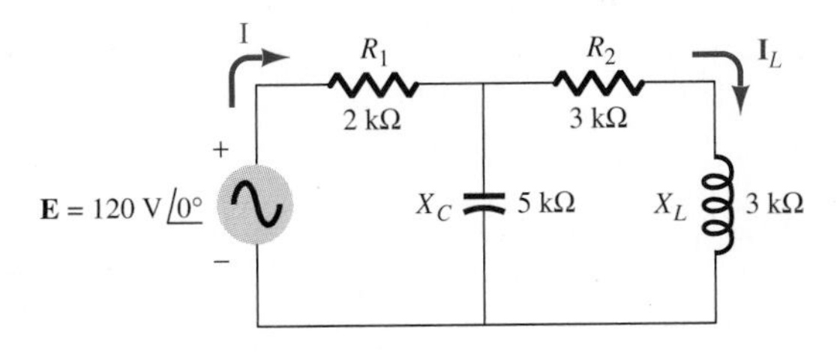

FIG. 5.77

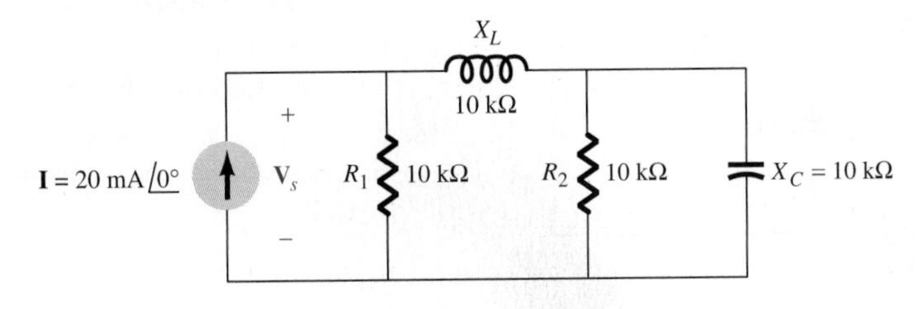

FIG. 5.78

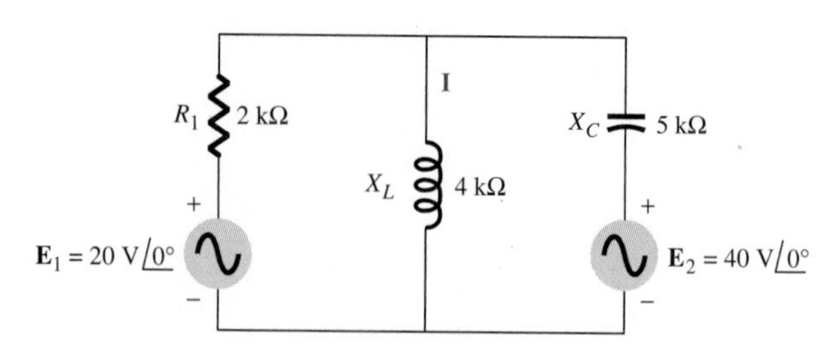

FIG. 5.79

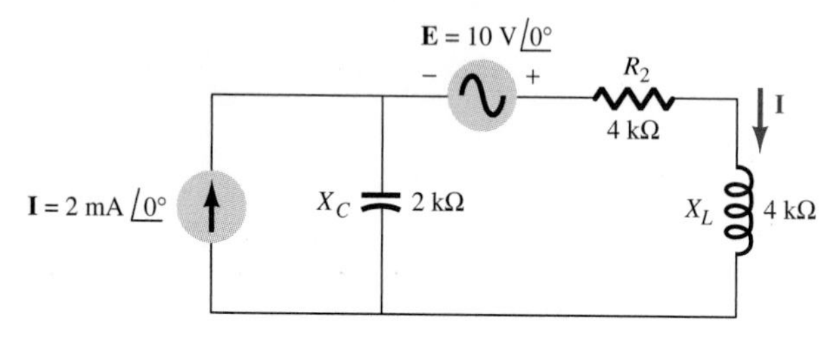

FIG. 5.80

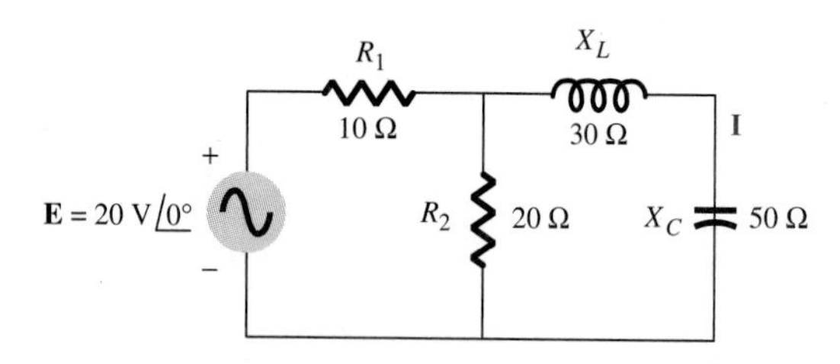

FIG. 5.81

SECTION 5.3 Multisource Ac Networks

3. Determine the current **I** for the network in Fig. 5.79 using superposition.
4. Determine the current **I** for the network in Fig. 5.80 using superposition.
5. Determine the current **I** in Fig. 5.79 using branch-current analysis.
6. Determine the current **I** in Fig. 5.81 using branch-current analysis.
7. Determine the current **I** in Fig. 5.79 by first converting both voltage sources to current sources and then combining the parallel current sources.
8. Convert the current source in Fig. 5.80 to a voltage source and find the current **I** by combining the series voltage sources and using Ohm's law.
9. **a.** Determine the current through the inductor of Fig. 5.79 using mesh analysis.
   b. Using the results for part (a) find the voltage across the inductor.
10. Determine the current **I** in Fig. 5.81 using mesh analysis.
11. Determine the current through resistor R_2 in Fig. 5.86.
12. Determine the voltage across the resistor R_2 in Fig. 5.86 using nodal analysis.
13. **a.** Determine the nodal voltages for the network in Fig. 5.82.
    b. Calculate the current through each impedance using the results for part (a).

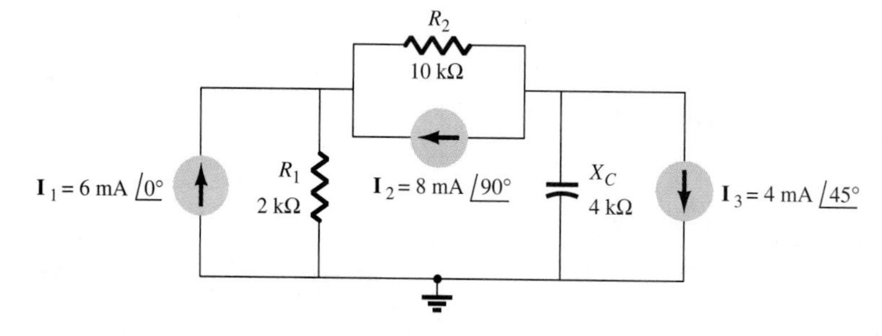

FIG. 5.82

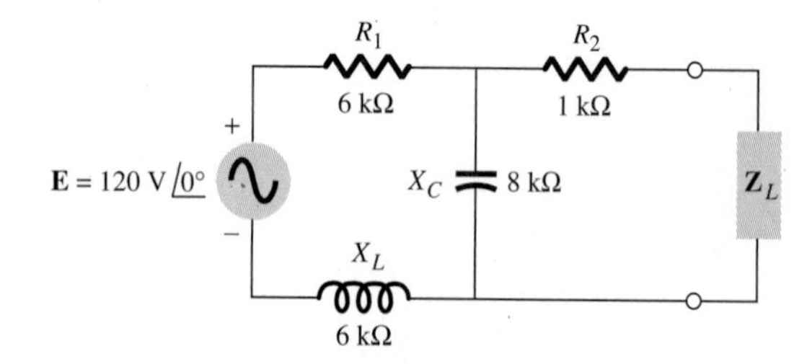

FIG. 5.83

SECTION 5.4 Network Theorems

14. Find the Thévenin equivalent circuit for the network to the left of the load $\mathbf{Z}_L$ in Fig. 5.83.
15. **a.** Determine $\mathbf{Z}_L$ for maximum power to $\mathbf{Z}_L$ in Fig. 5.83.
    b. Calculate the maximum power that can be delivered to $\mathbf{Z}_L$.

16. **a.** Find the Thévenin equivalent circuit for the network to the left of $X_C = 50\ \Omega$ in Fig. 5.81.

b. If X_C in Fig. 5.81 is replaced by a load $\mathbf{Z}_L$, determine R and X for the load so that the load will receive maximum power.

17. **a.** Find the Thévenin equivalent circuit for the network to the left of $\mathbf{Z}_L$ in Fig. 5.84.

b. Determine $\mathbf{Z}_L$ for maximum power to $\mathbf{Z}_L$.

c. Calculate the maximum power to $\mathbf{Z}_L$.

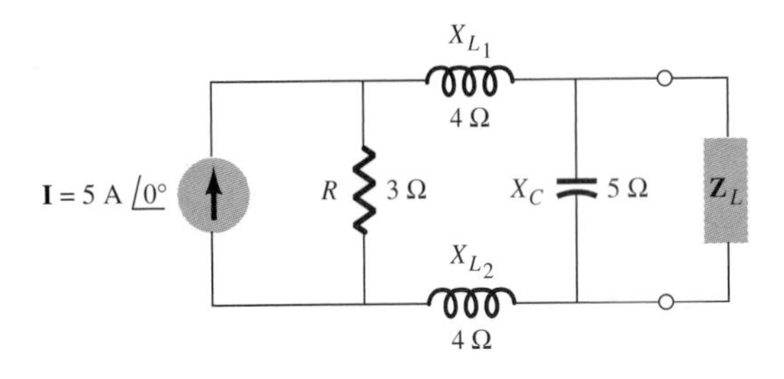

FIG. 5.84

SECTION 5.6 Δ-Y Conversions

18. Determine the total impedance of the network in Fig. 5.85.

19. Determine the current **I** for the network in Fig. 5.86.

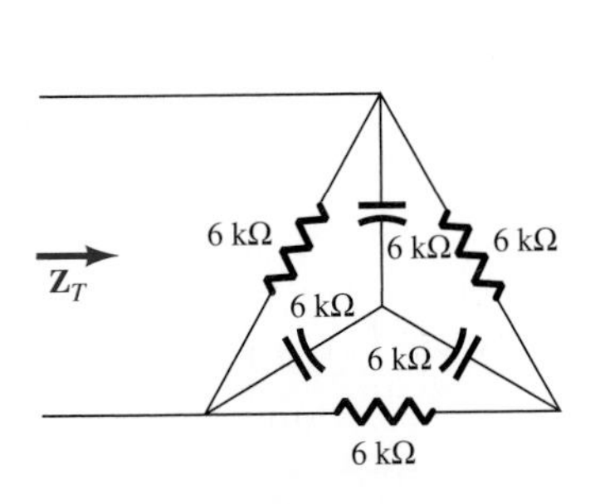

FIG. 5.85

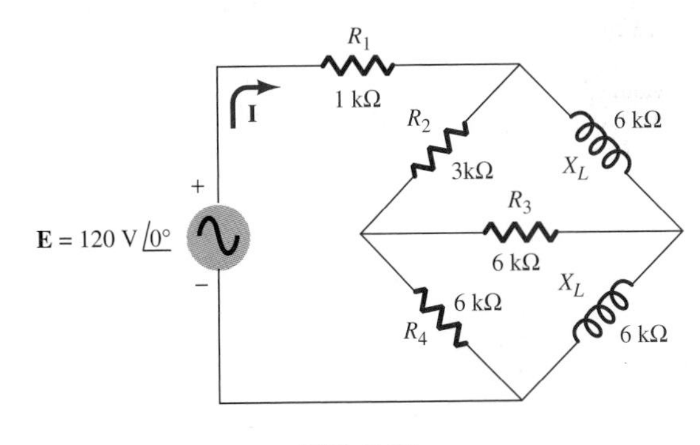

FIG. 5.86

SECTION 5.7 Bridge Configuration

20. **a.** Is the bridge in Fig. 5.87 balanced?

b. If not, which element could be changed to achieve this condition?

21. Derive the equations for the Maxwell bridge in Table 5.1 (b) by applying Eq. (5.8). That is, show that $L_x = CR_2R_3$ and $R_x = R_2R_3/R_1$ for balance conditions.

22. Repeat Problem 20 for the capacitance comparison bridge in Table 5.1 (c).

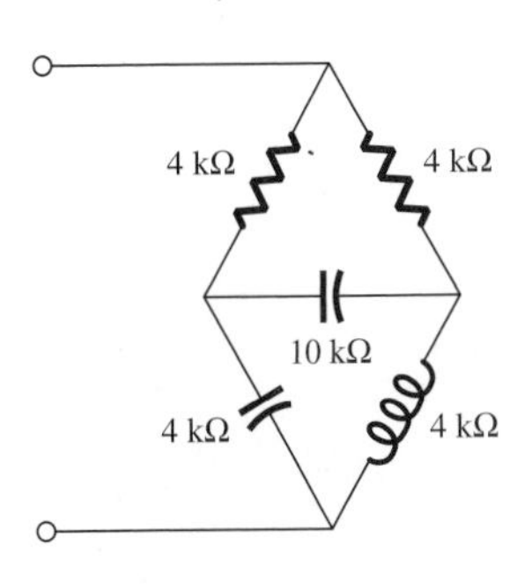

FIG. 5.87

SECTION 5.8 Power (Ac)

23. For the network in Fig. 5.88, determine:
   a. The current through each branch using Ohm's law.
   b. The total power dissipated (real power).
   c. The net reactive power.
   d. The total apparent power.
   e. The network power factor.

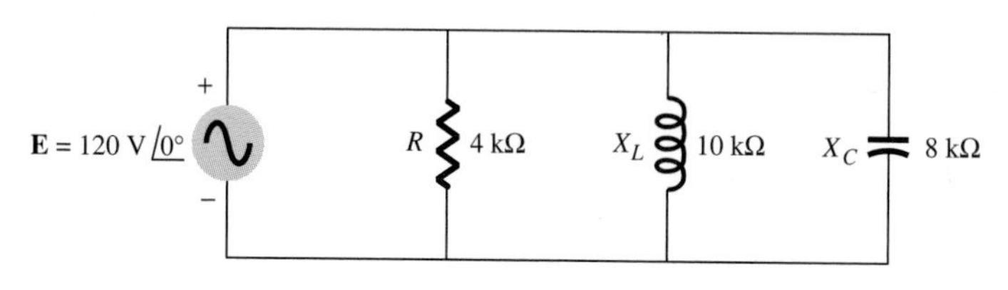

FIG. 5.88

24. For the system in Fig. 5.89, determine:
   a. P_T.
   b. Q_T.
   c. S_T.
   d. **I**.
   e. F_p.

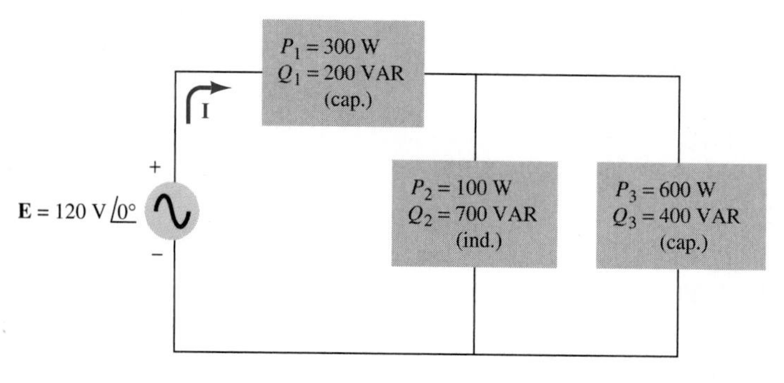

FIG. 5.89

25. Repeat Problem 23 for the network in Fig. 5.81.

26. A factory has the following loads: two 5-hp motors, 0.8 lagging F_p and 75% efficiency, and a 10-kW heating load, $F_p = 1$.
   a. Determine the current drain of the supply if $E = 220$ V at 60 Hz.
   b. Determine the parallel capacitance required to raise the power factor of the system to unity.
   c. Find the current drain with the added capacitance in part (b). Compare with part (a).

SECTION 5.9 Polyphase Systems

27. A three-phase, four-wire, Y–Y-connected system has a series *RL* impedance ($R = 4\ \text{k}\Omega$, $X_L = 6\ \text{k}\Omega$) in each phase of the load. If the line voltage is 200 V, determine:
a. The generated and load phase voltage.
b. The phase currents of the load.
c. The line current magnitudes.
d. The total apparent power delivered to the load.

28. Repeat Problem 26 for a three-phase, three-wire, Δ–Δ system.

29. Repeat Problem 26 for phase impedances of $\mathbf{Z} = 3\ \Omega - j12\ \Omega$ in each leg.

30. A three-phase, Y–Δ system has a parallel combination of $R = 6\ \text{k}\Omega$ and $X_L = 6\ \text{k}\Omega$ in each leg of the load. If the line voltage is 220 V, determine:
a. P_T.
b. Q_T.
c. S_T.
d. F_p.

SECTION 5.11 Tuned (Resonant) Networks

31. For the series *RLC* network in Fig. 5.90, determine:
a. X_C for resonance.
b. Q_s.
c. BW if $f_s = 5$ kHz.
d. The power delivered at resonance and the half-power frequencies.
e. L and C using the information in part (c).

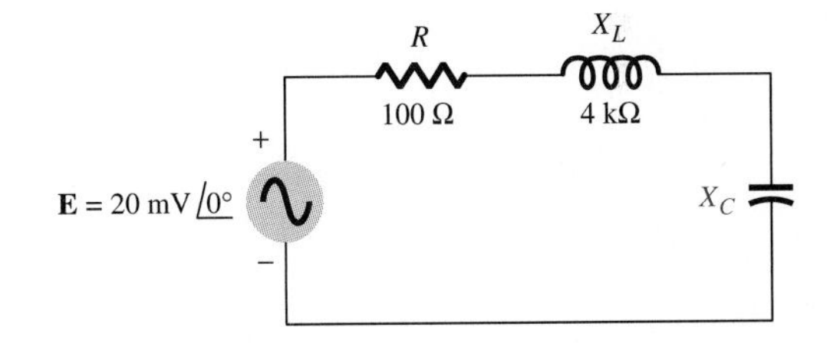

FIG. 5.90

32. For the series *RLC* circuit in Fig. 5.91, determine:
a. f_s.
b. X_L and X_C (at resonance).
c. Q_s.
d. BW.
e. f_1 and f_2.
f. P_{HPF}.

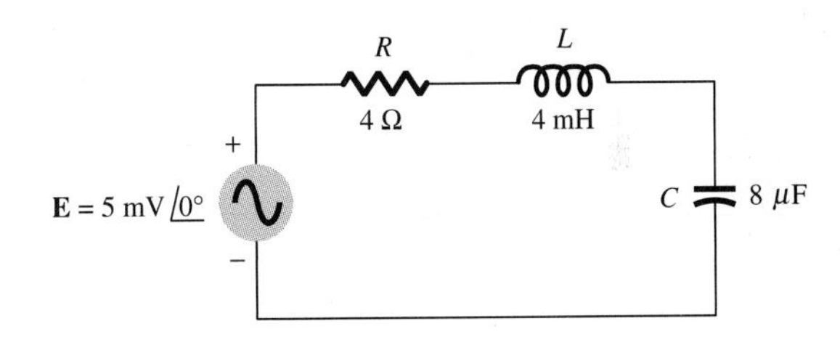

FIG. 5.91

33. For the parallel resonant circuit in Fig. 5.92, determine:
a. The value of X_C for resonance.
b. The impedance at resonance.
c. f_p if $C = 0.01\ \mu$F.
d. The cutoff frequencies.
e. The resonance curve for V_C.

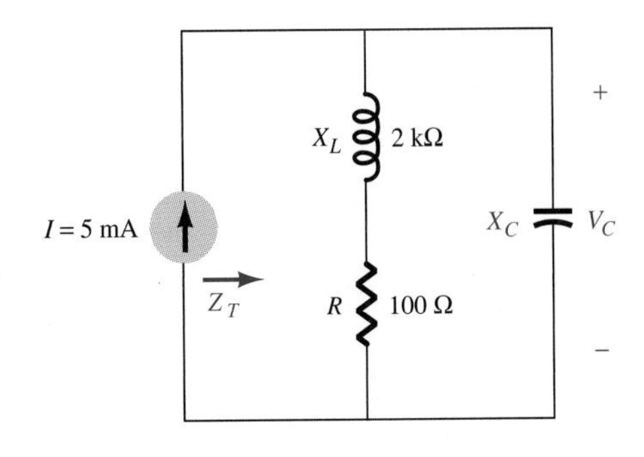

FIG. 5.92

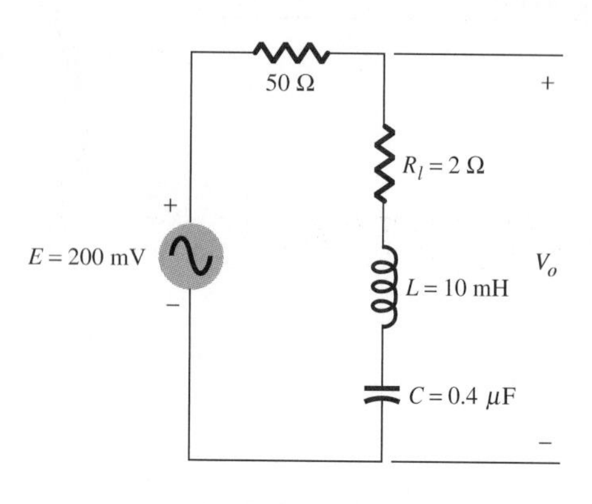

FIG. 5.93

SECTION 5.12 Filters

34. For the filter in Fig. 5.93:

a. Determine whether it is a pass-band or stop-band filter.
b. Calculate the resonant frequency.
c. Find the BW and cutoff frequencies.
d. Determine V_{min} at resonance.
e. Sketch the output waveform (versus frequency).

6 Electromagnetism

6.1 INTRODUCTION

The impact of electromagnetic effects on a multitude of engineering systems requires that a chapter be devoted to introducing the basic concepts and a few areas of application. Recording systems, electronic instrumentation, speakers and microphones, telephones, automobile ignition systems, computer hard disks, Hall effect sensors, magnetic reed switches, and magnetic resonance imaging (MRI) all depend on magnetic effects to function properly. Each will be described in some detail later in the chapter. The controlling variables of a magnetic circuit are so closely related to the variables of an electric circuit that the basic concepts of electric circuits can be easily related to magnetic circuits and serve as a guide through the analysis to follow.

Three different systems of units are often employed in the analysis and design of electromagnetic systems: *Système International* (SI), centimeter-gram-second (CGS), and English. Since the SI system has been defined as the system to be universally accepted as the standard, it will be employed throughout this chapter. A conversion table appears in Appendix C if the other units of measurement should be encountered. Whichever system is employed, it is absolutely necessary that each quantity in a particular equation have units of measurement that are in the same system of units. They cannot be interchanged unless the equation calls for it specifically. All the equations in this chapter will be in the SI system. If the units of measurement in a particular problem are all in the English or CGS system, simply convert each to the SI using the conversion table and substitute into the equations as they appear in this chapter.

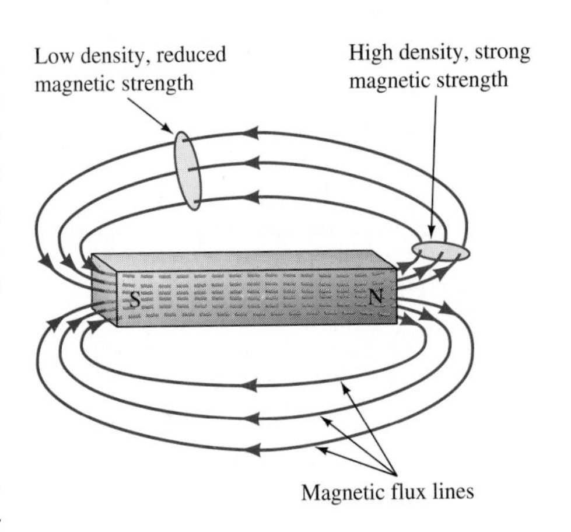

FIG. 6.1

6.2 BASIC PROPERTIES OF MAGNETISM

We have all, at one time or another, enjoyed experimenting with a horseshoe or *permanent* bar magnet, as shown in Fig. 6.1. When two north (or south) poles are brought close to each other, each makes every effort to avoid or repel the other. The reverse is true if opposite poles are brought together. In other words, *like poles repel, while unlike poles attract*. This phenomenon is demonstrated for each case in Fig. 6.2 using *magnetic flux lines,* whose density can indicate the strength of the magnetic field in a particular region. They are continuous lines that leave the north pole of a magnet, return to the south pole, and pass through the magnetic material. Flux lines characteristically strive to be as short as possible in their continuous path. In Fig. 6.2(a), therefore, the flux lines that pass directly from one magnet to the other tend to draw the two magnets together. For like poles a buffer action exists between the poles, as shown in Fig. 6.2(b).

Permanent magnets such as appear in Fig. 6.2 are constructed of a *ferromagnetic material* such as iron or steel that characteristically permits the setting up of flux lines through them with very little external pressure. Once magnetized, their retention properties are such that they exhibit magnetic properties after the external magnetizing agent is removed.

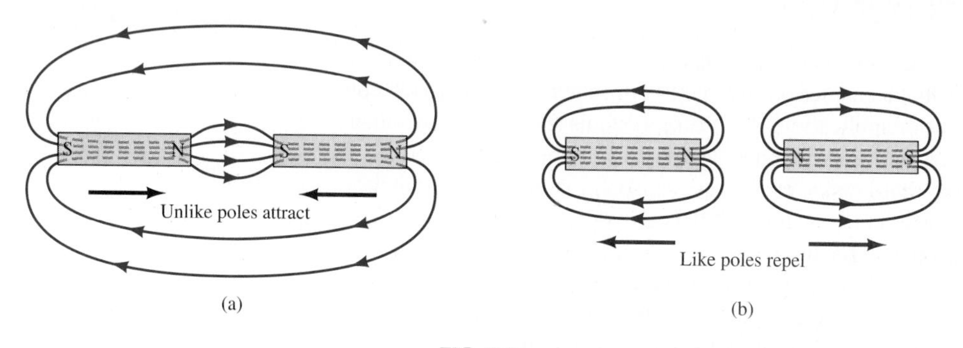

FIG. 6.2
(a) Attraction; (b) repulsion.

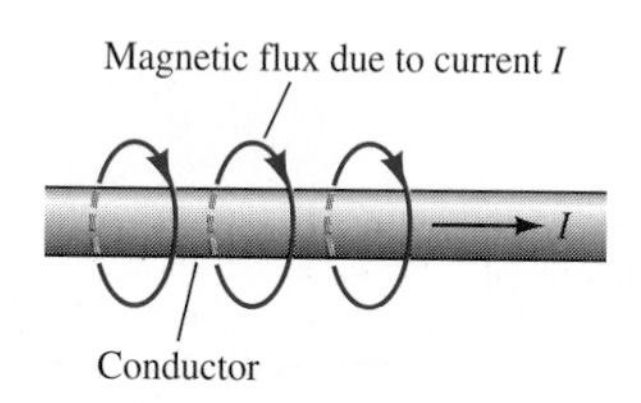

FIG. 6.3
Magnetic flux lines surrounding a current-carrying conductor.

The current through a conductor establishes a circular flux pattern around the conductor such as shown in Fig. 6.3. The direction of the flux lines can be determined by placing the thumb of the right hand in the direction of conventional current and noting the direction of the fingers. This determination is shown in Fig. 6.3. If we wrap a ferromagnetic material with a conducting wire as shown in Fig. 6.4(a), a strong magnetic field can be established through the core. An expanded view of three adjacent windings on the top of the core is shown in Fig. 6.4(b), with the resulting flux patterns for each loop as determined by the technique demonstrated in Fig. 6.3. Note that between the windings there appears to be a canceling effect on the net flux, while above and below the two windings there is a strengthening of the flux in each direction. The net result for the wrapped core in Fig. 6.4(a) is a north pole on the right, since the flux patterns are leaving, and a south pole on the left since the flux patterns are entering. The direction of the resultant flux can be determined by placing the fingers of the right hand in the direction of current for the wrappings, as shown in Fig. 6.5. The thumb then points in the direction of the resultant flux. This particular structure, which re-

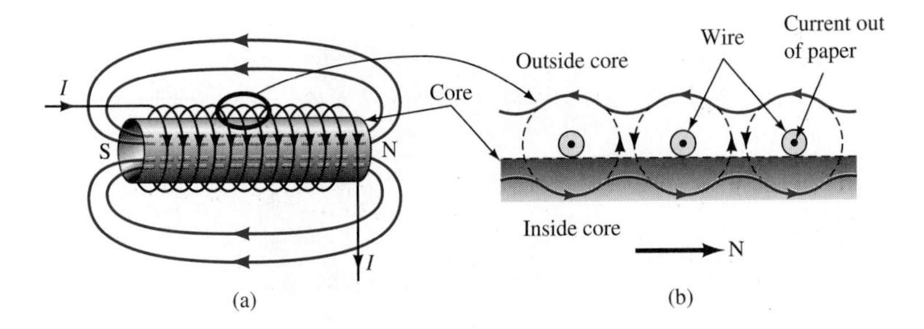

FIG. 6.4
Electromagnet: (a) complete structure; (b) three adjacent windings on the top of the core.

quires electric current to develop its magnetic characteristics, is called an *electromagnet.* The strength of the magnet can be increased by increasing the current through the conductors. Eventually, however, a condition called *saturation* is reached such that any additional current has very little effect on the strength of the magnet. Once the magnetizing current is removed, the magnetic characteristics of the ferromagnetic core decreases to a value determined only by the ability of that sample to retain the magnetic characteristics—a measure of its *retentivity*. Materials with high retentivities are used to make *permanent* magnets by simply subjecting them to a strong magnetic field such as developed by the coil in Fig. 6.4.

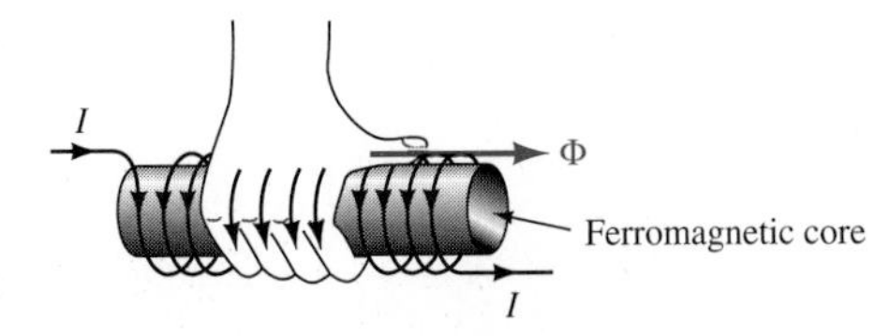

FIG. 6.5
Determining the direction of flux developed by an electromagnet.

The electromagnetic effect is employed in systems too numerous to possibly list here. However, we shall examine a few of the more interesting areas of application. A photograph and sectional drawing of a *magnetic* relay appear in Fig. 6.6. When the specified voltage is applied to the coil, the electromagnet is energized by the resulting current and draws the arm (armature) down away from the normally closed (NC) contacts to the normally open (NO) contacts. When the energizing voltage is removed, the spring draws the contact back to the NC terminal. In the nonenergized state a direct connection exists between contacts A and C of either figure. In the energized state a direct connection exists between contacts B and C.

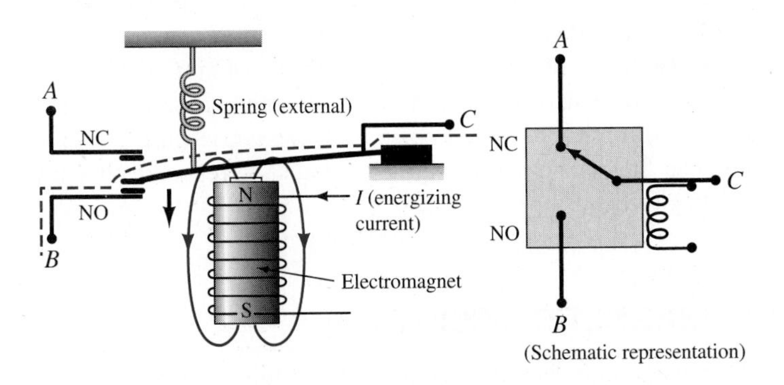

FIG. 6.6
Magnetic relay.

FIG. 6.7
Circuit breaker. (Courtesy of Thomas Sundheim, Inc.)

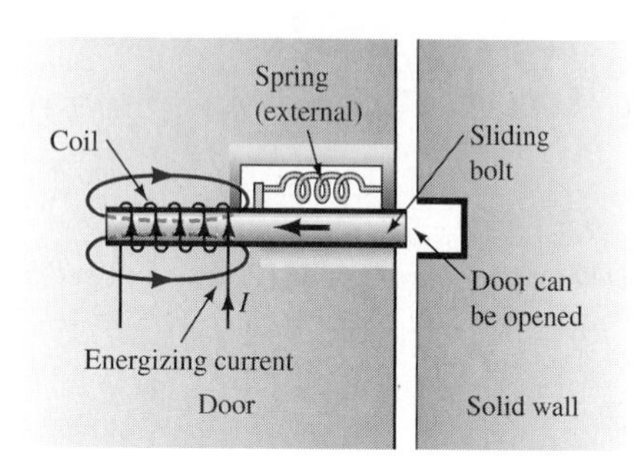

FIG. 6.8
Electromagnetically controlled lock.

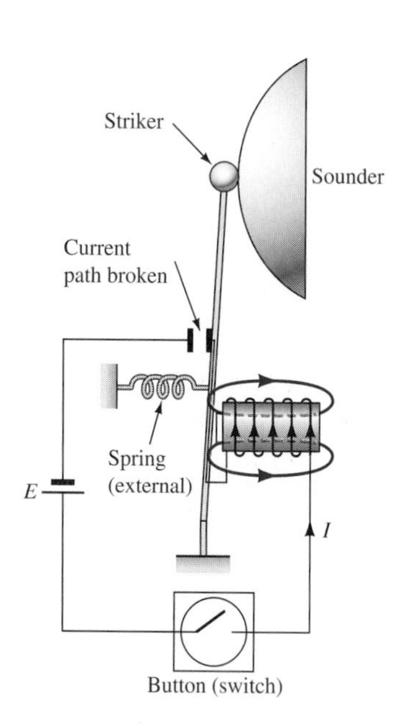

FIG. 6.9
Electric bell.

The internal construction of a circuit breaker appears in Fig. 6.7. The electromagnet at the bottom is sensitive to the service current passing through the breaker. When too high a level is reached, the movable arm at the bottom left is drawn to the electromagnet, and the circuit opens at the contacts to prevent damage due to excessive current. Although it seems that the basic construction of such a device would be quite simple, the additional complexity in appearance in Fig. 6.7 is there to ensure proper operation even when it is subject to shock, vibration, or thermal variations.

When the core inside the coil is not fixed but free to move, a whole new vista of applications of electromagnetic effects is possible. In this capacity the electromagnetic system is referred to as a *solenoid*. In Fig. 6.8, the locking bolt on the door can be controlled by energizing or deenergizing the coil. When the proper signal voltage is applied, the magnetizing effect draws the core back into the coil and permits the door to be opened. When deenergized, the spring draws the core back and locks the door.

The last application of the electromagnet to be described in this section is in the operation of an electric bell such as shown in Fig. 6.9. When the button is depressed, the circuit is completed and current flows through the coil, drawing the striker to the gong and sounding the signal, alarm, and so on. However, note that once the striker moves toward the sounder, the NC contacts are open, as shown in Fig. 6.9. The electromagnet is then deenergized and the striker is free to return under the action of the spring to the initial position. However, when it does return, current can flow again and the cycle continues to repeat itself until the controlling switch is opened.

6.3 MAGNETIC CIRCUITS

Let us now examine in more detail the controlling variables of the simple applications of magnetic effects introduced in the previous section. The magnetic flux established by the permanent magnet or through the electromagnetic core is defined by the Greek letter phi (ϕ) and measured in webers (Wb) in the SI system. Its properties are very similar to those of current in an electric circuit. In an electric circuit the current always seeks the path of least resistance. The flux in a magnetic system always seeks the path of least *reluctance* ($\mathcal{R}$). Ferromagnetic materials have very low values of reluctance, while substances such as air, wood, and glass have a very high reluctance. In other words, flux lines always strive to establish the shortest path possible through ferromagnetic materials such as iron and steel in the region of magnetic excitation. The reluctance of a sample of material is determined by the following equation, as depicted in Fig. 6.10:

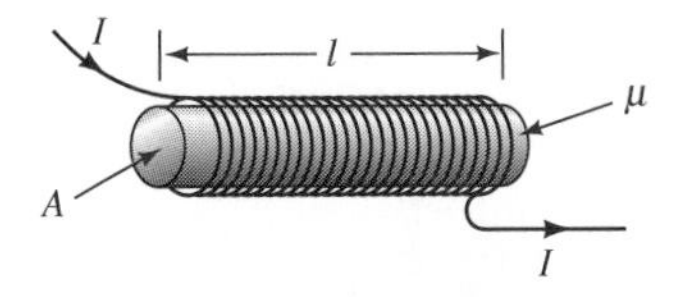

FIG. 6.10

$$\mathcal{R} = \frac{l}{\mu A} \tag{6.1}$$

where l = length in meters
A = area in square meters
μ = permeability
$\mathcal{R}$ = reluctance in rels or ampere-turns per weber (At/Wb)

The permeability μ is a measure of how easily flux lines can be set up in the material. The relative permeability is given by

$$\mu_r = \frac{\mu}{\mu_0} \tag{6.2}$$

where μ_r = relative permeability
μ = permeability
μ_0 = permeability of air

It is a measure of the quality of a magnetic material as compared with air. In the SI system the permeability of air is

$$\mu_0 = 4\pi \times 10^{-7}\ \text{Wb/Am} \tag{6.3}$$

For ferromagnetic materials, $\mu_r \geq 100$, while for air, wood, glass, and so on, μ_r is for all practical purposes 1. Values of μ_r are not tabulated, since they are dependent on the operating conditions of the magnetic circuit.

Based on Eq. (6.2), the permeability of a material is given by

$$\mu = \mu_r \mu_0 \tag{6.4}$$

In other words, once μ_r is known, the permeability of the material can be determined by multiplying by the permeability of air. For instance, if

$\mu_r = 2000$ (note that since it is a ratio of permeabilities, it is unitless) for a particular steel sample,

$$\mu = \mu_r\mu_0 = (2000)(4\pi \times 10^{-7}\ \text{Wb/Am})$$
$$= 8\pi \times 10^{-4}\ \text{Wb/Am}$$

The fact that magnetic flux lines always take the path of least reluctance can be put to good use in the shielding of instruments, electronic circuitry, and so on, that may be quite sensitive to magnetic effects. When the piece of equipment is surrounded by a *shield* of ferromagnetic material such as shown in Fig. 6.11, the magnetic flux will pass through the ferromagnetic material of the shield rather than disturb the sensitive instrumentation inside.

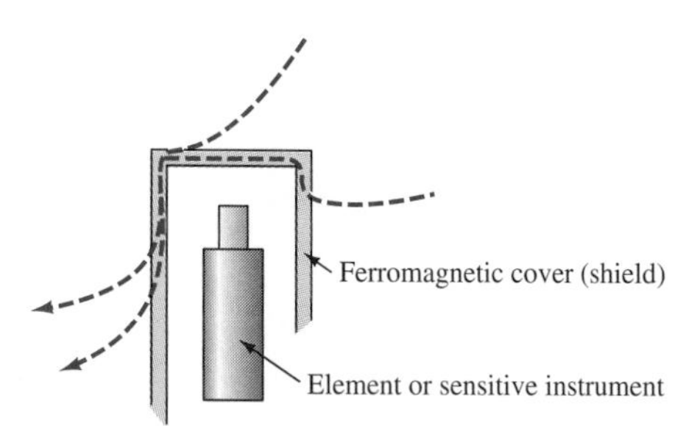

FIG. 6.11
Magnetic shield.

Series, parallel, and series–parallel configurations are also encountered in magnetic systems. However, the majority of the simpler applications of magnetic effects appear in the realm of *series magnetic circuits.* A great deal of literature is available for research into the other areas.

For a series magnetic circuit such as shown in Fig. 6.12, the flux (like the current in a series electric circuit) is the same throughout. The "pressure" (like the applied voltage of an electric circuit) that results in the flux in the magnetic circuit is determined by the product of the number of turns N in the coil and the current I through the coil. It is called the *magnetomotive force* (mmf) of the magnetic circuit and is given the symbol $\mathscr{F}$ with the units *ampere-turns* (At). That is,

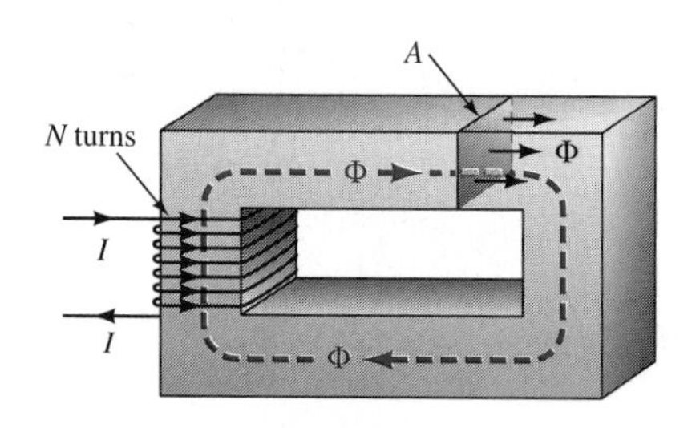

FIG. 6.12
Series magnetic circuit.

$$\boxed{\mathscr{F} = NI} \qquad \text{(ampere-turns)} \qquad \textbf{(6.5)}$$

For any series magnetic circuit the flux established by a particular mmf is determined by the reluctance of the core. The greater the opposition, the less the magnetic flux. The Ohm's law equivalent for magnetic circuits is therefore given by

$$\boxed{\phi = \frac{\mathscr{F}}{\mathscr{R}} = \frac{NI}{\mathscr{R}}} \qquad \text{(webers)} \qquad \textbf{(6.6)}$$

In Fig. 6.12 the direction of the resulting flux can be found in the same manner as for the electromagnet. Place the fingers of the right hand in the direction of the current through the turns of the coil, and the thumb of that hand will point in the direction of the flux.

The flux per unit area of the core (note Fig. 6.12) is called the *flux density*, has the symbol B, and is measured in teslas (T), where $1\ \text{T} = 1\ \text{Wb/m}^2$. It is determined by the equation

$$\boxed{B = \frac{\phi}{A}} \qquad \text{(teslas)} \qquad \textbf{(6.7)}$$

where B = teslas (T)
ϕ = webers
A = square meters

EXAMPLE 6.1 If the flux density in a circular core of 1-cm diameter is 0.4 T, find the flux in the core in webers.

Solution:

$$1\ \text{cm}\left(\frac{1\ \text{m}}{100\ \text{cm}}\right) = 0.01\ \text{m} = 10^{-2}\text{m}$$

$$\text{Area} = \frac{\pi d^2}{4} = \frac{(3.1416)(10^{-2}\ \text{m})^2}{4} = 7.85 \times 10^{-5}\ \text{m}^2$$

$$\phi = BA = (0.4\ \text{T})(7.85 \times 10^{-5}\ \text{m}^2) = \mathbf{3.14 \times 10^{-5}\ Wb}$$

Meters are available for measuring the flux density present. A photograph of a gaussmeter appears in Fig. 6.13. The conversion factor as provided in Appendix C is

$$10^4\ \text{gauss (G)} = 1\ \text{T} \qquad \textbf{(6.8)}$$

FIG. 6.13
Gauss meter. (Courtesy of LDJ Electronics, Inc.)

For example, if a reading of 5×10^2 G was obtained, the flux density in webers per square meter would be

$$(5 \times 10^3\ \text{G})\left(\frac{1\ \text{T}}{10^4\ \text{G}}\right) = 5 \times 10^{-1}\ \text{T} = 0.5\ \text{T}$$

The magnetomotive force *per unit length* required to establish a particular flux in a core is called the magnetizing force (H). In equation form,

$$H = \frac{\mathscr{F}}{l} = \frac{NI}{l} \qquad \text{(At/m)} \qquad \textbf{(6.9)}$$

The magnetizing force and the flux density are related in the following manner:

$$B = \mu H \qquad \textbf{(6.10)}$$

For a particular core sample with a specified reluctance, Eq. (6.6) reveals that an increase in the number of turns around the sample or the magnitude of the current through the coil results in greater flux in the core. If we start with an unmagnetized sample ($\phi = 0$ Wb) and increase the current from 0A (fixed $\mathcal{R}$ and N), a curve following the path appearing in the first quadrant in Fig. 6.14 will result (A to B). A further increase in current will result in a minimal increase in flux density because saturation has been established. In other words, all the magnetic alignment that can be established within the core has been attained, and any further increase in current through the coil will have little effect on the level of flux in the core.

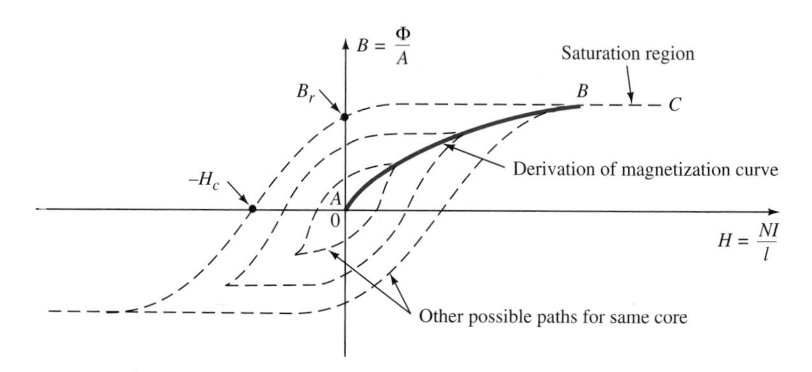

FIG. 6.14
Hysteresis curve (cast steel).

If the current is now reduced in magnitude, the curve will follow the path from B to B_r, the *retentivity* level of the magnetic sample. If the coil is now removed from the core, the magnetic material will retain a level of magnetic flux determined by $\phi = B_rA$. It is this retentivity that permits the establishment of *permanent magnets* (metallic samples with magnetic characteristics in the absence of a magnetizing force). If the direction of the current is reversed, the magnetic alignment will be reversed until saturation is established in the reverse flux direction. A reduction in magnitude and then in direction for the magnetizing current will then return the curve to point B, as shown in Fig. 6.14. For lower peak values of magnetizing current, the other curves in Fig. 6.14 will result.

Since a multitude of curves such as shown in Fig. 6.14 can be obtained simply by limiting the variables for each pass in the B-H curve, the singular curve in Fig. 6.15 was obtained from the B-H curves for application purposes. The curve for Fig. 6.15 is called the *normal magnetization curve*, while the cycle in Fig. 6.14 is called a B-H or *hysteresis curve*. Since curves of the type just described are quite important for application purposes, a sophisticated piece of equipment such as shown in Fig. 6.16 was devised to determine the B-H curve for magnetic samples. For a particular value of B the value of μ can be determined from Fig. 6.15 after determining the corresponding values of H and using Eq. (6.10).

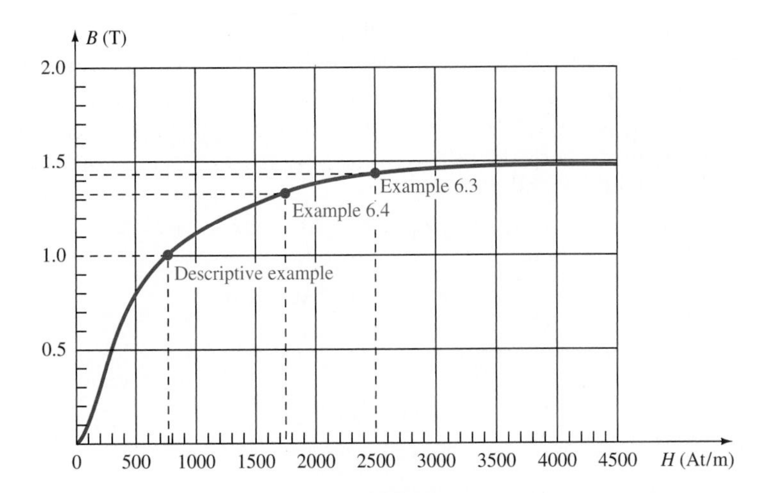

FIG. 6.15
Normal magnetism curve for cast steel.

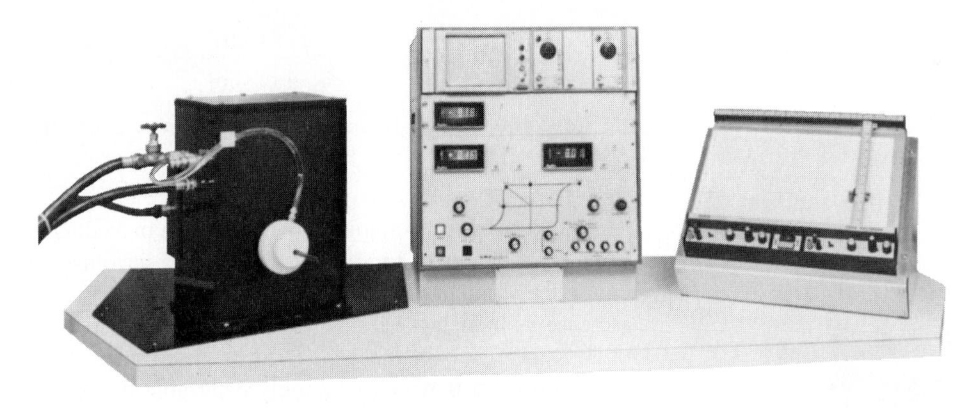

FIG. 6.16
B-H meter. (Courtesy of LDJ Electronics, Inc.)

It is particularly important to note that $\mu = B/H$, and since the curve relating B and H is not a straight line (Fig. 6.15), the value of μ varies from point to point. It is therefore impractical to use a particular value of μ for a magnetic core, since the level of H determines the level of μ and the resulting flux density. The analysis of magnetic circuits [unlike electric circuits where ρ of $R = \rho(l/A)$ can be treated as a constant] must therefore include reference to a magnetization curve to determine B or H from the other variable.

EXAMPLE 6.2

a. For the magnetic sample in Fig. 6.15, determine the maximum flux in a core having an area of 0.01 m^2.
b. Determine μ and μ_r for the conditions in part (a) if $H = 3500$ At/m.

c. Determine μ and μ_r at 250 At/m and compare the results with those for part (b).

Solution:

a. At saturation $B \cong 1.46$ T and

$$\phi_{max} = B_{sat}A = (1.46\text{ T})(10^{-2}\text{ m}^2) = \mathbf{1.46 \times 10^{-2}\ Wb}$$

b. $\mu = \dfrac{B}{H} = \dfrac{1.46\text{ T}}{3500\text{ At/m}} = \mathbf{4.17 \times 10^{-4}\ Wb/Am}$

with

$$\mu_r = \frac{\mu}{\mu_0} = \frac{4.17 \times 10^{-4}}{4\pi \times 10^{-7}} = \mathbf{332}$$

c. $\mu = \dfrac{B}{H} = \dfrac{0.5\text{ T}}{250\text{ At/m}} = \mathbf{2 \times 10^{-3}\ Wb/Am}$

with

$$\mu_r = \frac{\mu}{\mu_0} = \frac{2 \times 10^{-3}}{4\pi \times 10^{-7}} = \mathbf{1590}$$

revealing a much higher level of μ_r at lower magnetizing levels.

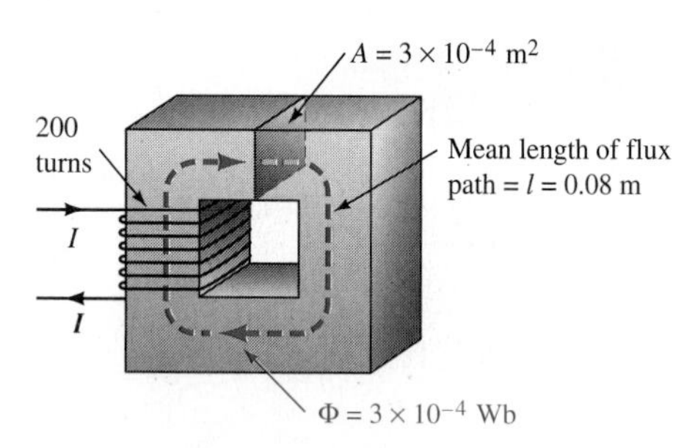

FIG. 6.17
Inductor.

As a descriptive example of the use of the preceding equations, consider the inductor in Fig. 6.17, which has a ferromagnetic core to increase the strength of the magnetic field linking the coil and thereby to increase its inductance level. The current I necessary to establish a flux of 3×10^{-4} Wb in the core is to be determined. It would be quite straightforward if we could simply determine the reluctance from the material and dimensions of the core and then plug them into the Ohm's law equation for magnetic circuits [Eq. (6.5)]. Unfortunately, however, as noted previously, the reluctance cannot be found, since the value of μ is determined by the conditions of operation of the magnetic circuit and cannot simply be read off a chart. Rather, we must first apply a law to the magnetic circuit that is very similar to Kirchhoff's voltage law for electric circuits. The law, called *Ampère's circuital law*, states that the algebraic sum of the mmfs around a closed path is zero. Application of the law requires that Eq. (6.9) be written in the form $\mathscr{F} = Hl$ so that the mmf drops across the branches of the magnetic circuit can be included.

For the *series* magnetic circuit in Fig. 6.17 the impressed mmf (like the applied voltage E of a series electric circuit) is NI, while the mmf drops (like the voltage drops across the resistive loads of the series circuit) is Hl. Based on Ampère's circuital law, the algebraic sum of the mmfs must equal zero, and

$$NI - Hl = 0$$

or

$$NI = Hl$$

which states in words that the impressed mmf must equal the mmf drops around a closed path.

The unknown quantities in the equation obtained include I and H. The latter quantity can be determined directly from Fig. 6.15 once B is determined. Therefore,

$$B = \frac{\phi}{A} = \frac{3 \times 10^{-4}\ \text{Wb}}{3 \times 10^{-4}\ \text{m}^2} = 1\ \text{T}$$

From the curve we find $H \cong 770$ At/m for the material used. Substituting into the equation above, we have

$$NI = Hl$$

$$(200\ \text{t})I = (770\ \text{At/m})(0.08\ \text{m})$$

and

$$I = \frac{61.6}{200}\ \text{A} = \mathbf{308\ mA}$$

The value of μ and μ_r can then be determined for the ferromagnetic material using Eq. (6.10),

$$\mu = \frac{B}{H} = \frac{1\text{T}}{770\ \text{At/m}} = \mathbf{1.30 \times 10^{-3}\ Wb/Am}$$

and Eq. (6.2),

$$\mu_r = \frac{\mu}{\mu_0} = \frac{1.30 \times 10^{-3}}{4\pi \times 10^{-7}} \cong \mathbf{1035}$$

The general technique for analyzing magnetic circuits has thus been described. In total, it differs only slightly, because of the need for a *B-H* curve, from the analysis of electric circuits. In this past example, the applied mmf was to be determined. For a simple series magnetic circuit such as the one just considered, if the impressed mmf was provided, the resulting flux could be determined by simply reversing the sequence of steps. However, for more complex networks such as series–parallel configurations, the flux cannot be determined directly but must be determined through a "cut and try again" technique. Incidentally, for series–parallel configurations, the analogue of Kirchhoff's current law is applicable here in the sense that the total magnetic flux entering a magnetic junction must equal that leaving. That is, for two parallel branches, $\phi_T = \phi_1 + \phi_2$, where ϕ_T is the net flux entering the junction.

EXAMPLE 6.3 Determine the flux in the core of the transformer in Fig. 6.18. Transformers are examined in detail in Section 6.4.

Solution: Since the second winding is left open, the current through the coil is zero and its mmf is zero. It does not affect the magnitude of the flux ϕ. Application of Ampère's circuital law gives

$$NI = H_{\text{core}} l_{\text{core}}$$

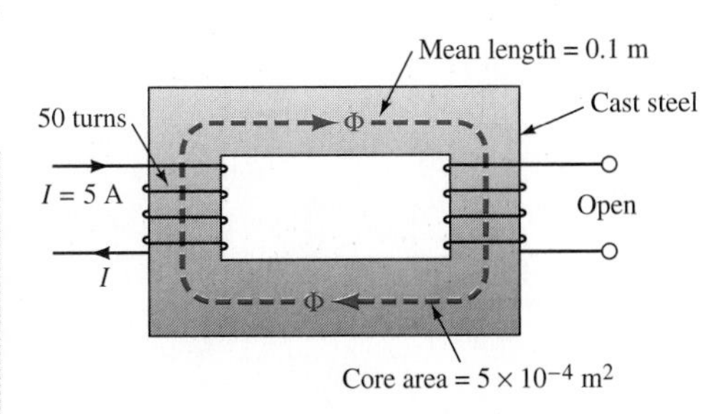

FIG. 6.18
Transformer (unloaded, $I_s = 0$ A).

Substitution yields

$$(50\text{t})(5\text{ A}) = H_{core}(0.1)$$

and

$$H_{core} = \frac{250\text{ At}}{0.1\text{ m}} = 2500\text{ At/m}$$

and from the curve in Fig. 6.15 we find that

$$B_{core} \cong 1.42\text{ T}$$

and

$$\begin{aligned}\phi_{core} &= B_{core}A_{core}\\ &= (1.42\text{ T})(5 \times 10^{-4}\text{ m}^2)\\ &= \mathbf{7.1 \times 10^{-4}\ Wb}\end{aligned}$$

In a number of applications involving magnetic effects, air gaps are encountered. In Fig. 6.19, the air gap is quite evident for the magnetic relay. For air gaps, the flux density and magnetizing force are related by

$$H_g = \frac{B_g}{\mu_0} = \frac{B_g}{4\pi \times 10^{-7}} = 7.96 \times 10^5 B_g \qquad \textbf{(6.11)}$$

The example to follow will demonstrate that even the smallest air gap with its high-reluctance characteristics requires the bulk of the impressed mmf to establish a flux of some magnitude in the core and across the gap. In fact, in most situations, the mmf required to establish the flux in the core can be ignored compared with that of the air gap when determining either the mmf required or the flux developed.

EXAMPLE 6.4 For the relay in Fig. 6.19, determine the current I required to establish a flux of 4×10^{-4} Wb in the core.

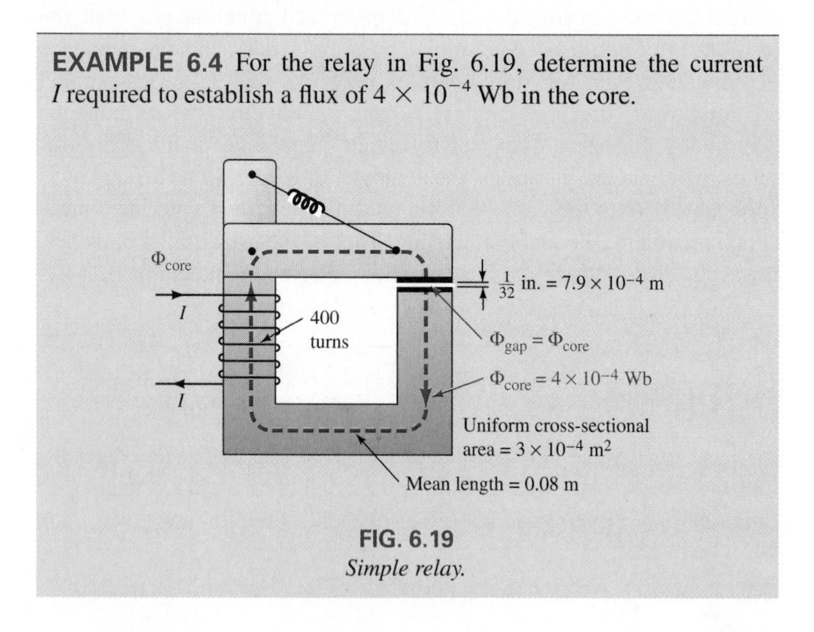

FIG. 6.19
Simple relay.

Solution: Ampère's circuital law results in

$$NI = H_{\text{core}}l_{\text{core}} + H_g l_g$$

an equation very similar to that obtained for a series electrical network where $E = V_1 + V_2$.

$$B_{\text{core}} = B_g = \frac{\phi}{A} = \frac{4 \times 10^{-4}\ \text{Wb}}{3 \times 10^{-4}\ \text{m}^2} = 1.33\ \text{T}$$

and from Fig. 6.15,

$$H_{\text{core}} \cong 1750\ \text{At/m}$$

$$H_g = 7.96 \times 10^5 B_g = (7.96 \times 10^5)(1.33) = 10.59 \times 10^5\ \text{At/m}$$

Substitution into the equation obtained above gives

$$(400\ \text{t})(I) = (1750\ \text{At/m})(0.08\ \text{m}) + (10.59 \times 10^5\ \text{At/m})(7.9 \times 10^{-4}\ \text{m})$$

$$400I = 140 + 836.61 \quad \text{(note the difference in magnitude between the two terms)}$$

$$400I = 976.61$$

and

$$I = \mathbf{2.44\ A}$$

Note the ratio of the mmf for the air gap as compared with that for the core: 836.61/140 ≅ 6:1, which certainly substantiates the statement made earlier, that is, to assume that NI ≅ Hglg is certainly a valid first approximation in this example.

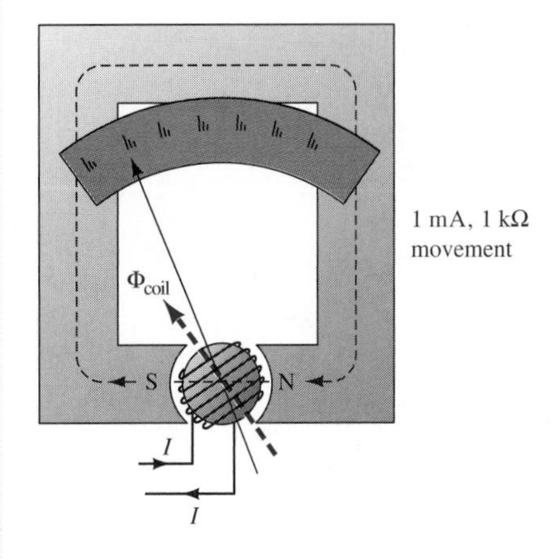

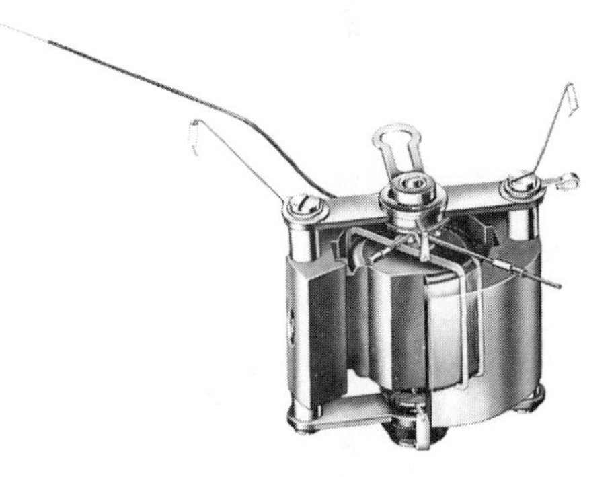

FIG. 6.20
D'Arsonval movement. (Courtesy of Weston Instruments.)

The basic construction of the D'Arsonval movement as used in both dc and ac instruments is shown in Fig. 6.20. Note that in this case both a permanent horseshoe magnet and an electromagnet are present with two series air gaps between the movable and stationary parts of the movement. The nameplate data for the movement appear in the figure. The 1-kΩ resistance is simply the dc resistance of the windings around the movable core. When 1 mA is passed through the movable core, a flux develops through the coil and core that reacts with the flux of the permanent magnet and causes the movable core to turn on its shaft. For the full 1-mA rated current the pointer attached to the movable core indicates a full-scale deflection. For lesser values of current the deflection is that much less on a linear (straight-line) basis.

Since the maximum current is 1 mA, how is an ammeter designed to read higher values such as 1 A? In Fig. 6.21 we find the internal circuitry for a 1-A meter using the movement in Fig. 6.20.

With 1 A applied (the maximum for that scale) we want the full deflection current of 1 mA through the movements as shown in the figure. The remaining current through R_{shunt} is then 1 A − 0.001 A = 0.999 A = 999 mA; the voltage across the 1-kΩ resistor is (1 kΩ)(1 mA) = 1 V, which is also across R_{shunt} since they are in parallel. Therefore,

$$R_{\text{shunt}} = \frac{V}{I} = \frac{1\ \text{V}}{0.999\ \text{A}} \cong 1.001\ \Omega \cong \mathbf{1\ \Omega}$$

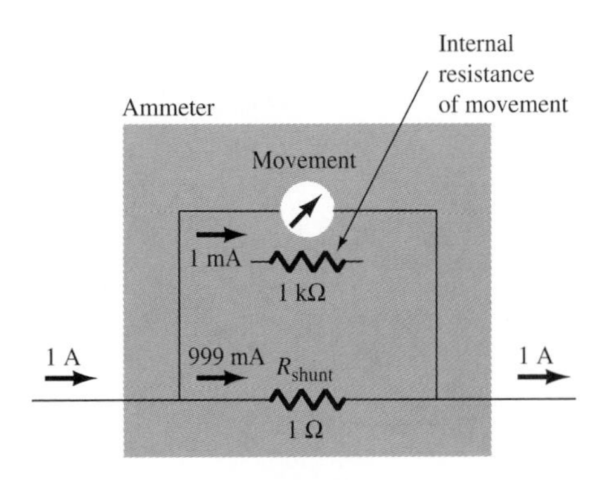

FIG. 6.21
Ammeter.

In other words, high current levels can be read by the sensitive movement simply by diverting that current above its maximum level to the parallel branch.

For voltmeters, a series resistor is added as shown in the 100-V voltmeter in Fig. 6.22 using the same movement. The maximum voltage that the movement can handle is 1 V. The remaining 99 V must appear across R_{series}. Therefore,

$$R_{\text{series}} = \frac{V}{I} = \frac{99\text{ V}}{1\text{ mA}} = \mathbf{99\ k\Omega}$$

Multirange ammeters and voltmeters can be designed using rotary switches that insert the proper shunt or series resistor into the network.

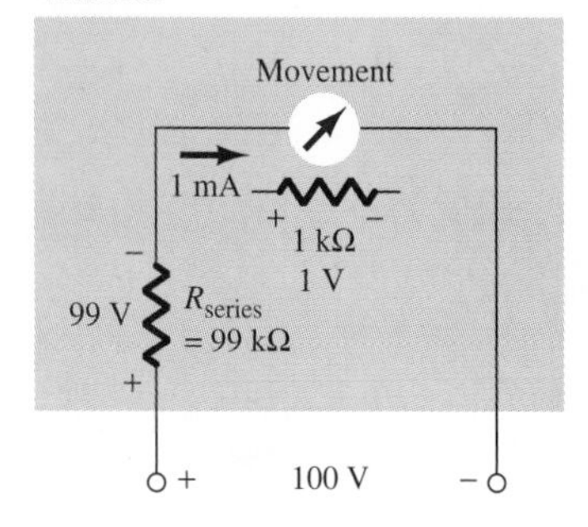

FIG. 6.22
Voltmeter.

6.4 TRANSFORMERS

The *transformer* is an electromagnetic device primarily designed to perform one of the three following functions:

1. Step up or step down the voltage or current.
2. Act as an impedance-matching device.
3. Isolate one portion of a network from another.

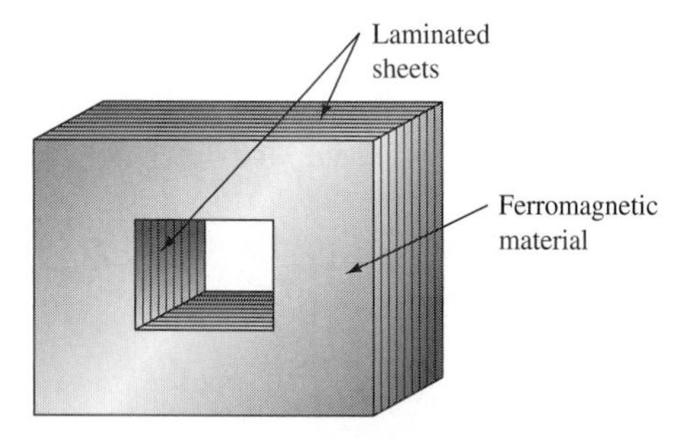

FIG. 6.23
Transformer core.

The core of a transformer, as shown in Fig. 6.23, is laminated (made of thin metallic sheets separated by an insulating material) to reduce the eddy current losses in the core. *Eddy currents* are small, circular currents that flow in a core induced by the changing flux through the core. The changing flux induces voltages in the core that set up these currents, which result in an additional I^2R loss. When the sheets of insulation are inserted, the resistance in the path of the currents is so high that their magnitude is reduced significantly.

A second loss in the core is called the *hysteresis* loss. It is a result of the reversing magnetic field, which causes the small magnetic alignments in the atoms of the material to reverse their polarities. This reversal motion results in friction and therefore heating loss in the core. It can be reduced significantly by introducing small amounts of silicon into the core.

Additional factors that make the actual transformer less than ideal include the resistance of the windings, the inductive effects of the windings, the stray capacitance between the windings, and so on. However, today's transformers are so close to ideal (95 to 99% level) that we usually assume (except for large power applications) that they are ideal. In addition, keep in mind that transformers are ac devices. They cannot perform the functions listed earlier for dc inputs. Although the terminology *dc transformer* does appear in catalogs, and so on, it is understood that an additional amount of circuitry is required to perform this function. As shown in Fig. 6.24, at least two windings are required. The winding to which the signal is applied is called the *primary,* while the winding to which the load is connected is called the *secondary.*

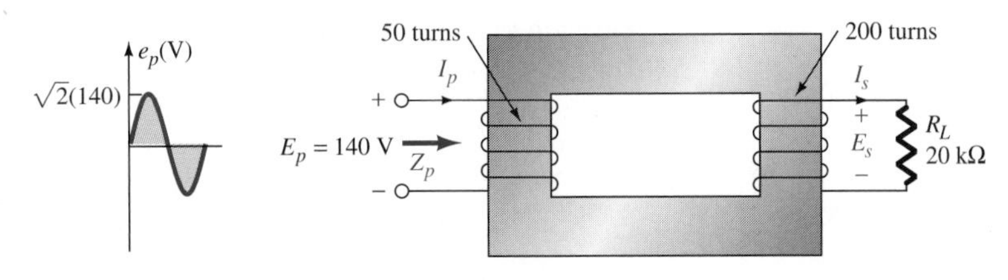

FIG. 6.24

The basic operation of the transformer is quite fundamental in nature. An ac voltage is applied to the primary that induces a sinusoidally varying ϕ in the core having the same frequency as the input signal. This sinusoidally varying flux in the core induces a voltage across the secondary as determined by Faraday's law of magnetic induction, introduced earlier: $e = N\,d\phi/dt$. The result is a sinusoidally induced voltage at the secondary that has the same frequency as the applied signal and a magnitude related to the input by the turns ratio of the coils. That is, the voltages are related by

$$\frac{E_p}{E_s} = \frac{N_p}{N_s} \tag{6.12}$$

where the indicated voltages are effective values. Solving for E_s, $E_s = (N_s/N_p)(E_p)$, we find that if the secondary turns are greater than the primary turns, the secondary voltage will be greater than the primary voltage by the ratio of the turns. A quantity called the *transformation ratio* is defined by

$$a = \frac{N_p}{N_s} \tag{6.13}$$

If a is less than 1, the secondary voltage will be greater than the primary voltage, and the device is a *step-up* transformer. If a is greater than 1, the reverse is true, and the device is a *step-down* transformer.

The currents of the ideal iron-core transformer are related by

$$\frac{I_s}{I_p} = \frac{N_p}{N_s} = a \tag{6.14}$$

Note that the currents are related by the inverse of the turns ratio, so that if the voltage increases, the current decreases, and vice versa.

Setting the voltage and current ratios equal to the same turns ratio results in

$$\frac{E_p}{E_s} = \frac{N_p}{N_s} = \frac{I_s}{I_p}$$

and since things equal to the same thing are equal to each other,

$$\frac{E_p}{E_s} = \frac{I_s}{I_p}$$

or

$$E_p I_p = E_s I_s \quad \textbf{(6.15)}$$

which states that the apparent power of the primary must equal the apparent power of the secondary for an ideal transformer. Or, for resistive loads, the power applied is equal to the power to the load.

EXAMPLE 6.5 For the transformer in Fig. 6.24, determine:
a. The transformation ratio.
b. The secondary voltage.
c. The secondary current.
d. The primary current.
e. The power to the load.
f. The power supplied by the source.

Solution:

a. $a = \frac{N_p}{N_s} = \frac{50}{200} = \mathbf{\frac{1}{4}}$ **(step-up transformer)**

b. $E_s = \left(\frac{N_s}{N_p}\right)E_p = \left(\frac{200}{50}\right)(140\text{ V}) = \mathbf{560\ V}$

c. $I_s = \frac{E_s}{R_L} = \frac{560\text{ V}}{20\text{ k}\Omega} = \mathbf{28\ mA}$

d. $I_p = \left(\frac{N_s}{N_p}\right)I_s = \left(\frac{200}{50}\right)(28\text{ mA}) = \mathbf{112\ mA}$

e. $P_L = E_s I_s = 560\text{ V}(28 \times 10^{-3}\text{ A}) = \mathbf{15.68\ W}$

f. $P_s = E_p I_p = (140\text{ V})(112 \times 10^{-3}\text{ A}) = \mathbf{15.68\ W} = P_L$

If we divide Eq. (6.12) by (6.14) and perform a few mathematical manipulations, we will obtain the following result:

$$\frac{E_p/E_s = a}{I_p/I_s = 1/a} \rightarrow \frac{E_p/I_p}{E_s/I_s} = a^2 \rightarrow \frac{Z_p}{Z_s} = a^2$$

and

$$Z_p = a^2 Z_s \qquad \textbf{(6.16)}$$

which in words states that the impedance "seen" at the primary of a transformer is the transformation ratio squared times the impedance connected to the secondary. Since the impedances are related by a constant (a^2), if a load is capacitive or inductive, it will also appear capacitive or inductive at the primary.

For Example 6.5,

$$Z_p = a^2 Z_s = \left(\frac{N_p}{N_s}\right)^2 R_L = \left(\frac{50}{200}\right)^2 20\text{ k}\Omega = \left(\frac{1}{16}\right) 20\text{ k}\Omega$$
$$= \mathbf{1.25\ k\Omega}$$

If impedance is to be transferred from the primary to the secondary, it is multiplied by the factor $1/a^2$.

Equation (6.16) can be used to its full advantage in application of the transformer as an *impedance-matching* device. Consider the following example.

EXAMPLE 6.6 Determine the turns ratio of the transformer in Fig. 6.25 required to ensure maximum power to the resistive load and calculate the maximum power.

Solution: The maximum power theorem states that maximum power is delivered to the load when its resistance equals the internal resistance R_{Th} of the supply. In this case R_L should be 800 Ω. As it stands, with $R_L = 8\ \Omega$, *if the transformer were not present,* the power to the load would be

$$P_L = I_L^2 R = \left(\frac{E}{R_{Th} + R_L}\right)^2 R_L = \left(\frac{120\text{ V}}{800\ \Omega + 8\ \Omega}\right)^2 (8\ \Omega) = (0.1485)^2(8)$$
$$= \mathbf{176.5\ mW}$$

However, if we insert the transformer and ensure that $Z_p = 800\ \Omega$, the load will "appear" to be equal to R_{Th}, and maximum power will be delivered. That is,

$$Z_p = 800\ \Omega = a^2 R_L = a2(8\ \Omega)$$

and

$$a^2 = \frac{800}{8} = 100$$

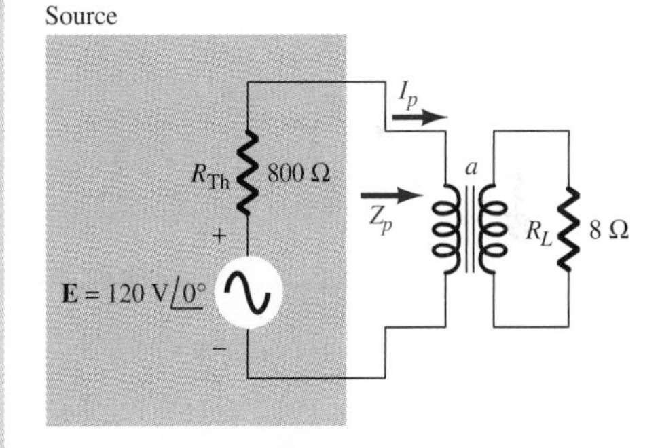

FIG. 6.25

or

$$a = \frac{N_p}{N_s} = \mathbf{10}$$

The primary current is then

$$I_p = \frac{120\text{ V}}{R_{\text{Th}} + Z_p} = \frac{120\text{ V}}{1600\ \Omega} = \mathbf{75\ mA}$$

and

$$I_s = aI_p = 10(75 \times 10^{-3}\text{ A}) = \mathbf{750\ mA}$$

with the power to the load determined by

$$P_L = I_s^2 R_L = (0.75\text{ A})^2 8\ \Omega = \mathbf{4.5\ W}$$

The increase over the power delivered without the transformer is

$$4.5{:}0.1765 \cong \mathbf{25.5{:}1}$$

The third application of the transformer listed in the introduction to this section was as an isolation device, which isolates a network insofar as dc and ground connectors are concerned but permits the passage of ac quantities. Recall the discussion of grounding in Section 5.10.

In Fig. 6.26 the supply and one end of R_2 have a ground in common with the oscilloscope. Before the scope is applied the peak value of the current through the circuit is

$$I_m = \frac{E}{R_T} = \frac{100\text{ V}}{100\ \Omega} = 1\text{ A}$$

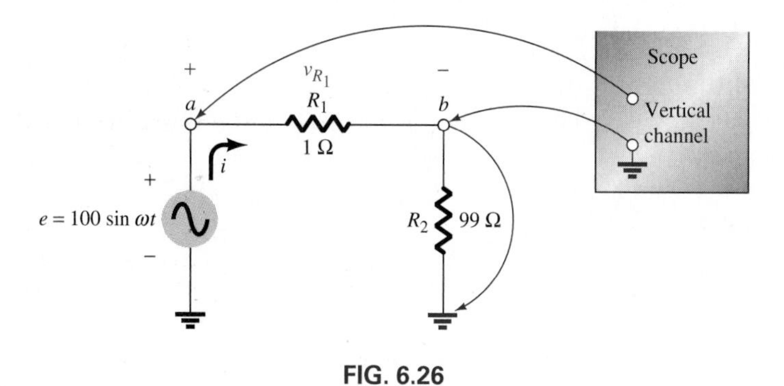

FIG. 6.26

If the scope is applied as shown, the scope will ground point b, setting both ends of the 99-Ω resistor to 0 Ω. The result is that the 99-Ω resistor will be "shorted out," and the total load of the circuit will be just R_1 or 1 Ω. The resulting peak current will then be

$$I_m = \frac{E}{R_T} = \frac{100\text{ V}}{1\ \Omega} = 100\text{ A}$$

which is probably of sufficient magnitude to cause dangerous side effects.

Such a situation can be avoided by placing a transformer between the oscilloscope and circuit as shown in Fig. 6.27 to remove any direct connection between the two and to remove the concern about grounding effects.

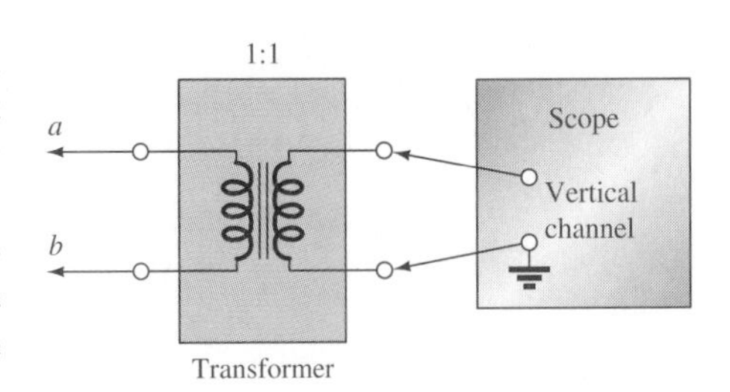

FIG. 6.27

Transformers are rated in kilovolt-amperes (kVA) rather than kilowatts (kW), for the reason indicated in Fig. 6.28. Even though I_s may be well above the rated value and can possibly cause severe damage to the secondary circuit, the wattmeter shows zero deflection (ideally speaking, no dc resistance is present in the system), since the load C is purely reactive. However, the apparent power rating of E_pI_p or E_sI_s sets a limit on the transformer currents for a particular voltage. In other words, an ammeter should be employed to determine maximum conditions for a fixed voltage rather than a wattmeter.

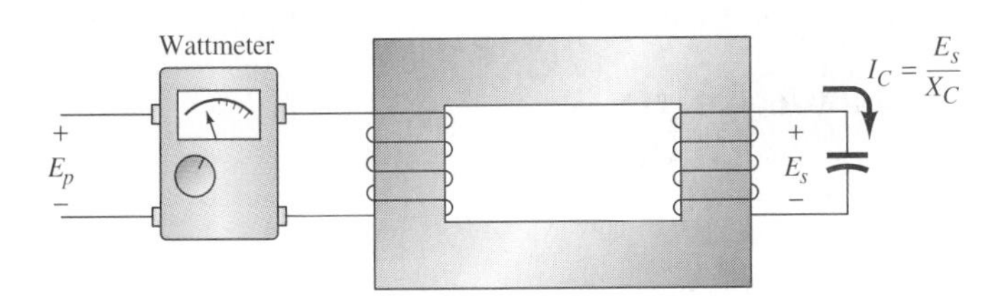

FIG. 6.28

In addition to the current and voltage rating, the frequency range of application is also specified for a transformer. For a particular transformer the output voltage versus frequency (on a log scale) may appear as shown in Fig. 6.29. At low frequencies, inductive effects in the primary circuit create a "shorting" effect that results in low levels of secondary voltage. At very high frequencies the stray capacitance introduces its shorting effect at both the primary and secondary, causing a "tail-off" in E_s. The peak in the curve near 10 kHz is due to a resonant effect established by the inductive and capacitive elements. Different iron-core transformers have different frequency response curves, requiring that E_s have a reasonable level at the frequency of application. All, however, rise and drop off at low and high frequencies, respectively, as shown in Fig. 6.29.

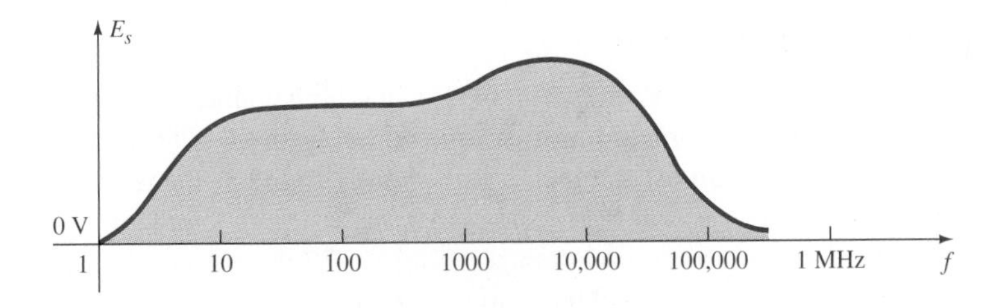

FIG. 6.29
Transformer frequency response.

At the expense of giving up the isolation characteristic, a two-circuit isolation transformer can be hooked up as an *autotransformer* to increase its kilovolt-ampere rating. For instance, the isolation transformer in Fig. 6.30(a) has a kVA rating of (500 V)(2 A) = 1 kVA. When used as an autotransformer, it has the general appearance of Fig. 6.30(b). Note that the secondary winding is connected so that its voltage is added to the primary voltage when the new secondary voltage is determined.

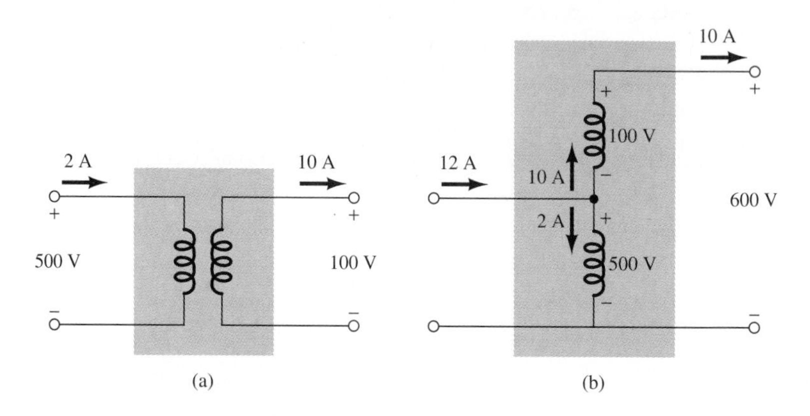

FIG. 6.30
Transformer connections: (a) two-circuit isolation configuration; (b) autotransformer.

The rated current for each winding is indicated next to each coil in Fig. 6.30(b). Applying Kirchhoff's current law, we find that the new primary current is the sum of the two, or 12 A, while that of the secondary remains at the 10-A level. The new kVA remains the same for the primary and secondary at (500 V)(12 A) = (600 V)(10 A) = 6 kVA—a level six times that obtained using the two-circuit configuration. An additional advantage of the autotransformer configuration is an improved efficiency rating. A portion of the energy can now be transferred to the output terminals directly rather than using solely transformer action. There are losses associated with the transfer of energy through transformer action that are not present using a direct conductive path.

6.5 OTHER AREAS OF APPLICATION

Recording Systems

Magnetic tape such as shown in Fig. 6.31 is in high demand today for the increasing number of recording instruments being used. The tape cartridge has become almost as much a part of daily living as the television or radio. The basic recording process is in actuality quite simple. For the single-channel recording head depicted in Fig. 6.32 the current direction through the coil determines which side of the recording head is north or south. In this case the resulting direction of the flux pattern is shown as established in the magnetic tape. The strength of the current in the coil

(a)

(b)

FIG. 6.31
Magnetic tape: (a) videocassette and recording cassette. (Courtesy of Maxell Corporation of America.) (b) manufacturing process. (Courtesy of Ampex Corporation.)

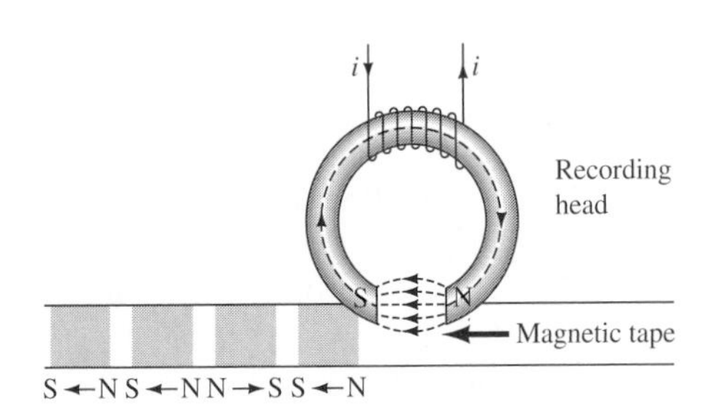

FIG. 6.32
Magnetic tape recording.

also affects the flux strength established in the head and therefore in the tape. A reversal in current direction reverses the magnetic polarization in the tape as also shown in the figure. The playback process employs a playback head similar in design to the recording head that is sensitive to the small permanent magnets established on the tape. As the tape passes under the reading head, a flux pattern is established in the playback core that induces a voltage pattern in the coil that can be translated into an audible sound.

Instruments

The *iron-vane* meter movement in Fig. 6.33 can read the effective value of any network without the additional circuitry required for the D'Arsonval movement. Two vanes are connected to a common hinge within a

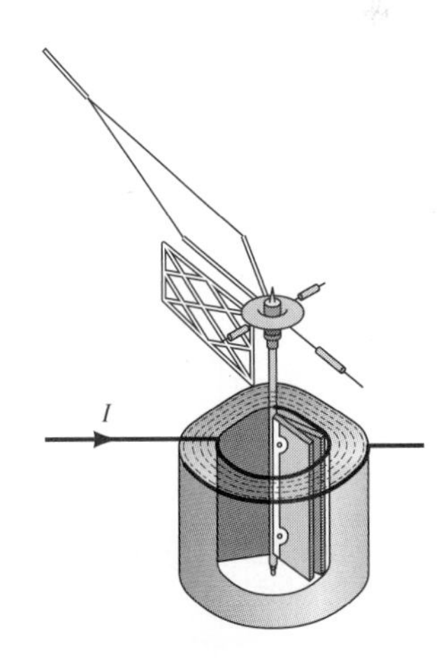

FIG. 6.33
Iron-vane movement. (Courtesy of Weston Instruments.)

coil that carries the current to be measured. The stronger the current, the stronger the opposite poles induced in the vanes and the greater the deflection. The scale is then properly calibrated to indicate the effective value.

The *electrodynamometer movement* in Fig. 6.34 is the most expensive of the three movements but usually the movement with the highest degree of accuracy and the only one used in the design of a wattmeter. It has a movable coil, as shown in the figure, that is free to rotate inside a fixed coil. Each coil senses the current, voltage, or power being measured, and the two fluxes react to develop a torque on the movable coil to which the pointer is attached. This movement also indicates the effective value of any waveform.

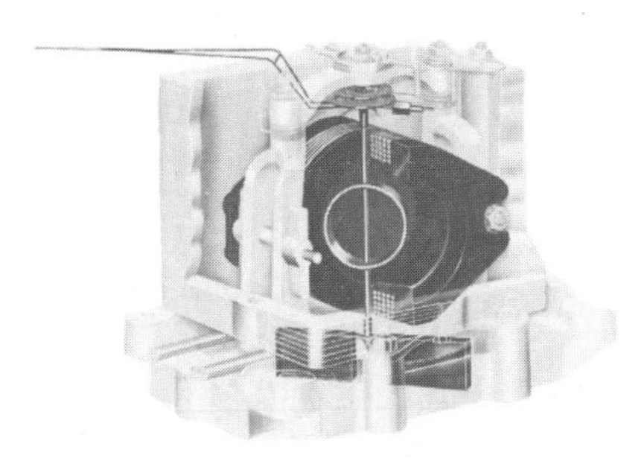

FIG. 6.34
Electrodynamometer movement. (Courtesy of Weston Instruments.)

The Amp-Clamp is an instrument that employs transformer effects to measure current without having to break the circuit. A typical commercially available unit appears in Fig. 6.35 that has 0-, 5-, 10-, 25-, 50-, 100-, and 250-A ac scales. Surrounding every current-carrying conductor is a magnetic field such as shown in Fig. 6.36. An ac current in the primary causes a varying flux pattern directly related to the magnitude of the current through the conductor. This is sensed by the "secondary" winding in the clamp of the instrument, and a reading is obtained that is calibrated directly to the induced voltage of the secondary.

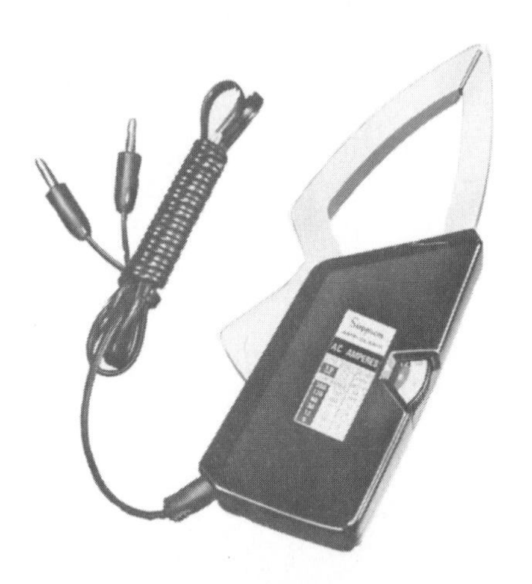

FIG. 6.35
Amp-Clamp®. (Courtesy of Simpson Electric Company.)

Speakers and Microphones

Electromagnetic effects are the moving force in the design of speakers such as shown in Fig. 6.37. The shape of the pulsating waveform of the input current is determined by the sound to be reproduced by the speaker at a high audio level. As the current peaks and returns to the valleys of the sound pattern the strength of the electromagnet varies in exactly the same manner. This causes the cone of the speaker to vibrate at a frequency directly proportional to the pulsating input. The higher the pitch of the sound pattern, the higher the oscillating frequency between the peaks and valleys and the higher the frequency of vibration of the cone.

A second design used more frequently in more expensive speaker systems appears in Fig. 6.38. In this case the permanent magnet is fixed,

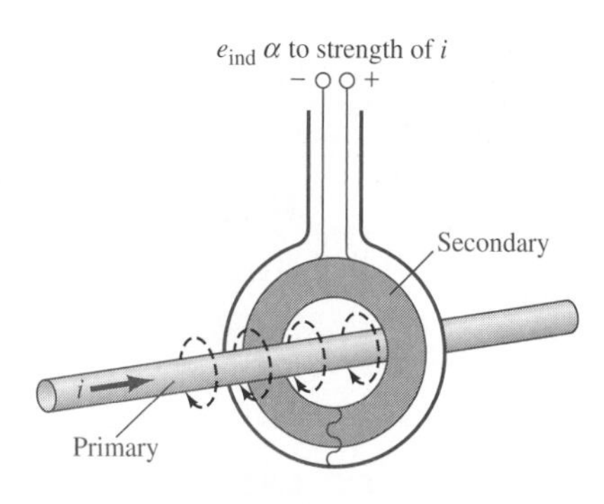

FIG. 6.36
Amp-Clamp® (internal view).

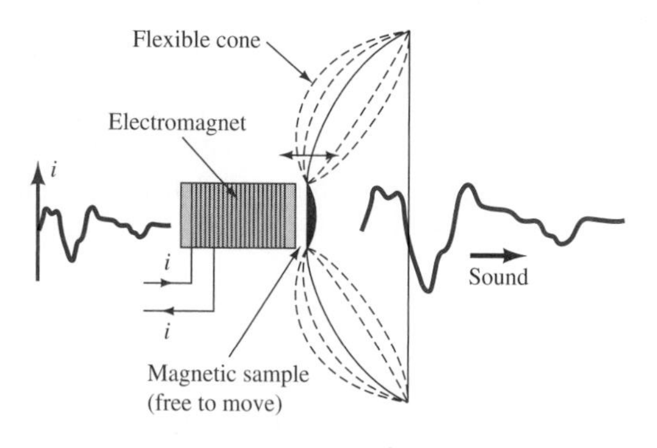

FIG. 6.37
Speaker.

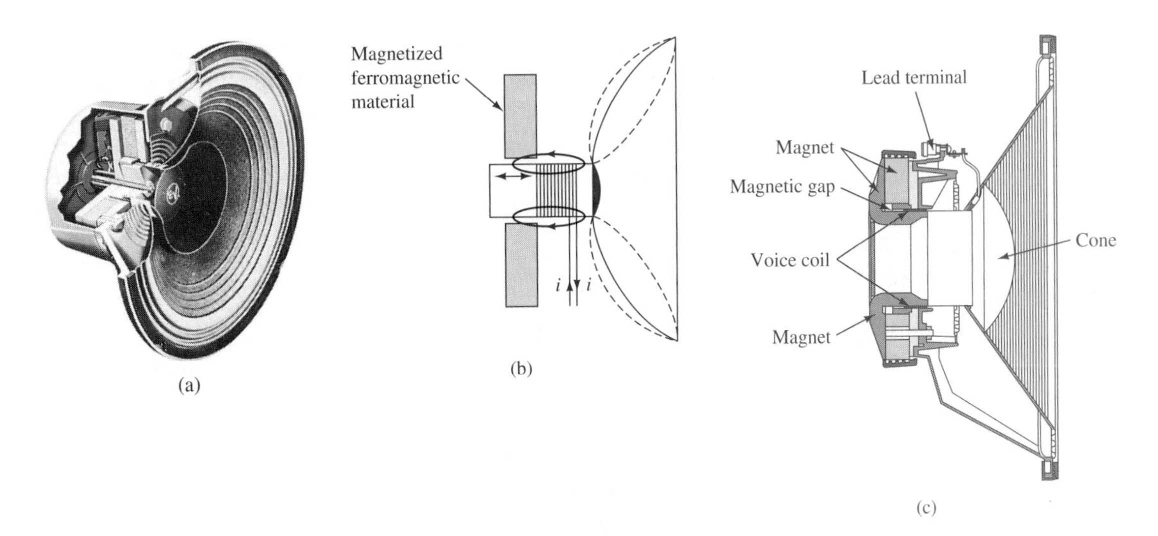

FIG. 6.38
Coaxial high-fidelity loudspeaker: (a) photograph; (b) basic operation; (c) cross section of actual unit. (Courtesy of Electro-Voice, Inc.)

and the input is applied to a movable core within the magnet, as shown in the figure. High peaking currents at the input produce a strong flux pattern in the voice coil, causing it to be drawn well into the flux pattern of the permanent magnet. As in the speaker of Fig. 6.37, the core then vibrates at a rate determined by the input and provides the audible sound.

Microphones such as those in Fig. 6.39 also employ electromagnetic effects. The sound to be reproduced at a higher audio level causes the core and attached moving coil to move within the magnetic field of the permanent magnet. Through Faraday's law ($e = N\,d\phi/dt$) a voltage is induced across the movable coil proportional to the speed with which it is

(a)

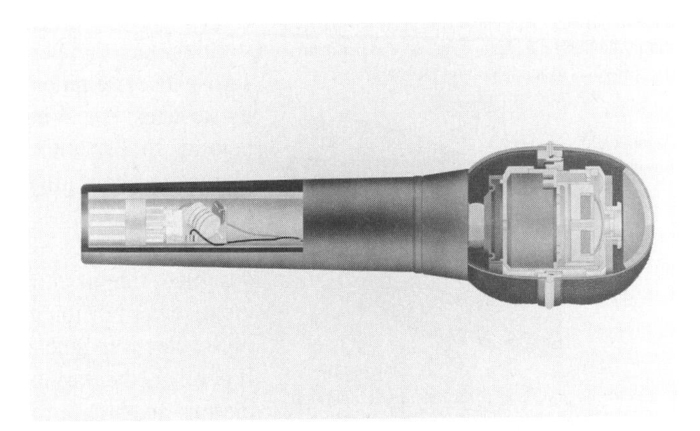
(b)

FIG. 6.39
Dynamic microphone: (a) external view; (b) internal view. (Courtesy of Electro-Voice, Inc.)

moving through the magnetic field. The resulting induced voltage pattern can then be amplified and reproduced at a much higher audio level through the use of speakers, as described earlier. Microphones of this type are the most frequently employed, although other types that use capacitive, carbon granular, and piezoelectric* effects are available. This particular design is commercially referred to as a *dynamic* microphone.

Telephone

Magnetic effects play a number of important roles in the telephone handset and base, as shown in Fig. 6.40. The piezo ringer replaced the clapper–gong unit of earlier years. *Piezo* is short for *piezoelectric*, a semiconductor material such as barium titanate that generates a voltage across its extremities when its shape is changed or pressure is applied to its surfaces. It can be used as a ringer because the piezoelectric crystal belongs to a special class of tranducers that are reversible. That is, applying an alternating signal to the opposite surfaces of a disc-shaped piezoelectric material will change its shape at the frequency of the applied signal. If the disc is attached to a thin-film surface and if properly designed, the vibrating surface will emit a ringing sound. The details of the receiver portion are shown in Fig. 6.41. The current through the coil of the electromagnet is controlled by the voice pattern to be reproduced. It controls the varying strength of the electromagnet and causes the vibrating diaphragm to vibrate at the same rate. The vibrating diaphragm causes the surrounding air to vibrate in the same manner, producing the audible sounds or voice of the sender.

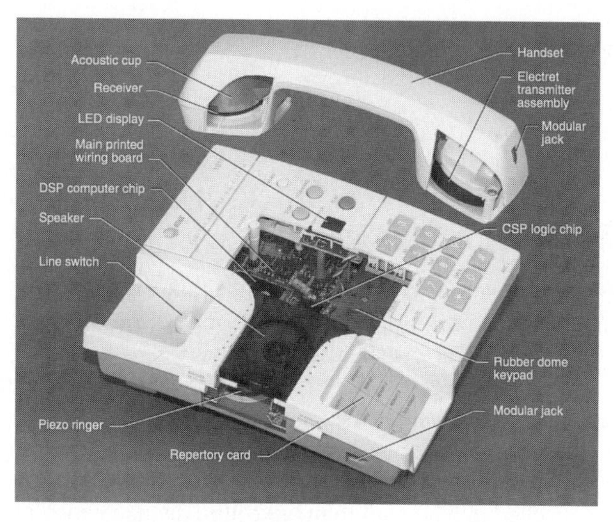

FIG. 6.40
Digital telephone answering set. (Courtesy of AT&T.)

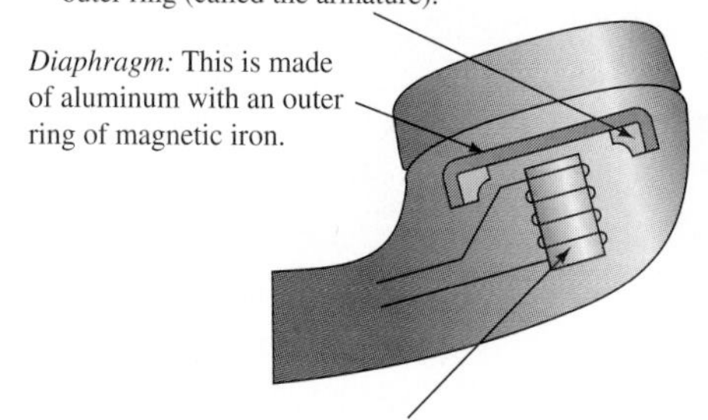

FIG. 6.41
Telephone receiver. (Courtesy of Western Electric Company.)

---

*Piezoelectricity is a small voltage generated by the exertion of pressure across certain crystals.

The transmitter or sending unit has the basic construction shown in Fig. 6.42. When the handpiece of the telephone is lifted, charge flows through the carbon granules. As you speak into the handset, the pressure on the diaphragm varies in accordance with your voice pattern. The greater the pressure on the diaphragm, the more compact the carbon granules and the less the resistance encountered for the current through the transmitter. The net result is a variation in current related directly to the voice pattern to be reproduced. The signal is then passed through the wires in the form of a current variation to the receiver set at the other end of the conversation. That charge can travel through the lines at the speed of light: 186,000 mi/s. This same principle can be employed in the design of carbon microphones, as indicated earlier in this section.

Diaphragm: This is a circular piece of very thin aluminum. Its outer edge is fixed firmly to a frame in the mouthpiece. But the rest of it can vibrate back and forth.

Carbon chamber: The center of the diaphragm has a small gold-plated brass dome which projects into this small chamber which contains small grains of carbon made from anthracite coal. When you are using the telephone, the carbon chamber is part of the circuit and charge flows through the carbon.

FIG. 6.42
Telephone transmitter. (Courtesy of Western Electric Company.)

Automobile Ignition System

The transformer introduced in Section 6.4 plays a vital role in the ignition system of an automobile, as demonstrated by the control system in Fig. 6.43. In present-day vehicles the ignition system is part of a much larger computer command control (CCC) system. The CCC system electronically controls a variety of engine functions to reduce exhaust emissions and provide good fuel economy.

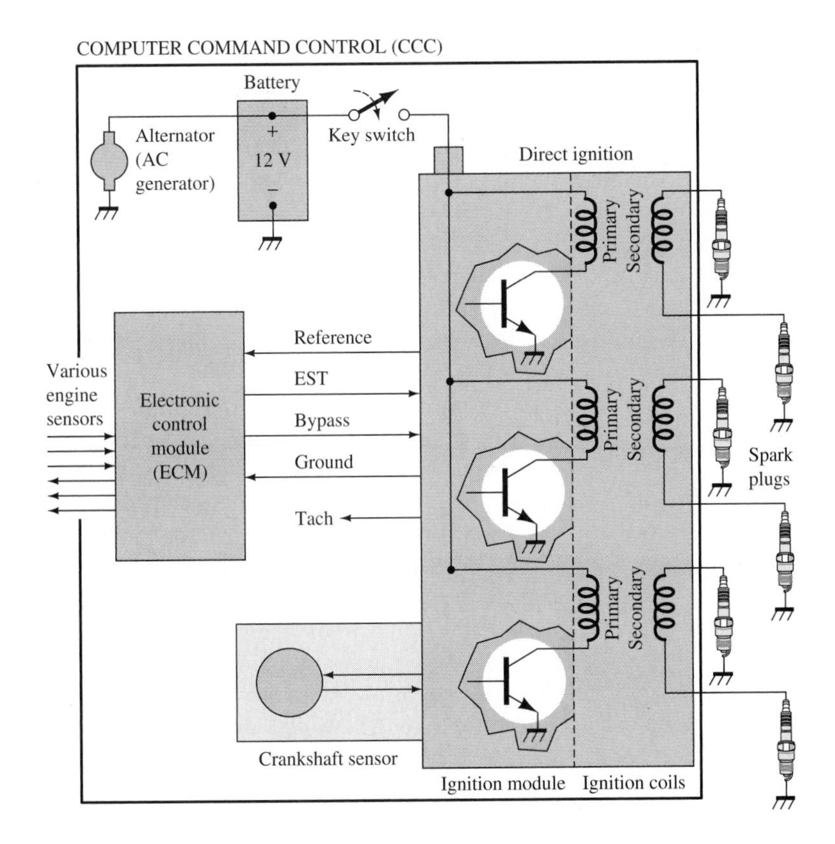

FIG. 6.43
Computer command control unit for a six-cylinder automobile.

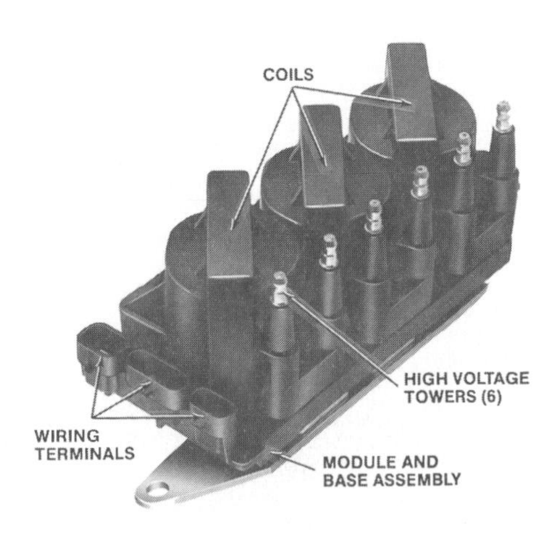

FIG. 6.44
Direct ignition assembly for the six-cylinder system in Fig. 6.43.

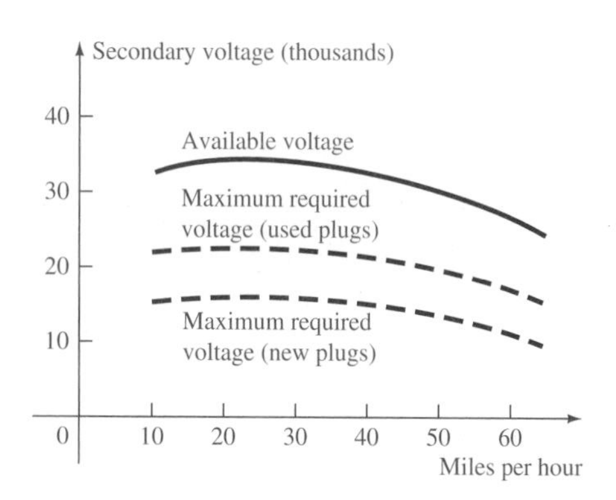

FIG. 6.45
Spark plug voltage versus speed for an automobile. (Courtesy of Delco-Remy.)

The heart of the command control system is the electronic control module (ECM), containing a sophisticated electronic system constructed using a number of complex integrated circuits. As indicated by the arrows, information flows to and from the module as dictated by the operation to be performed.

The alternator (ac generator) driven by a belt linked to the rotating crankshaft ensures through a rectifying system (to provide a dc charging level) that the battery is fully charged at 12 V. When the ignition switch is turned on, the direct ignition region of the CCC is also switched to the on state. The direct ignition assembly for a six-cylinder engine is shown in Fig. 6.44. The wiring terminals are used to carry information to and from the ECM and crankshaft sensors and to provide information to the tachometer. Each high-voltage terminal in Fig. 6.44 is connected directly to a spark plug as shown in Fig. 6.43. Note that in Figs. 6.43 and 6.44 a six-cylinder assembly has three coils with each secondary connected to two plugs. Each ignition coil (transformer) consists of an iron core with a primary of a few turns of heavy wire (heavier current level) and a secondary with many turns of fine wire (less current). Typically, the ratio between secondary and primary is 100:1. Recall that the voltage across a coil is sensitive to the *change* in current through the coil as defined by the equation $v_L = L\, di/dt$. In other words, the voltage across a coil is sensitive to the *rate of change* of current through a coil and not to the magnitude of the steady current level through the coil. When the ignition coil primary circuit is opened and closed, the current through the primary coil changes quite rapidly, causing a significant primary voltage of about 350 V. A turns ratio of 100 then results in a secondary voltage of about 35,000 V. Figure 6.45 reveals that the available voltage is about 35,000 V, with the maximum required voltage typically about 22,000 V. As the speed increases, the available and required voltage drops. Note also that older plugs require significantly more voltage (about 5000 V) because of the larger plug gap and rounding of the center electrode.

Note the transistor appearing between each primary winding and ground. The ignition module housed under the coils in Fig. 6.44 is another electronic IC system that interacts with the ECM unit and the crankshaft and controls the operation of the transistor. If all the signals received by the ignition module indicate a go status, it sends an appropriate signal to the ECM, which combines that input with a host of other sensor inputs such as temperature, barometric pressure, and rpm to determine whether a signal should be sent back to the ignition module to turn on the transistor. If the feedback is positive, the transistor is turned on and the collector current of the transistor is the primary current of the coil. Without the *on* transistor the primary winding would be left *floating*, removing the possibility of a primary current and ignition of the plugs. At the instant the transistor turns on, the primary current increases to a predetermined level. The controlling network then removes the base voltage, quickly turning off the transistor. The rapidly decaying primary current then induces a voltage in the secondary sufficient to fire the spark plug.

When the automobile is running at normal operating speeds, the bypass signal from the ECM to the ignition module directs the ignition module to establish a direct path from the EST (Electronic Signal Transmission) terminal to the base of the transistor, removing any control of the transistor by the ignition module and leaving it totally in the hands of the ECM unit. The status of the EST terminal is determined by the reference voltage pulse it receives from the ignition module and a number of other engine sensors. Under these conditions the ECM uses its complex electronic control system to continually *fine-tune* the engine to ensure minimum exhaust emissions and the highest fuel economy.

In a *four-cycle* gasoline engine, each piston goes through the four strokes of intake, compression, power, and exhaust during two complete revolutions of the crankshaft. For a six-cylinder engine half the pistons have a power stroke during each revolution of the crankshaft, resulting in three ignitions per revolution. At a typical rpm of 4500 there are 225 sparks per second:

$$4500\,\frac{\cancel{\text{rev}}}{\cancel{\text{min}}}\left(\frac{1\,\cancel{\text{min}}}{60\text{ s}}\right)\left(\frac{3\text{ sparks}}{\cancel{\text{rev}}}\right) = \mathbf{225\ sparks\ per\ second}$$

which is significant when you consider that in 1 h 810,000 ignition sparks occur.

Computer Hard Disks

The computer *hard disk* is a sealed unit in a computer that stores data on a magnetic coating applied to the surface of circular platters that spin like a record. The platters are constructed on a base of aluminum or glass (both nonferromagnetic), which makes them rigid, hence the term *hard disk*. Since the unit is sealed, the internal platters and components are inaccessible, and a “crash” (a term applied to the loss of data from a disk or the malfunction thereof) usually requires that the entire unit be replaced. Hard disks are currently available with diameters from 1⅓ in. to 5¼ in. with the 3½ in. the most popular for today’s desk- and laptop units. All hard disk drives are often referred to as *Winchester drives,* a term first applied in the 1960s to an IBM drive that had 30 MB [a byte is a series of binary bits (0s and 1s) representing a number, letter, or symbol] of fixed (nonaccessible) data storage and 30 MB of accessible data storage. The term *Winchester* was applied because the 30-30 data capacity matched the name of the popular 30-30 Winchester rifle.

The magnetic coating on the platters is called the *media* and is of either the oxide or the *thin-film* variety. The oxide coating is formed by first coating the platter with a gel containing iron oxide (ferromagnetic) particles. The disk is then spun at a very high speed to spread the material evenly across the surface of the platter. The resulting surface is then covered with a protective coating that is made as smooth as possible. The thin-film coating is very thin but durable, with a surface that is smooth and consistent throughout the disk area. In recent years the trend has been toward the thin-film coating because the read/write heads (to be described shortly) must travel closer to the surface of the platter, requiring a consistent coating thickness. Recent techniques have resulted in thin-film magnetic coatings as thin as one-millionth of an inch.

The information on a disk is stored around the disk in circular paths called *tracks*, with each track containing so many bits of information per inch. The product of the number of bits per inch and the number of tracks per inch is the *Areal* density of the disk, which provides an excellent quantity for comparison with early systems and reveals how far the field has progressed in recent years. In the 1950s the first drives had an Areal density of about 2 kbits/in.^{2} as compared to today's typical 1–2 Gbits/in.^{2}, an incredible achievement; consider 1,000,000,000,000 bits of information on an area the size of the face of your watch. *Electromagnetism* is the key element in the *writing* of information on the disk and the *reading* of information off the disk. In its simplest form the *write/read head* of a hard disk (or floppy disk) is a U-shaped electromagnet with an air gap that rides just above the surface of the disk as shown in Fig. 6.46.

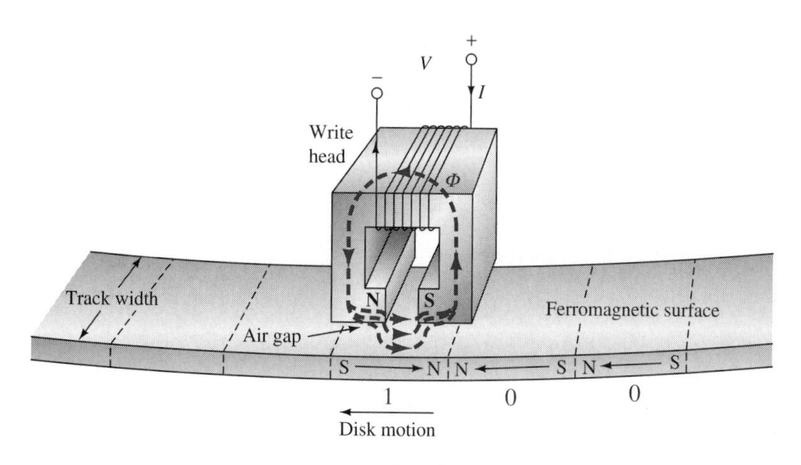

FIG. 6.46
Hard disk storage using a U-shaped electromagnet write head.

As the disk rotates, information in the form of a voltage with changing polarities is applied to the winding of the electromagnet. For our purposes we will associate a positive voltage level with a 1 level (of binary arithmetic) and a negative voltage level with a 0 level. Combinations of these 0 and 1 levels can be used to represent letters, numbers, or symbols. If energized as shown in Fig. 6.46 with a 1 level (positive voltage), the resulting magnetic flux pattern will have the direction shown in the core. When the flux pattern encounters the air gap of the core, it jumps to the magnetic material (since magnetic flux always seeks the path of least reluctance and air has a high reluctance) and establishes a flux pattern as shown on the disk until it reaches the other end of the core air gap, where it returns to the electromagnet and completes the path. As the head then moves to the next bit sector it leaves behind the magnetic flux pattern just established from the left to the right. The next bit sector has a 0 level input (negative voltage), which reverses the polarity of the applied voltage and the direction of the magnetic flux in the core of the head. The result is a flux pattern in the disk opposite that associated with a 1 level. The next bit of information is also a 0 level, resulting in the same pattern just generated. In total, therefore, information is stored on the disk in the form of small magnets whose polarity defines whether they represent a 0 or a 1. Once the data have been stored, we have to have some method of retrieving the information when desired. The first few hard disks used the same head for both the write and read functions. In Fig. 6.47(a) the

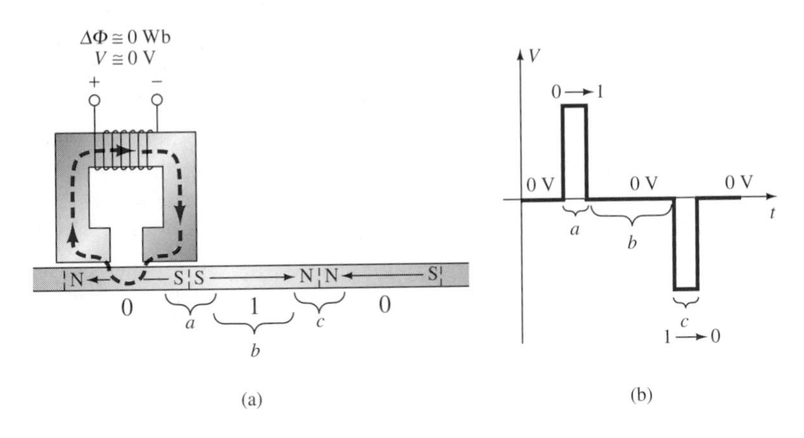

FIG. 6.47
Reading the information off a hard disk using a U-shaped electromagnet.

U-shaped electromagnet in the read mode simply picks up the flux pattern of the current bit of information. *Faraday's law of electromagnet induction* states that a voltage is induced across a coil if exposed to a changing magnetic field. The change in flux for the core in Fig. 6.47(a) is minimal as it passes over the induced bar magnet on the surface of the disk. A flux pattern is established in the core because of the bar magnet on the disk, but the lack of a significant change in flux level results in an induced voltage at the output terminals of the pickup of approximately 0 V, as shown in Fig. 6.47(b) for the readout waveform. A significant change in flux occurs when the head passes over the transition region, labeled *a* in Fig. 6.47(a). In this region the flux pattern changes from one direction to the other—a significant change in flux occurs in the core as it reverses direction, causing a measurable voltage to be generated across the terminals of the pickup coil as dictated by Faraday's law and indicated in Fig. 6.47(b). In region *b* of Fig. 6.47(a) there is no significant change in the flux pattern from one bit area to the next, and a voltage is not generated, as also revealed in the pulse train. However, when region *c* is reached, the change in flux is significant but opposite that occurring in region *a,* resulting in another pulse but of opposite polarity. In total, therefore, the output bits of information are in the form of pulses that have a shape totally different from the read signals but are certainly representative of the information being stored. In addition, note that the output is generated at the transition regions and not in the constant-flux region of the bit storage.

In the early years the use of the same head for the read and write functions was acceptable, but as the tracks became narrower and the seek time (the average time required to move from one track to another a random distance away) had to be reduced, it became increasingly difficult to construct the coil or core configuration in a manner that was sufficiently thin with minimum weight. In the late 1970s IBM introduced the *thin-film inductive head,* which was manufactured in much the same way as the small integrated circuits of today. The result is a head having a length typically less than $\frac{1}{10}$ in. and a height less than $\frac{1}{50}$ in., with minimum mass and high durability. The average seek time has dropped from the few hundred milliseconds for the first units to about 8 ms for current models. In addition, production methods have improved to the point that the head can "float" above the surface (to minimize damage to the disk) at a height of only 5 microinches, or 0.000005 in. For a typical hard speed of 3600 rpm (some are as high as 7200 rpm) and an average diameter of 1.75 in. for a 3.5-in. disk, the speed of the head over the track is about 38 mph. Scaling the floating height up to $\frac{1}{4}$ in. (multiplying by a factor of 50,000) the speed would increase to about 1.9×10^6 mph. In other words, the speed of the head over the surface of the platter is analogous to a mass traveling $\frac{1}{4}$ in. above a surface at 1.9 million miles per hour—quite a technical achievement and amazingly enough one that perhaps will be improved by a factor of 10 in the next decade. Incidentally, the speed of rotation of floppy disks is about one-tenth that of the hard disk, or 360 rpm. In addition, the head touches the magnetic surface of the floppy disk, limiting the storage life of the unit. The typical magnetizing force needed to lay down the magnetic orientation is 400 mA-turn (peak-to-peak). The result is a write current of only 40 mA for a 10-turn thin-film inductive head.

Although the thin-film inductive head could be used as a read head also, the *magnetoresistive* (MR) head has improved reading characteristics. The MR head depends on the fact that the resistance of a soft ferromagnetic conductor such as Permalloy is sensitive to changes in external magnetic fields. As the hard disk rotates, the changes in magnetic flux from the induced magnetized regions of the platter change the terminal resistance of the head. A constant current passed through the sensor displays a terminal voltage sensitive to the magnitude of the resistance. The result is output voltages with peak values in excess of 300 V, which exceeds that of typical inductive read heads by a factor of 2: or 3:1.

Further investigation will reveal that the best write head is of the thin-film inductive variety and that the optimum read head is of the MR variety. Each has particular design criteria for maximum performance, resulting in the increasingly common *dual-element head*, with each head containing separate conductive paths and different gap widths. The Areal density of the new hard disks will essentially require the dual-head assembly for optimum performance.

The preceding is clear evidence of the importance of magnetic effects in today's growing industrial, computer-oriented society. Although research continues to maximize the Areal density, it appears certain that the storage will remain magnetic for the write/read process and not be replaced by any of the growing alternatives such as the optic laser variety used so commonly in CD-ROMS.

A 3.5-in. full-height disk drive manufactured by the Micropolis Corporation that has a formatted capacity of 1.75 gigabytes (GB) with an average search time of 10 ms appears in Fig. 6.48.

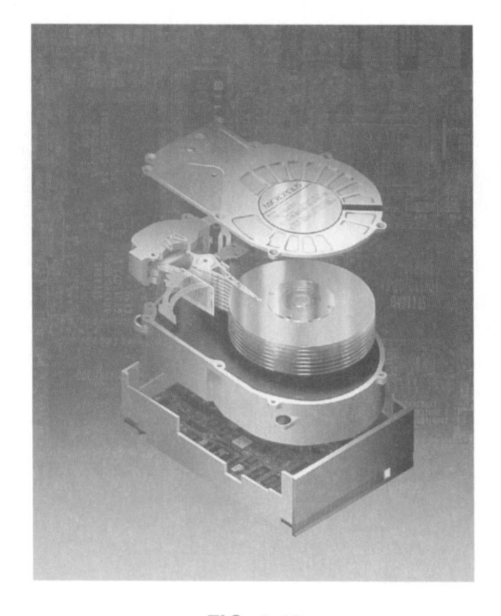

FIG. 6.48
A 3.5-in. hard-disk drive with a capacity of 1.75 GB with an average search time of 10 ms. (Courtesy of Micropolis Corporation.)

Hall Effect Sensor

The Hall effect sensor is a semiconductor device that generates an output voltage when exposed to a magnetic field. The basic construction consists of a slab of semiconductor material through which a current is passed, as shown in Fig. 6.49(a). If a magnetic field is applied as shown in the figure perpendicular to the direction of the current, a voltage V_H will be generated between the two terminals, as indicated in Fig. 6.49(a). The difference in potential is due to the separation of charge established by the Lorentz force first studied by Professor Hendrick Lorentz in the early eighteenth century. He found that electrons in a magnetic field are subjected to a force proportional to the velocity of the electrons through the field and the strength of the magnetic field. The direction of the force is determined by the left-hand rule. Simply place the index finger of the left hand in the direction of the magnetic field, with the second finger at right angles to the index finger in the direction of conventional current through the semiconductor material, as shown in Fig. 6.49(b). The thumb, if placed at right angles to the index finger, will indicate the direction of the force on the electrons. In Fig. 6.49(b) the force causes the electrons to accumulate in the bottom region of the semiconductor (connected to the negative terminal of the voltage V_H), leaving a net positive charge in the upper region of the material (connected to the positive terminal of V_H). The stronger the current or strength of the magnetic field, the greater the induced voltage V_H.

In essence, therefore, the Hall effect sensor can reveal the strength of a magnetic field or the level of current through a device if the other determining factor is held fixed. Two applications of the sensor are therefore apparent—to measure the strength of a magnetic field in the vicinity of a sensor (for an applied fixed current) and to measure the level of current through a sensor (with knowledge of the strength of the magnetic field linking the sensor). The gaussmeter in Fig. 6.13 employs a Hall effect sensor. Internal to the meter a fixed current is passed through the sensor with the voltage V_H indicating the relative strength of the field. Through amplification, calibration, and proper scaling the meter can display the relative strength in gauss.

The Hall effect sensor has a broad range of applications that are often quite interesting and innovative. The most widespread is as a trigger for an alarm system in large department stores where theft is often a difficult problem. A magnetic strip is attached to the merchandise that sounds an alarm when a customer passes through the exit gates without paying for the product. The sensor, control current, and monitoring system are housed in the exit fence and react to the presence of the magnetic field as the product leaves the store. When the product is paid for, the cashier removes the strip or demagnetizes the strip by applying a magnetizing force that reduces the residual magnetism in the strip essentially to zero.

The timing system of an automobile or the speed of a motor can be determined by placing tabs in the flywheel as shown in Fig. 6.50(a). In the absence of a tab there is a large air gap in the bottom of the series magnetic circuit, resulting in a small magnetic flux through the sensor at the top of the core. The output voltage, which is sensitive to the strength of the magnetic field, is relatively small. When the tab is present, the magnetic flux strength increases substantially, and the output voltage

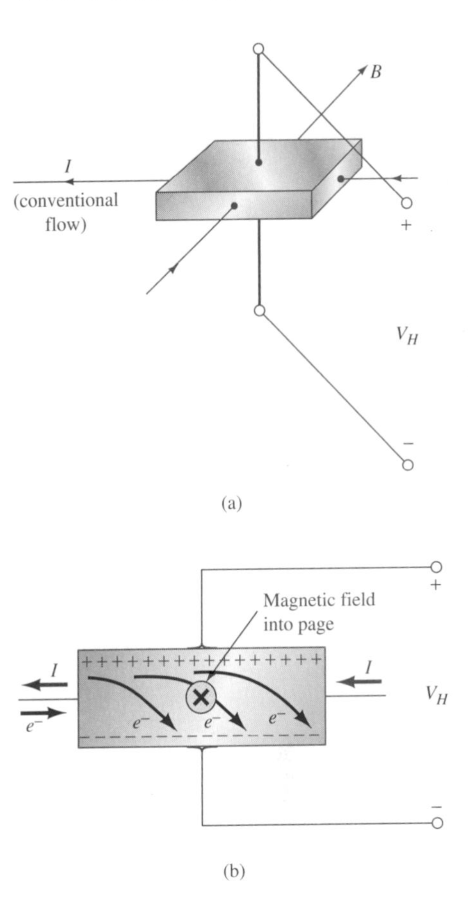

FIG. 6.49
Hall effect sensor: (a) orientation of controlling parameters; (b) effect on electron flow.

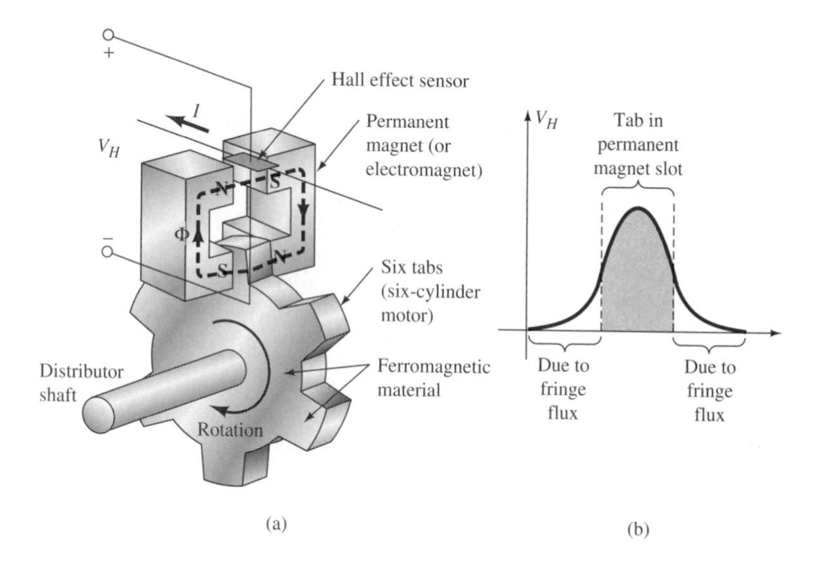

FIG. 6.50
Timing system of an automobile employing a Hall effect sensor: (a) system; (b) output voltage.

peaks as shown in Fig. 6.50(b). Knowledge of the diameter of the rotating wheel and the time between pulses reveals the speed of rotation.

The Hall effect sensor is also used to indicate the speed of a bicycle on a digital display conveniently mounted on the handlebars. As shown in Fig. 6.51(a), the sensor is mounted on the frame of the bike and a small permanent magnet is mounted on a spoke of the front wheel. The magnet must be carefully mounted to be sure that it passes over the

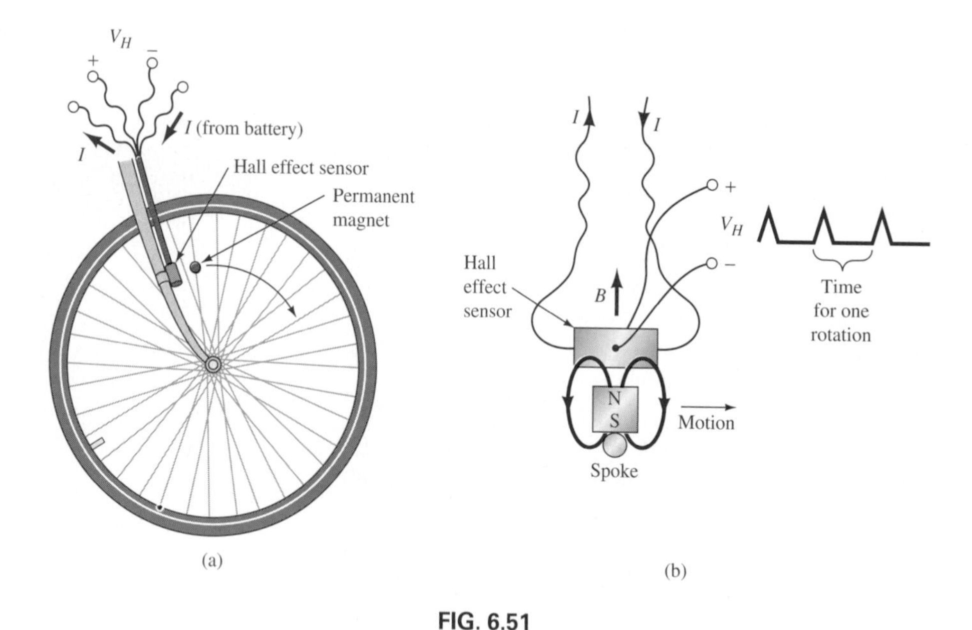

FIG. 6.51
Obtaining a speed indication for a bicycle using a Hall effect sensor: (a) mounting the components; (b) Hall effect response.

proper region of the sensor. When the magnet passes over the sensor, the flux pattern in Fig. 6.51(b) results and a voltage with a sharp peak is developed by the sensor. For a bicycle with a 26-in. diameter wheel, the circumference is about 82 in. Over 1 mi the number of rotations is

$$5280\ \text{ft}\left(\frac{12\ \text{in.}}{1\ \text{ft}}\right)\left(\frac{1\ \text{rotation}}{82\ \text{in.}}\right) \cong 773\ \text{rotations}$$

If the bicycle is traveling at 20 mph, an output pulse will occur at a rate of 4.29 per second. It is interesting to note that at a speed of 20 mph the wheel is rotating at more than 4 revolutions per second and the total number of rotations over 20 mi is 15,460.

Magnetic Reed Switch

One of the most frequently employed switches in alarm systems is the *magnetic reed switch,* shown in Fig. 6.52. There are two components of the reed switch—a permanent magnet embedded in one unit, which is normally connected to the movable element (door, window, and so on), and a reed switch in the other unit, which is connected to the electrical control circuit. The reed switch is constructed of two iron alloy (ferromagnetic) reeds in a hermetically sealed capsule. The cantilevered ends of the two reeds do not touch but are in very close proximity to each other. In the absence of a magnetic field the reeds remain separated. However, if a magnetic field is introduced, the reeds will be drawn to each other because flux lines seek the path of least reluctance and if possible exercise every alternative to establish the path of least reluctance. The situation is similar to placing a ferromagnetic bar close to the ends of a U-shaped magnet. The bar is drawn to the poles of the magnet, establishing a magnetic flux path without air gaps and minimum reluctance. In the open-circuit state the resistance between reeds is in excess of 100 MΩ, while in the on state it drops to less than 1 Ω.

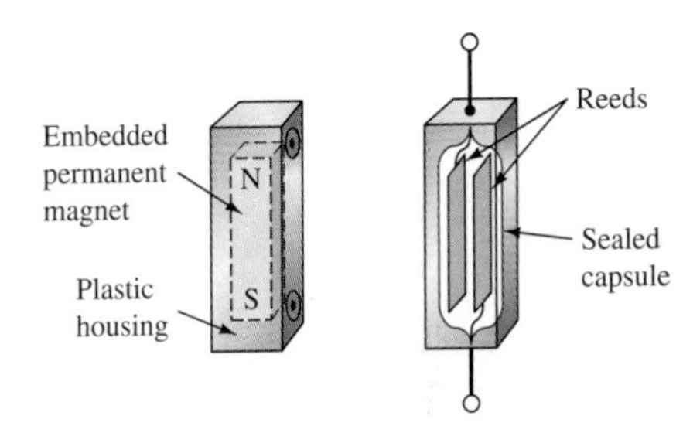

FIG. 6.52
Magnetic reed switch.

In Fig. 6.53 a reed switch has been placed on the fixed frame of a window and a magnet on the movable window unit. When the window is closed as shown in Fig. 6.53, the magnet and reed switch are sufficiently close to establish contact between the reeds, and a current is established through the reed switch to the control panel. In the armed state the alarm system accepts the resulting current flow as a normal secure response. If the window is opened, the magnet leaves the vicinity of the reed switch and the switch opens. The current through the switch is interrupted, and the alarm reacts appropriately.

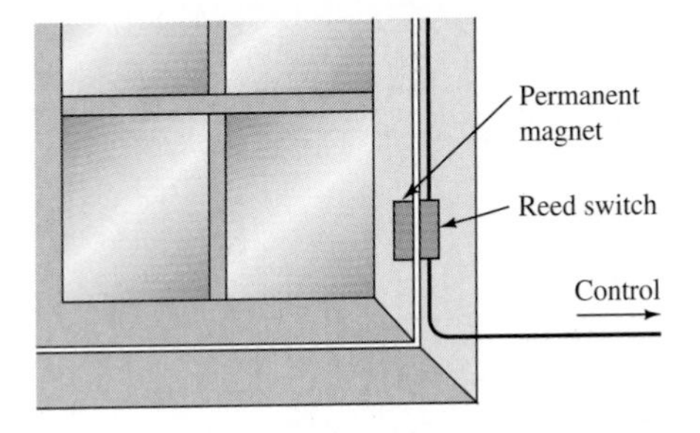

FIG. 6.53
Using a magnetic reed switch to monitor the state of a window.

One of the distinct advantages of the magnetic reed switch is that the proper operation of any switch can be checked with a portable magnetic element by simply bringing the magnet to the switch and noting the output response. There is no need to continually open and close windows and doors. In addition, the reed switch is hermetically enclosed so that oxidation and foreign objects cannot damage it, and the result is a unit that can effectively last indefinitely. Magnetic reed switches are also available in other shapes and sizes, allowing them to be concealed from obvious view. One is a circular variety that can be set into the edge of a door and door jamb, revealing only two small visible disks when the door is open.

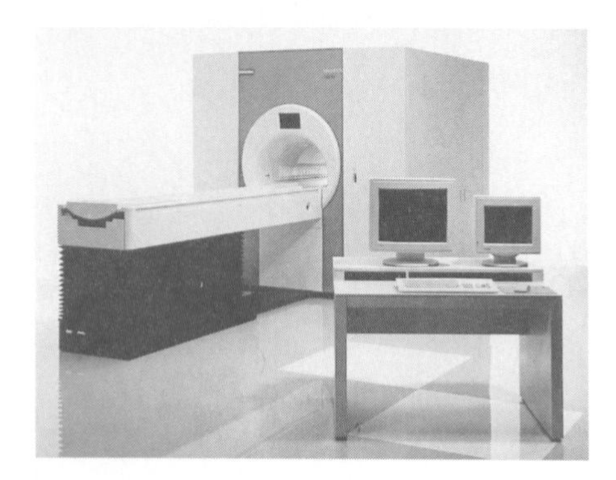

FIG. 6.54
Magnetic resonance imaging equipment. (Courtesy of Siemens Medical Systems, Inc.)

Magnetic Resonance Imaging

Magnetic resonance imaging [MRI, also called nuclear magnetic resonance (NMR)] is receiving more and more attention as scientists strive to improve the quality of the cross-sectional images of the body so useful in medical diagnosis and treatment. MRI does not expose the patient to potentially hazardous X-rays or injected contrast materials such as those employed to obtain computerized axial tomography (CAT) scans.

The three major components of an MRI system are a huge magnet that can weigh up to 100 tons, a table for transporting the patient into the circular hole in the magnet, and a control center, as shown in Fig. 6.54. The image is obtained by placing the patient in the tube to a precise depth depending on the cross section to be obtained and applying a strong magnetic field that causes the nuclei of certain atoms in the body to line up. Radio waves of different frequencies are then applied to the patient in the region of interest, and if the frequency of the wave matches the natural frequency of the atom, the nuclei are set into a state of resonance and absorb energy from the applied signal. When the signal is removed, the nuclei release the acquired energy in the form of weak but detectable signals. The strength and duration of the energy emission vary from one tissue of the body to another. The weak signals are then amplified, digitized, and translated to provide a cross-sectional image such as shown in Fig. 6.55.

MRI units are very expensive and therefore are not available at all locations. In recent years, however, their numbers have grown and one is available in almost every major community. For some patients the claustrophobic feeling they experience while in the circular tube is difficult to contend with. Today, however, a more open unit has been developed, as shown in Fig. 6.56, that alleviates most of this discomfort.

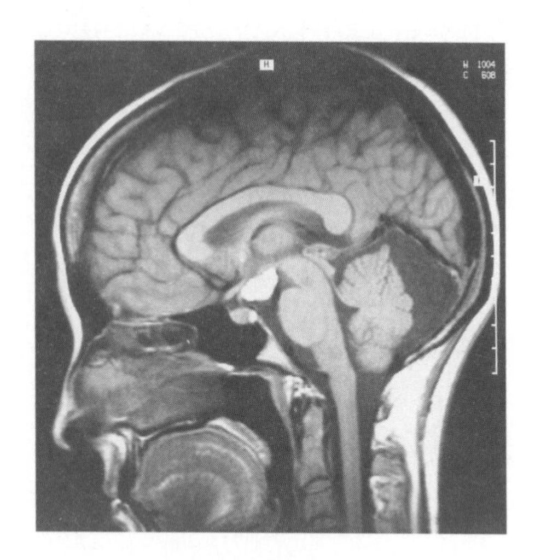

FIG. 6.55
Magnetic resonance image. (Courtesy of Siemens Medical Systems, Inc.)

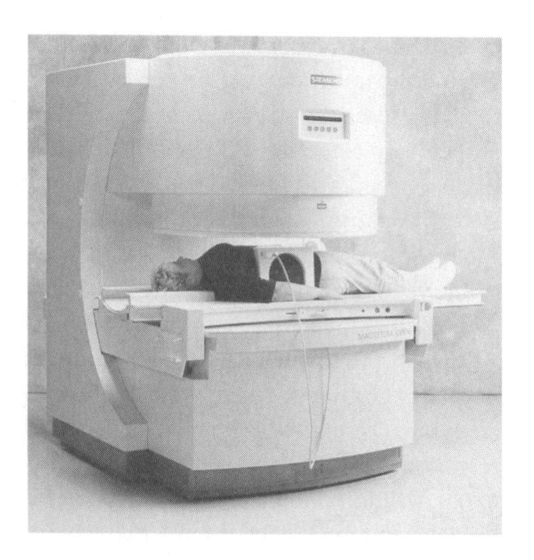

FIG. 6.56
Magnetic resonance imaging equipment (open variety). (Courtesy of Siemens Medical Systems, Inc.)

Patients who have metallic implants or pacemakers or those that have worked in industrial environments where minute ferromagnetic particles may have become lodged in open sensitive areas such as the eyes, nose, and so on, may have to use a CAT scan system because it does not employ magnetic effects. The attending physician is well trained in such areas of concern and will allay any unfounded fears or suggest alternative methods.

6.6 SUMMARY

- **Electromagnetism** is the setting up of **magnetic flux lines** through the application of **electric current.**
- **Opposite** magnetic poles **attract,** while **like** poles **repel.**
- The **denser** the magnetic flux lines (ϕ) the **stronger** the magnetic field.
- **Retentivity** is a measure of the ability of a ferromagnetic material to **hold on to** its magnetic properties.
- **Reluctance** ($\mathcal{R}$) is a measure of the ability of a material to **oppose** the setting up of magnetic flux lines in its core.
- **Magnetic flux lines always** seek the path of **least reluctance.**
- **Permeability** (μ) is a measure of the **ease** with which magnetic flux lines can be established in a ferromagnetic material.
- **Magnetomotive force** ($\mathcal{F}$) is the **dual** of voltage in an electric circuit. It is the applied **"pressure"** to establish flux lines in a magnetic circuit.
- The **magnetizing force** is applied **magnetomotive force per unit length** of the magnetic path.
- **Magnetic saturation** is magnetization to the point beyond which the application of additional magnetomotive forces will have **little or no effect** on the flux density in a ferromagnetic core.
- The **hysteresis curve** of a magnetic sample reveals the relationship between the **magnetic flux density** in a core sample and the level of **applied magnetizing force.**
- **Ampere's circuital law** states that the **algebraic sum** of the magnetomotive forces around a closed path is **zero.**
- **Air gaps** in magnetic paths have a **very high reluctance** characteristic and usually account for the largest share of the applied magnetizing force.
- **Transformers** can **step up** or **step down** the voltage or current of a network, act as an **impedance matching device,** or provide **isolation** between electrical systems.
- An **autotransformer** can **increase** the kVA rating of a transformer but with the consequence that the **isolation property is lost.**

Equations:

$$\mathcal{R} = \frac{\ell}{\mu A} \quad \text{(rels)}$$

$$\mu_r = \frac{\mu}{\mu_0} \quad \text{(unitless)}, \ \mu_0 = 4\pi \times 10^{-7} \text{ Wb/Am}$$

$$\mathscr{F} = NI \quad \text{(At)}$$

$$\phi = \frac{NI}{\mathscr{R}} \quad \text{(webers)}$$

$$B = \frac{\phi}{A} \quad \text{(teslas)}$$

$$10^4 \text{ gauss (G)} = 1 \text{ T}$$

$$H = \frac{\mathscr{F}}{\ell} = \frac{NI}{\ell} \quad \text{(At/m)}$$

$$B = \mu H$$

$$H_g = \frac{B_g}{\mu_0}$$

Transformer:

$$\frac{E_p}{E_s} = \frac{N_p}{N_s},\ a = \frac{N_p}{N_s},\ \frac{I_s}{I_p} = \frac{N_p}{N_s} = a$$

$$\frac{E_p}{E_s} = \frac{I_s}{I_p}$$

$$P_i = P_o,\ E_p I_p = E_s I_s$$

$$Z_p = a^2 Z_L$$

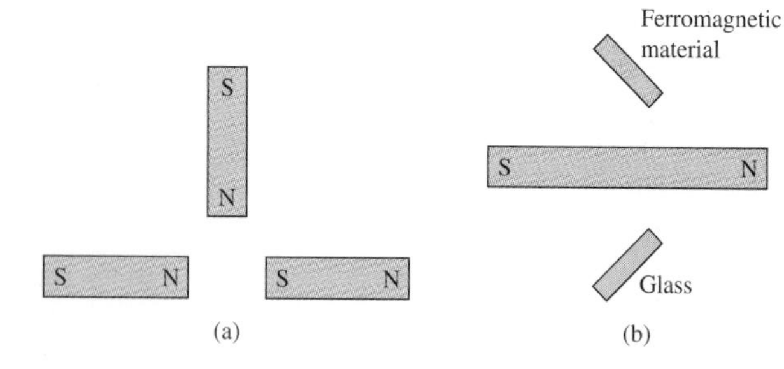

FIG. 6.57

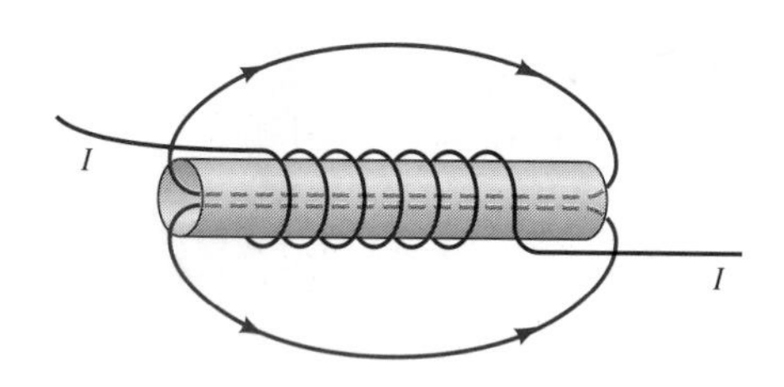

FIG. 6.58

PROBLEMS

SECTION 6.2 Basic Properties of Magnetism

1. Draw the flux distribution patterns, including relative strength and direction, for the magnetic arrangements of Fig. 6.57.

2. For the electromagnet in Fig. 6.58, determine the direction of I needed to establish the flux pattern, and label the induced north and south poles.

SECTION 6.3 Magnetic Circuits

3. A cast-steel core has a relative permeability of 1200. Find its permeability.

4. **a.** If the length of a magnetic core is increased for the same magnetomotive force, what will happen to the magnitude of the resulting flux?
   b. If the area of a magnetic core is doubled and the length reduced to one-third, what will be the effect on the resulting flux if the magnetomotive force is held constant?

5. **a.** For the system in Fig. 6.17, determine the current I if the area is doubled.
   b. Is the resulting current in part (a) half of that obtained in the descriptive example? Why not?

6. **a.** For the system in Fig. 6.17, determine the resulting flux if the current is reduced to 200 mA.
   b. Find the relative permeability of the core.

7. For the magnetic system in Fig. 6.59, determine:
   a. The magnetomotive force.
   b. The magnetizing force applied to the core.
   c. The flux density.
   d. The flux ϕ in the core.
8. Determine the current I necessary to establish the flux indicated in Fig. 6.60.
9. Determine the current I_1 necessary to establish a net flux $\phi = 5 \times 10^{-4}$ Wb in the transformer in Fig. 6.61.
10. Repeat Problem 7 if an air gap of 0.01 in. is cut through the core.
11. Repeat Problem 8 if an air gap of 250 μm is cut through the core.
12. Determine the current required to establish a flux of 5×10^{-4} Wb in the core of the transformer in Fig. 6.18.
13. If the air gap in Fig. 6.19 is doubled ($\frac{1}{16}$ in.), will the current required to establish the same flux increase by a factor of 2 also? Determine the resulting current and comment on the results.
14. Find the magnetic flux ϕ established in the series magnetic circuit in Fig. 6.62.
15. Using a 50-μA, 20,000-Ω movement, design:
   a. A 10-A ammeter.
   b. A 10-V voltmeter.

SECTION 6.4 Transformers

16. A transformer with a turns ratio $a = N_p/N_s = 12$ has a load of 2.2 kΩ applied to the secondary. If 120 V is applied to the primary, determine:
   a. The reflected impedance at the primary.
   b. The primary and secondary currents.
   c. The load voltage.
   d. The power to the load.
   e. The power supplied by source.
17. For a power transformer, $E_p = 120$ V, $E_s = 6000$ V, and $I_p = 20$ A:
   a. Determine the secondary current I_s.
   b. Calculate the turns ratio a.
   c. Is it a step-down or a step-up transformer?
   d. If $N_s = 100$ turns, $N_p = ?$
18. An inductive load $\mathbf{Z}_L = 4\Omega + j4\Omega$ is applied to a 120 V/6 V filament transformer.
   a. What is the magnitude of the secondary current?
   b. What is the power delivered to the load?
   c. What is the primary current?
19. A purely capacitive load $\mathbf{Z}_L = -j2$ is applied to a 120 V/6 V filament transformer with a 12 VA apparent power rating.
   a. Determine the magnitude of the secondary current.
   b. Calculate the power delivered to the primary and secondary.
   c. Do you expect the transformer to heat up if operating under these conditions?

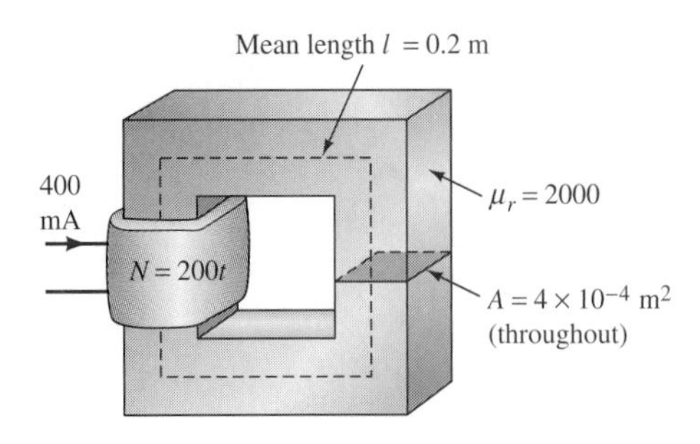

FIG. 6.59

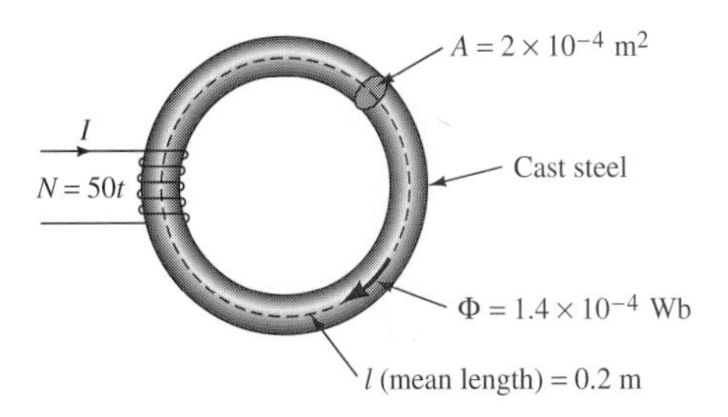

FIG. 6.60

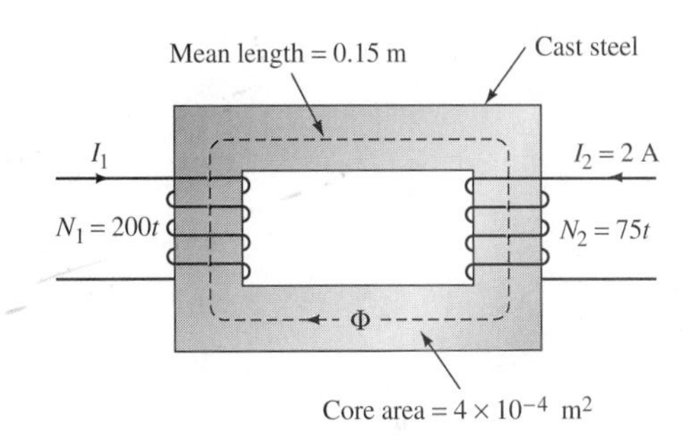

FIG. 6.61

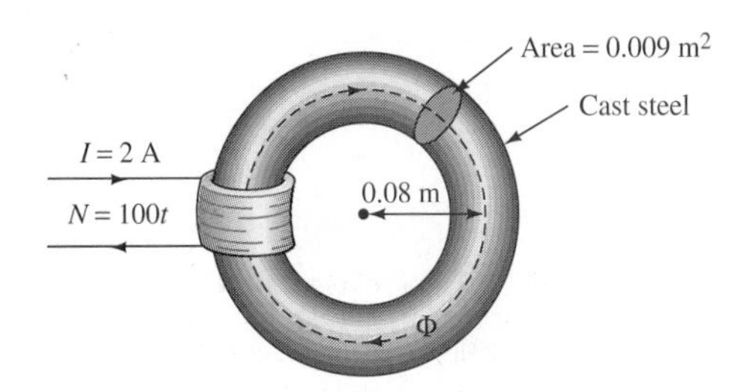

FIG. 6.62

20. **a.** If $E = 120$ V, $R_{Th} = 0.5$ kΩ, and $a = 5$ in the network in Fig. 6.25, determine the load value for maximum power to the load.

b. Determine the power to the load under these conditions.

21. A two-circuit transformer has the following measured quantities: $E_p = 240$ V, $I_p = 2$ A, $E_s = 40$ V, $I_s = 12$ A. Sketch the autotransformer connection if 280 output volts are desired across the load. What is the new volt-ampere rating of the system? Compare this level with that of the two-circuit configuration if the coils carry 2 A and 12 A, respectively.

7 Generators and Motors

7.1 INTRODUCTION

There are four broad areas of coverage to be considered under the given chapter heading: dc generators and motors and ac generators and motors. The basic construction operation and nameplate data along with an area of application or two will be provided for each subject area. The sole purpose of the chapter is to ensure that the reader has sufficient knowledge of each type of machine to choose which to use for the area of application. Since the English system of units is the most frequently applied to the quantities of interest, it will be employed throughout this chapter. Like most texts devoted solely to dc and ac machinery, we shall begin with dc generators and motors.

7.2 DC GENERATORS

Generators (whether dc or ac) are dynamos (energy-converting devices) that convert mechanical energy to electrical energy. As the shaft of the dc generator in Fig. 7.1 is turned at the nameplate speed, a dc voltage is generated at the two output terminals. The basic construction of the dynamo can be broken down into the two basic components appearing in Fig. 7.2: the stationary *stator* and the movable *rotor* or *armature*. The stationary part has an even number of poles (alternating north and south poles), which are either the permanent magnetic type or induced by a continuous winding around all the poles through which an energizing dc current is passed, as shown in the figure. The flux pattern established by the poles reflects the need for alternating field poles. The stator frame (also called the *yoke*) is constructed of a ferromagnetic material to permit a continuous flux path for the poles with minimum reluctance in its path. When the rotor (also constructed of a ferromagnetic material in the laminated form discussed for transformers) is inserted within the stator, the flux pattern can continue through the rotor with only the air gap between the two structures to severely limit the generated flux. The air gap is typically between $\frac{1}{32}$ and $\frac{1}{8}$ in. depending on the relative size of the machine.

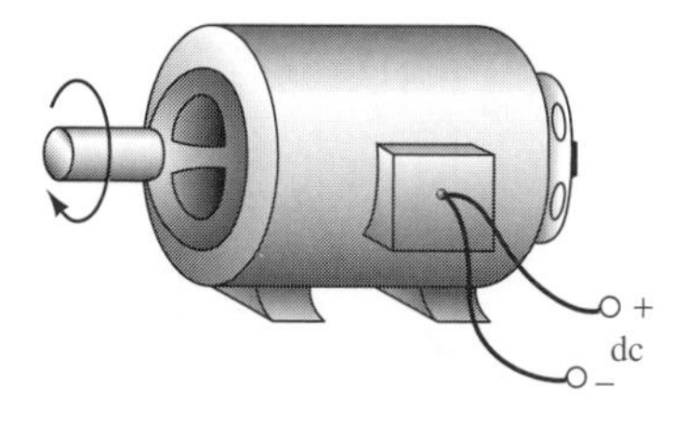

FIG. 7.1
Dc generator.

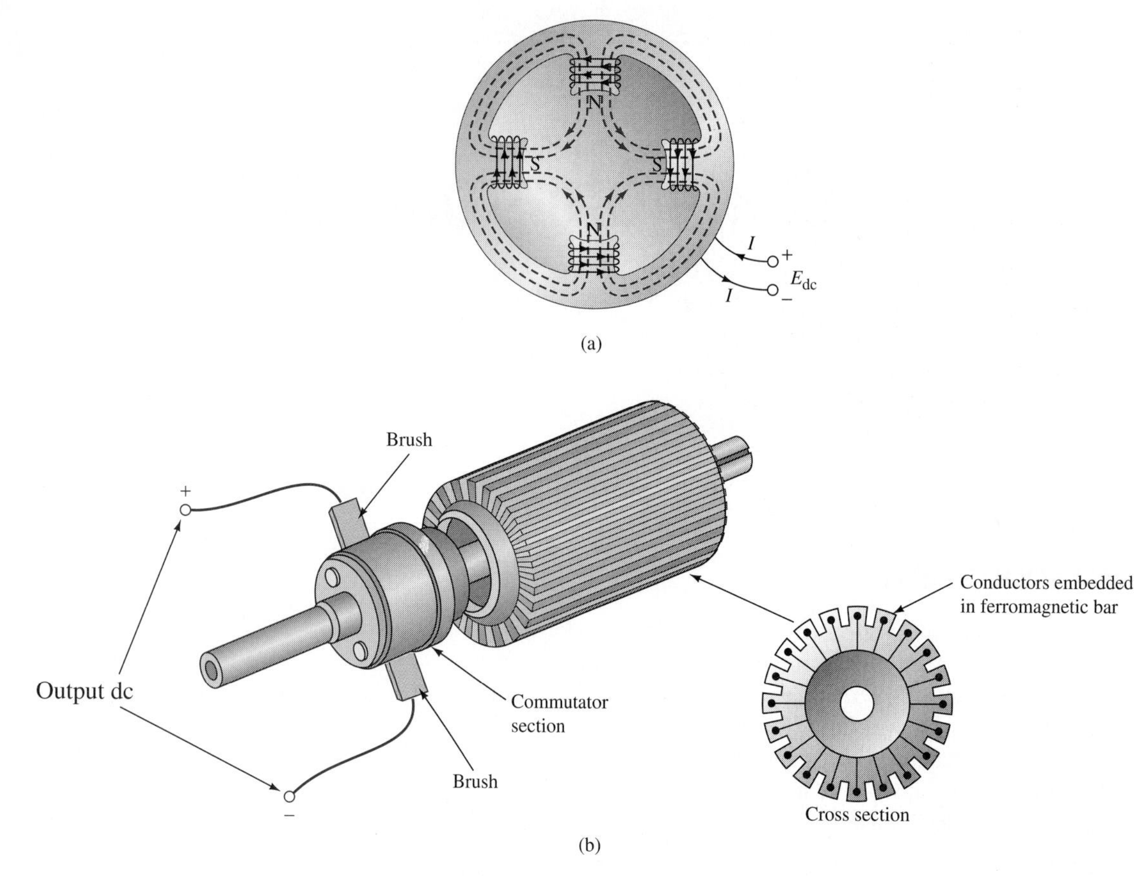

FIG. 7.2
Four-pole generator: (a) stator; (b) rotor.

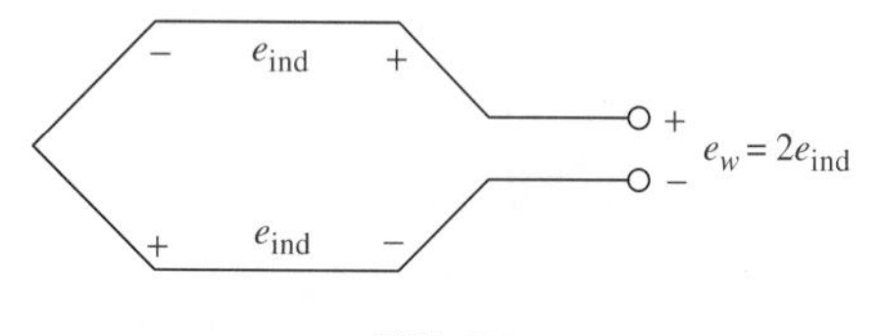

FIG. 7.3
Rotor winding.

Coils of the type depicted in Fig. 7.3 are set into the perimeter of the rotor and connected in a particular manner to develop the nameplate voltage and current rating. They have the shape shown in the figure because of the necessity to overlap the windings at each end of the rotor. There would be insufficient length in the coil for the overlapping if each coil were perfectly rectangular.

As the rotor turns, under some external force such as hydroelectric power, the conductors of each coil pass through the flux of the poles of the stator, and a voltage is induced across the conductors as determined by Faraday's law of electromagnetic induction: $e = N\,d\phi/dt$. The induced voltages of a turn add as shown in Fig. 7.3, and the series summation of the coils results in the generated voltage.

When the coils are connected as shown in Fig. 7.4, the rotor has a *lap winding* in which the windings *overlap* but the induced voltages are additive from one winding to the next. The polarity of the induced voltages is opposite for each conductor of a winding, since they appear under opposite poles of the stator. The only external connections are through the

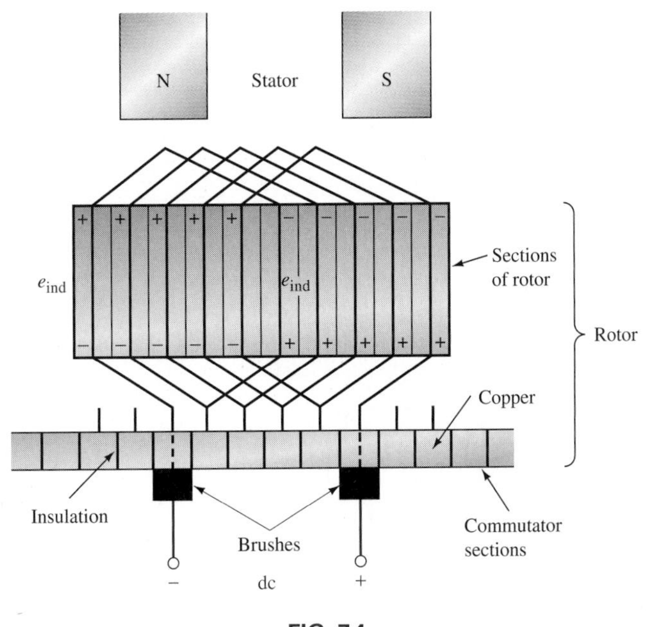

FIG. 7.4
Lap winding.

brushes and the two commutator sections, as shown in Fig. 7.4, resulting in an output dc voltage equal to the sum of the induced voltages from the − to + output terminals. The commutator is constructed of small parallel copper bars insulated from one another around the shaft of the rotor. As the brushes pass over the commutator sections there is some ripple in dc output, but it is usually ignored for most applications. Replacing the commutator section by a continuous conducting ring would result in an ac generator and a sinusoidal output voltage. Since carbon is a conductor, as a dc generator ages, carbon from the brushes can fill in the insulated sections between commutator sections. There is then a deterioration of output dc level until the carbon is removed from the grooves using a number of maintenance techniques.

A second method of connecting the coils appears in Fig. 7.5. For reasons obvious from the figure, the winding is referred to as a *wave winding*. The "wave" pattern extends around the rotor, with the output again pulled off the commutator sections. Note again how the position of the conductors is such that the voltage is additive as the wave winding is followed around the rotor.

Although not obvious from Fig. 7.4, the lap winding has *as many parallel paths of induced voltages as there are poles*. The current through the parallel paths is then additive to establish the rated output current. In general, therefore, the lap winding provides a low-voltage output with a high current rating. The wave winding has *two parallel paths of induced voltage no matter how many poles in the stator*, resulting in a higher output voltage at a lower rated current level.

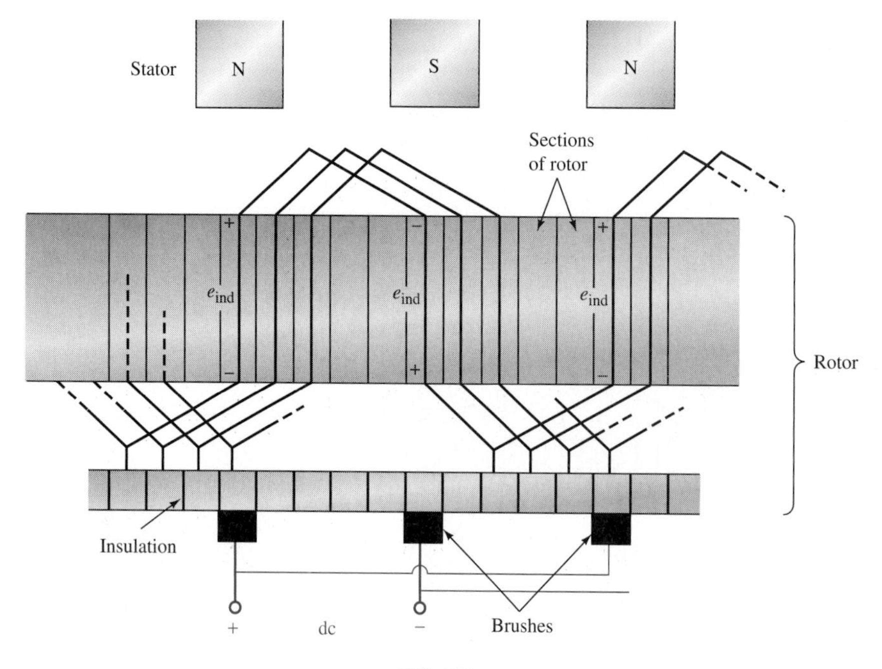

FIG. 7.5
Wave winding.

For each conductor in a lap or wave winding the voltage induced across each conductor is determined by

$$e_c = \frac{P\phi S}{60} \times 10^{-8} \quad \text{(volts)} \tag{7.1}$$

where P = number of poles
ϕ = flux per pole (lines)
S = rotational speed in rpm

For a rotor (or armature) of Z conductors with a parallel paths, the generated dc voltage is determined by

$$E_g = \left(\frac{Z}{a}\right)e_c$$

or

$$E_g = \left(\frac{Z}{a}\right)\frac{P\phi S}{60} \times 10^{-8} \quad \text{(volts)} \tag{7.2}$$

For lap windings $a = P$, and for wave windings $a = 2$, as revealed above.

EXAMPLE 7.1 The armature of a four-pole dc generator has 1200 conductors and rotates at a speed of 3600 rpm. If the flux per pole is 100 kilolines and the lap winding is employed, determine:

a. The induced voltage per conductor.
b. The generated voltage.

Solution:

a. Per Eq. (7.1),

$$e_c = \frac{P\phi S}{60} \times 10^{-8} = \frac{(4)(100 \times 10^3 \text{ W})(3600 \text{ rpm})}{60} \times 10^{-8}$$
$$= \mathbf{0.24\ V}$$

b. $$E_g = \left(\frac{Z}{a}\right)e_c$$
$$= \left(\frac{1200}{4}\right)(0.24 \text{ V}) = \mathbf{72\ V}$$

EXAMPLE 7.2 Repeat Example 7.1 for a wave winding.

Solution:

a. Same as Example 7.1: $e_c = \mathbf{0.24\ V}$
b. $a = 2$ and

$$E_g = \left(\frac{Z}{a}\right)e_c$$
$$= \left(\frac{1200}{2}\right)(0.24 \text{ V}) = \mathbf{144\ V}$$

EXAMPLE 7.3 If the rated output power of the generator in Examples 7.1 and 7.2 is 288 W, determine the rated current for a lap and a wave winding.

Solution: For a lap winding,

$$E_g = 72 \text{ V}$$

$$P_g = E_g I \quad \text{and} \quad I = \frac{P_g}{E_g} = \frac{288 \text{ W}}{72 \text{ V}} = \mathbf{4\ A}$$

For a wave winding,

$$E_g = 114 \text{ V}$$

$$I = \frac{P_g}{E_g} = \frac{288 \text{ W}}{144 \text{ V}} = \mathbf{2\ A}$$

That dc generators need a dc voltage to energize the field poles appears to be a contradiction, since dc voltages are required to develop the desired dc output. The generated levels, however, are much greater than those required to develop the desired output. Dc generators can be divided into two basic types: *separately excited* and *self-excited.* The labels imply that the separately excited generator requires a separate dc supply, while the self-excited generator obviously does not. In fact, in the latter case the generated voltage provides the necessary dc voltage to energize the field windings. This phenomenon is possible only because of an effect referred to as *residual magnetism,* which is the magnetism remaining after the current is removed from the magnetizing coil of an electromagnetic system. When the rotor first begins to turn, this residual flux in the core develops sufficient voltage across the field of the stator to result in a measurable field current. The result is an increase in the field flux and a simultaneous gain in induced voltage. This behavior is cyclic until rated generated voltage is reached and both the generated voltage and limiting field resistance reach a point of common satisfaction. Therefore, as long as the small generated voltage due to the residual flux appears directly across the field windings, the dc generator can build up to rated voltage without the need for a separate dc supply.

If we define as a constant

$$K = \left(\frac{Z}{a}\right)\frac{P}{60} \times 10^{-8}$$

then Eq. (7.2) becomes

$$E_g = K\phi S \tag{7.3}$$

The equation clearly reveals that the greater the speed of the rotor or magnitude of the flux developed by the poles, the greater the generated voltage.

In general, the *nameplate* on a dc generator includes three important pieces of data that must be carefully considered in choosing a machine to perform a particular task. They include the rated terminal voltage of the generator V_t; the rated line current I_L, indicating the maximum current that can be drawn from the supply; and finally, the speed of rotation of the shaft required to sustain the voltage and current rating.

The output power rating can be determined by the product of the rated output voltage and current:

$$P_{max} = V_t I_L \qquad \text{(watts)} \tag{7.4}$$

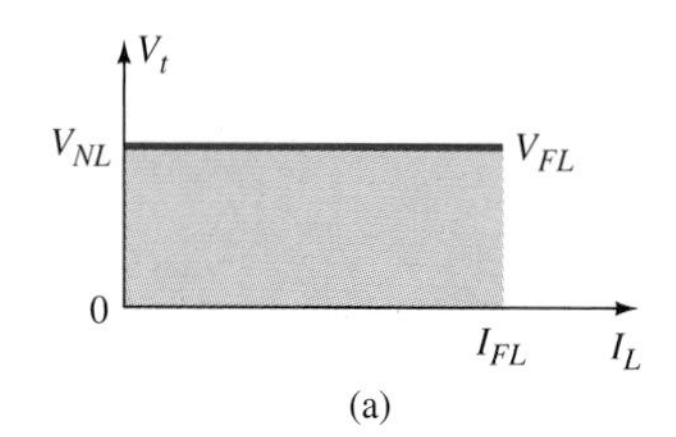

(a)

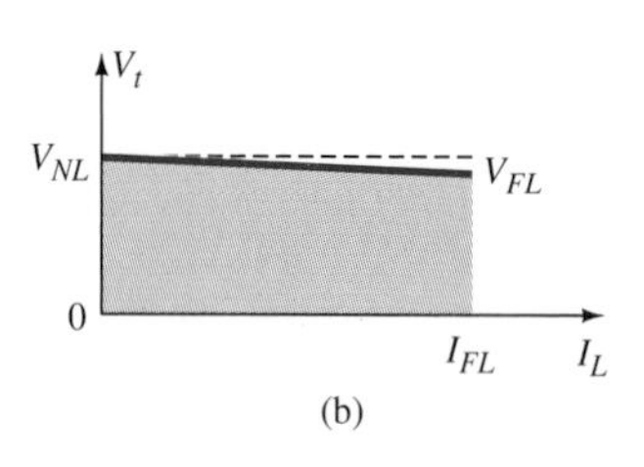

(b)

FIG. 7.6
Terminal voltage variation with load: (a) ideal; (b) practical.

For a generator, laboratory dc bench supply, or what have you, the *voltage regulation* (introduced earlier) is an excellent measure of the quality of the supply. The ideal situation appears in Fig. 7.6(a), in which the terminal voltage is *unaffected* by the current drawn by the load. In the practical world, however, every supply has some internal resistance, as introduced for simple dc batteries. As the load current increases there is an increase in the voltage drop across the internal resistance of the supply, and the terminal voltage naturally decreases, as shown in Fig. 7.6(b). Certainly, the closer the curve to the ideal, the better the supply for the broad range of applications. In equation form the *voltage regulation* (VR) is given by

$$\text{voltage regulation (VR\%)} = \frac{V_{NL} - V_{FL}}{V_{FL}} \times 100\% \qquad \textbf{(7.5)}$$

There is more than one type of dc generator commercially available. Thus far our interest has been in the *shunt* (parallel-type) generator that has its field coil (of the stator) in parallel with generated voltage, as shown in Fig. 7.7. The generated voltage E_g appears across the schematic representation of the rotor and brushes. The resistance R_A is the resistance of the rotating armature. The resistance R_F and the coil represent the field circuit equivalent. The circuit clearly shows that the generated voltage must be greater than the rated terminal voltage due to the drop across R_A. That is,

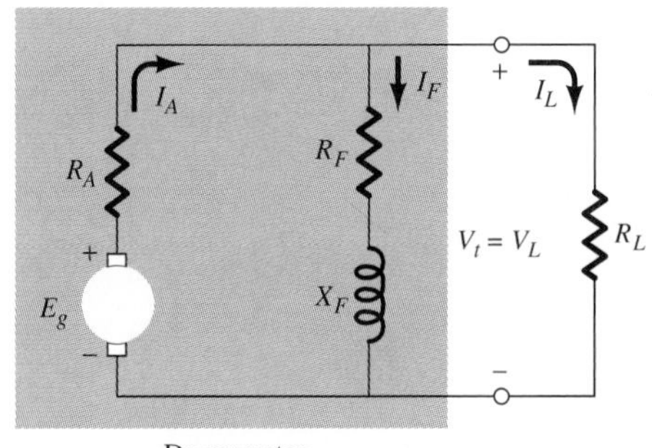

FIG. 7.7
Shunt generator.

$$V_t = E_g - I_A R_A \qquad \text{(volts)} \qquad \textbf{(7.6)}$$

The current I_A must be sufficiently high to guarantee the proper levels of I_F and I_L, with

$$I_A = I_F + I_L \qquad \textbf{(7.7)}$$

EXAMPLE 7.4 Determine the following for the 120-V shunt dc generator in Fig. 7.8.

a. I_L, I_F, I_A.
b. The voltage drop across R_A.
c. The resistance R_A.
d. The power to the load.

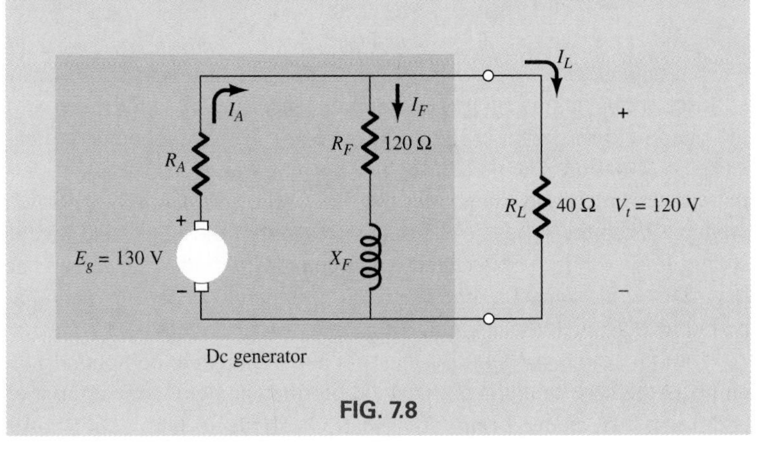

FIG. 7.8

Solution:

a. $I_L = \dfrac{120\text{ V}}{40\ \Omega} = \mathbf{3\ A},\qquad I_F = \dfrac{V_t}{R_F} = \dfrac{120\text{ V}}{120\ \Omega} = \mathbf{1A}$

$I_A = I_F + I_L = 1\text{ A} + 3\text{ A} = \mathbf{4\ A}$

b. Kirchhoff's voltage law: $Eg - V_{R_A} - V_t = 0$ or $V_{R_A} = E_g - V_t =$ $130\text{ V} - 120\text{ V} = \mathbf{10\ V}$

c. $R_A = \dfrac{V_{R_A}}{I_A} = \dfrac{10\text{ V}}{4\text{ A}} = \mathbf{2.5\ \Omega}$

d. $P_L = V_tI_L = (120\text{ V})(3\text{ A}) = \mathbf{360\ W}$

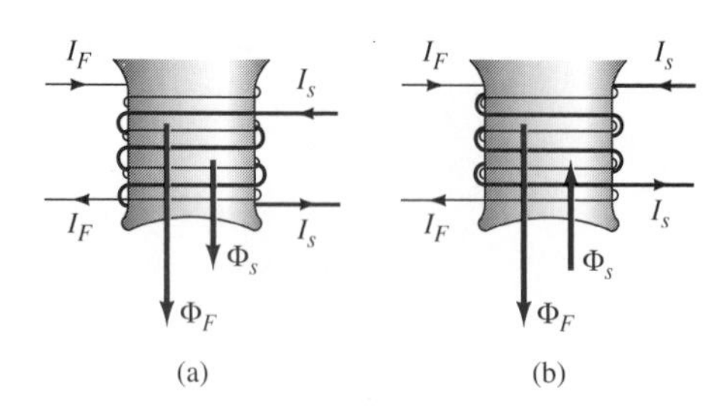

FIG. 7.9
Compound pole winding: (a) cumulative; (b) differential.

Another type of dc motor that has increased control of the terminal voltage is the *compound* dc generator, which has two separate insulated windings around the same field poles as shown in Fig. 7.9.

An electrical schematic of a compound configuration revealing the location of the additional winding is provided in Fig. 7.10. Note that the additional winding is placed in series with generated voltage and therefore has the label *series* windings. The current through the series winding that results in the additional flux is the armature current I_A. The turns of the series winding are far fewer than those of the main field shunt winding. The series winding is designed solely to act as a vernier (a fine control) on the output terminal voltage (V_t) and not the major provider of flux necessary for generating a sufficiently high voltage.

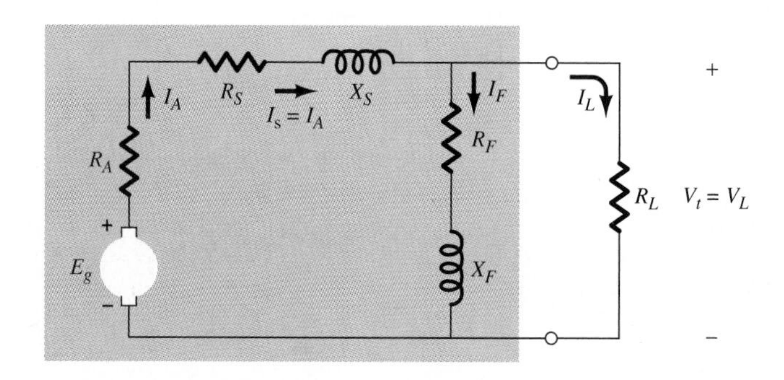

FIG. 7.10
Compound dc generator.

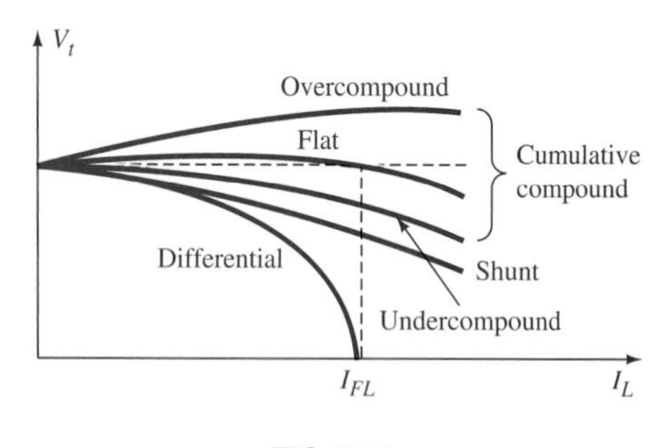

FIG. 7.11
Dc generator characteristics.

If the series field is connected so that its resulting flux aids the flux of the main field as shown in Fig. 7.9(a), we say we have a compound *cumulative* situation. The greater the line current drawn, the greater the armature current I_A and the greater the flux in the core and the generated voltage. When the series field flux opposes that of the shunt field flux as shown in Fig. 7.9(b), it is called a compound *differential* machine. The net effect with increasing line current is a reduction in net flux and generated voltage such as shown in the characteristics in Fig. 7.11.

Note that three possibilities exist for the cumulative compound machine. If the series field is allowed to contribute a significant amount of additional flux, an *overcompound* situation will result. A reduction in the effect of the series field will result in the *flat* response, and a still further reduction will result in the *undercompound* situation. When the series coil is bypassed, we return to the shunt field characteristics.

For the covercompound case the voltage regulation has a negative sign, since the full-load voltage is greater than the no-load voltage. Further, the flat response has 0% regulation and the undercompound a positive regulation. For most compound-wound machines a diverter (a variable resistor) is placed across the series coil to control the amount of current that passes through the field. For maximum resistance values the overcompound situation results, and for the 0-Ω setting all the armature current passes through the parallel path, and the series coil has absolutely no effect. Although one might wonder why the differential characteristic would be desirable at all, consider that in the far right region there is an almost constant current region for a variation in terminal voltage (like a current source). This characteristic is excellent for applications such as welding and series connection of loads where a fixed current must exist even if the load varies.

From Fig. 7.10 the following current and voltage equations can easily be developed:

$$I_A = I_F + I_L \qquad I_S = I_A \qquad I_F = \frac{V_t}{R_F} \tag{7.8}$$

and

$$V_t = E_g - I_A(R_A + R_S) \tag{7.9}$$

EXAMPLE 7.5 For the elements of the network in Fig. 7.10, $R_F = 100\ \Omega$, $R_A = 0.5\ \Omega$, $R_S = 1\ \Omega$, $V_t = 240$ V, and $R_L = 20\ \Omega$. Determine:

a. The rated line current.
b. The generated voltage.
c. The power to the load.
d. The power delivered by the armature.
e. The total power lost in the armature, series, and shunt field coils.

Solution:

a. $I_L = \dfrac{V_t}{R_L} = \dfrac{240\text{ V}}{20\ \Omega} = \mathbf{12\ A}$

b. $I_A = I_F + I_L = \dfrac{V_t}{R_F} + I_L = \dfrac{240\text{ V}}{100\ \Omega} + 12\text{ A} = 2.4\text{ A} + 12\text{ A} = 14.4\text{ A}$,

and $E_g = V_t + I_A(R_A + R_S) = 240\text{ V} + 14.4\text{ A}\ (0.5\ \Omega + 1\ \Omega)$
$= \mathbf{261.6\ V}$

c. $P_L = V_t I_L = (240\text{ V})(12\text{ A}) = \mathbf{2880\ W}$
d. $P = E_g I_A = (261.6\text{ V})(14.4\text{ A}) = \mathbf{3767.04\ W}$
e. $P_{\text{lost}} = P_i - P_o = 3767.04\text{ W} - 2880\text{ W} = \mathbf{887.04\ W}$

The third and last type of dc generator to be introduced is the series generator, which has the circuit model in Fig. 7.12. A single field coil is placed in series with the generated voltage rather than in the usual parallel position. At small levels of I_L, the current $I_S = I_L$ is small and the flux strength ϕ_S of the coil is low. The result is a small generated voltage, as seen in the characteristics in Fig. 7.13. As the line current increases, the

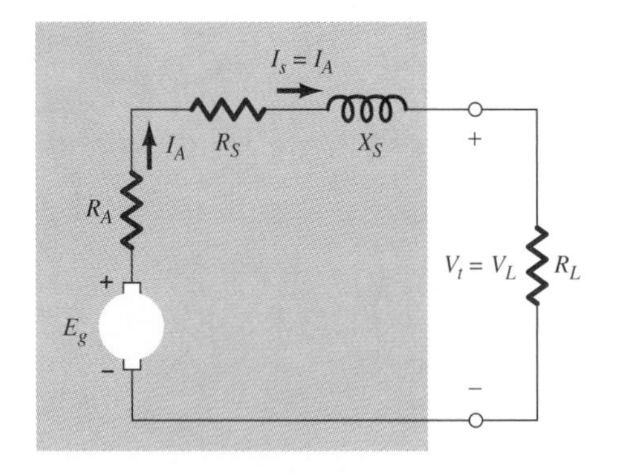

FIG. 7.12
Series dc generator.

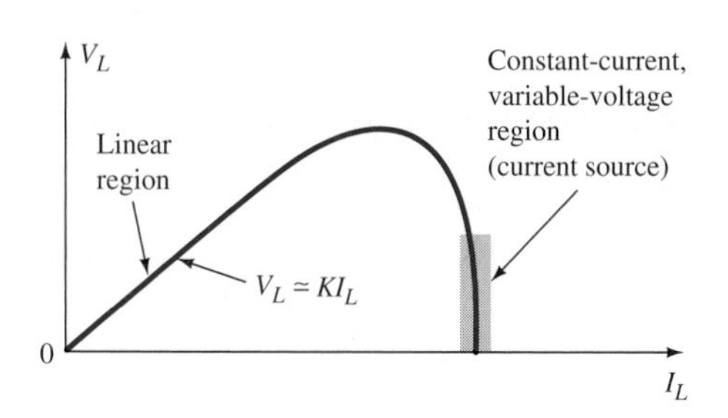

FIG. 7.13
Series dc generator characteristics.

field strength and generated voltage increase, as shown in the same figure. Because of an effect known as *armature reaction,* which can be researched in many available texts, the curve drops to zero as shown, and the areas of application indicated for the differential compound machine apply here also. One interesting application of the rising characteristic of the series generator is shown in Fig. 7.14, where it is being used as a *booster.* Some of the generated voltage V'_t is lost across the long transmission line from the generator to the load. The greater the current demand, the greater the loss across the lines. To compensate for this loss, a series generator is placed near the load as shown in the figure. Increasing line currents result in increasing generated voltage across the series

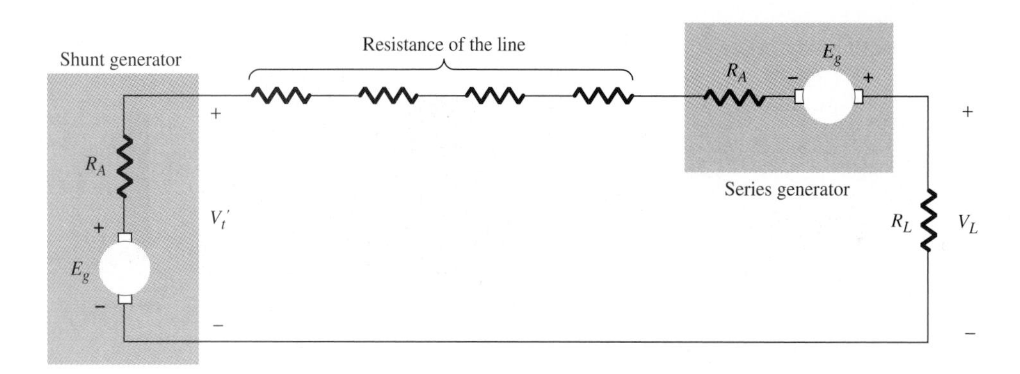

FIG. 7.14
Series generator in application as a booster.

booster (characteristics in Fig. 7.13), compensating directly for the increasing line losses so that V_t can be maintained at some relatively constant and sufficiently high magnitude.

For the series generator the following equations are applicable:

$$\boxed{I_L = I_S = I_A} \quad \text{(amperes)} \qquad \textbf{(7.10)}$$

$$\boxed{V_t = E_g - I_A(R_A + R_S)} \quad \text{(volts)} \qquad \textbf{(7.11)}$$

EXAMPLE 7.6 For the series dc generator in Fig. 7.12, if $V_t = 110$ V, $R_S = 2\ \Omega$, $R_A = 8\ \Omega$, and $E_g = 126$ V, determine:

a. I_L.
b. The power lost to R_A and R_S.
c. The efficiency of the system if $\eta = P_L/P_g \times 100\%$.

Solution:

a. Eq. (7.11):

$$I_A = \frac{E_g - V_t}{R_A + R_S} = \frac{126\ \text{V} - 110\ \text{V}}{8\ \Omega + 2\ \Omega} = \frac{16\ \text{V}}{10\ \Omega} = \mathbf{1.6\ A}$$

b. $P = I_A^2(R_A + R_S) = (1.6\ \text{A})^2(8\ \Omega + 2\ \Omega) = \mathbf{25.6\ W}$

c. $P_L = V_tI_L = (110\ \text{V})(1.6\ \text{A}) = 176\ \text{W}$
$P_g = E_gI_A = (126\ \text{V})(1.6\ \text{A}) = 201.6\ \text{W}$

$$\eta = \frac{P_L}{P_g} \times 100\% = \frac{176\ \text{W}}{201.6\ \text{W}} \times 100\% = \mathbf{87.3\%}$$

7.3 DC MOTORS

The motor is an electrical dynamo that can convert electrical energy into mechanical energy. In other words, as shown in Fig. 7.15, an applied voltage results in a turning shaft that can do mechanical work.

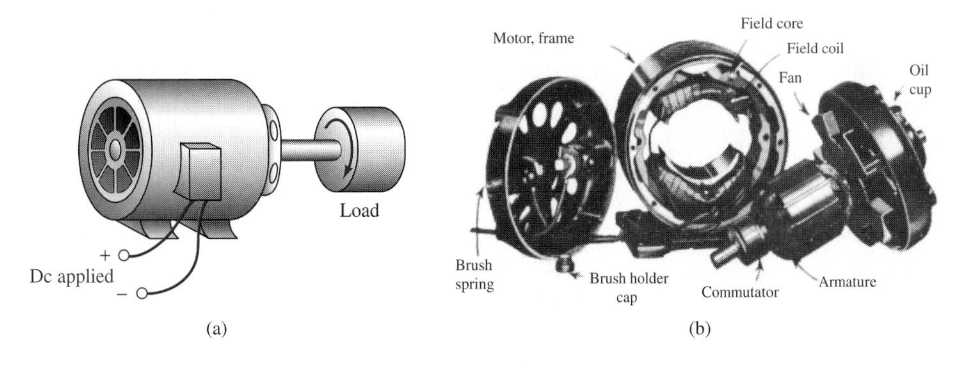

FIG. 7.15
Shunt-wound dc motor: (a) operation; (b) internal construction.

The network schematic of a dc motor, as shown in Fig. 7.16(a), is simply the reverse of that for a dc generator. As in the case of the generator, there are shunt, compound, and series dc motors. The shunt configuration appears in Fig. 7.16(a). The applied voltage results in a field current I_F and the necessary flux for the poles of the stator as indicated in Fig. 7.16(b). The applied voltage also results in an armature current I_A

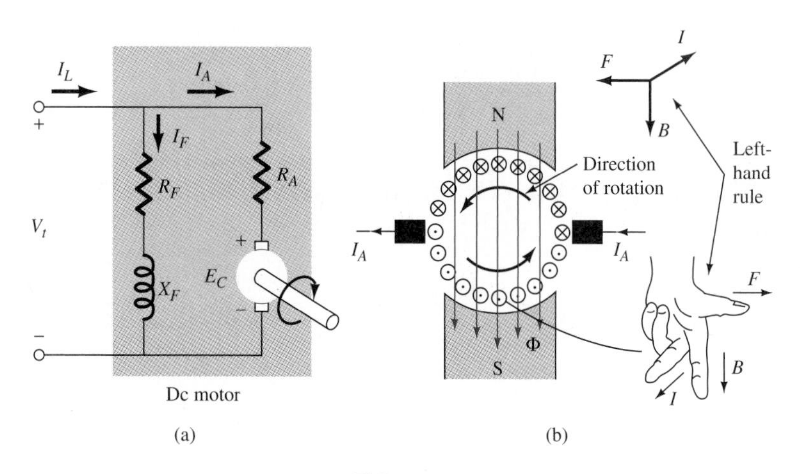

FIG. 7.16
Dc motor: (a) schematic; (b) internal construction.

through the conductors of the armature, as shown in the same figure. The current-carrying conductors in the magnetic field experience a force in a direction determined by the left-hand rule. Simply place the index finger in the direction of the magnetic flux and the second finger at right angles to the index finger in the direction of the current in the conductor of the rotor. The thumb, if placed at right angles to the index finger, will give the direction of the force on the conductor. The force direction for the conductors at the top and bottom of Fig. 7.16(b) are indicated. Note that they are aiding forces in the counterclockwise direction. The reversal in force direction in the lower half is due to the reversal in current direction. The force on the conductors can be determined by

$$F = kBIL \sin \theta \quad \text{(pounds)} \tag{7.12}$$

where F = pounds
B = lines per square inch
I = amperes
L = inches
$k = 8.85 \times 10^{-8}$
θ = angle between the magnetic field and current

which clearly indicates that the greater the strength of the magnetic field, conductor current, or length of the conductor *in* the magnetic field, the greater the force on each conductor and therefore on the rotating member (rotor) of the machine.

As the rotor turns, a voltage is induced across the rotating conductors of the rotor that opposes the applied voltage at the input terminals. It is labeled E_C, as shown in Fig. 7.16(a), with the subscript derived from the expression *counter-emf.* It is determined by an equation similar to that for E_g for the dc generator:

$$E_C = K\phi S \qquad \text{(volts)} \qquad \textbf{(7.13)}$$

revealing that the greater the speed of rotation or flux of the poles, the greater the counter-emf. For the shunt motor in Fig. 7.16, the following relationships result:

$$I_L = I_F + I_A \qquad \text{(amperes)} \qquad \textbf{(7.14)}$$

$$I_F = \frac{V_t}{R_F} \qquad \text{(amperes)} \qquad \textbf{(7.15)}$$

$$V_t = E_C + I_A R_A \qquad \text{(volts)} \qquad \textbf{(7.16)}$$

EXAMPLE 7.7 The dc shunt motor in Fig. 7.16(a) has the following component values:

$$V_t = 120\text{ V}, \qquad I_L = 8\text{ A}, \qquad R_F = 100\ \Omega, \qquad R_A = 1\ \Omega$$

Determine the counter-emf, E_C.

Solution:

$$I_F = \frac{V_t}{R_F} = \frac{120\text{ V}}{100\ \Omega} = 1.2\text{ A}$$

$$I_A = I_L - I_F = 8\text{ A} - 1.2\text{ A} = 6.8\text{ A}$$

and

$$E_C = V_t - I_A R_A = 120\text{ V} - (6.8\text{ A})(1\ \Omega) = \mathbf{113.2\ V}$$

A quantity of major importance to any type of motor is the *torque* the shaft can develop, since it determines the load the machine can efficiently handle. As depicted in Fig. 7.17 and given by

$$T = Fr \qquad \text{(ft-lb)} \qquad \textbf{(7.17)}$$

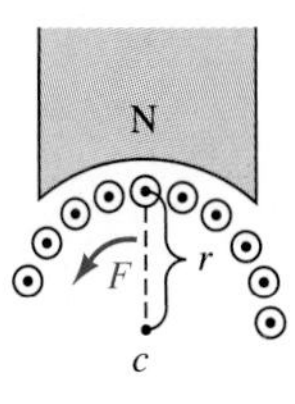

FIG. 7.17

the torque is directly proportional to the force developed on the conductors of the rotor and their distance from the center of the rotating shaft. In terms of the flux and armature current, torque is given by

$$T = K\phi I_A \qquad \text{(ft-lb)} \qquad \textbf{(7.18)}$$

clearly indicating that an increase in magnetic field strength or current through the conductors of the rotor increases the torque by the same amount.

The nameplate data of any dc motor include the input voltage, current rating, and power rating. The power rating given is always the rated output power of the motor. It is related to the torque on the output shaft by the equation

$$T = \frac{7.04P}{S} \qquad \text{(lb-ft)} \qquad \textbf{(7.19)}$$

where T = lb-ft
P = watts
S = rpm

Proportionally speaking, the following is true:

$$TS \propto P \qquad \textbf{(7.20)}$$

which says in words that for a particular horsepower machine there is a trade-off between torque and speed. In other words, in Eq. (7.20), for a *fixed* power P, the torque available can be increased, but only at the expense of speed. Many of us have experienced this effect with a simple $\frac{1}{4}$-in. drill. When the drilling becomes difficult, the drill may continue to perform but at a sharply reduced speed. If the torque of the rotating shaft is insufficient to handle the load, the drill may be damaged if pressured to complete the task. The solution may be to use a drill with a higher horsepower rating, such as the standard $\frac{1}{2}$-in. drill.

The efficiency of a dc motor (or generator) can be determined by the equation

$$\eta = \frac{P_o}{P_i} \times 100\% \qquad \text{(percent)} \qquad \textbf{(7.21)}$$

For the dc motor the output power is the nameplate rating, and the input power is determined by $P_i = V_t I_L$. For the generator the input power is the power delivered to the shaft, and the output power is $P_o = V_t I_L$.

EXAMPLE 7.8 The following nameplate data were provided for a dc shunt motor: 5 hp, 240 V, 18 A, 2400 rpm. Determine:

a. The output torque.
b. The efficiency of operation.

Solution:

a. $T = \dfrac{7.04P}{S} = \dfrac{7.04[(5)(746\text{ W})]}{2400\text{ rpm}} = \mathbf{10.94\text{ ft-lb}}$

b. $P_i = V_t I_L = (240\text{ V})(18\text{ A}) = 4320\text{ W}$

$\eta = \dfrac{P_o}{P_i} \times 100\% = \dfrac{(5)(746\text{ W})}{4320\text{ W}} \times 100\% = \mathbf{86.34\%}$

For the shunt motor the field flux remains fairly constant during load variations, since the applied voltage is directly across the field circuit. Replacement of ϕ with the constant K' in the equation for torque results in

$$T = K\phi I_A = \underbrace{KK'}_{K_T}I_A = K_T I_A \tag{7.22}$$

and we find that the torque varies directly as the armature current. Since I_A and I_L are usually quite close in magnitude, the graph of torque versus current includes the former, as shown in Fig. 7.18. The curve clearly indicates that as the torque demand on the motor increases, the current increases proportionally. This effect is noticed in the heating up of the handle as too heavy a load is put on the biting edge of the drill in the example described in the discussion of torque. The curve drops off slowly at higher currents because of an effect mentioned earlier called *armature reaction* that reduces the net flux of the field poles by introducing a flux component (developed by the rotating armature) that appears perpendicular to that of the main field winding. As determined by Eq. (7.18), a reduction in net flux most certainly reduces the available torque.

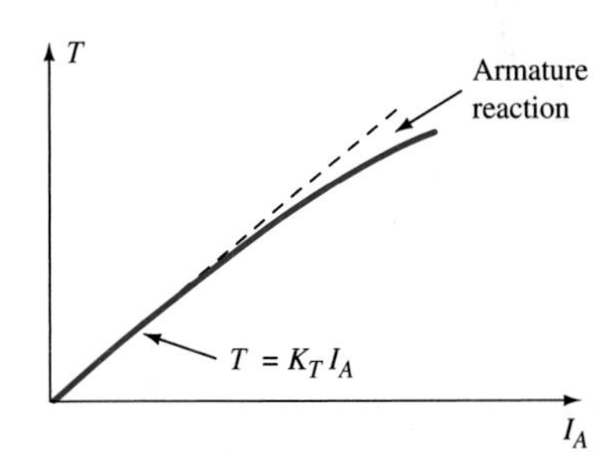

FIG. 7.18
Torque characteristics of a shunt motor.

A second curve of primary importance for dc motors is its speed versus I_A characteristics—in other words, how the speed varies with change in load. Equation (7.13) can be rewritten as

$$S = \frac{E_C}{K\phi} \quad \text{(rpm)} \tag{7.23}$$

Substituting Eq. (7.16), we have

$$S = \frac{V_t - I_A R_A}{K\phi} \quad \text{(rpm)} \tag{7.24}$$

For increasing values of I_A (for fixed V_t and R_A) the numerator obviously becomes smaller. The net flux ϕ also decreases slightly because of

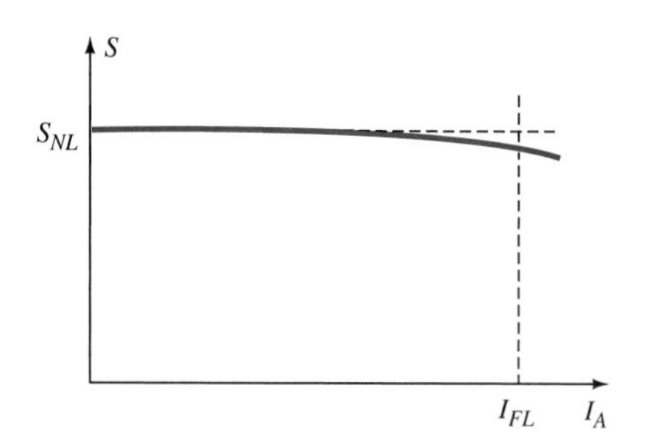

FIG. 7.19
Speed characteristics of a shunt motor.

armature reaction. However, on the whole the shunt motor is considered to be a fairly constant speed device as shown in Fig. 7.19. Therefore, applications that require a fairly constant speed for variations in load should probably employ a dc shunt-wound motor.

Speed control can be introduced by making R_F a variable resistor, since it will control I_F and therefore the flux ϕ in the denominator of the equation. Additional control can be gained by making R_A a variable quantity. Of course, variations in V_t will also affect the speed characteristics.

Dc motors, like dc generators, are also manufactured using a second series coil such as shown in the compound configuration in Fig. 7.20. The flux of the added series coil again acts as a vernier on the total magnetic flux of the stator poles. As in the dc generator, the series coil is wrapped around the same poles as the main field winding although totally insulated from the other winding. For the cumulative compound ($\phi_F + \phi_S$) machine the three conditions of overcompound, flat, and undercompound can be established depending on the amount of additional series flux. Since $T = K\phi I_A$, the reduction in net flux associated with changing from an overcompound to an undercompound situation results in the variation in curves indicated in Fig. 7.21. For the differential compound situation ($\phi_F - \phi_S$), the torque drops off more rapidly as shown in the same figure.

For the compound configuration the equation for speed becomes

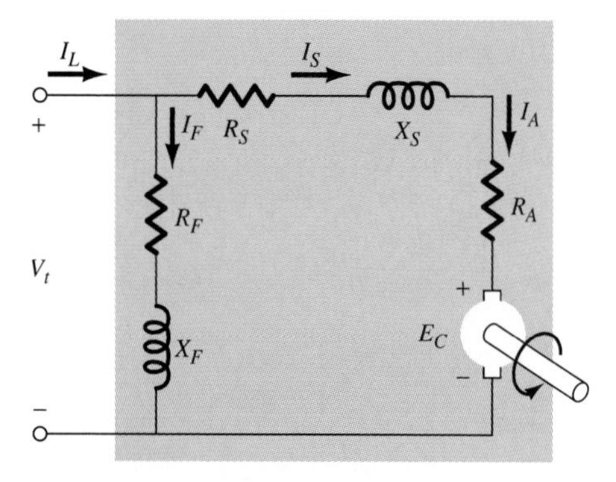

FIG. 7.20
Compound dc motor.

$$S = \frac{V_t - I_A(R_A + R_S)}{K(\phi_F \pm \phi_S)} \quad \text{(rpm)} \tag{7.25}$$

For the cumulative or differential case an increase in current (I_A) certainly causes a reduction in the numerator and a decrease in speed. However, ϕ_S, which appears in the denominator, is also dependent on the current I_A. For increasing I_A (and therefore ϕ_S) in the cumulative case the denominator continues to get larger and the speed decreases as shown by the set of curves in Fig. 7.22. For the differential case, however, the total flux decreases and the speed curve can rise as shown in the

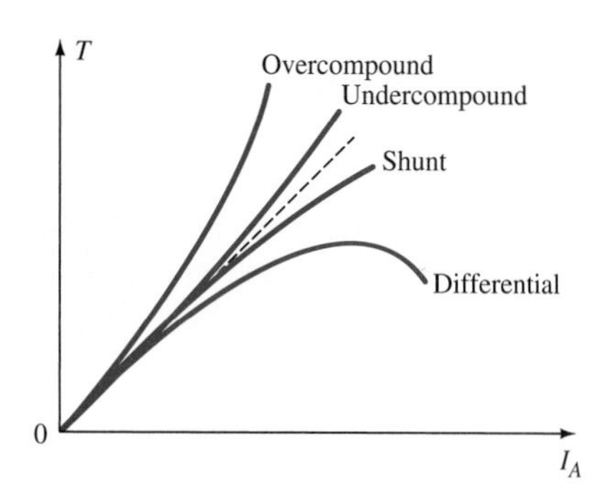

FIG. 7.21
Torque characteristics of the compound dc motor.

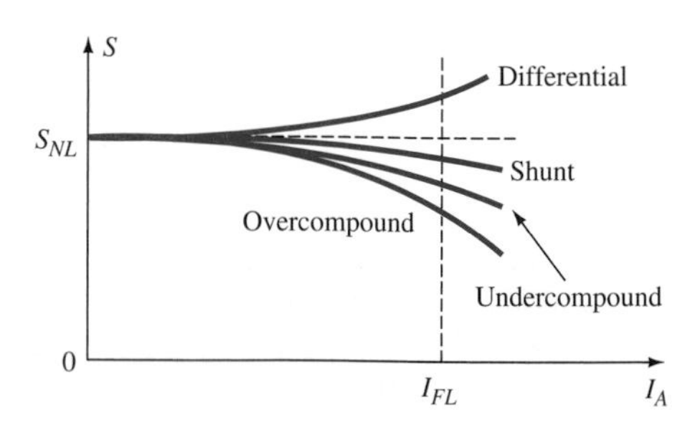

FIG. 7.22
Speed characteristics of the compound dc motor.

same figure. For cumulative or differential machines the effect of armature reaction is to decrease the net flux and cause an increase in speed.

EXAMPLE 7.9 For the compound dc motor in Fig. 7.20, $R_F = 100\ \Omega$, $R_A = 4\ \Omega$, $R_S = 1\ \Omega$, $E_C = 106$ V, and $V_t = 120$ V. Determine:

a. I_F.
b. I_S and I_A.
c. I_L.
d. The applied power.
e. P_o if $\eta = 78\%$.
f. The output torque if $S = 1200$ rpm.

Solution:

a. $I_F = \dfrac{V_t}{R_F} = \dfrac{120\text{ V}}{100\ \Omega} = \mathbf{1.2\ A}$

b. $I_S = I_A = \dfrac{V_t}{R_S + R_A} = \dfrac{120\text{ V}}{1\ \Omega + 4\ \Omega} = \dfrac{120\text{ V}}{5\ \Omega} = \mathbf{24\ A}$

c. $I_L = I_F + I_S = 1.2\text{ A} + 24\text{ A} = \mathbf{25.2\ A}$

d. $P_{\text{del}} = V_t I_L = (120\text{ V})(25.2\text{ A}) = \mathbf{3024\ W}$

e. $\eta = \dfrac{P_o}{P_i} \times 100\% \Rightarrow P_o = \dfrac{\eta P_i}{100} = \dfrac{(78)(3024\text{ W})}{100} = \mathbf{2358.72\ W}$

f. $T_o = \dfrac{7.14P_o}{S} = \dfrac{7.04(2358.72\text{ W})}{1200\text{ rpm}} = \mathbf{13.84\ ft\text{-}lb}$

Speed regulation (SR), defined by

$$\text{speed regulation (SR\%)} = \frac{S_{NL} - S_{FL}}{S_{FL}} \times 100\% \qquad \text{(percent)} \quad \mathbf{(7.26)}$$

is a measure (like voltage regulation) of how much the speed varies from no-load to full-load conditions. For the differential case, since the full-load value is greater than the no-load value, a negative result is obtained.

The series dc motor has the network configuration in Fig. 7.23. Since the line current $I_L = I_A$, the flux developed by the series coil is directly related to the I_A, and in equation form, $\phi = K_1 I_A$. If we substitute for ϕ in the torque equation, we find

$$T = K\phi I_A = \underbrace{K(K_1 I_A)}_{K_T} I_A = K_T I_A^2$$

indicating that the torque increases exponentially with the line or armature current. The resulting torque characteristic appears in Fig. 7.24. The speed characteristics of the series motor can be determined from

$$S = \frac{V_t - I_A(R_A + R_S)}{K\phi_S} \qquad \text{(rpm)} \quad \mathbf{(7.27)}$$

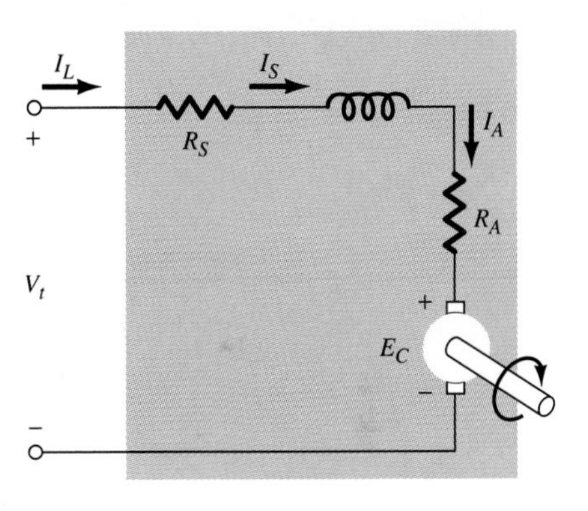

FIG. 7.23
Series dc motor.

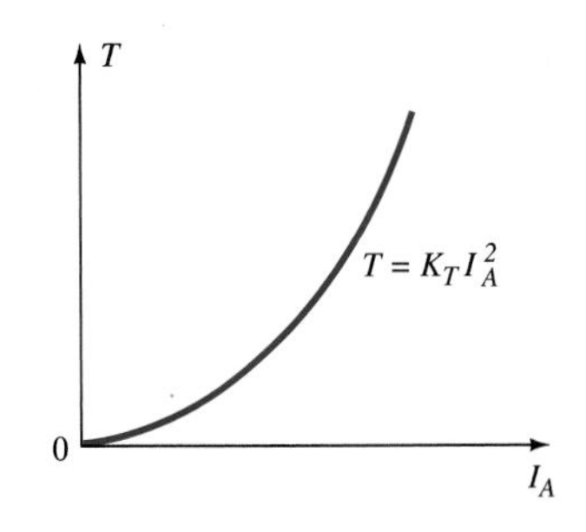

FIG. 7.24
Series dc motor torque characteristics.

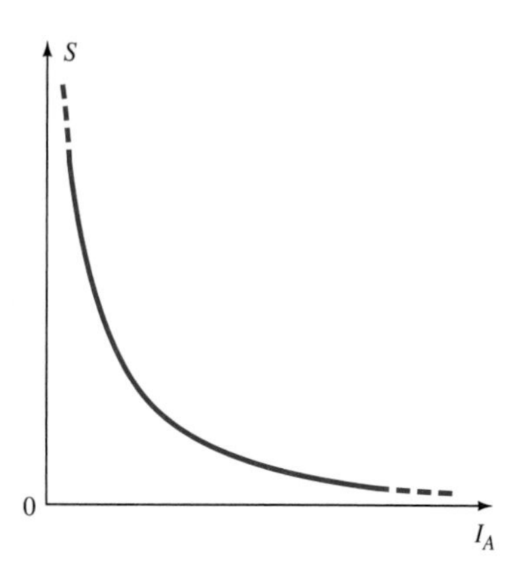

FIG. 7.25
Series dc motor speed characteristics.

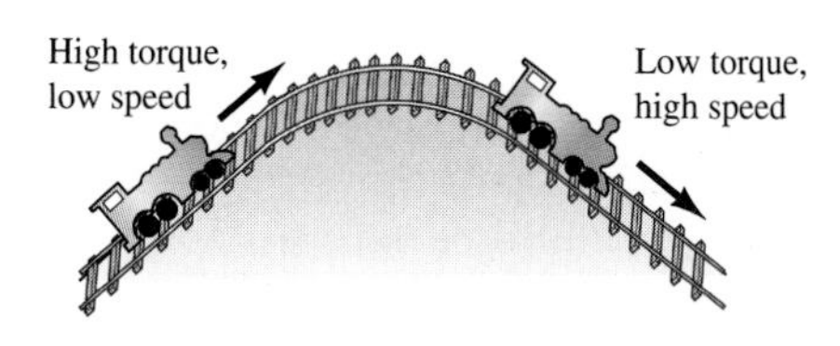

FIG. 7.26
Application of a series dc motor.

When $I_L = I_A = I_S = 0$ A, ϕ_{residual} is quite small, resulting in a very small magnitude for the denominator of the equation and a very high speed (which is usually excessive and must be limited). As $I_L = I_A$ increases, the flux increases, and the speed decreases toward zero, as indicated by the characteristics in Fig. 7.25.

One interesting application of the series motor is in the cog train climbing the very steep grade in Fig. 7.26. When the train is climbing the hill, a very high torque must be developed, although the speed can be quite low. This condition is described by high levels of I_A for each characteristic. When the train is going down the other side, the torque demand is significantly reduced, and the speed can increase substantially. This condition is described by low values of I_A on each curve.

The description of dc motors in this section implies that dc voltage can be applied directly across the field and armature circuits when starting the dynamo. In actuality, an additional resistive element must initially be inserted in series with the armature circuit at starting, since the counter-emf, $E_C = K\phi S$, is zero (the speed of the rotating shaft is initially zero). The entire applied voltage appears directly across R_A, since

$$V_t = E_C + I_A R_A = 0 + I_A R_A \qquad \text{(at starting)}$$

and

$$I_A = \frac{V_t}{R_A} = \frac{120\text{ V}}{1\ \Omega} = 120\text{ A} \qquad \text{(typically)}$$

Current levels of 120 A will most assuredly damage the armature circuit. This effect can easily be eliminated by inserting a starting resistance in series with RA at starting as shown in Fig. 7.27. The starting resistor can be designed to limit the starting current to perhaps 200% of rated line current. For short periods of time the machine can safely handle this level of current. As the machine builds up, the counter-emf E_C increases, and the voltage across the series combination of R_{st} and R_A decreases. The value of R_{st} can then be reduced in step with the increasing

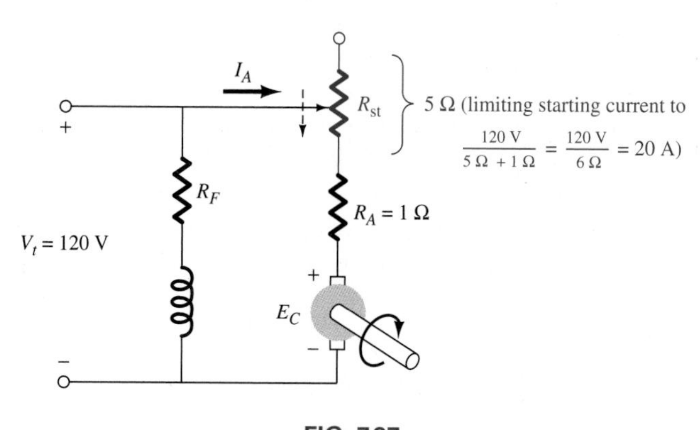

FIG. 7.27
Starting resistor.

level of E_C until running conditions are established and R_{st} can be eliminated altogether.

The basic design of a four-point starter appears in Fig. 7.28. The handle is manually moved through the contacts of the series resistors until the final position is reached where the energized magnet will hold the arm until the applied emf is removed. Note that in the final rest position the entire starting resistance is out of the armature circuit, while initially it is all in series with R_A.

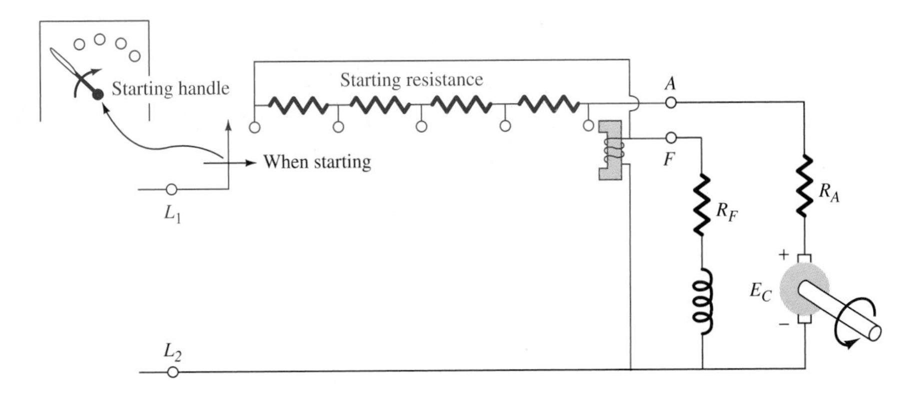

FIG. 7.28
Four-point starter.

Technological advances of recent years have resulted in a number of interesting innovations in the design of dc motors and generators. For instance, commutation can now be performed with an electronic package rather than the copper sections introduced earlier, which were a definite maintenance problem. There are now rotor designs in which the coils are wrapped near the surface so that the interior of the rotor is hollow. The result is a smaller rotating mass, with less inertia for braking and bearing design. Recently, printed circuit techniques have been employed to reduce the overall length of the machine.

All in all, however, the basic operation of the vast majority of commercially available dc generators and motors is as described in the previous sections. Only the techniques for performing certain functions are receiving some redesign attention.

7.4 AC GENERATORS

We shall now examine the basic construction and operation of single-phase and polyphase ac generators, which are also referred to as *alternators*. If the commutator sections of the dc generator in Fig. 7.1 were replaced with two slip rings (continuous circular bands) such as shown in Fig. 7.29, a sinusoidal voltage would result at the output terminals as shown. The slip rings are insulated from one another and represent the two ends of the continuous connections of conductors in the rotor. However, the vast majority of larger commercial units are not made in this

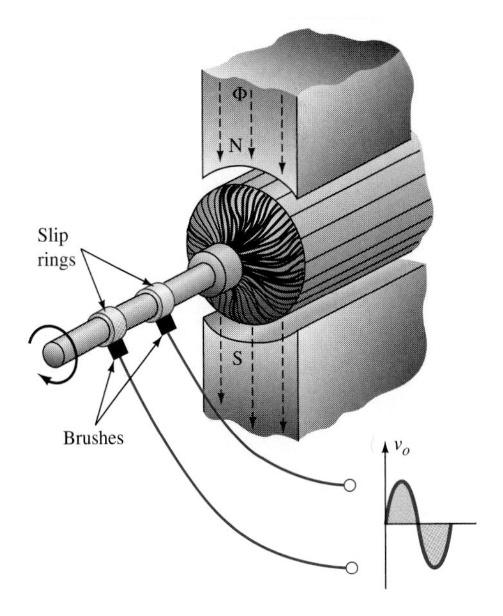

FIG. 7.29
Ac generator.

manner. Rather, the field poles become part of the rotating member, and the voltages are induced across the conductors set in the slots of the fixed stator, as shown in Fig. 7.30. The required current for the field poles of

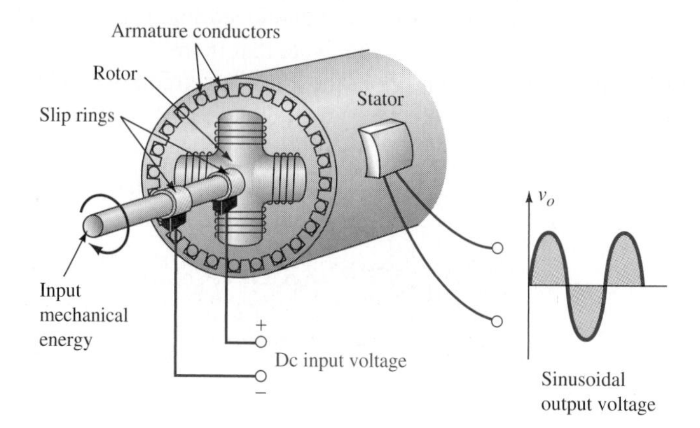

FIG. 7.30
Single-phase ac alternator.

the rotor is passed through the brushes to the slip rings. There are a number of reasons for this reversal in approach. The rotating mass in the center is considerably lighter, requiring less concern about such design problems as the end bearings and braking. In addition, if very large generated voltages were taken off the brushes of the construction in Fig. 7.29, heavy sparking and brush wear would quickly result. With the design in Fig. 7.30 the high voltages do not pass across a brush connector but are connected directly to the load. The dc voltages across the brushes (to energize the fields) are usually 240 V or less. For polyphase generators such as the three-phase type, four slip rings would have to appear on the shaft of the design in Fig. 7.29, requiring a longer frame for the dynamo, while for the other design only two slip rings would be required and an additional two wires would be required at the stator output.

The frequency of the sinusoidal voltage produced by an ac generator is determined by the number of poles on the rotor, the speed at which the rotor is turned, and a factor of 120, as dictated by

$$f = \frac{PN}{120} \qquad \text{(hertz)} \qquad \textbf{(7.28)}$$

where P = number of poles of rotor
N = speed of rotor in rpm

EXAMPLE 7.10 Determine the number of poles of the rotor of a three-phase generator if a frequency of 60 Hz is generated at a rotor speed of 3600 rpm.

Solution:

$$P = \frac{120f}{N} = \frac{120(60\text{ Hz})}{3600\text{ rpm}} = \frac{7200}{3600} = \mathbf{2\ poles}$$

Today's automobiles use an alternator rather than the dc generator of years past. Figure 7.31 is a cross-sectional view of a typical ac generator,

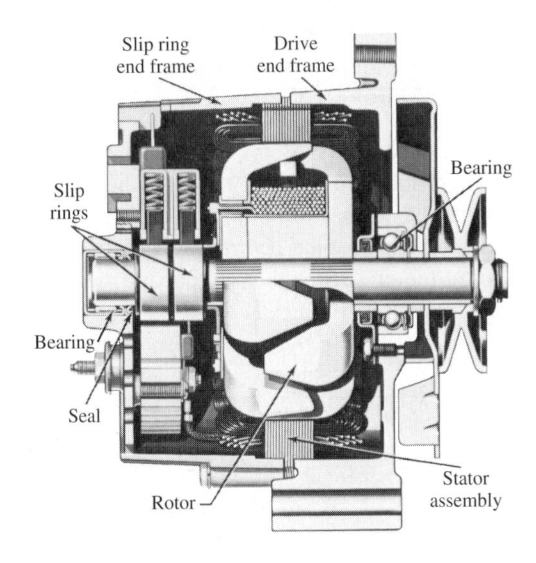

FIG. 7.31
Cross-sectional view of the Delcotron ac generator. (Courtesy of Delco-Remy, Division of General Motors.)

the Delcotron. The dc voltage to energize the field of the rotor is applied through the indicated brushes and slip rings. The output ac off the stator appears across the terminals at the bottom left of the figure. They are connected to a diode (a two-terminal electronic configuration, to be discussed in Chapter 8) that converts the generated ac to dc as required by the automobile electrical system.

Single-phase portable generators such as shown in Fig. 7.32 provide the necessary power in remote areas for electrical equipment, power tools, wildlife surveys, and so on. This particular unit can also supply dc power, and others are available that can provide the necessary 60 Hz for a cabin in a remote area.

The manner in which the windings of the stator are connected determines the number of phases to be generated by the alternator. In Fig. 7.33 the stator windings are connected such that three sinusoidal voltages (120° out of phase with each other) are produced by each rotation of the rotor. The frequency is the same for each phase as determined by Eq. (7.28).

The nameplate on a polyphase generator should include, at a minimum, the following information: three-phase, 10 kVA, 220 V, 3600 rpm,

FIG. 7.32
Single-phase portable generator. (Courtesy of Coleman Powermate, Inc.)

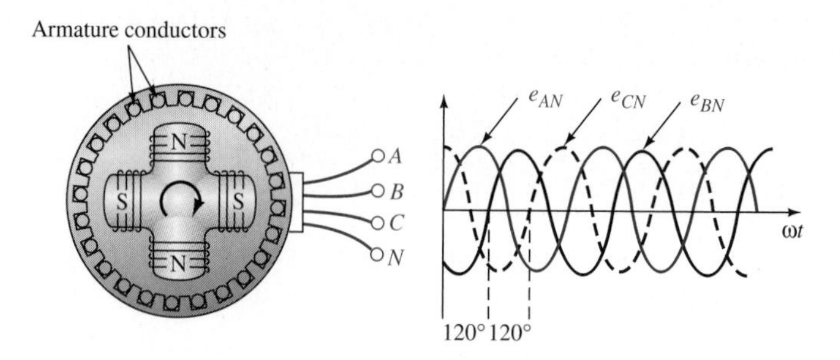

FIG. 7.33
Three-phase alternator.

Y-connected, 60 Hz. The "10 kVA" is the total apparent power rating of the generator as determined by $S_{a_T} = \sqrt{3}V_L I_L$, where V_L and I_L are defined by the specified Y connection of phases in Fig. 7.34. The "220 V" is always the line voltage (not the per-phase quantity), and the speed is the rpm of the rotor required through some outside means to develop the rated conditions (frequency, voltage, and so on).

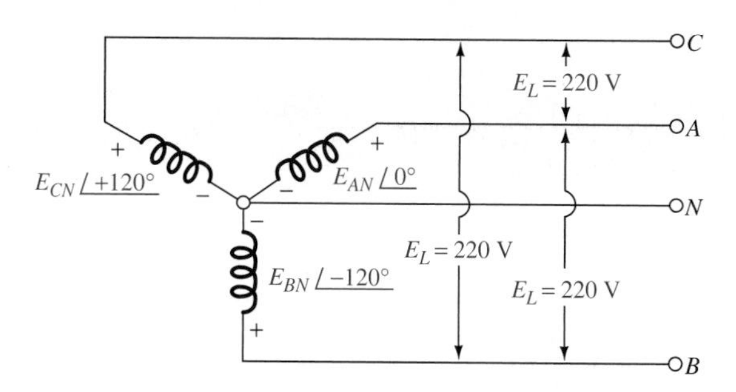

FIG. 7.34
Y-connected three-phase generator.

The voltage induced across each conductor of the stator can be determined by

$$E_C = 2.2\phi f \times 10^{-8} \quad \text{(volts)} \tag{7.29}$$

where ϕ = flux in lines
f = frequency generated

For a polyphase generator with Z stator conductors and P phases the generated phase voltage is given by

$$E_\phi = 2.22\frac{Z}{P}\phi f \times 10^{-8} \quad \text{(volts)} \tag{7.30}$$

EXAMPLE 7.11 Determine the voltage induced per phase of a three-phase ac alternator if the stator has 240 conductors, the flux is 1.5×10^6 lines, and the frequency is 60 Hz.

Solution:

$$E_\phi = 2.22 \frac{Z}{P} \phi f \times 10^{-8}$$

$$= 2.22 \left(\frac{240}{3}\right)(1.5 \times 10^6 \text{ lines})(60 \text{ Hz}) \times 10^{-8}$$

$$= \mathbf{159.84 \ V}$$

In our analysis of dc generators and motors we could ignore the inductive reactive elements (except for their internal resistances), since we simply assumed the dc short-circuit equivalent. For ac dynamos, however, the windings have a reactance dependent on the frequency generated that can measurably reduce the terminal voltage of the generator under loaded conditions. The network equivalent for one phase of a polyphase generator appears in Fig. 7.35(b) as derived from the three-phase, Y-connected

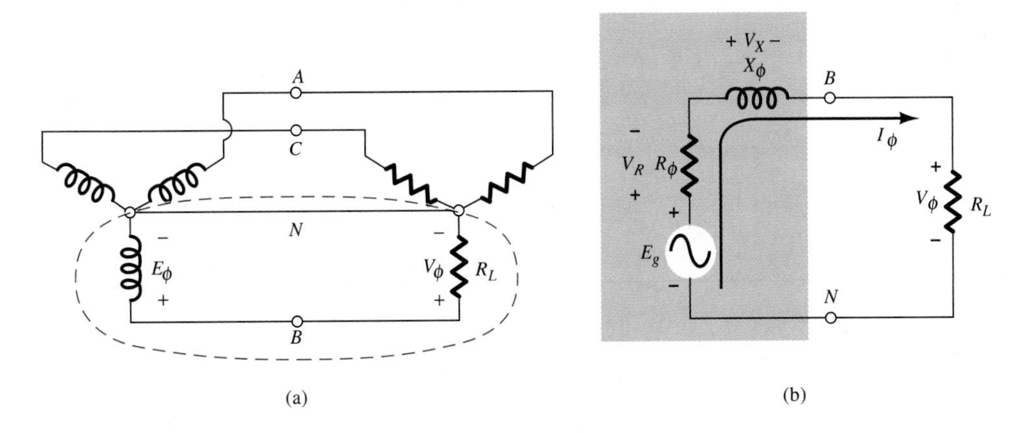

FIG. 7.35
Single-phase equivalent of a three-phase alternator.

generating system in Fig. 7.35(a). R_ϕ is the internal armature resistance, and X_ϕ is the reactance of that phase of the stator windings. The terminal voltage is then determined by the vector relationship

$$V_\phi = E_g - V_R - V_X \qquad \textbf{(7.31)}$$

or

$$V_\phi = E_g - I_\phi R_\phi - I_\phi X_\phi \qquad \textbf{(7.32)}$$

The vector diagram can quickly be determined by first drawing an arbitrary length current vector ($\mathbf{I}_\phi$) to act as a reference for the voltages of the closed loop. For a *resistive* load the voltage drop across the load ($\mathbf{V}_\phi$) is *in phase* with the current through it, and the vector for $\mathbf{V}_\phi$ is drawn parallel to that for $\mathbf{I}_\phi$. For $\mathbf{V}_R$ the same is true, and it is simply added to the end of the $\mathbf{V}_\phi$ vector in the same direction. $\mathbf{V}_X$ leads $\mathbf{I}_\phi$ by 90° and is therefore drawn vertically to $\mathbf{I}_\phi$. Through Eq. (7.31) we see that the generated phase voltage is determined by connecting the beginning of vector $\mathbf{V}_\phi + \mathbf{V}_R$ to the head of $\mathbf{V}_X$, as shown in Fig. 7.36. In equation form, the *magnitude* of $\mathbf{E}_g$ is determined by

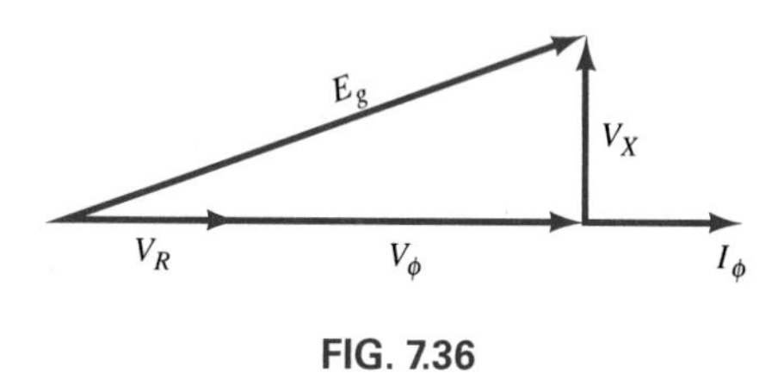

FIG. 7.36

$$E_g = \sqrt{(V_\phi + V_R)^2 + V_X^2} \quad \text{(volts)} \qquad \textbf{(7.33)}$$

EXAMPLE 7.12 A three-phase Y-connected generator has the following nameplate characteristics: 10 kVA, 220 V, $R_\phi = 0.5\ \Omega$, $X_\phi = 4\ \Omega$. Determine the voltage generated per phase to supply rated voltage to a unity F_p load (resistive).

Solution:

$$S_{a_T} = \sqrt{3}V_L I_L = 10{,}000\ \text{VA}$$

and

$$\sqrt{3}(220\ \text{V})I_L = 10{,}000\ \text{VA}$$

$$I_L = \frac{10{,}000\ \text{VA}}{(1.732)(220\ \text{V})} = \mathbf{26.24\ A}$$

For a Y-connected system (Fig. 7.34), $I_L = I_\phi$ and $I_\phi = 26.24$ A. In Fig. 7.37,

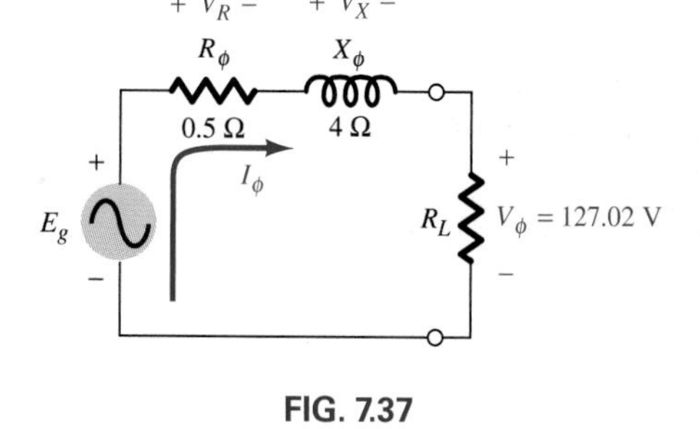

FIG. 7.37

$$V_R = I_\phi R_\phi = (26.24\ \text{A})(0.5\ \Omega) = 13.12\ \text{V}$$
$$V_X = I_\phi X_\phi = (26.24\ \text{A})(4\ \Omega) = 104.96\ \text{V}$$

and

$$V_\phi = \frac{V_L}{\sqrt{3}(\text{Y-connected})} = \frac{220\ \text{V}}{1.732} = 127.02\ \text{V}$$

$$\begin{aligned} E_g &= \sqrt{(V_\phi + V_R)^2 + (V_X)^2} \\ &= \sqrt{(127.02\ \text{V} + 13.12\ \text{V})^2 + (104.96\ \text{V})^2} \\ &= \mathbf{175.09\ V} \end{aligned}$$

In other words, in order to provide the rated 127.17 V at the load (resistive), the generator has to develop 175.72 V to compensate for the drops in potential across the internal resistance and reactance of each phase.

The analysis here was for a purely resistive load. If the load was inductive to any degree, E_g would increase, while for capacitive loads E_g

would decrease from that required for a purely resistive load, since the series capacitive and inductive reactances oppose each other and reduce the magnitude of the total impedance. In fact, Fig. 7.38 clearly indicates that the greater the leading of the power factor, the less the no-load voltage (E_g) required to meet rated conditions, although too low a leading F_p would result in very poor voltage regulation (variation in terminal voltage with load).

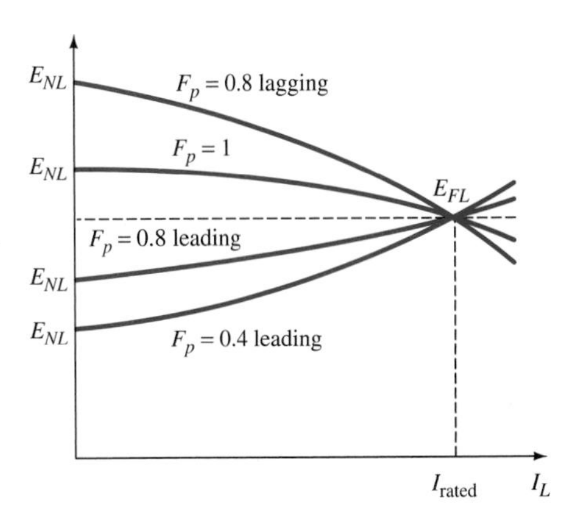

FIG. 7.38
Variation in terminal voltage for an ac alternator with change in power factor.

7.5 POLYPHASE INDUCTION AC MOTOR

The stator of a polyphase induction ac motor is constructed in much the same manner as that of the ac generator except that the voltage is applied to the stator, and the rotor turns the mechanical load.

The basic construction of the stator and rotor appears in Fig. 7.39 with a photograph of a commercially available unit. The conductors of the stator are connected in such a manner as to develop a resulting direction of flux such as shown in Fig. 7.39(a) when the three-phase ac voltage is applied. As the magnitude of the phase components varies in magnitude the conductors of the stator are connected such that the resulting flux direction appears to be rotating around the shaft as indicated in

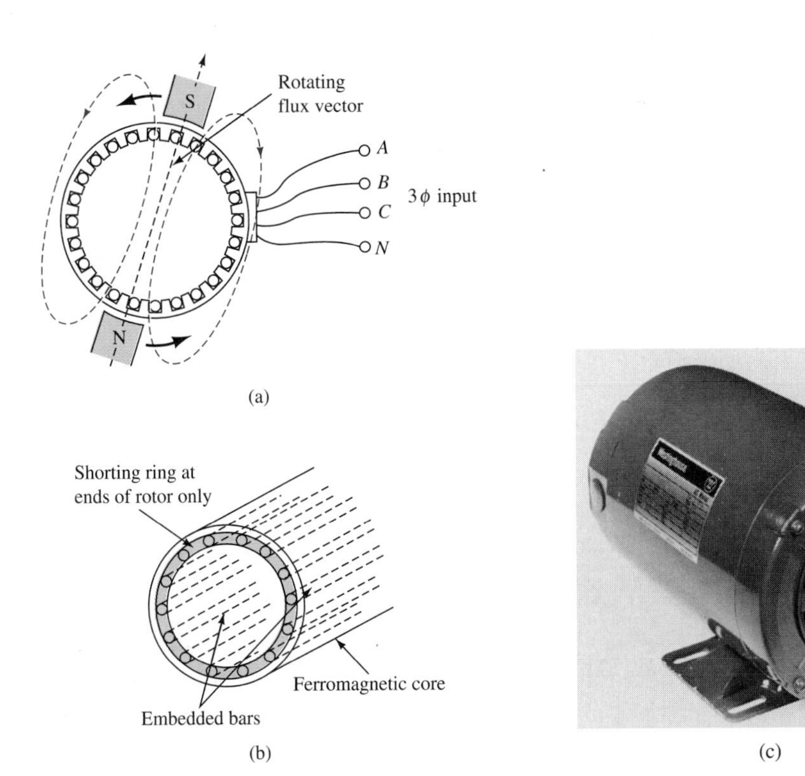

FIG. 7.39
Polyphase induction ac motor: (a) flux pattern; (b) squirrel-cage rotor; (c) typical unit. (Courtesy of Westinghouse Electric Corporation.)

Fig. 7.39(b). The rotating flux pattern is equivalent to having two opposite poles rotating around the frame of the machine. That is, the flux pattern is the same.

The speed with which this resultant flux rotates around the stator is determined by the number of poles around which the stator is wound (two for this example) which can be two, four, or more depending on the manner in which the stator windings are connected and on the frequency of the applied three-phase signal. In equation form, this *synchronous* speed is determined by

$$N_S = \frac{120f}{P} \quad \text{(rpm)} \tag{7.34}$$

where f = applied frequency
P = poles of stator
N_S = rpm

EXAMPLE 7.13 Determine the synchronous speed of a polyphase induction motor having a stator with four poles and an applied frequency of 60 Hz.

Solution:

$$N_S = \frac{120f}{P} = \frac{120(60)}{4} = \mathbf{1800\ rpm}$$

The construction of the rotor as indicated in Fig. 7.39(b) is nothing more than conducting rods embedded in a ferromagnetic core with shorting rings on each end. The bars are skewed to provide a more uniform torque. If the rods and shorting ring were removed from the ferromagnetic core, a cage would result, so called for no other reason than that its general size is about that of a squirrel cage. For this reason this particular motor is often referred to as a *squirrel-cage* induction motor.

The rotating flux of the stator cuts the embedded conductors in the rotor, induces a voltage across them, and establishes a current through the rotor conductors and shorting rings. The current-carrying conductors in a magnetic field experience a force in a direction determined by the left-hand rule described for dc motors. The torque developed is determined by the current through the conductors and the stator flux. That is,

$$T = K\phi I_C \quad \text{(lb-ft)} \tag{7.35}$$

where ϕ = flux in lines
I_C = conductor current
K = constant

Since current can be established only in the conductors when they are cut by the rotating stator flux, the speed (N) of the rotor can never equal the speed of the flux vector (N_S). If $N_S = N$, the flux vector cannot cut the conductors of the rotor, induce the voltage, and create the necessary current flow. The difference in speed between the stator flux and the speed of the rotor is called the *slip* and is measured in percent or rpm as determined by the equations

$$S = \frac{N_S - N}{N_S} \times 100\% \qquad \text{(percent)} \qquad \mathbf{(7.36)}$$

$$S = N_S - N \qquad \text{(rpm)} \qquad \mathbf{(7.37)}$$

In terms of N_S, P, and the slip, the speed of the rotor can be determined by

$$N = \frac{120f(1 - S)}{P} \qquad \text{(rpm)} \qquad \mathbf{(7.38)}$$

EXAMPLE 7.14 A four-pole, 60-Hz, three-phase induction motor has a full-load speed of 1725 rpm. Determine the slip in percent.

Solution:

$$N_S = \frac{120f}{P} = \frac{120(60)}{4} = 1800 \text{ rpm}$$

$$S = \frac{N_S - N}{N_S} \times 100\% = \frac{1800 - 1725}{1800} \times 100\% = \mathbf{4.167\%}$$

The starting torque of a three-phase induction motor can be shown to be related to the applied line voltage in the following manner:

$$T_{st} = KV_L^2 \qquad \text{(lb-ft)} \qquad \mathbf{(7.39)}$$

Equation (7.39) reveals that doubling the applied line voltage increases the starting torque by a factor of 4 (the squared value). This high starting torque is an excellent characteristic for use in pumps, compressors, industrial fans, and machine tools.

Under normal operating conditions it can be shown that the running torque is directly related to the slip. That is,

$$T_{\text{running}} = KS \qquad \text{(lb-ft)} \qquad \mathbf{(7.40)}$$

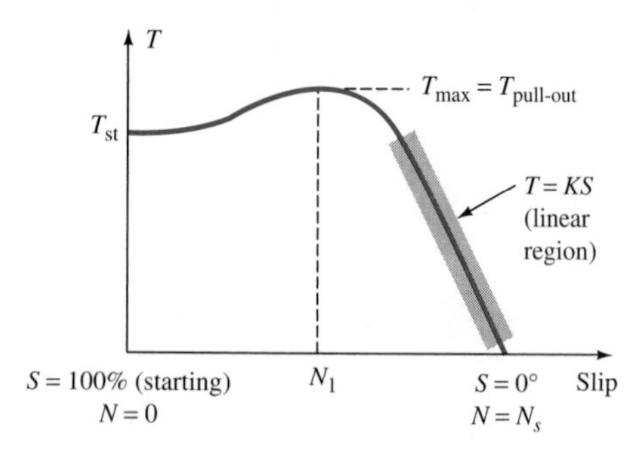

FIG. 7.40
Torque characteristics of a polyphase induction motor.

In other words, if the slip doubles (if the difference in speed between N_S and N increases by a factor of 2), the running torque increases by the same factor. The complete torque curve for a three-phase squirrel-cage induction motor is shown in Fig. 7.40. Note the straight-line region determined by Eq. (7.40). The pull-out torque is an indication of the maximum load the motor can handle at any speed. For this machine maximum torque is obtained at a rotor speed N_1.

The speed N of a motor can be determined directly using a tachometer. The slip can be determined by first calculating N_S from the provided data and substituting into Eq. (7.36).

The polyphase induction motor, unlike the dc motor, can be started with full-line voltage applied. The starting current may be 5 to 10 times the rated line current, but the machine is designed to handle these surge currents for short periods of time. As the machine builds up to rated speed the line current drops to rated value. For other loads on the same three-phase system the starting of the three-phase induction motor may result in momentary dimming of the lights, a slowing down of the machinery, and so on. If this reaction is objectionable, other techniques, such as using an autotransformer (Chapter 6), may have to be employed to reduce the voltage applied directly to the machine at starting.

Table 7.1 lists some commercially available 60-Hz, three-phase squirrel-cage induction motors. In Table 7.1 the length of the $\frac{1}{6}$-hp machine is $9\frac{13}{32}$ in. with a diameter of $5\frac{13}{16}$ in.

TABLE 7.1
Polyphase induction motors

hp	rpm	Volts	Amperes (I_L)
$\frac{1}{6}$	1725	240/480	0.9
$\frac{1}{4}$	1725	208	1.2
	1140	240/480	1.3
$\frac{1}{3}$	1725	208	1.8
$\frac{1}{2}$	3450	240/480	1.6
$\frac{3}{4}$	1725	208	3.6
1	3450	240/480	3.0
$1\frac{1}{2}$	1725	208	6.0
2	1725	208	6.8
3	3450	240/480	8.9
5	3450	208	15.5

Courtesy of General Electric Company.

For the 5-hp machine the length is $14\frac{3}{64}$ in. and the diameter is $6\frac{1}{2}$ in. In addition, the weight of the $\frac{1}{6}$-hp motor is only 16 lb, while that of the 5-hp machine is 44 lb—a large unit.

7.6 POLYPHASE SYNCHRONOUS MOTOR

Through the proper design of the rotor it is possible to have a polyphase motor that runs at synchronous speed (the speed of the rotating flux of the stator). It is then referred to as a *polyphase synchronous* motor. Basically, the stator has the same design as the squirrel-cage induction motor; however, the rotor is designed to have both the squirrel-cage structure and rotating field poles (the electromagnetic or permanent-magnet type). The electromagnetic field poles in the rotor can be energized through dc slip rings connected to a dc supply. There are also rotors of a very particular design, which will not be examined in this section, that actually have the field poles induced through a combination of the shape of the rotor and the stator flux.

After the motor is started, its speed is brought near synchronous speed by the torque developed by the induced current through the rods of the squirrel cage. Near synchronous speed, if the field poles in the rotor are energized, the rotor will jump forward and "lock in" with the rotating poles of the stator, as shown in Fig. 7.41. If like poles are encountered at the time the rotating rotor poles are energized, the rotor will slip back until unlike attracting poles are encountered. The slip has then dropped to zero, and the speed of the rotor is then equal to the synchronous value, namely,

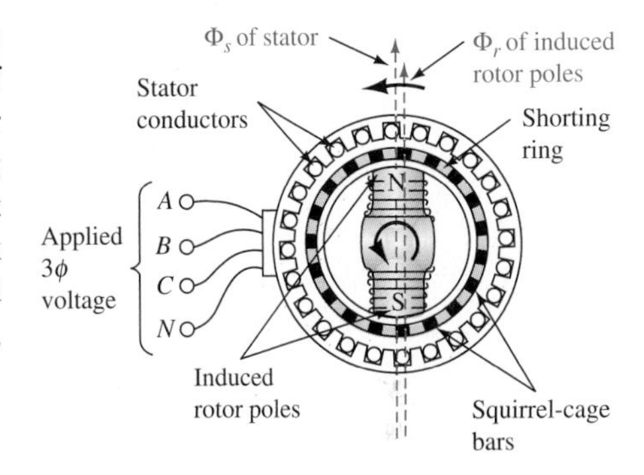

FIG. 7.41
Polyphase synchronous motor.

$$N = N_S = \frac{120f}{P} \quad \text{(rpm)} \tag{7.41}$$

where N is the speed of the rotor.

In an effort to bring the speed due to induced effects as close to synchronous speed as possible the design requires cutting the starting torque to significantly reduce the slip. In fact, it may be cut as much as 50% of full-load value—significantly different from the squirrel-cage type designed to have a very heavy starting torque for its areas of application.

For the polyphase synchronous motor with an external rotor field excitation (I_F) the change in terminal characteristics with field current can provide a very efficient method of cutting energy costs in a large industrial plant. At low values of I_F the synchronous machine has an inductive input impedance, and for high values of I_F it is capacitive. For each load condition there is a particular field current that results in a unity power factor. In general, therefore, no matter what the load on the shaft, there is a field current that can be applied to make the machine appear inductive, resistive, or capacitive.

The capacitive characteristics of the dynamo have important practical application as a *synchronous capacitor*. For a large industrial facility the vast majority of loads (machines, lighting, and so on) have inductive characteristics. As noted in Fig. 7.42 this results in an apparent power rating far beyond that necessary to supply the total kilowatts (real power) to the load. In large industrial plants, if the manufacturer is paying for the apparent power demand rather than the real power actually dissipated, the

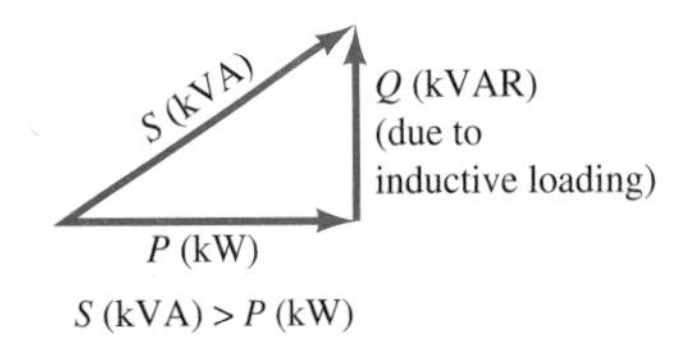

FIG. 7.42

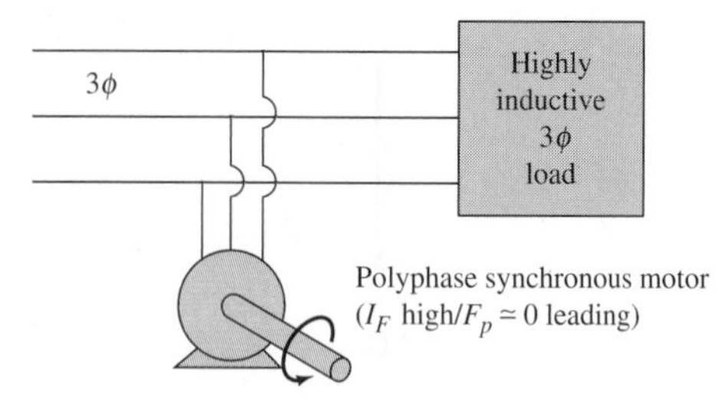

FIG. 7.43
Power factor correction employing a polyphase synchronous motor.

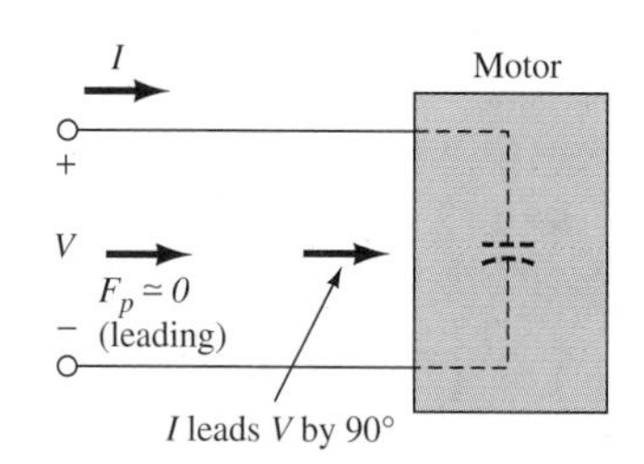

FIG. 7.44
Capacitive appearance of a polyphase synchronous motor with no load and high I_F.

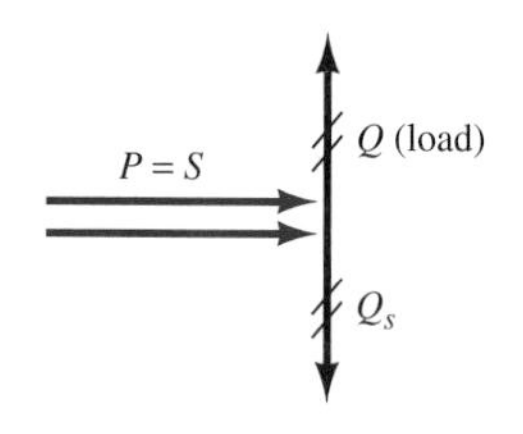

FIG. 7.45
Forcing a load power factor of unity (F_P = 1).

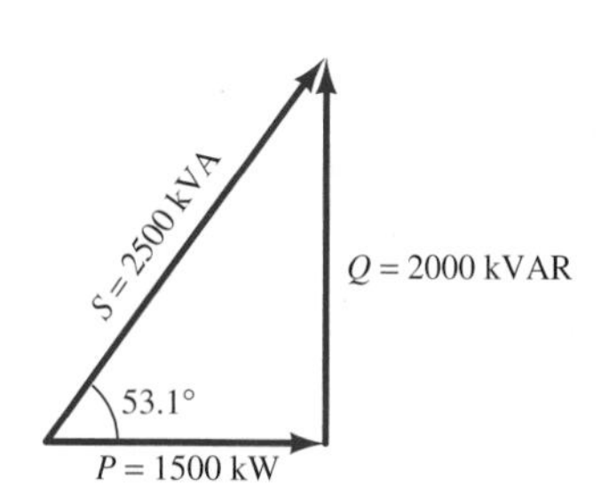

FIG. 7.46

introduction of a large reactive power (Q) component can be a very expensive item. In Fig. 7.43 a synchronous motor operating in its high field current region to establish the highly capacitive characteristics has been attached to a three-phase system with a highly inductive load. With no load on the shaft of the motor we can assume as a first approximation that applied voltage lags the line current of the synchronous motor by 90° if we apply a very high field current. For all practical purposes we can therefore assume for application sake that the synchronous motor is nothing more than a large capacitor, as shown in Fig. 7.44.

If we assume the machine to be purely capacitive, it will not have a real power component in its power triangle, and its apparent power rating will equal its reactive power component. That is,

Synchronous capacitor:
$$S_S = Q_S \tag{7.42}$$

For the power triangle in Fig. 7.42, if we choose a synchronous motor with a kVA rating equal to the inductive component of the factory load, the two reactive components will cancel as shown in Fig. 7.45, and the net apparent power (the cost element) will have its minimum value—simply equal to the real power dissipated. It certainly seems strange to apply a motor to a system simply to act as a capacitor and have absolutely no load on its shaft, but the practical world considers it a valid energy-conserving technique. Reduced benefits of the capacitive characteristics can still be derived if a load is applied to the shaft of the synchronous motor.

EXAMPLE 7.15 In Fig. 7.43, the total three-phase load of the industrial plant is 1500 kW at a lagging (inductive) power factor of 0.6.

a. Sketch the power triangle of the load.
b. Sketch the power triangle of a synchronous motor operating as a synchronous capacitor that can balance the reactive power component of the inductive load. Assume that the synchronous motor has pure capacitive characteristics and a zero power factor.

Solution:

a. $\theta = \cos^{-1} 0.6 = 53.13°$, $\sin 53.13° = 0.8$
   From $P = S\cos\theta$,
   $$S = \frac{P}{\cos\theta} = \frac{1500\text{ kW}}{0.6} = \mathbf{2500\ kVA}$$
   and from $Q = S\sin\theta$,
   $$Q = (2500 \times 10^3\text{ VA})(0.8) = \mathbf{2000\ kVAR}$$
   The power triangle for the inductive load appears as in Fig. 7.46.
b. To establish a unity power factor load characteristic the reactive power of the synchronous capacitor must equal the reactive power component of the load. Therefore,
   $$Q_S = Q_L = 2000\text{ kVAR}$$

and from Eq. (7.42),

$$S_S = Q_S = \mathbf{2000\ kVA}$$

The power triangle of the synchronous capacitor appears as in Fig. 7.47.

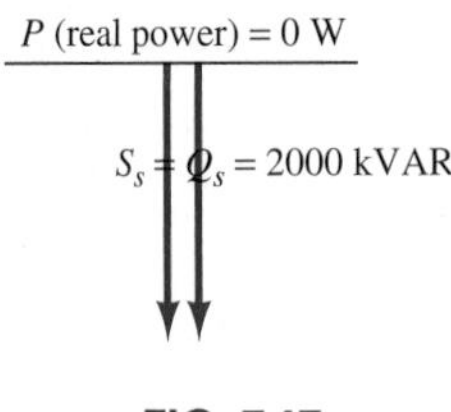

FIG. 7.47

A great deal of additional information on the construction and analysis of the synchronous machine under varying load conditions is readily available.

7.7 SINGLE-PHASE AC MOTORS

A single-phase outlet is typically all that is available in homes, educational institutions, many small businesses, office buildings, and so on. The single-phase ac motor is therefore the most frequently encountered. Since most single-phase outlets have an absolute maximum current rating of 20 A at 120 V, the majority of single-phase ac motors have power ratings of less than (20 A)(120 V) = 2400 W = 3.22 hp. The vast majority of single-phase ac motors are 1 hp or less. A few moments' reflection is all that is necessary to realize how many dynamos of this type appear in the home. Refrigerators, washing and drying machines, dishwashers, sewing machines, electric typewriters, electric heaters with fans, heating system circulators, and even electric shavers (plug-in) incorporate some type of single-phase ac motor. For most single-phase ac motors, the general characteristics such as starting torque, efficiency, and power factor are not as ideal as for a three-phase ac motor if compared on a relative horsepower scale. In fact, the efficiency of some single-phase ac motors can be as unbelievably low as 10%. However, since the power drawn during operation is usually quite small, the loss in watts can be tolerated. Of course, if we consider the number of machines of this type currently in use, perhaps a measure of reevaluation is required in light of the energy concerns we have today.

There are a wide variety of commercially available single-phase ac motors. However, the *series* or *universal, split phase, capacitive start, shaded pole,* and *hysteresis* types are the most frequently encountered. Each will be described briefly here with an application or two.

It is possible to simply apply an ac voltage to a dc shunt-wound motor and obtain an output torque on the shaft. In Fig. 7.48(a), the directions of I_F and I_A are shown for the positive pulse of the sinusoidal input. In the same figure a pole and rotor conductor current direction are shown that clearly indicate that (using the left-hand rule) the torque is clockwise. During the lower half of the sinusoidal input the current in the field and armature reverse, as indicated in Fig. 7.48(b); however, the torque is *still* clockwise, indicating that for the positive and negative regions of the sinusoidal input the torque is always in the clockwise (or counterclockwise) direction. A reversal in direction can be obtained by reversing either the field or armature circuit but obviously not both at the same time. The shunt motor configuration is seldom employed for ac application because of the reduction in output torque due to the resulting small main

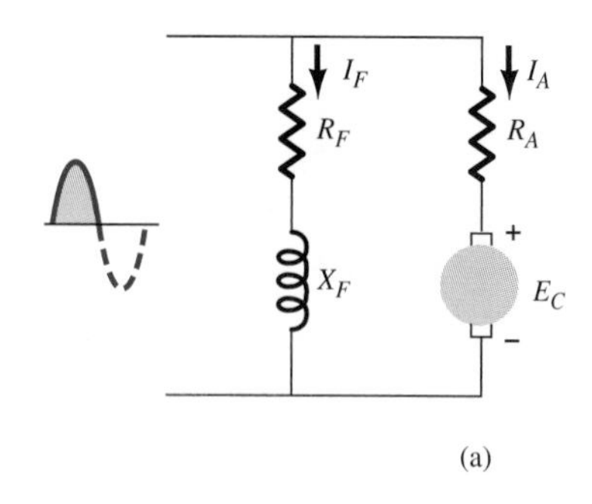

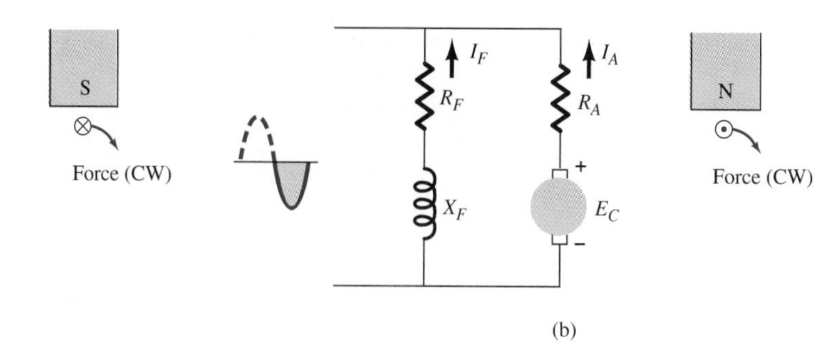

FIG. 7.48
Shunt single-phase ac motor.

field flux and the increased out-of-phase relationship between the field flux and the armature current. Both effects are introduced by the high inductive reactance of the main field winding, which is not present for dc conditions. The impedance of the field current is thereby increased substantially, and the field current (and therefore the field flux) is reduced to an ineffective level. This high reactance introduces a large phase shift between I_F and I_A, so that when I_A is its maximum, the flux is not, and the torque $T = K\phi I_A$ is substantially reduced.

FIG. 7.49
Single-phase series ac motor.

Series (Universal)

For reasons to be described, the series motor, called the *universal motor,* can display satisfactory characteristics when a single-phase ac voltage is applied. The basic elements of the machine appear in Fig. 7.49. Obviously, the field and armature currents always have the same direction, since they are in fact the same quantity ($I_S = I_A$). Thus, (as for the dc shunt motor) the torque always has the same direction even though an alternating emf is applied. To reverse direction either the series field coil *or* armature portion of the system must be reversed. Changing both will result in the same direction of rotation.

The effect of the inductance introduced by the turns of the series field is smaller because there are normally considerably fewer turns in the series field than in the shunt field. This is a necessity or there would be an intolerable voltage drop across the dc geometric resistance of the series coil, and an insufficient potential level would appear across the armature. The resulting field current and flux are therefore greater, and less phase shift exists between the pole strength and rotor conductor current. The result is a torque more comparable to an equivalent applied dc voltage and a more efficient system in general.

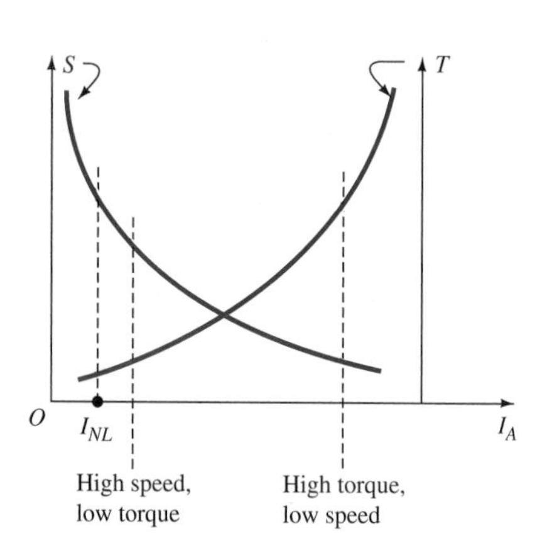

FIG. 7.50
Torque and speed characteristics of a single-phase series ac motor.

The characteristics of this *universal* motor are quite similar to those of the dc series motor, as shown in Fig. 7.50. Applications that require very high speed but need only low torques to meet the load work their design around the low-current region. Applications where very high torques are needed but speed is not a factor require the use of the high-current region. As noted on the curves, the current must not reach too low a level or the speed could increase to dangerous levels. It is therefore necessary to ensure with resistive elements or other design techniques that the minimum line current is the no-load value. There are numerous applications for this device in the home that make use of the high-speed, low-torque requirements: electric shavers, sewing machines, food mixers, vacuum cleaners, electric typewriters. The full range of its characteristics are employed in electric power tools (drills, and so on) where heavy loads require the high-torque, low-speed characteristic, and lighter loads, the high-speed, low-torque characteristic.

Keep in mind that for the preceding discussion, the design is basically dc in the sense that commutator sections and brushes are present, which results in increased maintenance and cost. Other designs called simply *series* ac motors are available that work only on ac applied signals with increased simplicity in design (only slip rings) and increased efficiencies. However, the term *universal* is employed to indicate the versatility of the machine in that it can operate on dc or ac applied voltages.

Split-Phase

The induction principle as applied in the three-phase squirrel-cage induction motor can also be applied to single-phase machines with one important modification: the single phase must be split to appear as two with a phase shift between them as close to 90° as possible—hence the terminology for this group of single-phase ac motors: *split phase*. The splitting of the phase is necessary to establish a rotating flux around the stator of the machine. If left as a single phase, the magnetic pole alignment would remain fixed. An attempt will not be made here to show why splitting the phase by 90° results in rotation of the induced poles about the stator, but we shall investigate how the phase can be split. There are many references for further investigation in this area. The speed of the rotating flux is determined by the same equation employed for the three-phase machine:

$$N_S = \frac{120f}{P} \quad \text{(rpm)} \tag{7.43}$$

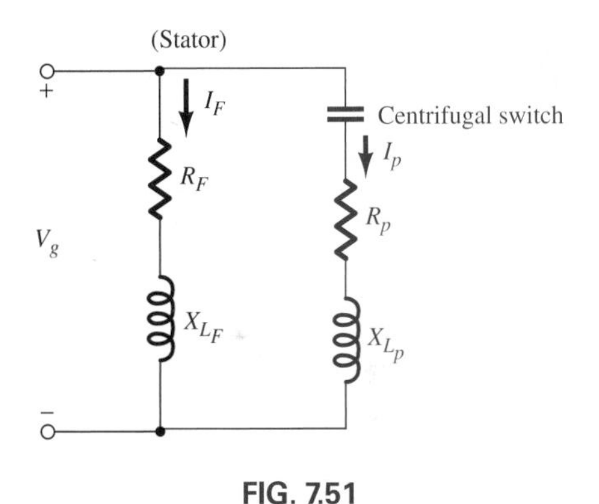

FIG. 7.51
Split-phase single-phase ac motor.

To split the phase, a second winding is added in parallel with the main field winding as shown in Fig. 7.51 that has considerably fewer turns than the main field winding. The additional winding therefore has a much lower inductance level and consequently lower inductive reactance at the applied frequency. This secondary winding is set in the same slots of the stator as the main field winding but is totally insulated from it.

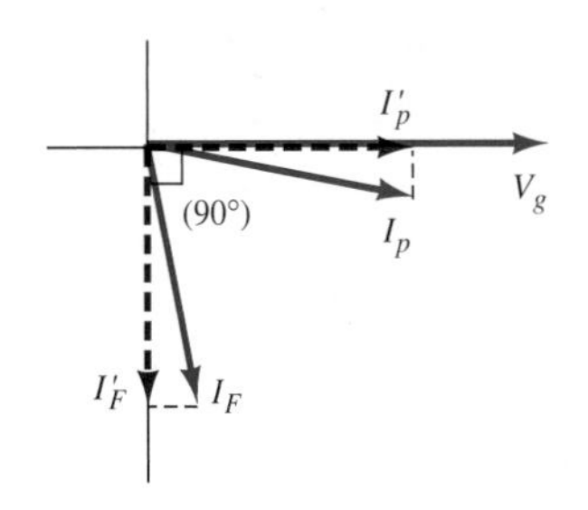

FIG. 7.52
Phasor diagram for split-phase single-phase ac motor.

The applied single-phase voltage V_g in Fig. 7.51 appears across both windings and is therefore our reference vector in the phasor diagram for the system in Fig. 7.52. At the applied frequency, the inductive reactance of the main field winding is very high compared with its series internal resistance, and I_F lags V_g by almost 90°, as shown in Fig. 7.52. In the parallel branch the inductive reactance is considerably less than its series resistance, and I_p lags V_g (since it still is an inductive branch lagging F_p)

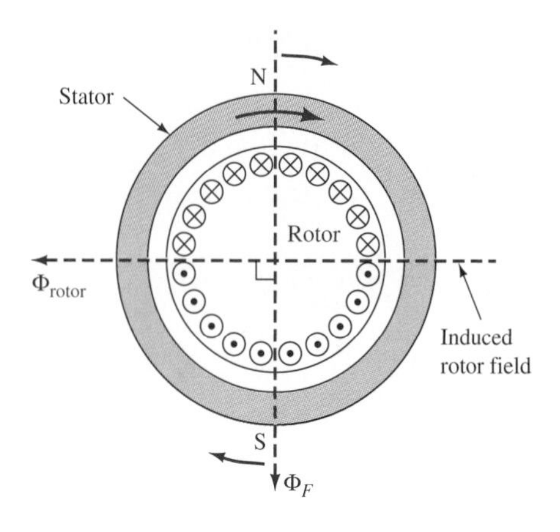

FIG. 7.53

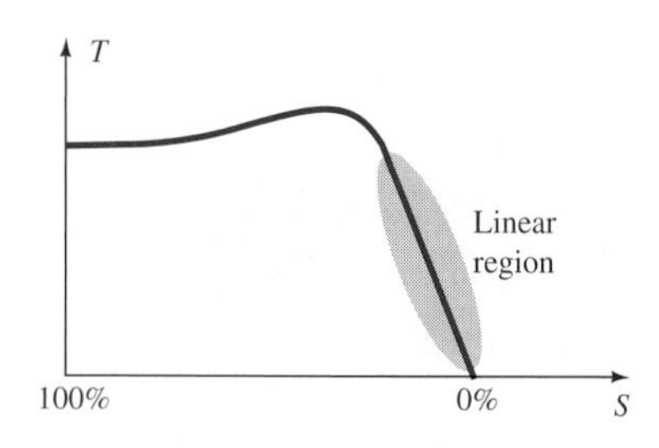

FIG. 7.54
Torque characteristics of a split-phase ac motor.

FIG. 7.55
Split-phase ac motor. (Courtesy of Westinghouse Electric Corporation.)

by only a few degrees, as shown in Fig. 7.52. Although I_F and I_p are not exactly 90° out of phase, they have components that are exactly 90° out of phase, as shown in the same figure. The flux generated by these component currents in their respective windings is also 90° out of phase, resulting in the split phase and a flux pattern that rotates around the stator, inducing currents in the conductors of the rotor and establishing the necessary torque to turn the shaft.

Near synchronous speed the necessary torque can be developed solely by the main field winding for reasons shown in Fig. 7.53. With the auxiliary field winding removed, the main field induces currents in the rotor in the direction shown in Fig. 7.53 when the rotor is rotating at or near synchronous speed. The flux pattern (ϕ_{rotor}) developed by these induced currents appears 90° out of phase with ϕ_F, resulting in a phase shift of 90° between the induced voltages across the rotor conductors and the resulting rotor current. The requirement for two flux vectors 90° out of phase to establish a resultant rotating flux vector is therefore present, and the motor can continue to turn without the auxiliary winding. In fact, the auxiliary winding cuts down on the performance characteristics of the motor and should be removed. In the neighborhood of 80% of synchronous speed, a centrifugal switch (a switch whose contacts open when rotating at a preset speed) opens and removes the auxiliary winding from the system.

The characteristics of the split-phase ac motor as depicted in Fig. 7.54 are quite similar to those of the polyphase induction motor, that is, it is a fairly constant speed device near synchronous speed, although its slip is more typically 5 to 6% rather than the 2 to 3% encountered for polyphase machines. The starting torque is typically about 150% of rated value at a starting current of 5 to 10 times the rated value. The torque characteristics are quite linear near the synchronous speed, with a direct relationship with the slip at these speeds. Since this type of motor is frequently used in washing machines, dishwashers, and oil-burner circulator motors, whenever they are engaged it is possible that a dimming of the lights in the household may be noticeable. The high initial surge current is often the cause of the circuit breaker "popping" if two or more appliances require this high starting current simultaneously on the same circuit in the home. To reverse the direction of rotation of this machine it is necessary only to reverse either one of the windings. A split-phase motor appears in Fig. 7.55.

Capacitive-Start

The 90° phase shift can also be obtained by placing a capacitor in series with an auxiliary winding such as shown in Fig. 7.56. At the applied frequency the capacitive reactance heavily outweighs the inductive reactance of the auxiliary winding, and so the added parallel branch is effectively capacitive in terminal impedance. The auxiliary winding is necessary to establish the out-of-phase flux resulting from the current through that branch. The phasor diagram of the system appears in Fig. 7.57 with the applied voltage the reference vector as before. The current I_F still lags the voltage across that branch because of the high inductance

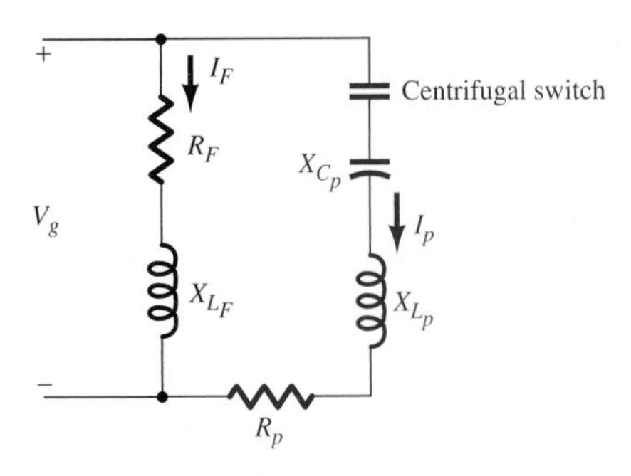

FIG. 7.56
Capacitive-start induction motor.

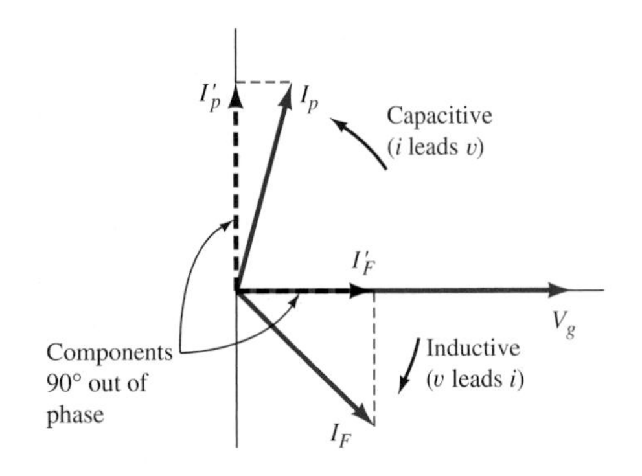

FIG. 7.57
Phasor diagram of capacitive-start induction motor.

of the main field winding. However, since the parallel branch is, in total, capacitive, the current I_p leads the applied voltage by the indicated angle. Again a phase shift of exactly 90° is not usually obtained, but components of I_F and I_p do have this phase relationship and establish the two flux components 90° out of phase. As before, a centrifugal switch opens the auxiliary branch at a speed close to synchronous.

The increased angle between the parallel branch currents results in improved starting characteristics over those of the split-phase inductive type. The starting torque can now be 2 or 3 times the rated value, but with starting currents only 2 or 3 times the rated value rather than the 5 to 10 times encountered for the split-phase inductive machine. In general, however, the capacitive start is more expensive than the split-phase variety. Machines of this type are not limited to the fractional horsepower variety in part because of the significantly reduced starting current. A machine with 5 to 7 hp is quite possible and relatively efficient.

The increased starting torque is desirable for applications involving pumps, compressors, air conditioners, freezers, refrigerators, and farm equipment. A typical commercial unit appears in Fig. 7.58.

FIG. 7.58
Capacitive-start ac induction motor. (Courtesy of Westinghouse Electric Corporation.)

Shaded-Pole

The last of the split-phase motors to be considered is the shaded pole, which acts on a principle quite different from the two just described. The poles of the stator are constructed as shown in Fig. 7.59. A small portion of the pole is separated from the main structure by the indicated cut and a winding of a limited number of turns (often a single heavy copper strap) around the smaller section called a *shading coil.* The effect of a shading coil is to introduce a time phase shift of 90° between the two poles (of the singular structure). This is accomplished through the current induced in the shading coil by the changing flux of the pole structure. An increasing flux strength through the pole induces a current in the strap that develops a flux opposing the increasing flux pattern in the

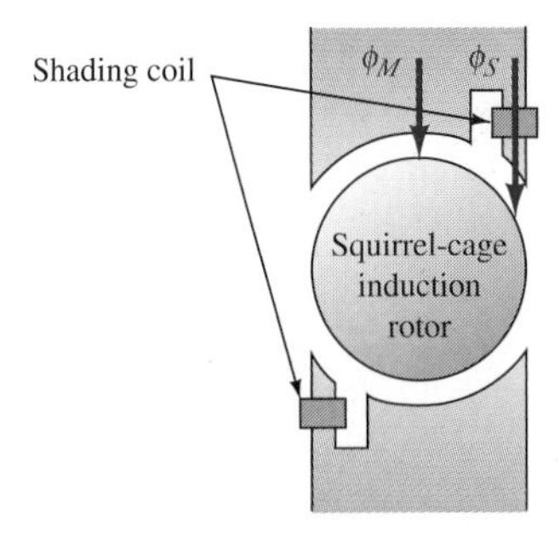

FIG. 7.59
Shaded-pole induction motor.

shaded region and severely reduces the flux through the shaded portion. Then, as the main field flux reaches its maximum, the rate of change of flux is a minimum and the induced current in the shaded coil is quite small. The resulting opposing flux of the shading coil is therefore quite small, and the flux strength of the shaded region increases. In other words, when the main field flux is changing at its highest rate, the shaded pole flux is a minimum, and when the main field flux is changing at its minimum rate, the shaded pole is a maximum. The result is a time delay between maximum flux conditions for each part of the pole, which makes it appear as if the flux pattern is moving clockwise around the stator as required to set the rotor in motion.

The single-phase shaded-pole induction motor is one of the least expensive to manufacture because of its relatively simple construction, but it is also one of the most rugged and finds application in fans, vending machines, movie projectors, and so forth. Its horsepower rating is typically less than $\frac{1}{25}$ hp. It must be limited to very small power applications, since its efficiency can be as low as 10%.

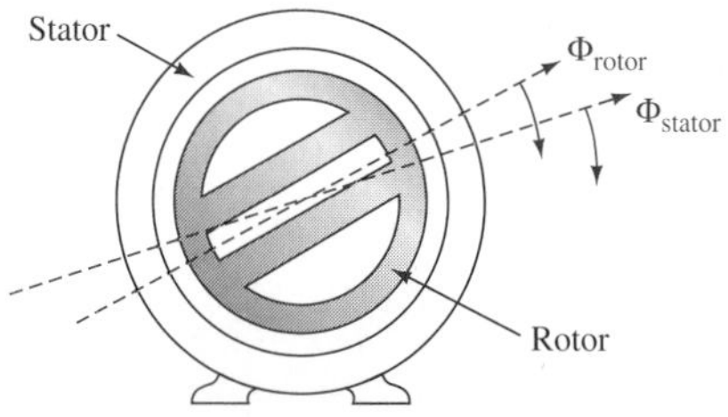

Stator
Φ_{rotor}
Φ_{stator}
Rotor

FIG. 7.60
Hysteresis ac motor.

Hysteresis

The principle of operation of the *hysteresis* motor is quite different from that of those just described. Although the rotating stator field can be established by a polyphase applied voltage, capacitor start, or shaded pole, the rotor as shown in Fig. 7.60 does not have salient (protruding) poles or dc excitation. It is made of very high retentivity steel (it holds onto the polarized flux pattern even when the applied mmf is removed) that has a large hysteresis loop. The large loop ensures a lag in time between removal of the applied mmf and the rotor's giving up its magnetic characteristics. The induced polarized flux pattern of the rotor tries to align itself with the rotating stator flux of the stator, causing a torque on the rotor. The lag between the two is due to the high-retentivity effect. When synchronous conditions are reached such that the speed of the rotor equals that of the stator, the eddy current torque is zero, leaving only the hysteresis-induced torque. The crossbars provide a low-reluctance path for the induced polarized flux in the rotor to establish a type of permanent magnet in the rotor that can lock in on the rotating stator flux. A commercially available unit appears in Fig. 7.61. The quiet running of the motor due to the absence of the salient poles and lack of windings and its constant-speed characteristic make it an excellent choice for applications such as clocks, magnetic tape drives, turntables, and movie projectors.

FIG. 7.61
Single-phase hysteresis ac motor. (Courtesy of Singer Company.)

7.8 STEPPER MOTORS

The tremendous growth of and interest in *stepper* motors during the last decade is due primarily to the growth of the computer industry, although the range of applications for the motor in the industrial community grows daily. Fundamentally, the stepper motor provides a method whereby discrete steps in rotation can be obtained that extend from less than 1° to 90°. It replaces the gears and slidewires used in the past, which were mechanically less reliable and slower. From experience we all realize that the motor employed by a record or tape player should provide a uniform torque at a fixed speed, starting and stopping on a low-frequency basis. A stepper motor permits quickly moving to a particular position on a disk or tape to obtain an important piece of data and then perhaps moving backward to obtain further information. We have all seen the large tape wheels on computer systems on TV or in the movies move back and forth as the system searches for "ultra" important data.

The operation of the stepper motor is best understood if a clear picture of the basic construction is first presented. The stator of the stepper motor in Fig. 7.62 has two separate coils, 1–1′ and 2–2′. Note the dashed

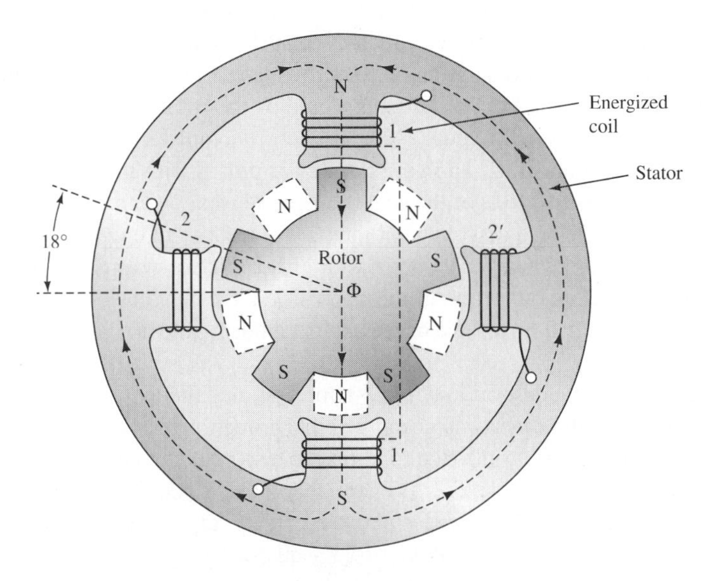

FIG. 7.62
Stepper motor with 1–1′ energized.

line between 1 and 1′ indicating that the current through 1 also appears in 1′ to establish alternate poles at the same instant. For the polarity of the applied dc voltage a north pole has been established at 1 and a south pole at 1′. Coil 2–2′ is unenergized for the moment, with no polarity, simply acting as an extension of the stator.

The rotor in Fig. 7.62 is the *permanent-magnet* type, resulting in a series of south poles at one end and a series of north poles at the other end, as shown in Figs. 7.62 and 7.63. Figure 7.63 is a simplified version of the actual construction specifically designed to explain the general principle of operation. The "face-on" appearance of Fig. 7.62 is a close reproduction, but the manner in which the permanent magnet is placed on the rotor in Fig. 7.63 has been modified for clarity.

In Fig. 7.62 note that the north poles at the other end of the rotor (7.63) are between the south poles of the rotor (actually, 36° separates the south and north poles as you face the rotor in Fig. 7.62). The north

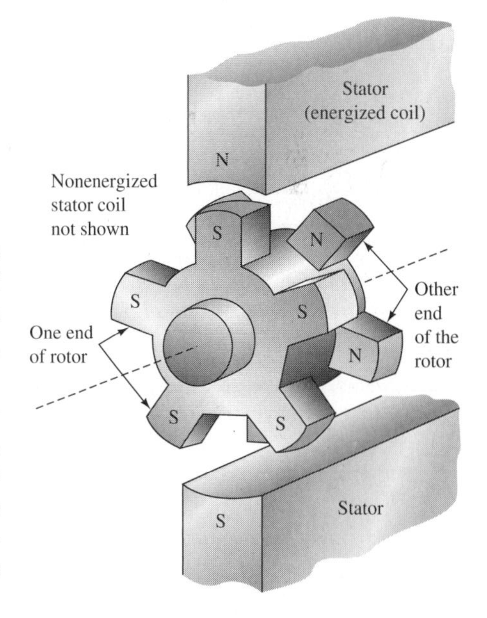

FIG. 7.63
Stepper motor permanent-magnet rotor construction.

pole at the bottom of the rotor then appears directly above the south pole of the stator, providing a direct flux path from the north pole of the stator through the top south pole of the rotor, leaving the bottom north pole of the rotor to the south pole of the stator. The rotor is therefore "locked in" the position shown in Fig. 7.62 as long as coil 1–1′ remains energized.

If coil 1–1′ is now deenergized and coil 2–2′ energized such that 2 becomes a north pole and 2′ a south pole, the south pole of the rotor closest to stator pole 2 will move in a counterclockwise (CCW) direction to lock in with stator pole 2. At the same instant the north pole of the rotor closest to stator pole 2′ also moves CCW to lock in with stator pole 2′. The actual CCW rotation is 18°, as shown in Fig. 7.62. As long as 2–2′ is energized in this manner, the rotor remains in its new position. Of course, if important data are on a disk between positions 1–1′ and 2–2′, these data are read off by the reading head on the disk. A further 18° shift can be obtained by energizing coil 1–1′ in such a manner as to place the north pole of the stator at 1′ and the south pole at 1 (keeping in mind the new location of the bottom left south pole of the rotor after the movement just established by coil 2–2′). In fact, a complete rotation of the rotor can be obtained by applying a series of pulses with the proper polarity to alternating coils of the stator. A reverse rotation can be obtained simply by applying a dc voltage of the proper polarity to the stator coils. In other words, when locked in position 2–2′, if we wanted to move clockwise back to our original starting position, the dc voltage applied to coil 1–1′ would simply have to reestablish a north pole at 1 and a south pole at 1′.

The preceding description of operation assumed that one coil was energized while the other was deenergized. Through proper design of the controlling network a half-step rotation can also be obtained. If coil 1–1′ remains energized while 2–2′ is energized, a CCW rotation of 18°/2 = 9° can be obtained. The degrees per step or half-step can be determined from the rotor construction if we first divide 360 by the number of rotor poles to obtain the angle between rotor poles:

$$\theta \text{ (between poles)} = \frac{360^\circ}{P_r \text{ (poles of rotor)}} \quad \textbf{(7.44)}$$

If we then divide by 2, we will have the angle between the north and south poles of the rotor:

$$\theta_p = \frac{\theta}{2} \quad \textbf{(7.45)}$$

Dividing by 2 again will provide the step angle:

$$\theta_s = \frac{\theta_p}{2} = \frac{\theta/2}{2} = \frac{\theta}{4} = \frac{360^\circ}{4P_r} \quad \textbf{(7.46)}$$

and

$$\theta_s = \frac{90°}{P_r}$$

For half-stepping,

$$\frac{\theta_s}{2} = \frac{45°}{P_r} \qquad \textbf{(7.47)}$$

Rewriting Eq. (7.46) as $P_r = 90°/\theta_s$, we find that P_r must be chosen such that P_r is a whole number. For $\theta_s = 18°$, $P_r = 90°/\theta_s = 90°/18° = 5$ (four stator poles):

$$\theta_s = 15° \qquad P_r = \frac{90°}{15°} = 6 \qquad \text{(4 stator poles)}$$

$$\theta_s = 9° \qquad P_r = \frac{90°}{9°} = 10 \qquad \text{(8 stator poles)}$$

$$\theta_s = 7.5° \qquad P_r = \frac{90°}{7.5°} = 12 \qquad \text{(10 stator poles)}$$

$$\theta_s = 1.8° \qquad P_r = \frac{90°}{1.8°} = 50 \qquad \text{(40 stator poles)}$$

Understandably, the controlling network must be quite sophisticated to provide the proper pulse and polarity for systems that may have as many as 40 stator poles. However, through proper logic design and integrated circuits (described in a later chapter), the size of such a controlling system can be minimized. The details of such a system are beyond the needs and scope of this book, but a growing list of references is now available on the subject.

There is a second type of stepper motor, called a *reluctance* type, that operates on much the same principle as the permanent-magnet type. The most obvious differences are the absence of a permanent magnet on the rotor and an increased number of stator poles. Instead, the ferromagnetic rotor appears the same as shown in Fig. 7.62, but it is not polarized and does not have the alternating north poles at the other end of the rotor. Additional stator coils establish rotation by ensuring that the reluctance path between stator and rotor is a minimum. Magnetic flux lines always strive to be as short as possible (take the path of least reluctance), thereby drawing magnetic materials together such as occurs when opposite poles of permanent magnets are placed close together. In Fig. 7.64, if coil 1–1′ is energized, the rotor will slip clockwise to lock in with the north and south poles to establish the shortest flux path for ϕ (path of least reluctance), causing the desired step in rotation.

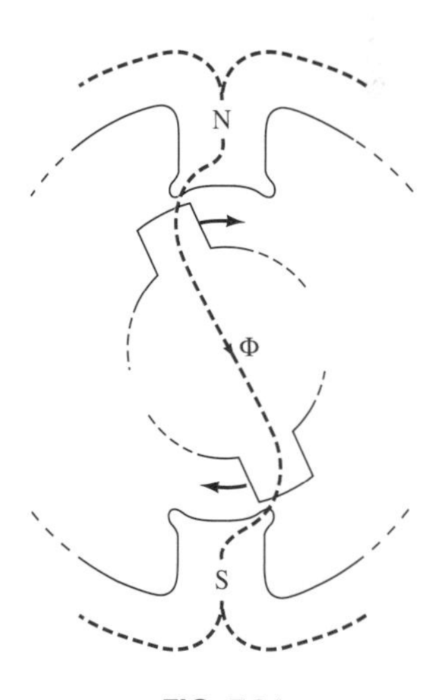

FIG. 7.64
Reluctance-type stepper motor.

7.9 SUMMARY

- **Generators** convert **mechanical energy into electrical energy,** while motors perform the reverse conversion.
- The **stationary** part of a dynamo is called the **stator;** the **rotating** member is called the **rotor** or armature.
- **Lap windings** on a dc generator have **many parallel paths** of induced voltage resulting in a low voltage output and high current rating, while wave-wound machines have a **high output voltage** and **lower current rating.**
- The **greater** the **speed** of the rotor or **magnitude** of a **magnetic flux** cut by the rotor the **greater** the induced voltage.
- The **voltage generated** by a dc generator is **always** more than the **terminal voltage** due to drop in voltage across the armature resistance.
- The **compound dc generator** has **two separate insulated windings** around the field poles. If the flux due to the series windings **aids** the main flux, the configuration is an **overcompound** one; if it **opposes** the main flux, it is **differential.**
- The **series dc generator** has a **constant-current characteristic** for a range in load voltage.
- **Current-carrying conductors** in a magnetic field experience a **force** in a direction determined by the **left-hand rule.**
- As the **rotor of a motor** turns, a voltage (called a **counter-emf**) is induced across the rotating conductors of the rotor that **opposes** the applied terminal voltage.
- An **increase** in the **magnetic field strength or current** through the conductors of the rotor will **increase** the torque of a dc motor.
- The **shunt dc motor** is considered a fairly **constant speed** dynamo.
- A **compound dc motor** offers a high level of **speed control** through the use of the series winding.
- The **torque** of a series dc motor **increases exponentially** with the **armature current,** but the **speed drops** dramatically with **increase** in available **torque.**
- An **ac generator or alternator** does not have the commutator section of a dc generator but rather has **continuous slip rings** for access to the generated voltage.
- In a **polyphase induction motor** the terminal voltage is applied to the **stator windings** as the rotor turns a mechanical load.
- The **difference** between the **synchronous speed** (stator flux rotational speed—the maximum possible) and the speed of the rotor is called the **slip.**
- A **polyphase synchronous motor** runs at **synchronous speed** by introducing a squirrel-cage structure and field poles in the rotor.
- The **most common motor** is the single-phase ac motor. Various types include the shunt, series, split-phase, capacitive-start, shaded-pole, and hysteresis.
- **Stepper motors** are used throughout the **computer industry** to provide quick access to particular sections of magnetic disks.

Equations:

Dc Generators:

$$e_c = \frac{P\phi S}{60} \times 10^{-8} \quad \text{(V)}$$

$$E_g = \left(\frac{Z}{a}\right)\frac{P\phi S}{60} \times 10^{-8} \quad \text{(V)}$$

$$E_g = K\phi S \text{ (V)}$$

$$P_{\max} = V_t I_L$$

$$\text{VR\%} = \frac{V_{NL} - V_{FL}}{V_{FL}} \times 100\%$$

$$V_t = E_g - I_A R_A$$

$$I_A = I_F + I_L$$

Compound Dc Generator:

$$I_A = I_F + I_L,\ I_S = I_A,\ I_F = \frac{V_t}{R_F}$$

$$V_t = E_g - I_A(R_A + R_S)$$

Series Dc Generator:

$$I_L = I_S = I_A$$

$$V_t = E_g - I_A(R_A + R_s)$$

Dc Motors:

$$F = 8.85 \times 10^{-8} \text{ BIL} \sin\theta$$

$$E_C = K\phi S \quad \text{(V)}$$

Shunt Dc Motor:

$$I_L = I_F + I_A$$

$$I_F = \frac{V_t}{R_F}$$

$$V_t = E_C + I_A R_A$$

$$T = Fr = K\phi I_A = \frac{7.04P}{S} \quad \text{(ft-lb)}$$

$$S = \frac{E_C}{K\phi} = \frac{V_t - I_A R_A}{K\phi} \quad \text{(rpm)}$$

Compound Dc Motor:

$$S = \frac{V_t - I_A(R_A + R_S)}{K(\phi_F \pm \phi_S)}$$

$$SR\% = \frac{S_{NL} - S_{FL}}{S_{FL}} \times 100\%$$

Series Dc Motor:

$$T = K_T I_A^2$$

$$S = \frac{V_t - I_A(R_A + R_S)}{K\phi S} \qquad \text{(rpm)}$$

Ac Generator:

$$f = \frac{PN}{120} \qquad \text{(Hz)}$$

$$E_C = 2.2\phi f \times 10^{-8} \qquad \text{(V)}$$

$$E_\phi = 2.22\frac{Z}{P}\phi f \times 10^{-8} \qquad \text{(V)}$$

$$E_g = \sqrt{(V_\phi + V_R)^2 + V_X^2} \qquad \text{(V)}$$

Ac Motors:

$$N_S = \frac{120f}{P}$$

$$T = K\phi I_C$$

$$S = N_S - N \qquad \text{(rpm)}$$

$$S\% = \frac{N_S - N}{N_S} \times 100\%$$

$$N = \frac{120f(1 - S)}{P}$$

$$T_{st} = KV_L^2$$

$$T_{\text{running}} = KS$$

Polyphase Synchronous Motor:

$$N = N_S = \frac{120f}{P}$$

Stepper Motor:

$$\theta \text{ (between poles)} = \frac{360°}{P_r \text{ (poles of rotor)}}$$

PROBLEMS

SECTION 7.2 Dc Generators

1. A four-pole dc generator with 200 conductors is driven at 3600 rpm. If the flux in pole is 100 kilolines, determine the average generated voltage if it is:
a. Lap-wound.
b. Wave-wound.

2. For the system in Problem 1, determine the voltage generated per conductor for:
a. Lap-wound.
b. Wave-wound.

3. If the power rating of the system in Problem 1 is 72 W, determine the current per conductor for:
   a. Lap-wound.
   b. Wave-wound.

4. A six-pole lap-wound dc generator generates 2 V per conductor at a speed of 1800 rpm. If the armature has 240 conductors, find the required flux per pole.

5. A dc generator develops 120 V at 1800 rpm. At what speed will it generate 140 V if all the other variables of the dynamo remain fixed?

6. If the flux (ϕ) of a 240-V, 2400-rpm generator is reduced by 25%, determine the new generated voltage. The speed remains fixed.

7. Determine the new generated voltage if the speed is doubled and the flux reduced by 20%. Before the change in conditions, the dc generator provided 120 V at 2400 rpm.

8. With no load applied, the output terminal voltage of a dc generator is 122 V. When fully loaded there is a 2.2-V drop across the internal resistances of the supply. What is the voltage regulation of the supply in percent?

9. A dc shunt generator ($R_F = 100\ \Omega$, $R_A = 2.5\ \Omega$) develops 120 V (E_g) when applied to a 20-Ω load. Determine:
   a. I_L, I_F, I_A.
   b. V_{R_A}.
   c. V_L.
   d. The power to the load.

10. For the cumulative compound dc generator in Fig. 7.10, where $R_F = 100\ \Omega$, $R_A = 1\ \Omega$, $R_S = 0.5\ \Omega$, $V_t = 126$ V, and $I_L = 4$ A, determine:
   a. I_F and I_A.
   b. V_{R_A} and V_{R_S}.
   c. E_g.
   d. The power to the load.
   e. The power generated.
   f. The power lost.

11. For a series dc generator with $R_S = 1\ \Omega$, $R_A = 2\ \Omega$, and $V_t = 60$ V that delivers 240 W to the load, determine:
   a. I_A, I_S, and I_L.
   b. E_g.

12. For a compound dc generator when $R_F = 120\ \Omega$, $R_A = 2\ \Omega$, $R_S = 1\ \Omega$, $V_t = 160$ V, and $R_L = 20\ \Omega$, determine the following for a flat response characteristic:
   a. I_F and I_A.
   b. E_g.
   c. P_g.
   d. P_L.
   e. $\eta\%$.

SECTION 7.3 Dc Motors

13. Determine the force on a conductor 6 in. long that is perpendicular to the field if the flux density is 50,000 lines/in.^{2} and the current is 6 A.

14. What is the effect on the counter-emf of a dc motor if the flux is increased by a factor of 2 and the speed is reduced by 5%?

15. The counter-emf generated by a dc shunt motor is 80 V at a speed of 1800 rpm and a flux of 120 kilolines. What is the counter-emf generated if the speed is 3600 rpm and the flux is reduced to 100 kilolines?

16. For the dc motor in Fig. 7.16, $R_F = 80\ \Omega$, $R_A = 2\ \Omega$, and the power supplied is 120 V at 6 A. Determine:
a. I_F and I_A.
b. The counter-emf.
c. The output power if the machine is 60% efficient.
d. The output torque if the speed is 1800 rpm.

17. For a dc motor with the nameplate data 2.4 hp, 120 V, 20 A, and 3600 rpm, determine:
a. The output power.
b. The input power.
c. The efficiency of operation.

18. Determine the torque on a 3-in.-diameter shaft of a dc shunt-wound motor if the force is 16 lb.

19. What is the torque developed by a 5-hp motor if the speed is:
a. 400 rpm.
b. 3600 rpm.

20. For the compound configuration in Fig. 7.20, $R_F = 100\ \Omega$, $R_A = 2\ \Omega$, $R_S = 0.5\ \Omega$, $E_C = 100$ V, and $V_t = 116$ V. Determine:
a. I_A and I_F.
b. I_L.
c. The input power.
d. The power out in horsepower if the machine is 70% efficient.

21. a. If the armature current is doubled, will the shunt, series, or compound differential machine have the greatest increase in torque?
b. Repeat part (a) for the effect on speed.

22. Determine the starting resistor necessary to limit the starting current to 150% of its rated 10-A level if the applied voltage is 120 V and $R_A = 2\ \Omega$.

23. For a series dc motor with $R_S = 1\ \Omega$, $R_A = 4\ \Omega$, $E_C = 106$ V, and $V_t = 120$ V, determine:
a. The output power in horsepower if the system is 72% efficient.
b. The resulting torque if the speed is 1800 rpm.
c. The resulting torque if the current is increased 100%.

SECTION 7.4 AC GENERATORS

24. Calculate the frequency generated by a six-pole alternator running at a speed of 1800 rpm.

25. Determine the line current and phase current of a three-phase ac generator with the following nameplate data: 5 kVA, 200 V, 1800 rpm, Y-connected, 60 Hz.

26. Determine the number of conductors per phase of a three-phase generator that can generate 199.88 V with $\phi = 3 \times 10^6$ lines and $f = 60$ Hz.

27. Determine the voltage generated per phase of a three-phase, Y-connected alternator if $R_\phi = 2\ \Omega$, $X_\phi = 3\ \Omega$, and the machine is rated 5 kVA at 208 V.

28. A three-phase, Y-connected alternator generates 180 V per phase. If the machine is rated 8 kVA at 212 V with $X_\phi = 4\ \Omega$, find R_ϕ if the load is resistive.

29. Repeat Problem 27 if the applied load is $\mathbf{Z}_L = 10 + j20$.

SECTION 7.5 POLYPHASE INDUCTION AC MOTOR

30. Calculate the number of poles of a 60-Hz polyphase induction motor that runs at a speed of 1200 rpm.

31. A polyphase induction motor has a slip of 2.5%. Determine its running speed if it has four poles and the applied frequency is 60 Hz.

32. Determine the new starting torque of a polyphase induction motor if the applied line voltage is increased by 200%.

33. What is the effect on the running torque if the slip drops to half of its initial value in the linear region (polyphase induction motor)?

SECTION 7.6 Polyphase Synchronous Motor

34. a. Determine the speed of a polyphase synchronous motor with four poles and an applied frequency of 60 Hz.
b. What is the slip in percent? (Think.)

35. In Fig. 7.43 the total three-phase load is 1200 kW at a lagging F_p of 0.6.
a. Sketch the power triangle of the load.
b. Determine the apparent power rating of a synchronous motor that will result in a unity power factor load.

SECTION 7.7 Single-Phase Ac Motors

36. a. List four types of single-phase ac motors and briefly describe (in your own words) their mode of operation.
b. Which of these has a constant-speed characteristic?

37. For the split-phase single-phase ac motor in Fig. 7.51, if $R_F = 10\ \Omega$, $X_{L_F} = 100\ \Omega$, $R_p = 60\ \Omega$, and $X_{L_f} = 5\ \Omega$, determine the quadrature components I'_p and I'_F in Fig. 7.52 if $\mathbf{V}_g = 120\ \text{V}\ \angle 0°$.

38. For the capacitive-start induction motor in Fig. 7.56, if $R_F = 60\ \Omega$, $X_{L_F} = 100\ \Omega$, $X_{C_P} = 120\ \Omega$, $X_{L_p} = 40\ \Omega$, and $R_p = 20\ \Omega$, determine the magnitude of the current I'_p that leads I_F by 90° if $\mathbf{V}_g = 120\ \text{V}\ \angle 0°$.

SECTION 7.8 Stepper Motors

39. a. Sketch the appearance of a PM stepper motor (like the one in Fig. 7.62) if $\theta_s = 30°$ and the number of stator poles is four.
b. Briefly describe how a CCW rotation of 30° and then 60° can be obtained (including the polarities of the stator poles).

40. If $\theta_s/2 = 0.9°$, how many poles must the rotor have?

41. Is $\theta_s = 12°$ a possibility for a *standard* PM stopper motor design? Why not?

8 Two-Terminal Electronic Devices

8.1 INTRODUCTION

In recent years the advertising media have added a touch of glamor to the electronics industry through their exposure of the new advances in miniaturization. Although laypersons may not be totally familiar with the terminal characteristics of the device, they have become familiar with some of the terminology and how it has effectively reduced the size of electronic systems from the calculator to the large-memory computer.

In this chapter a host of two-terminal semiconductor electronic devices will be introduced along with a variety of applications. The term *semiconductor* refers to those materials that have characteristics somewhere between those of a conductor and an insulator. The mathematics and physics of each device that determine its terminal behavior will not be covered in detail. Rather, a surface understanding of its operation and function will be the goal. Approximations will be used throughout the discussion to minimize the complexity that often surrounds a basically simple device when the complete complex mathematical relationships of the device are employed.

8.2 SEMICONDUCTOR DIODE CHARACTERISTICS

The *diode* is a two-terminal device that ideally behaves like an ordinary switch with the special condition that it can conduct in only one direction. It has an *on* state, in which it appears on an ideal basis to be simply a *short circuit* between its terminals, and an *off* state, in which its terminal characteristics are similar to those of an *open circuit*. The ideal characteristics of the device appear in Fig. 8.1 with the symbol for the semiconductor device. Note that for positive values of V_D ($V_D > 0$ V) the diode is in the short-circuit state ($R = 0\ \Omega$), and the current through it is limited by the network in which it appears. For the opposite polarity ($V_D < 0$ V), the diode is in the open-circuit state ($R = \infty\ \Omega$), and $I_D =$ 0 mA. For the semiconductor element, the diode current (conventional

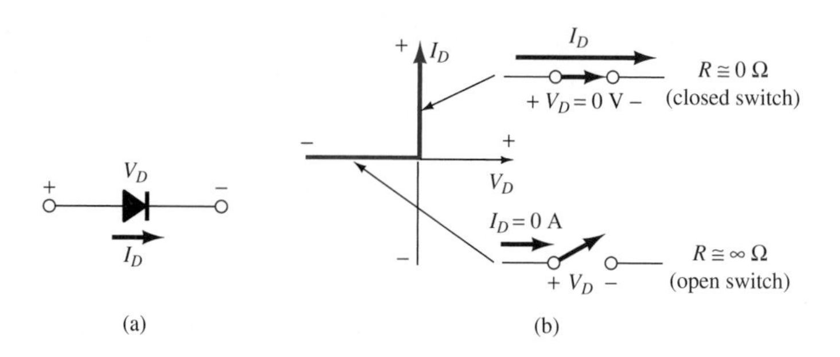

FIG. 8.1
Diodes: (a) symbol; (b) characteristics of the ideal device.

flow) can pass through the device only *in the direction of the arrow appearing in the symbol.*

The construction of the semiconductor diode appears in Fig. 8.2. Two semiconductor materials of opposite charge content are placed side by side. The *n*-type material is a semiconductor material such as *silicon* or *germanium* heavily *doped* with negative charges (electrons) as shown in the figure. The *p*-type material is the same semiconductor material heavily doped with other elements so that it will have an excess of positive regions called *holes*. The holes are in actuality the absence of a negative charge (electron). When the *n*-type and *p*-type materials are brought together in the construction process, a layer of positive ions is established at the boundary in the *n*-type material and a layer of negative ions at the boundary in the *p*-type material. When a forward-bias (a term for the application of a dc voltage that establishes the conduction state) voltage is applied as indicated by the polarities in Fig. 8.2, the ionic region at the boundary is reduced, and the negative carriers in the *n*-type material can overcome the remaining negative wall of positive ions and continue through to the positive applied potential. A similar statement of transition can be applied to the positive carriers of the *p*-type material. The region near the boundary defined by the positive and negative ions is referred to as the *depletion* region because of the lack of free carriers. Conduction is therefore established through the device, and, ideally, the short-circuit state is established.

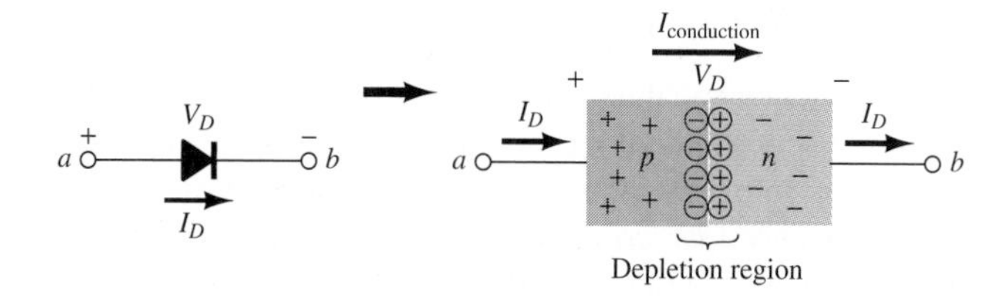

FIG. 8.2
Construction and forward-bias states of the semiconductor diode.

For the reverse polarity across the diode the negative charges of the *n*-type material are attracted to the positive potential, and the positive holes of the p-type material to the negative potential. The result is an increased depletion region between the n- and p-type materials that results in very few "free" charge carriers. Thus, the two possible states of a semiconductor diode have been described in a very superficial manner. The details of the construction and operation of this two-terminal device require many pages of a text devoted to electronic devices and systems.

If the actual characteristics of each device were perfectly ideal, the application of each in the practical world would be greatly simplified. However, as shown in Fig. 8.3, the actual characteristics of the device

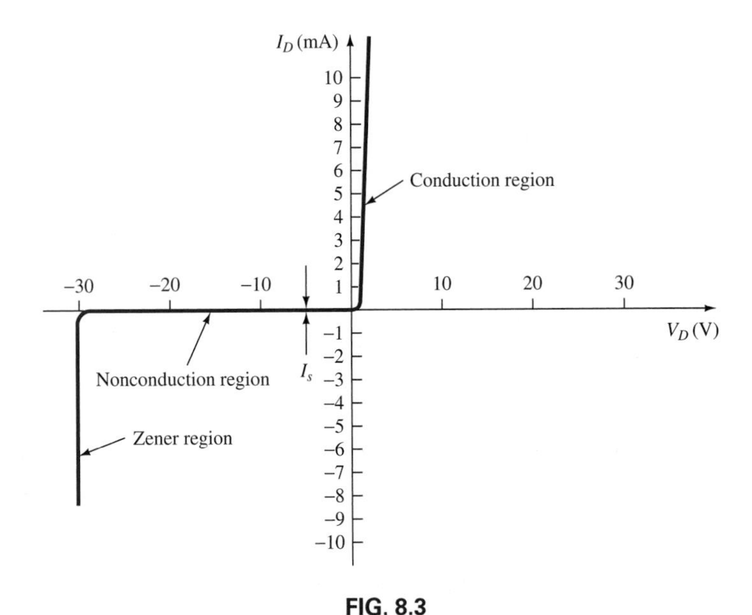

FIG. 8.3
Diode characteristics (continuous scale).

are not ideal, although it is also immediately obvious that except for large negative voltages the match with Fig. 8.1(b) is excellent. A change in scale, as shown in Fig. 8.4, permits a closer examination of the actual characteristics in the forward-bias region.

The fact that the characteristics are a curved line reveals that the resistance of the device changes from point to point. The curve of a fixed resistance value is a straight line through the origin. The steeper the slope of the curve, the less the terminal resistance of the device approaches the ideal 0 Ω at higher levels of diode current. The ac resistance of the diode in a specific region can be found using the equation

$$r_{ac} = r_d = \frac{\Delta V_D}{\Delta I_D} \tag{8.1}$$

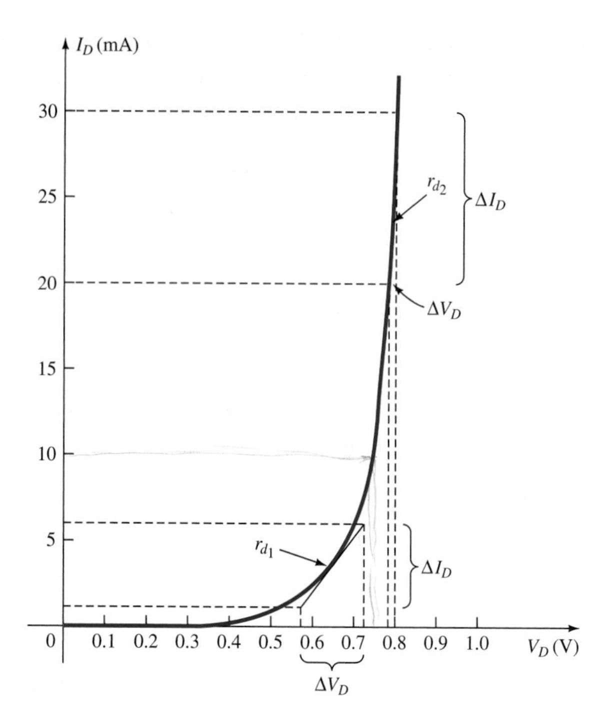

FIG. 8.4

Application of Eq. (8.1) to the characteristics in Fig. 8.4 for the regions indicated results in

$$r_{d_1} = \frac{\Delta V_D}{\Delta I_D} \cong \frac{(0.72 - 0.57)\text{V}}{(6 - 2)\text{ mA}} = \frac{0.15\text{ V}}{4\text{ mA}} = \mathbf{37.5\ \Omega}$$

$$r_{d_2} = \frac{\Delta V_D}{\Delta I_D} \cong \frac{(0.8 - 0.78)\text{ V}}{(30 - 20)\text{ mA}} = \frac{0.02\text{ V}}{10\text{ mA}} = \mathbf{2\ \Omega}$$

Note, as indicated earlier, that the resistance of the characteristics increases as the curve becomes more and more horizontal. Since for the vast majority of characteristic curves for electronic devices the current is plotted on the vertical axis and the voltage, on the horizontal axis, this generalization can be applied to the majority of characteristic curves.

It should be fairly obvious from the care with which Eq. (8.1) was applied that the ac resistance can be determined in this manner only from an expanded set of characteristics and a careful estimate of intersections with the axis. Fortunately, however, there is an alternative approach to determining the ac resistance of a diode that is more direct and provides results of practical value.

The method is based on the fact that

The derivative of a function at a point is equal to the slope of a tangent line drawn at that point.

If we could therefore find the derivative dV/dI at the point of interest on the curve in Fig. 8.4, we would have the ac resistance at that point, since $r_d = dV_d/dI_d$.

For the diode characteristics in Fig. 8.5, a general equation for the current at any particular voltage level except in the Zener region is given by

$$I = I_s(e^{kV/T_K} - 1) \tag{8.2}$$

where I_s = reverse saturation current (the diode current when reverse-biased)

$k = 11{,}600/\eta$, with $\eta = 1$ for Ge and 2 for Si for low levels of I_D and $\eta = 1$ for Ge and Si for higher levels of I_D

$T_K = T_C + 273°$ (T_K = kelvins, T_C = degrees Celsius)

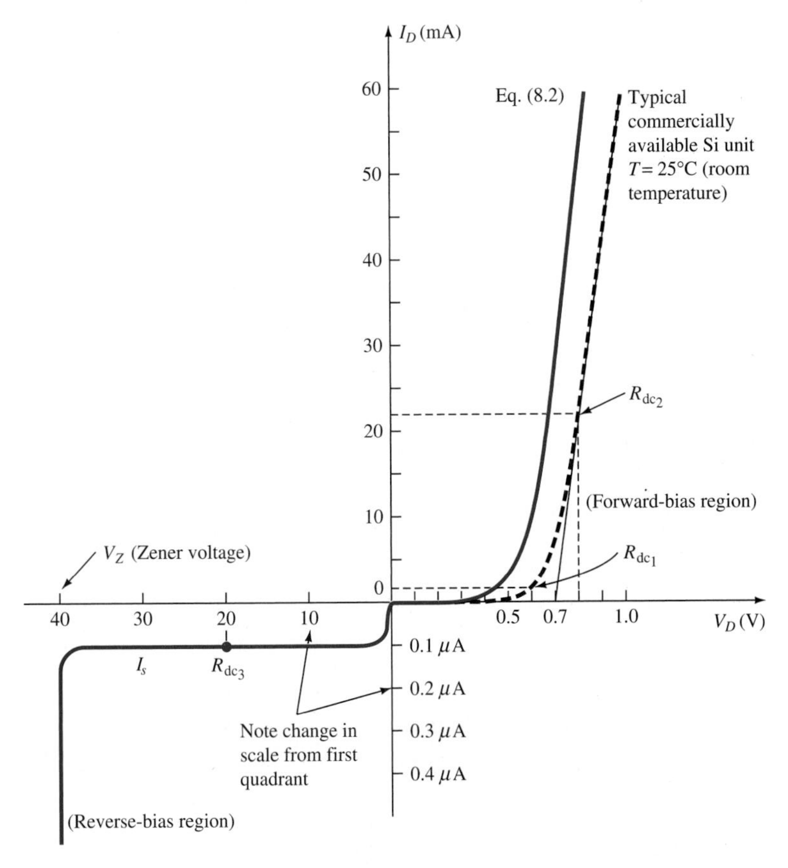

FIG. 8.5
Semiconductor diode (Si) characteristics.

If we find the derivative of the general equation for the semiconductor diode with respect to the applied forward bias and then invert the result, we will have an equation for the dynamic or ac resistance in that region.

That is, taking the derivative of Eq. (8.2) with respect to the applied bias results in

$$\frac{d}{dV} I = \frac{d}{dV} [I_s(e^{kV/T_K} - 1)]$$

and

$$\frac{dI}{dV} = \frac{k}{T_K}(I + I_s)$$

following a few basic operations of differential calculus. In general, $I \gg I_s$, and

$$\frac{dI}{dV} = \frac{k}{T_K} I$$

Substituting $\eta = 1$ for Ge and Si in the vertical-rise section of the characteristics, we obtain

$$k = \frac{11{,}600}{\eta} = \frac{11{,}600}{1} = 11{,}600$$

and at room temperature,

$$T_K = T_C + 273° = 25° + 273° = 298°$$

so that

$$\frac{k}{T_K} = \frac{11{,}600}{298} \cong 38.93$$

and

$$\frac{dI}{dV} = 38.93I$$

Inverting the result to define a resistance ratio ($R = V/I$) gives us

$$\frac{dV}{dI} \cong \frac{0.026}{I}$$

or

$$r_d = \frac{dV}{dI} = \frac{26 \text{ mV}}{I_D (\text{mA})} \quad \text{(ohms) (Ge, Si)} \qquad \textbf{(8.3)}$$

The significance of Eq. (8.3) must be clearly understood. It implies that the dynamic or ac resistance can be found by simply substituting the quiescent (dc) value of the diode current into the equation. There is no need to have the characteristics available or to worry about sketching tangent lines as defined by Eq. (8.1).

Although Eq. (8.3) is usually applied to obtain a working level for r_d, it fails to include the nonideal effects of the *bulk* resistance of the semiconductor device (simply the resistance of the semiconductor material) or the *contact* resistance due to the resistance introduced when the leads are attached. The effect of the nonideal resistive elements is clearly shown in Fig. 8.5. It shifts the curve to the right, since $v = iR$, and for the same i an increase in R increases the levels of v.

Equation 8.3 can be modified as follows to include these effects:

$$r'_d = \frac{26 \text{ mV}}{I_D \text{ (mA)}} + r_B \qquad \text{(ohms)} \qquad \textbf{(8.4)}$$

with r_B typically 0.1 Ω for high-power devices and 2 Ω for some low-power general-purpose diodes. As construction techniques improve, this additional factor will continue to decrease in importance until it can be completely ignored. For the purposes of this text and the accuracy desired, Eq. (8.3) will be employed whenever the ac resistance of a diode is desired. In most cases the resistances in series with the diode are sufficiently large to completely ignore the effects of r_B.

Before we leave the subject of resistance, it is often important to know the dc resistance (insensitivity to the slope of the curve) of a device at a particular operating point. The *dc* or *static* resistance of a diode is determined by

$$R_{\text{dc}} = \frac{V_D}{I_D} \qquad \text{(ohms)} \qquad \textbf{(8.5)}$$

At $V_D = 0.6$ V in Fig. 8.5 (a commercially available unit),

$$R_{\text{dc}_1} = \frac{V_D}{I_D} = \frac{0.6 \text{ V}}{2 \text{ mA}} = 0.3 \times 10^3 \ \Omega = \mathbf{300 \ \Omega}$$

while at $V_D = 0.8$ V in Fig. 8.5,

$$R_{\text{dc}_2} = \frac{V_D}{I_D} = \frac{0.8 \text{ V}}{22 \text{ mA}} = 0.036 \times 10^3 \ \Omega = \mathbf{36 \ \Omega}$$

and at $V_D = -20$ V,

$$R_{\text{dc}_3} = \frac{V_D}{I_D} = \frac{20 \text{ V}}{1 \ \mu\text{A}} = 20 \times 10^{+6} \ \Omega = \mathbf{20 \ M\Omega}$$

Note again how lower current levels correspond with increasing resistance levels for the semiconductor diode.

In summary, keep in mind that the static or dc resistance of a diode is determined solely by the point of operation, while the dynamic resistance is determined by the shape of the curve in the region of interest.

Note in Fig. 8.5 that the vertical rise in the characteristics does not occur until the forward-bias potential is about 0.7 V. For germanium diodes this potential is typically about 0.3 V. It might now be asked why germanium diodes are not used exclusively, since they appear to be closer to the ideal. The answer is that each device has a limited range of operation dependent on its construction and the materials used. In other words, for each diode there is a maximum current that can pass through the device in its conduction state before it reaches its burnout state and its characteristics are destroyed. This limiting current is specified for each manufactured device. It has been found that silicon material can handle much higher currents and power levels than can germanium, so silicon is used in much greater quantities even though its characteristics may appear less "ideal." The offshoot voltage indicates that the diode must be forward-biased more than this voltage before reaching the conduction state.

In the reverse-bias region in Fig. 8.5 note that the current is not 0 A, as required by the ideal situation. However, this leakage current, due to the minority carriers (electrons in the *p*-type material and holes in the *n*-type material) is usually quite small compared with the operating current level of the network and can be ignored. The sharp vertical drop in the characteristics at V_Z (called the *Zener potential*) is due to a breakdown in the ionic barrier at the junction at high applied potentials. In rather simplistic terms, the carriers in the *n*- and *p*-type materials are able to over-come the ionic barrier at the junction through their attraction to the high potentials applied at the opposite ends of the device. More will be said about the breakdown region in the paragraphs and sections that follow.

Through proper design the Zener region can usually be avoided and the reverse-bias current I_s ignored. The result is the straight-line approximation in Fig. 8.6(b) for the silicon diode in Fig. 8.6(a). If the diode resistance is sufficiently small compared with the other series resistive elements of the network, the slope of the forward-bias region in Fig. 8.6(b) can be ignored, resulting in the practical diode equivalent in Fig. 8.6(c). A circuit equivalent for the diode characteristics in Fig. 8.6(c) is provided in Fig. 8.7(b).

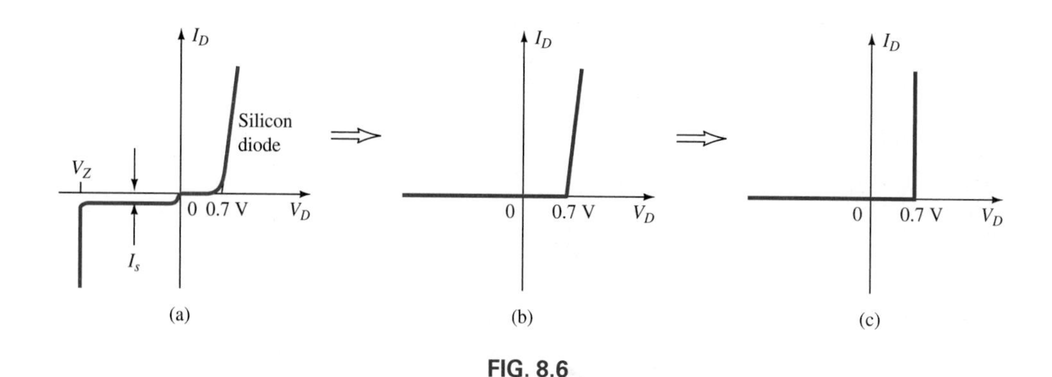

FIG. 8.6

FIG. 8.7
Diode practical equivalent.

It is important to understand that the 0.7-V battery in Fig. 8.7(b) is not an internal source of energy. Placing a voltmeter across an isolated diode would not result in a reading of 0.7 V. However, the circuit equivalent in Fig. 8.7(b) reveals that *at least* 0.7 V must be applied across the diode with the proper polarity (forward bias) before conduction is established. Any forward-bias voltage of less than 0.7 V will result in a nonconduction mode. In other words, the 0.7 V is the price that has to be paid to turn the device on.

A quick check of the condition of a diode can be made with most digital multimeters using a diode checking scale. A reading in the vicinity of 0.7 V (for silicon diodes) in the conduction direction (leads applied in a specific manner), and an O.L. indication in the other normally reveal a good device. Any other indication or combination reveals an open or shorted diode.

The ohmmeter section of a VOM can also be used to determine a diode's condition using the small dc battery voltage required in the ohmmeter design. Placement of the leads across the device in one direction should result in a high sensing current and a high ohmmeter indication (low resistance), while the reverse will indicate the opposite. A high or very low current (indication) for both directions indicates a damaged element. The maximum charge that can flow through the ohmmeter circuit even with the two leads shorted together is usually not high enough to be of concern in terms of damaging the diode during the measurement of the forward-bias state.

An instrument called a *curve tracer* can display the characteristics of the diode on a screen for comparison and checking purposes. A well-designed commercially available unit appears in Fig. 8.8. The characteristics on the scope are those of a transistor, to be described in a later section.

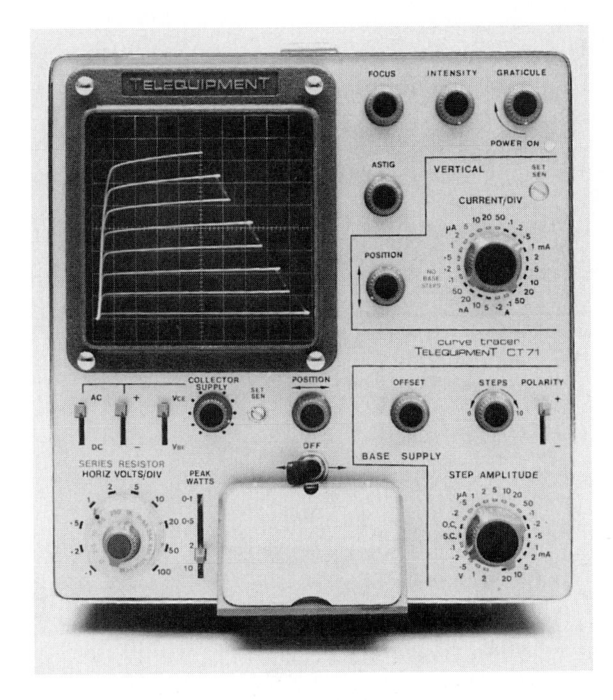

FIG. 8.8
Curve tracer. (Courtesy of Telequipment, Inc.)

As indicated earlier, there is a specific current rating for each diode that must not be exceeded. There are some semiconductor devices that can handle hundreds of amperes of current but require the use of heat sinks such as shown in Fig. 8.21(b) to draw the heat away from the device. A number of semiconductor diodes of different sizes and ratings appear in Fig. 8.9. The direction of diode conduction can be determined by a dot, band, small diode symbol, and so on, appearing on the casing of the device, as shown in Fig. 8.10.

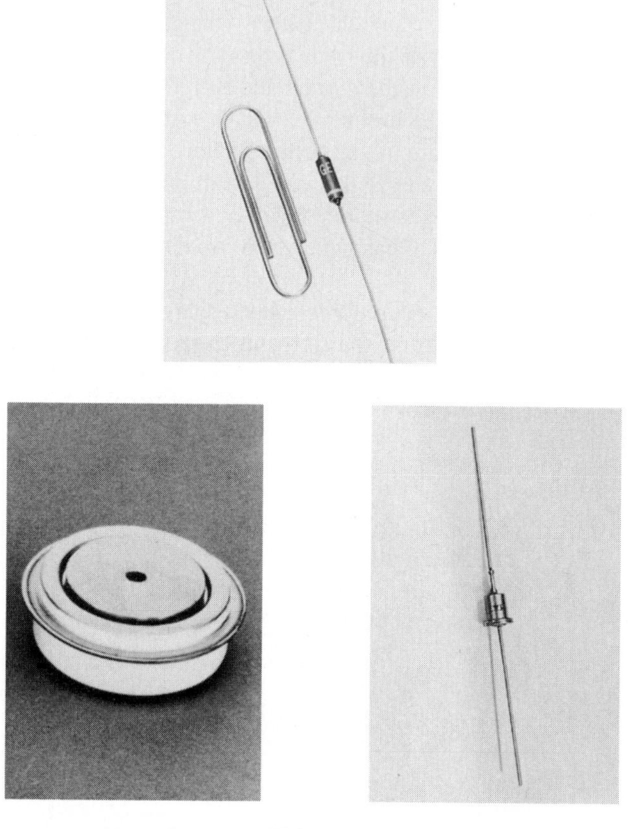

FIG. 8.9
Semiconductor diodes.

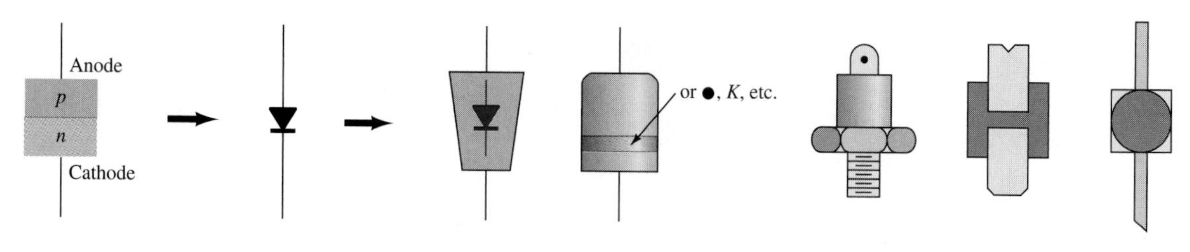

FIG. 8.10
Semiconductor diode notation.

8.3 SERIES DIODE CONFIGURATIONS WITH DC INPUTS

For the analysis to follow both the ideal and approximate models of the diode will be utilized. The symbol for the ideal diode will appear as shown in Fig. 8.11 with its characteristics. Keep in mind that the ideal diode is *on* if the network current has the same direction as the arrow in the symbol, or if the voltage across the diode has the polarity appearing in Fig. 8.11. If the current has the reverse direction or the voltage the opposite polarity, the diode is *off.*

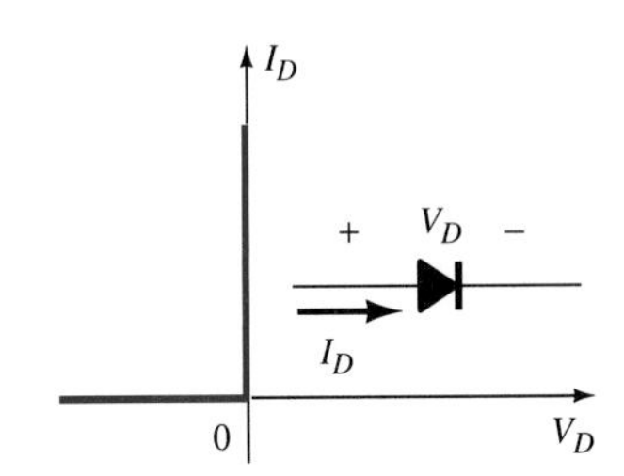

FIG. 8.11
Ideal diode and its characteristics.

The symbol for the approximate or practical model will appear as shown in Fig. 8.12 with the simple addition of Si. In this case be aware that the voltage across the diode must be at least 0.7 V with the polarity shown to be *on,* and the only direction of conduction continues to be in the direction of the arrow in the symbol. Any voltage less than 0.7 V across the diode, such as 0.5 V or −2 V, will result in the *off* condition.

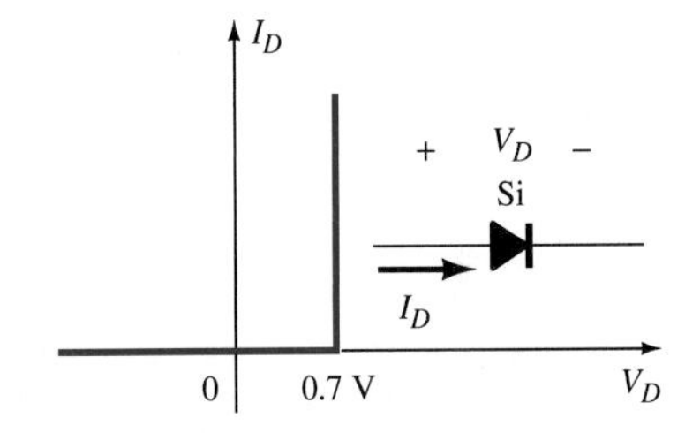

FIG. 8.12
Approximate characteristics and symbol for a silicon diode.

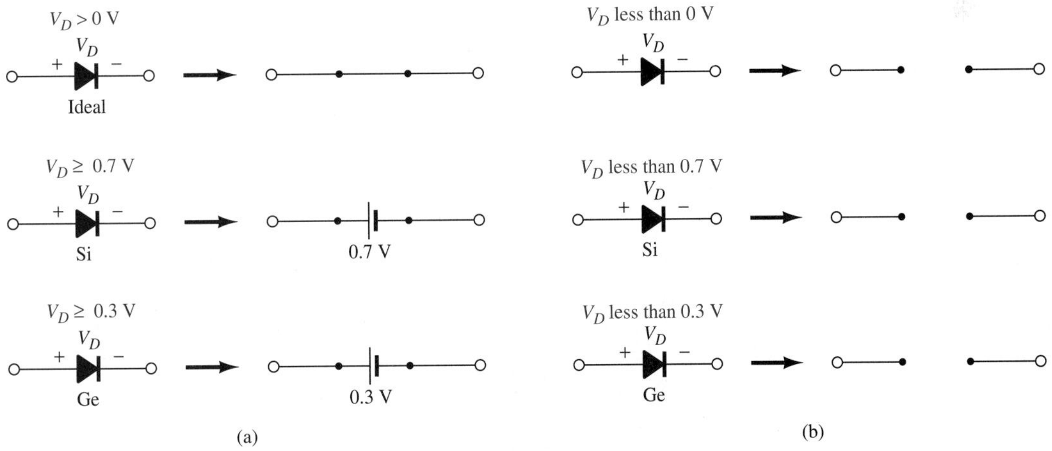

For the ideal diode the *on* condition will result in a short-circuit equivalent as shown in Fig. 8.13(a), with the *off* state resulting in the open-circuit equivalent as shown in the same figure. For the practical equivalent the *on* condition can be represented simply by including a 0.7-V drop across the diode or replacing it with a small battery of 0.7 V as shown in Fig. 8.13(b). Again, the *off* condition is represented by the open-circuit equivalent. For the practical diode always keep in mind that the 0.7 V is the price that must be paid to turn on the diode. The drop is always working against the source trying to turn the diode on. In addition, keep in mind that the 0.7 V appearing in the equivalent model is not an *independent* source of energy. An isolated diode on a table will not register 0.7 V if a voltmeter is connected across its terminals. Finally, the germanium diode has characteristics similar to those of the practical silicon diode except that the turn-on voltage is 0.3 V.

$V_D > 0$ V — Ideal
$V_D \geq 0.7$ V — Si — 0.7 V
$V_D \geq 0.3$ V — Ge — 0.3 V
(a)

V_D less than 0 V
V_D less than 0.7 V — Si
V_D less than 0.3 V — Ge
(b)

FIG. 8.13
On and off conditions of ideal and approximate diode models.

The magnitude and polarity of the applied forward bias has a pronounced effect on the behavior of the diode. However, what about the resulting current level? For the open-circuit situation, it is obviously 0 A. When the short-circuit state is appropriate, the *current is determined by the network to which the diode is connected.* Of course, it must be less than the maximum rating of the device, but ideally a forward-biased diode will have a fixed voltage drop across it (0 V for the ideal, 0.7 V for Si, and 0.3 V for Ge), with a current determined by the surrounding network.

Let us begin by determining the various voltages and the current level for the series dc configuration in Fig. 8.14. The first question of major importance involves the state of the diode. Should the short-circuit or the open-circuit state be assumed? For most situations, simply removing the diode from the picture entirely, as shown in Fig. 8.15, and determining the direction of the resulting conventional current provides the hint needed regarding the diode state. If the current direction has the same direction as the arrow in the diode symbol, the diode will be in the *on* state as long as the applied voltage is greater than the V_T back-bias voltage (0.7 V) of the diode.

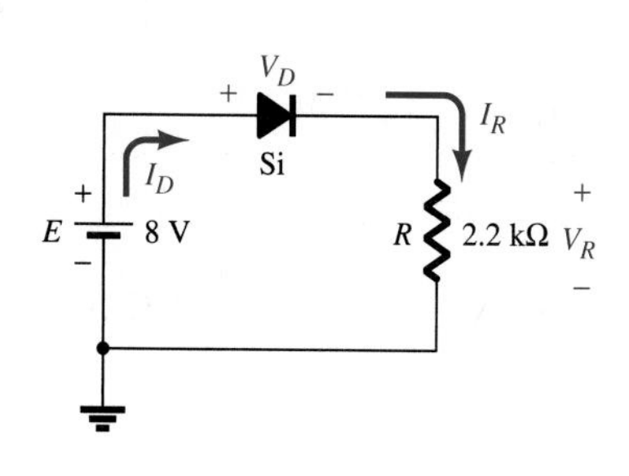

FIG. 8.14
Series diode configuration.

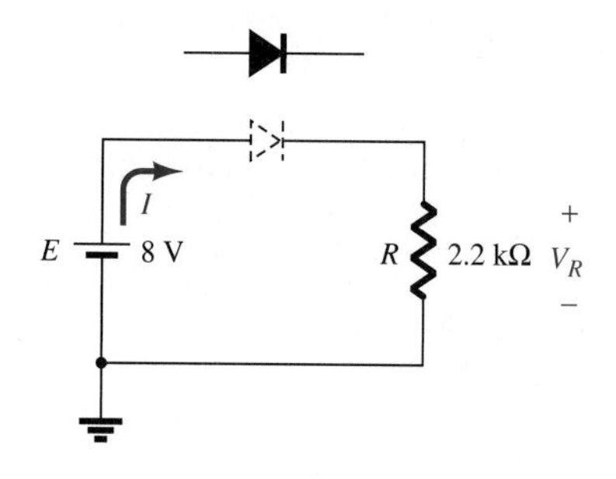

FIG. 8.15
Determining the state of the diode.

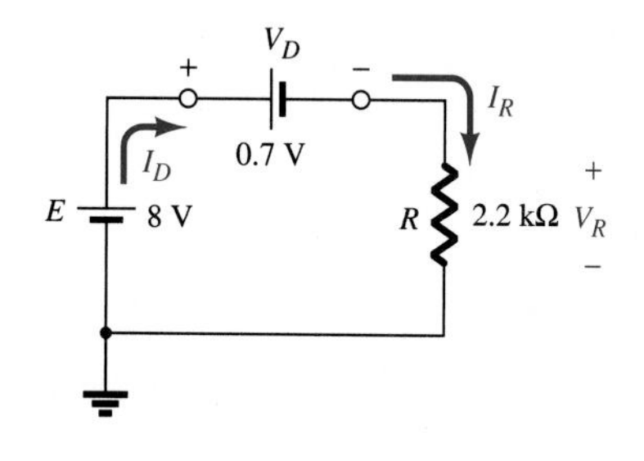

FIG. 8.16
Substituting the on*-state equivalent for the Si diode.*

Since $E > V_T$ and the established current has the same direction as the arrow in the diode symbol, the diode is *on* and the equivalent network in Fig. 8.16 is the result.

Clearly,

$$V_D = V_T = \mathbf{0.7\ V}$$

$$V_R = E - V_T = 8\ \text{V} - 0.7\ \text{V} = \mathbf{7.3\ V}$$

$$\text{and } I_D = I_R = \frac{V_R}{R} = \frac{7.3\ \text{V}}{2.2\ \text{k}\Omega} \cong \mathbf{3.32\ mA}$$

The point of operation on the approximate characteristics is shown in Fig. 8.17 with a dashed line representing the possible actual characteristics. Note that for any diode current of the approximate model the voltage V_D will be 0.7 V.

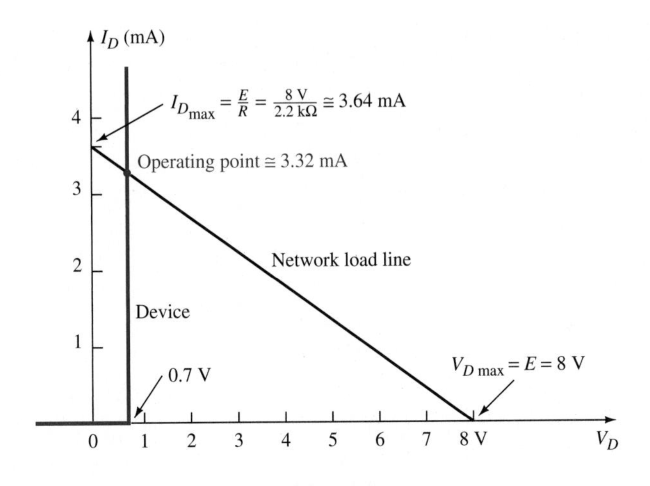

FIG. 8.17
Point of operation for the circuit in Fig. 8.16.

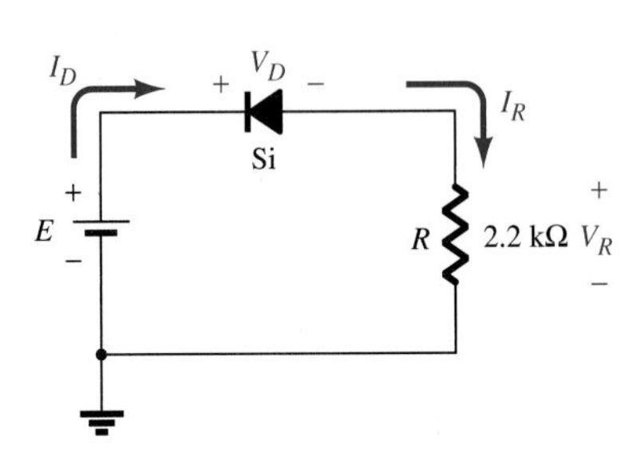

FIG. 8.18
Series diode circuit with reverse-biased diode.

Let us now repeat the analysis of Fig. 8.14 with the diode reversed as shown in Fig. 8.18.

Removing the diode, we find that the direction of I in Fig. 8.19 is opposite the arrow in the diode symbol, and the diode equivalent is the open circuit no matter which model is employed. The result is the network in Fig. 8.20, where $I_D = 0$ A due to the open circuit. Since $V_R = I_R R$, $V_R = (0)R = 0$ V. Applying Kirchhoff's voltage law around the closed loop, we have

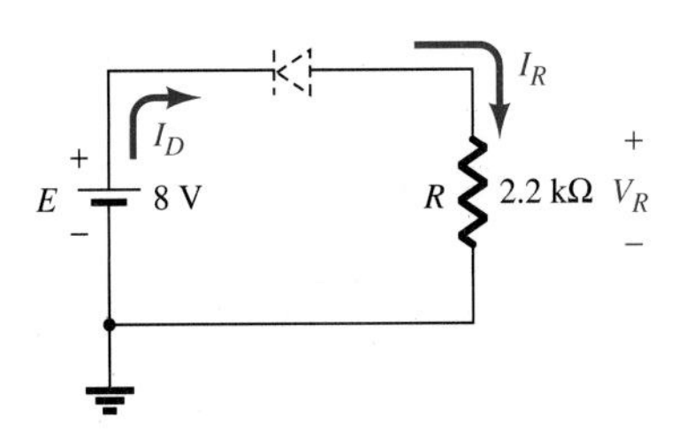

FIG. 8.19

$$E - V_D - V_R = 0$$

and

$$V_D = E - V_R = E - 0 = E = \mathbf{8\ V}$$

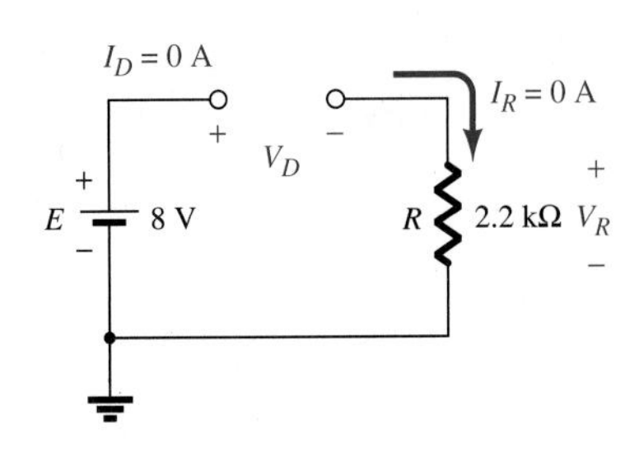

FIG. 8.20

In particular, note the high voltage across the diode even though it is in an *off* state. The current is zero, but the voltage is significant. For review purposes, keep the following in mind for the analysis to follow:

1. *An open circuit can have any voltage across its terminals, but the current is always 0 A.*
2. *A short circuit has a 0-V drop across its terminals, but the current is limited only by the surrounding network.*

In some of the examples to follow, the notation in Fig. 8.21 will be employed for the applied voltage. It is a common industry notation and one with which you should become very familiar.

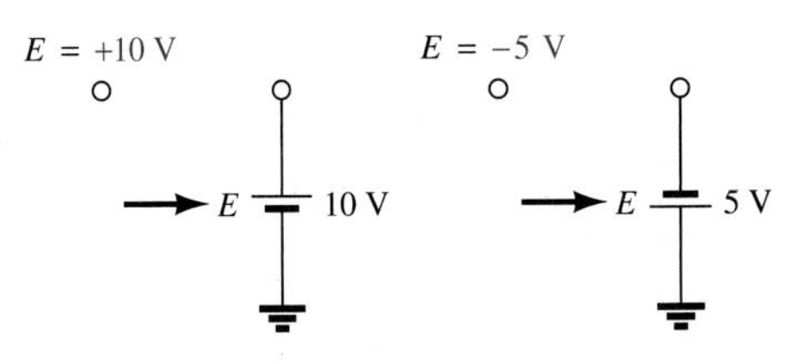

FIG. 8.21
Source notation.

EXAMPLE 8.1 Determine V_o and I_D for the series circuit in Fig. 8.22.

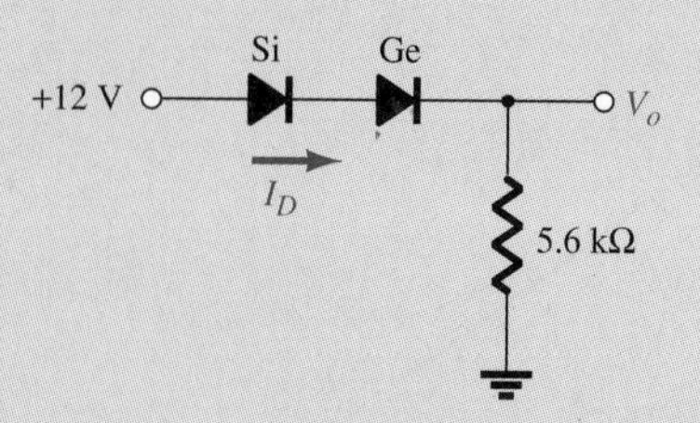

FIG. 8.22

Solution: The applied voltage ("pressure") is such that the resulting current has the same direction as the arrowheads in the symbols for both diodes, and the network in Fig. 8.23 results because $E = 12\text{ V} > (0.7\text{ V} + 0.3\text{ V}) = 1.0\text{ V}$. Note the redrawn supply of 12 V and the polarity of V_o across the 5.6-kΩ resistor. The resulting voltage

$$V_o = E - V_{T_1} - V_{T_2} = 12\text{ V} - 0.7\text{ V} - 0.3\text{ V} = \mathbf{11.0\text{ V}}$$

and

$$I_D = I_R = \frac{V_R}{R} = \frac{V_o}{R} = \frac{11\text{ V}}{5.6\text{ k}\Omega} \cong \mathbf{1.96\text{ mA}}$$

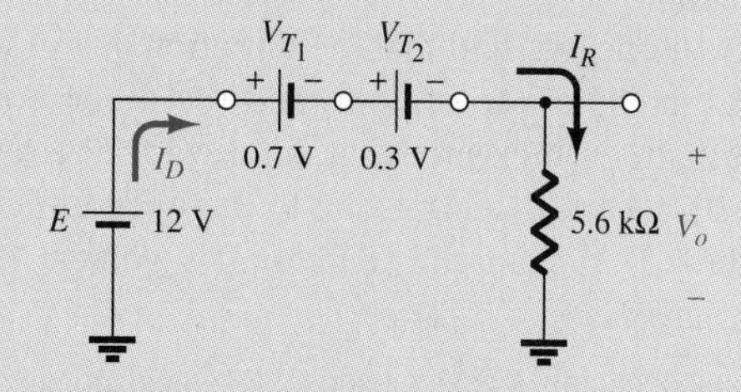

FIG. 8.23

EXAMPLE 8.2 Determine I, V_1, V_2, and V_o for the series dc configuration in Fig. 8.24.

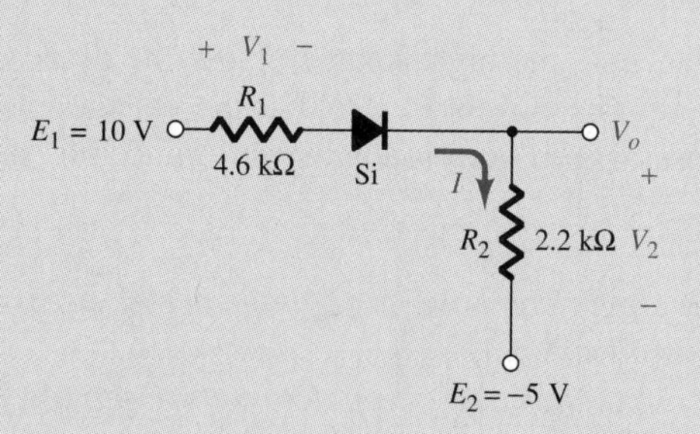

FIG. 8.24

Solution: The sources are drawn and the current direction determined as indicated in Fig. 8.25. The diode is in the *on* state and the alternative indication of an *on* diode is applied in Fig. 8.26.

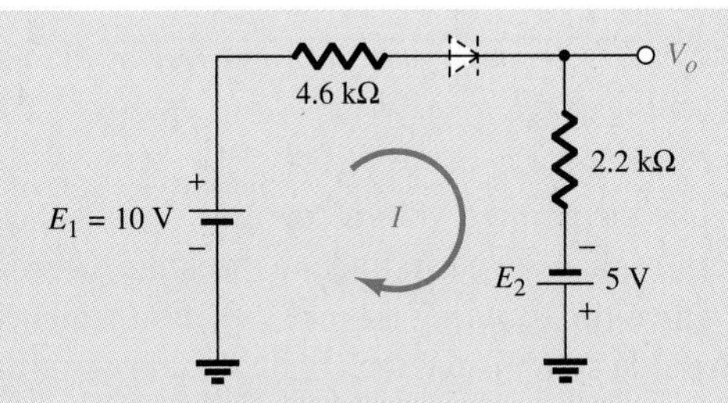

FIG. 8.25

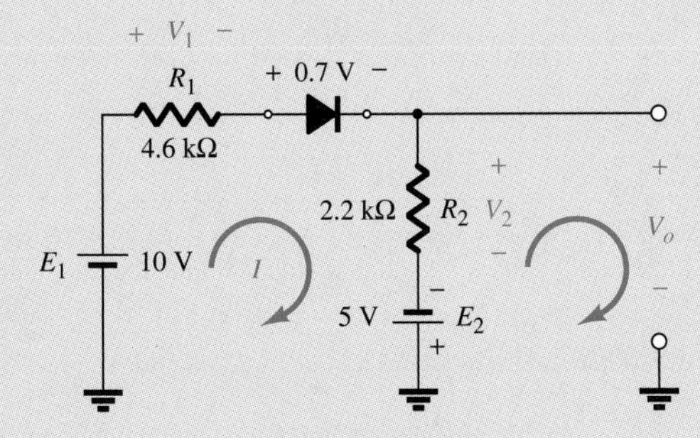

FIG. 8.26

$$I = \frac{E_1 + E_2 - V_D}{R_1 + R_2} = \frac{10\text{ V} + 5\text{ V} - 0.7\text{ V}}{4.6\text{ k}\Omega + 2.2\text{ k}\Omega} = \frac{14.3\text{ V}}{6.8\text{ k}\Omega}$$

$$\cong \mathbf{2.1\text{ mA}}$$

$$V_1 = IR_1 = (2.1\text{ mA})(4.6\text{ k}\Omega) = \mathbf{9.66\text{ V}}$$

$$V_2 = IR_2 = (2.1\text{ mA})(2.2\text{ k}\Omega) = \mathbf{4.62\text{ V}}$$

Applying Kirchhoff's voltage law to the output section in the clockwise direction, we have

$$-E_2 + V_2 - V_o = 0$$

and

$$V_o = V_2 - E_2 = 4.62\text{ V} - 5\text{ V} = \mathbf{-0.38\text{ V}}$$

The minus sign indicates that V_o has a polarity opposite that appearing in Fig. 8.26.

In this example it is particularly important to note that V_o includes the voltage V_2 and E_2 and is not just the voltage across R_2.

8.4 PARALLEL AND SERIES–PARALLEL CONFIGURATIONS

The methods applied in Section 8.3 can be extended to the analysis of parallel and series–parallel configurations. For each area of application, simply match the sequential series of steps applied to series diode configurations.

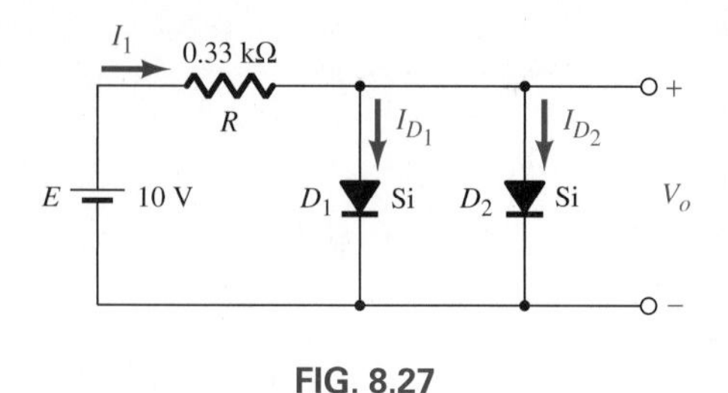

FIG. 8.27

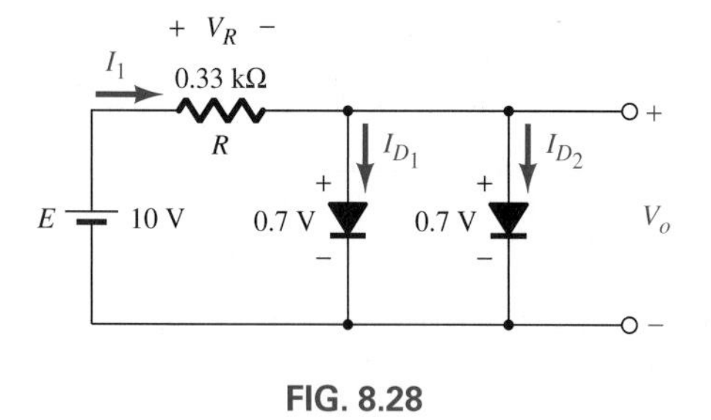

FIG. 8.28

EXAMPLE 8.3 Determine V_o, I_1, I_{D_1}, and I_{D_2} for the parallel diode configuration in Fig. 8.27.

Solution: For the applied voltage the pressure of the source is to establish a current through each diode in the same direction as shown in Fig. 8.28. Since the resulting current direction matches that of the arrow in each diode symbol and the applied voltage is greater than 0.7 V, both diodes are in the *on* state. The voltage across parallel elements is always the same, and

$$V_o = \mathbf{0.7\ V}$$

The current is

$$I_1 = \frac{V_R}{R} = \frac{E - V_D}{R} = \frac{10\text{ V} - 0.7\text{ V}}{0.33\text{ k}\Omega} = \mathbf{28.18\ mA}$$

Assuming diodes of similar characteristics, we have

$$I_{D_1} = I_{D_2} = \frac{I_1}{2} = \frac{28.18\text{ mA}}{2} = \mathbf{14.09\ mA}$$

Example 8.3 demonstrates one reason for placing diodes in parallel. If the current rating of the diodes in Fig. 8.27 was only 20 mA, a current of 28.18 mA would damage the device if it appeared alone in Fig. 8.27. Placing two in parallel limits the current to a safe value of 14.09 mA with the same terminal voltage.

EXAMPLE 8.4 Determine the current I for the network in Fig. 8.29.

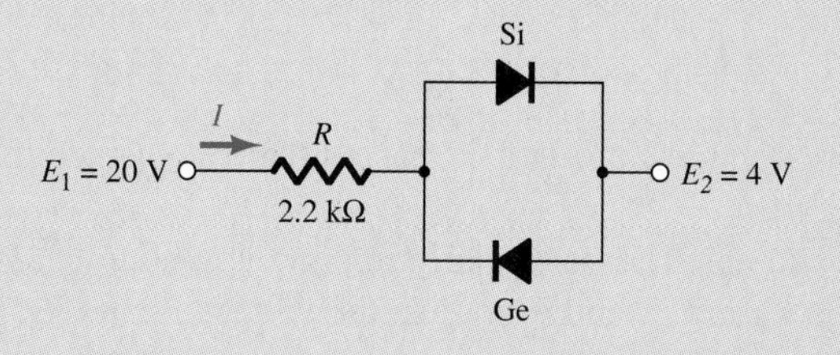

FIG. 8.29

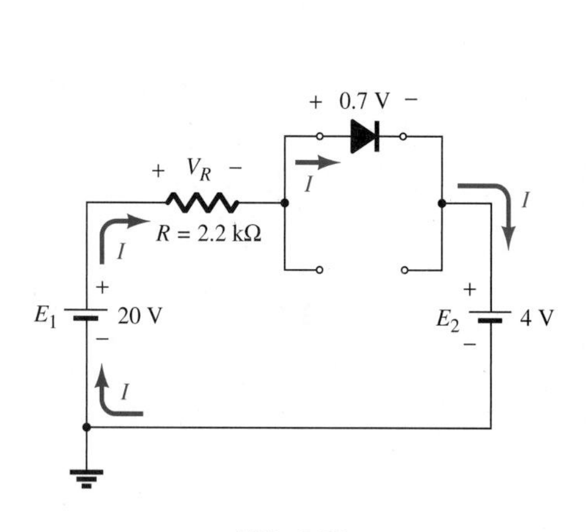

FIG. 8.30

Solution: Redrawing the network as shown in Fig. 8.30 reveals that the resulting current direction is such as to turn on the silicon diode and turn off the germanium diode. The resulting current I is then

$$I = \frac{E_1 - E_2 - V_D}{R} = \frac{20\text{ V} - 4\text{V} - 0.7\text{ V}}{2.2\text{ k}\Omega} \cong \mathbf{6.95\ mA}$$

8.5 AND/OR GATES

The tools of analysis are now at our disposal, and the opportunity to investigate a computer configuration is one that will demonstrate the range of applications of this relatively simple device. Our analysis will be limited to determining the voltage levels and will not include a detailed discussion of Boolean algebra or positive and negative logic.

The network to be analyzed in Example 8.7 is an OR gate for positive logic. That is, the 10-V level in Fig. 8.31 is assigned a 1 for Boolean algebra, while the 0-V input is assigned a 0. An OR gate is one whose output voltage level will be a 1 if either *or* both inputs are a 1. The output is a 0 if both inputs are at the 0 level.

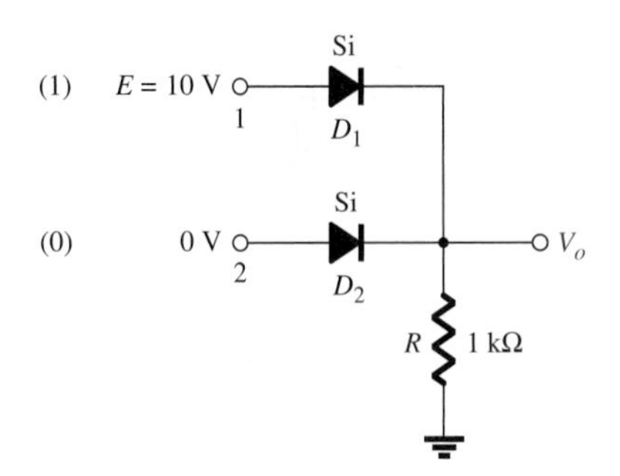

FIG. 8.31
Positive logic OR gate.

The analysis of AND/OR gates is made measurably easier by using the approximate equivalent for a diode rather than the ideal because we can stipulate that the voltage across the diode must be 0.7 V (0.3 V for Ge) positive for the silicon diode to switch to the *on* state.

In general, the best approach is simply to cultivate a "gut" feeling for the state of the diodes by noting the direction and the pressure established by the applied potentials. The analysis will then verify or negate your initial assumptions.

EXAMPLE 8.5 Determine V_o for the network in Fig. 8.31.

Solution: First, note that there is only one applied potential—10 V at terminal 1. Terminal 2 with a 0-V input is essentially at ground potential, as shown in the redrawn network in Fig. 8.32. Figure 8.32 "suggests" that D_1 is probably in the *on* state due to the applied 10 V, while D_2 with its "positive" side at 0 V is probably *off.* Assuming these states results in the configuration in Fig. 8.33.

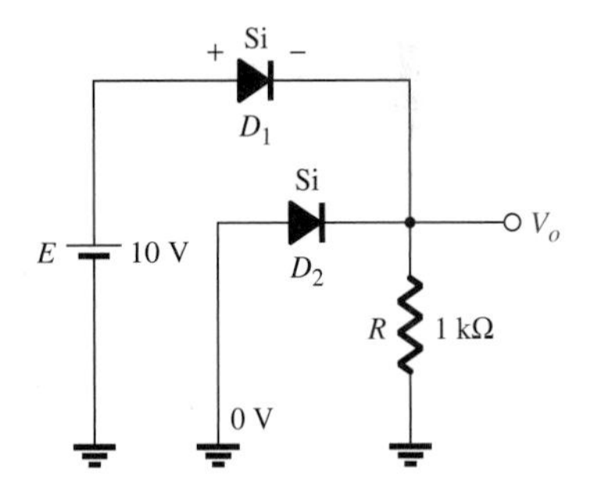

FIG. 8.32

The next step is simply to check that there is no contradiction to our assumptions. That is, note that the polarity across D_1 is such as to turn it on and the polarity across D_2 is such as to turn it off. For D_1 the *on* state establishes V_o at $V_o = E - V_D = 10\text{ V} - 0.7\text{ V} = \mathbf{9.3\text{ V}}$. With 9.3 V at the cathode (−) side of D_2 and 0 V at the anode (+) side, D_2 is definitely in the *off* state. The current direction and the resulting continuous path for conduction further confirm our assumption that D_1 is conducting. Our assumptions seem confirmed by the resulting voltages and current, and our initial analysis can be assumed to be correct. The output voltage level is not 10 V, as defined for an input of 1, but the 9.3 V is sufficiently large to be considered a 1 level. The output is therefore at a 1 level with only one input, which suggests that the gate is an OR gate. An analysis of the same network with two 10-V inputs will find both diodes being in the *on* state and an output of 9.3 V. A 0-V input at both inputs will not provide the 0.7 V required to turn the diodes on, and the output will be a 0 because of the 0-V output level. For the network in Fig. 8.33 the current level is determined by

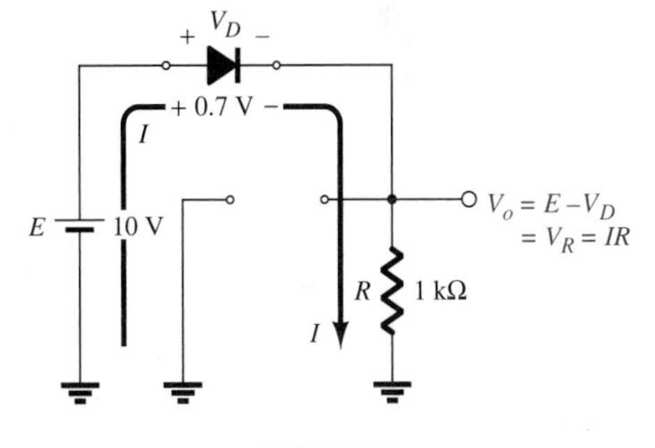

FIG. 8.33

$$I = \frac{E - V_D}{R} = \frac{10\text{ V} - 0.7\text{ V}}{1\text{ k}\Omega} = \mathbf{9.3\text{ mA}}$$

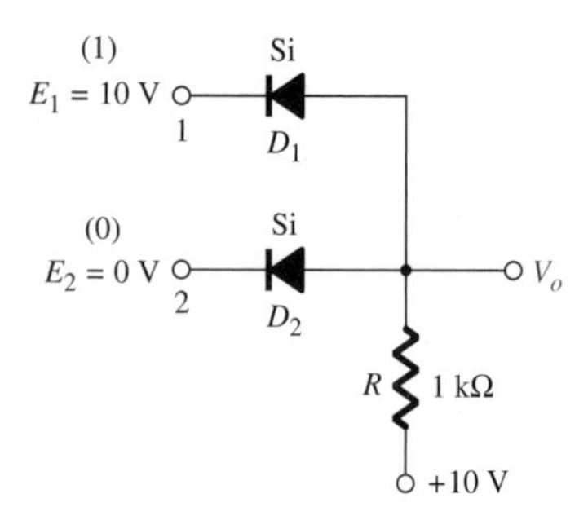

FIG. 8.34
Positive logic AND gate.

EXAMPLE 8.6 Determine the output level for the positive logic AND gate in Fig. 8.34.

Solution: Note in this case that an independent source appears in the grounded leg of the network. For reasons soon to become obvious it is chosen at the same level as the input logic level. The network is redrawn in Fig. 8.35 with our initial assumptions regarding the state of the diodes. With 10 V at the cathode side of D_1 it is assumed that D_1 is in the *off* state even though there is a 10-V source connected to the anode of D_1 through the resistor. However, recall that we mentioned in the introduction to this section that use of the approximate model will be an aid to the analysis. For D_1, where will the 0.7 V come from if the input and source voltages are at the same level and creating opposing pressures? D_2 is assumed to be in the *on* state because of the low voltage at the cathode side and the availability of the 10-V source through the 1-kΩ resistor.

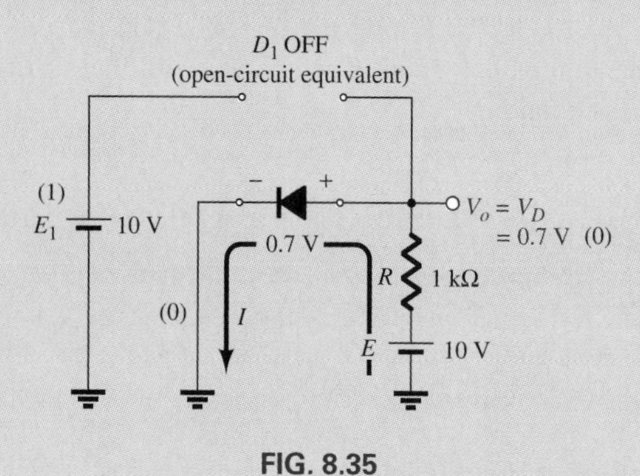

FIG. 8.35

For the network in Fig. 8.35 the voltage at V_o is **0.7 V** because of the forward-biased diode D_2. With 0.7 V at the anode of D_1 and 10 V at the cathode, D_1 is definitely in the *off* state. The current I will have the direction indicated in Fig. 8.35 and a magnitude equal to

$$I = \frac{E - V_D}{R} = \frac{10\text{ V} - 0.7\text{ V}}{1\text{ k}\Omega} = \mathbf{9.3\text{ mA}}$$

The state of the diodes is therefore confirmed, and our earlier analysis was correct. Although not 0 V as earlier defined for the 0 level, the output voltage is sufficiently small to be considered a 0 level. For the AND gate, therefore, a single input will result in a 0-level output. The state of the diodes for two inputs, or no inputs, will be examined in the problems at the end of the chapter.

8.6 HALF-WAVE RECTIFICATION

The diode analysis will now be expanded to include time-varying functions such as the sinusoidal waveform and the square wave. There is no question that the degree of difficulty will increase, but once a few fundamental operations are understood, the analysis will be fairly direct and follow a common thread.

The simplest of networks to examine with a time-varying signal appears in Fig. 8.36. For the moment we will use the ideal model to ensure that the approach is not clouded by additional mathematical complexity.

Over one full cycle, defined by the period T in Fig. 8.36, the average value of v_i (the algebraic sum of the areas above and below the axis) is zero. The circuit in Fig. 8.36, called a *half-wave rectifier,* generates a waveform v_o that has an average value of particular use in the ac-to-dc conversion process.

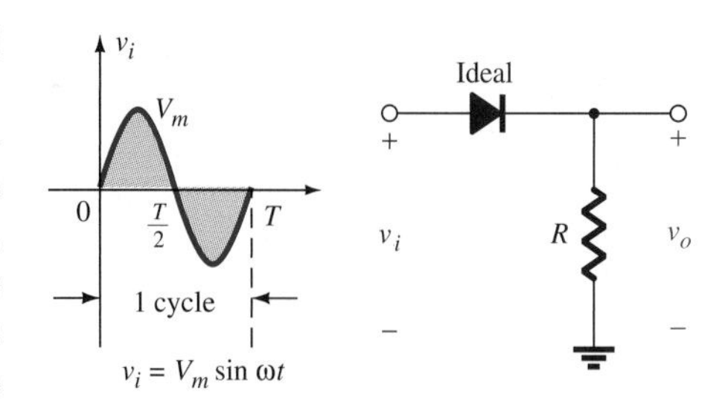

FIG. 8.36
Half-wave rectifier.

During the interval $t = 0 \rightarrow T/2$ in Fig. 8.36, the polarity of the input voltage v_i is defined as shown in Fig. 8.37. The result is the polarity shown across the diode, resulting in the short-circuit equivalent (for ideal diodes) appearing in the adjoining figure. The output is now connected directly to the input, with the result that for the period $0 \rightarrow T/2$, $v_o = v_i$.

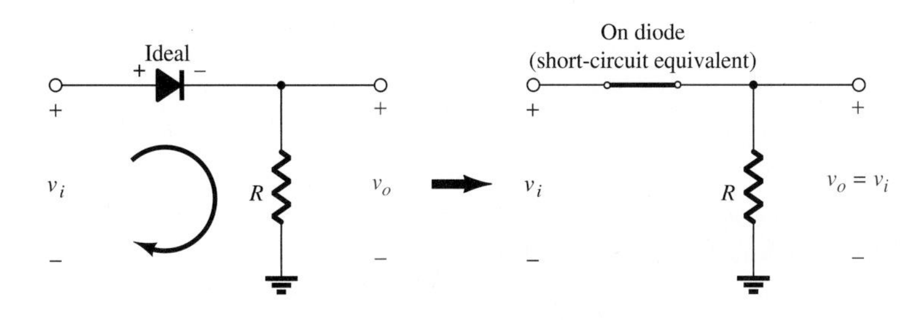

FIG. 8.37
Conduction region ($0 \rightarrow T/2$).

For the period $T/2 \rightarrow T$, the polarity of the input v_i is as shown in Fig. 8.38, and the resulting polarity across the ideal diode produces an *off* state with an open-circuit equivalent. The result is the absence of a path for charge to flow, and $v_o = iR = (0)R = 0$ V for the period $T/2 \rightarrow T$. The input v_i and the output v_o are presented together in Fig. 8.39 for comparison purposes. The output signal v_o now has a net positive area above the axis over a full period, and an average value is determined by

$$\text{average (dc value)} = 0.318V_m \tag{8.6}$$

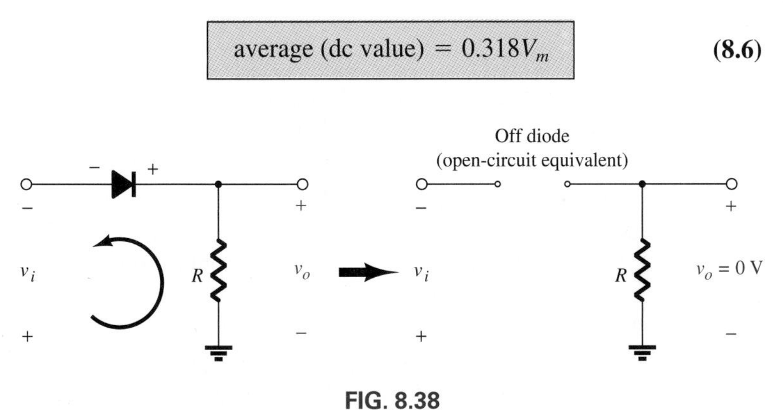

FIG. 8.38
Nonconduction region ($T/2 \rightarrow T$).

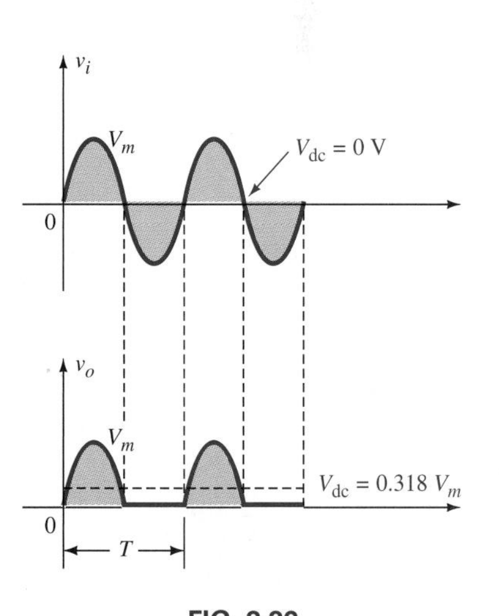

FIG. 8.39
Half-wave rectifier.

The process of removing half the input signal to establish a dc level is aptly called half-wave rectification. The term *rectification* comes from the use of the term *rectifier* for diodes employed in power supplies for the ac-to-dc conversion process. Later in this chapter the use of this pulsating voltage to establish a steady dc voltage will be demonstrated.

EXAMPLE 8.7 Sketch the output v_o and determine the dc level of the output for the half-wave rectifier in Fig. 8.40.

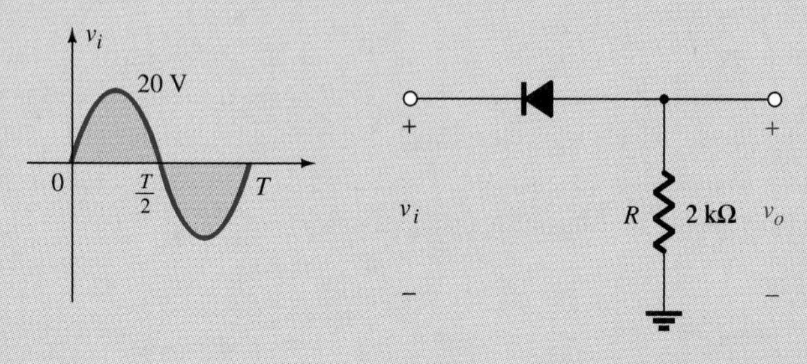

FIG. 8.40

Solution: In this situation the diode conducts during the negative part of the input, as shown in Fig. 8.41, and v_o appears as shown in the same figure. For the full period, the dc level is

$$V_{dc} = -0.318V_m = -0.318(20\ V) = \mathbf{-6.36\ V}$$

The negative sign indicates that the polarity of the output is the opposite of the defined polarity in Fig. 8.40.

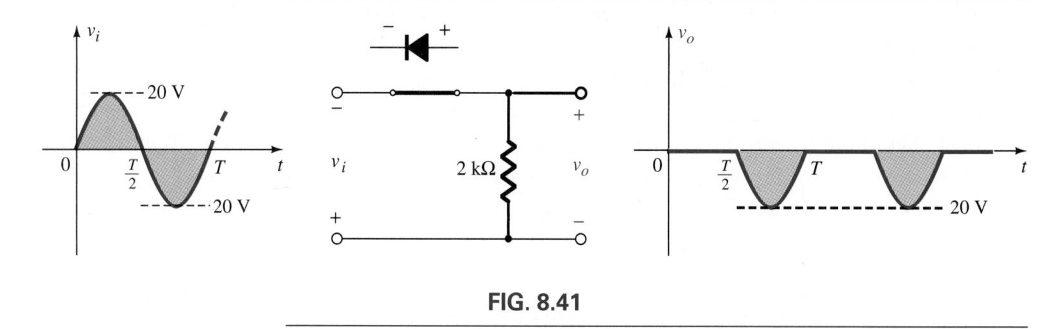

FIG. 8.41

The peak-inverse-voltage (PIV) rating of the diode is of primary importance in the design of rectification systems. Recall that it is voltage rating that must not be exceeded in the reverse-bias region or the diode will enter the Zener avalanche region. The required PIV rating for the half-wave rectifier can be determined from Fig. 8.42, which displays the reverse-biased diode in Fig. 8.36 with maximum applied voltage. Application of Kirchhoff's voltage law makes it fairly obvious that the PIV rating of the diode must equal or exceed the peak value of the applied voltage. Therefore,

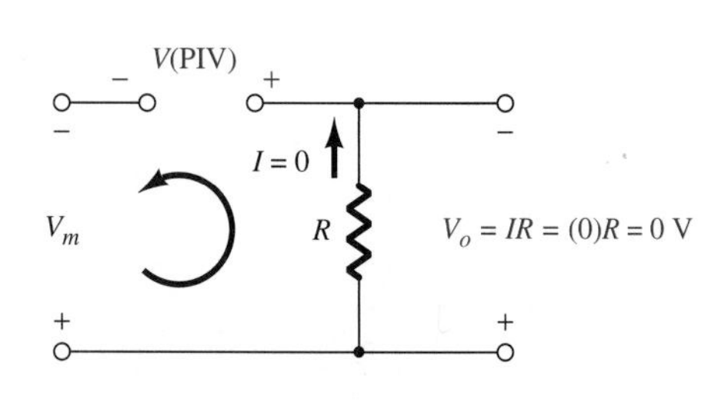

FIG. 8.42
Determining the required PIV rating for the half-wave rectifier.

$$\boxed{\text{PIV rating} = V_m} \quad \text{half-wave rectifier} \tag{8.7}$$

8.7 FULL-WAVE RECTIFICATION

The dc level obtained from a sinusoidal input can be improved 100% using a process called *full-wave rectification*. The most familiar network for performing such a function appears in Fig. 8.43 with its four diodes

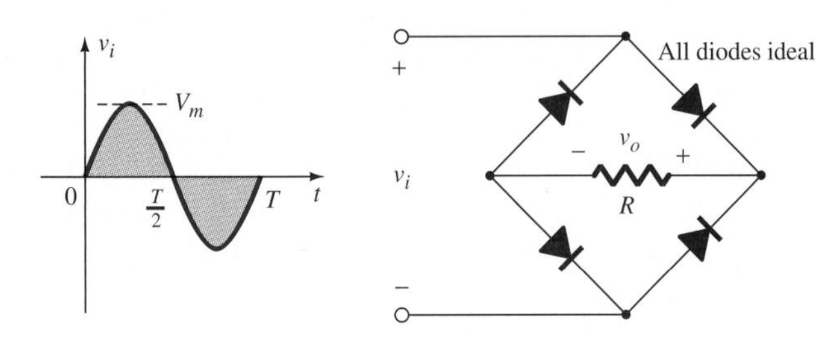

FIG. 8.43
Full-wave bridge rectifier.

in a *bridge* configuration. During the period $t = 0$ to $T/2$ the polarity of the input is as shown in Fig. 8.44. The resulting polarities across the ideal diodes are also shown in Fig. 8.44 to reveal that D_2 and D_3 are conducting while D_1 and D_4 are in the *off* state. The net result is the configuration in Fig. 8.45 with its indicated current and polarity across R. Since the diodes are ideal, the load voltage $v_o = v_i$, as shown in the same figure.

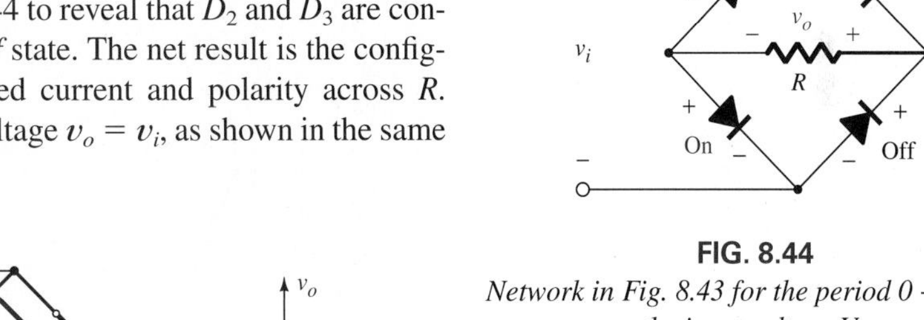

FIG. 8.44
Network in Fig. 8.43 for the period $0 \rightarrow T/2$ of the input voltage V_i.

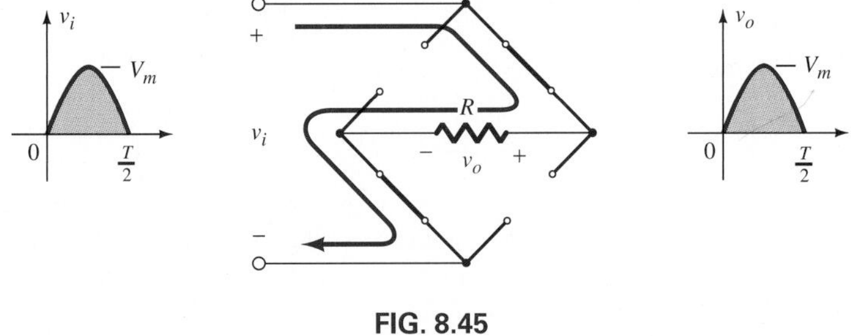

FIG. 8.45

For the negative region of the input the conducting diodes are D_1 and D_4, resulting in the configuration in Fig. 8.46. The important result is

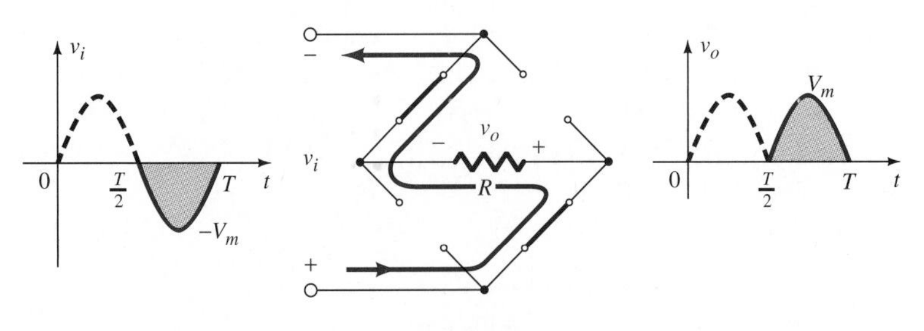

FIG. 8.46

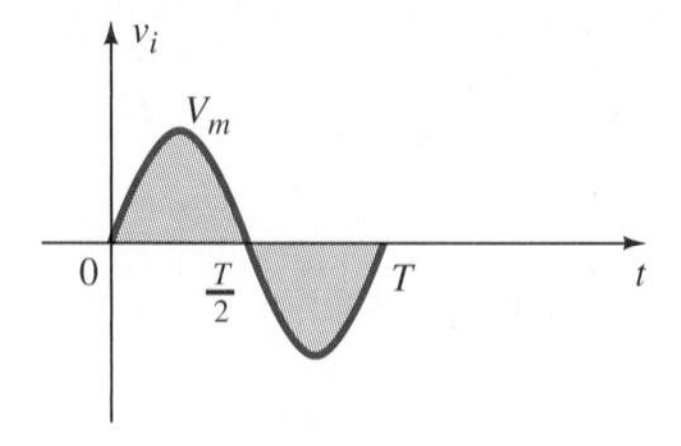

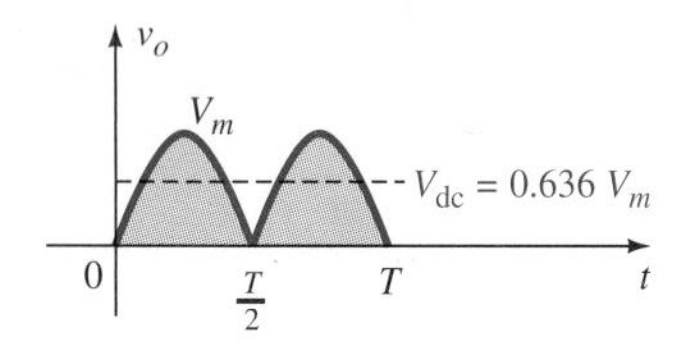

FIG. 8.47

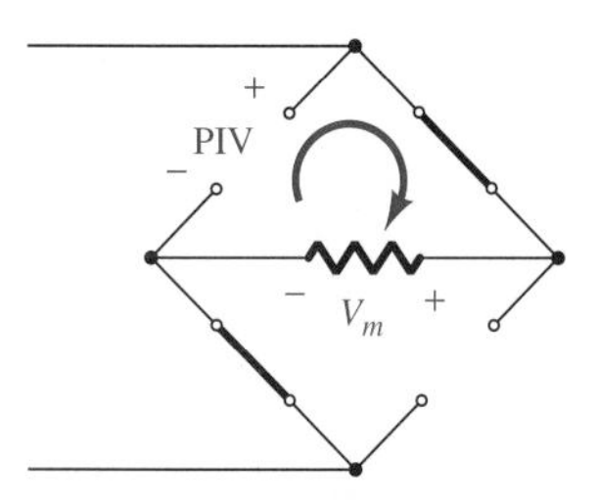

FIG. 8.48

that the polarity across the load resistor R is the same as in Fig. 8.44, establishing a second positive pulse, as shown in Fig. 8.46. Over one full cycle the input and output voltages appear as shown in Fig. 8.47.

Since the area above the axis for one full cycle is now twice that obtained for a half-wave system, the dc level has also been doubled and

$$\text{average (dc value)} = 0.636V_m \tag{8.8}$$

The required PIV of each diode (ideal) can be determined from Fig. 8.48 obtained at the peak of the positive region of the input signal. For the indicated loop the maximum voltage across R is V_m and

$$\text{PIV} = V_m \tag{8.9}$$

full-wave bridge rectifier

EXAMPLE 8.8 Determine the output waveform for the network in Fig. 8.49 and calculate the output dc level and the required PIV of each diode.

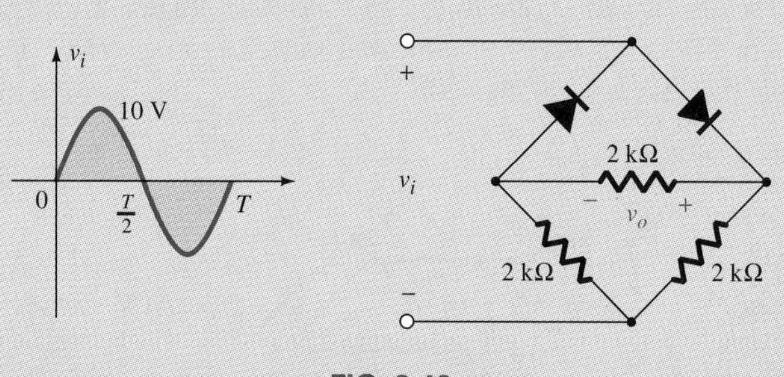

FIG. 8.49

Solution: The network will appear as shown in Fig. 8.50 for the positive region of the input voltage. Redrawing the network will result in the configuration in Fig. 8.51(a), where $v_o = \frac{1}{2}\,v_i$ or $V_{o(\text{max})} = \frac{1}{2}V_{i(\text{max})} = \frac{1}{2}(10\text{ V}) = 5\text{ V}$. For the negative part of the input the roles of the diodes will be interchanged, and v_o will appear as shown in Fig. 8.51(b).

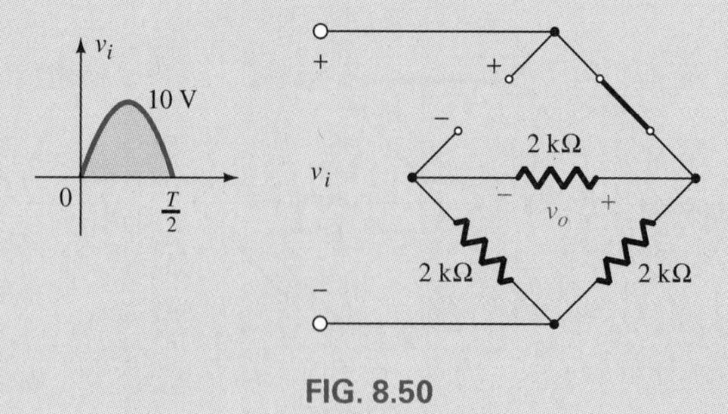

FIG. 8.50

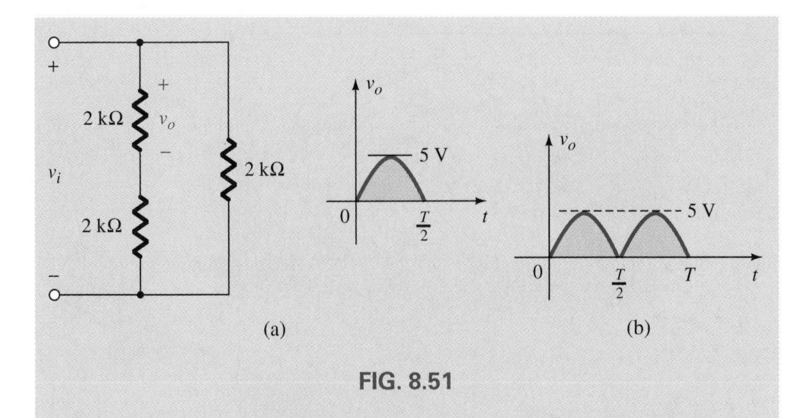

FIG. 8.51

The effect of removing two diodes from the bridge configuration is therefore to reduce the available dc level to

$$V_{dc} = 0.636(5\text{ V}) = \mathbf{3.18\text{ V}}$$

or that available from a half-wave rectifier with the same input. However, the PIV as determined from Fig. 8.48 is equal to the maximum voltage across R, which is 5 V or half of that required for a half-wave rectifier with the same input.

8.8 APPLICATION: 12-V CAR BATTERY CHARGER

Battery chargers are a common household piece of equpiment used to charge everything from small flashlight batteries to heavy-duty marine lead–acid batteries. Since all are plugged into a 120-V ac outlet such as found in the home, the basic construction of each is quite similar. In every charging system a *transformer* (Chapter 6) must be included to cut the ac voltage to a level appropriate for the dc level to be established. A *diode* (also called *rectifier*) arrangement must be included to convert the ac voltage that varies with time to a fixed dc level such as described in this chapter. Some dc chargers also include a *regulator* to provide an improved dc level (one that varies less with time or load). Since the car bettery charger is one of the most common, it will be described in the next few paragraphs.

The outside appearance and the internal construction of a Sears 6/2 AMP Manual Battery Charger are provided in Fig. 8.52. Note in Fig. 8.52(b) that the transformer (as in most chargers) takes up most of the internal space. The additional air space and the holes in the casing are there to ensure an outlet for the heat that will develop due to the resulting current levels.

The schematic of Fig. 8.53 includes all the basic components of the charger. Note first that the 120 V from the outlet is applied directly across the primary of the transformer. The charging rate of 6 A or 2 A is determined by the switch, which simply controls how many windings of the primary will be in the circuit for the chosen charging rate. If the battery is charging at the 2-A level, the full primary will be in the circuit, and the ratio of the turns in the primary to the turns in the secondary will

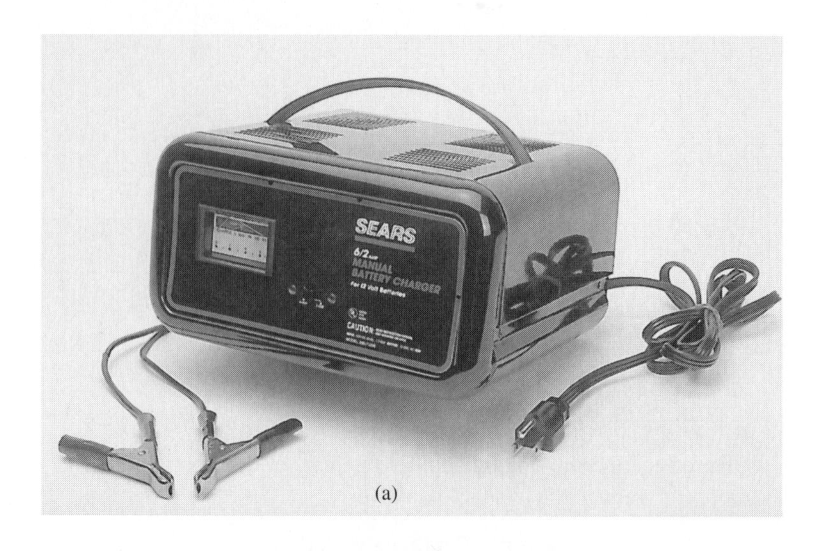

(a)

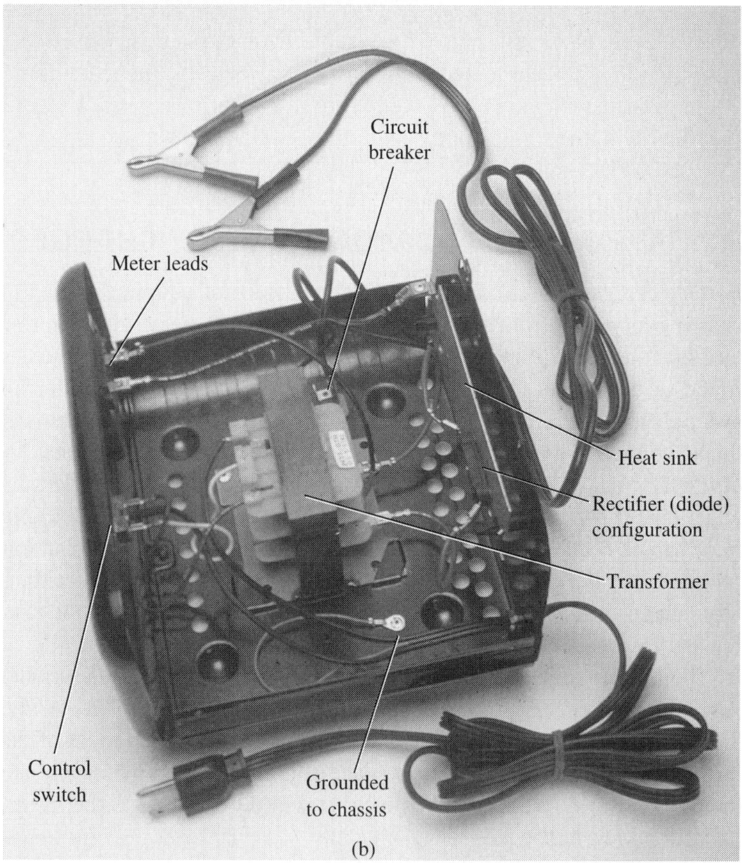

(b)

FIG. 8.52

Battery charger: (a) external appearance; (b) internal construction.

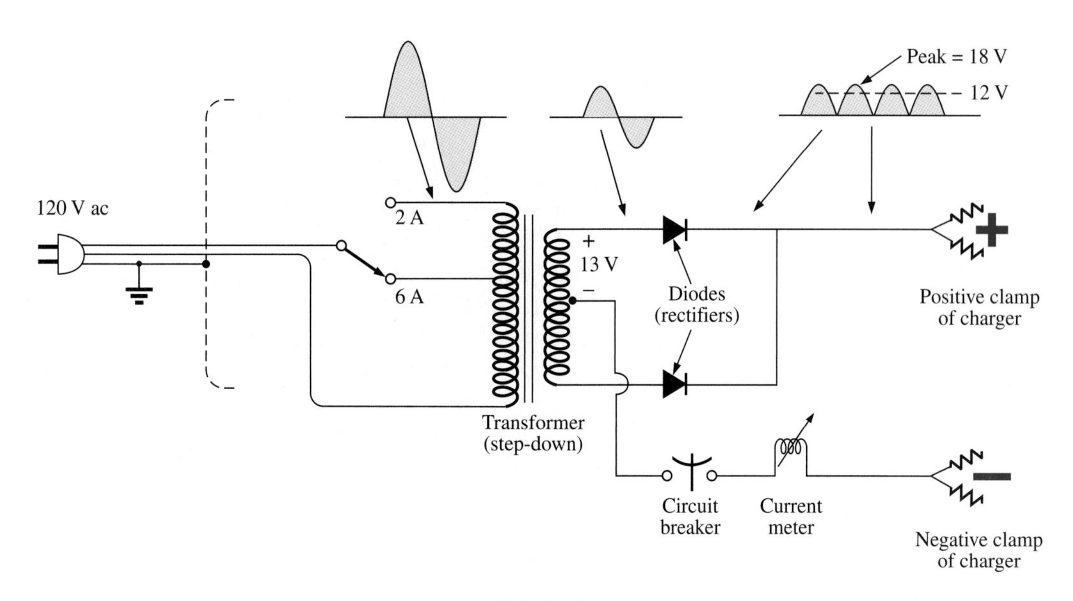

FIG. 8.53
Electrical schematic for the battery charger of Fig. 8.52.

be a maximum. If it is charging at the 6-A level, fewer turns of the primary are in the circuit, and the ratio drops. When you study transformers, you will find that the voltage at the primary and secondary is directly related to the *turns ratio*. If the ratio from primary to secondary drops, then the voltage drops also. The reverse effect occurs if the turns on the secondary exceed those on the primary.

The general appearance of the waveforms appears in Fig. 8.53 for the 6-A charging level. Note that so far, the ac voltage has the same wave shape across the primary and secondary. The only difference is in the peak value of the waveforms. Now the diodes take over and convert the ac waveform that has zero average value (the waveform above equals the waveform below) to one that has an average value (all above the axis), as shown in the same figure. For the moment simply recognize that diodes are semiconductor electronic devices that permit conventional current to flow through them in only one direction, and that direction is indicated by the arrow in the symbol. Even though the waveform resulting from the diode action has a pulsing appearance with a peak value of about 18 V, it will charge the 12-V battery whenever its voltage is greater than that of the battery, as shown by the shaded area. Below the 12-V level the battery cannot discharge back into the charging network because the diodes permit current flow in only one direction.

In particular, note in Fig. 8.52(b) the large plate that carries the current from the rectifier (diode) configuration to the positive terminal of the battery. Its primary purpose is to provide a *heat sink* (a place for the heat to be distributed to the surrounding air) for the diode configuration. Otherwise, the diodes would eventually melt down and self-destruct due to the resulting current levels. Each component of Fig 8.53 has been carefully labeled in Fig. 8.52(b) for reference.

When current is first applied to a battery at the 6-A charge rate, the current demand as indicated by the meter on the face of the instrument may rise to 7 A or almost 8 A. However, the level of current will decrease as the battery charges until it drops to a level of 2 A or 3 A. For units such as this that do not have an automatic shutoff, it is important to disconnect the charger when the current drops to the fully charged level; otherwise, the battery will become overcharged and may be damaged. A battery that is at its 50% level can take as long as 10 hours to charge, so don't expect it to be a 10-minute operation. In addition, if a battery is in very bad shape with a lower-than-normal voltage, the initial charging current may be too high for the design. To protect against such situations, the circuit breaker will open and stop the charging process. Because of the high current levels, it is important that the directions provided with the charger be carefully read and applied.

8.9 CLIPPERS

There are a variety of diode networks called *clippers* that have the ability to "clip" off a portion of the input signal without distorting the remaining part of the alternating waveform. The half-wave rectifier described in Section 8.6 is an example of the simplest form of diode clipper—one resistor and one diode. Depending on the orientation of the diode the positive or negative region of the input signal is clipped off.

There are two general categories of clippers: series and parallel. The series configuration is defined as one in which the diode is in series with the load, while the parallel variety has the diode in a branch parallel to the load.

The response of the series configuration in Fig. 8.54(a) to a variety of alternating waveforms is provided in Fig. 8.54(b). Although first intro-

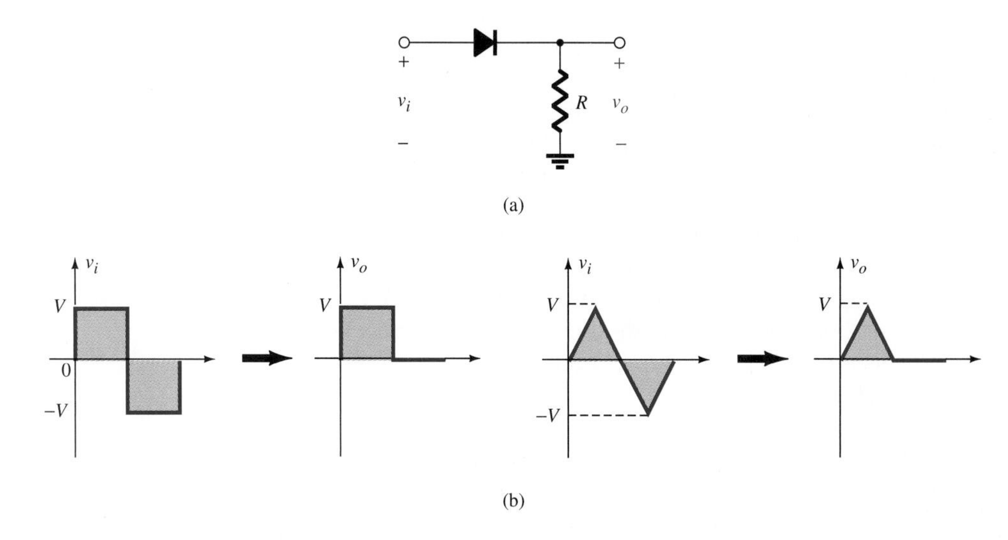

FIG. 8.54
Series clipper.

duced as a half-wave rectifier (for sinusoidal waveforms), a clipper has no boundaries on the type of signals that can be applied to it.

The addition of a dc supply such as shown in Fig. 8.55 can have a pronounced effect on the output of a clipper, as clearly revealed by the review table in Fig. 8.69

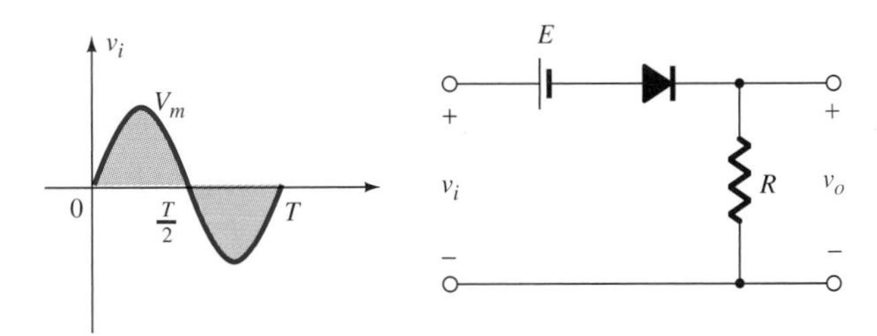

FIG. 8.55

There is no set procedure for analyzing networks like the one in Fig. 8.55, but there are a few thoughts to keep in mind as you work toward a solution.

1. *Make a mental sketch of the response of the network based on the direction of the diode and the applied voltage levels.* For the network in Fig. 8.55, the direction of the diode suggests that the signal v_i must be positive to turn it on. The dc supply further requires that the voltage v_i be greater than E volts to turn the diode on.

   The negative region of the input signal is pressuring the diode into the *off* state, supported further by the dc supply. In general, therefore, we can be quite sure the diode is an open circuit (*off* state) for the negative region of the input signal.
2. *Determine the applied voltage (transition voltage) that will cause a change in state for the diode.* For the ideal diode the condition $i_d = 0$ A or $v_d = 0$ V is employed to determine the level of v_i to effect a transition.

   Applying the condition $i_d = 0$ A at $v_d = 0$ V to the network in Fig. 8.55 results in the configuration in Fig. 8.56, where it is recognized that the level of v_i that will cause a transition in state is

$$v_i = E \tag{8.10}$$

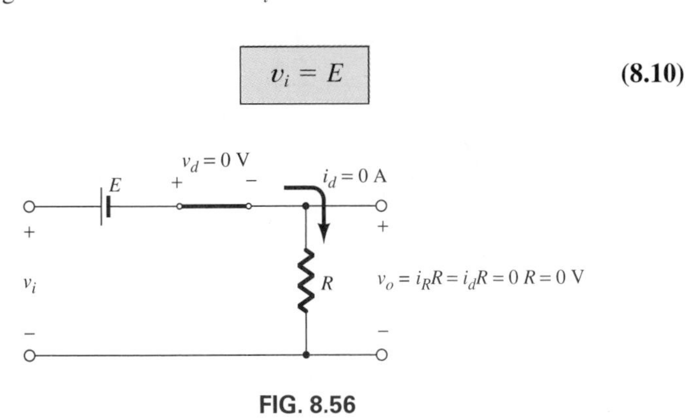

FIG. 8.56

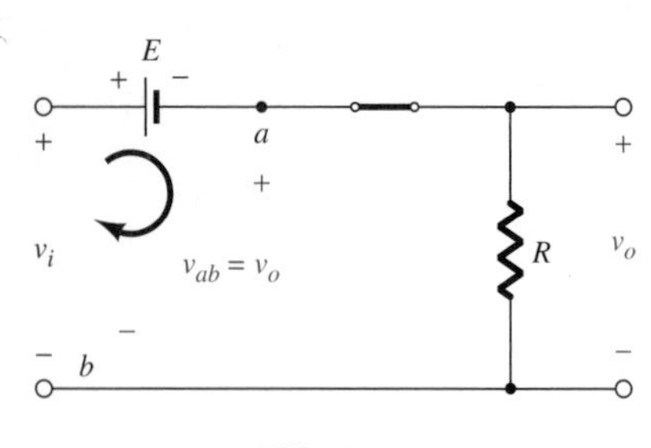

FIG. 8.57

For an input voltage greater than E volts the diode is in the short-circuit state, while for input voltages less than E volts it is in the open-circuit or *off* state.

3. *Be continually aware of the defined terminals and polarity of* v_o. When the diode is in the short-circuit state, such as shown in Fig. 8.57, the output voltage v_o is the same as from a to b, as noted in the figure. Applying Kirchhoff's voltage law, we have

$$v_i - E - v_o = 0 \qquad \text{(CW direction)}$$

and

$$\boxed{v_o = v_i - E} \tag{8.11}$$

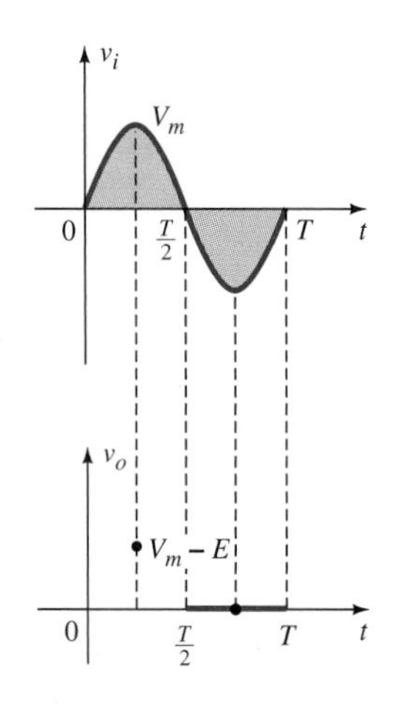

FIG. 8.58

4. *It can be helpful to sketch the input signal above the output as shown in Fig. 8.58 and determine the output at instantaneous values of the input.* It is then possible for the output voltage to be sketched from the resulting data points of v_o. Keep in mind that at an instantaneous value of v_i the input can be treated as a dc supply of that value and the corresponding dc value (the instantaneous value) of the output determined. For instance, at $v_i = V_m$ for the network in Fig. 8.55, the network to be analyzed appears in Fig. 8.59. For $V_m > E$ the diode is in the short-circuit state, and $V_o = V_m - E$, as shown in Fig. 8.60.

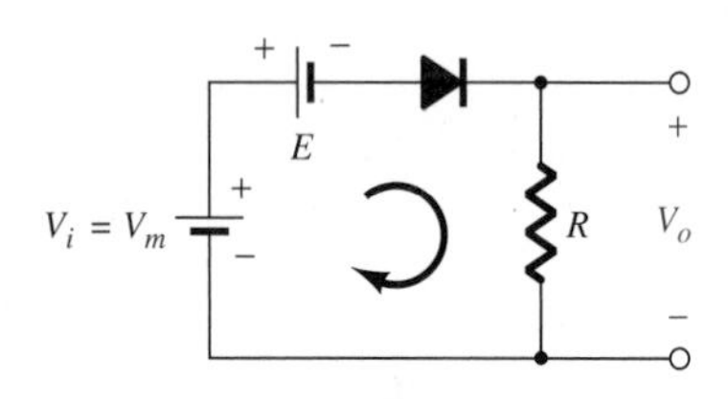

FIG. 8.59

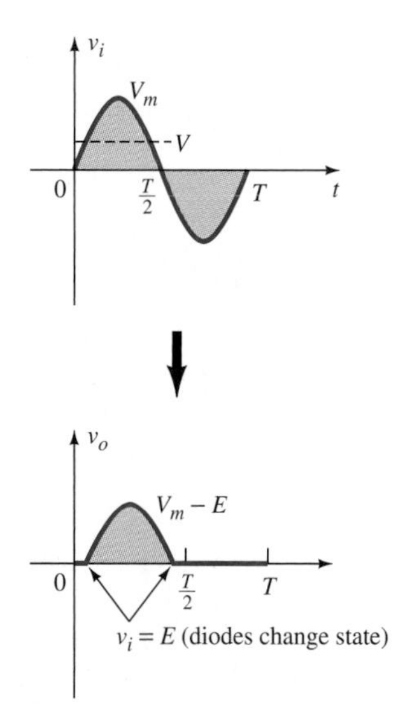

FIG. 8.60

EXAMPLE 8.9 Determine the output waveform for the network in Fig. 8.61.

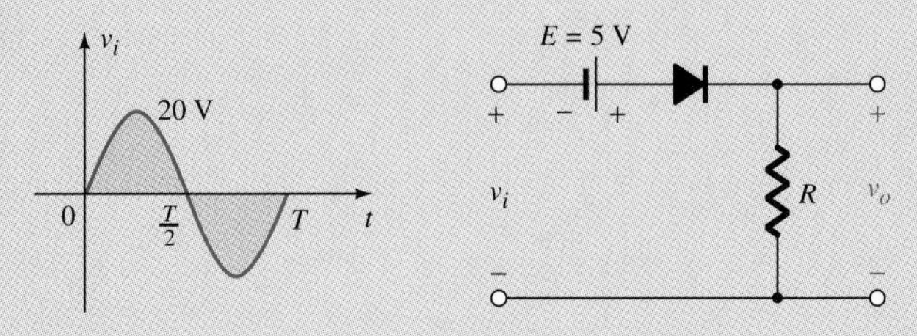

FIG. 8.61

Solution: Past experience suggests that the diode will be in the *on* state for the positive region of v_i—especially when we note the aiding effect of $E = 5$ V. The network will then appear as shown in Fig. 8.62, and $v_o = v_i + 5$ V. Substituting $i_d = 0$ A at $v_d = 0$ V for the transition voltage, we obtain the network in Fig. 8.63 and $v_i = -5$ V.

For voltages more negative than -5 V the diode will enter its open-circuit state, while for voltages more positive than -5 V the diode will be in the short-circuit state. The input and output voltages are shown in Fig. 8.64.

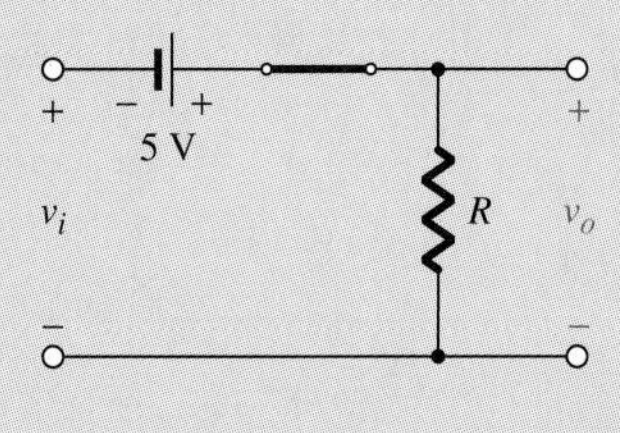

FIG. 8.62

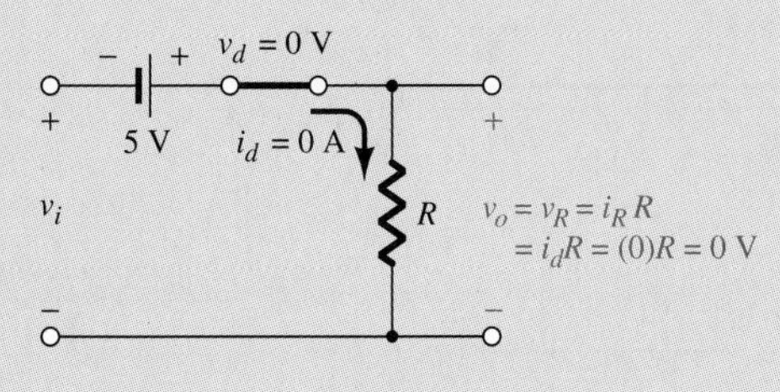

FIG. 8.63

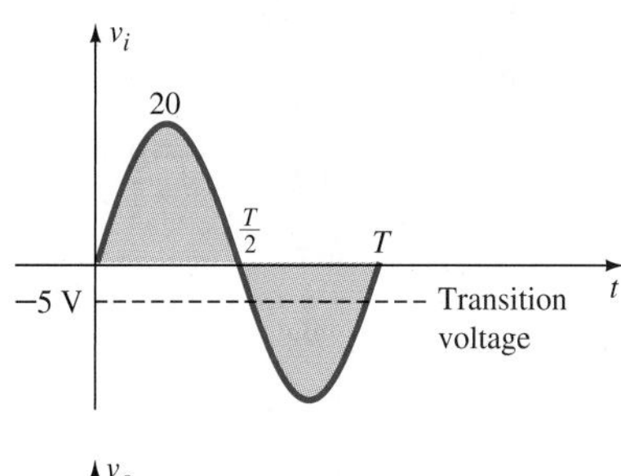

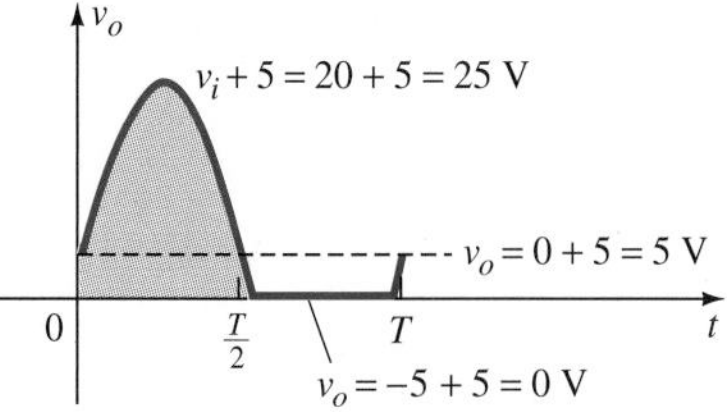

FIG. 8.64

Clipper networks with square-wave inputs are actually easier to analyze than those with sinusoidal inputs because only two levels have to be considered. In other words, the network can be analyzed as if it has two dc level inputs with the resulting output v_o plotted in the proper time frame.

EXAMPLE 8.10 Repeat Example 8.9 for the square-wave input in Fig. 8.65.

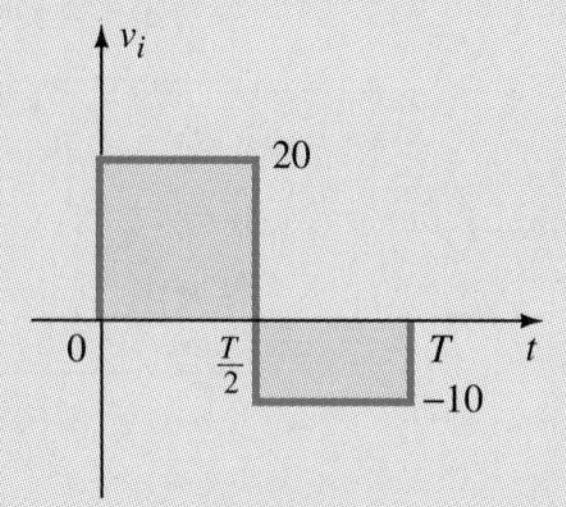

FIG. 8.65

Solution: For $V_i = 20$ V $(0 \rightarrow T/2)$ the network in Fig. 8.66 results. The diode is in the short-circuit state, and $V_o = 20\text{ V} + 5\text{ V} = 25$ V. For $v_i = -10$ V the network in Fig. 8.67 results, placing the diode in the *off* state and $v_o = i_R R = (0)R = 0$ V. The resulting output voltage appears in Fig. 8.68.

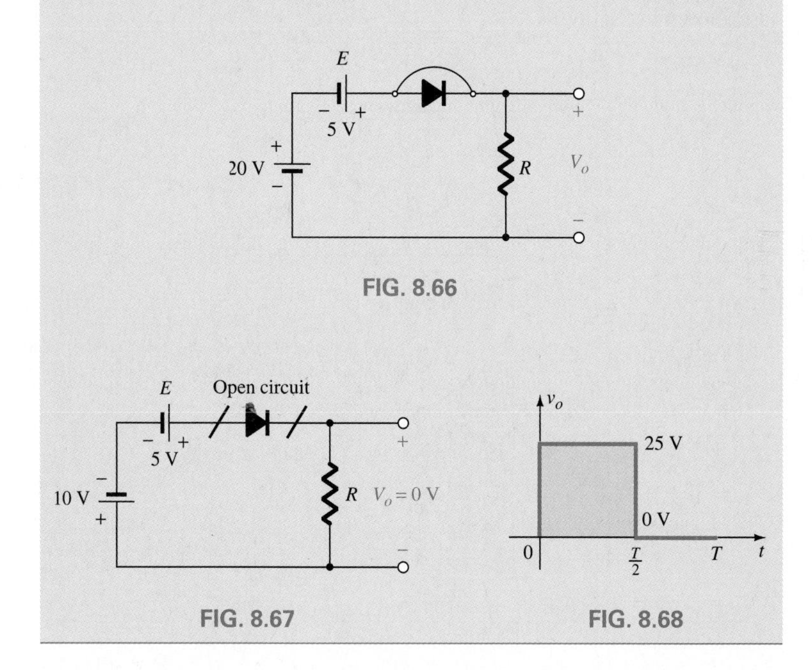

FIG. 8.66

FIG. 8.67

FIG. 8.68

Note in Example 8.10 that the clipper not only clipped off 5 V from the total swing but raised the dc level of the signal by 5 V.

A variety of series and parallel clippers with the resulting output for the sinusoidal input are provided in Fig. 8.69. In particular, note the response of the last configuration with its ability to clip off a positive and a negative section as determined by the magnitude of the dc supplies.

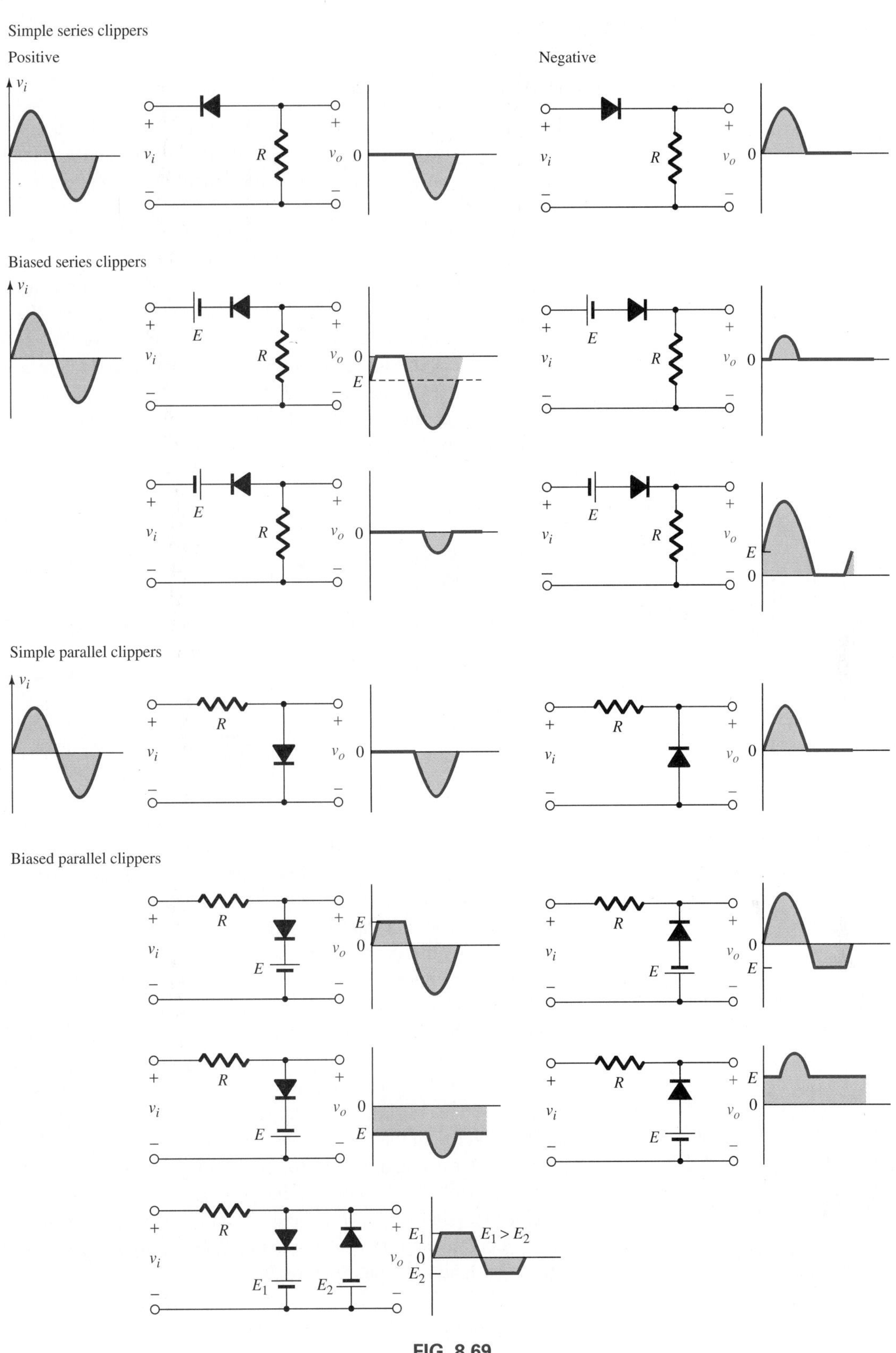

FIG. 8.69
Clipping circuits.

8.10 CLAMPERS

The *clamping* network is one that can "clamp" a signal to a different dc level. The network must have a capacitor, a diode, and a resistive element, but it can also employ an independent dc supply to introduce an additional shift.

The magnitude of R and C must be chosen such that the time constant $\tau = RC$ is large enough to ensure that the voltage across the capacitor does not discharge significantly during the interval the diode is nonconducting. Throughout the analysis we will assume that for all practical purposes the capacitor fully charges or discharges in five time constants.

The network in Fig. 8.70 can clamp the input signal to the zero level (for ideal diodes). The resistor R can be the load resistor or a parallel combination of the load resistor and a resistor designed to provide the desired level of R.

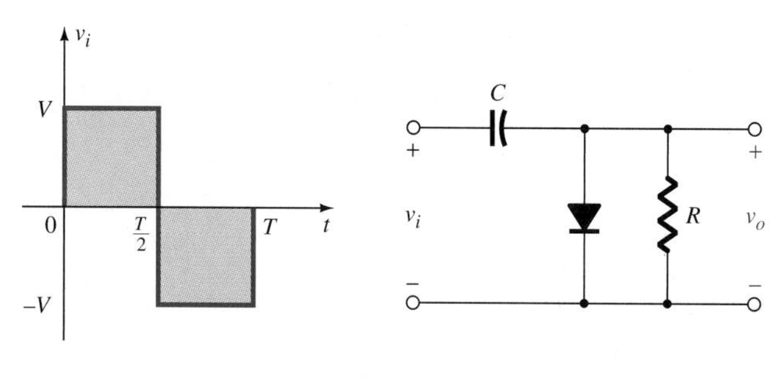

FIG. 8.70
Clamper.

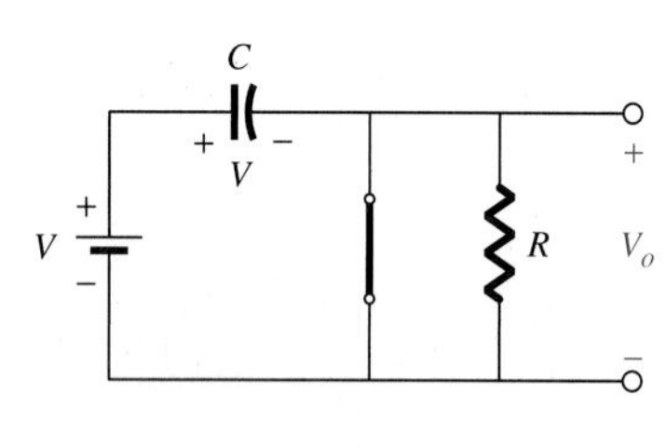

FIG. 8.71

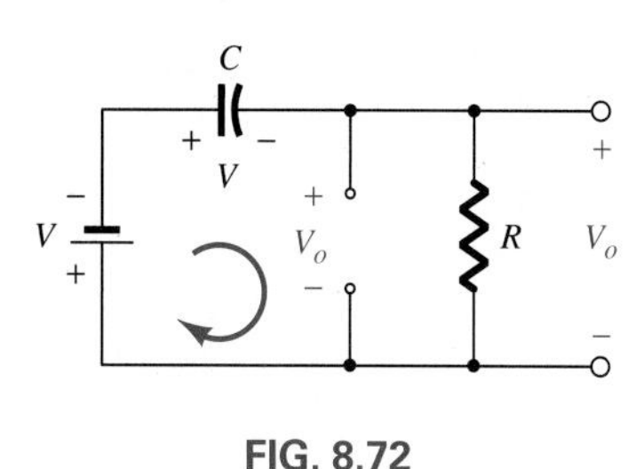

FIG. 8.72

During the interval $0 \rightarrow T/2$ the network appears as shown in Fig. 8.71, with the diode in the short-circuit state shorting out the effect of the resistor R. The resulting RC time constant is so small (R being determined by the inherent resistance of the network) that the capacitor charges to V volts very quickly. During this interval the output voltage is directly across the short circuit, and $v_o = 0$ V.

When the input switches to the $-V$ state, the network appears as shown in Fig. 8.72, with the open-circuit equivalent for the diode determined by the applied signal and stored voltage across the capacitor—both pressuring current through the diode from anode to cathode. Now that R is back in the network, the time constant determined by the RC product is sufficiently large to establish a discharge period 5τ much greater than the period $T/2 \rightarrow T$, and it can be assumed on an approximate basis that the capacitor holds onto all its charge and therefore voltage (since $V = Q/C$) during this period.

Since v_o is in parallel with the diode and resistor, it can also be drawn in the alternative position shown in Fig. 8.72. Applying Kirchhoff's voltage law around the input loop, we have

$$-V - V - V_o = 0$$

and

$$V_o = \mathbf{-2V}$$

The negative sign results from the fact that the polarity of $2V$ is the opposite of that defined for v_o. The resulting output waveform appears in Fig. 8.73 with the input signal. The output signal is clamped to 0 V for the interval 0 to $T/2$ but maintains the same total swing ($2V$) as the input.

For clamping networks, *the total swing of the output is equal to the total swing of the input signal*. This fact is an excellent checking tool for the result obtained.

In general, the following may be helpful when analyzing clamping networks:

1. Always start the analysis of clamping networks by *considering the part of the input signal that will forward-bias the diode*. This may require skipping an interval of the input signal (as demonstrated in an example to follow), but the analysis will not be extended unnecessarily.
2. During the period that the diode is in the short-circuit state, assume that the capacitor will charge up instantaneously (an approximation) to a level determined by the voltage across the capacitor in its equivalent open-circuit state.
3. Assume that during the period the diode is an open circuit (*off* state), the capacitor will hold onto all its charge and therefore voltage.
4. Throughout the analysis maintain a continual awareness of where v_o is defined to ensure that the proper levels for v_o are obtained in the analysis.
5. Keep in mind the general rule that the swing of the output must match the swing of the input signal.

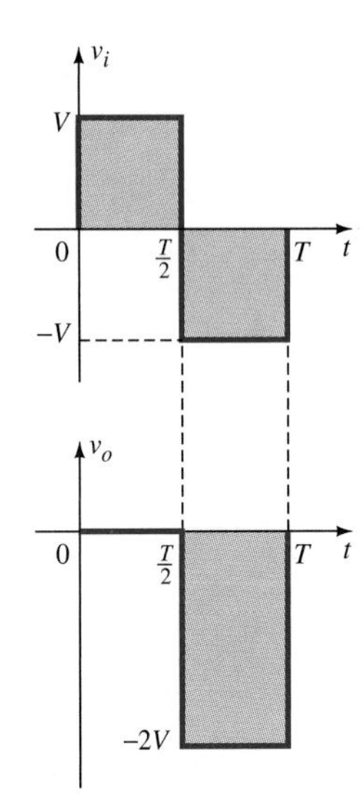

FIG. 8.73

EXAMPLE 8.11 Determine v_o for the network in Fig. 8.74 for the indicated input.

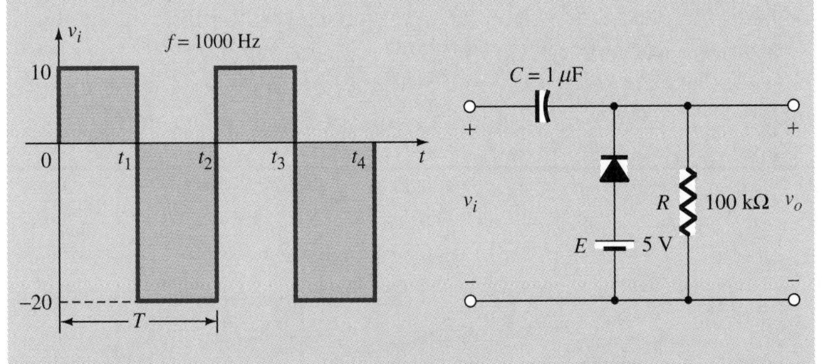

FIG. 8.74

Solution: Note that the frequency is 1000 Hz, resulting in a period of 1 ms and an interval of 0.5 ms between levels. The analysis will begin with the period $t_1 \rightarrow t_2$ of the input signal, since the diode is in its short-circuit state, as recommended by comment 1. For this interval the network appears as shown in Fig. 8.75.

The output is across R, but it is also directly across the 5-V battery if you follow the direct connection between the defined terminals

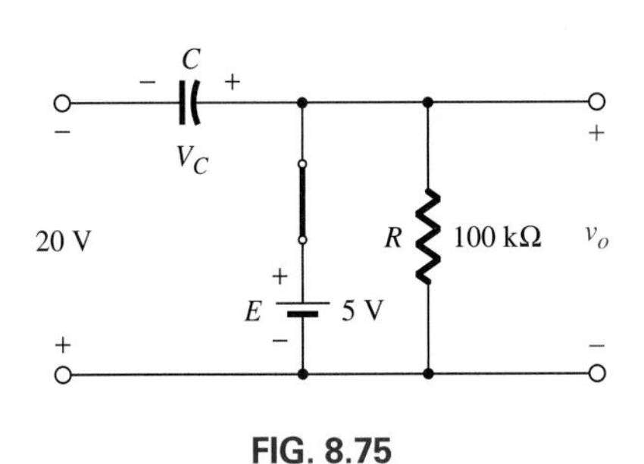

FIG. 8.75

for v_o and the battery terminals. The result is $v_o = 5$ V for this interval.

Application of Kirchhoff's voltage law around the input loop results in

$$-20\text{ V} + V_C - 5\text{ V} = 0$$

and

$$V_C = 25\text{ V}$$

The capacitor therefore charges up to 25 V, as stated in comment 2. In this case the resistor R is not shorted out by the diode, but a Thévenin equivalent circuit of that portion of the network that includes the battery and the resistor results in $R_{Th} = 0\ \Omega$ with $E_{Th} = V = 5$ V. For the period $t_2 \to t_3$ the network appears as shown in Fig. 8.76.

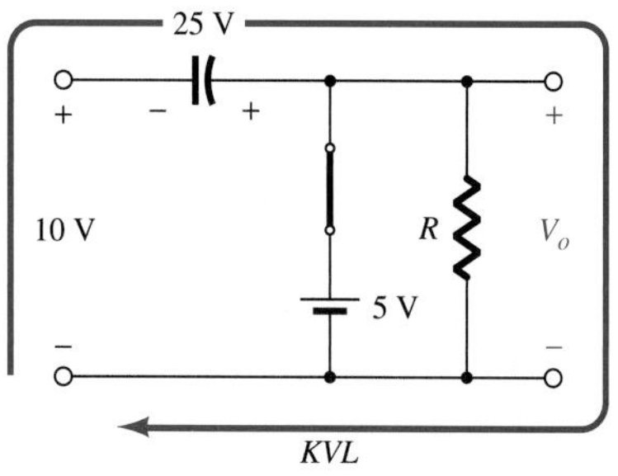

FIG. 8.76

The open-circuit equivalent for the diode removes the 5-V battery from having any effect on v_o, and application of Kirchhoff's voltage law around the outside loop of the network results in

$$+10\text{ V} + 25\text{ V} - V_o = 0$$

and

$$V_o = 35\text{ V}$$

The time constant of the discharging network in Fig. 8.76 is determined by the product RC and has the magnitude

$$\tau = RC = (100\text{ k}\Omega)(0.1\ \mu\text{F}) = 0.01\text{ s} = 10\text{ ms}$$

The total discharge time is therefore $5\tau = 5(10\text{ ms}) = 50$ ms.

Since the interval $t_2 \to t_3$ lasts for only 0.5 ms, it is certainly a good approximation to assume that the capacitor will hold its voltage during the discharge period between pulses of the input signal. The resulting output appears in Fig. 8.77 with the input signal. Note that the output swing of 30 V matches the input swing, as noted in comment 5. A number of clamping circuits and their effect on the input signal are shown in Fig. 8.78.

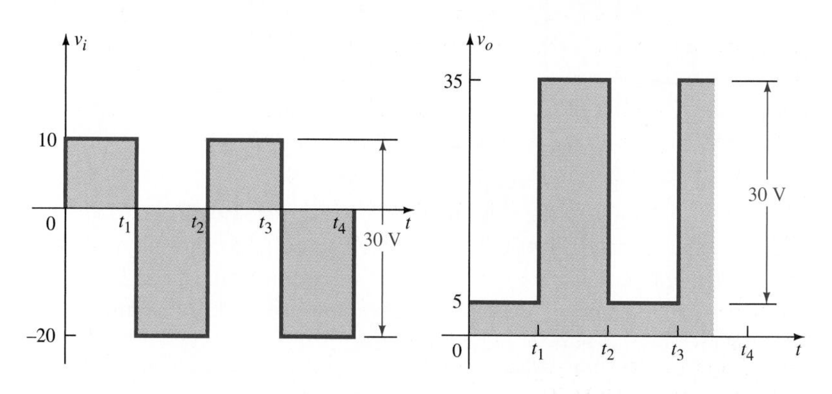

FIG. 8.77

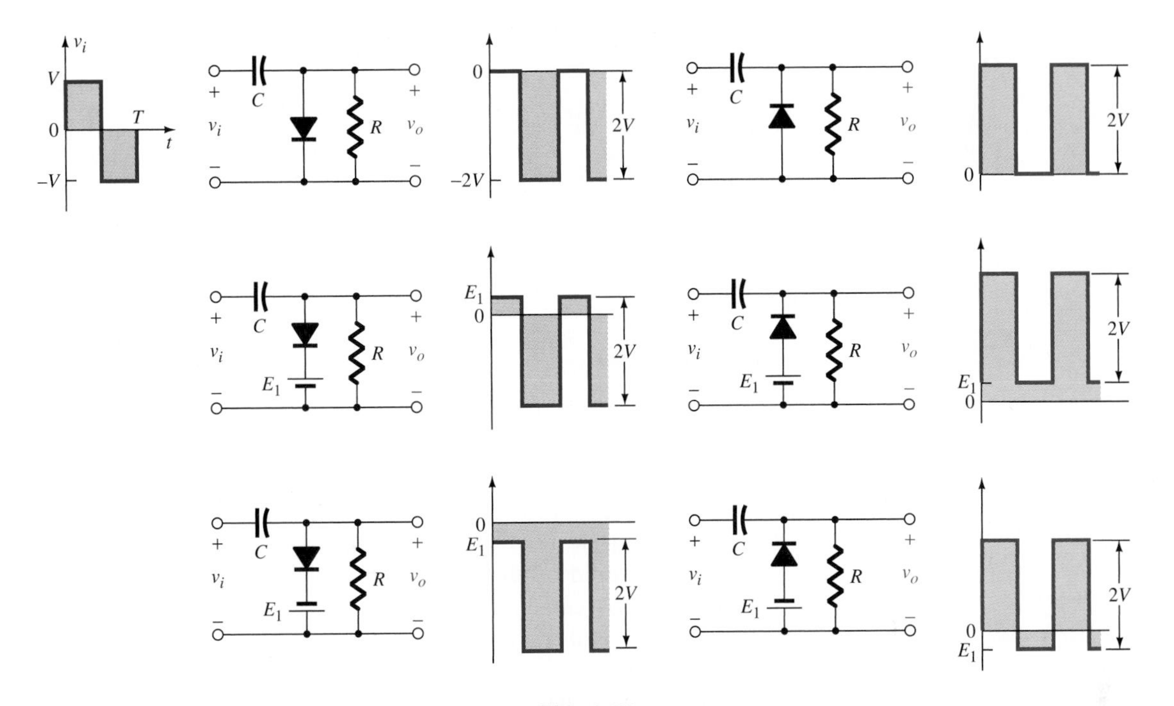

FIG. 8.78
Clamping circuits ($5\tau = 5RC >> T/2$).

8.11 ZENER DIODES

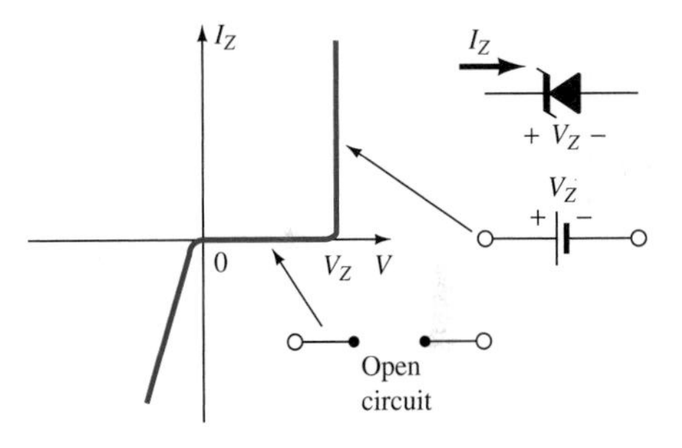

FIG. 8.79
Zener diode characteristics.

The Zener diode is a diode that takes full advantage of the avalanche region, as shown by its characteristics in Fig. 8.79. The symbol for the device and its defined polarities appear in the same figure. For V (with the polarity shown) less than V_Z but greater than 0 V, the device has the characteristics of an open circuit. For applied voltages $V \geq V_Z$ the diode conducts and assumes the short-circuit state indicated by the vertical line. Of course, in this state the short circuit must be associated with the required applied voltage V_Z, as shown in Fig. 8.79. For voltages less than $V = 0$ V the Zener diode has the characteristics of a forward-biased semiconductor diode. Note, in particular, that the direction of conduction for I_Z is opposite the direction of the arrow in the symbol. One frequently employed application of the Zener diode is as a reference potential for other points in the electrical or electronic system. In Fig. 8.80, for example, as long as the applied voltage is greater than 22 V, no matter how distorted the riding ripple may be, exactly 12 V appears at point 1 and 22 V at point 2, since each diode is in its *on* state ($V > V_Z$).

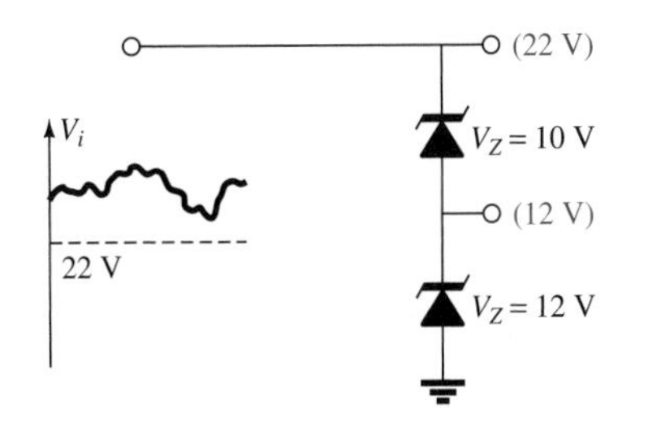

FIG. 8.80
Setting reference voltages using Zener diodes.

A simple square-wave generator can be developed using two back-to-back Zener diodes as shown in Fig. 8.81. For voltage levels in the positive region of the input signal that are greater than 10 V, Z_1 is reverse-biased and in the short-circuit equivalent region, while Z_2 is in the *on* state at its V_Z of 10 V. The remaining input voltage appears across the 5-kΩ series resistor. The result is a clipping of the top at 10 V. Similarly,

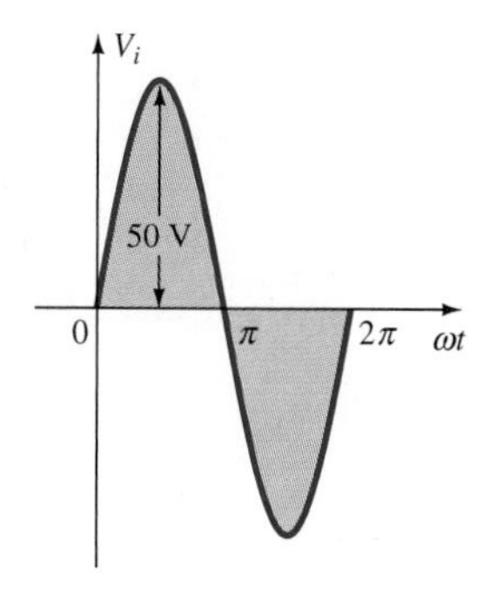

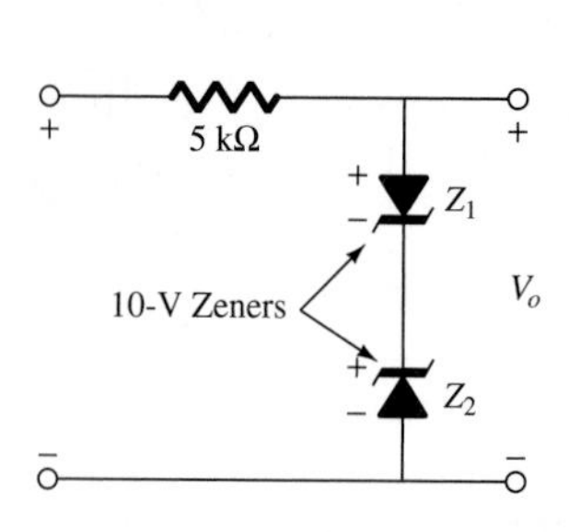

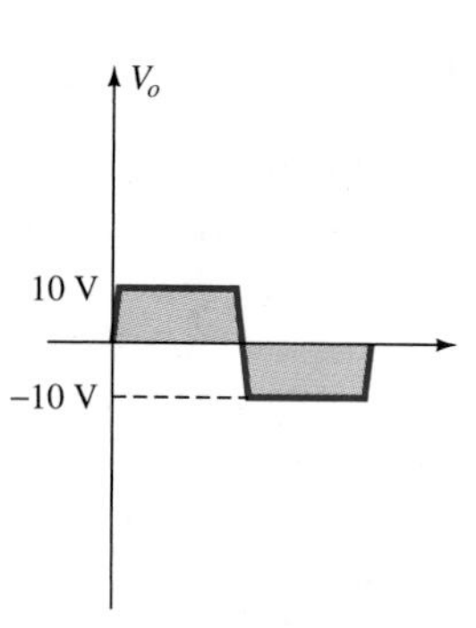

FIG. 8.81
Simple square-wave generator.

the negative portion is cut by 10 V with the state of the diodes reversed. The resulting square wave has only a slightly sloping size if the peak value of the sinusoid is sufficiently large compared with V_Z.

One of the most common applications of the Zener diode is in a *Zener regulator*, such as that shown in Fig. 8.82. The purpose of the Zener diode is to maintain V_L at a fixed level—specifically V_Z—even though E and R_L may vary in magnitude.

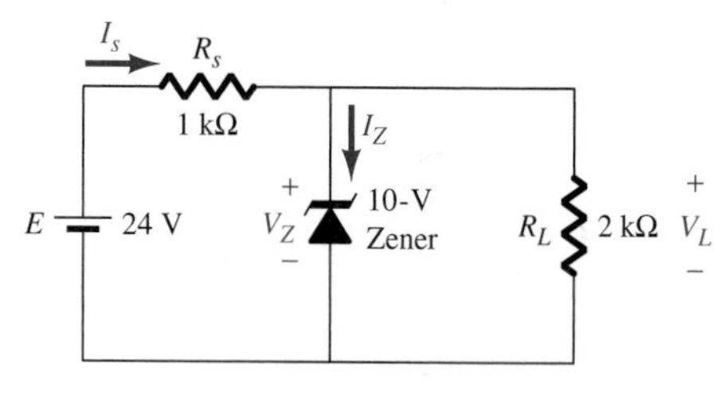

FIG. 8.82
Zener regulator.

The first step in an analysis of the circuit in Fig. 8.82 is to determine whether the Zener diode is in the *on* or the *off* state. The most direct method is simply to temporarily remove the diode as shown in Fig. 8.83 and calculate the voltage across the parallel load R_L. That is,

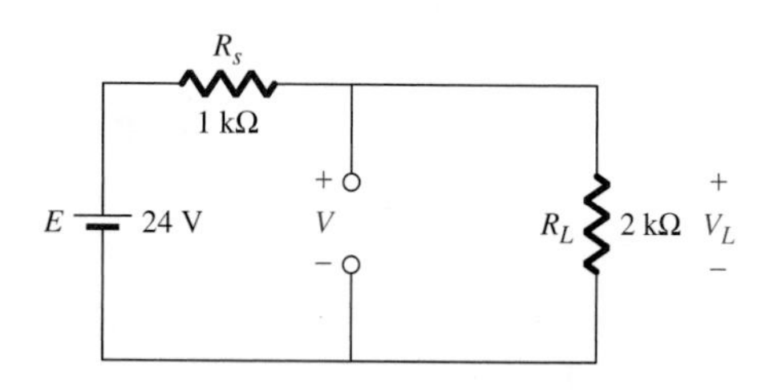

FIG. 8.83
Zener regulator in Fig. 8.82 following the removal of the Zener diode.

$$V = V_L = \frac{(2\text{ k}\Omega)(24\text{ V})}{2\text{ k}\Omega + 1\text{ k}\Omega} = 16\text{ V}$$

If $V_L \geq V_Z$ (as in this case), we can conclude that the Zener diode is in the *on* state because it would have turned on when $V_L = V_Z = 10$ V. The network can then be redrawn as shown in Fig. 8.84, with the circuit equivalent for a Zener diode. In other words, the sole purpose of Fig. 8.83 is to determine the state of the diode.

It should be obvious from Fig. 8.84 that $V_L = V_Z = 10$ V and $V_R = E - V_L = 24\text{ V} - 10\text{ V} = 14$ V, with $I_s = V_R/R = 14\text{ V}/1\text{ k}\Omega = 14$ mA, and $I_L = V_L/R_L = 10\text{ V}/2\text{ k}\Omega = 5$ mA. The magnitude of I_Z must be determined through an application of Kirchhoff's current law: $I_Z = I_s - I_L = 14\text{ mA} - 5\text{ mA} = 9$ mA. One must then check that the power delivered to the diode determined by

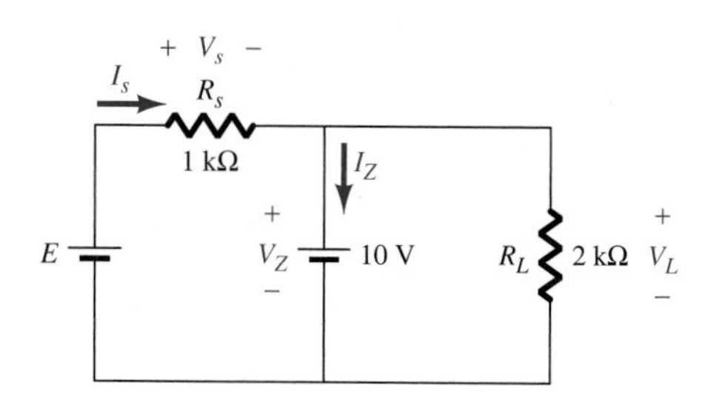

FIG. 8.84
The Zener regulator in Fig. 8.82 in the on state.

$$P = V_{ZM}I_{ZM} \qquad \text{(watts)} \qquad \textbf{(8.12)}$$

is less than the rated value for the device. In this case P (rated) = 100 mW, while $P = V_Z I_Z = (10\text{ V})(9\text{ mA}) = 90$ mW, placing the device within safe limits.

8.12 TUNNEL, VARICAP, AND POWER DIODES

The *tunnel* diode is a two-terminal semiconductor diode whose characteristics appear in Fig. 8.85. The characteristics of a typical semiconductor diode appear as a dashed line on the same set of characteristics for comparison purposes. The region shown as a heavy dark line is of particular interest, since it represents a *negative resistance characteristic*, which means that the current decreases with an increase in voltage rather than the reverse, as is true for most resistive loads. The label *tunnel* diode is derived from the tunneling action of the device at a particular potential level. Because the peak voltage that can be applied across the diode may be on the order of millivolts, the ohmmeter section of a multimeter should not be used for testing purposes, since the internal dc battery is typically a few volts or more.

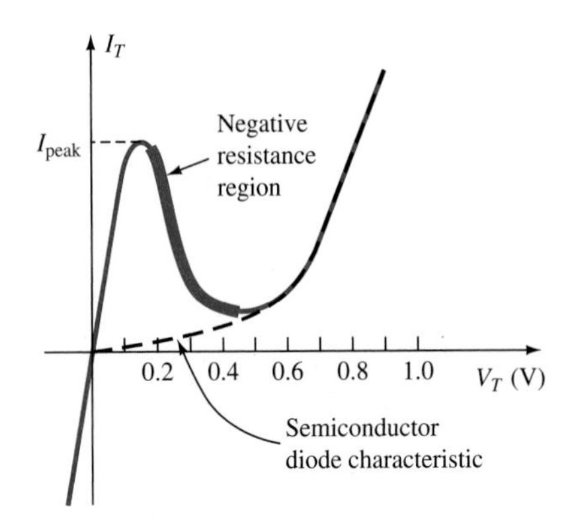

FIG. 8.85
Tunnel diode characteristics.

Like the junction diode described earlier, the tunnel diode is also a two-layer semiconductor device, as shown in Fig. 8.86 with a few of its most frequently employed symbols. The negative resistance has its application in oscillators (sine-wave generators), and its characteristics lend themselves readily to logic-switching applications in computers.

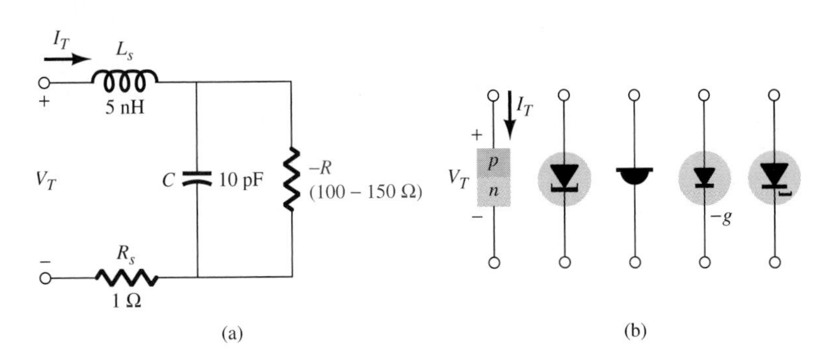

FIG. 8.86
Tunnel diode: (a) equivalent circuit; (b) symbols.

The *varicap* (VVC), or *varactor* diode as it is sometimes called, is also a two-layer semiconductor diode, but its usefulness lies in the fact that the capacitance between its terminal leads is dependent on the reverse-bias voltage applied across the device. In other words, it is a variable capacitance whose capacitance is dependent on the voltage applied across the device. In terms of the reverse-bias voltage, the capacitance is determined by

$$C_T = \frac{1}{k(V_T + V_r)^n} \quad \textbf{(8.13)}$$

where k = constant of semiconductor material
V_T = knee potential of diode characteristics
V_r = magnitude of the applied reverse-bias potential
$n = \frac{1}{2}$ for alloy junctions and $\frac{1}{3}$ for diffused junctions

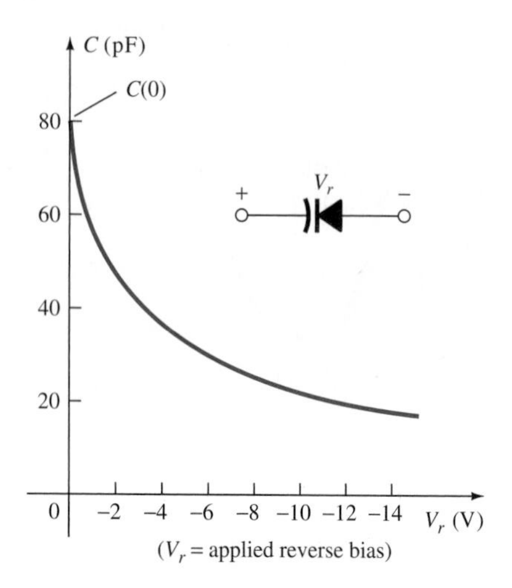

FIG. 8.87
Varicap characteristics: C(pF) versus V_r.

As the reverse-bias potential increases, the width of the depletion region increases, which in turn reduces the transition capacitance. The characteristics of a typical commercially available varicap diode appear in Fig. 8.87. Note the initial sharp decline in C_T with an increase in reverse bias. The normal range of V_r for VVC diodes is limited to about 20 V. In terms of the capacitance at the zero-bias condition $C(0)$, the capacitance as a function of V_r is given by

$$C_T(V_r) = \frac{C(0)}{(1 + |V_r/V_T|)^n} \tag{8.14}$$

The symbols most commonly used for the varicap diode and a first approximation for its equivalent circuit in the reverse-bias region are shown in Fig. 8.88. Since we are in the reverse-bias region, the resistance in the equivalent circuit is very large in magnitude—typically 1 MΩ or larger—while R_s, the geometric resistance of the diode, is, as indicated in Fig. 8.88, very small. The magnitude of C varies from about 2 pF to 100 pF depending on the varicap considered. To ensure that R_r is as large (for minimum leakage current) as possible, silicon is normally used in varicap diodes. The fact that the device is employed at very high frequencies requires that we include the inductance L_s even though it is measured in nanohenries. Recall that $X_L = 2\pi fL$, so a frequency of 10 GHz with $L_s = 1$ nH will result in $X_{L_s} = 2\pi fL = (6.28)(10^{10}$ Hz) $(10^{-9}$ H) $= 62.8\ \Omega$. There is obviously, therefore, a frequency limit associated with the use of each varicap diode.

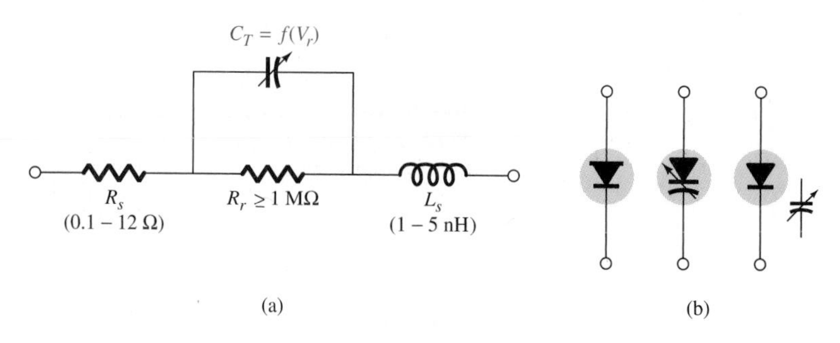

FIG. 8.88
Varicap diode: (a) equivalent circuit in the reverse-bias region; (b) symbols.

For the proper frequency range and a low value of R_s and X_{L_s} compared with the other series elements, the equivalent circuit for the varicap in Fig. 8.88(a) can be replaced with the variable capacitor alone.

The varicap diode finds application in control systems, tuning networks, variable filters, and fm systems, some of which will be described in a later chapter.

The last of the diodes to be introduced in this chapter is the power variety, which is similar in basic construction to the *pn* junction type first described but with certain important changes in its construction such as increased surface area at the junction to handle the heavy currents, almost exclusive use of silicon materials, and heat sinks to draw the heat away from the device. The increased size is noticeable in the few power diodes shown in Fig. 8.89 with their ratings. Consider the very heavy currents that can be handled by these devices of still relatively small size.

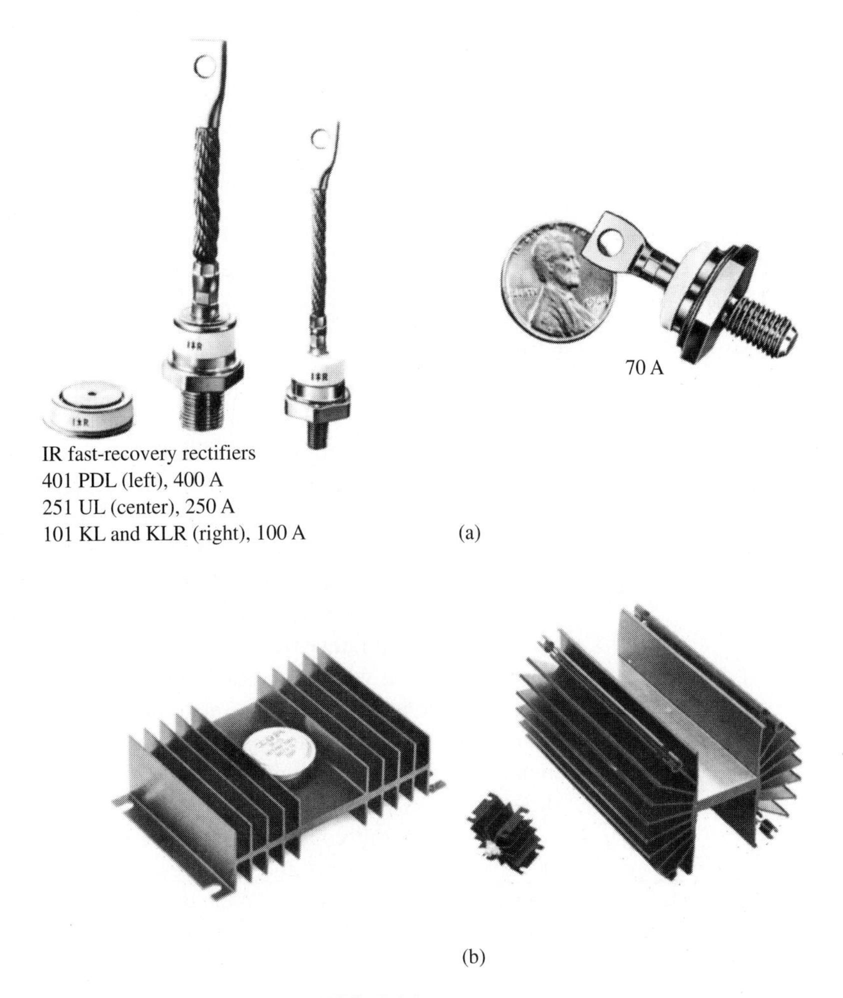

FIG. 8.89
(a) Power diodes; (b) heat sinks. (Courtesy of International Rectifier Corporation.)

8.13 SOLAR CELLS

In recent years there has been increasing interest in the solar cell as an alternative source of energy. When we consider that the power density received from the sun at sea level is about 100 mW/cm^2 (1 kW/m^2), it is certainly an energy source that requires further research and development to maximize the conversion efficiency from solar to electrical energy.

The basic construction of a silicon *pn* junction solar cell appears in Fig. 8.90. As shown in the top view, every effort is made to ensure that

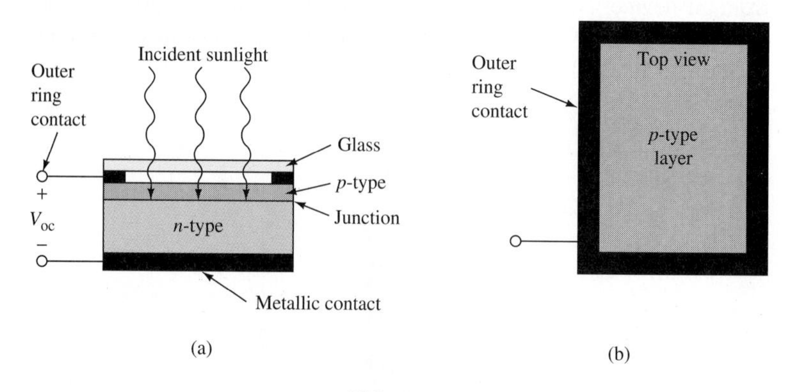

FIG. 8.90
Solar cell: (a) cross section; (b) top view.

the surface area perpendicular to the sun is a maximum. Also, note that the metallic conductor connected to the *p*-type material and the thickness of the *p*-type material are such that they ensure that a maximum number of photons, or packets of light energy, reach the junction. A photon in this region may collide with a valence electron and impart to it sufficient energy to leave the parent atom. The result is the generation of free electrons and holes. This phenomenon occurs on each side of the junction. In *p*-type material the newly generated electrons are minority carriers and move rather freely across the junction, as explained for the basic *pn* junction with no applied bias. A similar discussion is true for the holes generated in the *n*-type material. The result is an increase in the minority carrier flow, which is in the opposite direction of the conventional forward current of a *pn* junction. This increase in reverse current is shown in Fig. 8.91. Since $V = 0$ anywhere on the vertical axis and represents a short-circuit condition, the current at this intersection is called the *short-circuit current* and is represented by the notation I_{sc}. Under open-circuit conditions ($i_d = 0$) the *photovoltaic* voltage V_{oc} results. This is a logarithmic function of the illumination, as shown in Fig. 8.92. V_{oc} is the terminal voltage of a battery under no-load (open-circuit) conditions. Note, however, that in the same figure the short-circuit current is a linear function of the illumination. That is, it doubles for the same increase in illumination (f_{c_1} and $2f_{c_2}$ in Fig. 8.92), while the change in V_{oc} is less for this region. The major increase in V_{oc} occurs for lower-level increases in illumination. Eventually, a further increase in illumination will have very little effect on V_{oc}, although I_{sc} will increase, causing the power capabilities to increase.

Selenium and silicon are the most widely used materials for solar cells, although gallium arsenide, indium arsenide, and cadmium sulfide, among others, are also used. The wavelength of the incident light affects the response of the *pn* junction to the incident photons. Note in Fig. 8.93 how closely the selenium cell response curve matches that of the eye.

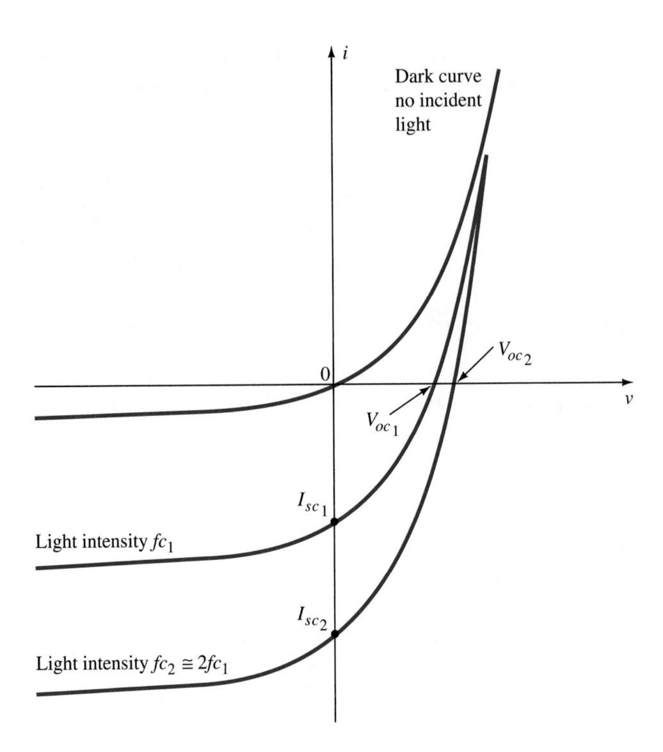

FIG. 8.91
Short-circuit current and open-circuit voltage versus light intensity for a solar cell.

This fact has widespread application in photographic equipment such as exposure meters and automatic exposure diaphragms. Silicon also overlaps the visible spectrum but has its peak at the 0.8-μm (8000-Å) wavelength, which is in the infrared region. In general, silicon has a higher conversion efficiency and greater stability and is less subject to fatigue. Both materials have excellent temperature characteristics. That is, they

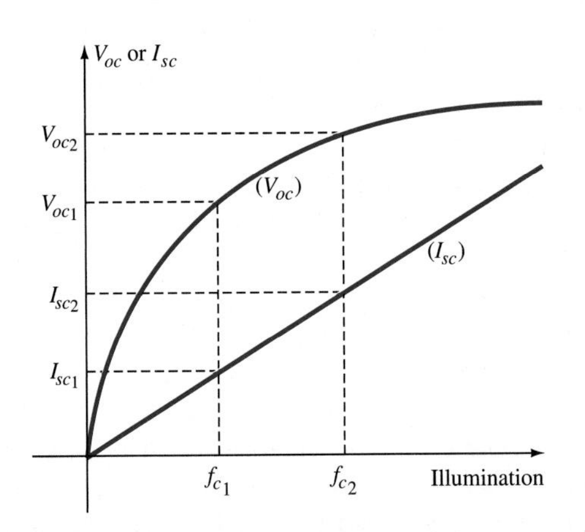

FIG. 8.92
V_{oc} and I_{sc} versus illumination for a solar cell.

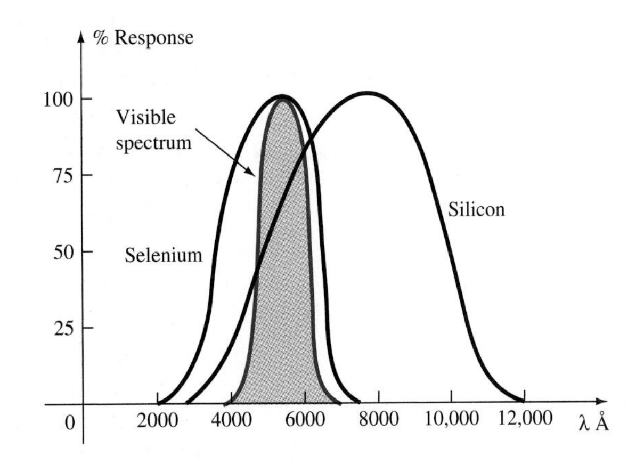

FIG. 8.93
Special response of Se, Si, and the naked eye.

can withstand extreme high or low temperatures without a significant drop-off in efficiency. Some typical solar cells, with their electrical characteristics, appear in Fig. 8.94.

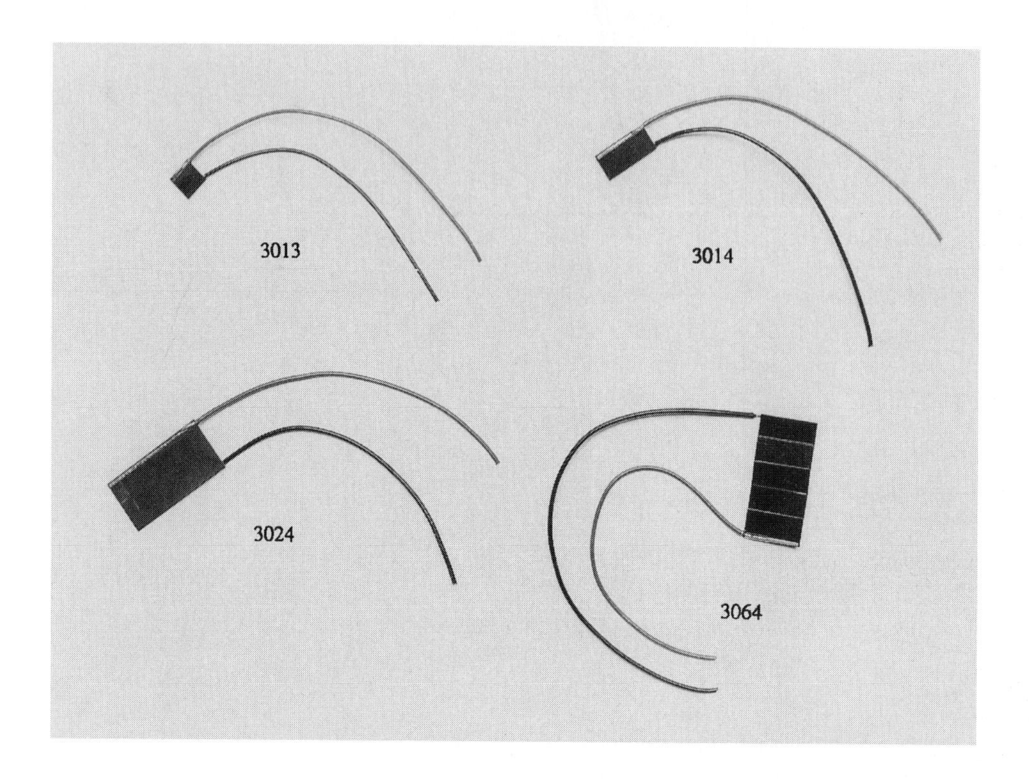

Electrical characteristics

Part no.	Active area	Test voltage	Minimum current test voltage
3013	0.032 in^2 (0.21 cm^2)	0.4 V	4.2 mA
3014	0.065 in^2 (0.42 cm^2)	0.4 V	8.4 mA
3024	0.29 in^2 (1.87 cm^2)	0.4 V	38 mA
3064	0.325 in^2 (2.1 cm^2)	2 V	8.4 mA

FIG. 8.94
Typical solar cells and their electrical characteristics. (Courtesy of EG&G VACTEC, Inc.)

It might be of interest to note that the Lockheed Missiles and Space Company was awarded a grant from the National Aeronautics and Space Administration to develop a massive solar-array wing for the space shuttle. The wing will measure 13.5 × 105 ft when extended and will contain 41 panels, each carrying 3060 silicon solar cells. The wing can generate a total of 12.5 kW of electrical power.

The efficiency of operation of a solar cell is determined by the electrical power output divided by the power provided by the light source. That is,

$$\eta = \frac{P_{o(\text{electrical})}}{P_{i(\text{light energy})}} \times 100\% = \frac{P_{\max(\text{device})}}{(\text{area in cm}^2)(100\text{ mW/cm}^2)} \times 100\% \tag{8.15}$$

Typical levels of efficiency range from 10 to 14%—a level that should improve measurably if the present interest continues. A typical set of output characteristics for silicon solar cells of 10% efficiency with an active area of 1 cm^2 appears in Fig. 8.95. Note the optimum power locus and the almost linear increase in output current with luminous flux for a fixed voltage.

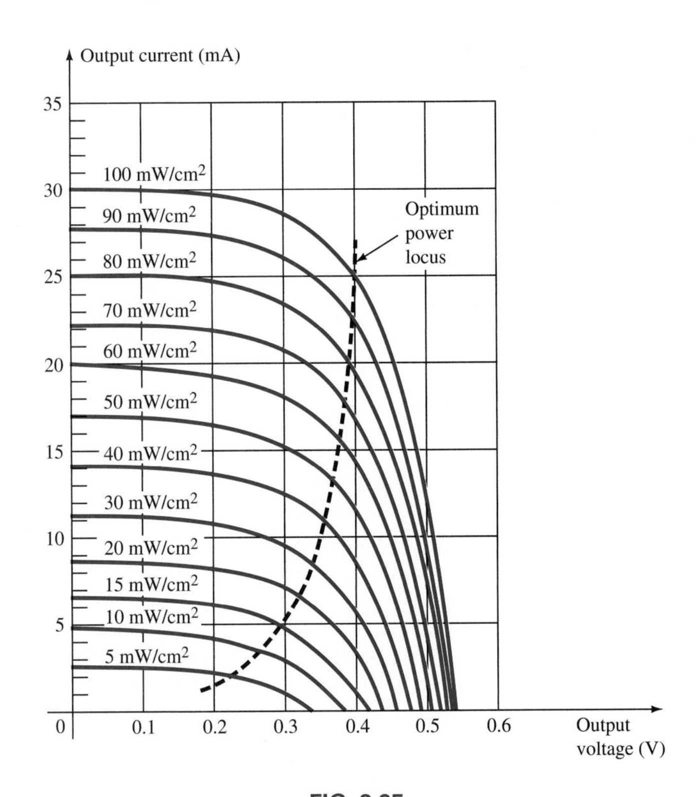

FIG. 8.95

Typical output characteristics for silicon cells of 10% efficiency having an active area of 1 cm^3. Cell temperature is 30°C.

8.14 THERMISTORS AND PHOTOCONDUCTIVE DEVICES

A *thermistor* is a two-terminal device whose terminal resistance is dependent on the body temperature of the device. It has a *negative temperature coefficient,* indicating that the resistance decreases with an increase in temperature rather than the reverse, as is typical for most commercial resistors introduced earlier. The characteristics of one such device appear in Fig. 8.96. Note the use of a log scale on the vertical axis to permit a broader range of resistance levels. At room temperature the terminal resistance is about 5 kΩ, while at 100°C (212°F—the boiling point of water) its resistance is only 100 Ω. There is therefore a 50-fold change in resistance for a temperature span of 80°C. The temperature of the device can be changed by either changing the current through the device or heating its surface through the surrounding medium or immersion. The symbol for the device is commercially accepted to be that appearing in Fig. 8.96. Thermistors come in all shapes and sizes, as indicated by the photograph in Fig. 8.97.

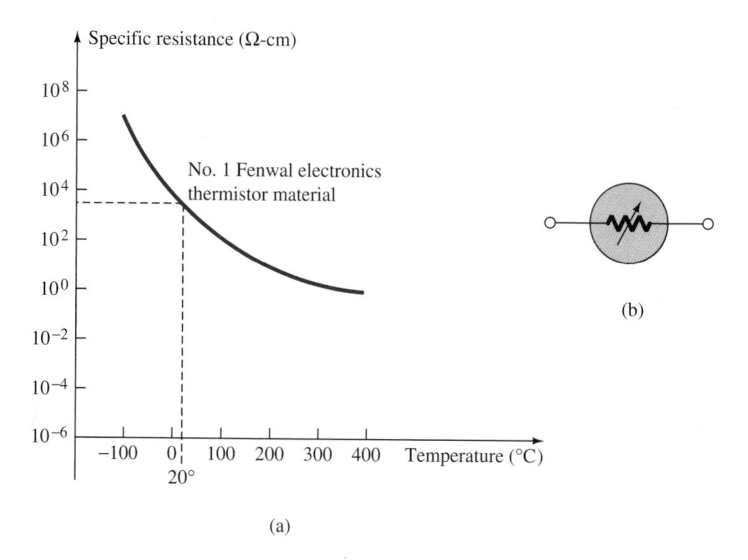

FIG. 8.96
Thermistor: (a) characteristics; (b) symbol.

Photoconductive devices react to the light incident on a particular surface of the element. As the light intensity increases as shown in Fig. 8.98, the resistance decreases, resulting in a negative resistance coefficient. The symbol for the device and a photograph of one commercially available unit appear in Fig. 8.99.

The *photodiode* is a two-terminal device whose current-versus-illumination characteristics (Fig. 8.100) closely resemble those of I_D versus V_D for a *pn* junction diode. The incident light is directed onto the junction between the semiconductor layers of the device by a lens in the cap of the diode (Fig. 8.101).

The energy conversion of a photodiode is reversed in the increasingly popular *light-emitting diodes* (LEDs) used for numerical displays in calculators, timepieces, instruments, and so on. In the process of *electroluminescence,* radiant light is emitted at a strength dependent on the current through the device, as shown by the plot in Fig. 8.102. The symbol for the device and commercially available units appear in Fig. 8.103.

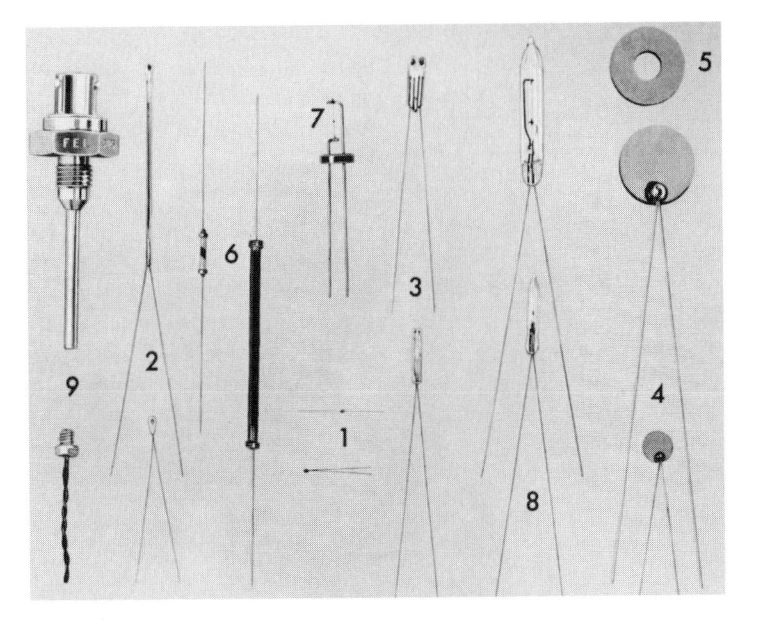

FIG. 8.97
Various types of thermistors: (1) beads; (2) glass probes; (c) iso-curve interchangeable probes and beads; (4) disks; (5) washers; (6) rods; (7) specially mounted beads; (8) vacuum- and gas-filled probes; (9) special probe assemblies.

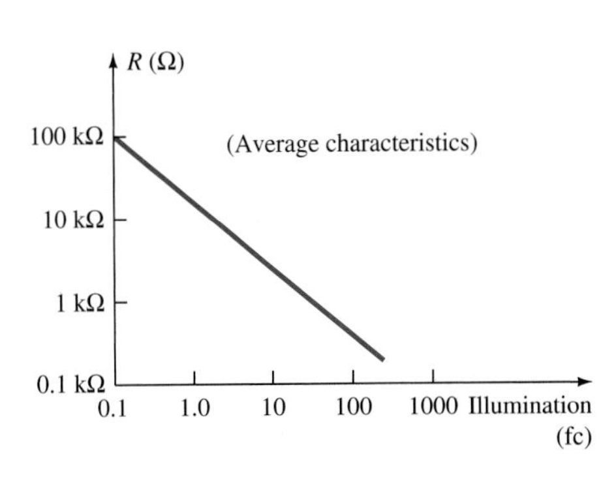

FIG. 8.98
Photoconductive cell terminal characteristics (GE type B425).

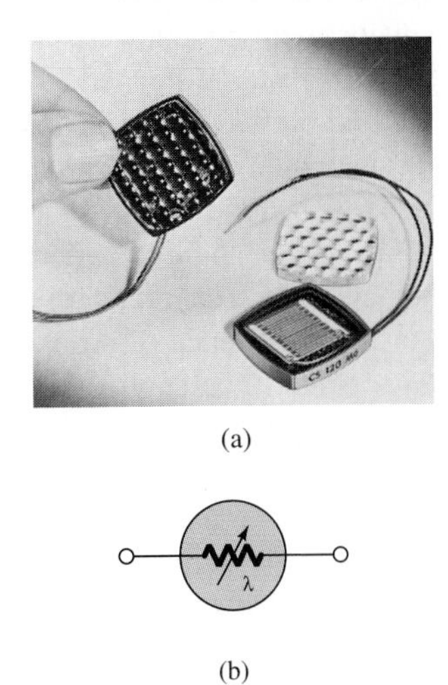

FIG. 8.99
Photoconductive cell: (a) type; (b) symbol. (Courtesy of International Rectifier Corporation.)

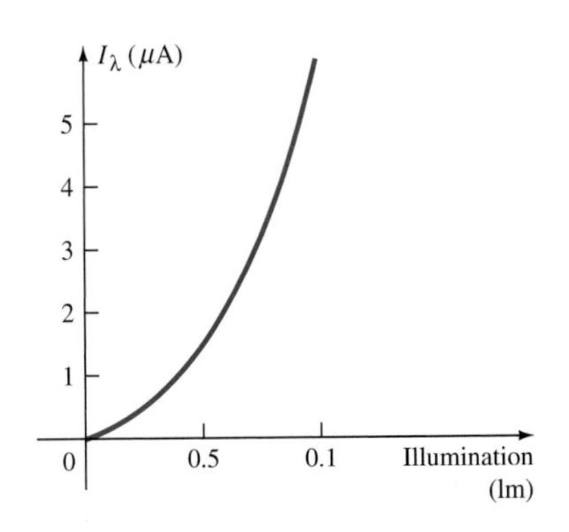

FIG. 8.100
Photodiode characteristics.

The *liquid-crystal display* (LCD) has the distinct advantage of having a lower power requirement than the LED. It is typically on the order of microwatts for the display, as compared with the same order of milliwatts for LEDs. It does, however, require an external or internal light source, is limited to a temperature range of about 0° to 60°C, and

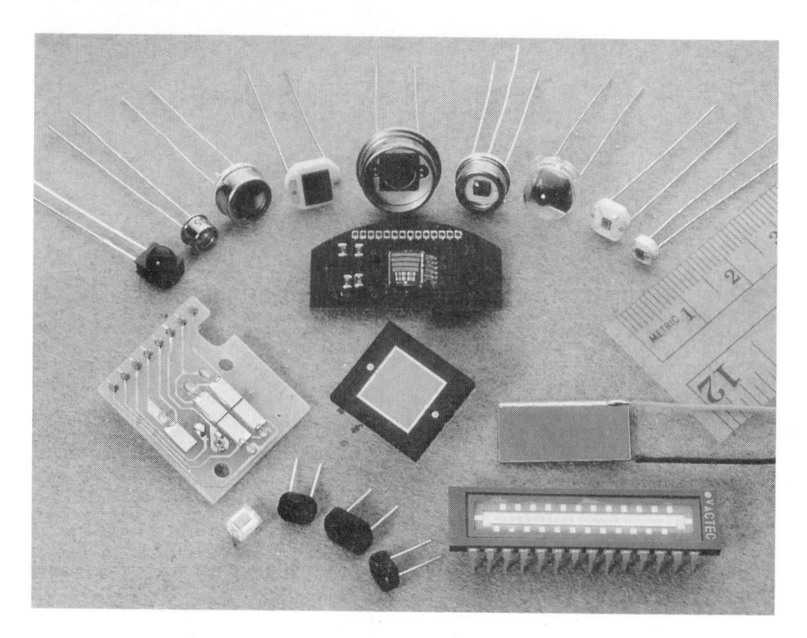

FIG. 8.101
Photodiodes. (Courtesy of EG&G VACTEC, Inc.)

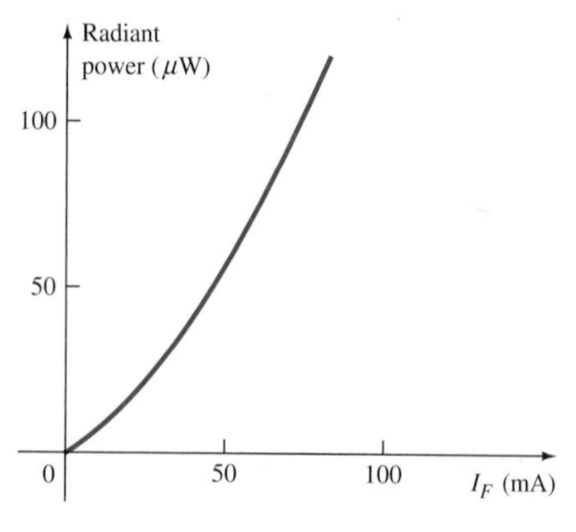

FIG. 8.102
LED characteristics.

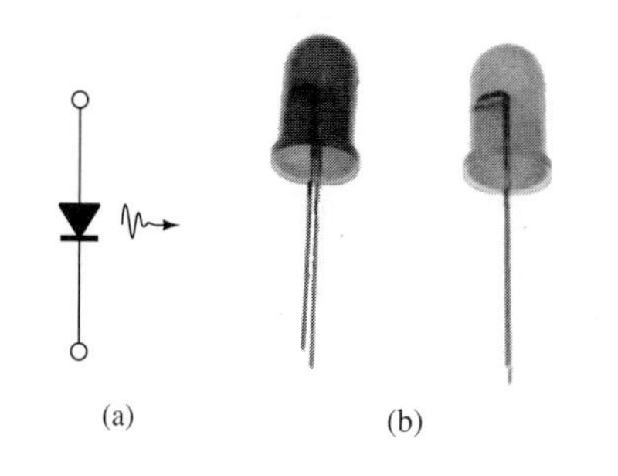

FIG. 8.103
LED: (a) symbol; (b) device appearance.

lifetime is an area of concern because LCDs can chemically degrade. The types receiving the major attention today are the dynamic-scattering and field-effect units. A liquid crystal is a material (normally organic for LCDs) that flows like a liquid but whose molecular structure has some properties normally associated with solids.

In both types introduced above, the liquid crystal is encapsulated in two parallel glass plates each of which has a pattern such as that in Fig. 8.104 applied to the glass surface in contact with the liquid crystal. In the dynamic-scattering units the absence of an applied voltage results in a clear display. When a potential is applied to terminals 3, 4, 5, 7, and 8 to display the number 2, that region of the liquid crystal has a voltage applied across it. The result is a change in the orientation of the atoms of the crystal in these regions, and the incident light is reflected in different directions. The energized regions take on a frosted appearance, which allows the numeric character to be distinguished from the remaining area of the display. The pressing of the button on a watch results in application of the required energizing potential to establish a readout.

The LCD does not generate its own light but depends on an external or internal source. Under dark conditions it is necessary for the unit to have its own internal light source either behind or to the side of the LCD. During the day, or in lighted areas, a reflector can be put behind the LCD to reflect the light back through the display for maximum intensity. For optimum operation, current watch manufacturers use a combination of the transmissive (an internal light source) and reflective modes called *transflective*.

The *field-effect* LCD has the same appearance and a thin layer of encapsulated liquid crystal, but its mode of operation is very different. Similar to the dynamic-scattering LCD, the field-effect LCD can be operated in the reflective or transmissive mode with an internal light source. The transmissive unit employs a polarizer that allows only vertical light patterns to pass through. The face of the display remains dark until a defined potential is applied to particular regions as described previously. The energized regions allow the vertical component of the entering light to pass through and lighten these regions of the display. The numeric or alphabetic characters are then distinguishable on the display.

The lifetime of LCD units is steadily increasing beyond the 10,000+ h limit. Since the color generated by LCD units is dependent on the source of illumination, there is a greater range of color choice.

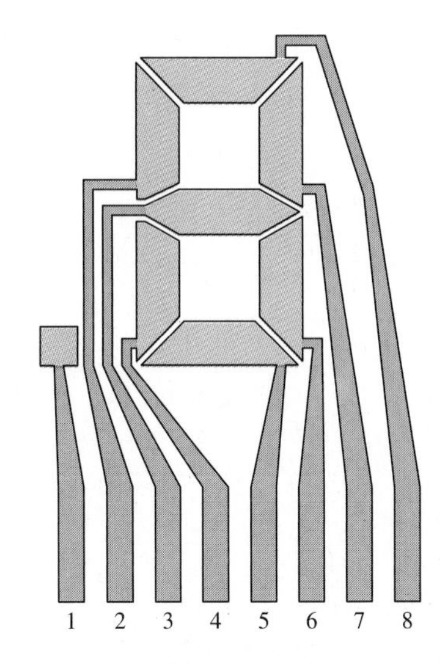

FIG. 8.104
LCD eight-segment digit display.

8.15 COMPUTER ANALYSIS

PSpice (Windows)

Use of PSpice for the network of Fig. 8.105 will result in the schematics screen as shown in Fig. 8.106 with the voltage at three nodes and the current at every point in the circuit.

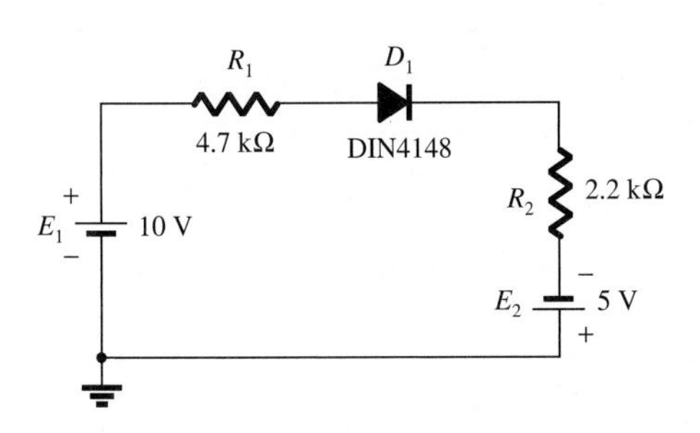

FIG. 8.105
Diode circuit to be analyzed using PSpice (Windows).

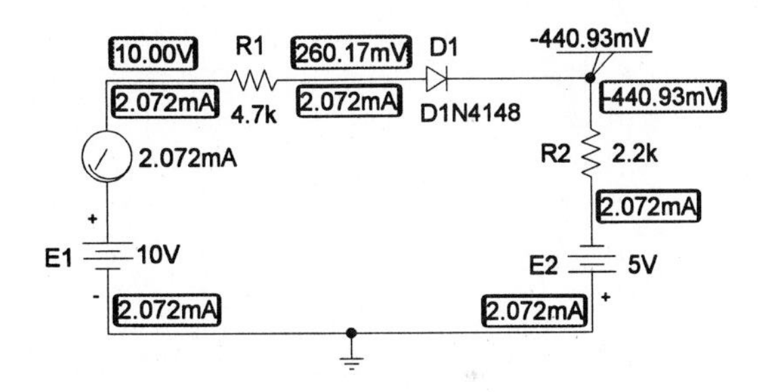

FIG. 8.106
Analyzing the circuit of Fig. 8.105 using PSpice (Windows).

For the network of Fig. 8.105 the current (clockwise, since both dc supplies are applying pressure in this direction) is determined by

$$I = \frac{10\text{ V} + 5\text{ V} - 0.7\text{ V}}{4.7\text{ k}\Omega + 2.2\text{ k}\Omega} = 2.072\text{ mA}$$

using a 0.7-V drop across the conducting silicon diode. The PSpice solution of Fig. 8.106 is an exact match for the current level.

The voltage drop across the conducting diode in Fig. 8.106 is

$$V_D = 260.17\text{ mV} - (-440.93\text{ mV}) \cong 0.7\text{ V}$$

which matches the assumed level for the longhand solution.

The voltage (to ground) to the right of the resistor R_1 is determined from the network of Fig. 8.105 in the following manner:

$$10\text{ V} - I(4.7\text{ k}\Omega) = 10\text{ V} - (2.072\text{ mA})(4.7\text{ k}\Omega) = 261.6\text{ mV}$$

which compares very well with the PSpice solution of 260.17 mV.

The characteristics of a D1N4148 silicon diode can be generated with PSpice using the simple cirucit of Fig. 8.107. Note in the figure that the initial source voltage is 0 V. Through an incrementing routine that will increase the voltage E from 0 V to 10 V in 0.1-V steps the voltage across the diode and the diode current can be found at a 100 different points to establish the data for the plot of Fig. 8.108.

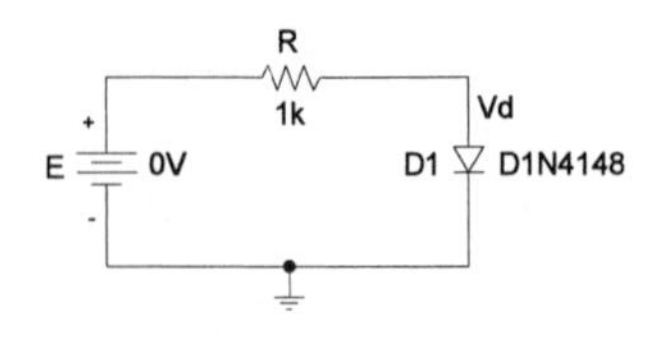

FIG. 8.107
Network for obtaining the characteristics of the D1N4148 diode.

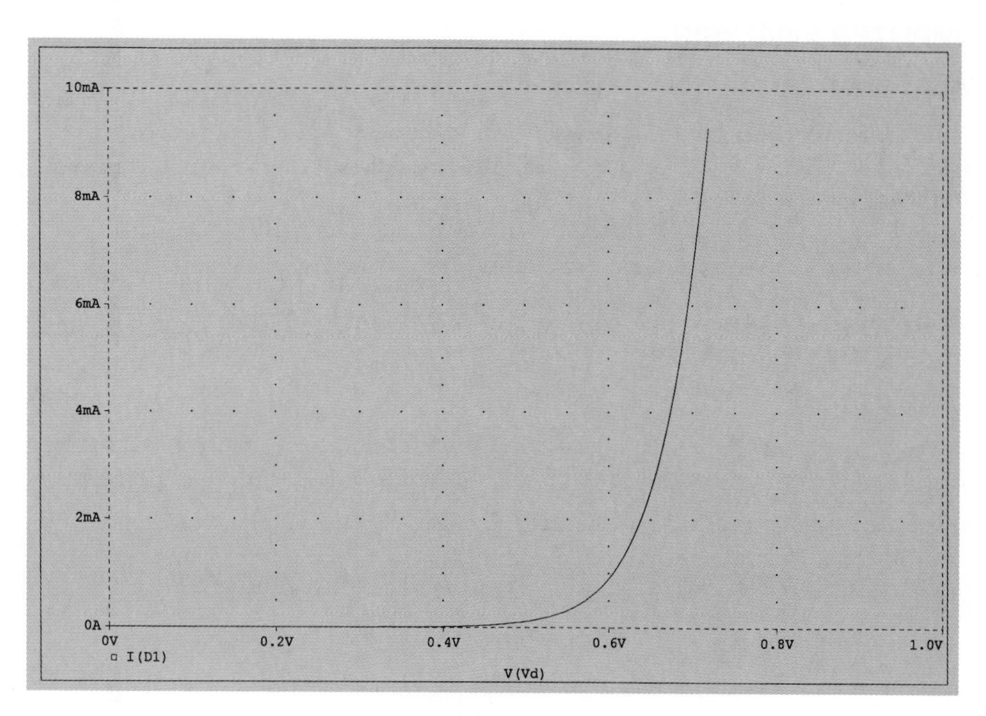

FIG. 8.108
Characteristics of the D1N4148 diode.

Note in Fig. 8.108 how the voltage across the diode rises from about 0.4 V to 0.72 V for different levels of diode current. At a diode voltage of 0.65 V the diode current is about 2.5 mA. Substitution of these levels into Fig. 8.107 will result in a required source voltage of

$$E = I_D R + V_D = (2.5\text{ mA})(1\text{ k}\Omega) + 0.65\text{ V} \cong 3.15\text{ V}$$

Electronics Workbench

Application of Electronics Workbench to the circuit of Fig. 8.105 will result in the display of Fig. 8.109. In this case, however, the interest was limited to the voltage at one node and the circuit current. As indicated in Fig. 8.109, the 1N4148 diode is also part of the Electronics Workbench library, resulting in measurements very close to those obtained using PSpice.

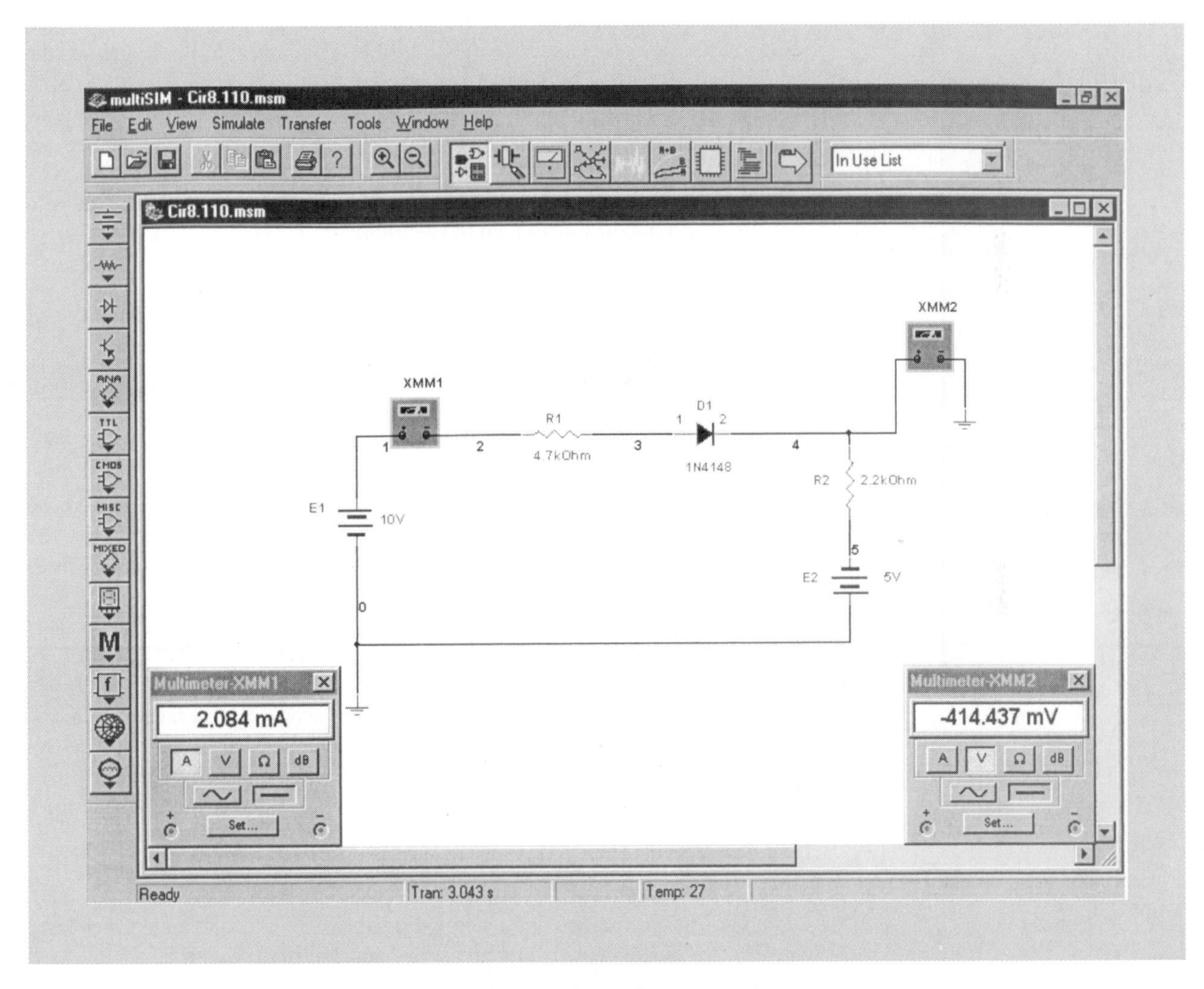

FIG. 8.109
Use of Electronics Workbench to determine the current and voltage from one end of the diode to ground for the circuit of Fig. 8.105.

8.16 SUMMARY

- The **diode** is a **two-terminal** electronic device that behaves much like an ordinary **switch** except that it can conduct in **only** one direction.
- The **silicon diode** is constructed of **two semiconductor materials** referred to as *n*-type and *p*-type. The ***n*-type** material is silicon material with an **excess of electrons,** while the ***p*-type** material has an **excess of holes.**
- The **depletion region** is a region **near the boundary** of the *n*- and *p*-type materials of a diode that is essentially **void of free carriers.**
- The **ideal diode** has **zero resistance** when biased in the forward-bias region and **infinite resistance** (open circuit) in the reverse-bias region.
- The **resistance of a diode** in the forward-bias region changes from point to point, with the **highest levels** found at **low voltages** and the **lowest** at **higher voltage levels.**
- The **ac resistance of a diode** is sensitive to the **shape** of the diode characteristics at the point of interest, while the dc resistance is determined **solely** by the voltage and current at the **same point.**
- The **derivative** of a function at a point is equal to the **slope of a tangent line drawn at the point.**
- The **ac resistance** of a diode can be found simply by dividing **26 mV** by the dc **diode current** at the point of interest.
- An **open circuit** can have **any voltage** across its terminals, but the **current** is always **0 A.**
- A **short circuit** has a **0-V drop** across its terminals, but the **current is limited** only by the surrounding network.
- A **half-wave rectifier** ensures that only **half** of the applied sinusoidal voltage reaches the load.
- A **full-wave rectifier flips** one portion of the sinusoidal voltage, so that both pulses appear across a load with the **same polarity.**
- A **clipper "clips" away** a portion of the applied signal on its way to the load.
- A **clamper "clamps"** an ac voltage to a **different dc level** but does not change the overall appearance of the waveform. The net swing of the applied signal and the load voltage is the same.
- The **time constant** of a clamping network must be **large enough** to ensure that the voltage across the capacitor **does not discharge significantly** during the interval the **diode is not conducting.**
- The **Zener region** of a diode has a fairly **fixed voltage level** for a significant range of diode current.
- A **tunnel diode** has a **negative-resistance** characteristic that is useful for oscillators and logic-switching applications.
- The **varicap** or **varactor diode** has a significant **capacitance** level that is sensitive to the applied **reverse-bias potential.**
- **Solar cells** have the ability to **convert sunlight** into **electrical energy** using a silicon *pn* junction.
- **Thermistors** are two-terminal semiconductor devices whose **resistance drops nonlinearly** with an **increase in temperature.**

- **LEDs** are diodes that **emit light** when a **forward current** of proper magnitude is passed through the device.
- **LCDs** are **liquid-crystal displays** that permit a display of numerical values through a **reorientation of the internal atoms** of the crystal when a voltage is applied.

Equations:

$$r_{ac} = r_d = \frac{\Delta V_D}{\Delta I_D} = \frac{26 \text{ mV}}{I_D \text{ (mA)}} \quad (\Omega)$$

$$I_D = I_s(e^{kV/T_K} - 1) \quad \text{(A)}$$

$$R_{dc} = \frac{V_D}{I_D} \quad (\Omega)$$

Half-Wave Rectified Waveform

average $= 0.318\ V_m$; PIV rating $= V_m$

Full-Wave Rectified Waveform

average $= 0.636\ V_m$; PIV rating $= V_m$

Zener Diode

$$P = V_{ZM} I_{ZM} \quad \text{(W)}$$

Varactor Diode

$$C_T = \frac{1}{k(V_T + V_r)^n}$$

$$C_T(V_r) = \frac{C(0)}{(1 + |V_r/V_T|)^n}$$

Solar Cell

$$\eta\% = \frac{P_{max}}{(\text{area in cm}^2)(100 \text{ mW/cm}^2)} \times 100\%$$

PROBLEMS

SECTION 8.2 Semiconductor Diode Characteristics

1. Describe in your own words the characteristics of the *ideal* diode and how they determine the *on* and *off* state of the device. That is, describe why the short-circuit and open-circuit equivalents are appropriate.

2. What is the one important difference between the characteristics of a simple switch and those of an ideal diode? (*Hint:* Consider the conduction state and the direction of charge flow.)

3. Describe in your own words the condition established by a forward- and reverse-bias condition on a *pn* junction diode and how it affects the resulting current.

4. Describe how you can remember the forward- and reverse-bias states of the *pn* junction diode. That is, how can you remember which potential (positive or negative) is applied to which terminal?

5. **a.** Determine the dc resistance of the silicon diode in Fig. 8.4 using Eq. (8.1) at 10 and 30 mA.
   b. Determine the ac resistance of the silicon diode in Fig. 8.4 using Eq. (8.3) at 10 and 30 mA.
   c. Discuss the results obtained for parts (a) and (b).

6. Determine the dc resistance of the silicon diode in Fig. 8.4 at $I_D = 10$ and 30 mA and compare this value with the results for Problem 5.

7. **a.** For the practical diode characteristics in Fig. 8.6(c), determine the current I_D at $V_D = +0.3$ V and -0.5 V.
   b. For the same diode characteristics determine V_D if $I_D =$ 1 mA and 3.6 mA.

8. Determine the dc resistance of the commercial diode in Fig. 8.5 at:
   a. $V_D = -5$ V and -20 V.
   b. $I_D = 5$ mA.
   c. $V_D = 0.7$ V.
   d. $I_D = 50$ mA.

9. Using Eq. (8.2), determine the diode current at 20°C for a silicon diode with $I_s = 50$ nA and an applied forward bias of 0.6 V.

10. Repeat Problem 9 for $T = 100$°C (the boiling point of water). Assume that I_s has increased to 1 μA.

11. **a.** If the ohmmeter section of an ohmmeter provides a high resistance indication in both directions (alternate lead connections) for a silicon diode, what is the condition of the diode and what is the possible reason?
    b. If the diode-checking position of a DMM provides an O.L. indication in both directions, what is the condition of the diode? Provide a possible reason for the readings. How does it match the results obtained for part (a)?
    c. Is it possible for a silicon diode to have a low resistance reading (VOM) or 0.7-V level in both directions? If not, why not?

SECTION 8.3 Series Diode Configurations with Dc Inputs

12. Determine V_o and I_D for the networks in Fig. 8.110.

13. Determine V_o and I_D for the networks in Fig. 8.111.

14. Determine V_{o_1} and V_{o_2} for the networks in Fig. 8.112.

SECTION 8.4 Parallel and Series–Parallel Configurations

15. Determine V_o and I_D for the networks in Fig. 8.113.

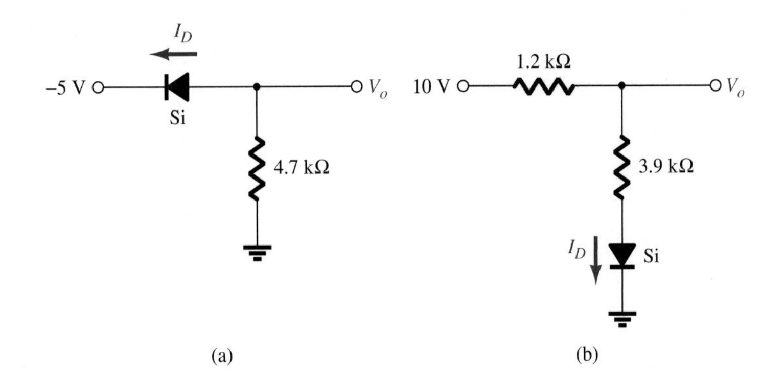

FIG. 8.110

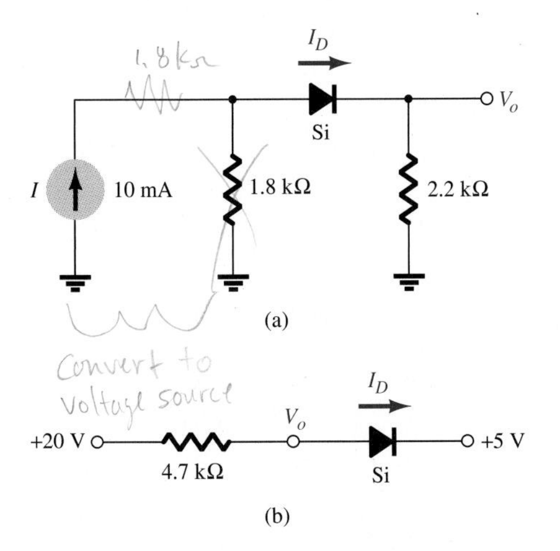

FIG. 8.111

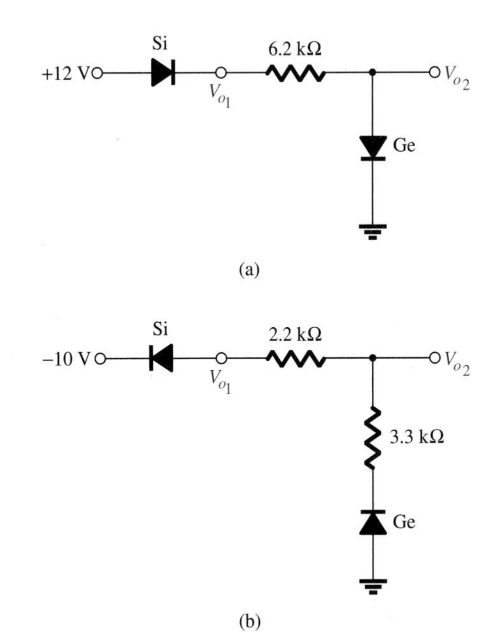

FIG. 8.112

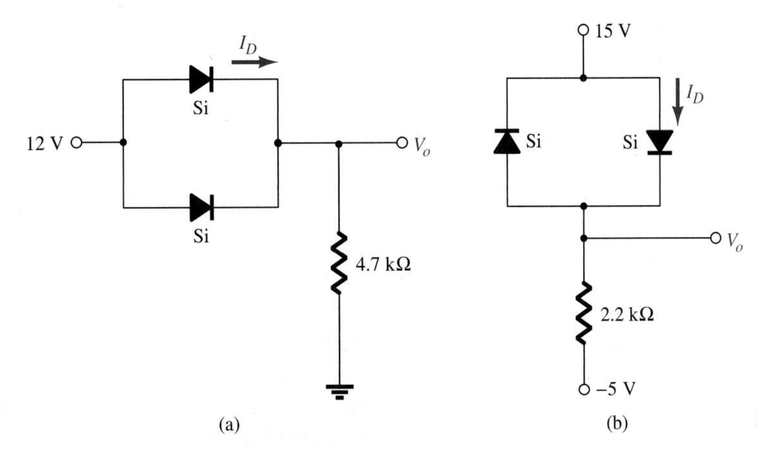

FIG. 8.113

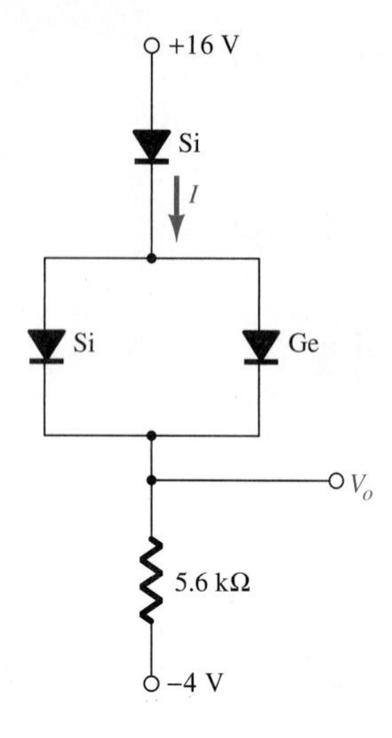

FIG. 8.114

16. Determine V_o and I for the network in Fig. 8.114.

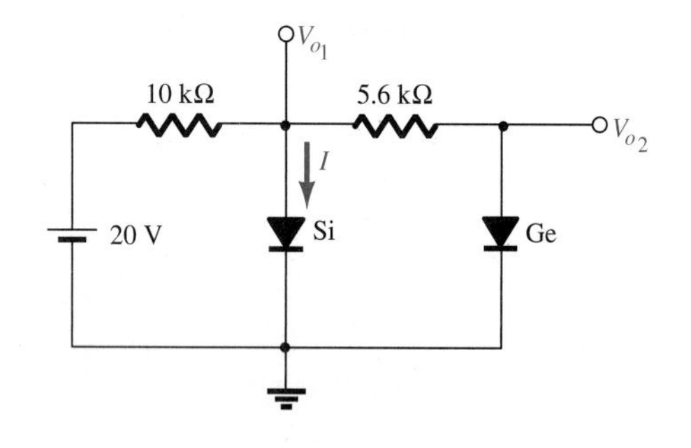

FIG. 8.115

17. Determine V_{o_1}, V_{o_2}, and I for the network in Fig. 8.115.

18. Determine V_o and I_D for the network in Fig. 8.116.

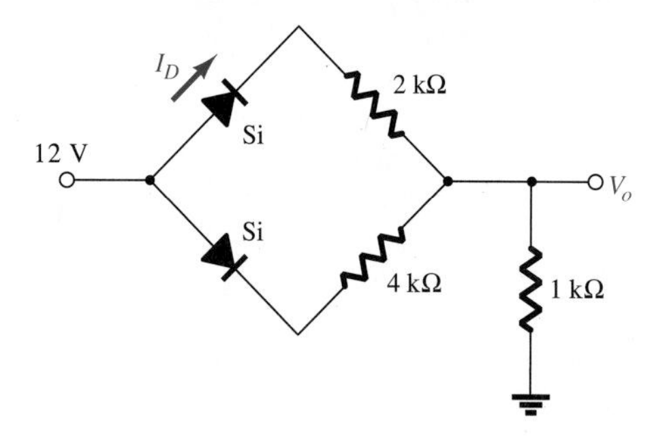

FIG. 8.116

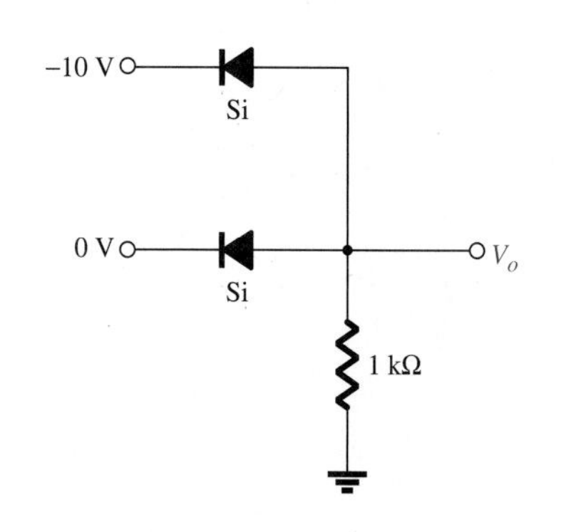

FIG. 8.117

SECTION 8.5 AND/OR GATES

19. Determine V_o for the network in Fig. 8.31 with 0 V on both inputs.

20. Determine V_o for the network in Fig. 8.31 with 10 V on both inputs.

21. Determine V_o for the network in Fig. 8.34 with 0 V on both inputs.

22. Determine V_o for the network in Fig. 8.34 with 10 V on both inputs.

23. Determine V_o for the negative logic OR gate in Fig. 8.117.

24. Determine V_o for the negative logic AND gate in Fig. 8.118.

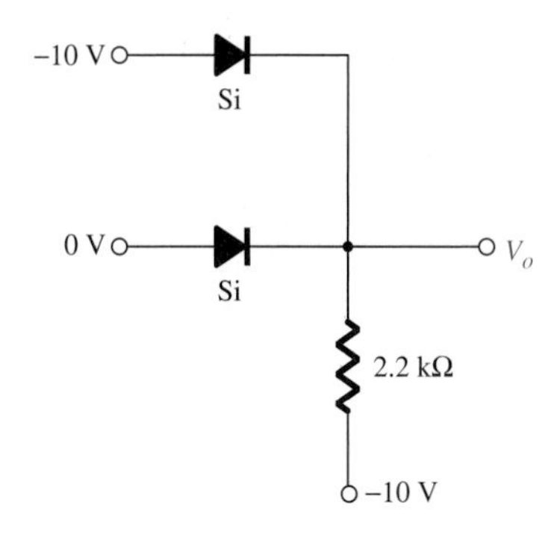

FIG. 8.118

SECTION 8.6 Half-Wave Rectification

25. Assuming an ideal diode, sketch v_i, v_d, and i_d for the half-wave rectifier in Fig. 8.119. The input is a sinusoidal waveform with a frequency of 60 Hz.

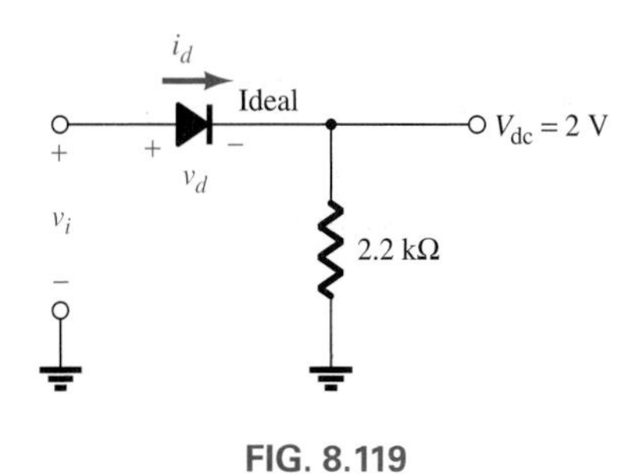

FIG. 8.119

26. Repeat Problem 25 with a 6.8-kΩ load applied as shown in Fig. 8.120. Also sketch v_L and i_L.

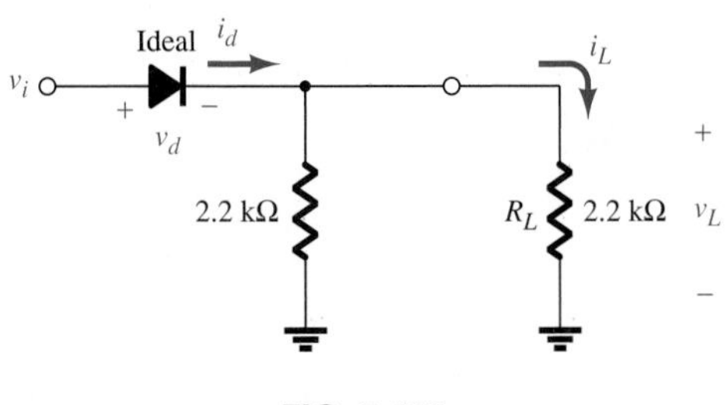

FIG. 8.120

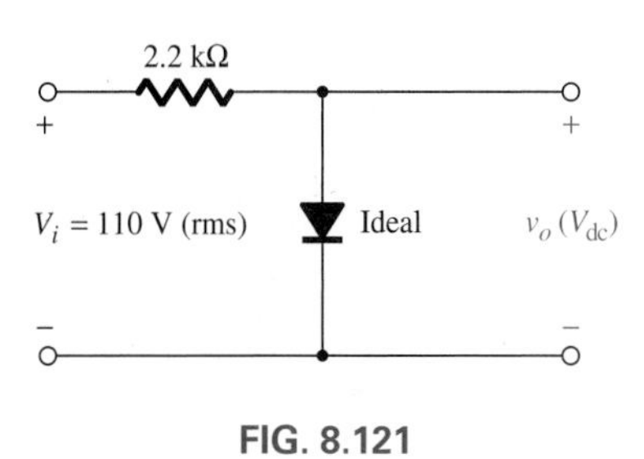

FIG. 8.121

27. For the network in Fig. 8.121 sketch v_o and determine V_{dc}.

28. For the network in Fig. 8.122 sketch v_o and i_R.

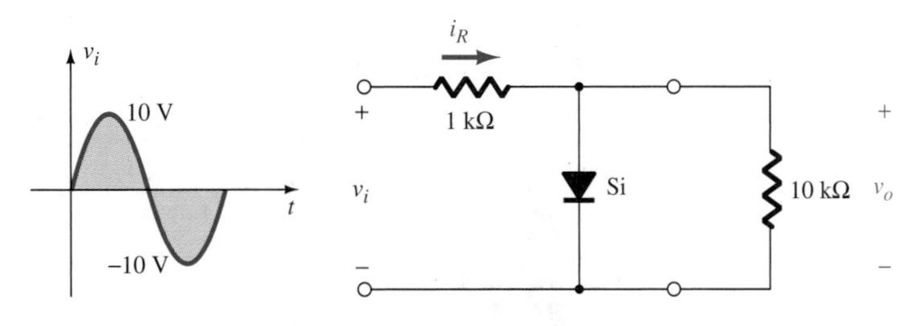

FIG. 8.122

29. **a.** Given $P_{max} = 14$ mW for each diode in Fig. 8.123, determine the maximum current rating of each diode.

b. Determine I_{max} for $V_{i_{max}} = 160$ V.

c. Determine the current through each diode for $V_m = 160$ V.

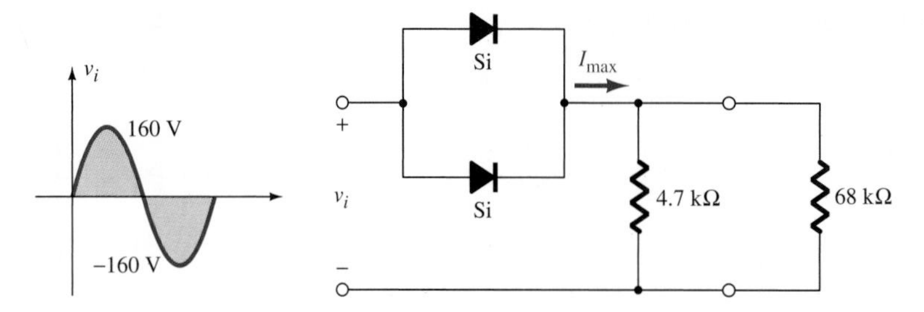

FIG. 8.123

d. Is the current determined in part (c) less than the maximum rating determined in part (a)?

e. If only one diode is present, determine the diode current and compare it with the maximum rating.

SECTION 8.7 Full-Wave Rectification

30. A full-wave bridge rectifier with a 120-V (rms) sinusoidal input has a load resistor of 1 kΩ.

a. If silicon diodes are employed, what is the dc voltage available at the load?

b. Determine the required PIV rating of each diode.

c. Find the maximum current through each diode during conduction.

d. What is the required power rating of each diode?

31. Determine v_o and the required PIV rating of each diode for the configuration in Fig. 8.124.

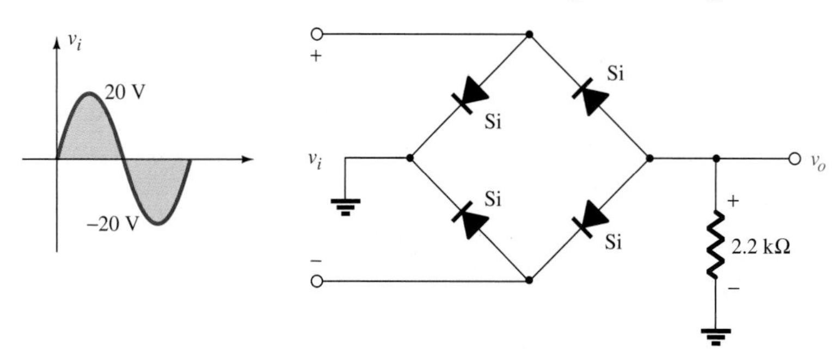

FIG. 8.124

32. Sketch v_o for the network in Fig. 8.125 and determine the dc voltage available.

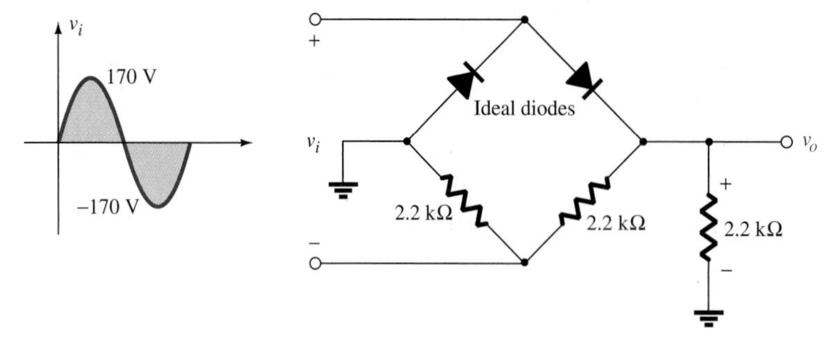

FIG. 8.125

SECTION 8.9 Clippers

33. Determine v_o for each network in Fig. 8.126 for the input shown.

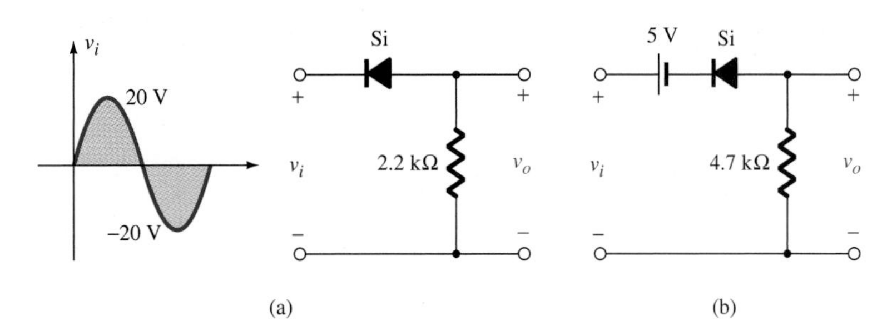

FIG. 8.126

34. Determine v_o for each network in Fig. 8.127 for the input shown.

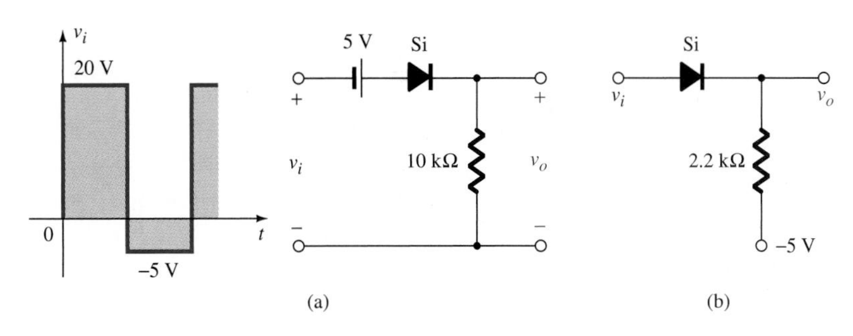

FIG. 8.127

35. Sketch i_R for the network in Fig. 8.128 for the input shown.

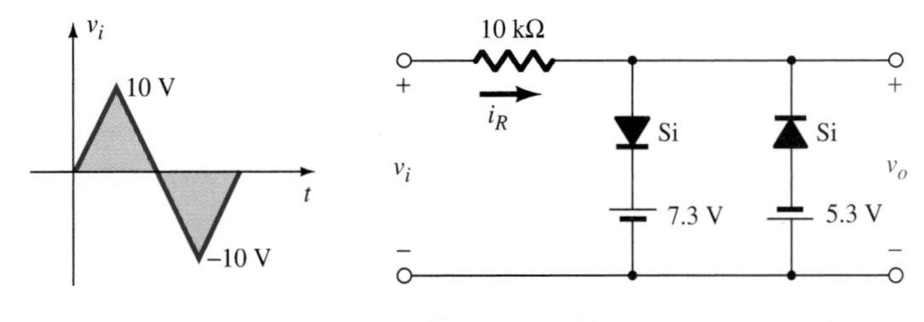

FIG. 8.128

SECTION 8.10 Clampers

36. Sketch v_o for each network in Fig. 8.129 for the input shown.

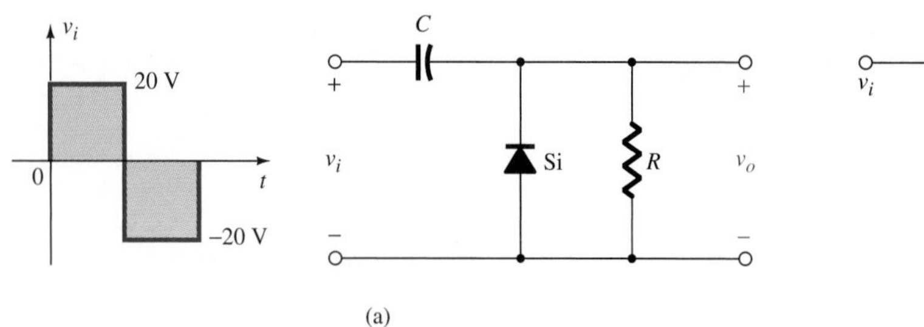

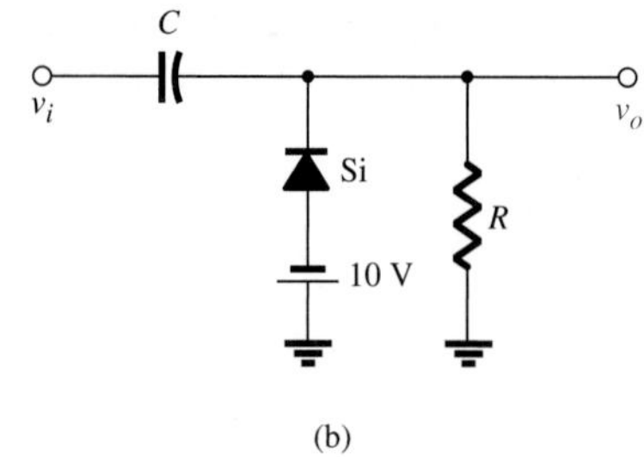

FIG. 8.129

37. For the network in Fig. 8.130:
- **a.** Calculate 5τ.
- **b.** Compare 5τ with half the period of the applied signal.
- **c.** Sketch v_o.

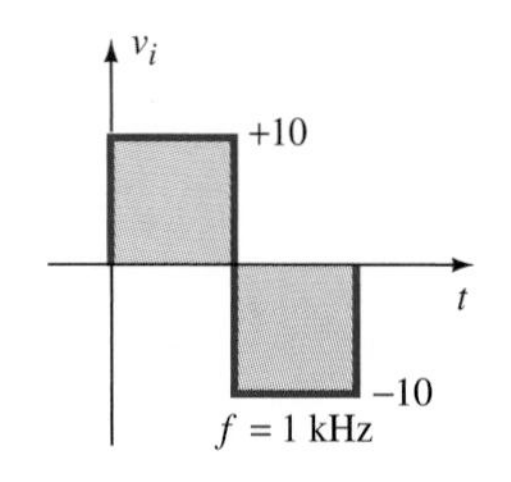

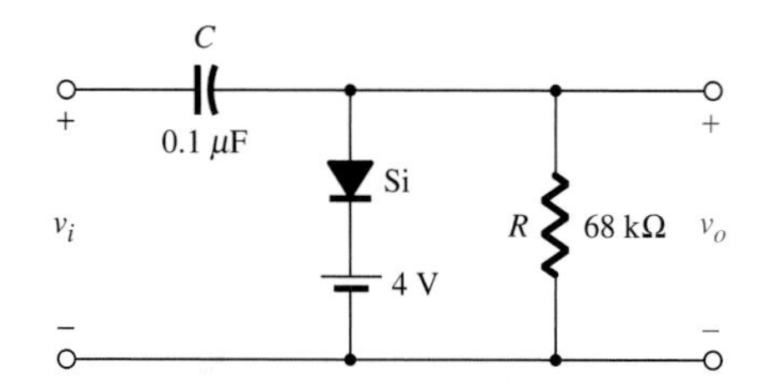

FIG. 8.130

38. Using ideal diodes, design a clamper to perform the function indicated in Fig. 8.131.

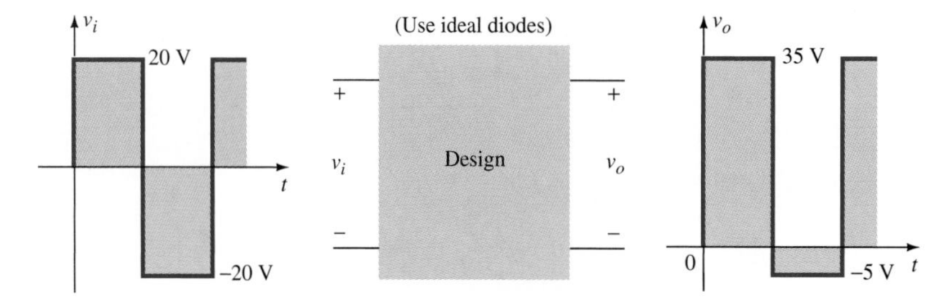

FIG. 8.131

SECTION 8.11 Zener Diodes

39. Sketch the output of the network in Fig. 8.81 if the input is a 50-V square wave. Repeat for a 5-V square wave.

40. **a.** Determine V_L, I_L, I_Z, and I_R for the network in Fig. 8.132 if $R_L = 270\ \Omega$.
- **b.** Repeat part (a) for $R_L = 470\ \Omega$.
- **c.** Determine the value of R_L that will establish maximum power conditions for the Zener diode.
- **d.** Determine the minimum value of R_L required to ensure that the Zener diode is in the *on* state.

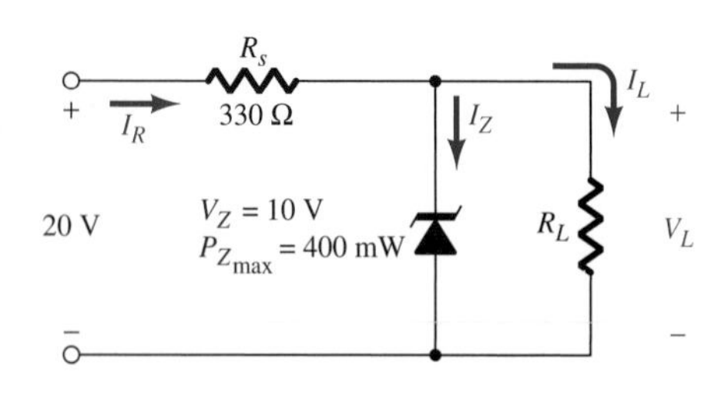

FIG. 8.132

41. **a.** Design the network in Fig. 8.133 to maintain V_L at 12 V for a load variation (I_L) from 0 to 200 mA. That is, determine R_s and V_Z.
b. Determine $P_{Z\max}$ for the Zener diode in part (a).

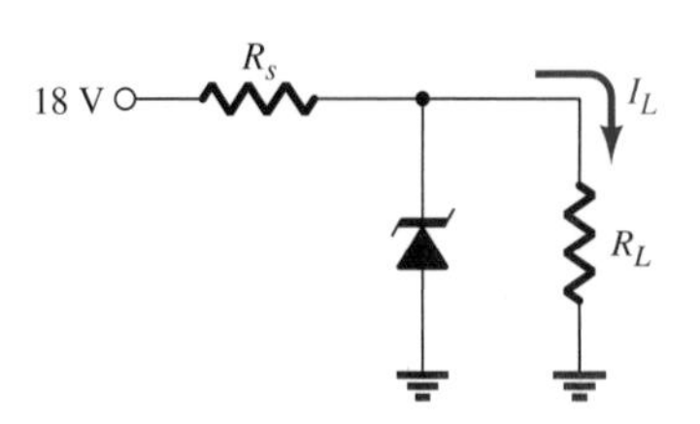

FIG. 8.133

42. For the network in Fig. 8.134, determine the range of V_i that will maintain V_L at 8 V and not exceed the maximum power rating of the Zener diode.

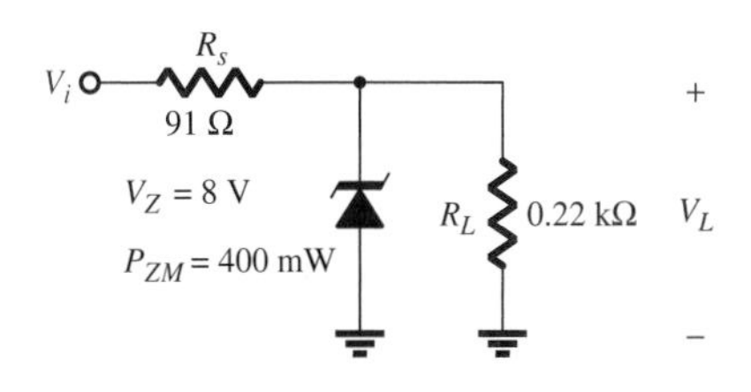

FIG. 8.134

43. Design a voltage regulator that will maintain an output voltage of 20 V across a 1-kΩ load with an input that varies between 30 and 50 V. That is, determine the proper value of R_s and the maximum current I_{ZM}.

SECTION 8.12 Tunnel, Varicap, and Power Diodes

44. If $I_P = 2$ mA in Fig. 8.85, what is the approximate negative resistance for the diode for the range $V_T = 0.2$ to 0.3 V?

45. If $P_{\max} = 20$ mV for the tunnel diode in Fig. 8.85, what is the maximum forward current if V_T is assumed to remain fixed at 0.7 V?

46. Note in the equivalent circuit in Fig. 8.86 that the capacitor appears in parallel with the negative resistance. Determine the reactance of the capacitor at 1 MHz and 100 MHz if $C = 10$ pF, and determine the total impedance of the parallel combination (with $R = -120\ \Omega$) at each frequency. Is the magnitude of the inductive reactance anything to be overly concerned about at either of these frequencies if $L_s = 5$ nH?

47. **a.** Determine the transition capacitance of a diffused junction varicap diode at a reverse potential of 4.2 V if $C(0) = 80$ pF and $V_T = 0.7$ V.
b. From the information in part (a), determine the constant k in Eq. (8.13).

48. **a.** For the equivalent model in Fig. 8.88 if $R_s = 10\ \Omega$, $C_T = 100$ pF, $R_r = 2$ MΩ, and $L_s = 2$ nH, determine the total impedance in rectangular form at:
1. $f = 1$ kHz.
2. $f = 1$ MHz.
3. $f = 1$ GHz.
b. For the frequencies in parts (a) to (c), is the varicap diode essentially capacitive (compare X_C to R_T), or have other elements come into play?

SECTION 8.13 Solar Cells

49. A solar cell 1 × 2 cm has a conversion efficiency of 9%. Determine the maximum power rating of the device.

50. If the power rating of a solar cell is determined on a very rough scale by the product $V_{oc}I_{sc}$, is the greatest rate of increase obtained at lower or higher levels of illumination? Explain your reasoning.

51. **a.** Referring to Fig. 8.95, determine the power density required to establish a current of 24 mA at an output voltage of 0.25 V.

b. Why is 100 mW/cm^2 the maximum power density in Fig. 8.95?

c. Determine the output current per square centimeter if the power is 40 mW/cm^2 and the output voltage is 0.3 V.

52. **a.** Sketch a curve of output current versus density at an output voltage of 0.15 V using the characteristics in Fig. 8.95.

b. Sketch a curve of output voltage versus power density at a current of 19 mA.

c. Is either of the curves from part (a) or (b) linear within the limits of the maximum power limitation?

SECTION 8.14 Thermistors and Photoconductive Devices

53. Find the specific resistance of the thermistor in Fig. 8.96 at 0°C. How does this value compare with its resistance at 100°C? Note the use of a log scale.

54. For the thermistor in Fig. 8.96, determine the dynamic rate of change in specific resistance with temperature at $T = 20$°C. How does this result compare with the value determined at $T = 300$°C? From the results, determine whether the greatest change in resistance per unit change in temperature occurs at lower or higher temperature levels. Note the vertical log scale.

55. Using the information provided in Fig. 8.96, determine the total resistance of a 2-cm length of material having a perpendicular surface area of 1 cm^2 at a temperature of 0°C. Note the vertical log scale.

56. Determine the resistance of the photoconductive cell in Fig. 8.98 at 0.5 and 50 fc. Did the relative change in resistance compare closely with the relative change in illumination?

57. Find the resistance of the photodiode in Fig. 8.100 at an illumination of 0.5 and 0.8 lm ($V = 40$ mV). Are increased levels of illumination associated with higher or lower levels of resistance?

58. Referring to Fig. 8.104, determine which terminals must be energized to display the number 7.

59. In your own words, describe the basic operation of an LCD.

60. Discuss the relative differences in mode of operation between an LED and an LCD display.

61. What are the relative advantages and disadvantages of an LCD display compared with an LED display?

9 Transistors and Other Important Electronic Devices

9.1 INTRODUCTION

In the previous chapter the discussion was limited to electronic devices with two accessible terminals. We shall now turn our attention to electronic devices with at least three terminals. In general, all amplifiers (electronic devices that increase the ac level of the applied signal) have at least three terminals—the third one controls the action between the other two. Some are referred to as *current-controlled amplifiers*, while others are referred to as *voltage-controlled devices*. If a current at the control terminal determines the voltage or current of the other two terminals, it is called a current-controlled device. If a voltage at the control terminal defines the response, it is called a voltage-controlled amplifier.

The first to be described is a current-controlled BJT (bipolar junction transistor) transistor, which was the first developed at Bell Laboratories in 1945. The construction will be discussed first, followed by the dc relationships, and finally the ac response. Next, the FET (field effect transistor) will be introduced and discussed following the same pattern. Finally, other important devices such as the unijunction transistor, SCR, DIAC, and TRIAC, and opto-isolators will be examined.

Take particular note as you progress through the chapter that superposition is applicable to the operation of each device. That is, the dc analysis can be determined totally separately from the ac response. The final result of the behavior at a particular point in the network is then simply the sum of the dc and ac responses. It is also important to note that the dc conditions will define the ac response. If the dc design is poor, the ac response will suffer. The conditions go hand-in-hand, even though each can be analyzed separately.

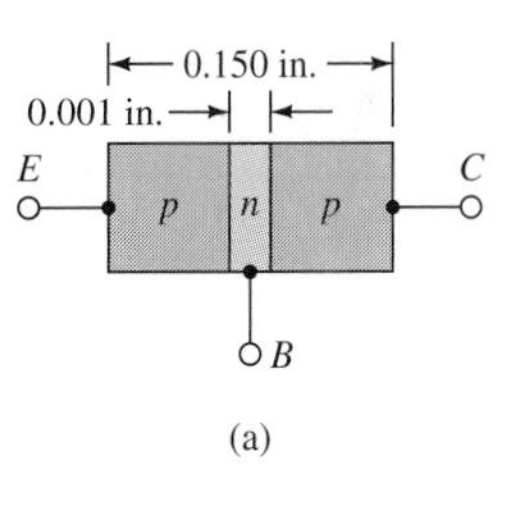

(a)

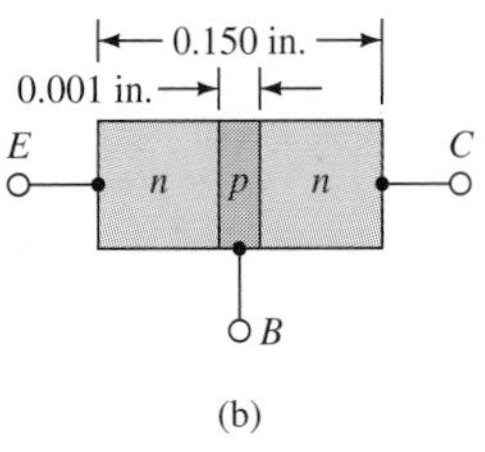

(b)

FIG. 9.1
Types of transistors: (a) pnp; (b) npn.

9.2 BIPOLAR JUNCTION TRANSISTORS

A *transistor* is a three-terminal device primarily used to amplify (increase in magnitude) the time-varying portion of the applied signal. Figure 9.1 reveals that the transistor is constructed of three alternating layers of semiconductor material. Note that the sandwiched layer is significantly thinner than the other two. In general this permits a "punching through" action for the carriers passing between the collector and emitter terminals. The term *emitter* for one terminal was chosen because it is the source of the majority carrier flow—it emits the necessary carrier flow. The *collector* terminal was so named because it "collects" the major part of the majority carrier flow. The term *base* was chosen because it was the base of the original structure. You will find in the characteristics to follow that the magnitude of the collector (C) and emitter (E) currents is essentially equal (on an approximate basis), while the base (B) current is significantly smaller in magnitude. The alternating layers provide the common commercial name for each: *npn* and *pnp*. The *npn* and *pnp* transistors can be used in one of three configurations. Each configuration has terminal characteristics that make it particularly useful for specific applications.

The normal operation of a transistor requires that one *pn* junction be forward-biased and the other be reverse-biased, as shown in Fig. 9.2.

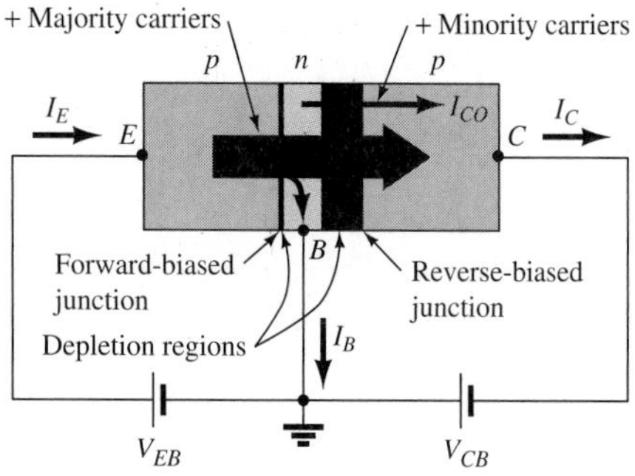

FIG. 9.2
Majority and minority carrier flow of pnp transistor.

The resulting depletion regions are depicted by the vertical black regions at each junction—much larger for the reverse-biased region. Note that the major conventional flow is from emitter to collector terminal, with I_B being of a much smaller magnitude. Applying Kirchhoff's current law to the three-terminal device, we find

$$I_E = I_C + I_B \tag{9.1}$$

Since I_C is typically 100 to 250 times larger than I_B, the following approximation is frequently employed:

$$I_E \cong I_C \tag{9.2}$$

The *common-base* configuration appears in Fig. 9.3 with the symbol for the *pnp* device. The terminology comes from the fact that the *base* terminal is common to both the emitter and collector terminals. In Fig. 9.3 note that two separate dc supplies are required to properly bias the

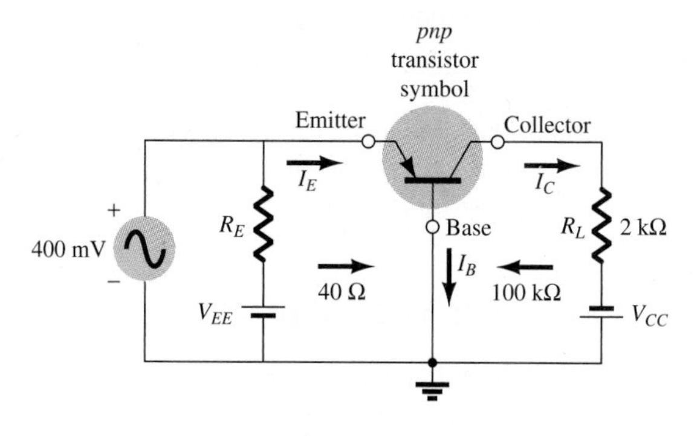

FIG. 9.3

device. In addition, a resistive element appears in the input and output circuit for further biasing control, a topic to be introduced in a later chapter. In Fig. 9.3 the emitter–base junction is forward-biased, accounting for the low input resistance of 40 Ω for this configuration. The collector–base junction is reverse-biased, accounting for the high output impedance of 100 kΩ. Since the input junction is forward-biased, the dc voltage from base to emitter is typically about 0.7 V for a silicon device transistor (as encountered for the forward-biased silicon diode). When the 400-mV ac signal is applied, the magnitude of the ac component of the emitter current as determined by Ohm's law is

$$I_E = \frac{400 \text{ mV}}{40 \ \Omega} = 10 \text{ mA}$$

From Eq. (9.2), $I_C \cong I_E = 10$ mA and the output ac voltage across R_L is

$$V_o = I_C R_L = (10 \text{ mA})(2 \text{ k}\Omega) = 20 \text{ V}$$

with a resulting voltage (amplification) gain:

$$A_v = \frac{V_o}{V_i} = \frac{20 \text{ V}}{0.4 \text{ V}} = \mathbf{50}$$

In the preceding example (as for all transistor networks) there was a *transfer* of the input current from a low- to a high-*resistance* circuit. Hence, the derivation of the name of the device:

*trans*fer + re*sistor* = transistor

In Fig. 9.4 the symbol and biasing arrangements for both the *npn* and *pnp* devices are provided. Note that the arrow in the transistor symbol is always associated with the emitter leg (and provides the direction of actual flow), while the heavy bar identifies the base terminal. The arrow is

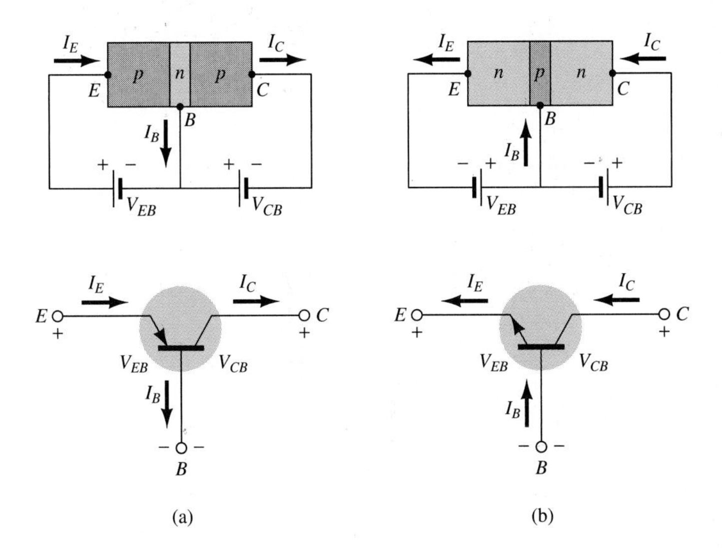

FIG. 9.4
Notation and symbols used with the common-base configuration: (a) pnp transistor; (b) npn transistor.

out (from the base) for *npn* (*n*ot *p*ointing i*n*) transistors and *in* for *pnp* (*p*ointing i*n*) devices. The indicated current directions in Fig. 9.4 are the actual flow directions, while the polarities in the lower figure of each set define the standard notation for each configuration.

Two sets of characteristics are required to fully describe a three-terminal device: an input and an output set. For the *pnp* transistor the characteristics in Fig. 9.5 apply. It is very seldom that the input or emitter–base characteristics of a transistor are available. Rather, the group of curves are looked upon as a single curve for silicon with a rise at 0.7 V. The slope of the rise is normally ignored, and the base-to-emitter voltage is assumed to be constant at 0.7 V (0.3 V for germanium) as shown in Fig. 9.5(b). From this point on, therefore, the following approximation will be applied: α

$$|V_{BE}| \cong 0.7 \text{ V} \tag{9.3}$$

The negative sign for v_{CB} in Fig. 9.5(a) simply indicates that the actual polarity of the voltage is the opposite of that defined in Fig. 9.4.

The collector characteristics in Fig. 9.5(a) clearly indicate that the previous statement that $I_C \cong I_E$ is valid. Simply follow any horizontal line directly over to the collector current axis to verify it. The characteristics also indicate that V_{CB} has very little effect on the magnitude of I_E, since the curves are essentially horizontal. The input current I_E is therefore determined by the input circuit.

The collector and emitter currents are related by the *short-circuit amplification factor*, called *alpha* (α). It is defined by

$$\alpha = \frac{I_C}{I_E} \tag{9.4}$$

and is typically in the range 0.98 to 0.995, which is very close to the unity factor employed in Eq. (9.2).

The base and collector currents are related by a factor called *beta* (β) in the following manner:

$$\beta = \frac{I_C}{I_B} \tag{9.5}$$

so that

$$I_C = \beta I_B \tag{9.6}$$

a relationship of important practical value.

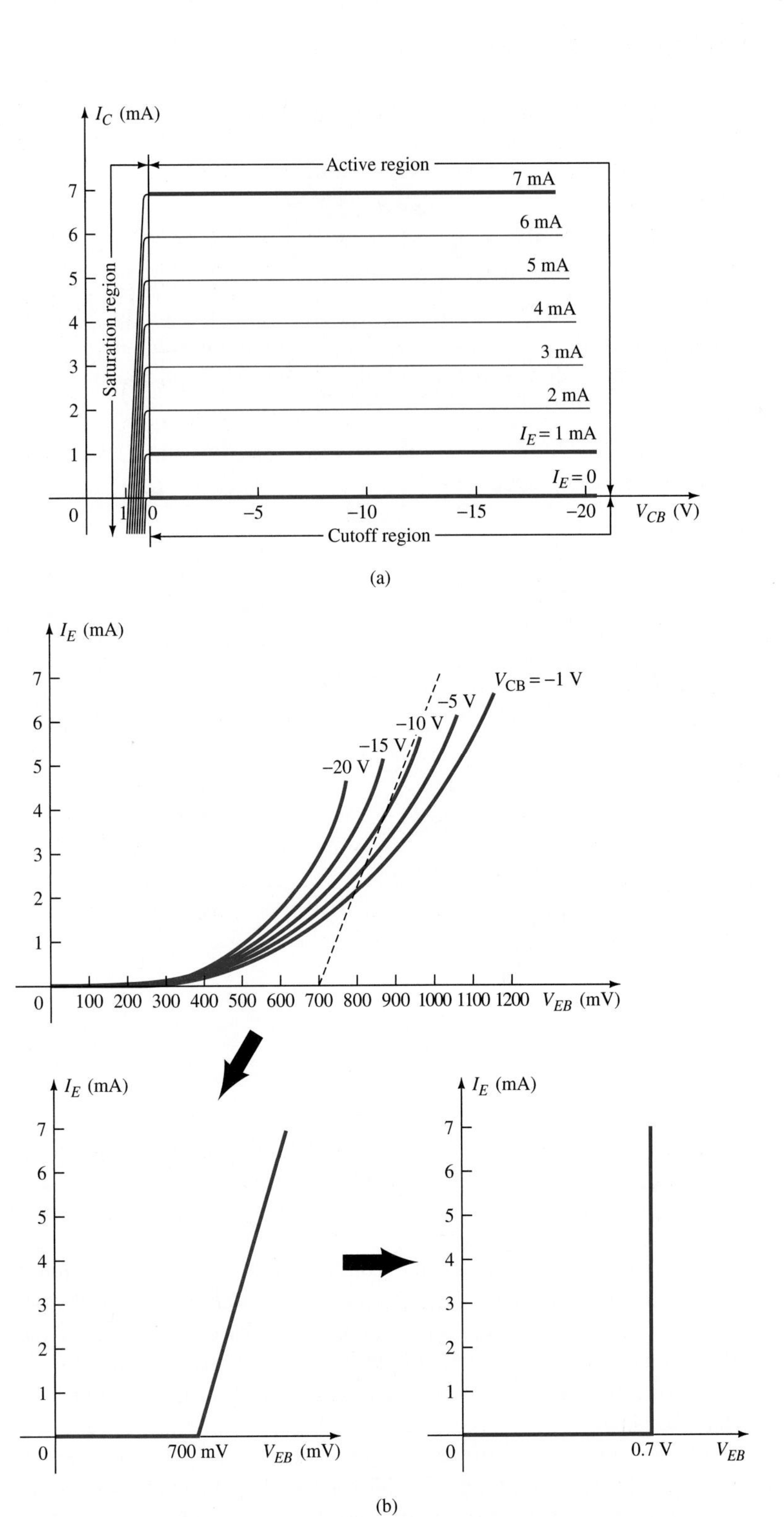

FIG. 9.5
Characteristics of a pnp transistor in the common-base configuration: (a) collector characteristics; (b) emitter characteristics.

The terms α and β are further related by

$$\beta = \frac{\alpha}{1 - \alpha} \tag{9.7}$$

A transistor configuration that appears more frequently than the common-base or common-collector is called the *common-emitter* configuration. It is shown in Fig. 9.6 for the *pnp* and *npn* transistors. It

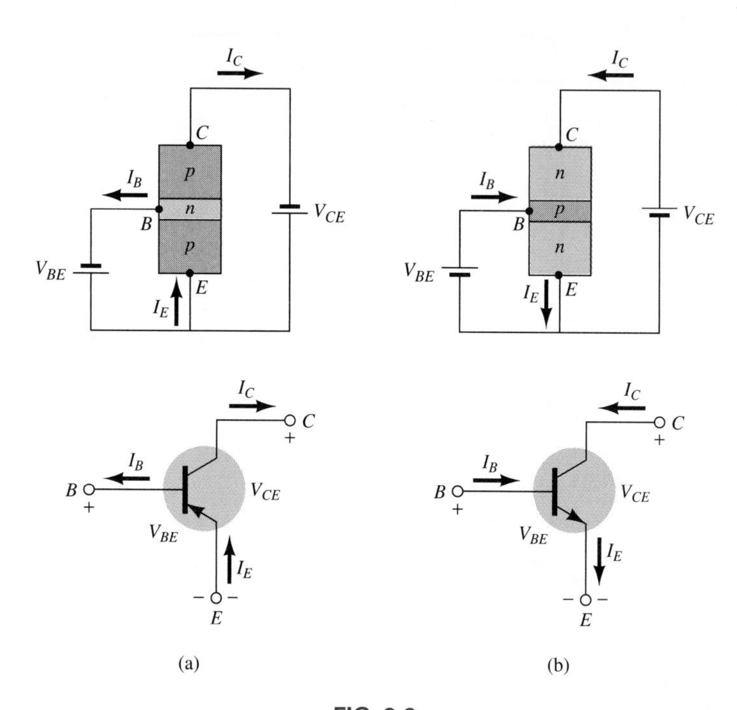

FIG. 9.6
Notation and symbols used with the common-emitter configuration: (a) pnp transistor; (b) npn transistor.

is important to note that for the common-base and common-emitter configurations a change from a *pnp* to an *npn* transistor requires that the dc biasing be reversed. As in the common-base configuration, the base-emitter junction is forward-biased, and we shall assume for dc conditions that $V_{BE} \cong 0.7$ V and remove the need for the input or base characteristics.

The collector characteristics are quite useful, however, and appear in Fig. 9.7 for the *npn* transistor. For this set note that the input base current determines the horizontal line to be employed and therefore the resulting collector current for a particular value of V_{CE} (the collector-to-emitter potential). Since the output collector current is related directly to the input base current by the beta factor, the transistor is considered a *current-controlled* amplifying device.

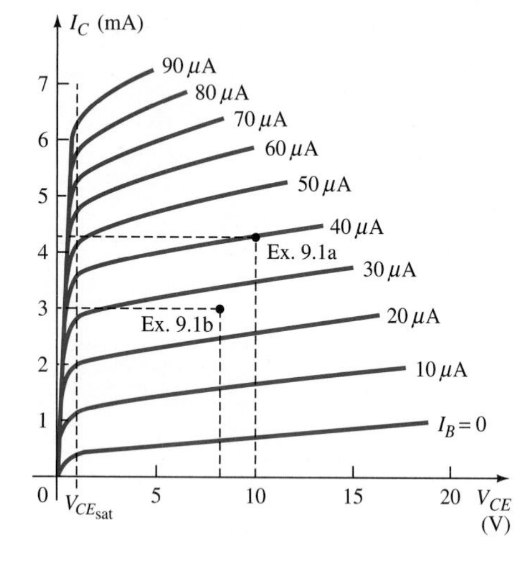

FIG. 9.7
Collector characteristics of an npn transistor in the common-emitter configuration.

EXAMPLE 9.1 For the characteristics in Fig. 9.7:

a. Determine I_C if $I_B = 40\ \mu\text{A}$ and $V_{CE} = 10$ V.
b. Determine I_B if $I_C = 3$ mA and $V_{CE} = 8$ V.
c. Calculate β for the results for part (a).
d. Calculate β for the results for part (b).
e. Discuss the results for parts (c) and (d).

Solution:

a. From the characteristics $I_C = \mathbf{4.3\ mA}$
b. For a situation such as encountered here where the intersection of the two given points is between two I_B curves, an equal number of divisions must be superimposed between $I_B = 20\ \mu\text{A}$ and $I_B = 30$ μA to give

$$I_B \cong \mathbf{27\ \mu A}$$

c. $\beta = \dfrac{I_C}{I_B} = \dfrac{4.3\text{ mA}}{40\ \mu\text{A}} = \mathbf{107.5}$

d. $\beta = \dfrac{I_C}{I_B} = \dfrac{3\text{ mA}}{27\mu\text{A}} = \mathbf{111.1}$

e. The beta relationship between I_C and I_B is dependent on the point of operation of the device. In some regions the dc amplification factor is greater than in others.

The remaining common-collector configuration normally appears as shown in Fig. 9.8 rather as one might expect with the collector terminal below the emitter of the transistor. In the configuration shown it is referred to as the *emitter follower,* which is employed for impedance-matching purposes, to be discussed in a later chapter in more detail. Briefly, however, the ac output is normally taken off the emitter terminal (the collector is at ground potential), which *follows* the signal applied to the base potential in both phase and magnitude. The base–emitter junction is forward-biased, and $V_{BE} \cong 0.7$ V (dc) as before. The output characteristics are essentially those of the common emitter since the vertical scale of I_E is related to that of the common-emitter characteristics by $I_E \cong I_C$.

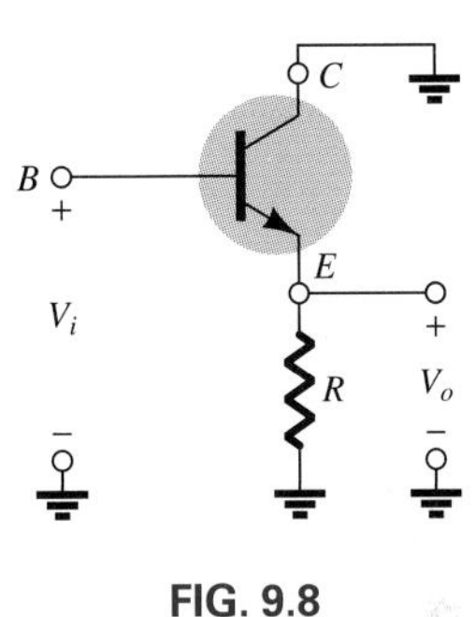

FIG. 9.8
Common-collector configuration used for impedance-matching purposes.

For each transistor configuration, there are maximum ratings of importance: maximum collector-to-emitter voltage (V_{CE}), maximum collector current (I_C), and maximum power dissipation $P_D = V_{CE}I_C$. Each of these ratings appears on the collector characteristics in Fig. 9.9. The maximum collector voltage and current are simply lines drawn at these values on the characteristics. The power is obtained by simply picking a value for V_{CE} (or I_C) and solving for I_C (or V_{CE}, respectively) using the maximum power rating. For example, if we pick $V_{CE} = 10$ V, then $I_C = P/V_{CE} = 30\text{ mW}/10\text{ V} = 3$ mA, and an intersection on the power curve is obtained as shown in Fig. 9.9. Other points can be found and the maximum power curve obtained. Points of operation must now be chosen within this power curve. To avoid distorted (nonlinear) outputs, values of V_{CE} less than the saturation value ($V_{CE_{\text{sat}}}$) must not be chosen. Further, the region below $I_B = 0\ \mu$A is the cutoff region for the device. The

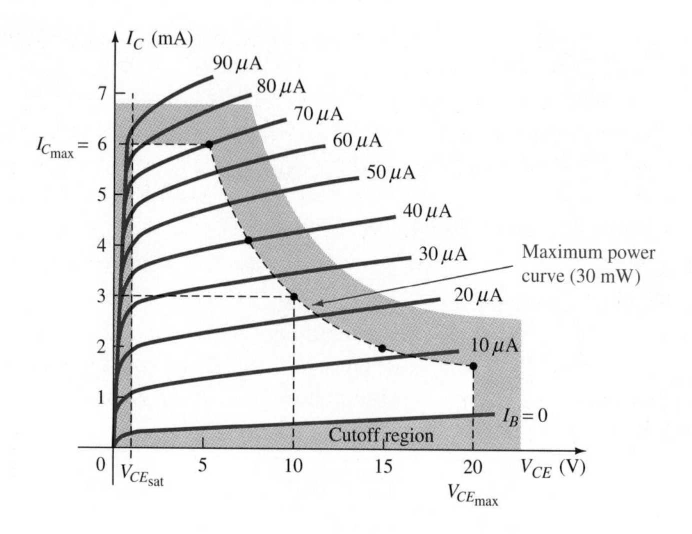

FIG. 9.9
Region of operation for amplification purposes.

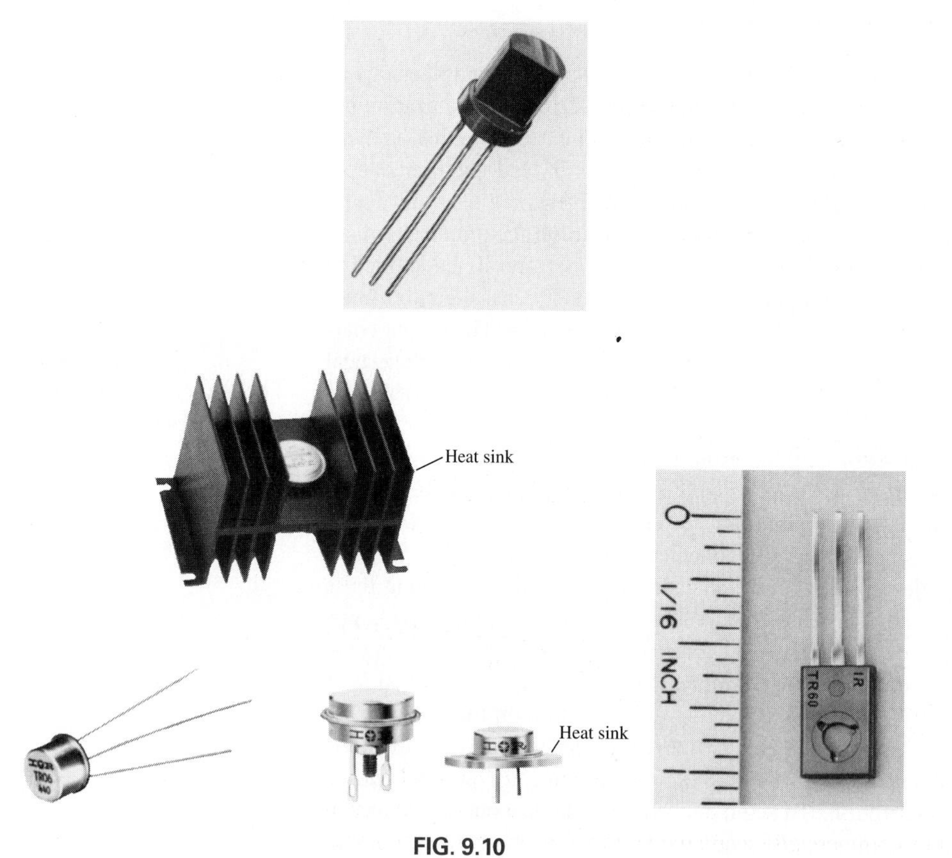

FIG. 9.10
Various types of transistors. (Courtesy of International Rectifier Corporation.)

remaining unshaded region must be employed if a nondistorted amplified version of the input is to be obtained.

A variety of commercially available transistors appears in Fig. 9.10. For increasing wattage ratings the casing must be redesigned to withstand the heat, or heat sinks must be applied as shown for two transistors in Fig. 9.10. For transistors connected directly to a chassis the chassis itself behaves as a heat sink. The terminal identifications of some appear in Fig. 9.11, although every effort is now being employed to label each terminal directly on the casing.

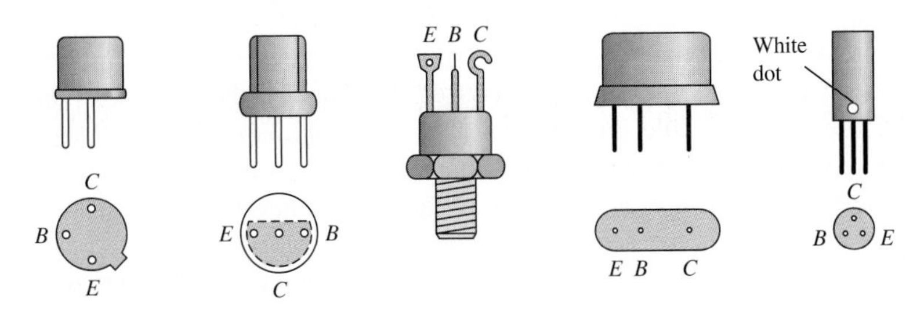

FIG. 9.11
Transistor terminal identification.

9.3 BIPOLAR JUNCTION TRANSISTOR AMPLIFIER CIRCUITS

By definition, an amplifier is a system that can increase (amplify) the magnitude of the applied signal. All electronic amplifiers must have at least three terminals, such as a BJT with the driving signal employing one terminal and the amplified output appearing at the other two. The size of the applied signal determines whether small- or large-signal methods of analysis will be employed—both methods will be covered in the sections and chapters to follow.

Amplifiers are further characterized by

1. *Type:* voltage, current, or power.
2. *Operating class:*
   Class A—output conduction for the full 360° of the applied signal.
   Class B—only 180° cycle conduction for the output waveform.
   Class AB—output conduction for 180° to 360°.
   Class C—conduction for less than 180°.
3. *Coupling method between stages:* capacitor-coupled, direct-coupled, transformer-coupled.
4. *Range of frequencies:* low frequency, audiofrequency, high frequency, intermediate frequency, ultrasonic frequency.
5. *Area of application:* such as RF amplifier, audio amplifier, or microphone amplifier.

We begin the discussion with an introduction to the ideal small-signal current-controlled amplifier to set the stage for comparison with the practical or commercially available variety. The discussion of large-signal amplifiers will be left for a later chapter.

Bipolar junction transistors are used in such a wide variety of applications that it would be impractical and in fact impossible to touch on each area with sufficient detail to have any redeeming value. Rather, we will examine the general procedure for analyzing small-signal BJT transistor amplifiers in sufficient detail to ensure a carryover to most areas of application. There is no defining line between the realms of small- and large-signal applications. Small-signal methods are used when the applied signal is relatively small compared with the scale of the parameters of the characteristics. When the applied signal results in a significant sweep across the characteristics of the device, large-signal methods must be applied.

There are two major components of the analysis of any BJT amplifier configuration: the dc and ac responses. Fortunately, the *superposition* theorem is applicable, and the dc and ac analyses can be performed separately. The *total solution* is then simply the sum of the two responses. The application of a dc voltage to establish an *operating point* on the characteristics of the device is referred to as the *biasing* operation. Not only does the biasing circuit *turn on* the device but it establishes a region of the characteristics to be employed in the amplification process. The absence of a biasing arrangement would leave the device in the *off state,* and the amplification process could not be performed. The biasing arrangement establishes a fixed point on the characteristics around which the applied signal will operate to establish a voltage, current, or power gain. In addition, keep in mind that when a transistor is used as an amplifier, the output ac power is greater than the applied signal power. This is possible (as we recall the conservation of energy criteria) because some of the applied dc power is being converted to the ac domain. In other words, the transistor acts like a power converter, converting dc power from the dc supply to signal power.

The operating region of a BJT transistor is limited to the region within the shaded area in Fig. 9.12. The maximum current ($I_{C_{max}}$) and voltage ($V_{CE_{max}}$) must not be exceeded, with the curved area between the two limits defined by the maximum power dissipation level of the device. Low values of voltage along the vertical axis must be avoided to prevent undue distortion of the amplified signal. In addition, the region below $I_B = 0\ \mu A$ is not part of the linear amplification region because the applied signal loses control of the shape of the output signal.

The absence of a biasing arrangement establishes point A on the characteristics as the operating point. The result is $I_C = 0$ mA and $V_{CE} = 0$ V, and the device is off because we realize from the previous chapter that the base–emitter junction must be forward-biased by about 0.7 V. The BJT would *turn on* only when the applied signal was sufficient to meet this condition, leaving a portion of the applied signal without an amplified output response. The output would also be a distorted version of the applied signal.

The application of a biasing arrangement that would establish point B as the operating point would limit the swing of the output voltage V_{CE} and current I_C to the region between B and the vertical and horizontal axes, respectively. Incidentally, the operating point is also called the *quiescent point* or Q point because it is a point that does not change with

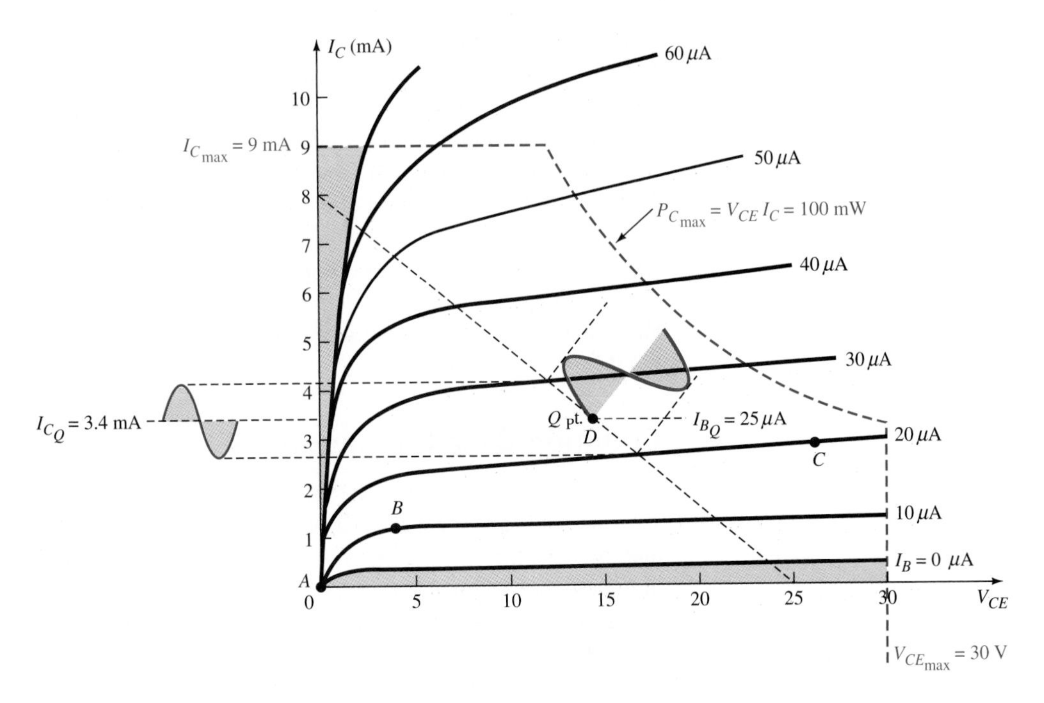

FIG. 9.12
Defining the operating region of a transistor.

time (still, unmoving by definition). Point C is also a poor choice because the swing around the fixed Q point is limited by the adjacent maximum voltage line and the power curve.

The most suitable operating point is point D because it permits a fairly equal voltage swing to either side (horizontally) before reaching the power curve or the distorted region at low voltages. The biasing current is also equally distant from the power curve to the minimum current. It is also important to realize that the spacing between I_B lines in this region is fairly constant. The result will be an output signal having most of the characteristics of the applied signal. If the spacing varies drastically from one side of the Q point to the other, it will be reflected in a distortion of the output signal. This is another reason why operating points at low or high voltages or currents should be avoided.

Once the biasing point is established by the resulting bias current and voltage of the BJT configuration, the applied ac signal causes a swing from the Q point along a *load line* defined by the chosen network. If we assume a load line defined by the dashed line in Fig. 9.12, an applied ac signal causes the current I_B to change value above and below the Q point value. If we assume a variation from $I_B = 20\ \mu\text{A}$ to $30\ \mu\text{A}$ as shown by the superimposed signal on the load line, the result is a change in collector current from $I_C = 2.7$ mA to $I_C = 4.1$ mA as shown along the vertical axis. Using the peak-to-peak values of I_B and I_C, we find that the undistorted ac gain is

$$A_i = \frac{I_o}{I_i} = \frac{I_{C(p\text{-}p)}}{I_{B(p\text{-}p)}} = \frac{1.4\text{ mA}}{10\ \mu\text{A}} = \mathbf{140}$$

The BJT has three basic configurations, each with its own special characteristics. They are referred to as the *common-base* (CB), *common-emitter* (CE), and *common collector* (CC) *configurations*, and each is defined by the terminal that is common to both the input and output sides of the amplifier (and usually the terminal closest to ground potential). Each configuration defines a Q point, a load line, specific levels of input and output impedances, and a typical ac gain. As mentioned earlier, superposition is applicable, resulting in a dc analysis followed by an ac review. In fact, the dc analysis defines the parameters to be used in the ac analysis. We will now investigate, in some detail, the most common of the BJT configurations.

Common-Base Configuration

A complete common-base BJT amplifier is provided in Fig. 9.13. The capacitors C_s and C_C are essentially open circuits for direct current, thereby isolating the dc levels of the amplifier from the source and loading networks. The resistors R_E and R_C and the voltage sources V_{EE} and V_{CC} establish a dc operating point on the device characteristics. The input and output signals of the amplifier are V_i and V_o, respectively, with the applied signal being V_s with a series internal resistance R_s. The transistor is of the *npn* variety with the input applied from emitter to base and the output from collector to base. The base terminal is common to both input and output levels and is connected to ground potential, resulting in the *common-base* terminology for this configuration.

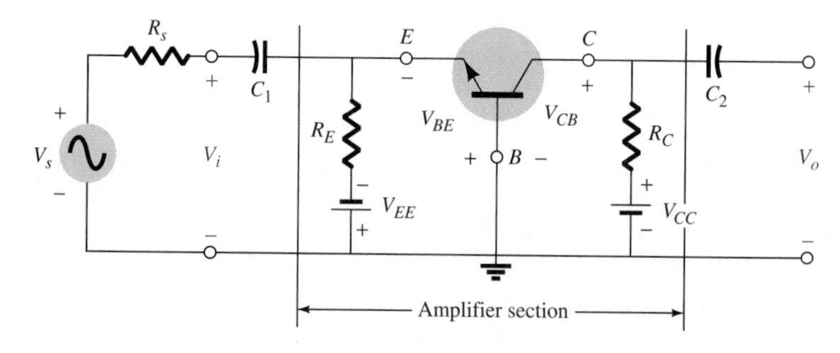

FIG. 9.13
Common-base amplifier.

Our analysis will begin with a determination of the dc levels because one of the resulting quantities will define an important parameter of the ac configuration. Recall that the superposition theorem is applicable, permitting a separate determination of the dc and ac responses.

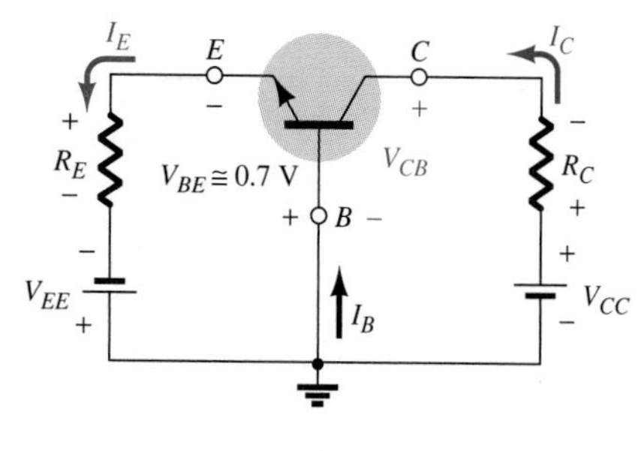

FIG. 9.14
Dc equivalent of Fig. 9.13.

Dc Bias

Since the two capacitors C_s and C_C isolate the dc levels of the amplifier, the dc analysis of the BJT configuration can be limited to the amplifier section appearing in Fig. 9.14.

The utilization of any BJT transistor as an amplifier requires that the *base-to-emitter junction be forward-biased and the base-to-collector junction be reversed-biased.* This is true for any BJT configuration to be utilized for amplification purposes. For the silicon transistor the dc base-to-emitter voltage must be about 0.7 V to turn the transistor *on*. This being the case, whenever we analyze a BJT amplifier, we will immediately assume $V_{BE} = 0.7$ V. Even though the base-to-emitter voltage varies slightly (in the practical device) with a change in current level, we will assume for all the dc analyses that

$$\boxed{V_{BE} = 0.7 \text{ V}}_{\,\textit{on}\text{ transistor}} \tag{9.8}$$

Our ability to immediately assume $V_{BE} = 0.7$ V for an *on* transistor permits a determination of all the required dc levels of the BJT transistor configuration. For instance, the current I_E in Fig. 9.14 is determined using Ohm's law in the following manner:

$$I_E = \frac{V_{R_E}}{R_E}$$

but Kirchhoff's voltage law reveals that around the input circuit,

$$V_{R_E} = V_{EE} - V_{BE}$$

so that

$$\boxed{I_E = \frac{V_{EE} - V_{BE}}{R_E}} \tag{9.9}$$

The equation reveals that I_E can be set by the values of V_{EE} and R_E. Usually, V_{EE} is restricted to some available battery or supply used with the rest of the circuitry, leaving the setting of I_E to the chosen R_E. Having set the value of I_E, we have also set the level of I_C, since $I_C = \alpha I_E$, where α is typically between 0.9 and 1 and usually above 0.95. This being the case, the following approximation is usually applied:

$$\boxed{I_C \cong I_E} \tag{9.10}$$

The voltage drop across the resistor R_C can then be determined by applying Ohm's law:

$$\boxed{V_{R_C} = I_C R_C} \tag{9.11}$$

with the voltage from collector to base determined by Kirchhoff's voltage law:

$$\boxed{V_{CB} = V_{CC} - I_C R_C} \tag{9.12}$$

Since V_{CC} is usually already set by the available circuit supply voltage and I_C is determined by the biasing of the base–emitter junction, the value of V_{CB} is primarily set by the chosen value of R_C.

EXAMPLE 9.2 For the network in Fig. 9.13, if $V_{CC} = 12$ V, $V_{EE} =$ 3 V, $R_E = 2.4$ kΩ, and $R_C = 4.7$ kΩ:

a. Determine I_{E_Q} (where the subscript Q indicates a quiescent or dc level).
b. Find I_{C_Q}.
c. Calculate V_{CB}.

Solution:

a. Since $V_{BE} = 0.7$ V for an *on* transistor,

$$\text{Eq. (9.9): } I_E = \frac{V_{EE} - 0.7\text{ V}}{R_E} = \frac{3\text{ V} - 0.7\text{ V}}{2.4\text{ k}\Omega} = \mathbf{0.958\text{ mA}}$$

b. Eq. (9.10): $I_{C_Q} \cong I_{E_Q} = \mathbf{0.958\text{ mA}}$
c. Eq. (9.11): $V_{R_C} = I_C R_C = (0.958\text{ mA})(4.7\text{ k}\Omega) = \mathbf{4.5\text{ V}}$
d. Eq. (9.12): $V_{CB} = V_{CC} - I_C R_C = 12\text{ V} - 4.5\text{ V} = \mathbf{7.5\text{ V}}$

The transistor, therefore, is biased to operate at $I_C \cong 1$ mA and $V_{CB} = 7.5$ V, which should be an operating point within the linear operating region of the chosen transistor. Note that the resulting levels were primarily set by the dc supplies and the chosen resistors and not by the transistor (except for $V_{BE} = 0.7$ V). The transistor chosen, however, determines the levels of I_C and V_{CB} that will ensure that the chosen operating point is in the linear amplification region.

Ac Analysis

Having provided the necessary dc bias so that the transistor operates as an amplifier in its linear operating region, we shall next investigate the circuit and device factors that determine the ac voltage amplification and other important ac circuit values. These include circuit ac input impedance, output impedance, and ac voltage and current gains.

Figure 9.15 shows the capacitors replaced by short circuits, as the capacitors used usually have very much smaller ac impedance than the other circuit elements for the frequency range of interest. The dc supplies are replaced by short-circuit equivalents, since they do not affect the ac swing of the output ac voltage V_o. The result is the reduced ac equivalent in Fig. 9.16. The ac input voltage is applied directly across the transistor emitter–base, which "looks like" a forward-biased diode of resistance r_e. The value of this base–emitter resistance is dependent on the dc emitter current as determined by Eq. (8.3)

$$r_e = \frac{26\text{ mV}}{I_E(\text{dc})} \tag{9.13}$$

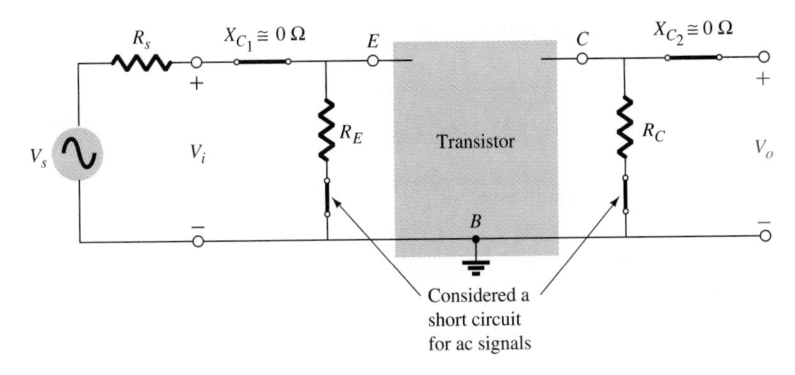

FIG. 9.15
Ac equivalent of Fig. 9.13.

Take particular note that an ac parameter is determined by a dc quantity. Changing a parameter of the dc biasing configuration may alter the magnitude of r_e also.

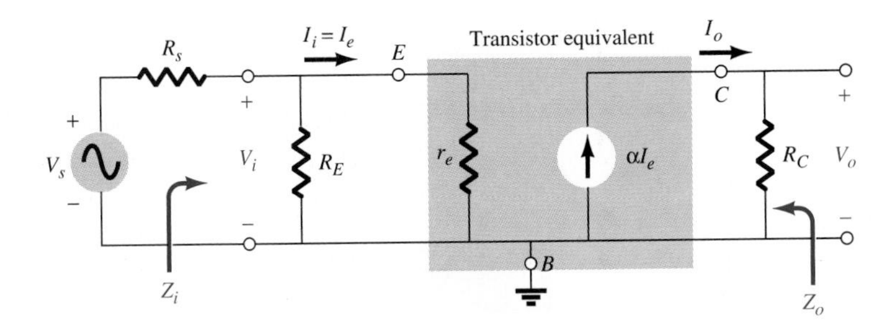

FIG. 9.16
Substituting the transistor equivalent model in the network of Fig. 9.15.

The transistor equivalent in Fig. 9.16 also reveals that the output current of the transistor amplifier is αI_e, which is approximately I_C since $\alpha \cong 1$. In addition, note that for a common-base configuration $Z_i = R_E \| r_e$. In this case r_o is assumed to be sufficiently large compared with R_C to be considered an open circuit and not appear in the equivalent circuit. The result is that $R_o = R_C \| r_o \cong R_C$.

The input ac current of the transistor is determined by

$$I_i = I_e = \frac{V_i}{r_e} \tag{9.14}$$

and

$$I_o = \alpha I_e \cong I_C \tag{9.15}$$

The output voltage is therefore

$$V_o = I_C R_C = I_e R_C \tag{9.16}$$

with a transistor voltage gain of

$$A_v = \frac{V_o}{V_i}$$

Substituting $V_o = I_e R_C$ and $V_i = I_e r_e$ [from Eq. (9.14)] we have,

$$A_v = \frac{V_o}{V_i} = \frac{I_e R_C}{I_e r_e}$$

or

$$A_v = \frac{R_C}{r_e} \tag{9.17}$$

The current gain is defined by

$$A_i = \frac{I_o}{I_i} = \frac{\alpha I_e}{I_e}$$

and

$$A_i = \alpha \cong 1 \tag{9.18}$$

In general, a common-base transistor amplifier has a low input impedance, a high output impedance, and a good voltage gain. It is therefore an excellent amplifier if driven by a low-impedance source with an applied load relatively close to, or less than, the biasing resistor R_C.

EXAMPLE 9.3 Continue the analysis of the network in Example 9.2 by finding the following ac parameters:

a. Input impedance.
b. Transistor voltage gain $A_v = V_o/V_i$.
c. Transistor current gain $A_i = I_o/I_i$.
d. Total voltage gain $A_{v_T} = V_o/V_s$ if $R_s = 20\ \Omega$.

Solution:

a. Eq. (9.13): $r_e = \dfrac{26\text{ mV}}{I_{E_Q}} = \dfrac{26\text{ mV}}{0.958\text{ mA}} = 27.14\ \Omega$

and $Z_i = R_E \parallel r_e = 2.4\text{ k}\Omega \parallel 27.14\ \Omega = \mathbf{26.84\ \Omega}$

b. Eq. (9.17): $A_v = \dfrac{R_C}{r_e} = \dfrac{4.7\text{ k}\Omega}{27.14\ \Omega} = \mathbf{173.18}$

c. Eq. (9.18): $A_i \cong \mathbf{1}$

d. Voltage-divider rule:

$$V_i = \frac{R'V_s}{R' + R_s} = \frac{(26.84\ \Omega)V_s}{26.84\ \Omega + 20\ \Omega} = 0.573\ V_s \text{ or } \frac{V_i}{V_s} = 0.573$$

$$\text{and } A_{v_T} = \frac{V_o}{V_s} = \frac{V_o}{V_i} \cdot \frac{V_i}{V_s} = (173.18)(0.573) = \mathbf{99.23}$$

Common-Emitter Configuration

The common-emitter configuration is the most common of the transistor configurations because of its excellent voltage, current, and power gain characteristics. The simplest of the common-emitter configurations is the *fixed-bias* arrangement shown in Fig. 9.17.

Although the network in Fig. 9.17 can operate properly and amplify small ac signals, there are other versions with improved characteristics that are used more frequently, but this network is an excellent starting point because of its relative simplicity.

An input ac signal V_i is coupled through capacitor C_s, which blocks any dc signal component of the input from affecting the present amplifier circuit (and vice versa). The amplified ac signal from the BJT collector is similarly coupled out through capacitor C_C. The values of resistors R_B and R_C are important in setting the dc bias of the BJT. They also affect the ac amplification of the circuit, as we shall shortly see. The ac input signal varies the base current set by the dc bias. This variation in base current then causes a variation in the collector current from the level set by the dc bias circuit and subsequently results in a variation of the voltage at the collector—the output voltage.

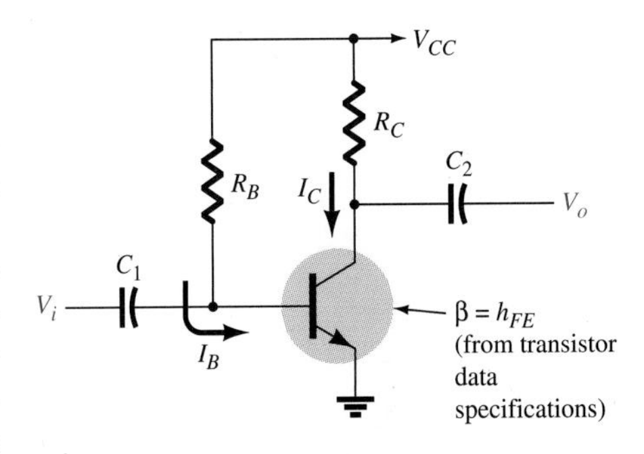

FIG. 9.17
BJT common-emitter amplifier.

Dc Bias

Figure 9.18 includes those elements of the circuit that establish the dc bias voltages and currents. In this circuit configuration the base current is *fixed* (hence the name *fixed-bias configuration*) or set to a specific value by resistor R_B and supply voltage V_{CC}. Notice that the dc base bias current I_B is set by the fixed voltage across R_B, which is the difference between the supply voltage and the base–emitter voltage drop when forward-biased—0.7 V, in general. We can thus calculate I_B as

$$I_B = \frac{V_{CC} - V_{BE}}{R_B} \qquad \textbf{(9.19)}$$

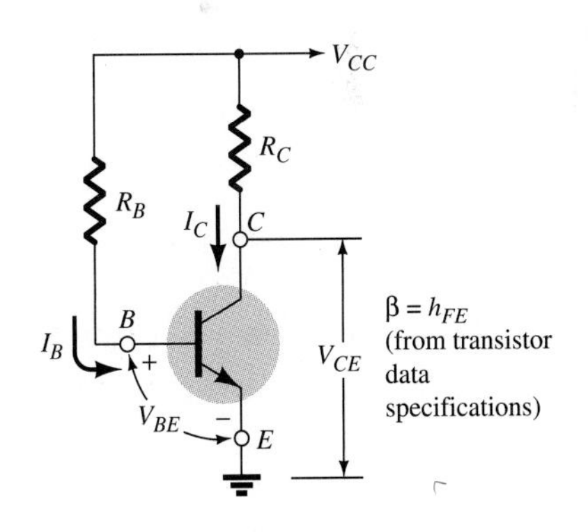

FIG. 9.18
Dc equivalent of Fig. 9.17.

Since the collector dc bias current is β times the base current, the collector current I_C is

$$I_C = \beta I_B \qquad \textbf{(9.20)}$$

It is important to realize that although the value of V_{BE} for a forward-biased base–emitter remains fairly constant, the value of β may vary considerably from the typical value specified for a particular type of transistor. This dependence of the value of I_C on β, a device parameter that cannot be set to an exact value in manufacturing a very large number of transistors, is the practical reason for the limited use of the circuit in Fig. 9.18.

Applying Kirchhoff's voltage law, we have

$$V_{CE} + V_{R_C} - V_{CC} = 0$$

and

$$V_{CE} = V_{CC} - V_{R_C}$$

but

$$V_{R_C} = I_C R_C$$

and

$$\boxed{V_{CE} = V_{CC} - I_C R_C} \qquad \textbf{(9.21)}$$

The values of R_B and R_C, as well as the transistor parameter β, have a pronounced effect on the operating point defined by I_{C_Q} and V_{CE_Q}. Since a package of transistors with the same label can have a variety of values of β (a range of β is specified on the specification sheet) it is necessary to adjust, say, R_B for each transistor used to obtain the same operating point. This individual adjustment of parameters to obtain the same operating point is undesirable when manufacturing a slew of the same configuration. The result is other configurations to be described shortly that are less sensitive to the value of β.

EXAMPLE 9.4 For the network in Fig. 9.17, if $V_{CC} = 9$ V, $R_B = 410$ kΩ, $R_C = 2.4$ kΩ, and $\beta_{dc} = h_{FE} = 100$:

a. Determine I_B.
b. Find I_{C_Q}.
c. Calculate V_{CE_Q}.

Solution:

a. For the *on* transistor $V_{BE} = 0.7$ V, and

$$I_B = \frac{V_{CC} - V_{BE}}{R_B} = \frac{9\text{ V} - 0.7\text{ V}}{410\text{ k}\Omega} = \mathbf{20.24\ \mu A}$$

b. $I_{C_Q} = \beta_{I_{B_Q}} = 100(20.24\ \mu\text{A}) = \mathbf{2.02\ mA}$

c. $V_{CE_Q} = V_{CC} - I_C R_C = 9\text{ V} - (2.02\text{ mA})(2.4\text{ k}\Omega) = 9\text{ V} - 4.85\text{ V}$ $= \mathbf{4.15\ V}$

Ac Analysis

The fact that the emitter terminal is now the common terminal between input and output circuits changes the ac equivalent circuit from that used for the common-base configuration. We still have the r_e equivalent between base and emitter employed for the common-base configuration as shown in Fig. 9.19(a), but now the output terminals are between collector and emitter, resulting in the controlled source βI_b between collector and base terminals.

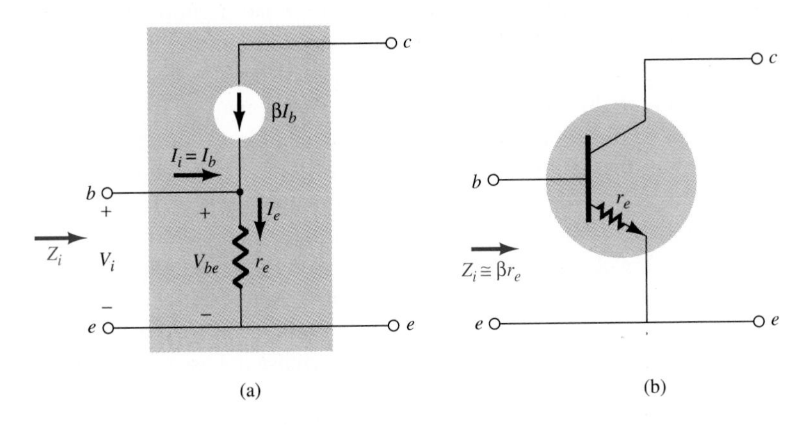

FIG. 9.19

Using the equivalent model in Fig. 9.19(a), we can calculate the input impedance as

$$Z_i = \frac{V_i}{I_i} = \frac{V_{be}}{I_b}$$

with

$$V_{be} = I_e r_e$$

and

$$I_e = (\beta + 1)I_b \cong \beta I_b$$

so that

$$V_{be} = I_e r_e = \beta I_b r_e$$

and

$$Z_i = \frac{V_{be}}{I_b} = \frac{\beta I_b r_e}{I_b}$$

so that

$$Z_i \cong \beta r_e \tag{9.22}$$

In essence, Eq. (9.22) states that the input impedance for the common-emitter amplifier in Fig. 9.17 is beta times the value of r_e. In other words, a resistive element in the emitter leg as shown in Fig. 9.19(b) is reflected into the input circuit by a multiplying factor β. Since β is often 100 or larger, the input impedance to the common-emitter amplifier is always considerably more than that to a common-base configuration. Typical values of Z_i for the common-emitter configuration range from a few hundred ohms to the kilohm range, with maximums of about 6 to 7 kΩ.

Since r_e is determined by $r_e = 26 \text{ mV}/I_E$ (dc), we find again that the dc biasing arrangement of the transistor determines an important ac parameter. If we use the fact that the input impedance is βr_e, that the collector current is βI_b, and that the output impedance is r_o, the equivalent model in Fig. 9.20 can be an effective tool in the analysis to follow and in fact will be the ac equivalent model employed for all common-emitter amplifiers in the ac domain. The output resistance r_o is the impedance *seen* at the output terminals with the input signal set to zero. For most applications we shall assume that its magnitude is sufficiently large to be ignored. In particular, note that the controlled source of the common-emitter equivalent circuit is pointing toward ground rather than away from ground, as in the common-base configuration. This is a result of the transistor characteristics that show that an increase in base current due to an increase in the applied swing results in a negative excursion for the output voltage v_{ce}. A careful review of Fig. 9.12 will clarify the need to include the current source as shown in Fig. 9.20. The effect of the minus sign in the gain is a 180° phase shift between input and output voltages. That is, as the input signal rises in magnitude the output signal becomes increasingly negative, and vice versa.

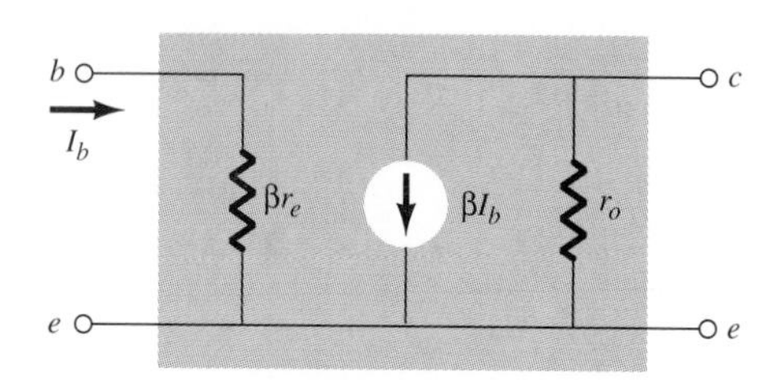

FIG. 9.20
Transistor common-emitter ac equivalent model.

The ac equivalent for the network in Fig. 9.17 is provided in Fig. 9.21, based on the model in Fig. 9.20. Note that the reactance of both capacitors is sufficiently small at the frequency of interest to be considered a short circuit. In addition, the dc source is set to zero (replaced by a short-circuit equivalent) since we are now only interested in the ac levels of the network.

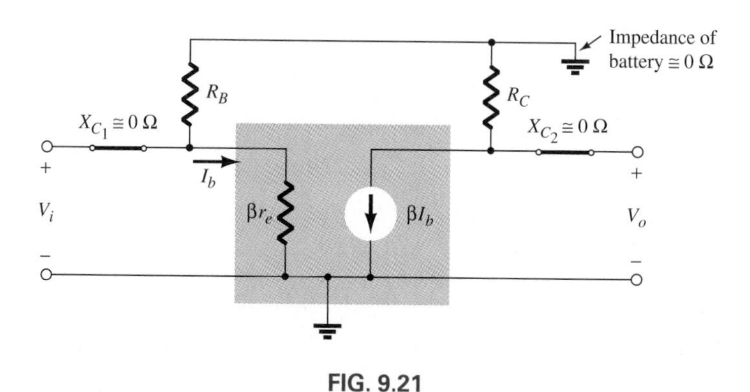

FIG. 9.21
Network of Fig. 9.18 in the ac domain with the transistor equivalent network.

Bringing R_B and R_C down to the transistor ground results in the ac equivalent network in Fig. 9.22. The resistors R_B and βr_e are now in parallel, resulting in an input impedance:

$$Z_i = R_B \parallel \beta r_e \tag{9.23}$$

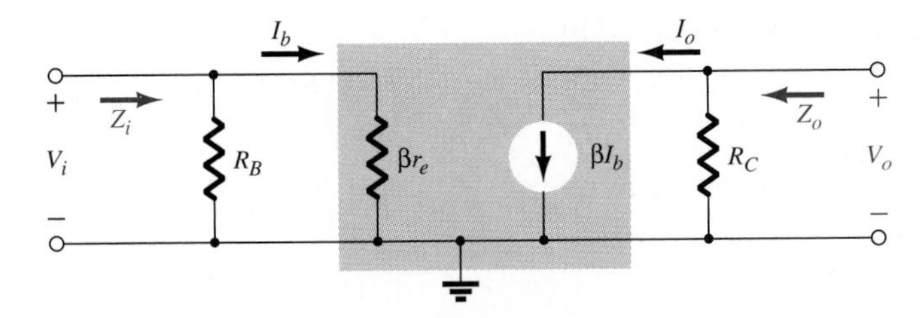

FIG. 9.22
Network of Fig. 9.21 redrawn.

Since r_o is considered sufficiently large compared with R_C to be ignored, the output impedance is

$$Z_o \cong R_C \tag{9.24}$$

Using the defined polarity of V_o, we see that the equation for the output voltage must have a negative sign because the direction of the current source creates a voltage across R_C that is the opposite of the defined V_o. The result is

$$V_o = -I_C R_C = -\beta I_b R_C$$

and for the input circuit,

$$V_i = I_b \beta r_e$$

The voltage gain is therefore

$$A_v = \frac{V_o}{V_i} = \frac{-\beta \cancel{I_b} R_C}{\cancel{I_b} \beta r_e}$$

with

$$A_v = -\frac{R_C}{r_e} \tag{9.25}$$

The current gain is affected by the division of I_i at the input circuit. Using the current-divider rule, we have

$$I_b = \frac{R_B I_i}{R_B + \beta r_e} \quad \text{or} \quad \frac{I_b}{I_i} = \frac{R_B}{R_B + \beta r_e}$$

The current gain is therefore

$$A_i = \frac{I_o}{I_i} = \left(\frac{I_o}{I_b}\right)\left(\frac{I_b}{I_i}\right) = \beta\left(\frac{R_B}{R_B + \beta r_e}\right)$$

and

$$A_i = \frac{\beta R_B}{R_B + \beta r_e} \tag{9.26}$$

The power gain defined by

$$A_p = \frac{P_o}{P_i} = \frac{V_o I_o}{V_i I_i} = \left(\frac{V_o}{V_i}\right)\left(\frac{I_o}{I_i}\right)$$

is therefore determined by

$$\boxed{A_p = A_v A_i} \tag{9.27}$$

EXAMPLE 9.5 For the network in Fig. 9.17, if the ac and dc beta values are equal, determine:

a. Input impedance Z_i.
b. Voltage gain A_v.
c. Total voltage gain A_{v_T} if a source with an internal resistance of 1 kΩ is applied.
d. Current gain A_i.
e. Power gain A_p.

Solution:

a. $Z_i = R_B \| \beta r_e$

with $r_e = \dfrac{26\text{ mV}}{I_E(\text{dc})} = \dfrac{26\text{ mV}}{2.02\text{ mA}}$ (from Example 9.4)
$= 12.87\ \Omega$

so that $Z_i = (410\text{ k}\Omega) \| [100(12.87\ \Omega)] = 410\text{ k}\Omega \| 1287\ \Omega$
$= \mathbf{1.28\ k\Omega}$

b. $A_v = -\dfrac{R_C}{r_e} = -\dfrac{2.4\text{ k}\Omega}{12.87\ \Omega} = \mathbf{-186.48}$

c. $V_i = \dfrac{R_i V_s}{R_i + R_s} = \dfrac{1.28\text{ k}\Omega\ V_s}{1.28\text{ k}\Omega + 1\text{ k}\Omega} = 0.561\ V_s$ or $\dfrac{V_i}{V_s} = 0.561$

$A_{v_T} = \dfrac{V_o}{V_s} = \dfrac{V_o}{V_i} \cdot \dfrac{V_i}{V_s} = A_v \dfrac{V_i}{V_s} = (-186.48)(0.561) = \mathbf{-104.62}$

d. $A_i = \dfrac{\beta R_B}{R_B + \beta r_e} = \dfrac{(100)(410\text{ k}\Omega)}{410\text{ k}\Omega + 1287\ \Omega} = \mathbf{99.69}$

e. $A_p = A_v \cdot A_i = (186.48)(99.69) = \mathbf{18.59 \times 10^3}$

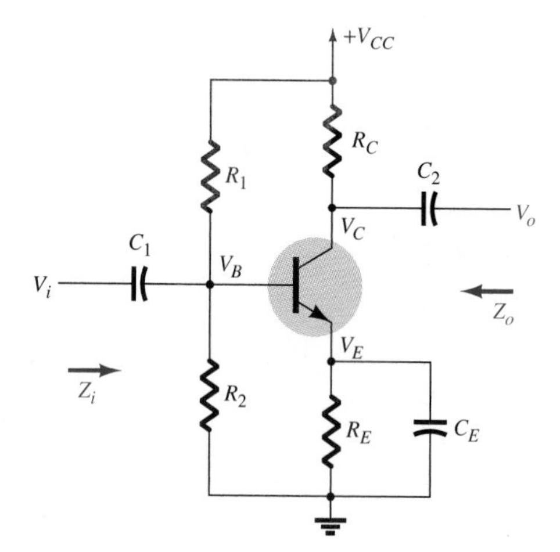

FIG. 9.23
Voltage divider BJT small-signal amplifier.

Voltage-Divider Configuration

A more practical dc bias configuration for small-signal amplification is that in Fig. 9.23. We will find that the ac gain is less sensitive to the value of β and more stable in the sense that temperature variations will have less impact on the network parameters. The analysis begins with the dc bias voltages and currents followed by a complete ac analysis.

Dc Bias

Our analysis begins with a look at the input circuit with the transistor configuration simply replaced by its input resistance. Recall that in the previous description the ac input resistance to the common-emitter transistor was beta times the resistance r_e—a reflection of the resistance r_e from the output circuit to the input circuit by multiplication by a factor beta. This same reflection of impedance in the ac domain can be applied to the dc configuration, resulting in an input impedance Z_i for the network in Fig. 9.23 of beta times R_E as shown in Fig. 9.24(a). That is,

$$Z_i \cong \beta R_E \tag{9.28}$$

For most voltage-divider configurations the impedance Z_i of the transistor is much larger than the parallel resistor R_2 in Fig. 9.24(b), resulting

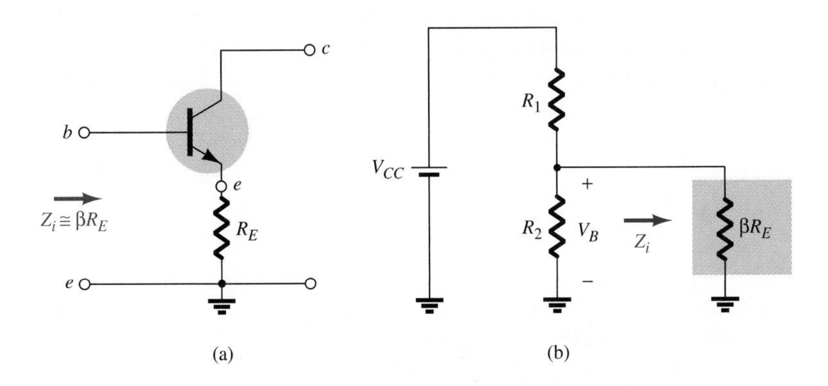

FIG. 9.24
(a) Reflection of R_E to the input circuit; (b) determining V_B.

in a situation in which it is a good approximation to ignore the effects of Z_i and determine the voltage V_B simply by using the resistor R_2 and the voltage-divider rule.

That is,

$$V_B \cong \frac{R_2 V_{CC}}{R_2 + R_1} \tag{9.29}$$

Once V_B is known, the voltage V_E for the dc bias network in Fig. 9.25 can be determined as follows because the voltage V_{BE} will be 0.7 V for an *on* transistor.

$$V_E = V_B - V_{BE} \tag{9.30}$$

The emitter current can then be determined using Ohm's law:

$$I_E = \frac{V_E}{R_E} \tag{9.31}$$

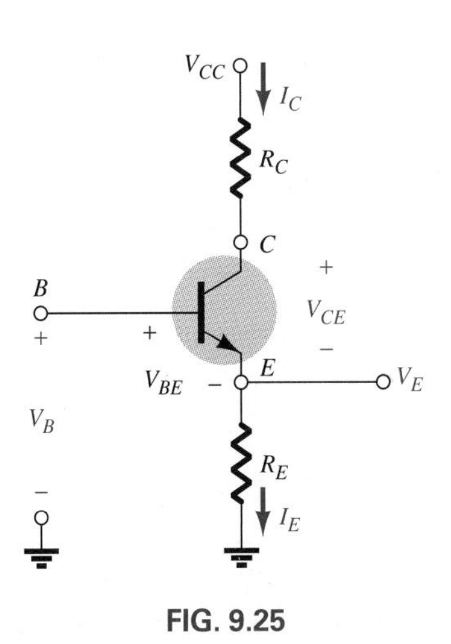

FIG. 9.25

The collector current is then available as follows:

$$I_C \cong I_E \tag{9.32}$$

and the collector-to-emitter voltage is

$$V_{CE} = V_{CC} - I_C R_C \tag{9.33}$$

A review of the preceding equations reveals that beta is absent from all the resulting equations, suggesting that beta can vary within the same *lot* of transistors without affecting the resulting dc operation and the ac parameters it may determine—a less sensitive situation. Of course, if the condition $\beta R_E \gg R_2$ is not satisfied by the chosen configuration, the preceding equations will have to be modified to include the effect of beta, and the improved stability and reduced level of sensitivity will be partially lost.

EXAMPLE 9.6 For the network in Fig. 9.23, if $V_{CC} = 20$ V, $R_1 = 40$ kΩ, $R_2 = 3.9$ kΩ, $R_C = 4.7$ kΩ, $R_E = 0.82$ kΩ, $C_s = 10$ μF, $C_E = 20$ μF, $C_C = 1$ μF, and $\beta = 200$:

a. Find V_B.
b. Determine I_{C_Q}.
c. Calculate V_{CE_Q}.

Solution:

a. $V_B = \dfrac{R_2 V_{CC}}{R_2 + R_1} = \dfrac{(3.9\text{ k}\Omega)(20\text{ V})}{3.9\text{ k}\Omega + 40\text{ k}\Omega} \cong \mathbf{1.78\ V}$

b. $V_E = V_B - V_{BE} = 1.78\text{ V} - 0.7\text{ V} = 1.08\text{ V}$

and $I_{E_Q} = \dfrac{V_E}{R_E} = \dfrac{1.08\text{ V}}{0.82\text{ k}\Omega} = 1.32\text{ mA}$, with $I_{C_Q} \cong I_{E_Q} = \mathbf{1.32\ mA}$

c. $V_{CE_Q} = V_{CC} - I_C(R_C + R_E) = 20\text{ V} - (1.32\text{ mA})(5.52\text{ k}\Omega) \cong \mathbf{12.71\ V}$

Ac Analysis

Substituting short circuits for the capacitors at the frequency of interest (assume $f = 10$ kHz) and replacing the dc supply with a direct connection to ground gives us the configuration in Fig. 9.26. To demonstrate the validity of replacing the capacitors with short-circuit equivalents, we calculate the reactance of the 10 μF at a frequency of 10 kHz using the equation $X_C = 1/2\pi fC$ and obtain the result 1.6 Ω. When you consider the typical resistive value of any component of the network, the 1.6 Ω can certainly be replaced as a short circuit as a good first approximation. Note that for this configuration replacing the capacitor C_E with a short-circuit equivalent *shorts out* the effect of the resistor R_E in the ac domain.

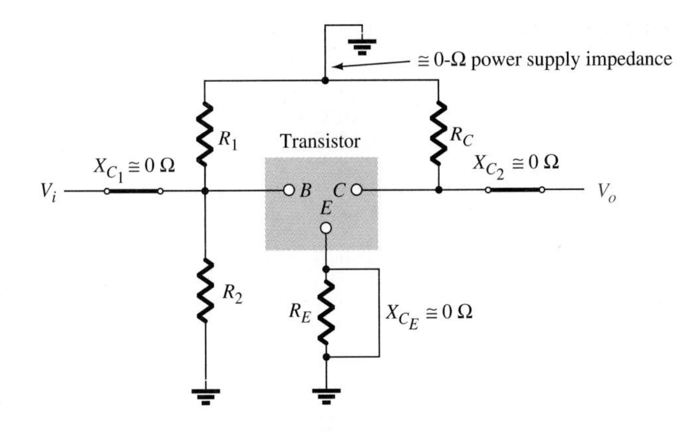

FIG. 9.26
Ac equivalent of network in Fig. 9.23.

In other words, the resistor R_E is important for the dc biasing of the device and establishing a desirable level of stabilization, but it has a negative impact on the ac operation.

Redrawing the network in Fig. 9.26 results in the ac equivalent network in Fig. 9.27.

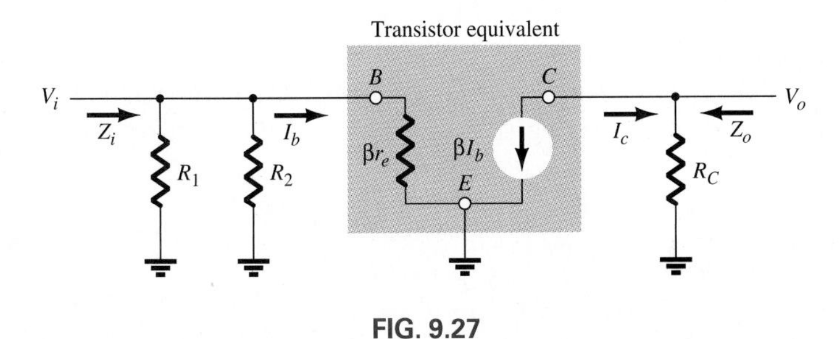

FIG. 9.27
Network of Fig. 9.26 with the transistor equivalent model.

The resistors R_1 and R_2 now drop down in parallel, and R_C is connected between the collector terminal and ground. The common-emitter configuration results in the same transistor equivalent network employed for the fixed-bias configuration as shown in the same figure. Note the similarities between the network in Figs. 9.27 and 9.22 for the fixed-bias configuration. The result is a set of equations similar to those obtained for the fixed-bias configuration.

The input impedance is now determined by the three parallel resistors connected from the base to ground. That is,

$$Z_i = R_1 \| R_2 \| \beta r_e \tag{9.34}$$

In the absence of an r_o parameter for the transistor, the output impedance is determined by

$$Z_o = R_C \tag{9.35}$$

The dynamic resistance r_e is still determined by

$$r_e = \frac{26\text{ mV}}{I_E(\text{dc})} \tag{9.36}$$

which results in the following equation for the voltage gain using the analysis described in detail in the previous section.

$$A_v = -\frac{R_C}{r_e} \tag{9.37}$$

The expressions for current gain and power gain will also be similar to those obtained for the fixed-bias configuration.

EXAMPLE 9.7 Using the parameters and results for Example 9.6 determine the following ac parameters for the network in Fig. 9.23.

a. Input impedance Z_i.
b. Output impedance Z_o.
c. Voltage gain A_v.
d. Output voltage if $V_i = 25$ mV (rms).
e. Current gain A_i.
f. Power gain A_p.

Solution:

a. $Z_i = R_1 \| R_2 \| \beta r_e$ with $r_e = \dfrac{26\text{ mV}}{I_E(\text{dc})} = \dfrac{26\text{ mV}}{1.32\text{ mA}} \cong 19.7\ \Omega$

$Z_i = 40\text{ k}\Omega \| 3.9\text{ k}\Omega \| [200(19.7\ \Omega)] = 3.55\text{ k}\Omega \| 3.94\text{ k}\Omega \cong \mathbf{1.87\ k\Omega}$

b. $Z_o = R_C = \mathbf{4.7\ k\Omega}$

c. $A_v = -\dfrac{R_C}{r_e} = \dfrac{-4.7\text{ k}\Omega}{19.7\ \Omega} = \mathbf{-238.6}$

d. $A_v = -\dfrac{V_o}{V_i} \Rightarrow V_o = -A_v v_i = -(-238.6)(25\text{ mV}) = \mathbf{5.97\ V}$

e. $A_i = \dfrac{\beta(R_1 \| R_2)}{(R_1 \| R_2) + \beta r_e} = \dfrac{200(3.55\text{ k}\Omega)}{3.55\text{ k}\Omega + 3.94\text{ k}\Omega} \cong \mathbf{94.8}$

f. $A_P = A_v A_i = (238.6)(94.8) = \mathbf{22.62 \times 10^3}$

Emitter-Follower Configuration

Another popular configuration is the common-collector or emitter-follower network in Fig. 9.28 with the output taken from the emitter terminal. Because the base-to-emitter resistance is relatively low in the ac domain, the applied signal at the base and the output signal will be relatively close in magnitude. In fact, the ac emitter voltage will be slightly smaller than the base voltage, and the two will be *in phase*. Since the output voltage is slightly less than the applied signal, the emitter-follower configuration is certainly not suitable for voltage amplification purposes. Although it often has a current gain close to the magnitude of beta, its primary application is as an impedance-matching device. That is, the input impedance is quite large because of the βR_E reflection in the ac domain, and the output impedance is quite small. For signals with a high internal source resistance, placing an emitter-follower network between the applied signal and the small-signal amplifier ensures that most of the applied signal will reach the input terminals of the amplifier and not be lost across the internal resistance of the source. The low output impedance ensures that most of the output signal of the emitter-follower will appear across the input terminals of the following amplifier. The name *emitter-follower* was chosen because the voltage at the emitter *follows* the applied voltage at the base as revealed by the *in-phase* relationship between input and output voltages noted above.

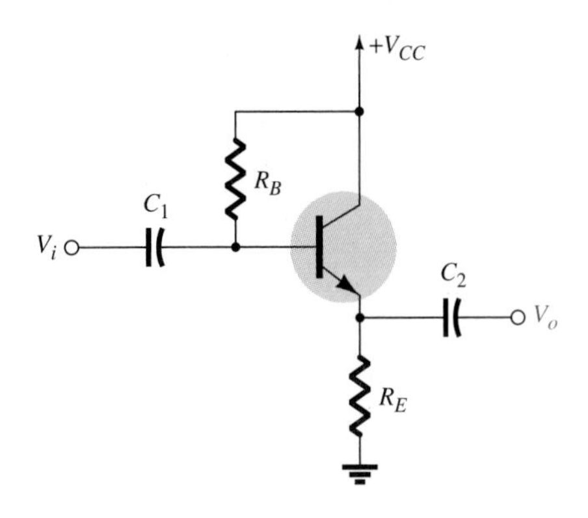

FIG. 9.28
Emitter-follower configuration.

Dc Bias

The dc configuration appears in Fig. 9.29, revealing a situation in which the emitter resistance is again reflected to the input circuit by a factor beta. In other words, for the input circuit the voltages present the V_{CC} and V_{BE} with V_{BE} always working against the applied source. The total resistance of the input circuit is the base resistance R_B and the reflected resistance βR_E. Applying Ohm's law, we have

$$I_B = \frac{V}{R} = \frac{V_{CC} - V_{BE}}{R_B + Z_i}$$

and

$$I_{B_Q} = \frac{V_{CC} - V_{BE}}{R_B + \beta R_E} \tag{9.38}$$

FIG. 9.29
Dc equivalent of Fig. 9.28.

The collector and emitter current are then determined by

$$I_{C_Q} = \beta I_{B_Q} \cong I_{E_Q} \tag{9.39}$$

and the collector-to-emitter voltage is

$$V_{CE_Q} = V_{CC} - I_{C_Q}R_E \tag{9.40}$$

Ac Analysis

The ac equivalent in Fig. 9.30 reveals that the emitter resistor is now part of the network rather than being shorted out by a bypass capacitor as in the voltage-divider configuration.

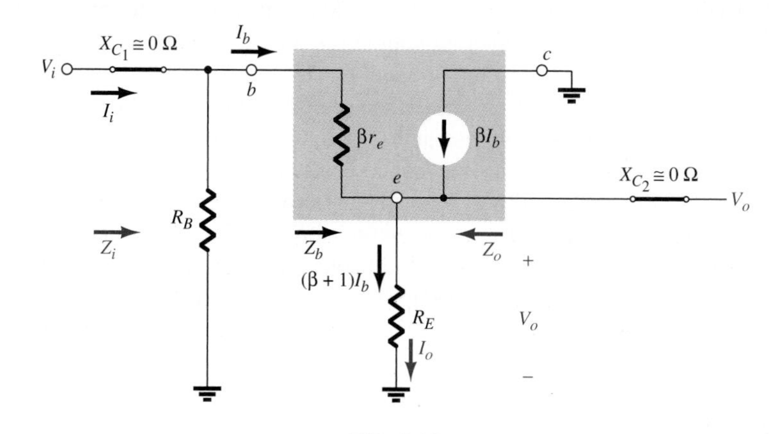

FIG. 9.30
Ac equivalent of the network in Fig. 9.28.

The voltage amplification can be determined by first finding the output voltage in the form

$$V_o = I_oR_E = I_eR_E = (\beta + 1)I_bR_E \cong \beta I_bR_E$$

The applied voltage can be written as

$$V_i = I_b\beta r_e + (\beta + 1)I_bR_E \cong I_b\beta r_e + \beta I_bR_E = \beta I_b(r_e + R_E)$$

resulting in the amplification factor

$$A_v = \frac{V_o}{V_i} = \frac{\beta I_bR_E}{\beta I_b(r_e + R_E)}$$

and

$$A_v = \frac{V_o}{V_i} = \frac{R_E}{r_e + R_E} = \frac{1}{1 + r_e/R_E} \tag{9.41}$$

Equation (9.41) clearly reveals that the voltage amplification factor is less than 1 because the ratio r_e/R_E has some magnitude. The ratio, however, is usually so small that the following approximation is normally applied to the emitter-follower configuration:

$$A_v \cong 1 \quad \textbf{(9.42)}$$

The input impedance can be found by first finding the impedance Z_b defined in Fig. 9.30.

$$Z_b = \beta r_e + (\beta + 1)R_E \cong \beta r_e + \beta R_E = \beta(r_e + R_E) \cong \beta R_E$$

and since R_B is in parallel with the impedance Z_b,

$$Z_i = R_B \parallel Z_b$$

and

$$Z_i = R_B \| \beta R_E \quad \textbf{(9.43)}$$

We have used the fact that an emitter resistor is reflected to the input circuit by multiplying by the factor beta. Working in reverse, however, we can reflect any resistance in the base circuit to the emitter output circuit by dividing the resistance by the same beta factor. For the emitter terminal, therefore, the ac network *seen* looking back into the transistor appears as shown in Fig. 9.31. Setting the source to zero as required to define the output impedance results in $Z_o = R_E \parallel r_e$, but since $R_E \gg r_e$

$$Z_o \cong r_e \quad \textbf{(9.44)}$$

which is quite low compared with most small-signal amplifiers. If the internal resistance of the applied signal is part of the input circuit, it will also be reflected to the output circuit by dividing by the factor beta.

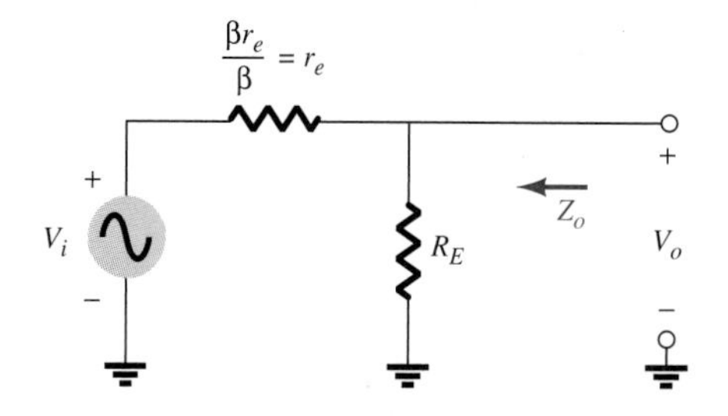

FIG. 9.31
Determining Z_o.

EXAMPLE 9.8 If the parameters of the network in Fig. 9.28 are $R_B = 1\text{ M}\Omega$, $R_E = 5\text{ k}\Omega$, $V_{CC} = 12\text{ V}$, and $\beta = 120$:

a. Find the dc levels of I_{B_Q}, I_{C_Q}, and I_{E_Q}.
b. Determine V_{CE_Q}.
c. Calculate Z_i.
d. Find A_v.
e. Calculate Z_o.

Solution:

a. $I_{B_Q} = \dfrac{V_{CC} - V_{BE}}{R_B + \beta R_E} = \dfrac{12\text{ V} - 0.7\text{ V}}{1\text{ M}\Omega + (120)(5\text{ k}\Omega)} = \dfrac{11.3\text{ V}}{1.6\text{ M}\Omega} = \mathbf{7.06\ \mu A}$

$I_{C_Q} = \beta I_{B_Q} = (120)(7.06\ \mu\text{A}) = \mathbf{0.847\ mA}$

$I_{E_Q} \cong I_{C_Q} = \mathbf{0.847\ mA}$

b. $V_{CE_Q} = V_{CC} - I_E R_E$
$= 12\text{ V} - (0.847\text{ mA})(5\text{ k}\Omega)$
$= 12\text{ V} - 4.24\text{ V} = \mathbf{7.76\text{ V}}$

c. $Z_i \cong R_B \| \beta R_E = 1\text{ M}\Omega \| (120)(5\text{ k}\Omega) = 1\text{ M}\Omega \| 600\text{ k}\Omega = \mathbf{375\text{ k}\Omega}$

d. $r_e = \dfrac{26\text{ mV}}{I_{E_Q}} = \dfrac{26\text{ mV}}{0.847\text{ mA}} = 30.7\ \Omega$

$A_v = \dfrac{1}{1 + r_e/R_E} = \dfrac{1}{1 + 30.7\ \Omega/5\text{ k}\Omega} = \dfrac{1}{1.006} = \mathbf{0.994} \cong \mathbf{1}$

e. $Z_o = R_E \| r_e = 5\text{ k}\Omega \| 30.7\ \Omega = \mathbf{30.5\ \Omega} \cong r_e$

Loading Effects

The application of a load such as another stage to a small-signal amplifier has the effect of reducing the overall gain of the system. In other words, the no-load gain is always the maximum attainable. Since it is normally tied to the output terminal through a capacitive element, the applied load does not affect the dc biasing arrangement of the amplifier, appearing only in the ac equivalent model. In Fig. 9.32 a load was intro-

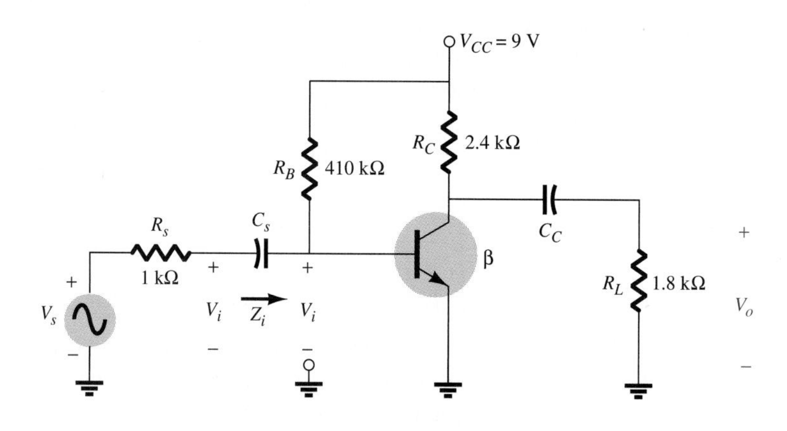

FIG. 9.32
Small-signal amplifier with applied load.

duced along with a signal source having an internal resistance R_s. The ac equivalent in Fig. 9.33 will demonstrate the detrimental effect of the

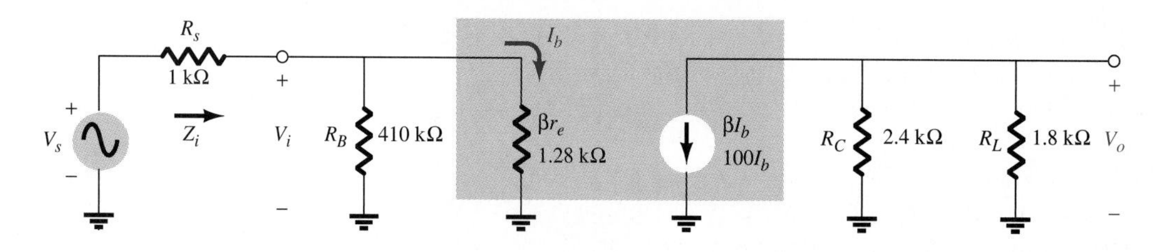

FIG. 9.33
Ac equivalent of the network in Fig. 9.32.

applied load and internal resistance of the source on the overall voltage gain.

The output voltage for the network in Fig. 9.33,

$$V_o = -(\beta I_b)(R_C \| R_L)$$

is now reduced because R_C and R_L are now in parallel. The addition of a resistor in parallel with any resistor always results in a reduced terminal resistance.

Substituting, we find

$$V_o = -(100\, I_b)(2.4\text{ k}\Omega \,\|\, 1.8\text{ k}\Omega) = -102.86 \times 10^3\, I_b$$

and

$$V_i = I_b Z_i = I_b(R_B \| \beta r_e) = I_b(410\text{ k}\Omega \,\|\, 1.28\text{ k}\Omega) \cong 1.28 \times 10^3\, I_b$$

with

$$A_v = \frac{V_o}{V_i} = \frac{-102.86 \times 10^3\, I_b}{1.28 \times 10^3\, I_b} \cong \mathbf{-80}$$

compared with $A_v \cong -186$ (Example 9.5) without R_L.

An alternative solution is to simply modify Eq. (9.25) as follows:

$$A_v = -\frac{R_C \| R_L}{r_e} = \frac{-1.029 \times 10^3}{12.87\ \Omega} \cong \mathbf{-80}$$

At the input circuit,

$$V_i = \frac{Z_i V_s}{Z_i + R_s}$$

clearly demonstrating a loss in applied voltage to the input terminals of the amplifier due to the drop across R_s.

Substituting, we have

$$V_i = \frac{1.28\text{ k}\Omega\ V_i}{1.28\text{ k}\Omega + 1\text{ k}\Omega}$$

and $V_i \cong 0.56\ V_s$, or a 44% loss across R_s.

The overall gain is

$$A_{v_T} = \frac{V_o}{V_s} = \frac{V_o}{V_i} \cdot \frac{V_i}{V_s} = (-80)(0.56) = \mathbf{-44.8}$$

If $R_s = 0\ \Omega$ and R_L was not applied, the overall gain would increase to -186.48 which is a significant increase over the calculated value.

The effect of R_s and R_L is therefore significant and must be considered in any design sequence.

Although assumed to be sufficiently large to be ignored for most applications, the effect of the output impedance of the amplifier (r_o) is also to reduce the overall gain. It often appears directly in parallel with the applied load and has the same effect on the total gain as demonstrated above.

Amplifier Notation

Since the most important characteristics of any amplifier are its gain, input impedance, and output impedance, the notation in Fig. 9.34(a) is often applied to represent the amplifier in package (system) form.

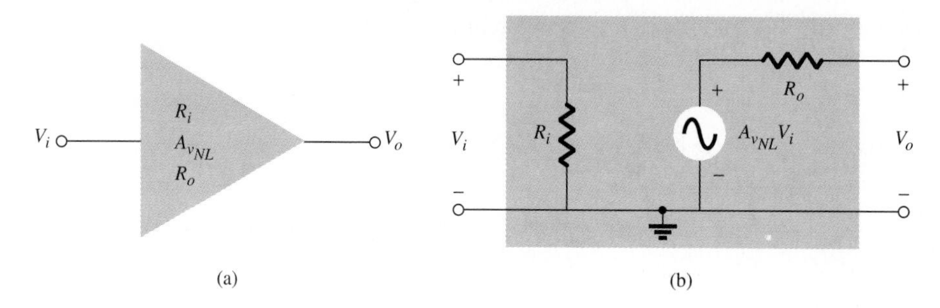

FIG. 9.34
Amplifier notation.

Note that the no-load (*NL* subscript) voltage gain is provided to permit its use with any applied load. The output impedance R_o is not simply r_o of the transistor but includes specific elements of the biasing arrangement. A network representation of the important parameters is provided in Fig. 9.34(b). In the absence of an applied load the output circuit is incomplete with $I_L = 0$ mA and $V_{R_o} = 0$ V. The result is $V_L = A_{v_{NL}}V_i$, and the amplifier gain is simply the no-load value $A_{v_{NL}}$.

For the voltage-divider configuration in Example 9.7 the equivalent models appear as shown in Fig. 9.35.

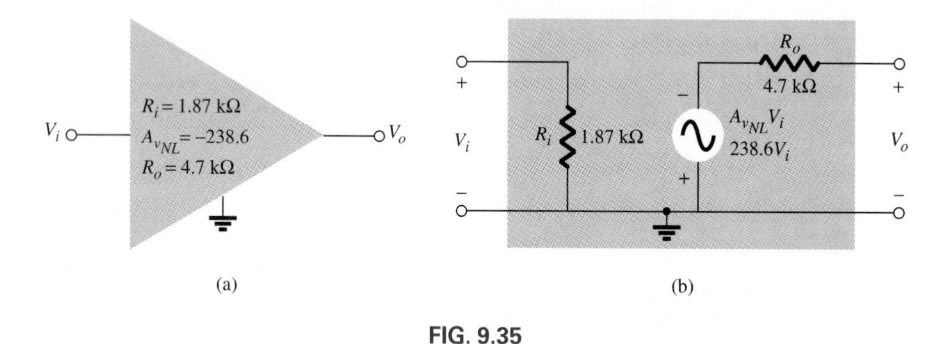

FIG. 9.35
System notation for the transistor amplifier in Example 9.7.

Figure 9.35(b) is clear evidence that the application of a load results in a reduced gain. The applied load completes the output circuit, resulting in a loss of voltage across R_o. Figure 9.35(b) also reveals that the larger the applied load, the higher the level of V_o (as determined by the voltage-divider rule) and the higher the gain. In other words, as the load increases in magnitude and approaches the open-circuit equivalent the gain approaches the no-load gain (the maximum). Note also in the same figure that the negative sign associated with the no-load gain is included by simply reversing the polarity of the supply.

The models in Fig. 9.35 can be applied to any small-signal amplifier, not simply BJT amplifiers. In general, they reveal that R_i should be as large as possible to capture a major part of the applied signal and reduce the net loss across the internal resistance of the supply. In addition, R_o should be as small as possible to maximize the voltage across the load and the resulting gain. Both equivalents will appear in the sections and chapters to follow.

Hybrid Parameters

Throughout the ac analysis of the BJT amplifier we used parameters such as r_e defined by the dc operation of the device. Specification sheets, however, are not aware of the application of the device or its dc biasing arrangement but would like to provide typical ac operating parameters. Normally, therefore, a typical collector current and collector–emitter voltage are chosen, and the parameters are provided for this region of the characteristics. However, rather than provide the level of r_e or βr_e, specification sheets provide a set of *hybrid parameters*. The term *hybrid* is used because a mix of units of measurement is applied to the parameter set. The hybrid equivalent model of a common-emitter amplifier is provided in Fig. 9.36(a).

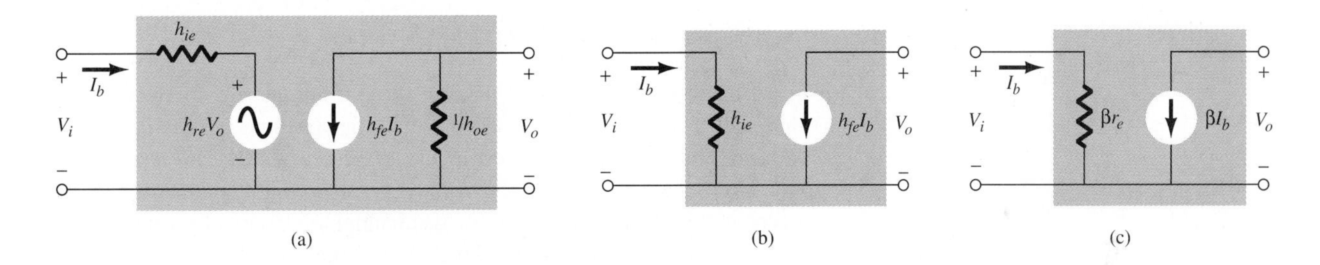

FIG. 9.36
Hybrid equivalent model of a common-emitter amplifier: (a) complete model; (b) reduced model; (c) equivalent r_e model.

The parameter h_{r_e} is a feedback parameter that takes into account that the input circuit is sensitive to the level of the output voltage. However, the level of $h_{r_e}V_o$ is usually sufficiently small compared with V_i to be ignored and is therefore left out of the reduced version in Fig. 9.36(b). The factor $1/h_{oe}$ provides the output impedance of the amplifier, which we normally assume to be large enough to be ignored for most applications and therefore does not appear in the reduced version of the same figure. In general, however,

$$r_o = \frac{1}{h_{oe}} \qquad \textbf{(9.45)}$$

Figure 9.36(b) should be somewhat familiar in that it has the same format as the equivalent model in Fig. 9.36(c), which we used in our recent analysis of the BJT ac configurations.

Comparing Fig. 9.36(b) and (c), we see that the relationships between the hybrid parameters and those defined by the biasing arranagement are

$$\beta_{\text{ac}} = h_{fe} \tag{9.46}$$

$$\beta r_e = h_{ie} \tag{9.47}$$

with

$$r_e = \frac{h_{ie}}{h_{fe}} \tag{9.48}$$

The level of r_e probably will not match the value calculated using the dc emitter current of the actual biasing arrangement, but it does provide a value for comparison with other available transistors and permits analysis of a network without concern for the dc levels of operation.

9.4 FIELD-EFFECT TRANSISTORS

In the last couple of decades, another three-terminal semiconductor device, called a *field-effect transistor* (FET), has become increasingly important. The basic construction in Fig. 9.37 reveals that an n-channel FET has a channel between the *drain* (D) and *source* (S) constructed of n-type material. Above and below the channel are layers of p-type material connected to the same bias level. The voltage V_{GS} between gate (G)

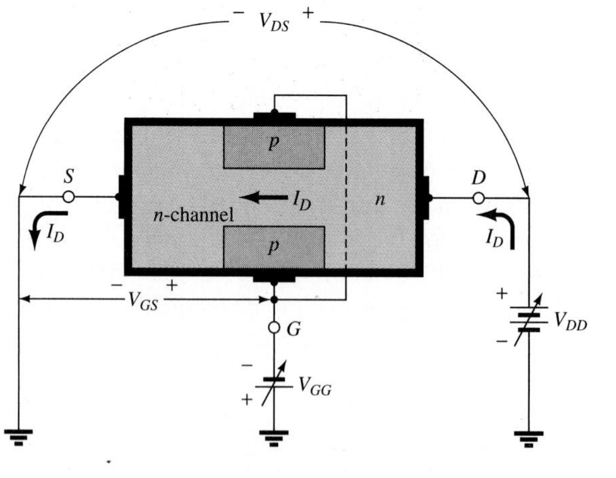

FIG. 9.37

and source (S) terminals is negative to establish a reverse-biased *pn* junction at the upper and lower edges of the channel, with a depletion region controlled by the strength of V_{GG}.

With $V_{GS} = V_{GG} = 0$ V, the effect of the gate can essentially be ignored and V_{DD} ($= V_{DS}$) will establish a healthy current I_D from drain to source (labeled I_{DSS} on specification sheets). Electrons at the source (S) and in the n-type channel will be highly attracted to the drain (D) and the applied positive potential. As V_{GG} increases in magnitude (V_{GS} more and more negative) the depletion region grows as shown in Fig. 9.38(a), reducing the width of the channel between depletion regions. Recall that

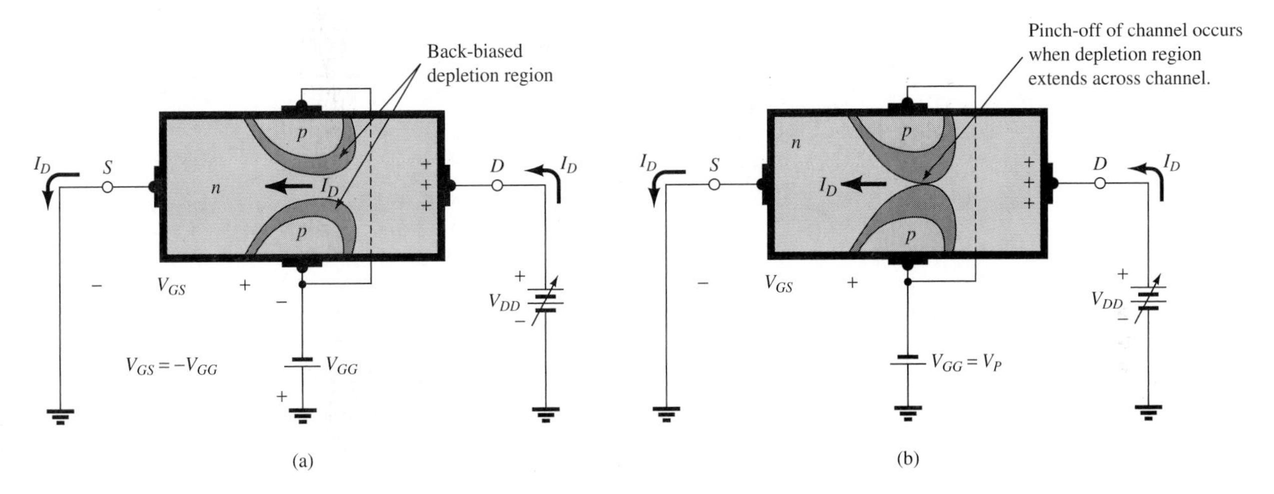

FIG. 9.38
(a) Reduced n-channel width; (b) pinch-off.

the depletion region cannot support charge flow because of the lack of "free carriers." As the channel width decreases, the current I_D naturally decreases, as demonstrated by the characteristics in Fig. 9.39. If V_{GG} continues to be made more and more negative, the channel will eventually be "pinched off" as shown in Fig. 9.38(b) and I_D will drop to insignificant levels ($\cong$0 mA). For the characteristics in Fig. 9.39, $V_{GS} = V_P$ (pinch-off level) is approximately -6 V.

In terms of the quantities often made available in data manuals and on specification sheets, the drain current of a junction field-effect transistor (JFET) is related to the gate-to-source voltage in the following manner:

$$I_D = I_{DSS}\left(1 - \frac{V_{GS}}{V_P}\right)^2 \tag{9.49}$$

In Fig. 9.39 if we follow the curve for $V_{GS} = 0$ V directly over to the I_D axis, we find that I_{DSS} for this transistor is 7.6 mA. For the same device $V_P = -6$ V. If we now plot the curve defined by Eq. (9.49), the

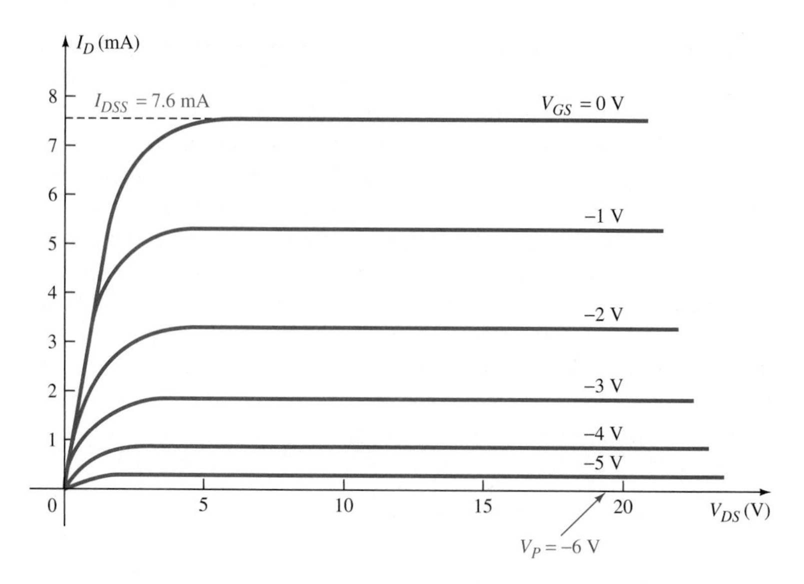

FIG. 9.39
n-channel JFET characteristics with I_{DSS} = 7.6 mA and V_P = −6 V.

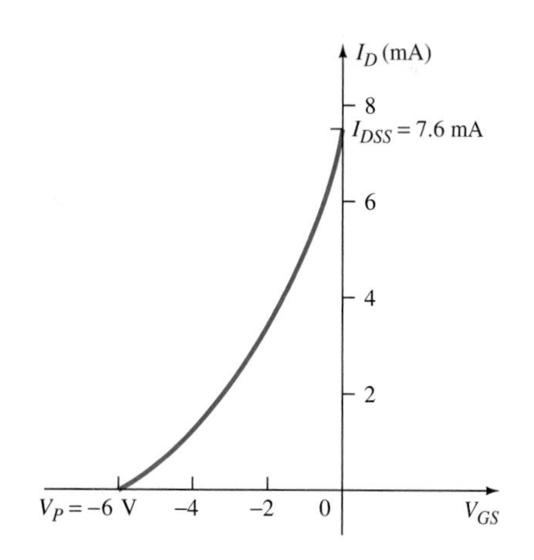

FIG. 9.40
JFET transfer characteristics for the device in Fig. 9.39.

transfer characteristics in Fig. 9.40 will result. The term *transfer* is employed because the curve relates an output quantity I_D to an input quantity V_{GS}. Note the absence of V_{DS} in Eq. (9.49). This is a direct result of the fact that V_{GS} curves are quite flat and values of V_{DS} have little effect on the drain current level except at very low values of V_{DS}. The transfer characteristics in Fig. 9.40 find extended use in the design of JFET amplifiers.

The *n*-type and *p*-type materials in Fig. 9.37 can be reversed, and a *p*-channel JFET will result, which requires a reversal in all biasing levels in the network and the characteristics. The symbol and basic biasing arrangement for both *n*-channel and *p*-channel JFETs are provided in Fig. 9.41. Although the basic construction of some commercially available FETs may differ slightly from that appearing in Fig. 9.37, the mode of operation is basically the same for each.

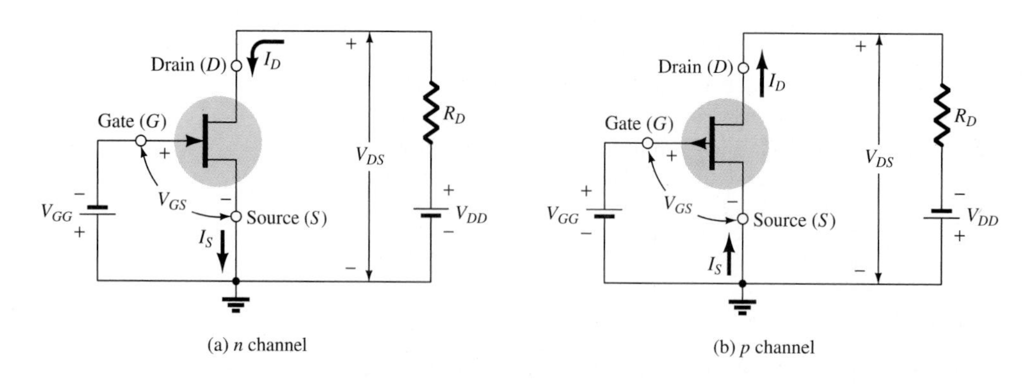

FIG. 9.41
Basic biasing arrangements for (a) n-channel and (b) p-channel JFETs.

Another type of field-effect transistor is the metal-oxide–silicon-field-effect transmitter (MOSFET). The basic construction of MOSFET devices is somewhat more complex than that of JFET devices, but a channel still appears whose width is controlled by the gate potential.

The *enhancement*-type MOSFET has the symbol and characteristics shown in Fig. 9.42 for the *p*-channel type. For the *n*-type the arrow is reversed and the polarities are reversed on the characteristics. The basic biasing arrangement is the same as for the JFET. The term *enhancement* is derived from the fact that with *no bias* on the gate ($V_{GS} = 0$ V) there is no channel for conduction between drain and source terminals and $I_D = 0$ A, as shown in Fig. 9.42. For increasing negative voltages on the gate the *p*-type channel materializes (the channel is enhanced) and increases in size so that I_D can increase in magnitude (see Fig. 9.42 for increasing values of V_{GS}). The broken line in the symbol reflects the fact that the channel does not exist without a gate bias.

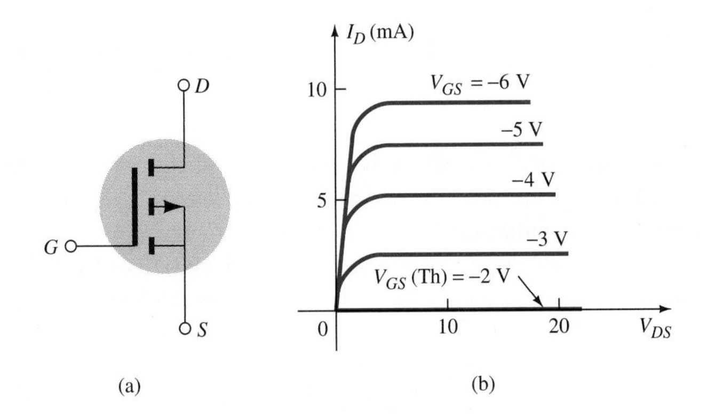

FIG. 9.42
Enhancement-type p-channel MOSFET with $V_{GS(Th)} = -2$ V: (a) symbol; (b) characteristics.

The symbol and characteristics for the *depletion-type n*-channel MOSFET appear in Fig. 9.43. Again the arrow is reversed and the polarities on the characteristics are changed for a *p*-channel device. In this case a channel is present with $V_{GS} = 0$ V, so that a measurable charge flows, as shown in Fig. 9.43. For increasing negative voltages on the gate the *n*-channel is reduced in width and the drain current diminishes. For increasing positive voltages the reverse is true. The continuous bar in the symbol connected to the gate terminal reflects that a channel is present for $V_{GS} = 0$ V. The biasing arrangement for this device is again the same as that employed for the JFET.

Amplifier Circuits

We will now examine the junction FET (JFET) and metal-oxide-silicon FET (MOSFET) small-signal amplifier configurations. The FET has a number of advantages over the BJT as an amplifying device. One is the

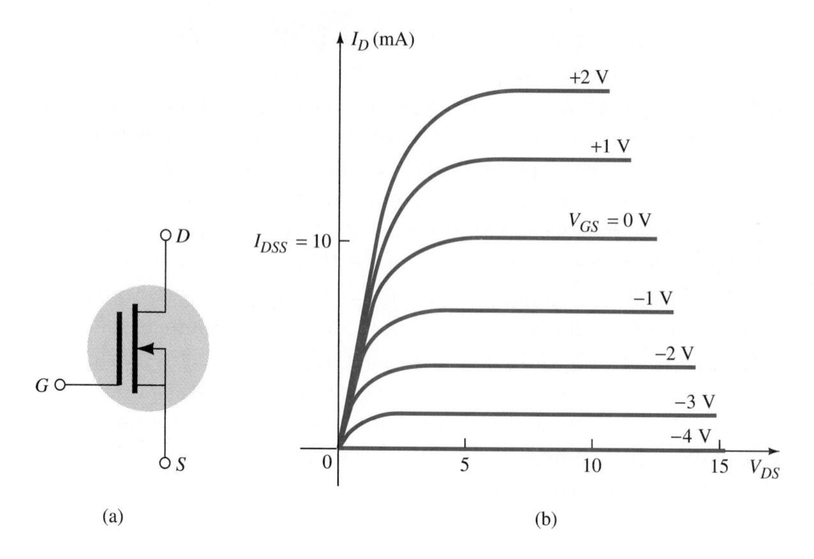

FIG. 9.43
Depletion-type n-channel MOSFET with $I_{DSS} = 10$ mA and $V_P = -4$ V: (a) symbol; (b) characteristics.

very high input impedance, which is usually considered to be infinite, resulting in the open-circuit equivalent at the input terminals. Another is the smaller size (smaller than the BJT) for manufacturing large-scale integrated circuits and the relative simplicity of constructing FET devices in the monolithic IC structure. MOSFETs are more common in IC structures, while JFETs see more use as discrete devices. Our analysis will begin with the JFET structure and the parameters that control the ac response. We will again find that the biasing arrangement determines the ac parameters employed in the small-signal analysis.

JFET Transfer Characteristics

The dc analysis of BJTs employed the beta factor, which was considered constant for the region of analysis. The operating relationship between an output and an input quantity of a JFET, however, is a nonlinear one requiring a slightly more complex mechanism for determining the dc response. There are two directions we can take with this nonlinear characteristic: one is a mathematical approach requiring iteration techniques because of the nonlinear function, and the other is a graphical approach using a *transfer* characteristic derived from the *drain* characteristics. The latter approach is the more popular method and will be described using the drain characteristics in Fig. 9.44.

Two device parameters of particular importance are the values determined from the $V_{GS} = 0$ V curve of the characteristic. Recall that it is possible to pass current through the JFET channel until pinch-off occurs. On the drain characteristic pinch-off occurs when the characteristic curve bends to the horizontal, showing that there is no further drain current (above a saturation value dependent on the value V_{GS}) with increased V_{DS} voltage. This pinch-off condition is shown as the knee of the

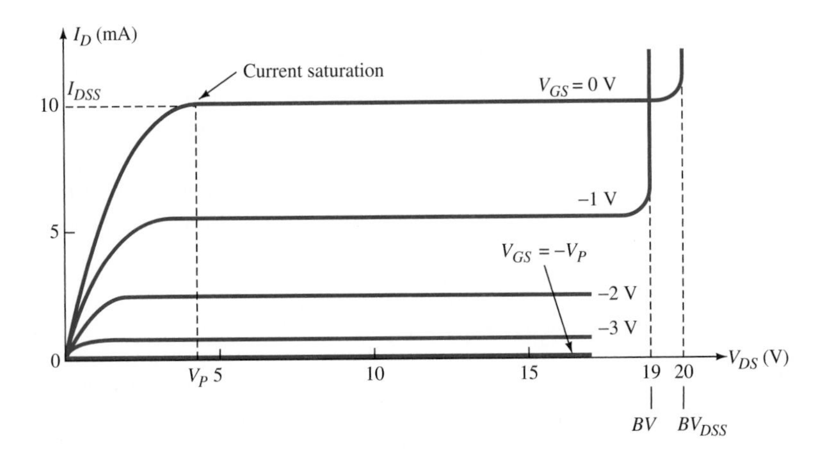

FIG. 9.44
FET drain characteristics.

curve, and for the $V_{GS} = 0$ V curve it is equal to the device pinch-off voltage V_P. It can be seen that at more negative gate–source voltages (negative for an *n*-channel or positive for a *p*-channel JFET) pinch-off occurs at lower values of V_{DS}, since the gate–source already has started to pinch off as a result of the gate–source bias voltage. The value V_P of interest, however, is that voltage required to fully pinch off the channel, and it can be obtained as either the knee of the curve for $V_{GS} = 0$ V or more clearly as the zero-current condition of the drain characteristics, as indicated at $I_D = 0$ mA in Fig. 9.44.

A second important device parameter is the value of drain saturation current at $V_{GS} = 0$ V, called I_{DSS} (current in *d*rain to *s*ource for *s*aturated condition, at $V_{GS} = 0$ V). It is the maximum drain current for the chosen JFET device. The drain current decreases from this value as the gate–source voltage becomes more negative (for an *n*-channel JFET), as shown in Fig. 9.44.

For the JFET the drain current and gate-to-source voltage are related by Eq. 9.48 repeated here for convenience. The equation is commonly referred to as Shockley's equation.

$$I_D = I_{DSS}\left(1 - \frac{V_{GS}}{V_P}\right)^2$$

I_{DSS} and V_P are as defined in Fig. 9.44. Note that the output quantity I_D is related to the input control variable V_{GS} by the square operation which results in a nonlinear relationship between the two, as noted earlier in this section. Also take note that the drain current is now controlled by input voltage rather than input current as occurred for the BJT. The JFET is therefore a *voltage-controlled* device rather than a *current-controlled* device as noted for the BJT. Plotting Eq. (9.49) results in the transfer characteristics in Fig. 9.45. Note that when V_{GS} = V, $I_D = I_{DSS}$, as defined in Fig. 9.44. In addition, when $V_{GS} = V_P$, the current drops to 0 mA. Choosing various levels of V_{GS} and solving for the level of drain current permits a complete drawing of the curve as shown in Fig. 9.45.

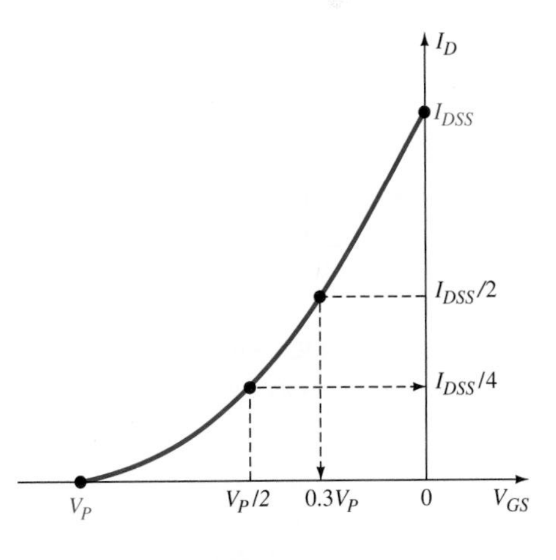

FIG. 9.45
The JFET transfer characteristics.

A fairly accurate rendering of the curve can be obtained from I_{DSS} and V_P if we recognize that when $V_{GS} = V_P/2$, $I_D = I_{DSS}/4$. In addition, when $I_D = I_{DSS}/2$, the voltage $V_{GS} \cong 0.3\ V_P$. Plotting the points determined by I_{DSS}, V_P, and the two just described permits a fairly accurate rendering of the transfer characteristics in a minimum amount of time. Keep in mind that the transfer characteristics are key to the analysis of JFET networks and have to be generated for the dc and ac analyses of any configuration.

Fixed-Bias Configuration

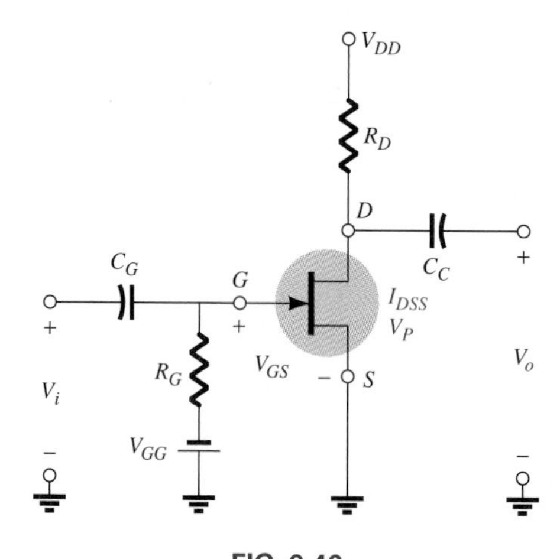

FIG. 9.46
Fixed-bias JFET configuration.

The first JFET configuration to be analyzed is the fixed-bias configuration in Fig. 9.46 requiring two dc supplies, V_{GG} and V_{DD}.

For the provided levels of I_{DSS} and V_P the transfer characteristics appear as shown in Fig. 9.47. Note that the four points described above for plotting the transfer characteristics were employed in the rendering of the curve. For the network in Fig. 9.46 we can ignore the effects of R_G because $I_G = 0$ mA for the dc configuration of a JFET amplifier. The result is $V_{R_G} = I_G R_G = 0$ V, permitting removal of R_G from the configuration and replacing it with a short-circuit equivalent. The resistor R_G will play an important role in the ac analysis to follow. Applying Kirchhoff's voltage law around the input circuit now results in

$$V_{GS} = -V_{GG} \tag{9.50}$$

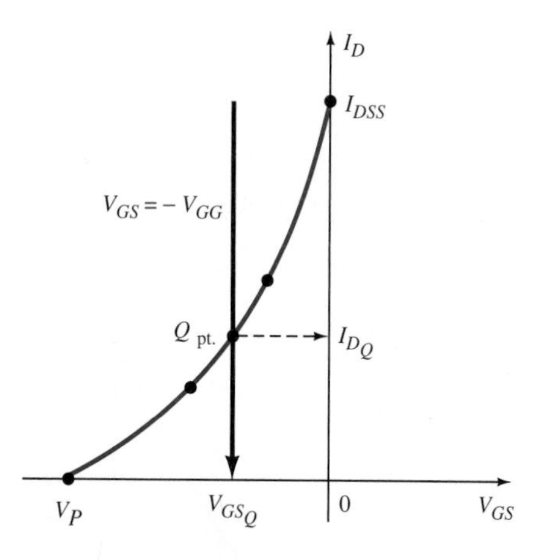

FIG. 9.47
Finding the Q point for the network in Fig. 9.46.

Since V_{GG} is a constant irrespective of the value of I_D, a straight vertical line representing the input circuit can be superimposed on Fig. 9.47. The intersection of the vertical line and the Shockley curve defines the point of operation (I_{D_Q} and V_{GS_Q}) for this network. Be aware that I_{D_Q} and V_{GS_Q} are the values one would obtain with a meter with the network in full operation.

Once I_{D_Q} and V_{GS_Q} are known, any other quantity of the dc configuration can be found using basic circuit analysis techniques. For instance, the voltage V_{DS} is determined by

$$V_{DS} = V_{DD} - I_{D_Q} R_D \tag{9.51}$$

The fact that V_{GS} is set at $-V_{GG}$ (no matter the level of I_D) has resulted in the term *fixed-bias configuration*. The requirement of two separate dc supplies limits the practical application of this configuration.

Ac Operation

For BJTs the dc operating conditions determined the value of r_e for the ac analysis. For JFETs the quiescent point determines the *transconductance parameter* g_m appearing in the small-signal equivalent circuit in

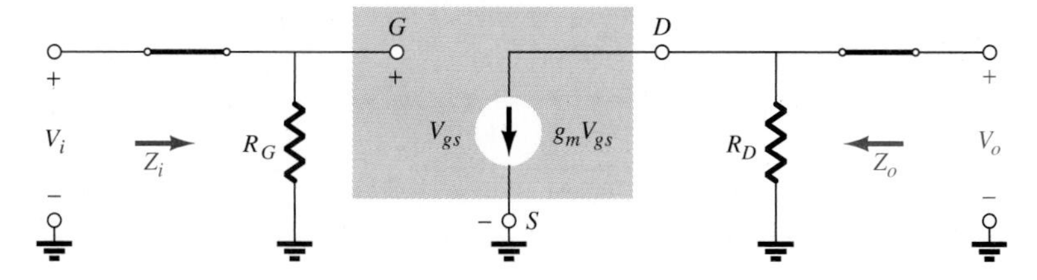

FIG. 9.48
Ac equivalent of Fig. 9.46.

Fig. 9.48. The magnitude of g_m is actually the slope of the transfer characteristic at the point of operation as defined by the differential equation $g_m = dI_D/dV_{GS}$. However, it can also be determined by the following equation, which includes the dc level of the gate-to-source voltage.

$$g_m = \frac{2I_{DSS}}{|V_P|}\left(1 - \frac{V_{GS_Q}}{V_P}\right) \qquad \textbf{(9.52)}$$

Note that in the ac equivalent in Fig. 9.48 the magnitude of the source is now controlled by a voltage present on the input side of the amplifier, specifically V_{gs}. The open circuit between gate and source clearly indicates that the input impedance is considered to be infinite, and the absence of r_o in the equivalent circuit also reveals that r_o is considered infinite for most applications. Figure 9.48 clearly shows the need for R_G in the network configuration. In its absence a short would be present from gate to source, and V_{gs} would be 0 V. In essence, it drops the input impedance from a level of infinite value to its value as determined by

$$Z_i = R_G \qquad \textbf{(9.53)}$$

In addition,

$$Z_o = R_D \qquad \textbf{(9.54)}$$

The capacitors present have been replaced by their short-circuit equivalent, resulting in V_i directly across R_G, and V_o across R_D. Using Ohm's law, we find that

$$V_o = -g_m V_{gs} R_D$$

and because of their parallel arrangement,

$$V_{gs} = V_i$$

The resulting ac gain is therefore

$$A_v = \frac{V_o}{V_i} = \frac{-g_m V_{gs} R_D}{V_{gs}}$$

and

$$A_v = -g_m R_D \tag{9.55}$$

Since the gate current (dc or ac) is considered to be 0 A, the current gain is an undefined element of JFET amplifiers—they are strictly voltage amplifiers.

EXAMPLE 9.9 For the fixed-bias JFET amplifier in Fig. 9.46 determine the following if $I_{DSS} = 10$ mA, $V_P = -4$ V, $V_{GG} = 1$ V, $V_{DD} = 15$ V, $R_G = 1$ MΩ, and $R_D = 1.5$ kΩ:

a. The transfer characteristics.
b. The operating point.
c. The level of V_{GS_Q} and I_{D_Q}.
d. The transconductance parameter g_m.
e. Z_i.
f. Z_o.
g. A_v.

Solution:

a. and b. See Fig. 9.49.

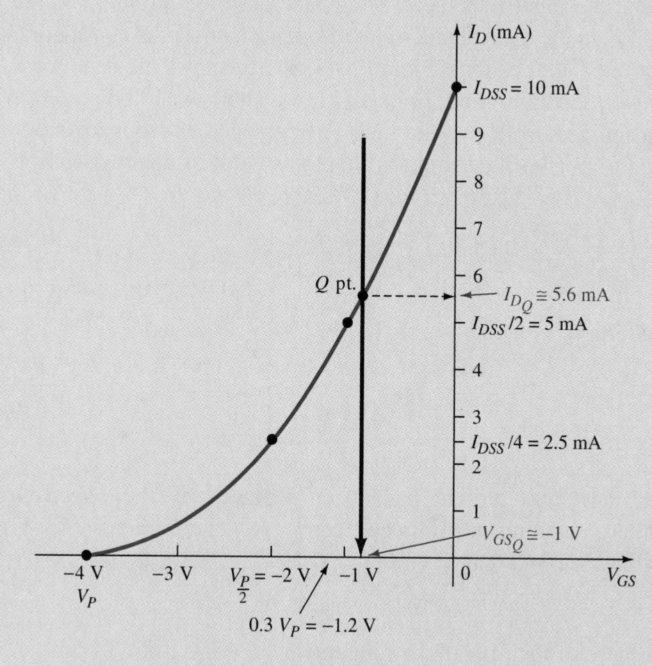

FIG. 9.49
Transfer characteristics for Example 9.9.

c. $V_{GS_Q} = \mathbf{-1\ V}$ and $I_{D_Q} = \mathbf{5.6\ mA}$

d. $g_m = \frac{2I_{DSS}}{|V_P|}\left(1 - \frac{V_{GS_Q}}{V_P}\right) = \frac{2(10\text{ mA})}{4\text{ V}}\left(1 - \frac{-1\text{ V}}{-4\text{ V}}\right) = (5\text{ mS})(0.75)$
$= \mathbf{3.75\ mS}$

e. $Z_i = R_G = \mathbf{1\ M\Omega}$

f. $Z_o = R_D = \mathbf{1.5\ k\Omega}$

g. $A_v = -g_m R_D = -(3.75\text{ mS})(1.5\text{ k}\Omega)$
$= \mathbf{-5.63}$

In the preceding example it is clear from the typical values employed that the gain is significantly less than typically encountered for BJT amplifiers. Keep in mind, however, that Z_i is considerably greater for a FET than for a BJT, which ensures that the majority of the applied signal will reach the input terminals of the amplifier and compensate for some of the difference when compared with a BJT amplifier. Note also the minus sign associated with the gain, revealing that, as in the BJT in the common-emitter configuration, there is a 180° phase shift between input and output quantities for the fixed-bias JFET amplifier.

Self-Bias Configuration

A more practical version of an FET amplifier stage is the *self-bias* configuration in Fig. 9.50 requiring only one independent supply. Note that a resistor R_S has been added to the configuration to establish the necessary dc bias configuration.

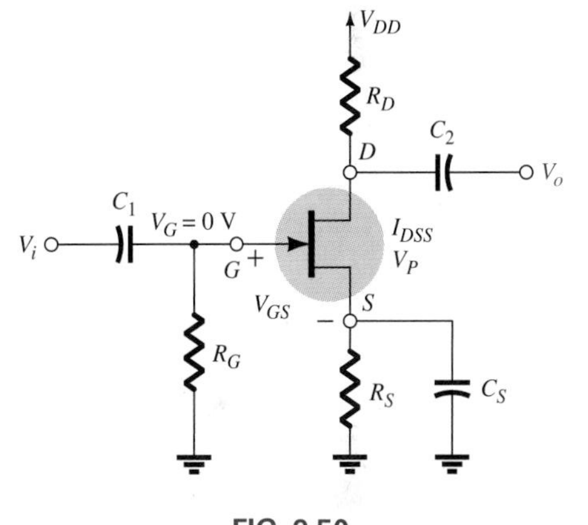

FIG. 9.50
Self-bias JFET configuration.

Dc Bias

Once the transfer characteristics are drawn using the device parameters I_{DSS} and V_P, a network characteristic must be introduced that will intersect the transfer characteristics and define the operating point. Under dc conditions all the capacitors are open circuits and the resistor R_G can be replaced by a short-circuit equivalent, since the voltage across R_G must be zero, as defined by $I_G = 0$ A and $V_{R_G} = I_G R_G$. Applying Kirchhoff's voltage law around the input circuit, we have

$$V_{GS} = -I_D R_S \tag{9.56}$$

This network equation can be plotted on the same graph as the transfer characteristics because the variables of the equation are the same as the axes of the graph. Since $V_{GS} = 0$ V when $I_D = 0$ mA, the origin is always a point on the straight line defined by Eq. (9.56). The other point required to draw the straight line can be obtained by choosing an appropriate level of I_D and solving for the level of V_{GS} as defined by R_S. The result is a load line as shown in Fig. 9.51 that defines the operating point at the point of intersection with the transfer characteristics.

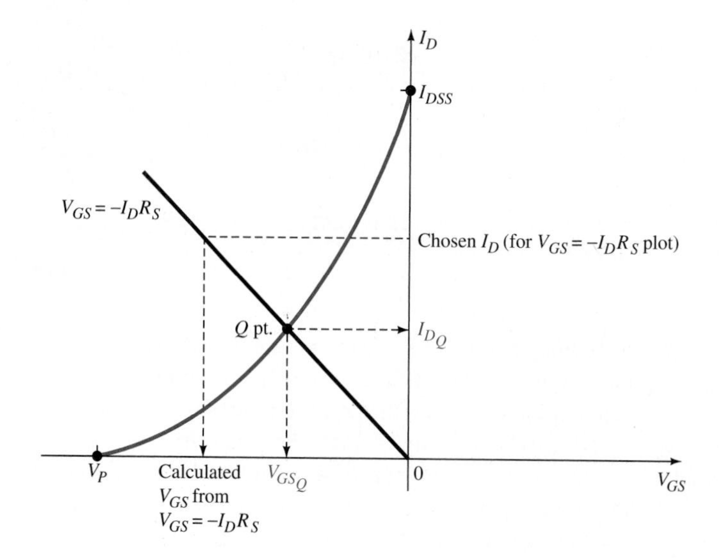

FIG. 9.51
Determining the Q point of a self-bias JFET configuration.

As with the fixed-bias configuration, once the level of I_{D_Q} is known, the other voltage levels of the network such as V_{DS} and V_S can be determined:

$$V_{DS} = V_{DD} - I_{D_Q}(R_D + R_S) \tag{9.57}$$

$$V_S = I_{D_Q} R_S \tag{9.58}$$

Ac Analysis

Since the capacitor C_S has a sufficiently small reactance at the frequency of interest, it can be considered a short circuit, removing the effect of R_S for the ac analysis. The resulting ac network therefore has the same configuration as obtained for the fixed-bias configuration, producing the same equations for Z_i, Z_o, and A_v. That is,

$$Z_i = R_G \tag{9.59}$$

$$Z_o = R_D \tag{9.60}$$

$$A_v = -g_m R_D \tag{9.61}$$

The FET transconductance g_m can still be determined using Eq. (9.52).

EXAMPLE 9.10 For the self-bias configuration in Fig. 9.50 determine the following if $I_{DSS} = 12$ mA, $V_P = -6$ V, $V_{DD} = 22$ V, $R_G = 1\ \text{M}\Omega$, $R_D = 2.7\ \text{k}\Omega$, and $R_S = 0.5\ \text{k}\Omega$:

a. The transfer characteristics.
b. The operating point.
c. The level of I_{D_Q} and V_{GS_Q}.
d. The transconductance parameter g_m.
e. Z_i.
f. Z_o.
g. A_v.

Solution:

a. and b. See Fig. 9.52.

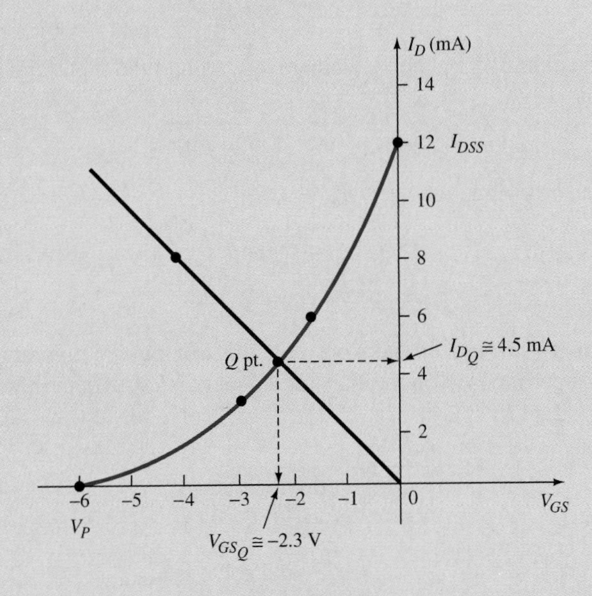

FIG. 9.52

c. $I_{D_Q} \cong \mathbf{4.5\ mA}$ and $V_{GS_Q} \cong \mathbf{-2.3\ V}$

d. $g_m = \dfrac{2I_{DSS}}{|V_P|}\left(1 - \dfrac{V_{GS_Q}}{V_P}\right) = \dfrac{2(12\text{ mA})}{6}\left(1 - \dfrac{-2.3\text{ V}}{-6\text{ V}}\right) = \mathbf{2.47\ mS}$

e. $Z_i \cong \mathbf{1\ M\Omega}$

f. $Z_o \cong \mathbf{2.7\ k\Omega}$

g. $A_v = -g_m R_D = -(2.47\text{ mS})(2.7\text{ k}\Omega) = \mathbf{-6.67}$

If the source resistance was not bypassed by a capacitor in the ac domain, the voltage gain would be divided by the factor $1 + g_m R_S$, which, in the preceding example, would result in a voltage gain of -2.98—a significant reduction. If the 0.5-kΩ source resistance is partially bypassed (two series resistors, one bypassed and the other not), the voltage gain can be set between the lowest value of -2.98 and the maximum of -6.67.

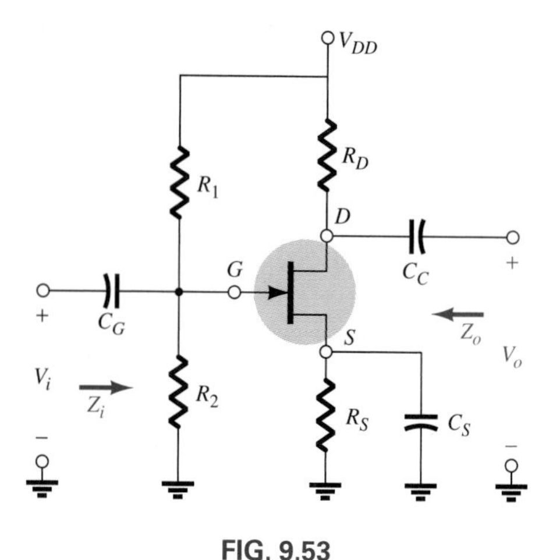

FIG. 9.53
Voltage-divider JFET bias configuration.

Voltage-Divider JFET Bias Configuration

The last JFET configuration to be described in detail is the voltage-divider bias arrangement in Fig. 9.53. The resistance R_S continues to have a bypass capacitor with the output off the drain conection.

Dc Bias

Since $I_G = 0$ A, the voltage from gate to ground is simply the voltage across the resistor R_2, which can be found using the voltage-divider rule as follows:

$$V_G = V_{R_2} = \frac{R_2 V_{DD}}{R_1 + R_2} \tag{9.62}$$

Application of Kirchhoff's voltage law around the input circuit then results in

$$+V_G - V_{GS} - V_{R_S} = 0$$

which can be written as

$$V_{GS} = V_G - I_D R_S \tag{9.63}$$

Equation (9.63) is the network equation for this configuration that must be superimposed on the transfer characteristics to determine the Q point. If we set $I_D = 0$ mA In Eq. (9.63) and solve for V_{GS}, we have

$$V_{GS} = V_G \quad (I_D = 0\text{ mA}) \tag{9.64}$$

If we set $V_{GS} = 0$ V in Eq. (9.63) and solve for I_D,

$$I_D = \frac{V_G}{R_S} \quad (V_{GS} = 0\text{ V}) \tag{9.65}$$

Since $I_D = 0$ mA defines the horizontal axis in Fig. 9.54, and $V_{GS} =$ 0 V defines the vertical axis in the same figure, we now have two points on the characteristics to plot the straight line defined by Eq. (9.63). The intersection of the two curves defines the Q point for the network.

As with the earlier configurations, once I_{D_Q} and V_{GS_Q} are known, all the dc levels of the network can be found, such as

$$V_{DS_Q} = V_{DD} - I_{D_Q}(R_D + R_S) \tag{9.66}$$

$$V_{R_S} = V_G - V_{GS_Q} = I_{D_Q} R_S \tag{9.67}$$

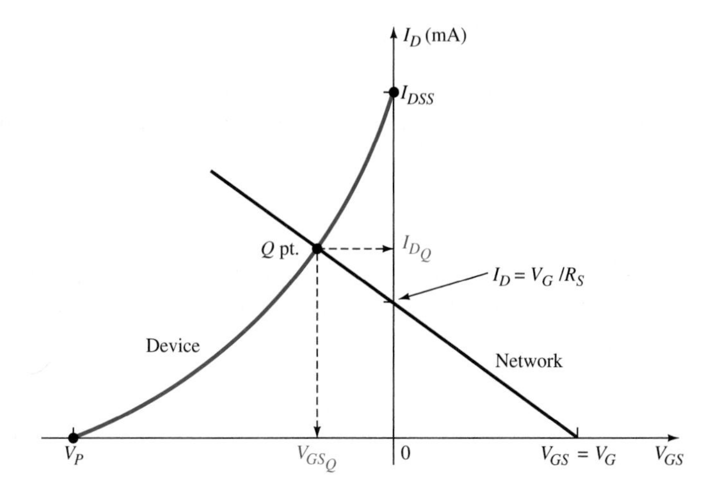

FIG. 9.54
Determining the Q point for the network in Fig. 9.53.

Ac Analysis

Substitution of the ac equivalent model for the JFET results in the network in Fig. 9.55 after we rearrange the elements in the practical manner.

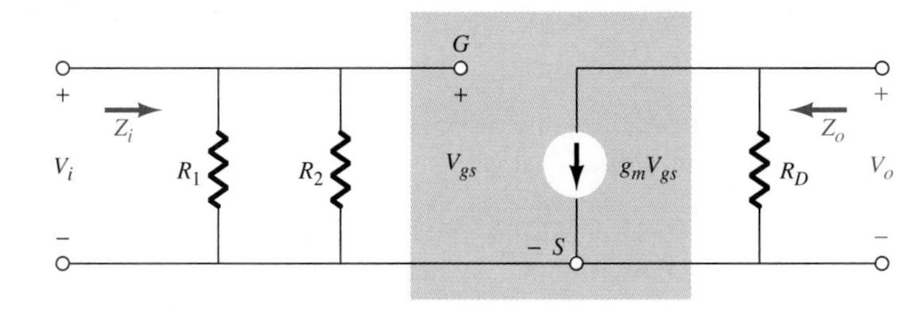

FIG. 9.55
Ac equivalent for the network in Fig. 9.53.

Note the absence of the resistor R_S because of the shorting effect of C_S and the fact that R_1 and R_2 now are in parallel with the input gate-to-source terminals of the JFET. The input impedance is now

$$Z_i = R_1 \| R_2 \quad \textbf{(9.68)}$$

with an output impedance

$$Z_o = R_D \quad \textbf{(9.69)}$$

The ac gain has the same format as earlier networks. That is,

$$A_v = -g_m R_D \tag{9.70}$$

Loading Effects of R_L and R_s

Both R_L and R_s have a negative effect on the gain of a JFET amplifier. However, for JFET amplifiers the input impedance is usually so large compared with the source resistance that for most applications we can assume $v_i = v_s$. The output impedance, however, is usually determined by finite resistor values in the biasing arrangement and has an impact on the gain. Consider the voltage-divider network in Fig. 9.56 with a source and load resistor.

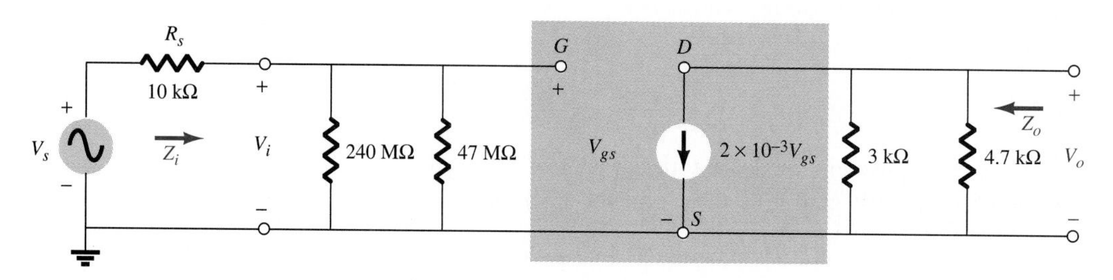

FIG. 9.56

Using the voltage-divider rule, we can determine the signal appearing at the input to the amplifier. That is,

$$V_i = \frac{Z_i V_s}{Z_i + R_s} = \frac{(240\ \text{M}\Omega \| 47\ \text{M}\Omega)V_s}{(240\ \text{M}\Omega \| 47\ \text{M}\Omega) + 10\ \text{k}\Omega} = \frac{(39.3\ \text{M}\Omega)(V_s)}{39.3\ \text{M}\Omega + 10\ \text{k}\Omega}$$

and

$$V_i = 0.9997\ V_s \cong V_s$$

The output voltage is now

$$\begin{aligned} V_o = -g_m V_{gs}(R_D \| R_L) &= -2\ \text{mS}(V_{gs})(3\ \text{k}\Omega \| 4.7\ \text{k}\Omega) \\ &= (-2\ \text{mS})(V_{gs})(1.83\ \text{k}\Omega) \\ &= -3.66\ V_{gs} \end{aligned}$$

resulting in an overall gain of

$$A_{v_T} = \frac{V_o}{V_s} = \frac{V_o}{V_i} \cdot \frac{V_i}{V_s} = \frac{V_o}{V_{gs}} \cdot \frac{V_i}{V_s} = (-3.66)(1) = \mathbf{-3.66}$$

which is measurably less than the no-load, no-source impedance level of **−6.**

MOSFET (Enhancement-Type) Transfer Characteristics

A typical enhancement-type MOSFET device (introduced in Chapter 8) has the characteristics in Fig. 9.57(a). The voltage $V_{GS(\text{Th})}$ is the threshold voltage between gate and source that initiates a measurable level of drain current. Any level of V_{GS} less than the threshold level results in a current of effectively 0 mA. For the characteristics in Fig. 9.57(a) it appears that the threshold level is about 4 V. As the gate-to-source voltage is increased above this value the drain current increases in a nonlinear fashion, as revealed by the increasing distance between the grid lines for V_{GS}. An equation relating the resulting drain current for different levels of V_{GS} is

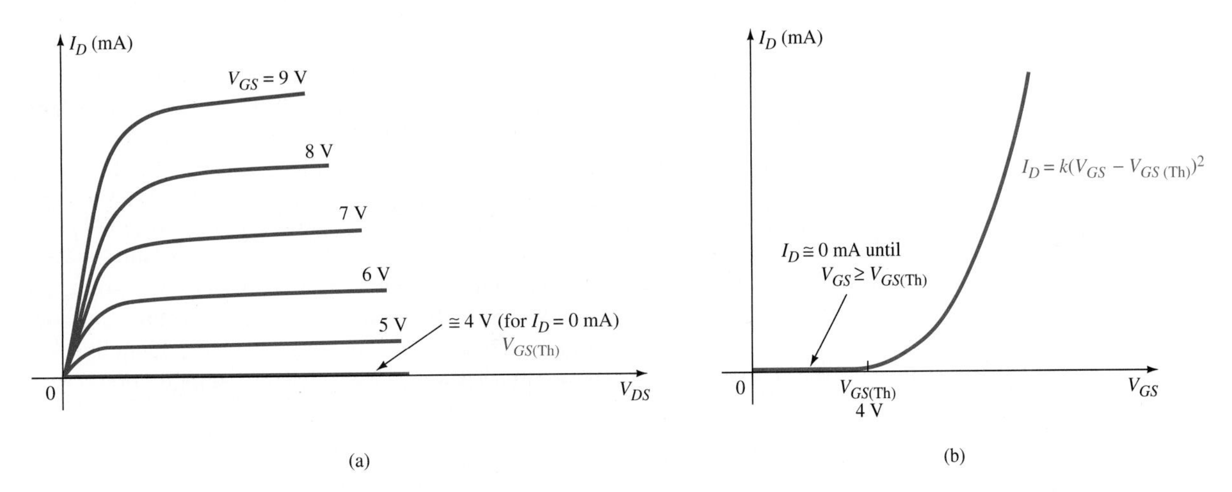

FIG. 9.57
MOSFET: (a) drain characteristics; (b) transfer characteristics.

$$I_D = k(V_{GS} - V_{GS(\text{Th})})^2 \qquad \textbf{(9.71)}$$

Since it relates an output quantity (I_D) to an input-controlling variable (V_{GS}), Eq. (9.71) can be used to plot the transfer characteristics in Fig. 9.57(b). The constant k is determined by the device physics and geometry and is typically 0.1 to 0.5 mA/V^2. The quantities I_{DSS} and V_P are not defined for MOSFET devices. The quantity k is seldom provided on specification sheets. Rather, a typical operating point is provided along with the threshold voltage from which the constant k can be determined, as will be demonstrated in the example to follow.

Let us begin our analysis with the practical MOSFET amplifier configuration in Fig. 9.58. Note in particular that the gate is connected directly to the drain through a resistance R_G.

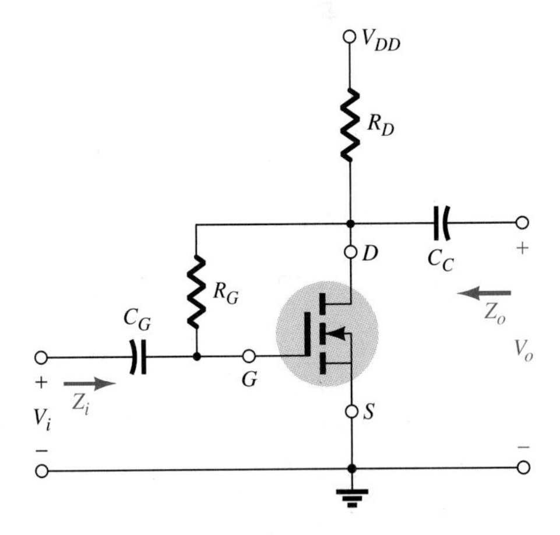

FIG. 9.58
Practical MOSFET amplifier.

Dc Bias

The dc operating point is determined in much the same manner as applied to JFET networks. First a transfer characteristic is drawn, and then a network load line is superimposed on the same graph with the intersection of the two defining the Q point. The load line is defined by the output side of the configuration. That is,

$$V_{DS} = V_{DD} - I_D R_D \tag{9.72}$$

The only remaining problem is that the two variables in Eq. (9.72) are I_D and V_{DS}, while those in Eq. (9.71) are I_D and V_{GS}. In order for both to be plotted on the same axes they must have the same two variables—hence the reason for the direct connection from gate to drain for this configuration. Since $I_G = 0$ mA, the dc voltage across R_G is 0 V, permitting removal of the element from the dc equivalent network and replacement of it with a short-circuit connection. The result is that for this configuration,

$$V_{DS} = V_{GS} \tag{9.73}$$

Substitution of Eq. (9.73) into Eq. (9.72) results in two equations with the same variables. Plotting Eq. (9.72) on the same graph is not a difficult task when one simply realizes that if we set $I_D = 0$ in Eq. (9.72), we can calculate the intersection of the straight line with the horizontal axis, since this is the only region in which $I_D = 0$ mA. For Eq. (9.72) the following results:

$$V_{DS} = V_{GS} = V_{DD} - I_D R_D \Big|_{(I_D = 0\text{ mA})} \tag{9.74}$$

If we now set V_{GS} to zero in Eq. (9.72), we will find the intersection with the vertical axis because $V_{GS} = 0$ V anywhere along this line. Substitution of $V_{GS} = 0$ V into Eq. (9.72) results in

$$V_{DS} = V_{GS} = 0 = V_{DD} - I_D R_D$$

and

$$I_D = \frac{V_{DD}}{R_D} \Big|_{(V_{GS} = 0\text{ V})}$$

With the two intersections the straight line defined by Eq. (9.72) can be drawn, as will be demonstrated in the example to follow.

As with JFETs, the intersection of the two curves defines the Q point and the resulting levels of I_{D_Q} and V_{GS_Q}, which for this configuration also leave us with V_{DS_Q}.

Ac Analysis

The ac model for the enhancement-type MOSFET has a transconductance factor g_m (or g_{fs}), which can be obtained from the manufacturer's standard specifications, by direct measurement using appropriate instrumentation, or by employing the following approximate calculation using the dc operating conditions:

$$g_m = 2k(V_{GS_Q} - V_{GS(\mathrm{Th})}) \tag{9.75}$$

The voltage $V_{GS(\mathrm{Th})}$ is again the device threshold voltage and V_{GS_Q} is the Q-point value of the gate-to-source voltage.

The ac model for the network in Fig. 9.58 is provided in Fig. 9.59, resulting in the following equations (to be left as an exercise for the reader).

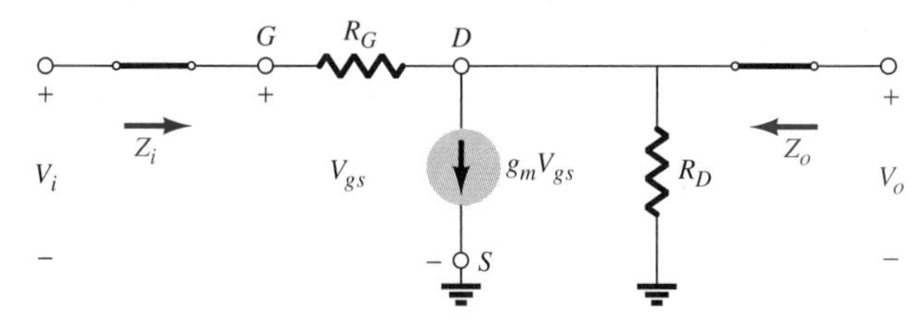

FIG. 9.59
Ac equivalent of the network in Fig. 9.58.

$$Z_i \cong \frac{R_G}{1 + g_mR_D} \tag{9.76}$$

$$Z_o \cong R_D \tag{9.77}$$

$$A_v = -g_mR_D \tag{9.78}$$

EXAMPLE 9.11 For $R_G = 10\ \mathrm{M\Omega}$, $R_D = 3\ \mathrm{k\Omega}$, $V_{DD} = 12\ \mathrm{V}$, $V_{GS(\mathrm{Th})} = 2.5\ \mathrm{V}$, $I_{D_{on}} = 2\ \mathrm{mA}$, and $V_{GS_{on}} = 5\ \mathrm{V}$ for the network in Fig. 9.58, determine:

a. Z_o.
b. A_v.
c. Z_i.

Solution:

a. $Z_o = R_D = \mathbf{3\ k\Omega}$

b. Determining k from the given operating condition gives

$$I_D = k(V_{GS} - V_{GS(\text{Th})})^2$$
$$2\text{ mA} = k(5\text{ V} - 2.5\text{ V})^2 = k(2.5\text{ V})^2$$
$$k = \frac{2\text{ mA}}{(2.5\text{ V})^2} = 0.32 \times 10^{-3}\text{ A/V}^2$$

In plotting the transfer characteristics, we already know that $I_D = 2$ mA when $V_{GS} = 5$ V. Substitution of various values of V_{GS} into Eq. (9.71) provides the corresponding level of I_D.

$$I_D = 0.32 \times 10^{-3}(V_{GS} - V_{GS(\text{Th})})^2$$
$$\text{For } V_{GS} = 4\text{ V}, I_D = 0.32 \times 10^{-3}(4 - 2.5)^2 = 0.72\text{ mA}$$
$$\text{For } V_{GS} = 6\text{ V}, I_D = 0.32 \times 10^{-3}(6 - 2.5)^2 = 3.92\text{ mA}$$

All three points are plotted in Fig. 9.60 along with the point defined by $V_{GS(\text{Th})}$.

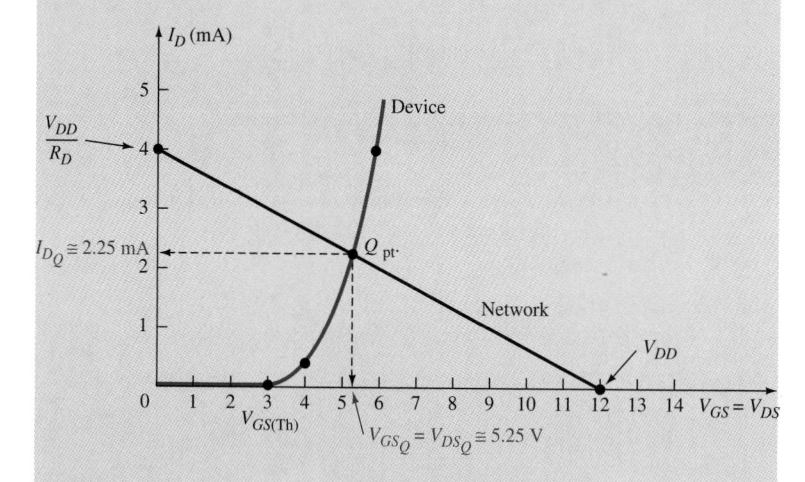

FIG. 9.60
Determining the Q point for Example 9.11.

Using Eq. (9.72), we obtain

$$V_{GS} = V_{DS} = V_{DD} - I_D R_D$$

When

$$I_D = 0\text{ mA}, V_{GS} = V_{DD} = 12\text{ V}$$

and when $V_{GS} = 0$ V,

$$I_D = \frac{V_{DD}}{R_D} = \frac{12\text{ V}}{3\text{ k}\Omega} = 4\text{ mA}$$

Drawing the load line on the graph of Fig. 9.60 will result in a Q point of

$$I_{D_Q} \cong \mathbf{2.25\ mA} \quad \text{and} \quad V_{GS_Q} = V_{DS_Q} \cong \mathbf{5.25\ V}$$

Solving for g_m, we have

$$\begin{aligned} g_m &= 2k(V_{GS_Q} - V_{GS(\text{Th})}) \\ &= 2(0.32 \times 10^{-3})(5.25 - 2.5) \\ &= 1.76\ \text{mS} \end{aligned}$$

resulting in the following gain:

$$\begin{aligned} A_v &= -g_m R_D \\ &= -(1.76\ \text{mS})(3\ \text{k}\Omega) \\ &= \mathbf{-5.28} \end{aligned}$$

c.

$$\begin{aligned} Z_i &= \frac{R_G}{1+g_m R_D} = \frac{10\ \text{M}\Omega}{1+(1.76\ \text{mS})(3\ \text{k}\Omega)} \\ &= \frac{10\ \text{M}\Omega}{1+5.28} = \frac{10\ \text{M}\Omega}{6.28} = \mathbf{1.59\ M\Omega} \end{aligned}$$

Depletion-Type MOSFETs

The analysis of depletion-type MOSFETs is the same as that applied to JFETs. I_{DSS} and V_P are provided, from which the transfer characteristics are drawn and the Q point determined. The value of g_m can then be determined using the same equation applied to JFETs followed by the use of an ac equivalent network that looks like the one employed for JFETs.

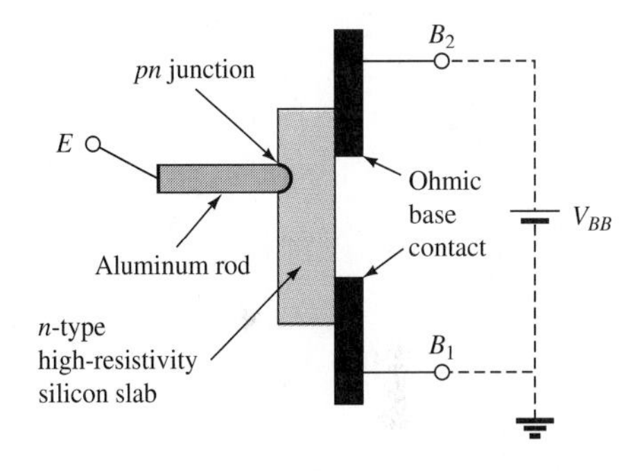

FIG. 9.61
Basic construction of unijunction transistor.

9.5 UNIJUNCTION TRANSISTOR

The *unijunction* transistor (UJT) is a three-terminal device having the basic construction in Fig. 9.61. Two contacts B_1 and B_2 are connected to a slab of *n*-type silicon material with a *pn* junction formed on the other side between an aluminum rod and the *n*-type material. The single *pn* junction is the reason for the prefix *uni-*. As shown in the sketch, the rod is closer to the B_2 connection, resulting in a different slab resistance between the rod and each terminal. However, the resistance between the rod and B_1 is quite sensitive to the emitter current of the network and may vary from 5 kΩ down to 50 Ω depending on the magnitude of the emitter current I_E. An equivalent network for the UJT appears in Fig. 9.62 with the resistance just described appearing as a variable element. With $I_E = 0$ A the total resistance R_{BB} is defined by

$$R_{BB} = R_{B_1} + R_{B_2}\big|_{I_E=0} \tag{9.79}$$

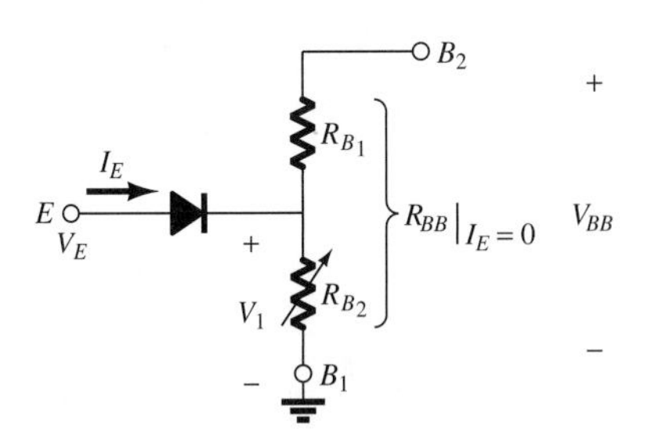

FIG. 9.62
Network equivalent for the unijunction transistor.

The diode in the equivalent circuit reveals one very important characteristic of the UJT. For the values of applied emitter voltage V_E not greater than V_1 by the forward-bias voltage of the diode ($V_F \cong 0.7$ V) the diode is reverse-biased and the input characteristics of the device are those of an open circuit. However, when sufficient voltage is applied at the emitter terminal, the diode is forward-biased. The major portion of I_E then flows through R_{B_1}, since this resistance decreases rapidly with the current I_E, and current always seeks the path of least resistance. The voltage V_E then decreases with increasing values of I_E until a minimum V_E is reached. Further increase in I_E is then linked with increasing values of R_{B_1}, causing the voltage V_1 to rise. The result is a back-biased diode, and the open-circuit input condition returns.

The characteristics of the device appear in Fig. 9.63 with V_P representing the voltage V_E necessary to initiate conduction in the diode. V_V represents the minimum voltage for which the diode is held in the conducting state. Beyond this point the increasing resistance is encountered until the back-biased condition for the diode results.

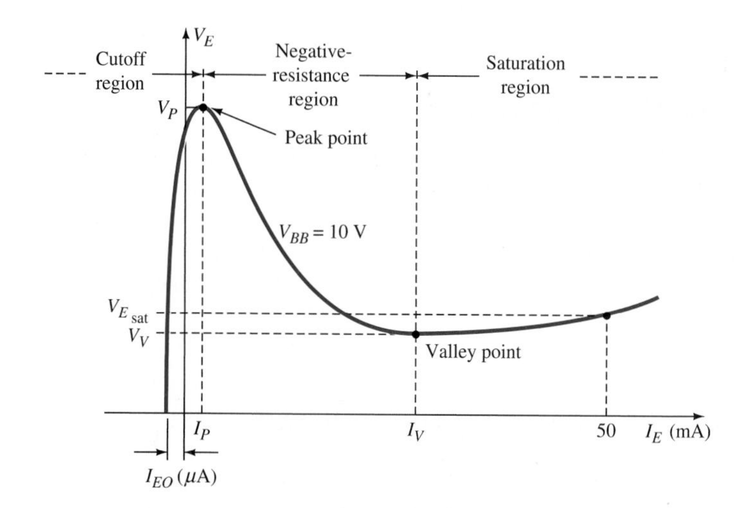

FIG. 9.63
UJT characteristics for a fixed $V_{BB} = 10$ V.

Note how the resistance decreases (negative resistance coefficient) with increasing values of I_E from V_P to V_V. This negative resistance region has application in areas such as oscillators (ac signal generators), to be described in a later chapter.

For increasing values of V_{BB} the curve rises, and increasing values of V_P are required to *fire* the device. This rise in the characteristics is noted in the characteristics in Fig. 9.64. Note, however, the same relative value of V_V for each curve.

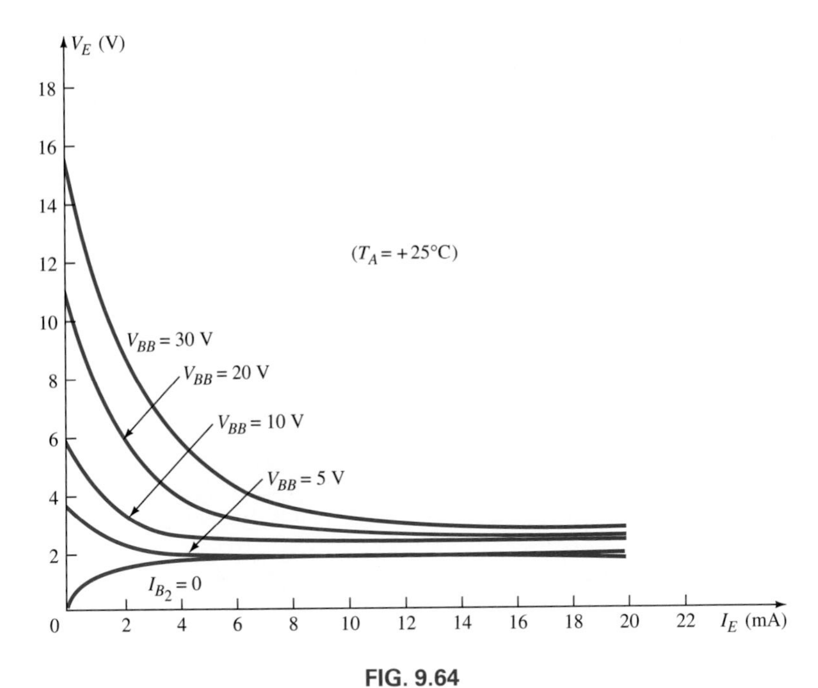

FIG. 9.64
Typical static emitter characteristics curves for a UJT.

The voltage V_1 is determined by a simple voltage-divider-rule equation when the diode is reverse-biased. That is,

$$V_1 = \frac{R_{B_1}}{R_{B_1}+R_{B_2}} V_{BB} \big|_{I_E=0} = \eta V_{BB} \big|_{I_E=0} \tag{9.80}$$

and

$$\eta = \frac{R_{B_1}}{R_{B_1}+R_{B_2}} \tag{9.81}$$

The quantity η (Greek lowercase letter eta) is typically available on specification sheets.

The *firing* potential is then determined by

$$V_P = \eta V_{BB} + V_F \tag{9.82}$$

A photograph of a commercially available UJT appears in Fig. 9.65 with its characteristics and terminal identification.

Absolute maximum ratings (25°C):

Power dissipation	300 mW
RMS emitter current	50 mA
Peak emitter current	2 A
Emitter reverse voltage	30 V
Interbase voltage	35 V
Operating temperature range	–65°C to +125°C
Storage temperature range	–65°C to +150°C

Electrical characteristics (25°C):

Characteristic	Symbol	Min.	Typ.	Max.
Intrinsic standoff ratio ($V_{BB} = 10$ V)	η	0.56	0.65	0.75
Interbase resistance ($V_{BB} = 3$ V, $I_E = 0$)	R_{BB}	4.7	7	9.1
Emitter saturation voltage ($V_{BB} = 10$ V, $I_E = 50$ mA)	$V_{E\,(sat)}$		2	
Emitter reverse current ($V_{BB} = 30$ V, $I_{B1} = 0$)	I_{EO}		0.05	12
Peak point emitter current ($V_{BB} = 25$ V)	I_P		0.4	5
Valley point current ($V_{BB} = 20$ V, $R_{B2} = 100\ \Omega$)	I_V	4	6	

(a) (b) (c)

B_1 B_2 E

FIG. 9.65
UJT: (a) appearance; (b) specification sheet; (c) terminal identification. (Courtesy of General Electric Company.)

EXAMPLE 9.12 For the UJT in Fig. 9.65, determine:

a. V_1.
b. V_P.
c. The relative magnitude of R_{B_1} as compared with R_{B_2} for $V_{BB} = 10$ V, $I_E = 0$, and typical values ($V_F = 0.67$ V).

Solution:

a. $V_1 = \eta V_{BB} = (0.65)(10\text{ V}) = \mathbf{6.5\ V}$

b. $V_P = \eta V_{BB} + V_F = 6.5\text{ V} + 0.67\text{ V} = \mathbf{7.17\ V}$

c. $\eta = \dfrac{R_{B_1}}{R_{B_1} + R_{B_2}} \rightarrow 0.65 = \dfrac{R_{B_1}}{R_{B_1} + R_{B_2}}$

and

$$\begin{aligned} 0.65R_{B_1} + 0.65R_{B_2} &= R_{B_1} \\ 0.35R_{B_1} &= 0.65R_{B_2} \\ R_{B_1} &= \mathbf{1.86}\boldsymbol{R_{B_2}} \end{aligned}$$

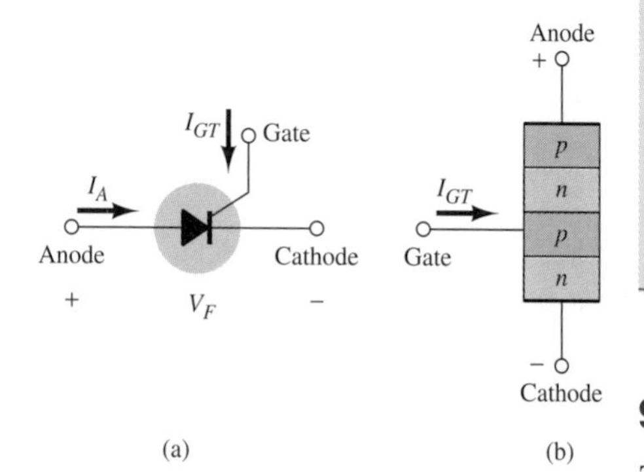

FIG. 9.66
SCR: (a) symbol; (b) basic construction.

9.6 SILICON-CONTROLLED RECTIFIER

The silicon-controlled rectifier (SCR) is a four-layer semiconductor device having the basic construction and symbol appearing in Fig. 9.66. Even though it has four layers, the SCR has only three external terminals, although the other four-layer semiconductor devices such as the

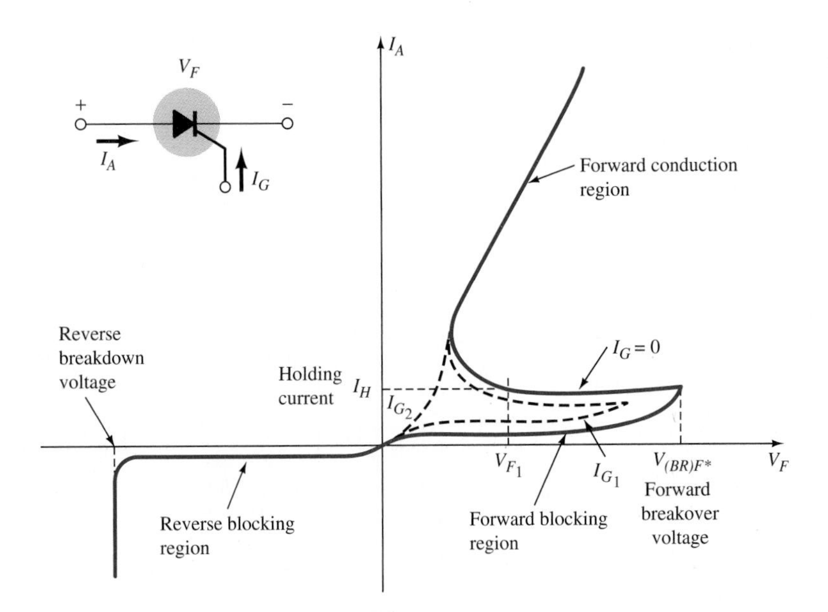

FIG. 9.67
SCR characteristics.

silicon-controlled switch (SCS) and gate turn-off switch (GTO) have four available connections. Priorities do not permit a detailed explanation of the latter two devices, but a number of references are available on each.

As the name implies, the SCR is a *s*ilicon-*c*ontrolled *r*ectifier whose state (open- or short-circuit equivalent) is controlled by a third terminal called the *gate*. In other words, it is not sufficient to forward-bias the diode by greater than the V_F ($\cong 0.7$ V) for the typical semiconductor diode. The gate current, as indicated on the characteristics in Fig. 9.67, determines the required firing voltage necessary to bring the device into the conduction state. For $I_G = 0$, V_F must be increased to at least $V_{(BR)F^*}$ before the short-circuit state between anode and cathode results. For increasing values of I_G, the voltage required to fire the device decreases until at I_{G_2} the diode appears to have the characteristics of the two-terminal junction diode. We can look at the behavior of the SCR in another light if we consider a fixed bias of say V_{F_1} (see Fig. 9.67) across the device. For conduction, even though V_{F_1} is greater than the required voltage for a typical semiconductor diode, the current must be increased to I_G before the short-circuit state is set. The holding current (I_H) is that value of current below which the SCR switches from the conduction to the blocking region. The characteristics of this device have widespread application in control systems in which a particular function is not to be performed until a gate-connected operation has occurred. A few common areas of application include relay controls, regulated power supplies, motor controls, battery chargers, and heater controls.

There are a wide variety of SCRs commercially available today—from milliwatts to megawatts with currents as high as 1000 A. A set of important gate characteristics for a top-hat SCR appears in Fig. 9.68. For

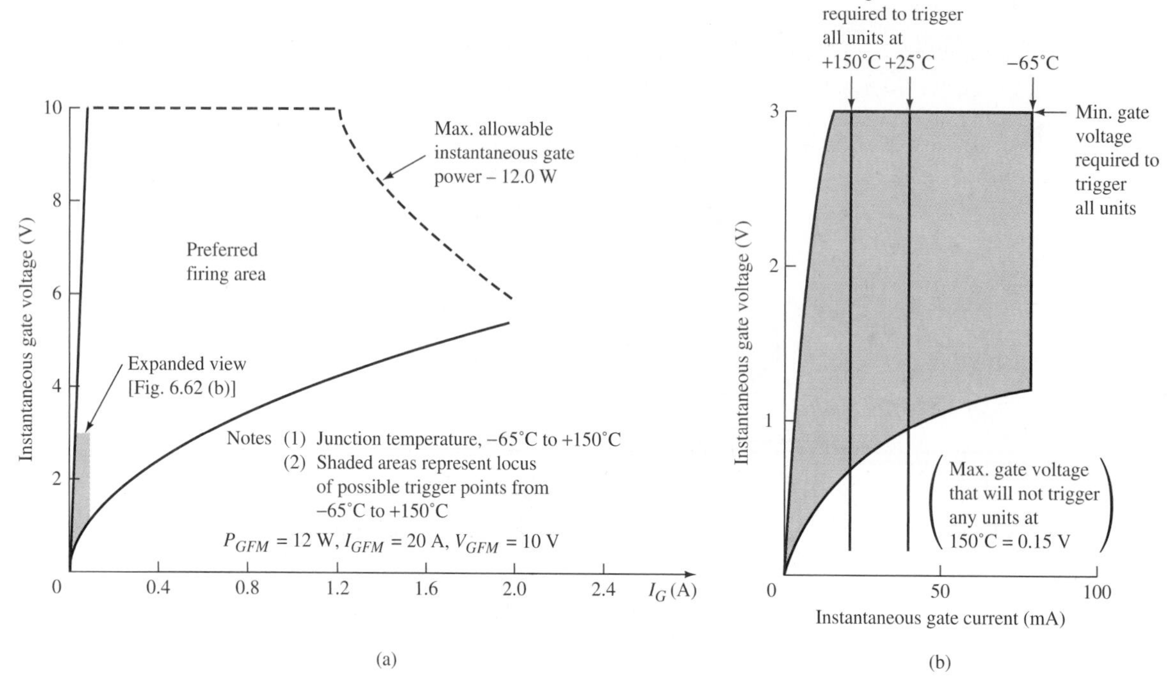

FIG. 9.68
SCR gate characteristics (GE series C38).

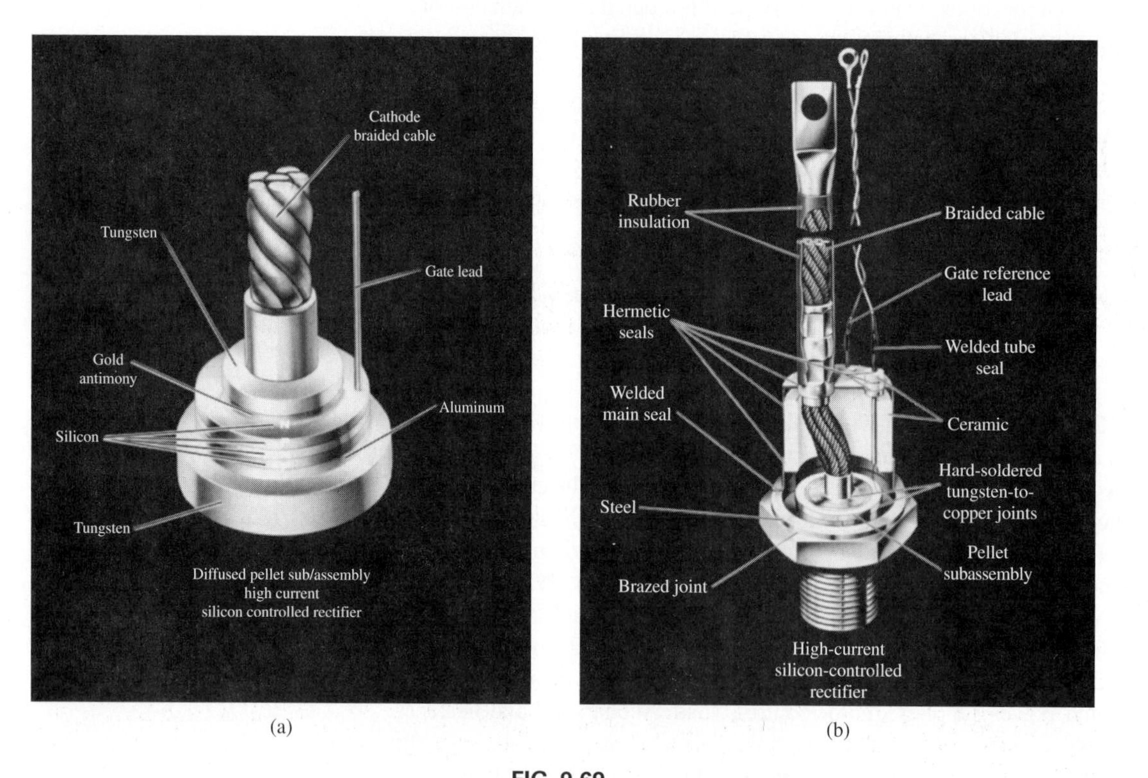

FIG. 9.69
(a) Alloy-diffused SCR pellet; (b) thermal fatigue–free SCR construction.
(Courtesy of General Electric Company.)

a particular gate current, the range of preferred firing potentials is indicated. For instance, a gate current of 1.2 A or greater requires a firing potential of 4.2 → 10 V. As indicated, the maximum gate current is 20 A, the maximum gate voltage is 10 V, and the maximum power dissipation ($P_G = I_G V_G$) is 12 W.

The internal construction of the device appears in Fig. 9.69. Note the very heavy appearance of the device to handle the high currents of this particular SCR. Other types appear in Fig. 9.70 with their terminal identification.

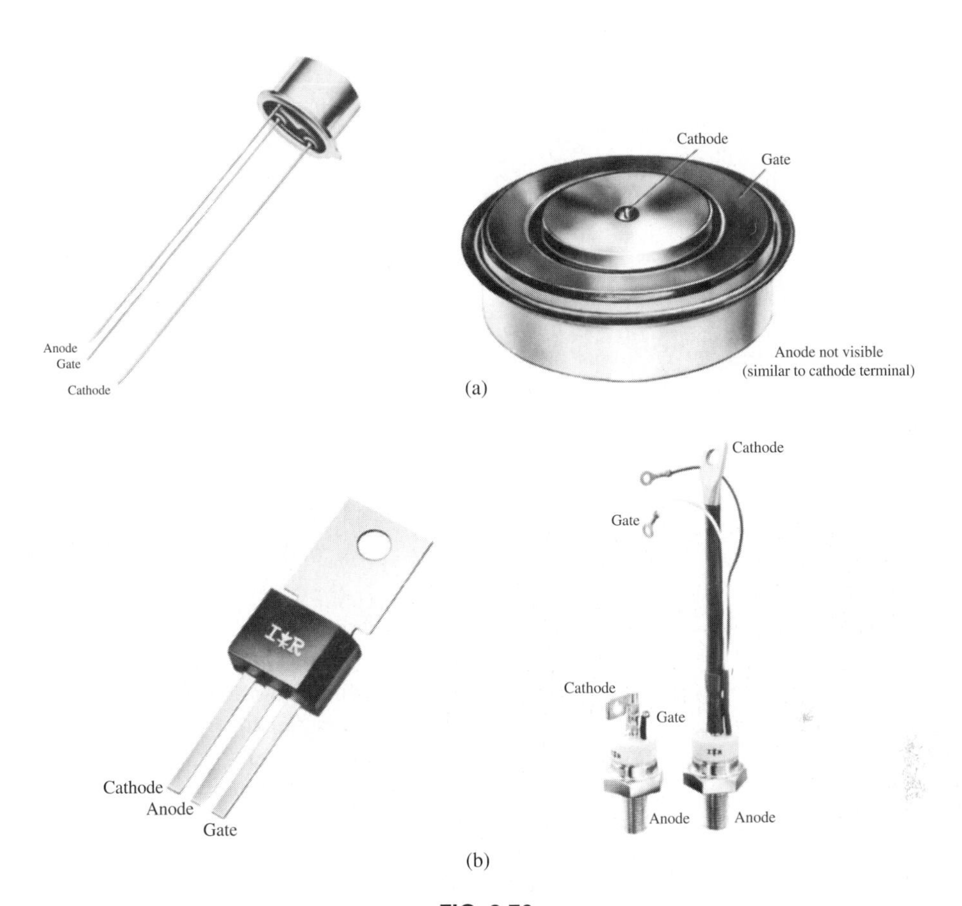

FIG. 9.70
SCR case construction and terminal identification. [(a) Courtesy of General Electric Company; (b) Courtesy of International Rectifier Corporation, Inc.]

9.7 DIAC AND TRIAC

The diac is a two-terminal, five-layer, semiconductor device constructed as shown in Fig. 9.71. Note that the characteristics in the first and third quadrants are somewhat similar to that obtained for the SCR in the first quadrant. For either voltage polarity across the device in Fig. 9.72 there is a potential V_{BR} above which the voltage decreases with increases in current until the short-circuit equivalent region is approached.

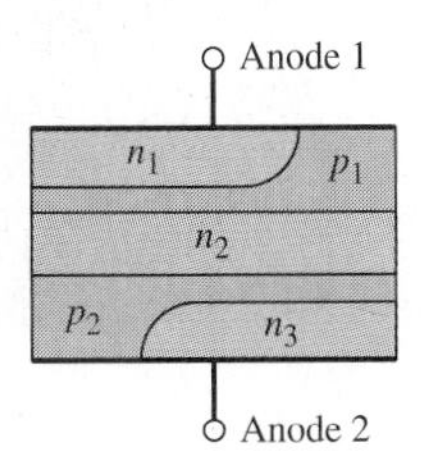

FIG. 9.71
DIAC construction.

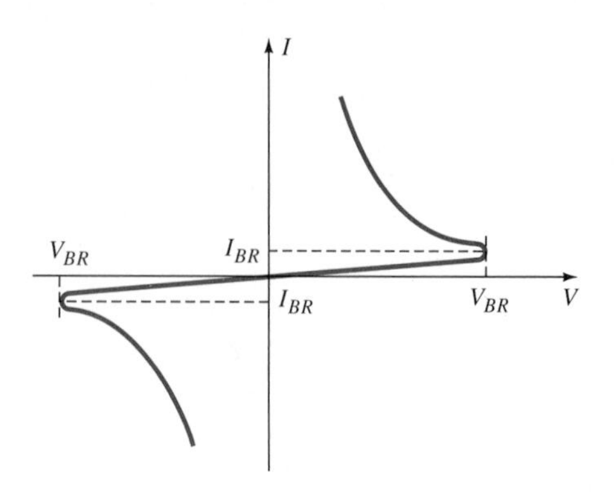

FIG. 9.72
DIAC characteristics.

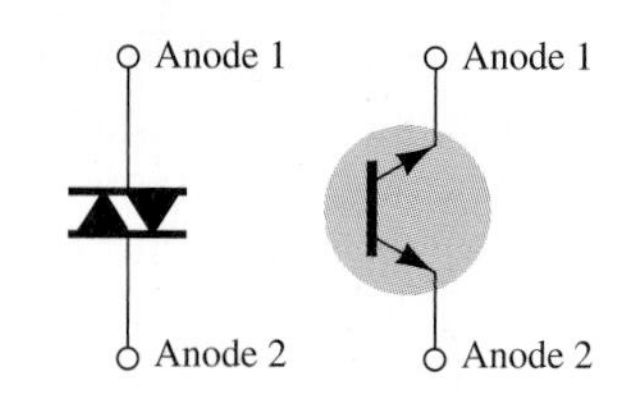

FIG. 9.73
DIAC symbols and appearance.

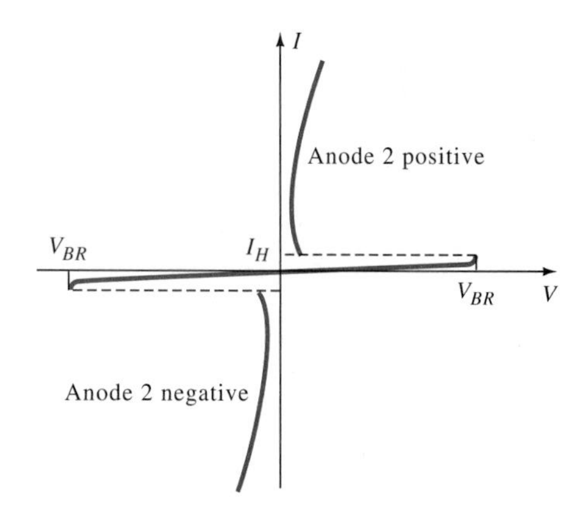

FIG. 9.74
TRIAC characteristics.

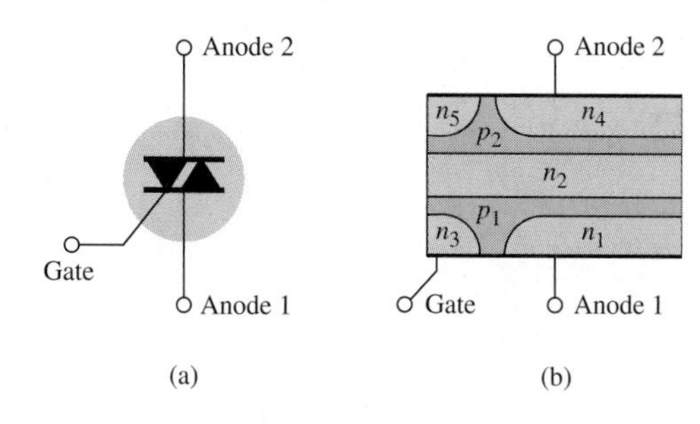

FIG. 9.75
TRIAC: (a) symbol; (b) construction.

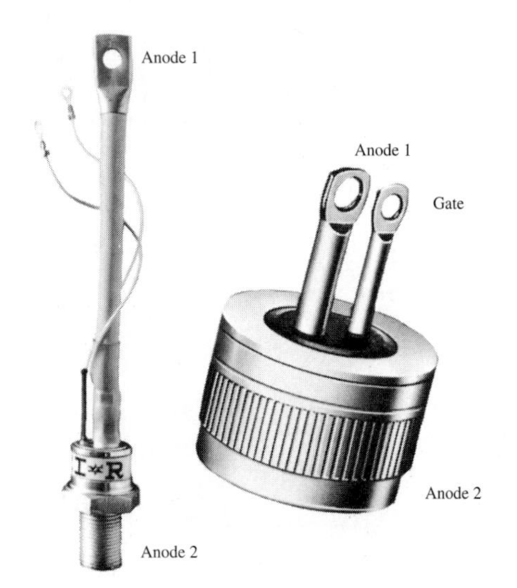

FIG. 9.76
TRIACs.

Two widely accepted symbols for the device appear in Fig. 9.73.

The triac has a set of characteristics very similar to those of the diac, and it has, in addition, a gate terminal that can control the state of the device in either direction. Note in Fig. 9.74 the presence of a holding current I_H not present for the diac. The symbol for and construction of the device appear in Fig. 9.75. Actual devices appear in Fig. 9.76.

9.8 OPTO-ISOLATORS

The *opto-isolator* is a package that contains both an infrared LED and a photodetector such as a silicon diode transistor, or SCR. The wavelength response of each device is tailored to be as close to identical as possible to permit the highest measure of coupling possible. In Fig. 9.77 two possible chip configurations are shown, with a photograph of each. There is a transparent insulating cap between the set of elements embedded in the structure (not visible) to permit the passage of light. They are designed with response times so small that they can be used to transmit data in the megahertz range.

The typical optoelectronic characteristic curves for each channel are provided in Figs. 9.78 to 9.80. Note the very pronounced effect of tem-

ISO-LIT 1

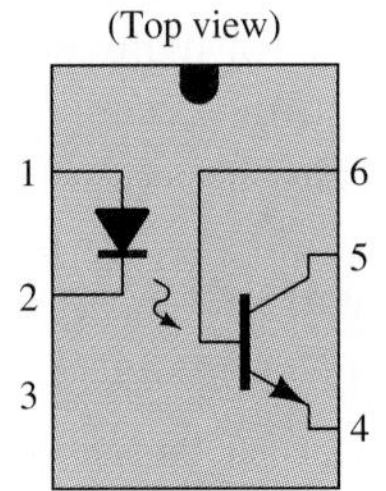

LED chip on pin 2
PT chip on pin 5

Pin no.	Function
1	anode
2	cathode
3	nc
4	emitter
5	collector
6	base

ISO-LIT Q1

(Top view)

Pin no.	Function
1	anode
2	cathode
3	cathode
4	anode
5	anode
6	cathode
7	cathode
8	anode
9	emitter
10	collector
11	collector
12	emitter
13	emitter
14	collector
15	collector
16	emitter

FIG. 9.77
Two Litronix opto-isolators. (Courtesy of Litronix, Inc.)

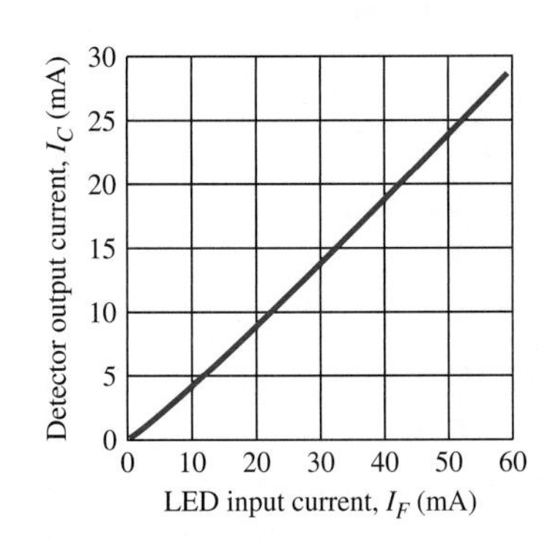

FIG. 9.78
Transfer characteristics.

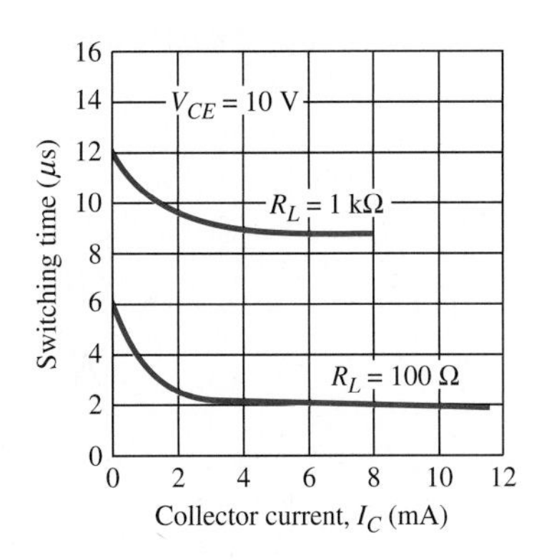

FIG. 9.79
Switching time versus collector current.

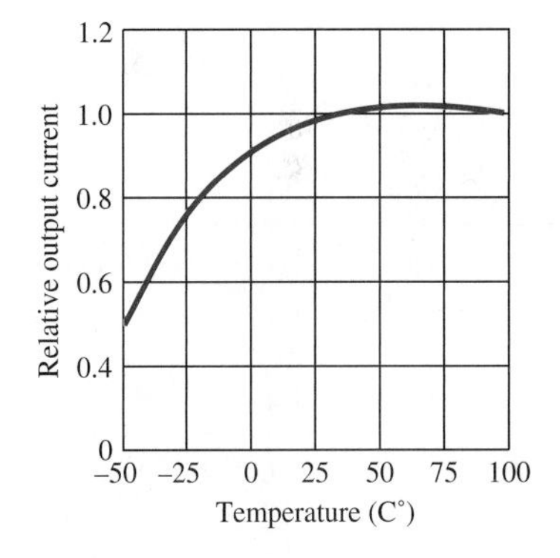

FIG. 9.80
Relative output versus temperature.

perature on the output current at low temperatures but the fairly level response at or above room temperature (25°C). The transfer characteristics in Fig. 9.78 compare the input LED current (which establishes the luminous flux) to the resulting collector current of the output transistor (whose base current is determined by the incident flux). It is interesting to note in Fig. 9.79 that the switching time of an opto-isolator decreases with increased current, while for many devices it is exactly the reverse. Consider that the switching time is only 2 μs for a collector current of 6 mA and a load R_L of 100 Ω. The relative output versus temperature appears in Fig. 9.80.

The schematic representations for a photodiode, phototransistor, and photo-SCR opto-isolator appear in Fig. 9.81.

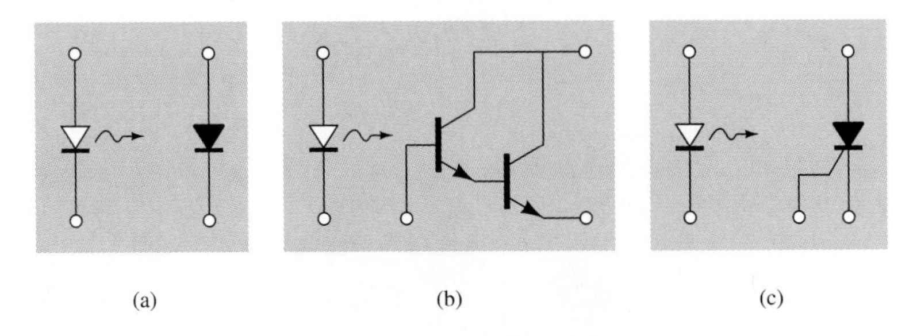

FIG. 9.81
Opto-isolators: (a) photodiode; (b) phototransistor; (c) photo-SCR.

9.9 PROGRAMMABLE UNIJUNCTION TRANSISTOR

Although there is a similarity in name, the construction and mode of operation of the programmable unijunction transistor (PUT) are quite different from those of the unijunction transistor. The similarity of the *I-V* characteristics and applications of each prompted the choice of names.

As indicated in Fig. 9.82, the PUT is a four-layer *pnpn* device with a gate connected directly to the sandwiched *n*-type layer. The symbol for the device and the basic biasing arrangement appears in Fig. 9.83. As the

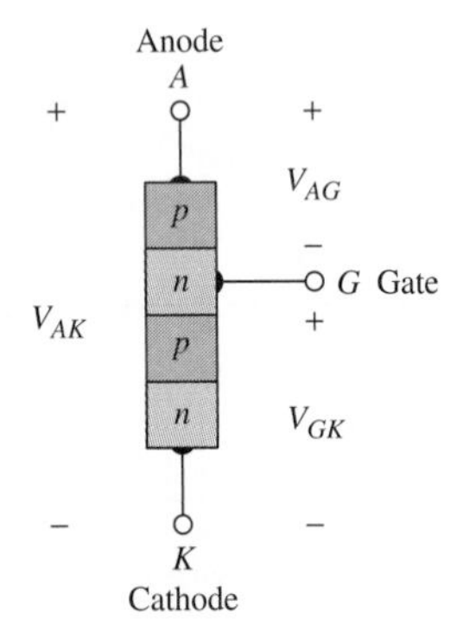

FIG. 9.82
Programmable UJT.

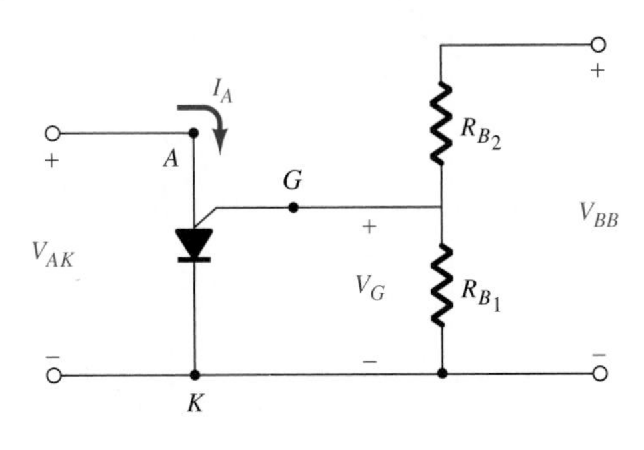

FIG. 9.83
Basic biasing arrangement for the PUT.

symbol suggests, it is essentially an SCR with a control mechanism that permits duplication of the characteristics of the typical SCR. The term programmable is applied because R_{BB}, η, and V_P as defined for the UJT can be controlled through the resistors R_{B_1}, R_{B_2}, and the supply voltage

V_{BB}. Note in Fig. 9.83 that through an application of the voltage-divider rule, when $I_G = 0$ A,

$$V_G = \frac{R_{B_1}}{R_{B_1} + R_{B_2}} V_{BB} = \eta V_{BB} \qquad \textbf{(9.83)}$$

where

$$\eta = \frac{R_{B_1}}{R_{B_1} + R_{B_2}}$$

as defined for the UJT.

The characteristics of the device are shown in Fig. 9.84. as noted on the diagram, the *off* state (I low, V between 0 and V_P) and the *on* state ($I \geq I_V$, $V \geq V_V$) are separated by the unstable region, as in the case of the UJT. That is, the device cannot stay in the unstable state—it simply shifts to either the *off* or the *on* stable state.

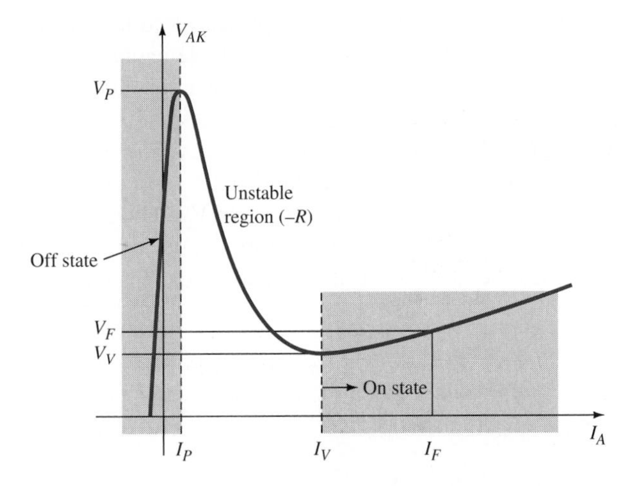

FIG. 9.84
PUT characteristics.

The firing potential (V_P) or voltage necessary to "fire" the device is given by

$$V_P = \eta V_{BB} + V_D \qquad \textbf{(9.84)}$$

as defined for the UJT. However, V_P represents the voltage drop V_{AG} in Fig. 9.82 (the forward voltage drop across the conducting diode). For silicon V_D is typically 0.7 V. Therefore,

$$V_P = \eta V_{BB} + V_D = \eta V_{BB} + V_{AG}$$

$$V_P = \eta V_{BB} + 0.7 \qquad \text{(silicon)} \qquad \textbf{(9.85)}$$

We noted in Eq. (9.83), however, that $V_G = \eta V_{BB}$, with the result that

$$V_P = V_G + 0.7 \qquad \text{(silicon)} \qquad \textbf{(9.86)}$$

Recall that for the UJT both R_{B_1} and R_{B_2} represent the bulk resistance and ohmic base contacts of the device—both inaccessible. We noted previously that R_{B_1} and R_{B_2} are external to the device, permitting an adjustment of η and hence V_G. In other words, a measure of control on the level of V_P is required to turn the device on.

Although the characteristics of the PUT and UJT are similar, the peak and valley currents of the PUT are typically lower than those of a similarly rated UJT. In addition, the minimum operating voltage is also less for a PUT.

The basic operation of the device can be reviewed through reference to Fig. 9.84. A device in the *off* state will not change state until the voltage V_P as defined by V_G and V_D is reached. The level of current until I_P is reached is very low, resulting in an open-circuit equivalent, since $R = V$ (high)/I (low) will result in a high resistance level. When V_P is reached, the device switches through the unstable region to the *on* state, where the voltage is lower but the current higher, resulting in a terminal resistance $R = V$(low)/I(high) which is quite small, representing the short-circuit equivalent on an approximate basis. The device therefore switches from essentially an open-circuit to a short-circuit state at a point determined by the choice of R_{B_1}, R_{B_2}, and V_{BB}. Once the device is in the *on* state, the removal of V_G will not turn the device *off*. The level of voltage V_{AK} must be dropped sufficiently to reduce the current below a holding level.

EXAMPLE 9.13 Determine R_{B_1} and V_{BB} for a silicon PUT if it is determined that η should be 0.8, $V_P = 10.3$ V, and $R_{B_2} = 5$ kΩ.

Solution: From Eq. (9.83),

$$\eta = \frac{R_{B_1}}{R_{B_1} + R_{B_2}} = 0.8$$
$$R_{B_1} = 0.8(R_{B_1} + R_{B_2})$$
$$0.2R_{B_1} = 0.8R_{B_2}$$
$$R_{B_1} = 4R_{B_2}$$
$$R_{B_1} = 4(5\text{ k}\Omega) = \mathbf{20\text{ k}\Omega}$$

From Eq. (9.84),

$$V_P = \eta V_{BB} + V_D$$
$$10.3\text{ V} = (0.8)(V_{BB}) + 0.7\text{ V}$$
$$9.6\text{ V} = 0.8V_{BB}$$
$$V_{BB} = \mathbf{12\text{ V}}$$

9.10 APPLICATION: HOUSEHOLD DIMMER SWITCH

Inductors can be found in a wide variety of common electronic circuits in the home. The typical household dimmer uses an inductor to protect the other components and the applied load from "rush" currents—currents that increase at very high rates and often to excessively high levels. The inductor will essentially "choke" the exponential rise in current and maintain safe operating levels. Recall that the current through an inductor cannot change instantaneously. This feature is particularly important for dimmers, since they are most commonly used to control the light intensity of an incandescent lamp. At "turn on," the resistance of incandescent lamps is typically very low, and relatively high currents may flow for short periods of time until the filament of the bulb heats up. The inductor is also effective in blocking high-frequency noise (RFI) generated by the switching action of the triac in the dimmer. A capacitor is also normally included from line to neutral to prevent any voltage spikes from affecting the operation of the dimmer and the applied load (lamp, etc.) and to assist with the suppression of RFI disturbances.

A photograph of one of the most common dimmers is provided in Fig. 9.85(a), with an internal view shown in Fig. 9.85(b). The basic components of most commercially available dimmers appear in the schematic of Fig. 9.85(c). In this design, a 14.5-μH inductor is used in the "choking" capacity described above, with a 0.068-μF capacitor for the "bypass" operation. Note the size of the inductor with its heavy wire and large ferromagnetic core and the relatively large size of the two 0.068-μF capacitors. Both suggest that they are designed to absorb high-energy disturbances.

The general operation of the dimmer is shown in Fig. 9.86. The controlling network is in series with the lamp and essentially acts as an impedance that can vary between very low and very high levels: very low impedance levels resembling a short circuit so that the majority of the applied voltage appears across the lamp [Fig. 9.86(a)] and very high impedances approaching open circuit where very little voltage appears across the lamp [Fig. 9.86(b)]. Intermediate levels of impedance control the terminal voltage of the bulb accordingly. For instance, if the controlling network has a very high impedance (open-circuit equivalent) through half the cycle, as shown in Fig. 9.86(c), the brightness of the bulb will be less than full voltage but not 50% due to the nonlinear relationship between the brightness of a bulb and the applied voltage.

The controlling knob, slide, or whatever other method is used on the face of the switch to control the light intensity is connected directly to the rheostat in the branch parallel to the Triac. Its setting determines when the voltage across the capacitor reaches a sufficiently high level to turn on the Diac (a bidirectional diode) and establish a voltage at the gate (G) of the Triac to turn it on. When it does, it establishes a very low resistance path from the anode (A) to the cathode (K), and the

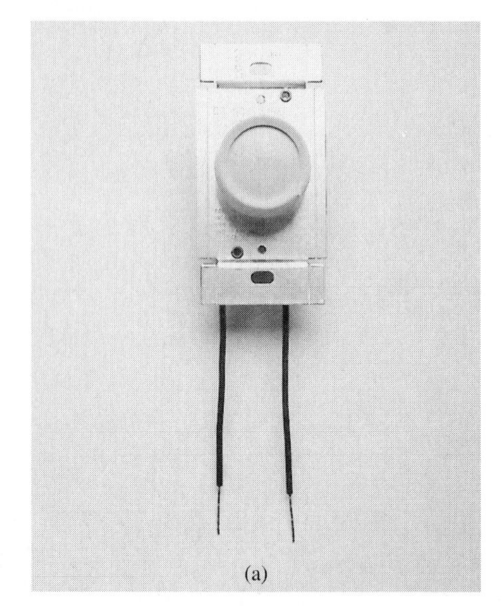

(a)

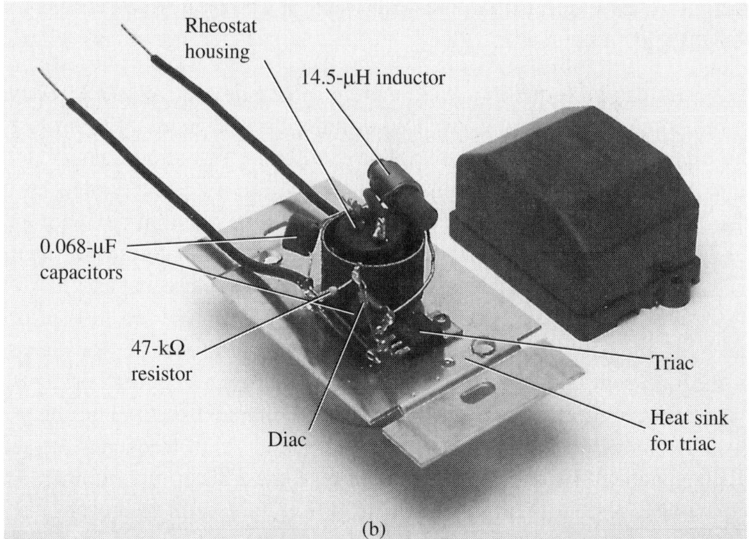

(b)

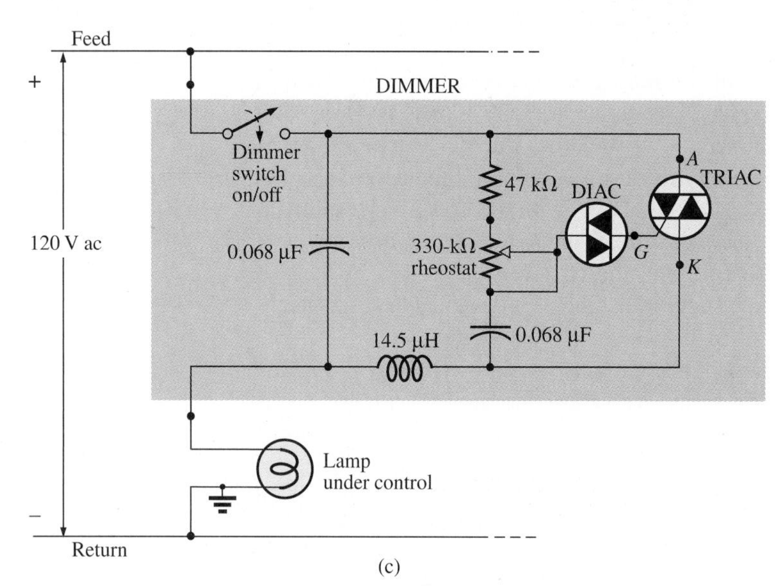

(c)

FIG. 9.85
Dimmer control: (a) external appearance; (b) internal construction; (c) schematic.

applied voltage appears directly across the lamp. A more detailed explanation of this operation will appear in a later chapter following the examination of some important concepts for ac networks. During the period the triac is off, its terminal resistance between anode and cathode is very high and can be approximated by an open circuit. During this period the applied voltage does not reach the load (lamp). During such intervals the impedance of the parallel branch containing the rheostat,

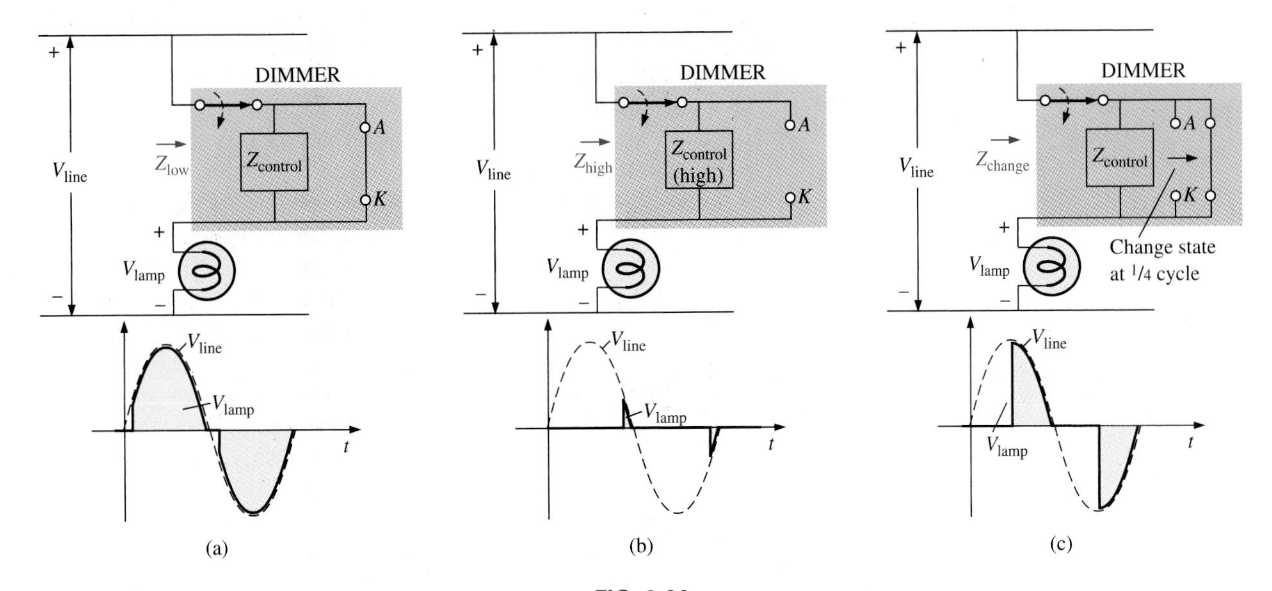

FIG. 9.86
Basic operation of the dimmer of Fig. 9.85: (a) full voltage to the lamp; (b) approaching the cutoff point for the bulb; (c) reduced illumination of the lamp.

fixed resistor, and capacitor is sufficiently high compared with the load that it can also be ignored, completing the open-circuit equivalent in series with the load. Note the placement of the elements in the photograph of Fig. 9.85 and that the metal plate to which the triac is connected is actually a heat sink for the device. The on/off switch is in the same housing as the rheostat. The total design is certainly well planned to maintain a relatively small size for the dimmer.

Since the effort here is simply to control the amount of power getting to the load, the question is often asked, Why don't we simply use a rheostat in series with the lamp? The question is best answered by examining Fig. 9.87, which shows a rather simple network with a rheostat in series with the lamp. At full wattage, a 60-W bulb on a 120-V line theoretically has an internal resistance of $R = V^2/P$ (from the equation $P = V^2/R$) $= (120\text{ V})^2/60\text{ W} = 240\ \Omega$. Although the resistance is sensitive to the applied voltage, we shall assume this level for the following calculations. If we consider the case where the rheostat is set for the same level as the bulb, as shown in Fig. 9.87, there will be 60 V across the rheostat and the bulb. The power to each element will then be $P = V^2/R = (60\text{ V})^2/240\ \Omega = 15\text{ W}$. The bulb will certainly be quite dim, but the rheostat inside the dimmer switch will be dissipating 15 W of power on a continuous basis. When you consider the size of a 2-W potentiometer in your laboratory, you can imagine the size rheostat you would need for 15 W, not to mention the purchase cost, although the biggest concern would probably be all the heat developed in the walls of the house. You would certainly be paying for electric power that would not be performing a useful function. Also, if you had four

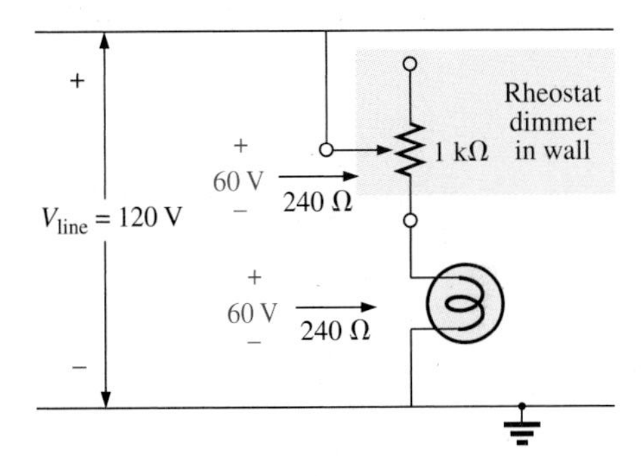

FIG. 9.87
Direct rheostat control of the brightness of a 60-W bulb.

dimmers set at the same level, you would actually be wasting sufficient power to fully light another 60-W bulb.

On occasion, especially when the lights are set very low by the dimmer, a faint "singing" can sometimes be heard from the light bulb. This effect sometimes occurs when the conduction period of the dimmer is very small. The short, repetitive voltage pulse applied to the bulb will set the bulb into a condition that can be likened to a resonance (Chapter 5) state. The short pulses are just enough to heat up the filament and its supporting structures, and then the pulses are removed to allow the filament to cool down again for a longer period of time. This repetitive heating and cooling cycle can set the filament in motion, and the "singing" can be heard in a quiet environment. Incidentally, the longer the filament, the louder the "singing." A further condition for this effect is that the filament be in the shape of a coil and not a straight wire so that the "slinky" effect can develop.

9.11 COMPUTER ANALYSIS

PSpice (Windows)

A voltage-divider network with a silicon Q2N2222 transistor was analyzed using PSpice in Fig. 9.88. When the analysis was run all the biasing dc levels appeared on the viewpoint symbols as shown. Note in

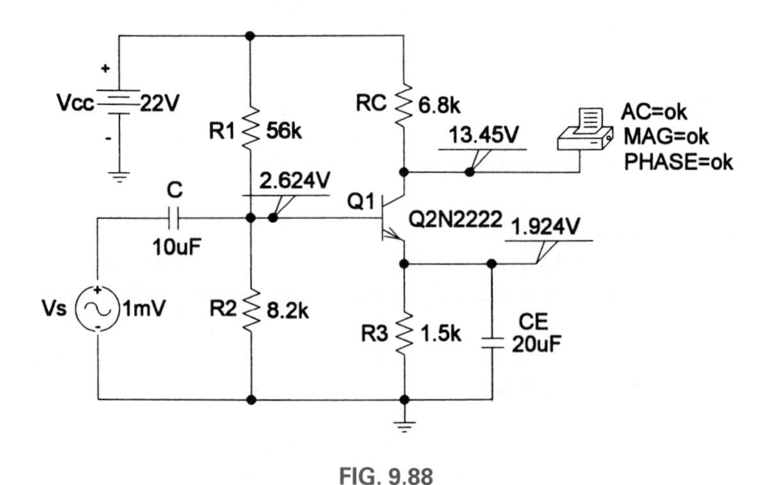

FIG. 9.88

Using PSpice Windows to analyze a voltage-divider transistor network.

particular that the base-to-emitter voltage of the transistor is 0.7 V, and the base voltage is set at 2.624 V by the voltage-divider biasing arrangement of resistors R_1 and R_2. To obtain the ac response the printer symbol must be added at the collector of the transistor, and the various quantities listed must be given the ok sign. The output file will provide a detailed list of information about the network plus the ac response. Given that the input is set at 1 mV the provided output level of 0.296 V = 296 mV reveals that the magnitude of the gain is 296. The phase angle associated with the voltage at the collector was also provided as $-178°$ or essentially $-180°$ to substantiate that the basic transistor amplifier introduces a phase shift of 180° between output and input.

Through a plotting routine the waveforms of the applied signal and the output voltage can be displayed using PSpice as shown in Fig. 9.89. Note that the dc levels associated with each are also indicated, and the 180° phase shift between the waveforms is quite obvious. Be aware that the vertical scale for the collector voltage is the far left scale and for the applied voltage, the adjoining scale.

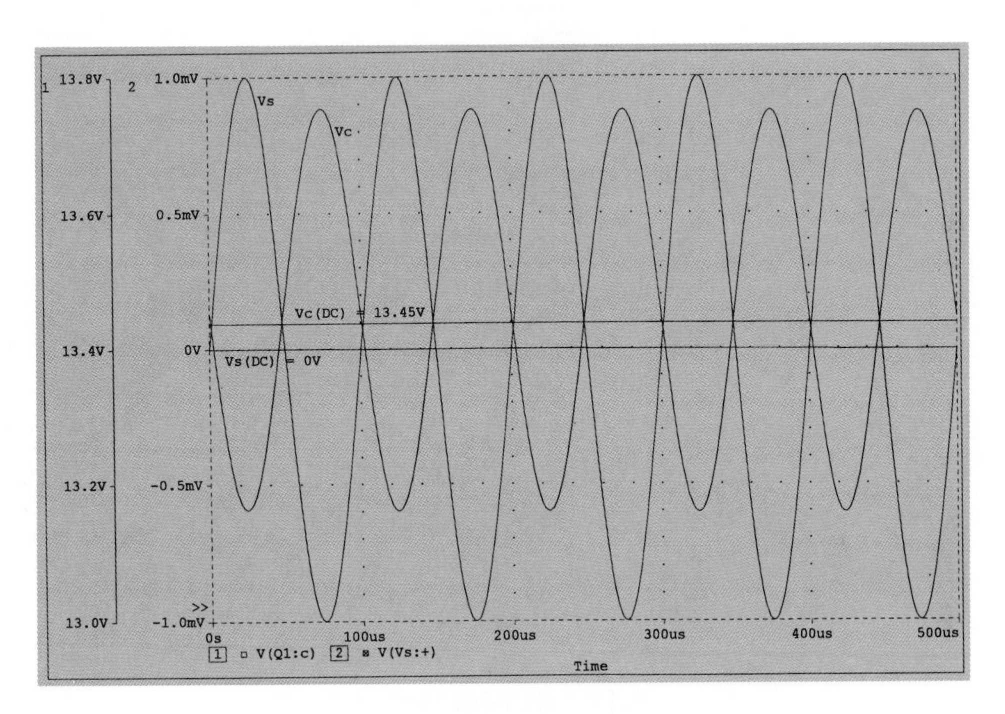

FIG. 9.89

The voltages v_C and v_s for the network of Fig. 9.88

Electronics Workbench

The FET network of Fig. 9.90 was investigated using Electronics Workbench. The dc levels at the gate and drain terminals were requested along with the ac level at the load to determine the ac gain of the amplifier. Both dc levels were as expected, and the ac gain was about 5.2.

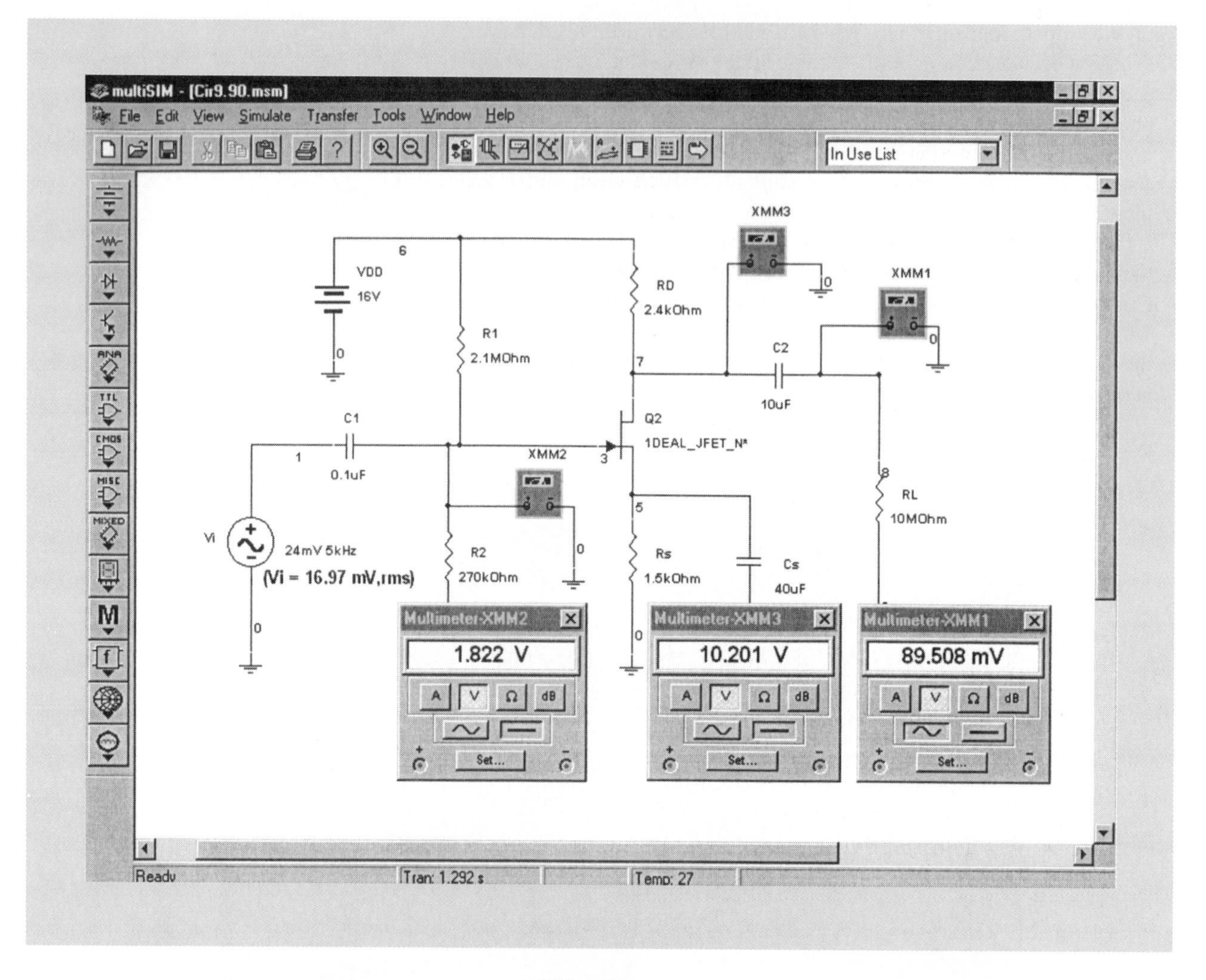

FIG. 9.90

Determining the dc and ac levels for a JFET network using Electronics Workbench.

9.12 SUMMARY

- The **normal operation** of a transistor requires that **one *pn* junction be forward-biased** and the **other reverse-biased.**
- The **BJT transistor** is a **current-controlled device** because the current at the base is reflected into the output collector–emitter branch by a multiplying factor beta.
- The **base-to-emitter voltage** of an ***on*** silicon transistor is about **0.7 V.**
- The **three configurations** for a transistor amplifier are the **common-emitter, common-base,** and **common-collector.**
- The **maximum power curve** for a transistor is determined by the **product** of the **collector-to-emitter voltage** and the **collector current.**

- Every **transistor amplifier network** must have a dc biasing arrangement to establish a **dc operating point** in the **linear region** of the characteristics.
- **Superposition** applies to the analysis of transistor amplifiers. That is, the **dc analysis** can be done **completely independently** of the **ac analysis,** and once the dc analysis is complete, the ac analysis can be performed independently of the dc conditions. However, it must be noted that some of the parameters used in the ac analysis may have been determined by the dc operating conditions.
- A **field-effect transistor** (FET) is a **voltage-controlled device** that has a **non-linear reslationship** between the **drain current** and the applied **gate-to-source voltage.** For a BJT it is usually a good approximation to assume that the factor beta, which relates the input current and load current, is fairly constant. However, for FETs a nonlinear relationship exists between the applied gate-to-source voltage and the drain current that requires the continuing use of Shockley's equation.
- The **analysis** of **depletion-type MOSFETs** is essentially the **same** as that applied to **JFETs;** however, the analysis of enhancement-type MOSFETs is quite different.
- The **characteristics** of **unijunction transistors** are **quite different** from those of **BJTs or FETs.** In fact, they have a **negative resistance region** that is useful in oscillators.
- A **silicon-controlled rectifier** (SCR) is, as the name implies, a **diode** whose **state** is **controlled by a third terminal called a gate.** The gate current determines what voltage must be applied across the diode to establish conduction.
- A **diac** is a **two-terminal, five-layer semiconductor device** that can **conduct in either direction** when sufficient voltage is applied across the device.
- A **triac** has **characteristics** very **similar** to those of a **diac** with the addition of a **third terminal** to determine when the device **will conduct in either direction.**
- An **opto-isolator** is a package containing both an **infrared LED** and a **photodetector.** It permits the control of a secondary circuit without a direct connection.
- A **programmable unijunction transistor** is essentially an **SCR** with a **control mechanism** that permits duplication of the **characteristics of the typical UJT** with **control** over the important parameters of the UJT.

Equations:

BJTs:

$$I_E = I_C + I_B$$
$$I_E \cong I_C$$
$$|V_{BE}| \cong 0.7\ \text{V}$$
$$\alpha = \frac{I_C}{I_E}$$

$$I_C = \beta I_B$$
$$\beta = \frac{\alpha}{1 - \alpha}$$
$$P_{\text{max}} = V_{CE} I_C$$
$$A_p = A_v A_i$$

Unbypassed Emitter Resistor:

$$Z_i \cong \beta R_E$$

Hybrid Parameters:

$$\beta_{ac} = h_{fe}$$

FETs:

$$I_D = I_{DSS}\left(1 - \frac{V_{GS}}{V_P}\right)^2$$

MOSFET:

$$I_D = k(V_{GS} - V_{GS(\text{Th})})^2$$
$$g_m = 2k(V_{GS_Q} - V_{GS(\text{Th})})$$

UJTs:

$$R_{BB} = R_{B_1} + R_{B_2}\big|_{I_E=0}$$
$$\eta = \frac{R_{B_1}}{R_{B_1} + R_{B_2}}$$
$$V_P = \eta V_{BB} + V_F$$

PUTs:

$$V_G = \frac{R_{B_1}}{R_{B_1} + R_{B_2}} V_{BB} = \eta V_{BB}$$
$$V_P = \eta V_{BB} + V_D$$

PROBLEMS

SECTION 9.2 Bipolar Junction Transistors

1. Determine the voltage gain (A_v) for the transistor configuration in Fig. 9.3 if $V_i = 100$ mV, $Z_i = 25\ \Omega$, and $R_L = 2\ \text{k}\Omega$.

2. Using the characteristics in Fig. 9.5, determine:
a. I_C if $I_E = 2$ mA and $V_{CB} = -10$ V.
b. I_C if $V_{EB} = 400$ mV and $V_{CB} = -10$ V.
c. V_{EB} if $I_C = 4$ mA and $V_{CB} = -5$ V.

3. For a particular transistor with $\alpha = 0.99$ and $I_C = 4$ mA, determine I_E and I_B.

4. Using the characteristics in Fig. 9.7:
a. Determine I_C if $I_B = 40\ \mu$A and $V_{CE} = 7.5$ V.
b. Find I_B if $I_C = 2$ mA and $V_{CE} = 5$ V.

c. Find I_E at $I_B = 20\ \mu A$ and $V_{CE} = 10$ V.
d. Determine β for the results for part (a).
e. Determine α for the results for part (a).

5. Sketch the power curve for $P_{C\max} = 20$ mW in Fig. 9.7 if $I_{C\max} = 6$ mA and $V_{CE\max} = 15$ V.

SECTION 9.3 Bipolar Junction Transistor Amplifier Circuits

6. Calculate the dc bias currents I_E and I_C and the voltage V_{CB} for the CB circuit in Fig. 9.91 for $R_E = 1\ k\Omega$.

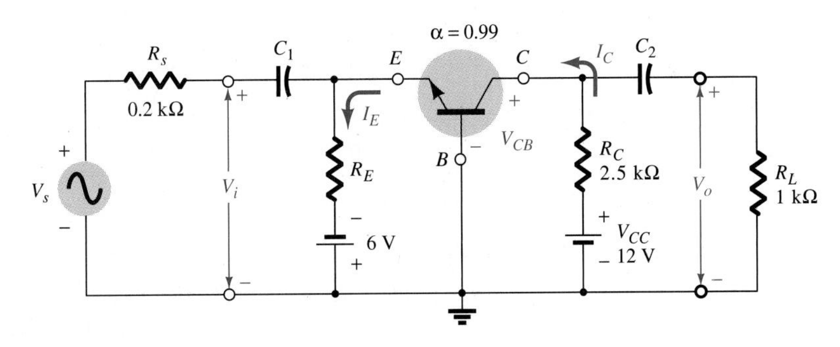

FIG. 9.91

7. If R_E is selected so that $I_{E_Q} = 3$ mA in the CB circuit in Fig. 9.91, calculate the voltage V_{CB_Q}. Determine the required value of R_E.

8. Calculate the ac output voltage V_o of the CB circuit in Fig. 9.91 for an input voltage V_s of 10 mV (rms) for the dc bias condition established in Problem 6.

9. For the common-base configuration in Fig. 9.92:
a. Determine I_E.
b. Determine r_e.
c. Calculate Z_i.
d. Find $A_v = V_o/V_i$.
e. Find $A_{v_T} = V_o/V_s$.
f. Calculate Z_o.

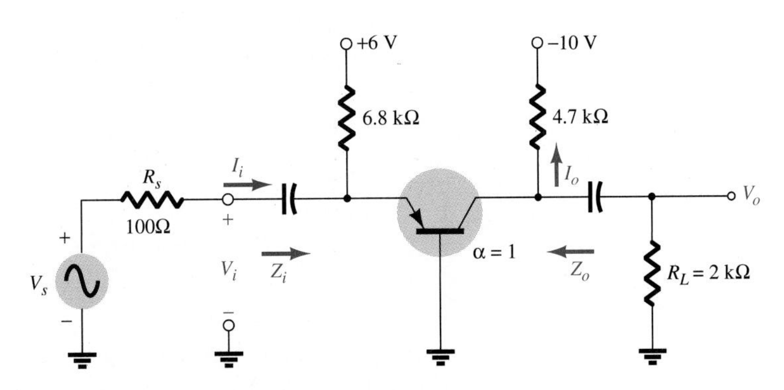

FIG. 9.92

10. Determine the bias currents I_{B_Q} and I_{C_Q} for the CE circuit in Fig. 9.93 when $R_B = 240\text{ k}\Omega$.

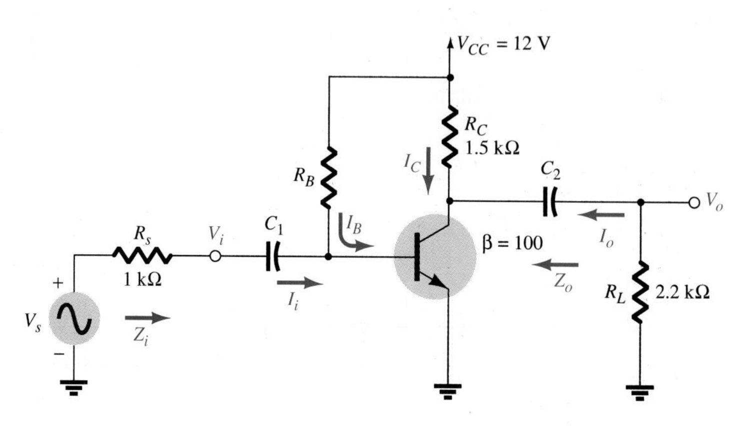

FIG. 9.93

11. If R_B is selected to give $I_{B_Q} = 50\mu\text{A}$, calculate the resulting bias voltage V_{CE_Q} (Fig. 9.93). Determine the required value of R_B.

12. For the network in Fig. 9.93:

a. Determine I_E.
b. Calculate r_e.
c. Find Z_i and Z_o.
d. Calculate $A_v = V_o/V_i$.
e. Find $A_{v_T} = V_o/V_s$.
f. Determine $A_i = I_o/I_i$.

13. For the network in Fig. 9.94:

a. Calculate I_B, I_c, and I_E.
b. Determine r_e.
c. Find Z_i and Z_o.
d. Calculate $A_v = V_o/V_i$.
e. Find $A_{v_T} = V_o/V_s$.
f. Calculate $A_i = I_o/I_i$.

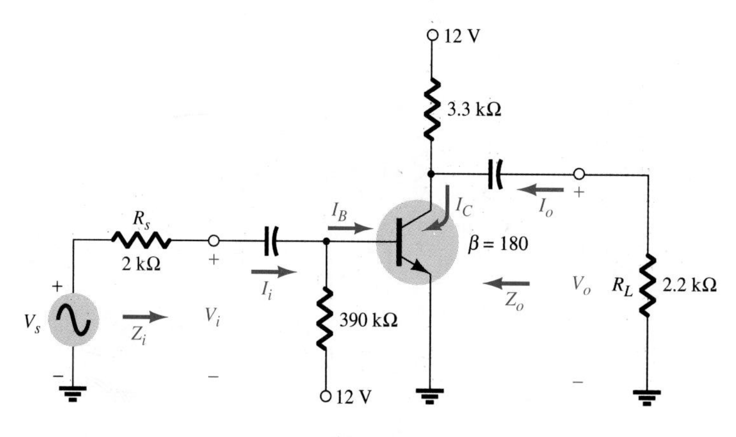

FIG. 9.94

14. For the network in Fig. 9.95:

a. Determine V_{B_Q}.
b. Calculate I_{C_Q} and V_{CE_Q}.
c. Find r_e.
d. Calculate $A_v = V_o/V_i$.
e. Find $A_{v_T} = V_o/V_s$.
f. Determine A_p.

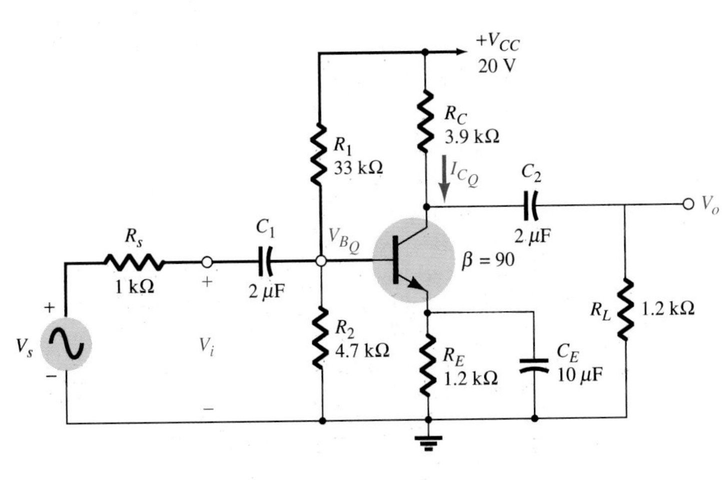

FIG. 9.95

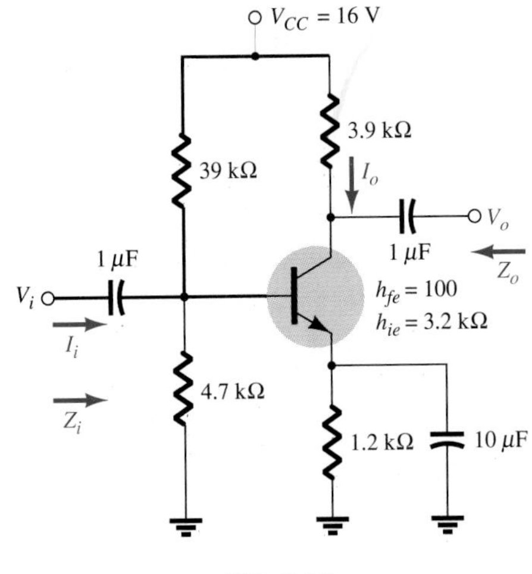

FIG. 9.96

15. For the voltage-divider configuration in Fig. 9.96 provided with the hybrid parameters:

a. Determine Z_i and Z_o.
b. Find $A_v = V_o/V_i$.
c. Calculate $A_i = I_o/I_i$.

16. For the emitter-follower in Fig. 9.97:

a. Determine I_{B_Q}, I_{C_Q}, and I_{E_Q}.
b. Calculate r_e.
c. Find h_{ie} and h_{fe}.
d. Find V_{CE_Q}.
e. Determine $A_v = V_o/V_i$ and $A_i = I_o/I_i$.
f. Calculate $A_{v_T} = V_o/V_s$.
g. Find Z_i and Z_o.

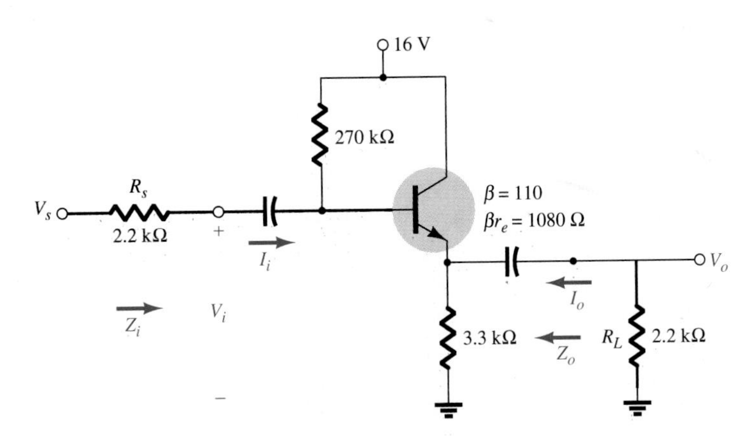

FIG. 9.97

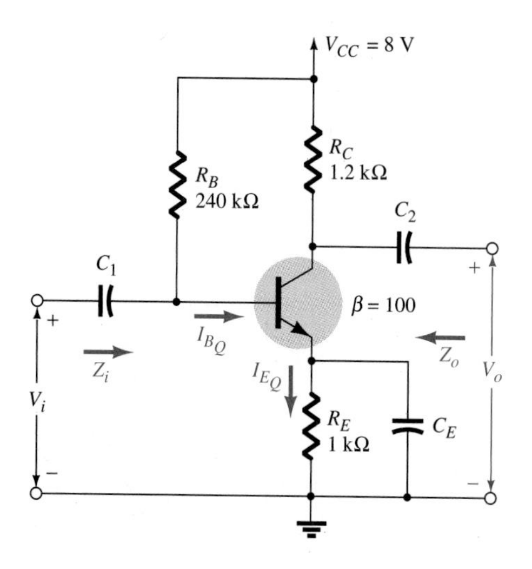

FIG. 9.98

17. The network in Fig. 9.98 was not analyzed in detail in Section 9.3. However, the required equations can be established using an approach very similar to that employed in the text.

a. Develop an equation for the current I_{B_Q} and then solve for I_{B_Q}.

b. Determine the current I_{E_Q} and then the resistance r_e.

c. Sketch the ac equivalent circuit and find the input and output impedances Z_i and Z_o.

d. Find the overall voltage gain $A_v = V_o/V_i$.

SECTION 9.4 Field-Effect Transistors

18. Determine the drain current I_D for the JFET in Fig. 9.39 at:

a. $V_{GS} = -3$ V, $V_{DS} = 15$ V.

b. $V_{GS} = -1$ V, $V_{DS} = 7$ V.

What conclusion can be drawn from the results?

19. Using Eq. (9.48), determine the drain current of JFET at $V_{GS} = -2$ V with $V_P = -8$ V and $I_{DSS} = 8$ mA.

20. Using Fig. 9.40, determine the drain current at $V_{GS} = -3$ V and -1 V and compare it with the results for Problem 18.

21. Describe (in your own words, from memory) the difference between a depletion-, an enhancement-, and an enhancement–depletion-type FET.

22. For the fixed-bias configuration in Fig. 9.99:

a. Find V_{GS_Q} and I_{D_Q}.

b. Find the input and output impedances Z_i and Z_o.

c. Calculate the voltage gain $A_v = V_o/V_i$.

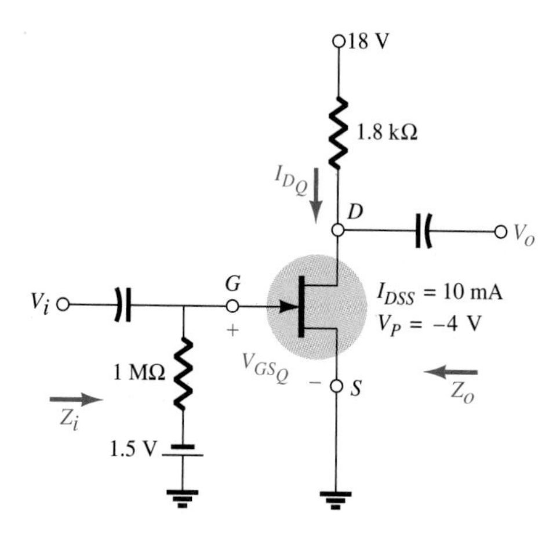

FIG. 9.99

23. For the self-bias JFET configuration in Fig. 9.100:

a. Find V_{GS_Q} and I_{D_Q}.

b. Find the input and output impedances Z_i and Z_o.

c. Calculate the voltage gain $A_v = V_o/V_i$.

d. Calculate the voltage gain $A_{v_T} = V_o/V_s$.

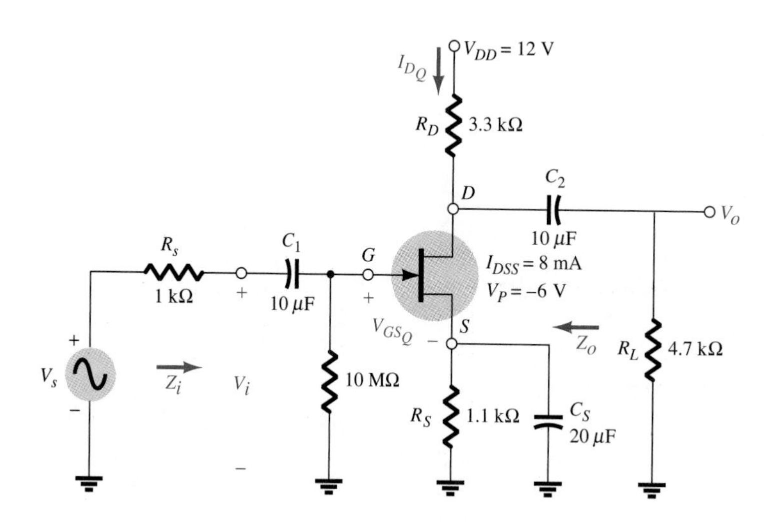

FIG. 9.100

24. For the voltage-divider JFET transistor configuration in Fig. 9.101:

a. Find V_{GS_Q} and I_{D_Q}.
b. Determine the input and output impedances Z_i and Z_o.
c. Find the voltage gain $A_v = V_o/V_i$.
d. Calculate the voltage gain $A_{v_T} = V_o/V_s$.

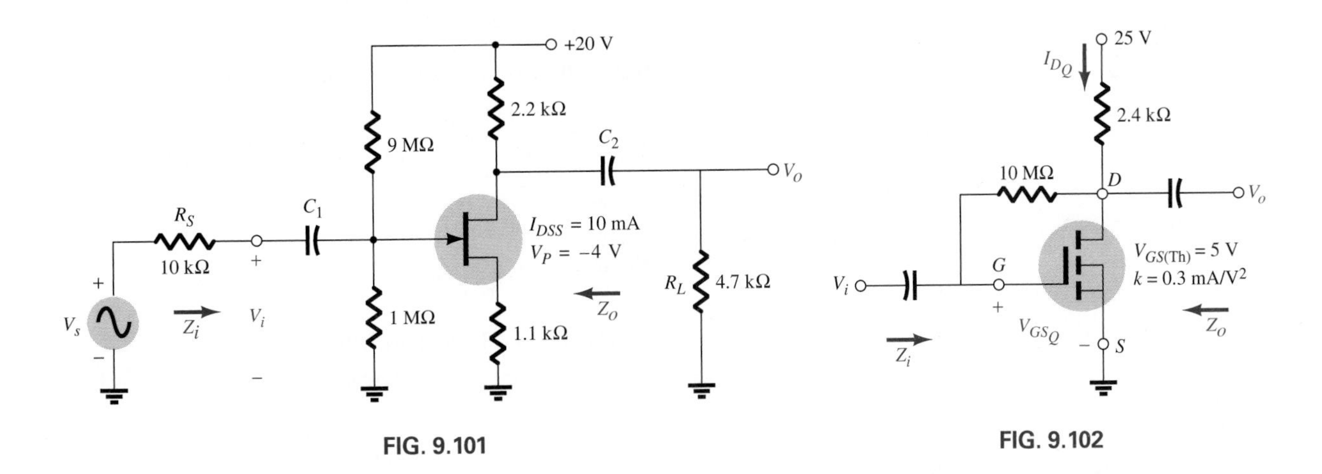

FIG. 9.101

FIG. 9.102

25. For the enhancement-type MOSFET configuration in Fig. 9.102:

a. Find V_{GS_Q} and I_{D_Q}.
b. Calculate the input and output impedances Z_i and Z_o.
c. Find the voltage gain $A_v = V_o/V_i$.

26. For the depletion-type MOSFET configuration in Fig. 9.103:

a. Find V_{GS_Q} and I_{D_Q}.
b. Calculate the input and output impedances Z_i and Z_o.
c. Find the voltage gain $A_v = V_o/V_i$.
d. Find the voltage gain $A_{v_T} = V_o/V_s$.

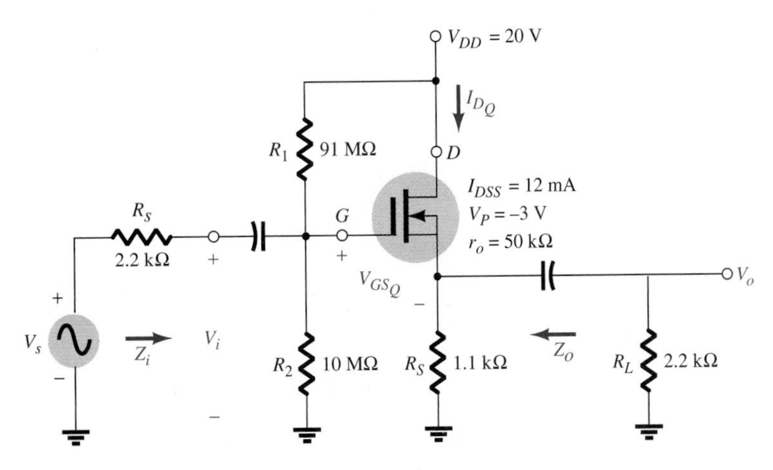

FIG. 9.103

SECTION 9.5 Unijunction Transistor

27. Determine the following for a UJT with $\eta = 0.62$, $V_{BB} =$ 20 V, $V_F = 0.7$ V, and $R_{BB} = 10$ kΩ.
a. V_1.
b. V_P.
c. R_{B_1}.

28. a. Using Fig. 9.64, determine V_P for $V_{BB} = 5$ and 30 V.
b. Determine I_E at $V_{BB} = 15$ V and $V_E = 4$ V.

SECTION 9.6 Silicon-Controlled Rectifier

29. Are the following statements true or false?
a. For increasing values of gate current, the required firing potential V_F across an SCR will increase.
b. For increasing gate current, at a fixed V_F, the holding current will decrease.

30. Determine the range of firing potentials (using Fig. 9.68) for a gate current of 1.2 A.

SECTION 9.7 DIAC and TRIAC

31. Describe the basic difference between the diac and triac semiconductor devices.

SECTION 9.8 Opto-Isolators

32. Using Fig. 9.78, determine the rate of change of output current per milliampere change in LED input current for the linear region.

33. Using Fig. 9.79, compare the switching times at $I_C = 6$ mA for a 1-kΩ and a 100-Ω load. Is the ratio of switching times the same as the ratio of load levels?

34. Using Fig. 9.80, compare the levels of output current at the boiling and freezing points of water.

SECTION 9.9 Programmable Unijunction Transistor

35. Calculate R_{B_2} and V_{BB} for a silicon PUT if $\eta = 0.7$, $V_P =$ 11.9 V, and $R_{B_1} = 15$ kΩ.

36. Calculate η and V_P if $R_{B_1} = 20$ kΩ, $R_{B_2} = 6$ kΩ, $V_{BB} =$ 12 V, and $V_D = 0.7$ V.

Operational Amplifiers (Op-Amps) 10

10.1 OP-AMP BASICS

θθThree important characteristics of an amplifier are

1. Voltage gain
2. Input impedance
3. Output impedance

An ideal amplifier would be one having infinite voltage gain, infinite input impedance, and zero output impedance. An operational amplifier is a real-world attempt to match these characteristics. Figure 10.1 provides some amplifier characteristic values.

Another important characteristic of amplifiers is the phase relationship between the input and output signals. The op-amp circuit has the unique ability to be able to provide an output that is in phase or out of phase with an applied input signal. As shown in Fig. 10.2, an op-amp provides two inputs, one resulting in an output that is in phase with the input, and the other providing an output that is out of phase with the input. Op-amp circuits are provided as integrated circuits (ICs) with one to four op-amps in a single IC. Figure 10.3 shows a 741 IC, providing a

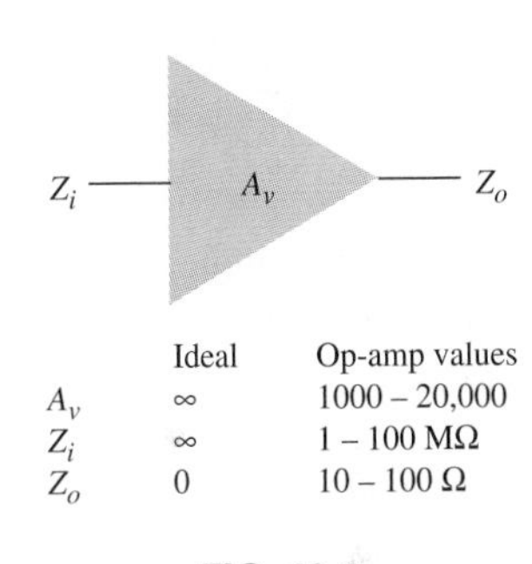

FIG. 10.1
Amplifier characteristics.

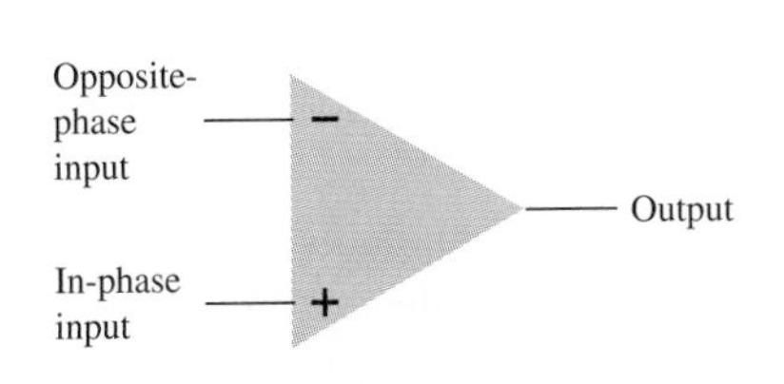

FIG. 10.2
Op-amp showing in-phase and opposite-phase inputs.

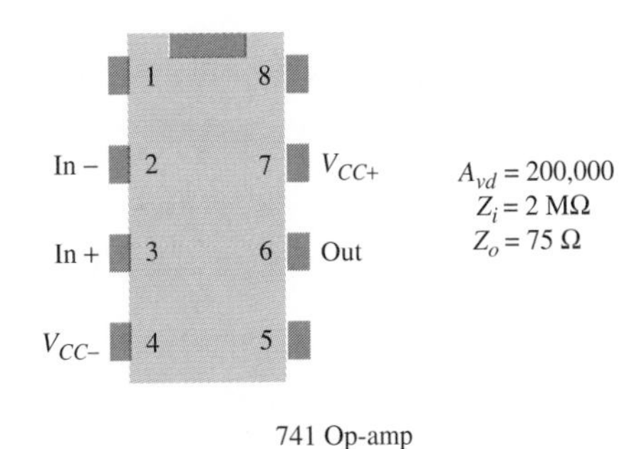

FIG. 10.3
741 op-amp pinouts and sample specifications.

description of the signals at specific pins and showing a selection of circuit characteristics.

The dc supply is connected between pin 7 (positive) and pin 4 (negative). Pin 4 can be connected to ground with pin 7 connected to the positive side of the voltage supply. An in-phase signal is then applied to pin 3, the opposite-phase signal or ground is applied to pin 2, and the output is taken from pin 6.

As indicated in Fig. 10.3, this single-stage op-amp has an input impedance of about 2 MΩ between pins 2 and 3, an output impedance of about 75 Ω from pin 6 to ground, and a differential voltage gain of about 200,000 between the input at pins 3 to 2 and the output at pin 6.

10.2 OP-AMP CIRCUITS

Constant-Gain Op-Amp Circuit

While an op-amp has a very high voltage gain, it typically ranges in value from one IC to another. Although the precise voltage gain of an IC is not well defined, a specific gain can be obtained using precision resistors in a circuit connection as in Fig. 10.4.

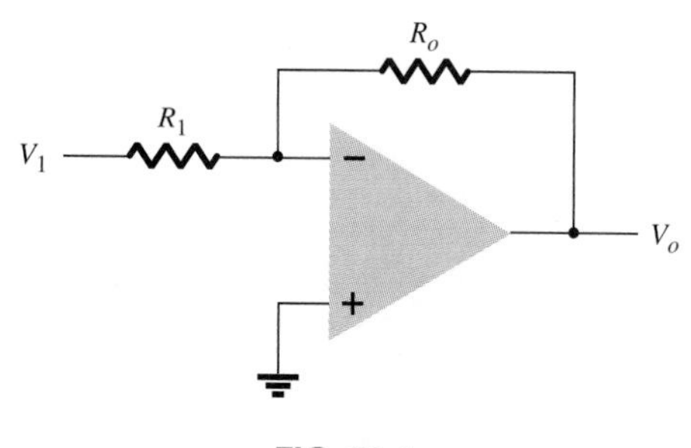

FIG. 10.4
Constant-gain op-amp circuit connection.

Notice that the input V_1 is applied through resistor R_1 to the inverting input with the output connected back to the inverting input through resistor R_o. The noninverting input is used as reference, which in this case is connected to ground.

If the op-amp is ideal, then the resulting voltage gain can be shown to be

$$A_v = \frac{V_o}{V_1} = -\frac{R_o}{R_1} \quad \textbf{(10.1)}$$

EXAMPLE 10.1 Calculate the voltage gain of the circuit in Fig. 10.4 for resistor components $R_1 = 10\ \text{k}\Omega$ and $R_o = 150\ \text{k}\Omega$.

Solution:

$$A_v = \frac{V_o}{V_1} = -\frac{R_o}{R_1} = -\frac{150\ \text{k}\Omega}{10\ \text{k}\Omega} = \mathbf{-15}$$

The amplifier has a voltage gain of precisely 15, with the negative sign indicating a phase shift of 180° between input and output.

Virtual Ground

The analysis of the constant gain amplifier in Fig. 10.4 and other op-amp configurations is simplified by using the concept of *virtual ground.* For an op-amp with a finite output voltage limited by the supply voltage, say 10 V, and a very high voltage gain, say 200,000, the input voltage is very small—virtually 0 V or ground, hence *virtual ground.* However, the input impedance of the op-amp is very high, say 2 MΩ, so that very little current is drawn by the op-amp. Thus, the input acts as virtual ground for voltage but as an open circuit for current.*

Figure 10.5 depicts this condition of signals. Using this model of the op-amp operation, we can write an equation for the current I as follows:

$$I = \frac{V_1}{R_1} = -\frac{V_o}{R_o}$$

which can be solved for V_o/V_1:

$$\frac{V_o}{V_1} = -\frac{R_o}{R_1}$$

which agrees with the result for Eq. (10.1).

While virtual ground can be used to analyze an op-amp circuit more easily, it depends on the op-amp's having very high voltage gain and high input impedance compared with the other circuit resistor components. Since this is typically true for an op-amp circuit, the resulting voltage gain equation provides numerical answers that are very close to those obtained using the actual circuitry.

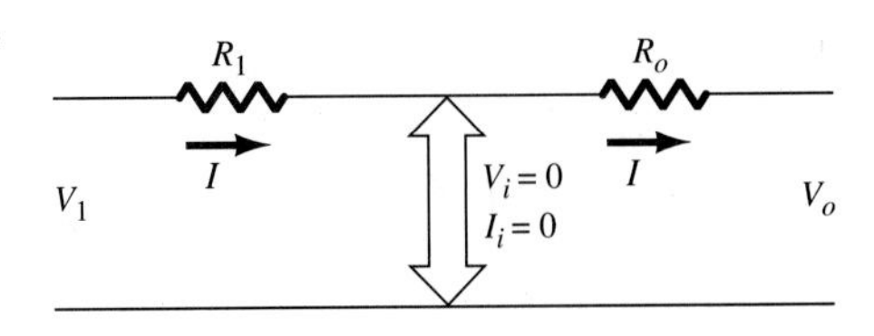

FIG. 10.5
Virtual ground in an op-amp circuit.

Noninverting Op-Amp Circuit

Figure 10.6 shows the connection of an op-amp as a noninverting amplifier. Using the virtual ground analysis ($V_i \approx 0$), we have

$$V_i = V_1 - V_o\left(\frac{R_1}{R_1 + R_o}\right) \approx 0$$

$$V_1 = V_o \frac{R_1}{R_1 + R_o}$$

$$\boxed{\frac{V_o}{V_1} = \frac{R_1 + R_o}{R_1} = 1 + \frac{R_o}{R_1}} \qquad \textbf{(10.2)}$$

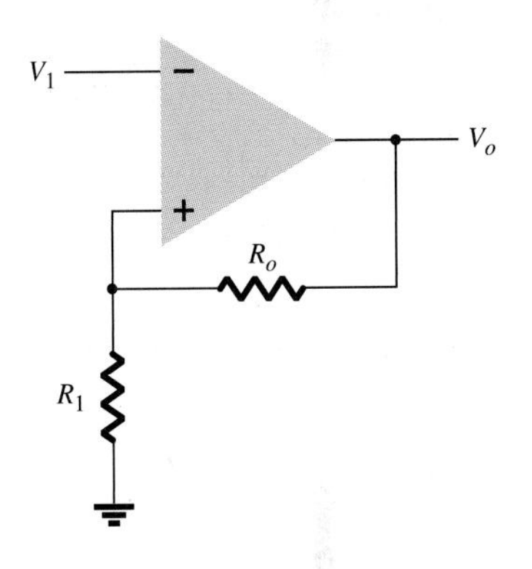

FIG. 10.6
Noninverting op-amp.

EXAMPLE 10.2 Calculate the voltage gain of the circuit in Fig. 10.6 with $R_o = 15\ \text{k}\Omega$ and $R_1 = 3\ \text{k}\Omega$.

Solution:
Using Eq. (10.2), we have

$$\frac{V_o}{V_1} = 1 + \frac{R_o}{R_1} = 1 + \frac{15\ \text{k}\Omega}{3\ \text{k}\Omega} = 1 + 5 = \mathbf{6}$$

---

*While V_i is very small, or ideally zero, it is not exactly zero, so that the very small input voltage V_i times the very large op-amp amplifier gain provides the finite output voltage V_o. Similarly, the very small input current, I_i, is negligible compared with I_1 and I_0, but is not exactly zero. Virtual ground provides a mathematical model for more easily analyzing an op-amp circuit.

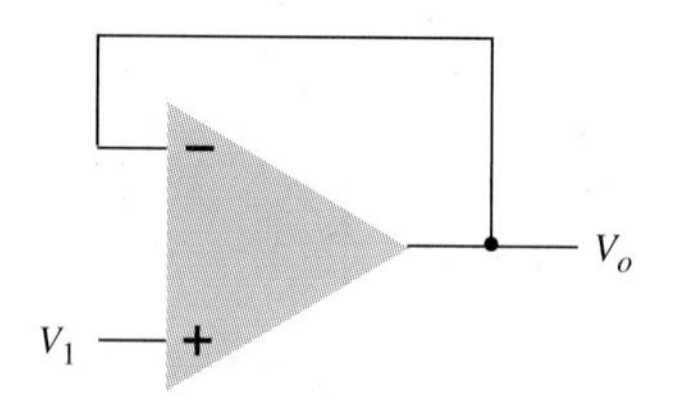

FIG. 10.7
Unity-gain connection.

Unity-Gain Op-Amp Circuit

The circuit in Fig. 10.7 shows the connection of an op-amp as a unity-gain amplifier (amplifier with a gain of 1). Using virtual ground analysis, we have

$$V_i = V_1 - V_o \approx 0$$

so that

$$V_1 = V_o$$

and

$$\boxed{\frac{V_o}{V_1} = 1} \tag{10.3}$$

The amplifier has a gain that is very nearly exactly 1, with no phase inversion. This type of amplifier is very useful as a buffer amplifier, providing an output that is exactly the same as the input but from a low-impedance output that can drive a number of loads. A unity follower circuit can thus be used to buffer a signal, providing a very high input impedance that essentially does not load the input signal and providing the same output signal (both magnitude and phase) from a low-impedance output that can be connected to one or more loads.

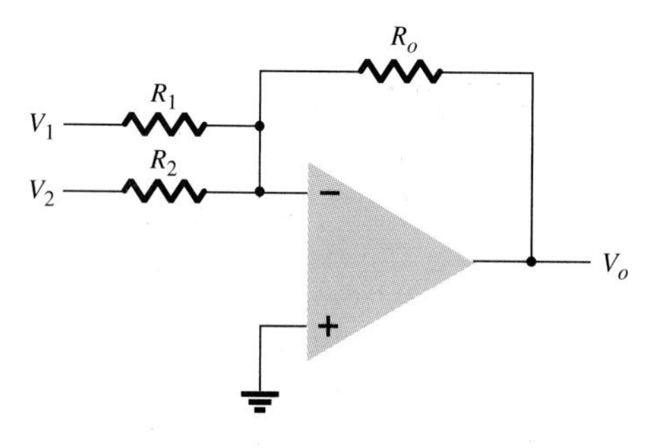

FIG. 10.8
Two-input summing circuit.

10.3 OP-AMP SUMMING CIRCUIT

While the basic circuits covered in Section 10.2 are quite popular, the connection of an op-amp that provides adding (or subtracting two or more signals) is also quite useful. Figure 10.8 shows the connection of an op-amp to sum two inputs. Using the virtual ground approach, with $V_i \approx$ 0 V, we see that the sum of the currents provides for

$$\frac{V_1}{R_1} + \frac{V_2}{R_2} = -\frac{V_o}{R_o}$$

which can be solved for V_o

$$\boxed{V_o = -\left(\frac{R_o}{R_1}V_1 + \frac{R_o}{R_2}V_2\right)} \tag{10.4}$$

EXAMPLE 10.3 Calculate the output voltage for the circuit in Fig. 10.8 for values of $R_o = 20\ \text{k}\Omega$, $R_1 = 1\ \text{k}\Omega$, $R_2 = 5\ \text{k}\Omega$, $V_1 = 0.5$ V, and $V_2 = 0.2$ V.

Solution
Using Eq. (10.4),

$$V_o = -\left(\frac{20\text{ k}\Omega}{1\text{ k}\Omega}0.5\text{ V} + \frac{20\text{ k}\Omega}{5\text{ k}\Omega}0.2\text{ V}\right) = -[20(0.5\text{ V}) + 4(0.2\text{ V})]$$
$$= -(10\text{ V} + 0.8\text{ V}) = \mathbf{-10.8\text{ V}}$$

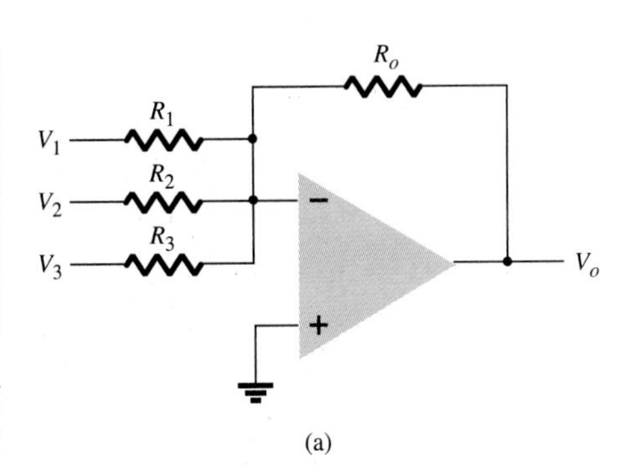

(a)

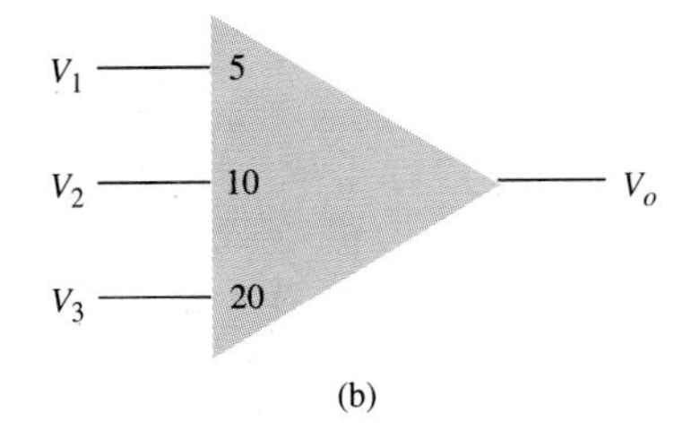

(b)

FIG. 10.9
(a) Three-input op-amp summing circuit; (b) simplified summing op-amp representation.

Figure 10.9(a) shows an op-amp summing circuit with three inputs of varied gain. Using the approach in Eq. (10.4), we calculate the gains for each of the three inputs to be

V_1 input amplified by $R_o/R_1 = 100\text{ k}\Omega/20\text{ k}\Omega = 5$
V_2 input amplified by $R_o/R_2 = 100\text{ k}\Omega/10\text{ k}\Omega = 10$
V_3 input amplified by $R_o/R_3 = 100\text{ k}\Omega/5\text{ k}\Omega = 20$

A simplified form of the op-amp summing circuit is shown in Fig. 10.9(b).

10.4 OP-AMP INTEGRATOR

When a capacitor is connected from output to input, as in Fig. 10.10(a), the circuit operates as an integrator. As such, the output is mathematically the integral function of the input signal. Thus a step input results in a ramp output, a linear or ramp input results in a squared output, and so on.

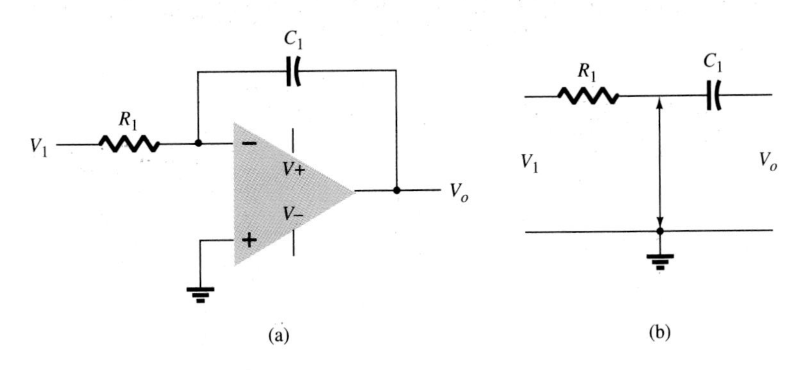

FIG. 10.10
Op-amp integrator: (a) circuit; (b) virtual connection.

The mathematical expression for the output of the circuit can be obtained using the virtual circuit shown in Fig. 10.10(b).

$$I_o = -I_1$$

Using Laplace notation ($s = -jw$), we have

$$\frac{V_o}{1/sC} = -\frac{V_1}{R}$$

$$\frac{V_o}{V_1} = -\frac{1}{sRC}$$

which translates in the time domain into

$$v_o(t) = -\frac{1}{RC}\int dt$$

so that

$$\boxed{v_o(t) = -\frac{1}{RC}\int v_1(t)\, dt} \tag{10.5}$$

EXAMPLE 10.4 Draw the output waveform for the input and circuit in Fig. 10.11(a).

Solution:
Using Eq. (10.5), we obtain

$$\begin{aligned} v_o(t) &= -\frac{1}{(100\ \text{k}\Omega)(0.5\ \mu\text{F})}\int (2V)\, dt \\ &= \frac{-2}{(100\ \text{k}\Omega)(0.5\ \mu\text{F})}t \\ &= -\frac{2}{50 \times 10^{-3}}t = \mathbf{-40}\boldsymbol{t} \end{aligned}$$

The output waveform is then a negative-going ramp voltage at a slope of -40, as shown in Fig. 10.11(b).

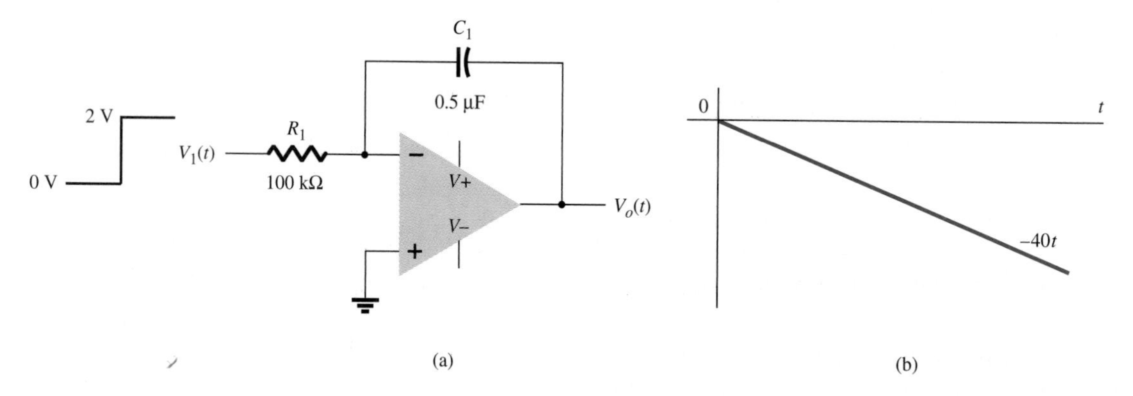

FIG. 10.11
Circuit for Example 10.4: (a) circuit; (b) output waveform.

The integrator circuit is useful for providing the integration operation as used in analog computers and also for shaping signals. Example 10.4 showed how the circuit converts a step input voltage into a ramp output voltage.

10.5 OP-AMP ICs

Op-amps are provided in IC packages with one to four op-amps in a single IC unit. The internal circuit can be made using only bipolar transistors, only FETs, or a combination of these in a Bi-FET or Bi-MOSFET circuit. Figure 10.12(a) shows a typical IC unit, the μ741, in an eight-pin package. The IC holds a single op-amp, as shown in Fig. 10.12(b).

A partial specification sheet listing for the μ741 IC is provided in Table 10.1. A number of the more important op-amp characteristics are discussed next.

TABLE 10.1
Selected μ741 op-amp characteristics

Parameter	Minimum	Typ.	Maximum	Units
Input resistance	0.3	2.0		MΩ
Common-mode rejection ratio	70	90		dB
Large-signal voltage gain	20,000	200,000		
Output resistance		75		Ω
Power consumption		50	85	mW
Slew rate		0.5		V/μs

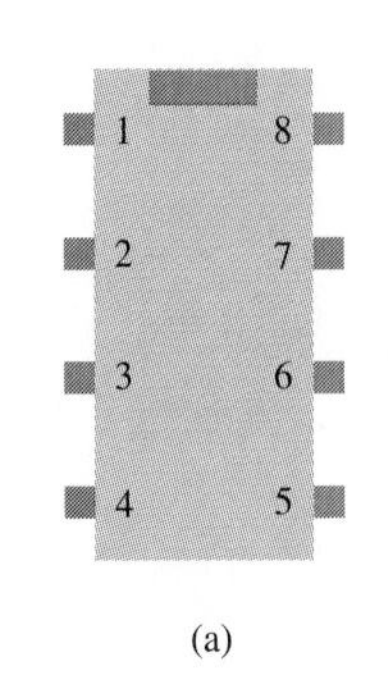

(a)

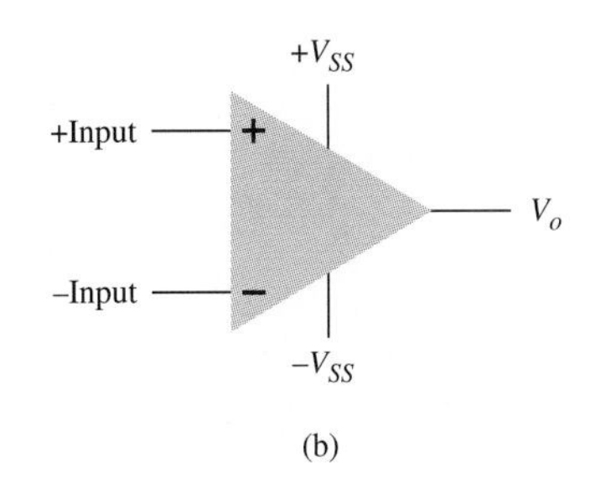

(b)

FIG. 10.12
μ741 op-amp: (a) IC unit; (b) circuit symbol.

Power Supply Terminals

Op-amp terminals marked $+V_{SS}$ and $-V_{SS}$ provide supply voltage to the entire IC circuit. Two usual connections are shown in Fig. 10.13.

In Fig. 10.13(a) dual voltage supplies are connected so that the output can swing equally as high as +10 V and as low as −10 V. Actually, the rail voltage or maximum voltage value the output can attain is typically about 0.5 to 1 V less than the supply voltage. Figure 10.13(b) shows a connection in which the output can swing as high as +10 V and as low as 0 V.

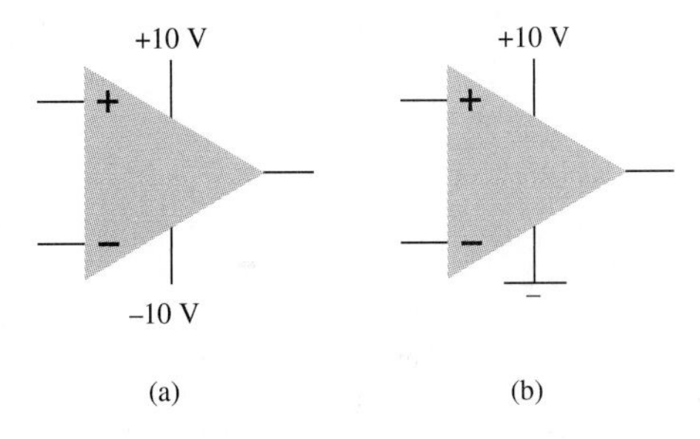

FIG. 10.13
Power supply connections: (a) dual supply; (b) single supply.

Output Terminal

The single output terminal, supplying a single-ended output, provides a voltage whose maximum value is limited by the supply voltage(s). The output voltage is provided at an output resistance that is shown in Table 10.1 to be

$$R_o = 75\ \Omega$$

Input Terminals

Two input terminals are provided. A signal applied to the positive input with respect to ground results in an output in phase with the input signal. A signal applied to the negative input with respect to ground results in an output that is the opposite of the phase of the input signal.

An input signal can also be connected between the + and − input terminals (a differential input) with the amplified output V_o.

The input impedance between either input and ground is typically very large. From Table 10.1 the value of R_i is seen to be

$$R_i = 2\ \text{M}\Omega$$

Open-Loop Voltage Gain

Table 10.1 shows the open-loop voltage gain to be

$$A_{OL} = 200{,}000$$

While an ideal op-amp gain is infinite, values are typically 100,000 to 1,000,000.

10.6 OP-AMP APPLICATIONS

Because the op-amp is such a versatile device there are numerous circuit applications that use it. This section mentions just a few of the more popular op-amp applications.

Gain-of-100 Amplifier

Figure 10.14 shows an amplifier providing a gain of exactly 100. For an op-amp having $A_{OL} = 200{,}000$, $R_1 = 5\ \text{k}\Omega$, and $R_o = 500\ \text{k}\Omega$, the actual gain will be very near the ideal value of 100 in the present example.

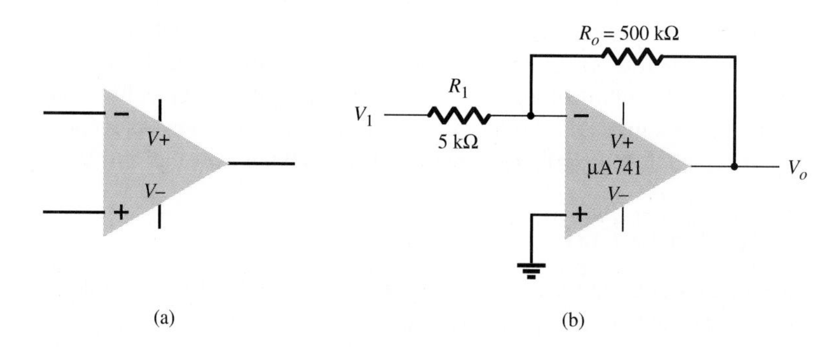

FIG. 10.14
Gain-of-100 amplifier.

Distribution Amplifier

Figure 10.15 shows a connection that provides outputs of 2, 5, and 10 times the input voltage. The connection shown thus provides distribution of a signal, with outputs available at a number of amplified levels. Since op-amp circuits are used, these gain values in practice are very close to the values listed.

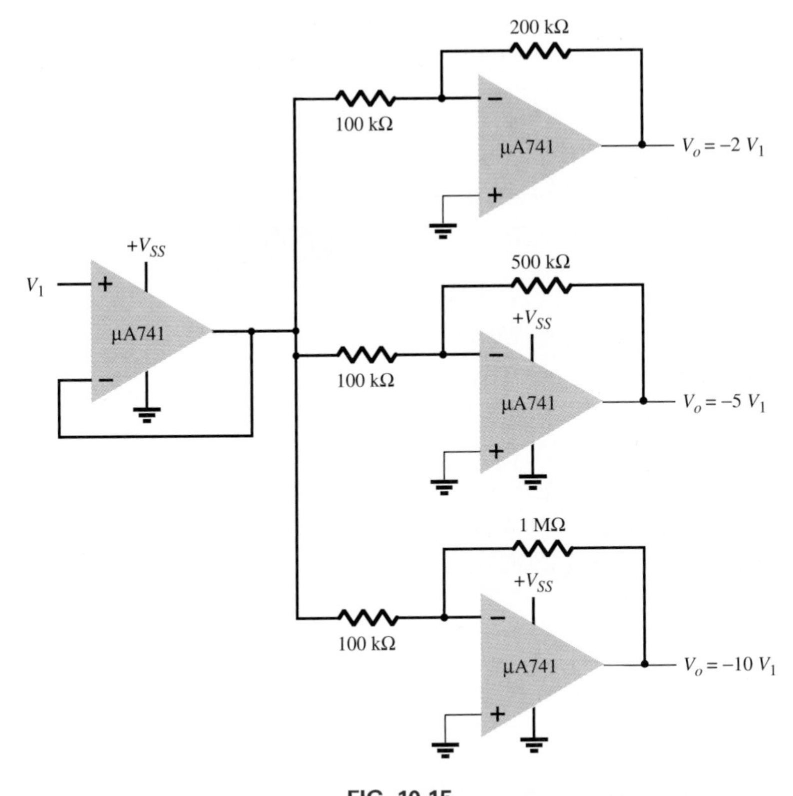

FIG. 10.15
Distribution amplifier.

10.7 COMPUTER ANALYSIS

PSpice Examples

PSpice provides op-amp models for many of the more popular units. Using them for engineering analysis gives results that compare favorably with those obtained using actual op-amp units. The version provided for educational use has only a few actual op-amp models. Figure 10.16 shows a μA741 op-amp obtained from the list of available parts. As shown, the positive and negative inputs are pins 3 and 2, respectively. The output is from pin 6, and supply voltages are connected to pins 4 and 7. The input impedance is very high, as determined by the components described in the model of the device. These provide for an input impedance of about 1 MΩ and a voltage gain of about 200,000.

You will see from the following examples that the results obtained using the "real" op-amp give answers that are close to those calcuated using "ideal" values.

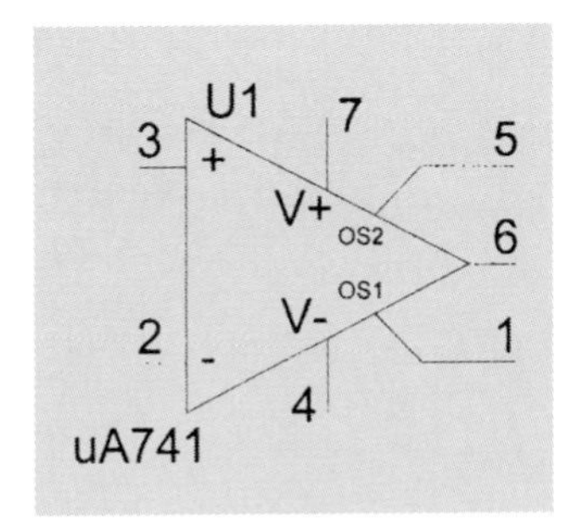

FIG. 10.16
μA741 op-amp using PSpice.

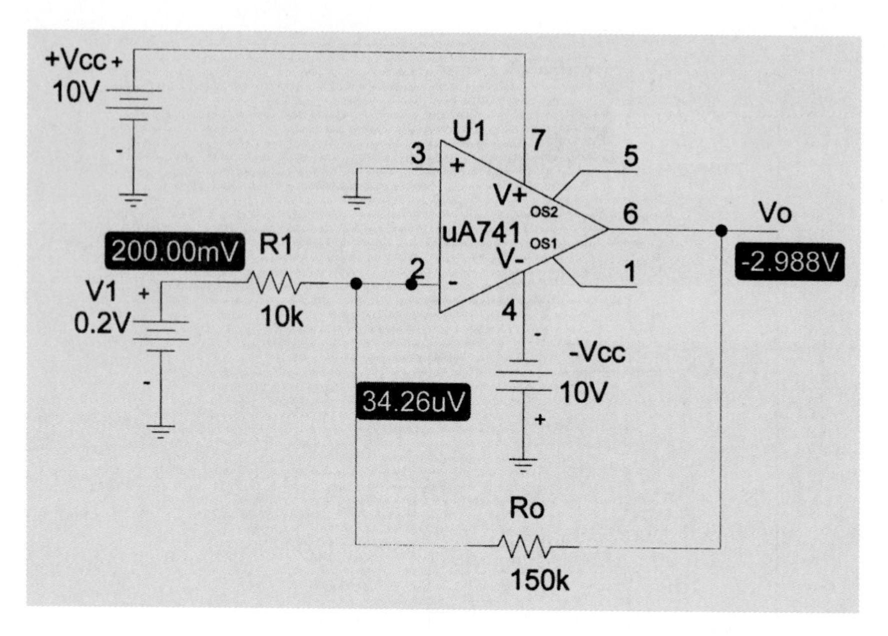

FIG. 10.17
Inverting constant-gain op-amp.

PSpice EXAMPLE 10.1 An inverting constant-gain voltage amplifier is shown in Fig. 10.17. The circuit resistors are $R_o = 150\text{ k}\Omega$, and $R_1 = 10\text{k}\Omega$.

The ideal voltage gain is calculated using Eq. (10.1):

$$A_v = -\frac{R_o}{R_1} = -\frac{150\text{ k}\Omega}{10\text{ k}\Omega} = \mathbf{-15}$$

Figure 10.17 shows that with an input of 0.2 V dc, the amplifier provides an output of −3 V dc (−2.988 V in the actual analysis), showing the gain to have a magnitude of 15 with a phase inversion indicated by the minus sign. Notice that the input to pin 2, which ideally is 0, is close to that at 34.26 μV.

PSpice EXAMPLE 10.2 A noninverting constant-gain voltage amplifier is shown in Fig. 10.18. The circuit contains values providing for a gain as calculated using Eq. (10.2):

$$A_v = 1 + \frac{R_o}{R_1} = 1 + \frac{50\text{ k}\Omega}{10\text{ k}\Omega} = 6$$

With an input of 1 V dc, the amplifier provides an output of 6 V dc, showing the gain to have a magnitude of 6 with no phase inversion indicated by the positive output voltage. Notice that the input of 1 V dc to pin 3 forces the op-amp output to be 6 V dc, so that the voltage at pin 2 is 1 V dc—provided that the voltage between pins 3 and 2 is very nearly 0 V dc.

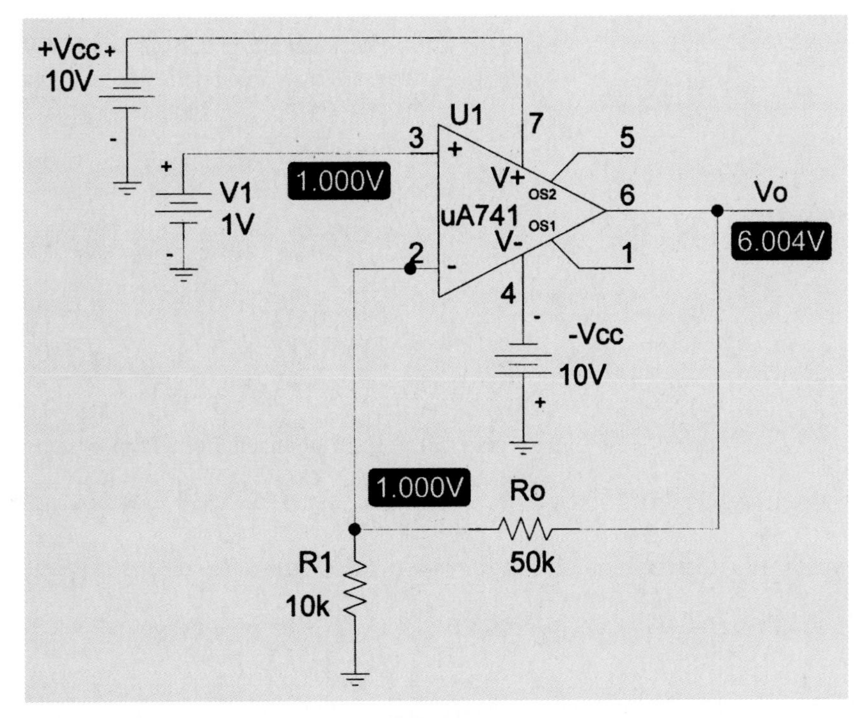

FIG. 10.18
Noninverting constant-gain op-amp.

PSpice EXAMPLE 10.3 A unity-gain voltage amplifier is shown in Fig. 10.19. With an input of 1 V dc, the amplifier provides an output of 1 V dc, showing the overall gain to be 1, with no phase inversion.

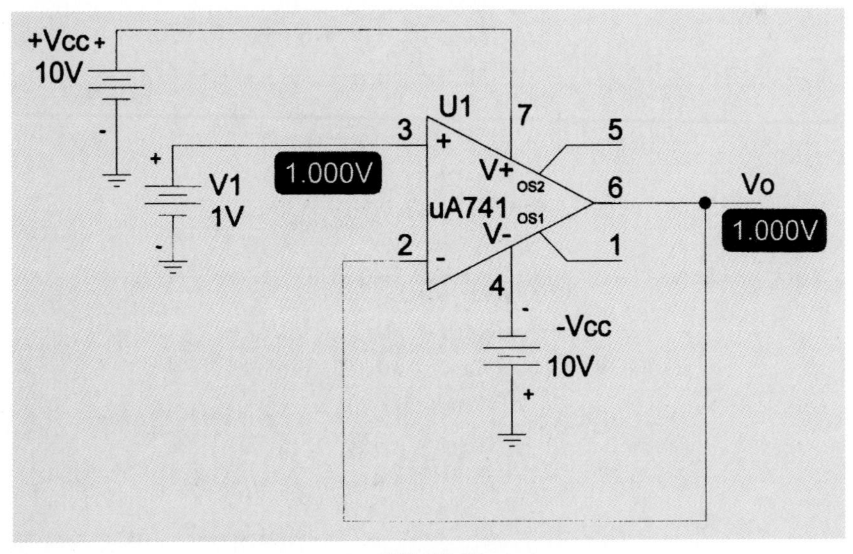

FIG. 10.19
Unity-gain amplifier.

Since the voltage between pins 3 and 2 is nearly 0 V, the output at pin 6 follows—is the same as—that applied to the input at pin 3.

PSpice EXAMPLE 10.4 A summing amplifier is shown in Fig. 10.20. The output is calculated using Eq. (10.4):

$$A_v = -\left[\left(\frac{R_o}{R_1}\right)V_1 + \left(\frac{R_o}{R_2}\right)V_2\right]$$

$$= -\left[\left(\frac{20\text{ k}\Omega}{1\text{ k}\Omega}\right)(0.2\text{ V}) + \left(\frac{20\text{ k}\Omega}{5\text{ k}\Omega}\right)(0.5\text{ V})\right]$$

$$= -[4\text{ V} + 2\text{ V}] = \mathbf{-6\text{ V}}$$

The practical voltage obtained from PSpice analysis is seen to be −5.997, which is very close to the ideal value.

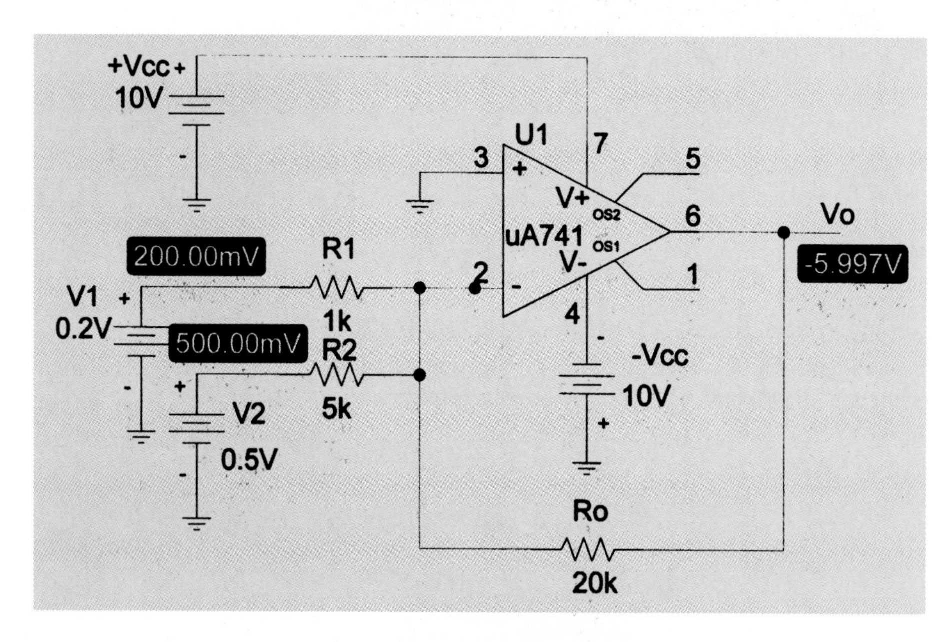

FIG. 10.20
Summing amplifier.

PSpice EXAMPLE 10.5 An integrating amplifier is shown in Fig. 10.21(a). The input is a step voltage having a value of 0.5 V. The output is then calculated using Eq. (10.5):

$$v_o(t) = \frac{1}{R_1C_1}\int v_1(t)$$

$$= \frac{1}{(100\text{ k}\Omega)(0.5\ \mu\text{F})}\int (0.5)$$

$$= (-20)(0.5t) = \mathbf{-10t}$$

Notice that a step input voltage results in an output that drops linearly at a rate of $-10t$, or 10 V/s. The probe output showing this result is in Fig. 10.21(b).

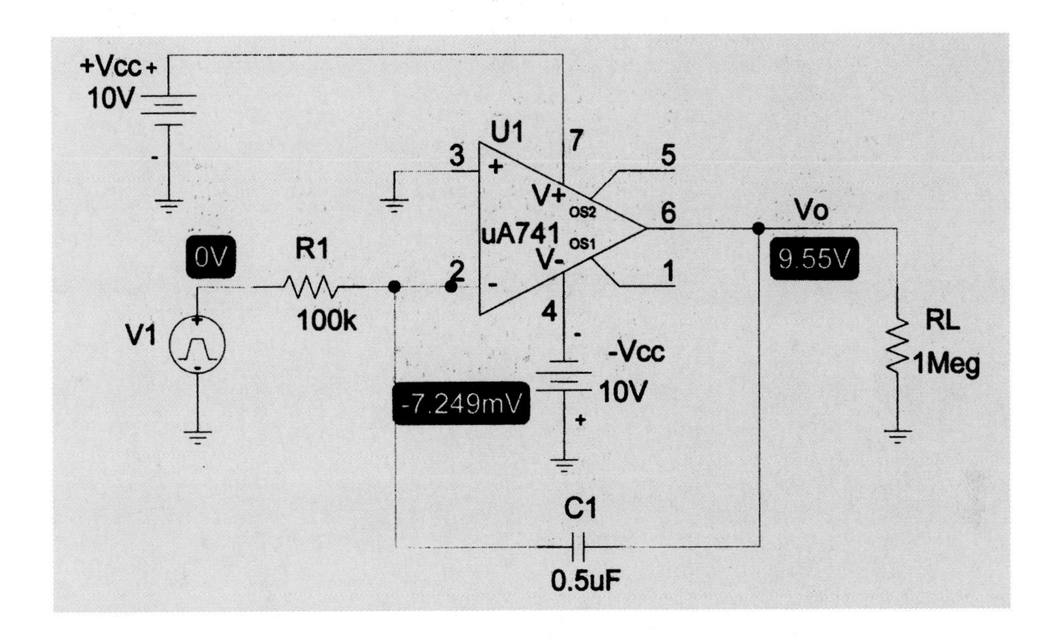

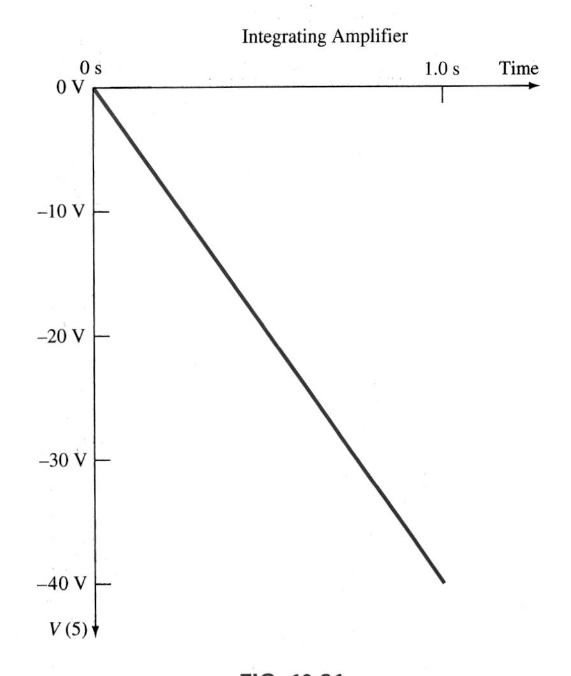

FIG. 10.21
(a) Integrator circuit; (b) PROBE output for PSpice Example 10.5.

Electronic Workbench Example

An example of an EWB op-amp circuit of an inverting constant-gain amplifier—the same as that in Fig. 10.17—is shown in Fig. 10.22. The use of a packaged μ741 op-amp, resistors, supply voltages, and a multimeter is similar to the setup using PSpice. The result of the EWB simulation shows a meter reading of -2.972 V, which is slightly less than the value of -2.988 V obtained in PSpice Example 10.1.

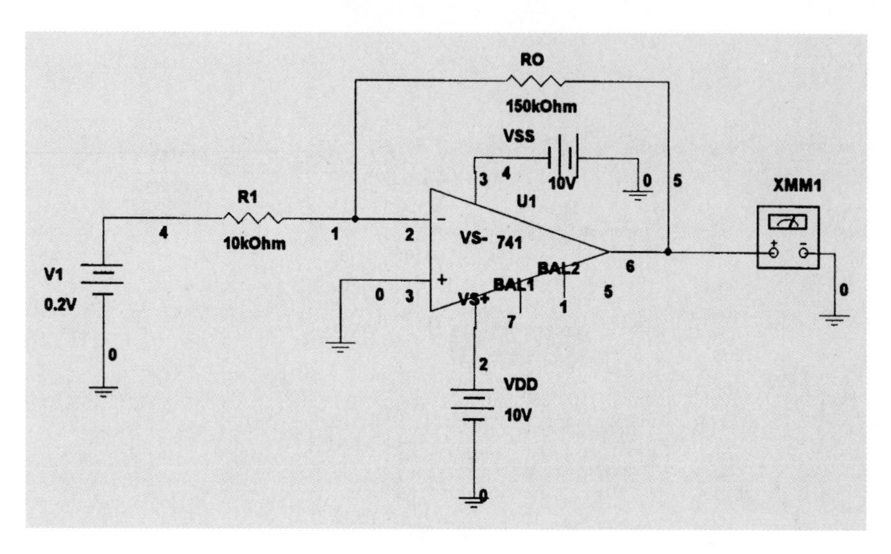

FIG. 10.22
EWB setup of inverting constant-gain amplifier.

10.8 SUMMARY

- Op-amps have the following basic features:
  Very high input impedance (ideally infinite)
  Very high voltage gain (ideally infinite)
  Low output impedance (ideally zero)
- Virtual ground is the concept that the voltage between positive and negative terminals is virtually 0 V.

Equations:

Constant-Gain Op-Amp Circuit

$$A_v = -\frac{R_o}{R_1}$$

Op-Amp Summing Circuit

$$A_v = -\left(\frac{R_o}{R_1}V_1 + \frac{R_o}{R_2}V_2\right)$$

Op-Amp Integrator Circuit

$$v_o(t) = \frac{-1}{RC}\int v_1(t)\,dt$$

PROBLEMS

SECTION 10.2 Op-Amp Circuits

1. Calculate the voltage gain of the circuit in Fig. 10.23.

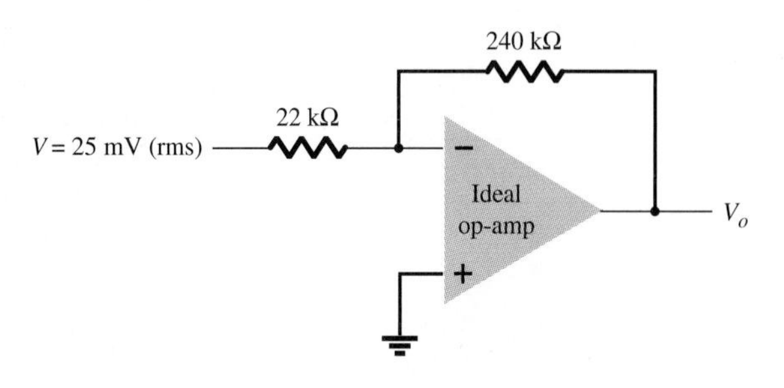

FIG. 10.23
Op-amp circuit.

2. Sketch the input V_i and output V_o waveforms for the circuit in Fig. 10.24.

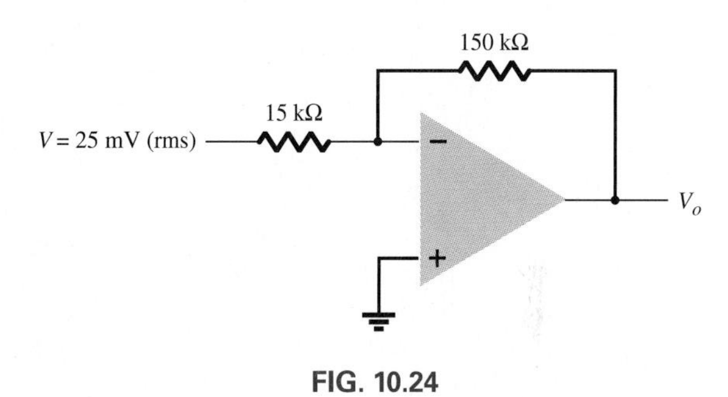

FIG. 10.24

3. Determine the value of R_i required to provide an output of $V_o = -3$ V (rms) in the circuit in Fig. 10.25.

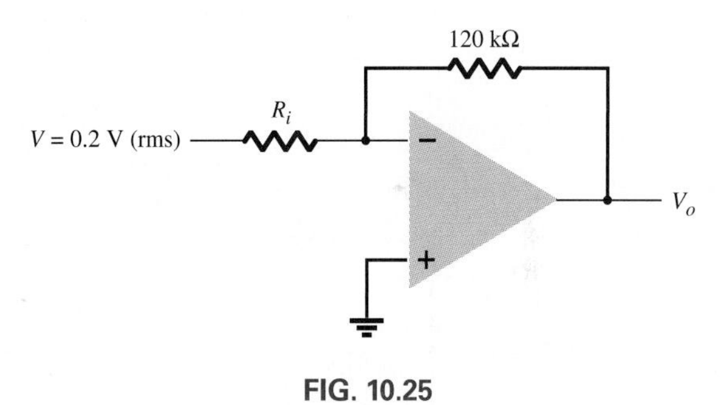

FIG. 10.25

4. Determine the value of V_i in the circuit in Fig. 10.26.

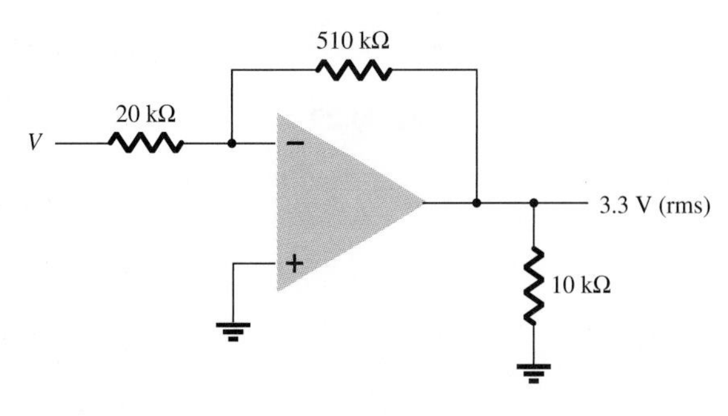

FIG. 10.26

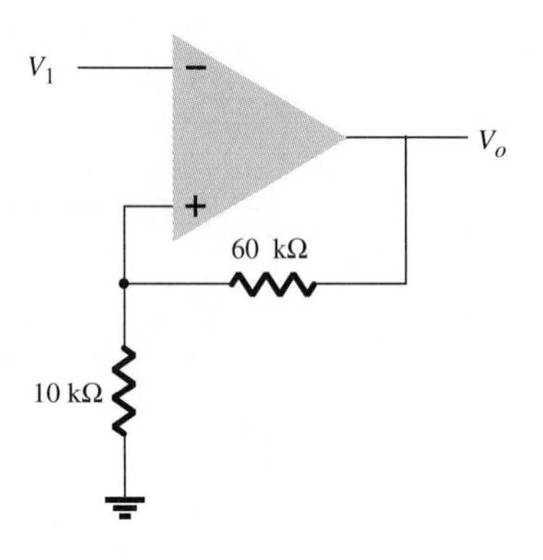

FIG. 10.27

5. Calculate the voltage gain of the circuit in Fig. 10.27.
6. Determine V_1 for the circuit in Fig. 10.27 if $V_o = 1.4$ V (rms).
7. Draw the input waveform V_1 and output waveforms V_{o1} and V_{o2} for the circuit in Fig. 10.28.

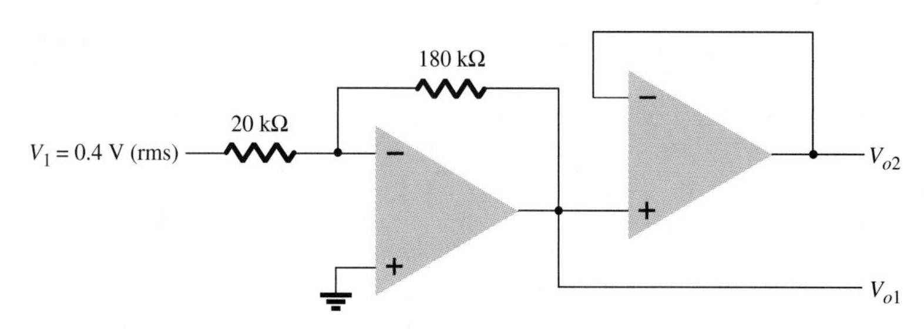

FIG. 10.28

SECTION 10.3 Op-Amp Summing Circuit

8. Calculate the output voltage V_o for the circuit in Fig. 10.29 with $R_1 = 3.3$ k, $R_2 = 1.2$ k, $R_3 = 2.4$ k, and $R_o = 30$ k. Input voltages are $V_1 = 0.5$ V, $V_2 = 0.3$ V, and $V_3 = 0.1$ V.

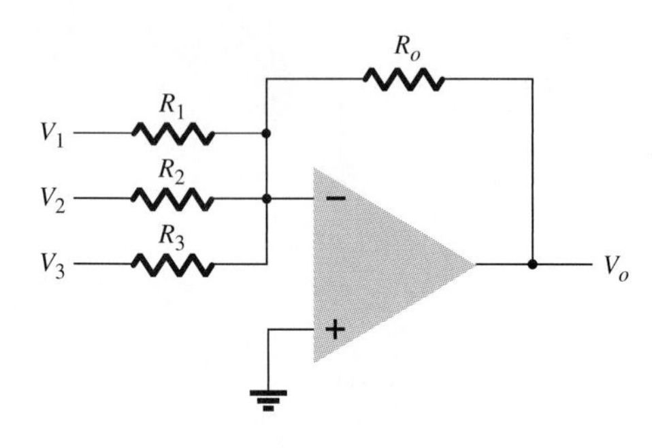

FIG. 10.29

9. Determine the value of V_o for the circuit in Fig. 10.29 with $R_1 = R_2 = R_3 = 10$ kΩ, and $V_1 = 0.25$ mV, $V_2 = 0.15$ mV, and $V_3 = 0.2$ mV. $R_o = 30$ kΩ.
10. Determine the value of R_o in the circuit in Fig. 10.29 to provide a gain of -100 using $R_1 = R_2 = R_3 = 10$ kΩ, and $V_1 = V_2 = V_3$.

11. Calculate the value of R_1 in the circuit in Fig. 10.29 to provide an output of -10 V dc with circuit values of $R_2 =$ 20 kΩ, $R_3 = 3$ kΩ $V_1 = 0.5$ V dc, $V_1 = V_3 = 0.2$ V dc and $R_o = 100$ kΩ.

12. Sketch inputs V_1, V_2, and V_3 and output V_o waveforms for the circuit in Fig. 10.29 with $V_1 = 0.1$ V rms, $V_2 = 0.2$ V rms, $V_3 = 0.3$ V rms, $R_1 = 1$ kΩ, $R_2 = 2$ kΩ, $R_3 = 3$ kΩ and $R_o = 6$ kΩ.

SECTION 10.4 Op-Amp Integrator

13. Sketch the input V_1 and output V_o waveform for the circuit in Fig. 10.30 for $V_1 = 0.4$ V rms sinusoidal at frequency of 10 kHz.

14. Calculate the value of capacitor C_1 in the circuit in Fig. 10.30 to provide a positively sloped ramp voltage for a step input of -0.5 V ($R_1 = 51$ kΩ).

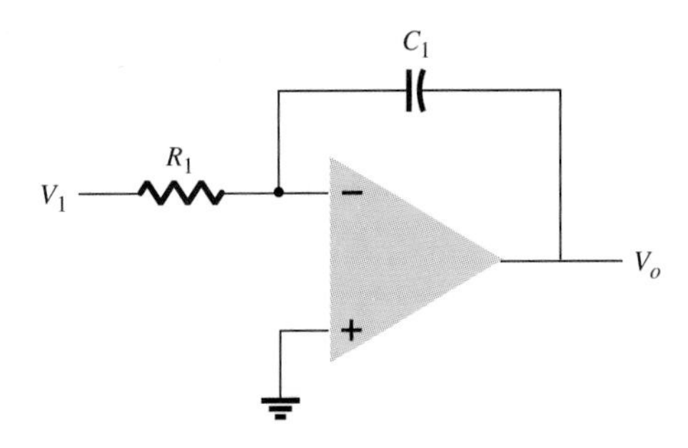

FIG. 10.30
Circuit for Problems 13 and 14.

SECTION 10.7 Computer Analysis

15. Use a PSpice circuit to calculate V_o for the circuit in Fig. 10.23.

16. Use a PSpice circuit to calculate V_{o1} and V_{o2} in the circuit in Fig. 10.28.

17. Use a PSpice circuit to calculate the output V_o for the circuit in Fig. 10.8.

18. Use a PSpice circuit to provide the input and output waveforms for the circuit in Problem 13. Use PROBE to obtain the waveforms.

11 Multistage and Large-Signal Amplifiers

11.1 MULTISTAGE AMPLIFIERS

It should be apparent that a single amplifier stage is not sufficient for most purposes. When a number of stages are connected, consideration must be paid to the method of coupling these stages (e.g., capacitor coupling, transformer coupling, direct coupling) to the loading effect of each stage, to the frequency range of the signal being amplified, and to the amount of signal distortion resulting.

RC Coupling

The most common type of coupling for non-integrated-circuit amplifiers is capacitor coupling (also called *RC* coupling), as previously indicated in single-stage circuits. Figure 11.1 shows a two-stage FET amplifier. A few considerations must be made in this two-stage (or more) connection that are not fully apparent when analyzing one stage at a time. First, the

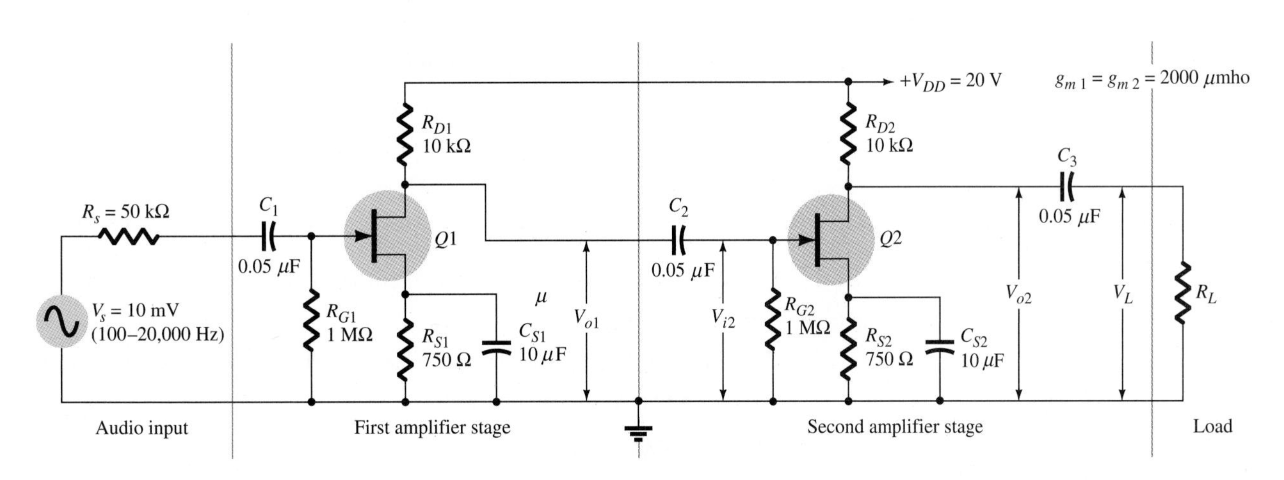

FIG. 11.1
Two-stage RC-coupled FET amplifier.

loading of the source by the first stage, of the first stage by the second stage, and of the second stage by the load must all be considered. In the present FET amplifier, however, these loadings are quite minimal and can be neglected, as will be shown. The voltage gain of the two amplifier stages is the *product* of the gains of each stage, as will also be shown.

For a practical signal source, such as that from an audio unit, a signal is obtained of typically, say, 10 mV (in the frequency range 100 to 16,000 Hz—most of the audio range) from a source having an impedance of 50 kΩ, typically. This signal is obtained from a pickup when not connected to any circuit (or load). In the present circuit the input impedance of the first FET stage acts as a voltage divider of the unloaded signal (V_s). Figure 11.2(a) shows the resulting input voltage V_{i1} reduced from the unloaded source value V_s by the voltage divider of R_s and the first FET stage input impedance—in this case approximately $R_{G1} = 1\ \text{M}\Omega$. The resulting input signal driving the first stage is thus

$$V_{i1} = \frac{R_{G1}}{R_{G1} + R_s} V_s = \frac{1\ \text{M}\Omega}{1\ \text{M}\Omega + 50\ \text{k}\Omega}(10\ \text{mV}) = 0.95(10\ \text{mV}) = 9.5\ \text{mV}$$

It seems apparent that little loading reduction of the input occurs for this FET amplifier stage, which may not be true in BJT circuits, as we saw in Chapter 9.

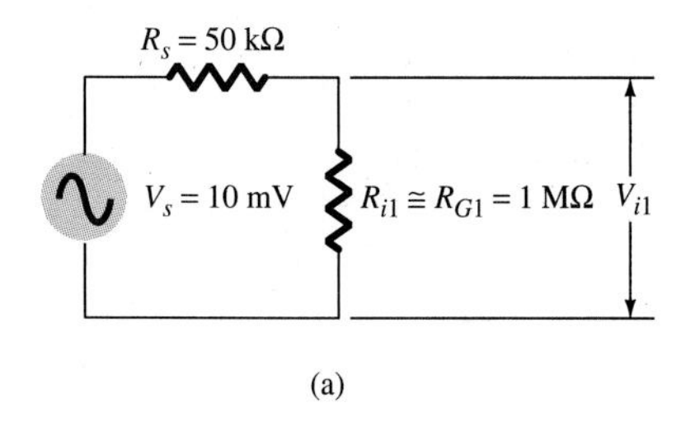

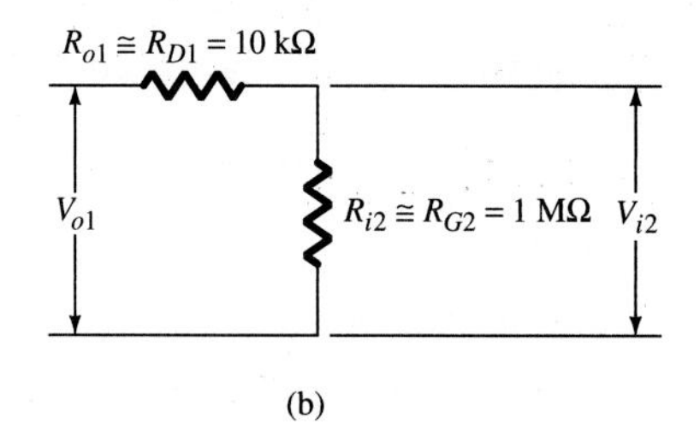

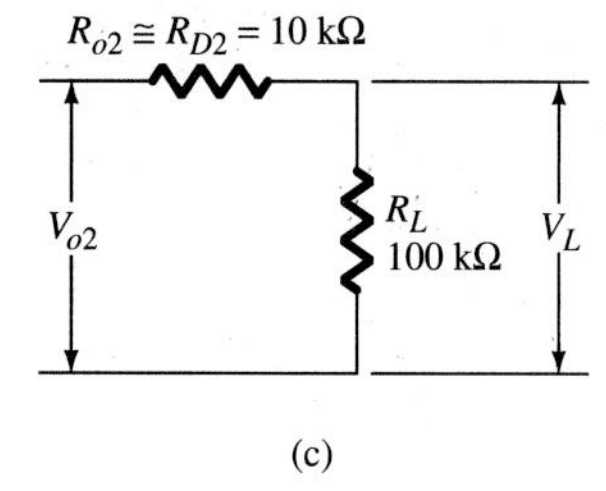

FIG. 11.2
Loading effects in a multistage amplifier: (a) input stage; (b) second stage; (c) load.

The voltage gain of stage 1 is approximately

$$A_{v1} = -g_m R_{D1} = -(2000 \times 10^{-6})(10 \times 10^3) = -20$$

(neglecting the output impedance of FET Q_1 and also the 1-MΩ input impedance of stage 2), and so

$$V_{o1} = V_{i2} = A_{v1} V_{i1} = -20(9.5\ \text{mV}) = -190\ \text{mV}$$

Using identical stages for this example, we can calculate the output of stage 2 to be

$$\begin{aligned} V_{o2} &= A_{v2} V_{i2} = -(g_m R_{D2}) V_{i2} = -(2000 \times 10^{-6})(10 \times 10^3)(-190\ \text{mV}) \\ &= 3800\ \text{mV} = 3.8\ \text{V} \end{aligned}$$

The overall gain of the two stages is

$$A_v = \frac{V_{o2}}{V_{i1}} = \frac{3.8\ \text{V}}{9.5\ \text{mV}} = 400$$

which is the same as the product of the stage gains:

$$A_v = A_{v1}A_{v2} = 20 \times 20 = 400$$

The overall circuit gain, however, should include any loading effects from unloaded source voltage V_s to the loaded output voltage V_L. The loading of stage 1 is seen in Fig. 11.2(b) to reduce V_{o1} to

$$V_{o1} = V_{i2} = \frac{R_{i2}}{R_{o1} + R_{i2}} V_{o1} = \frac{1\ \text{M}\Omega}{1\ \text{M}\Omega + 10\ \text{k}\Omega}(-190\ \text{mV}) = -188\ \text{mV}$$

Including this loading effect gives an output of

$$V_{o2} = A_{v2}V_{i2} = -20(-188\ \text{mV}) = 3.76\ \text{V}$$

If this output was connected to another amplifier stage or other circuit, then the effect of this load would be to further reduce the output voltage. With a load $R_L = 100\ \text{k}\Omega$ the load voltage delivered by the two-stage amplifier would be

$$V_L = \frac{R_L}{R_L + R_{o2}} V_{o2} = \frac{100\ \text{k}\Omega}{100\ \text{k}\Omega + 10\ \text{k}\Omega}(3.76\ \text{V}) = 3.42\ \text{V}$$

We see here that starting with two separate amplifier stages each having gains of 20 results in an overall circuit gain that is less than the "ideal" value of 400 due to the loading between stages and the loading of the source and load, so that in this case an overall gain of only 342 is achieved. It should be noted that the effects of loading are real considerations that must be taken into account when connecting amplifier stages together or when connecting source or load circuits to an amplifier. Suitable results can still be obtained, and high gains can easily be achieved, if required, by adding additional amplifier stages. In fact, with IC circuits, multistage amplifier units are readily available for a great variety of uses.

Differential Amplifier

An emitter-coupled differential amplifier is shown in Fig. 11.3 using BJTs. The two-transistor circuit makes up, essentially, one amplifier stage. Direct coupling of outputs V_{o1} and V_{o2} as input to another differential amplifier is usually made, and the frequency range of such a circuit is from dc signals up to some fairly high frequency (typically, in the 100-MHz range). This circuit arrangement not only affords a very wide frequency operating range including dc operation but also provides a flexible use of input and output using opposite polarity signals. Most importantly, it is a means of improving the signal-to-noise ratio by rejecting the same-polarity (noise) components while amplifying the opposite-polarity (signal) components. By amplifying one while reducing the other, the circuit achieves a common-mode rejection ratio of considerable value—typically 10^5 (as will be demonstrated).

Figure 11.4 shows a single input (V_{i1}) and the resulting output V_o. The input to the base of $Q1$ (V_{i1}) in Fig. 11.4(a) is amplified and inverted, resulting in the signal shown at output V_o. The signal V_{i2} shown in Fig. 11.5 produces a signal across R_E that results in the output at V_o—this

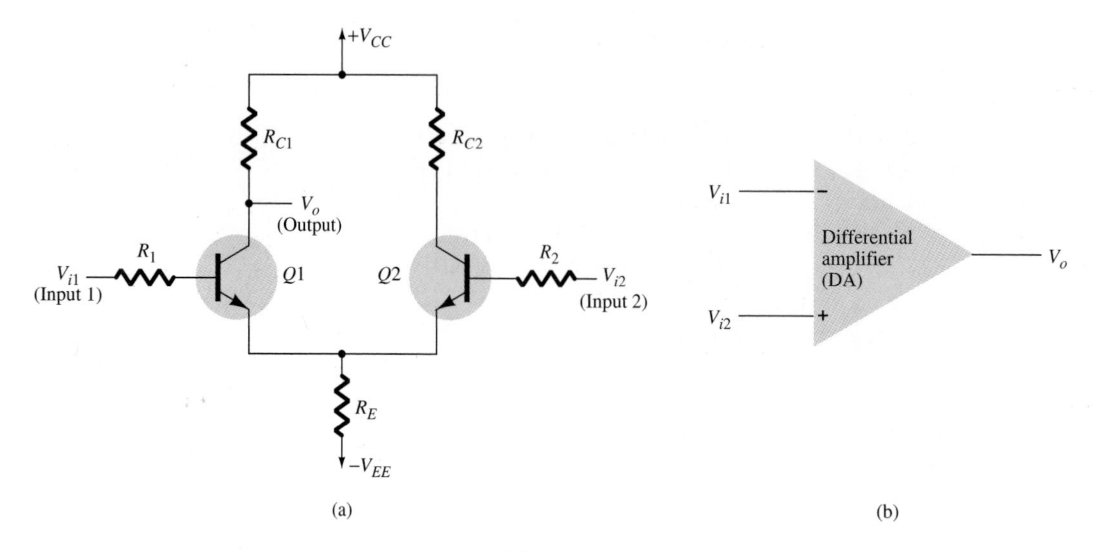

FIG. 11.3
Differential amplifier circuit: (a) circuit; (b) block symbol.

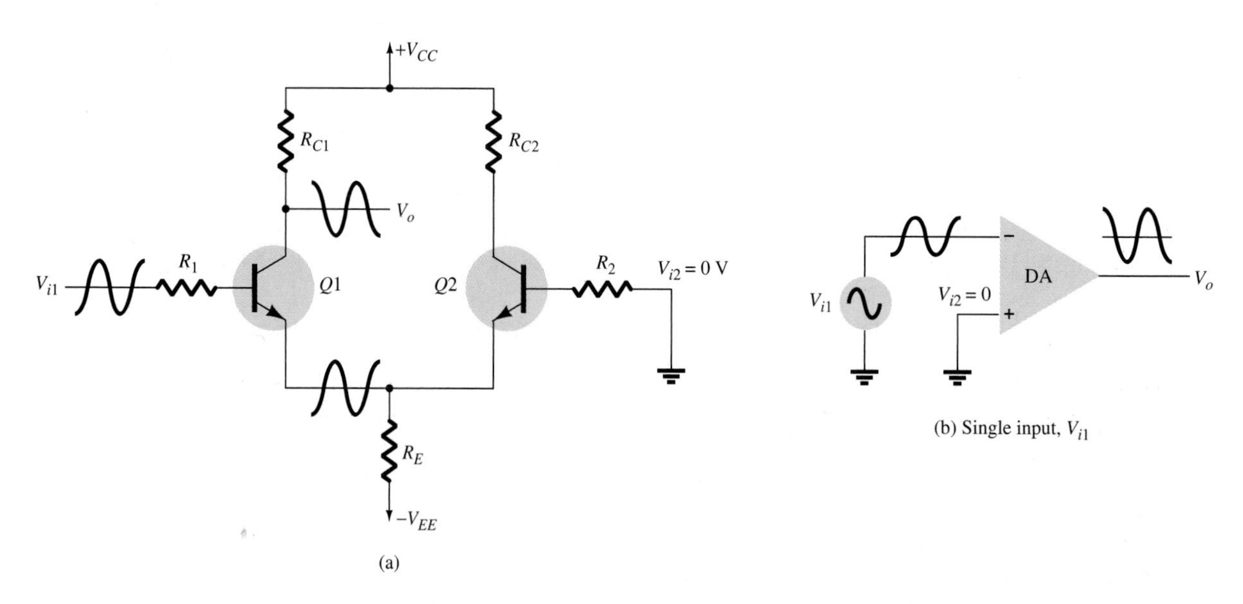

FIG. 11.4
Inverting operation of differential amplifier: (a) circuit; (b) block symbol.

signal being in phase with the input (V_{i2}). Thus the input V_{i1} can be marked as inverting (−), and that at V_{i2} as noninverting (+).

Common-Mode Rejection

A most important feature of differential amplifiers is their common-mode rejection: the ability to amplify opposite-polarity signals at the inputs while greatly attenuating or rejecting same-polarity inputs. Figure 11.6 shows the action of applying the same signal to both inputs and

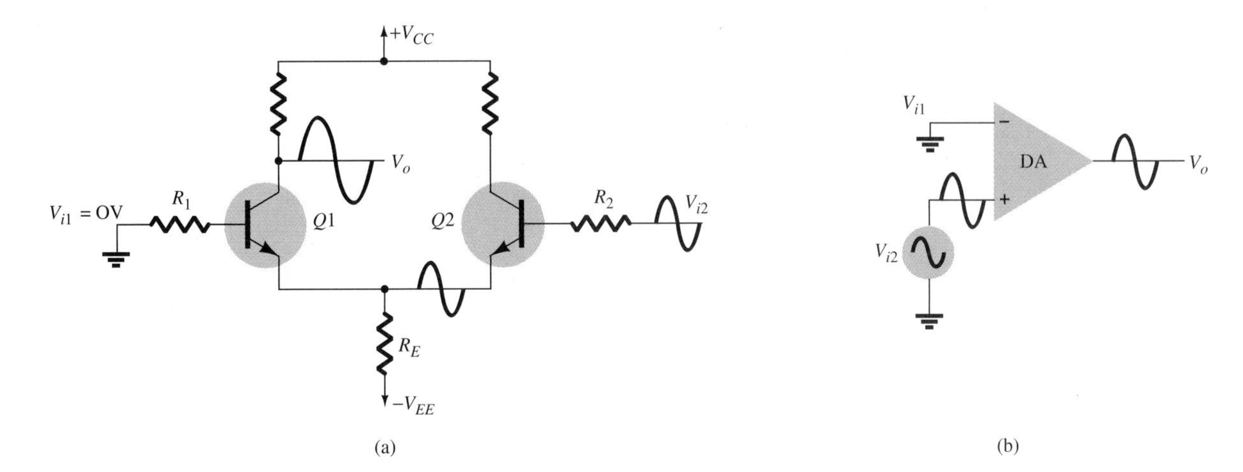

FIG. 11.5
Noninverting operation of differential amplifier: (a) circuit; (b) block symbol.

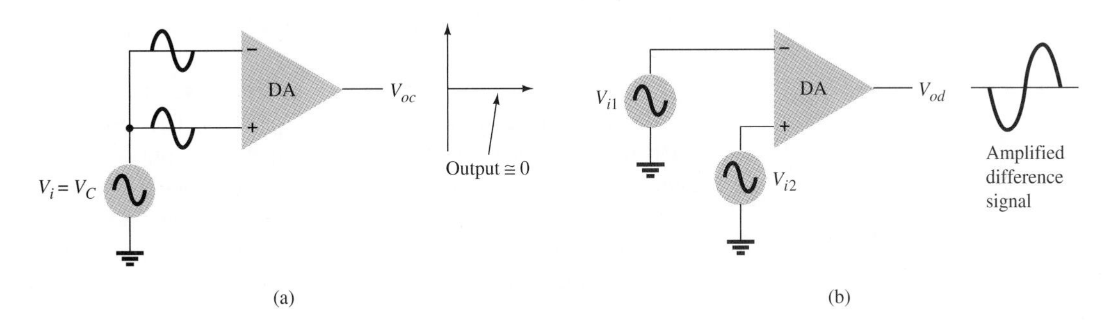

FIG. 11.6
Common-mode rejection of difference amplifier.

a difference signal between inputs. It should be apparent from Fig. 11.4(a) that applying the same input signal to transistor stages $Q1$ and $Q2$ produces opposite-polarity output signals that cancel out only if the gains of each stage are identical. In a practical circuit the gains of both sides of the difference amplifier are close enough so that only a small output signal results. For opposite-polarity inputs the amplified signals due to both inputs add at each output and result in a large amplified output signal.

The *common-mode rejection* (CMR) of a differential amplifier circuit is defined as the ratio of the differential gain to common gain, which is typically a very large number:

$$\text{common-mode rejection ratio (CMRR)} \equiv \frac{A_d}{A_c} \qquad \textbf{(11.1a)}$$

The common-mode rejection is generally given in decibels.

$$\text{CMR} = 20 \log \frac{A_d}{A_c} \text{ dB} \qquad \textbf{(11.1b)}$$

EXAMPLE 11.1 Calculate the CMR (in decibels) for a CMRR of 10,000.

Solution:

$$\text{CMR} = 20 \log \frac{A_d}{A_c} \text{ dB} = 20 \log(10{,}000) \text{ dB} = 20(4) \text{ dB} = \mathbf{80 \text{ dB}}$$

EXAMPLE 11.2 Calculate the ratio A_d/A_c for an amplifier with CMR = 120 dB.

Solution:

$$120 \text{ dB} = 20 \log \frac{A_d}{A_c} \text{ dB}$$

$$\frac{A_d}{A_c} = \text{antilog} \frac{120}{20} = \text{antilog } 6 = \mathbf{10^6}$$

A better version of differential amplifier uses a constant-current source instead of a large resistor, R_E (Fig. 11.7). The resulting value of CMR for this circuit version is quite good, being typically 90 to 100 dB. The action of constant-current stage $Q3$ is to provide a flexible adjustment of dc bias currents while acting as an open circuit (or high resistance) to an ac signal. The Zener diode $D1$ and resistor R_2 set the fixed value of emitter and collector current, which is held constant so that the

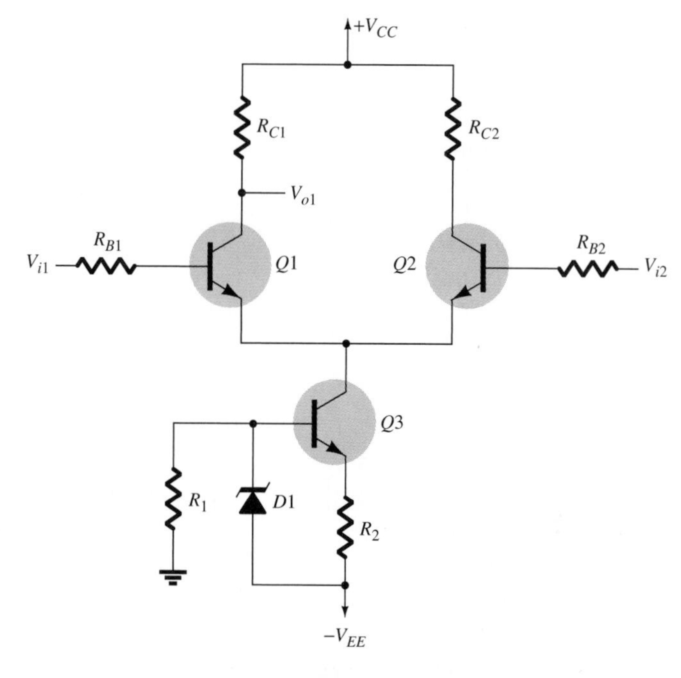

FIG. 11.7
Differential-amplifier circuit with constant-current source.

ac signals effectively "see" an open circuit or high impedance looking into stage $Q3$.

Op-Amp

A practical example of a differential amplifier is the 347 op-amp in Fig. 11.8. As shown in the symbol in Fig. 11.8(a), the differential-amplifier unit has two inputs, with one input inverting and the other noninverting. Two supply voltages are used to provide $+V$ and $-V$. The IC pin numbers are circled, the 347 being a 14-pin unit. One op-amp of the four in the quad op-amp IC is detailed in Fig. 11.8(b). A number of features in this circuit are superior to those in Fig. 11.7. The present circuit has JFET transistors at the input to provide high input resistance, current sources to provide large voltage gain, and complementary output bipolar transistors to provide low output resistance.

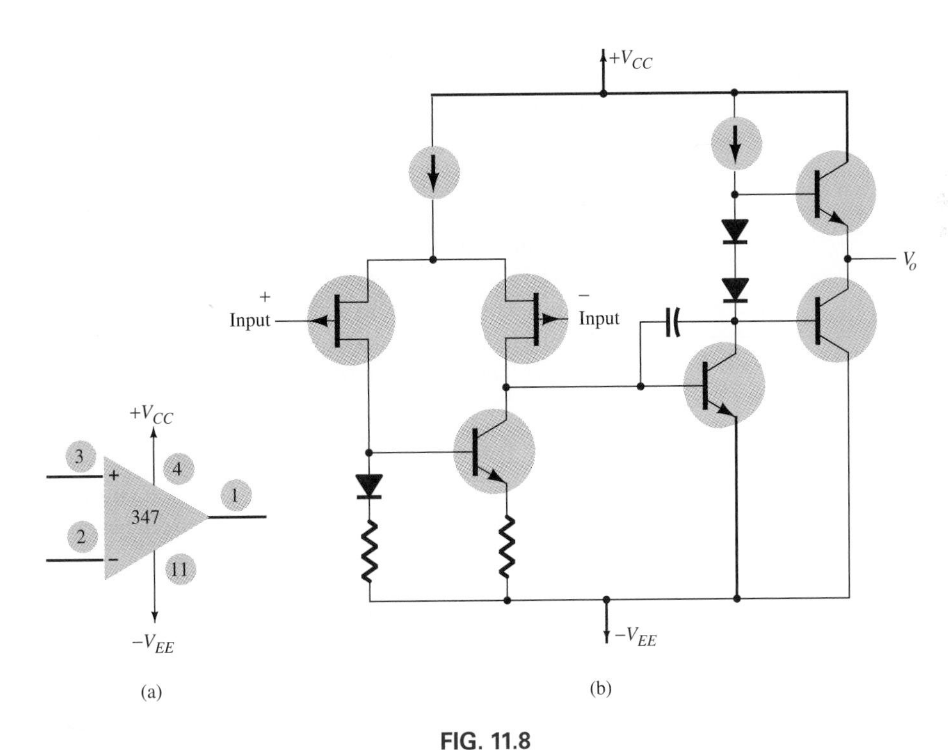

FIG. 11.8
347 op-amp: (a) symbol (one of four op-amps in IC); (b) circuit.

As an example, the input resistance (R_{in}) is typically 10^{12} Ω, the large signal voltage gain (A_{vol}) is typically 100 V/mV or 100,000 (100 dB), and the common-mode rejection is typically 100 dB.

An example of an amplifier with gain fixed at 100 is shown in Fig. 11.9. The amplifier gain is set by the resistor elements R_i and R_o* as long

*Theoretical analysis of op-amp gain for various circuit configurations is covered more fully in Chapter 13.

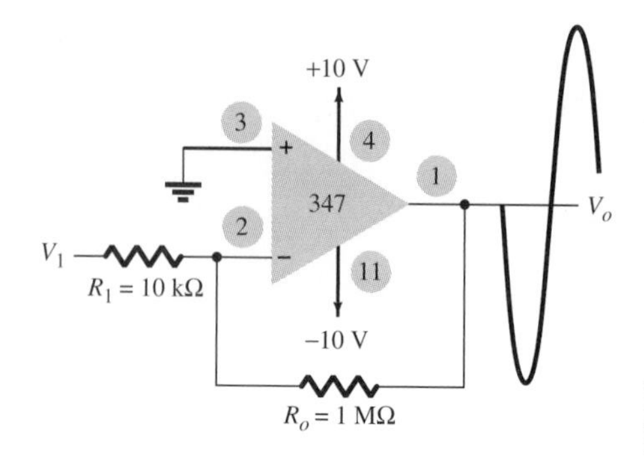

$$A_v = \frac{V_o}{V_1} = -\frac{R_o}{R_1} = -\frac{1\ \text{M}\Omega}{10\ \text{k}\Omega} = -100$$

FIG. 11.9
Constant-gain amplifier using 347 op-amp.

as that value is about 100 times less than A_{vol} of the op-amp. The output voltage can vary up to 1 to 2 V of the supply voltage, around ±8 V using the ±10-V supply as shown in Fig. 11.9.

$$A_v = \frac{V_o}{V_1} = -\frac{R_o}{R_1} \tag{11.2}$$

EXAMPLE 11.3 Determine the output voltage for the amplifier in Fig. 11.9 with $V_1 = 12$ mV, $R_o = 1.5$ MΩ, and $R_1 = 10$ kΩ.

Solution: Using Eq. (11.2), we have

$$A_v = \frac{V_o}{V_1} = -\frac{R_o}{R_1} = -\frac{1.5 \times 10^6}{10 \times 10^3} = -150$$

The output voltage is then

$$V_o = A_v V_1 = (-150)(12\ \text{mV}) = \mathbf{1.8\ V}$$

[As long as V_o is less than about ±8 V using the ±10-V supply voltages, the output varies as determined using Eq. (11.2).]

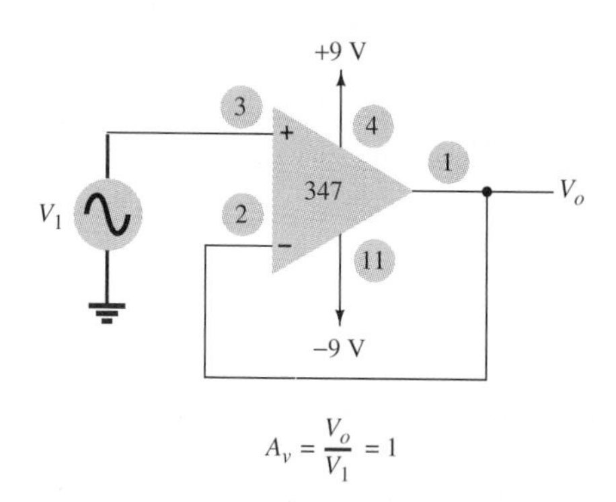

$$A_v = \frac{V_o}{V_1} = 1$$

FIG. 11.10
Unity-gain amplifier using 347 op-amp.

Another connection using an op-amp is shown in Fig. 11.10. This unity-gain amplifier has a noninverting gain of 1, providing very high input resistance and low output resistance. Essentially the circuit allows coupling of a signal from a high-resistance transducer so that the signal is available to other electronic circuits from a low-resistance source while not loading the transducer.

Multistage Op-Amps

A number of op-amp stages can be used in a variety of multistage arrangements. Figure 11.11 provides the pin connections for the 347 quad

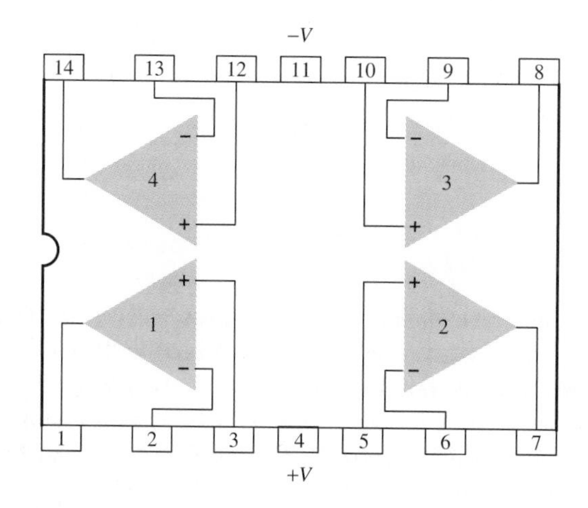

FIG. 11.11
Pin connections of 347 op-amp.

op-amp IC. Applying the supply voltages to pins 4 and 11 powers all four stages inside the single IC package.

A connection showing two op-amps in a multistage arrangement is provided in Fig. 11.12. The first stage has a gain of

$$A_{v1} = \frac{1\ \text{M}\Omega}{10\ \text{k}\Omega} = -100$$

The voltage gain of stage 2 is

$$A_{v2} = -\frac{2\ \text{M}\Omega}{10\ \text{k}\Omega} = -200$$

The overall amplifier gain is then

$$A_v = \frac{V_o}{V_1} = A_{v1}A_{v2} = (-100)(-200) = +20{,}000$$

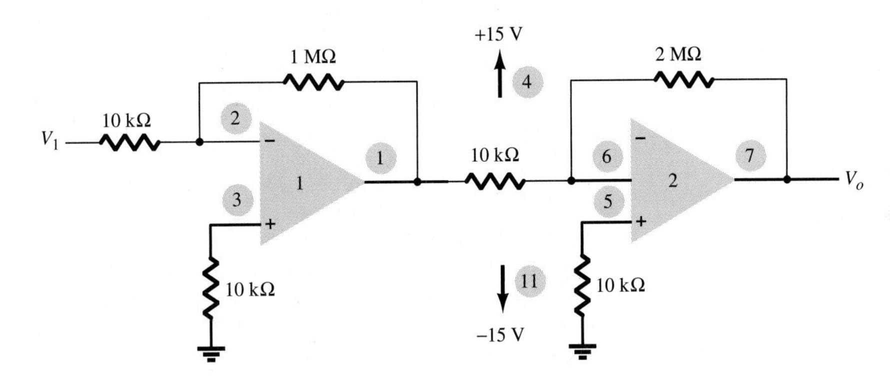

FIG. 11.12
Multistage amplifier using 347 op-amp.

EXAMPLE 11.4 Calculate the output voltage of the circuit in Fig. 11.12 for an input of 18 μV.

Solution: The output voltage is

$$V_o = A_vV_1 = (+20{,}000)(18 \times 10^{-6}) = \mathbf{0.36\ V}$$

11.2 LOGARITHMS AND DECIBEL MEASUREMENTS

In addition to the usual linear scale used to measure distance, length, or as a scale on a graph, scientific measurements also use a logarithmic scale. The logarithmic scale is particularly important in measurements covering a very wide range, such as sound and light measurements. A logarithmic measurement may be made using either base 10 (common) or base e (natural) logarithmic numbers. These two bases differ only by a fixed constant, as shown. A few examples of linear and logarithmic numbers are provided in Table 11.1. A number of observations can readily be made.

TABLE 11.1
Sample of linear and logarithmic numbers

Number, x	$\log_{10}(x)$	$\log_e(x)$
1	0.000	0.000
2	0.301	0.693
3	0.477	1.099
4	0.602	1.386
5	0.699	1.609
10	1.000	2.303
20	1.301	2.996
30	1.477	3.401
40	1.602	3.689
50	1.699	3.912
100	2.000	4.605
200	2.301	5.298
300	2.477	5.704
400	2.602	5.991
500	2.699	6.215
1000	3.000	6.908

$\log_{10}(10^n) = n$ e.g., $\log_{10}(100) = \log_{10}(10^2) = 2$

$\log_{10}(m \times 10^n) = \log_{10}(m) + n$ e.g., $\log_{10}(2000) = \log_{10}(2) + \log(1000)$
$= 0.301 + 3 = 3.301$

The ratio $\log_e(x)/\log_{10}(x)$ is a constant value of 2.3 (e.g., $\log_e(2)/\log_{10}(2) = 0.693/0.301 = 2.3$, and $\log_e(500)/\log_{10}(500) = 6.215/2.699 = 2.3$.

The decibel unit is derived from the logarithmic function. A number of different decibel definitions are used, most of them directed to a particular type or area of measurement. For our study of amplifier gain, the definition of voltage gain is

$$A_v = \frac{V_o}{V_i} \quad \text{(linear voltage gain)}$$

$$A_v\,(\text{dB}) = 20 \log_{10}\left(\frac{V_o}{V_i}\right) \quad \text{(voltage gain in decibels)}$$

EXAMPLE 11.5 Calculate the voltage gain (V_o/V_i) and dB voltage gain for an amplifier with input V_i = 2.5 mV (rms) and output 3.4 V (rms).

Solution:

$$A_v = \frac{V_o}{V_i} = \frac{3.4\text{ V (rms)}}{2.5\text{ mV (rms)}} = 1.36 \times 10^3 = \mathbf{1360}$$

$$A_v(\text{dB}) = 20 \log_{10}\left(\frac{V_o}{V_i}\right) = 20 \log_{10}(1.36 \times 10^3) = \mathbf{62.67\ dB}$$

EXAMPLE 11.6 What output voltage results from applying an input signal of 2 mV to an amplifier with a 55-dB voltage gain?

Solution:

$$A_v\,(\text{dB}) = 20 \log_{10}\left(\frac{V_o}{V_i}\right)$$

$$55\text{ dB} = 20 \log_{10}\left(\frac{V_o}{V_i}\right)$$

$$\log_{10}\left(\frac{V_o}{V_i}\right) = \frac{55}{20} = 2.75$$

$$\frac{V_o}{V_i} = \text{antilog}(2.75) = 5.62 \times 10^2 = 562.3$$

$$V_o = 562.3V_i = 562.3(2\text{ mV}) = \mathbf{1.12\ V}$$

[An antilog may be obtained on a calculator using the 10^x function, for example, antilog(2.75) = $10^{2.75}$ = 562.34.]

EXAMPLE 11.7 What input voltage is needed to provide an output voltage of 2 V with an amplifier having a 72-dB voltage gain?

Solution:

$$A_v\text{ (dB)} = 20\log_{10}\left(\frac{V_o}{V_i}\right)$$

$$\frac{V_o}{V_i} = \text{antilog}\left(\frac{72}{20}\right) = 3981.1$$

$$V_i = \frac{V_o}{3981.1} = \frac{2\text{ V}}{3981.1} = \mathbf{0.5\text{ mV}}$$

Whereas amplifier gains in series are multiplied, the same gains in decibel units are added, which is one reason why decibel measurements are often easier to handle. That is,

$$A_{vT} = A_{v1}A_{v2}A_{v3}\cdot\cdot\cdot$$

$$A_{vT}\text{(dB)} = A_{v1}\text{(dB)} + A_{v2}\text{(dB)} + A_{v3}\text{(dB)} + \cdot\cdot\cdot$$

EXAMPLE 11.8 Calculate the overall voltage gain and decibel voltage gain for the circuit in Fig. 11.13.

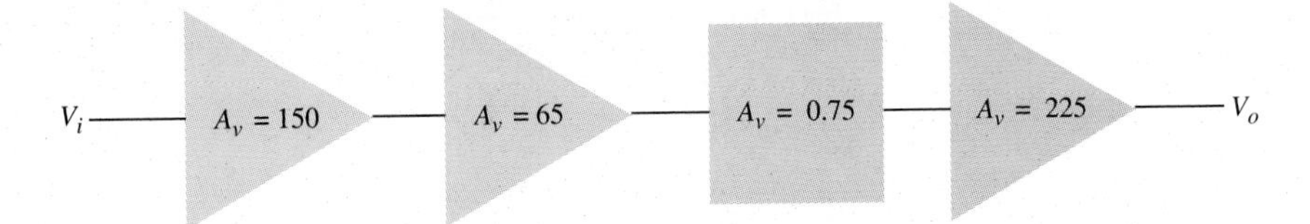

FIG. 11.13

Solution:

$$A_{vT} = A_{v1}\cdot A_{v2}\cdot A_{v3}\cdot A_{v4} = (150)(65)(0.75)(225) = \mathbf{1{,}645{,}312.5}$$

$$\begin{aligned} A_v\text{(dB)} &= 20\log_{10}A_{v1} + 20\log_{10}A_{v2} + 20\log_{10}A_{v3} + 20\log_{10}A_{v4} \\ &= 20\log_{10}(150) + 20\log_{10}(65) + 20\log_{10}(0.75) + 20\log_{10}(225) \\ &= 43.52\text{ dB} + 36.26\text{ dB} - 2.5\text{ dB} + 47.04\text{ dB} = 124.32\text{ dB} \end{aligned}$$

[Note that $20\log_{10}A_{vT} = 20\log_{10}(1{,}645{,}312.5) = \mathbf{124.32\text{ dB}}$.]

EXAMPLE 11.9 What overall decibel gain is provided by the circuit in Fig. 11.14?

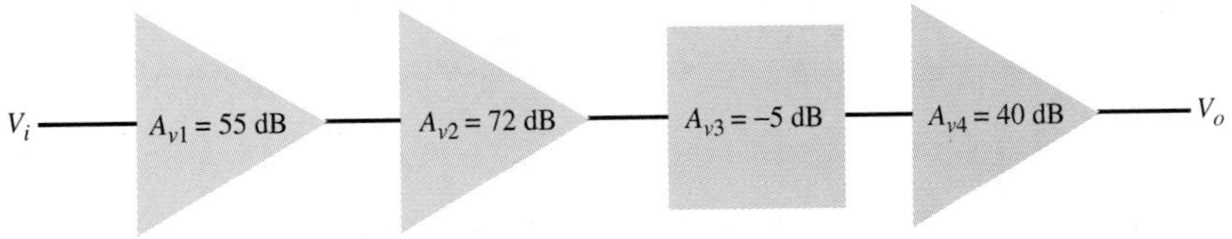

FIG. 11.14

Solution: The overall voltage gain is

$$A_{vT}(\text{dB}) = 55\text{ dB} + 72\text{ dB} - 5\text{ dB} + 40\text{ dB} = \mathbf{162\text{ dB}}$$

11.3 FREQUENCY CONSIDERATIONS

The frequency of the signal applied to an amplifier circuit can have considerable effect on the resulting gain of the circuit. Up to now the signal frequency has not been included in the circuit operation, or the frequency has had no detrimental effect—this condition being the midfrequency range. The gain of a single amplifier stage decreases at high frequency—the value at which the gain decrease begins is different for

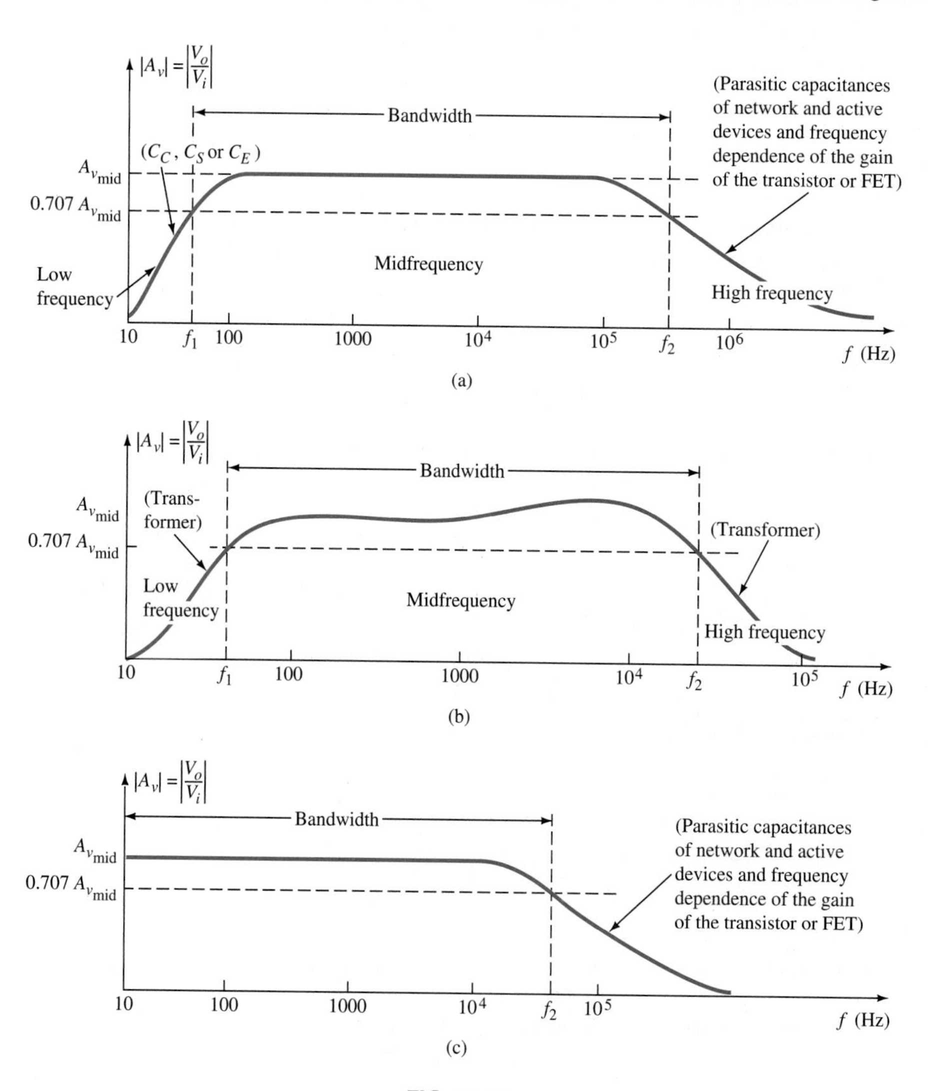

FIG. 11.15
Gain versus frequency for: (a) RC-coupled amplifiers; (b) transformer-coupled amplifiers; (c) direct-coupled amplifiers.

different transistors and circuits. When stages are coupled by capacitor or transformer elements, the overall circuit gain is greatly affected at both low and high frequencies. An example of how gain varies with frequency for a few types of coupling methods is shown in Fig. 11.15. Note that the horizontal axis is a logarithmic scale to permit a plot extending from the low- to the high-frequency regions. For each plot, a low-, high-, and midfrequency region has been defined. In addition, the primary reasons for the decrease in gain at low and high frequencies have also been indicated within the parentheses. For the *RC*-coupled amplifier the decrease at low frequencies is due to the increasing reactance of the coupling capacitor, while its upper frequency limit is determined by either parasitic capacitive elements of the network and the active device (transistor) or the frequency dependence of the gain of the active device. Notice that while a dc-coupled amplifier such as a differential amplifier does not have decreased gain at low frequencies, it does show gain reduction at high frequencies due to stray wiring capacitance and the frequency dependence of the transistor devices themselves.

The approximately constant gain for most of the indicated frequency range is referred to as the *midfrequency*. Accepted practice is to regard the condition at which the gain is 0.707 of the midfrequency gain (A_{mid}) as the point to describe either low- or high-frequency cutoff. In decibel units the gain decreases 3 dB when it is reduced $0.707A_{\text{mid}}$, and the upper and lower 3-dB frequencies are those at which the gain has dropped 3 dB (or to 0.707 of A_{mid}). The lower 3-dB frequency (f_1) for *RC*-coupled amplifiers [Fig. 11.15(a)] is determined mainly by coupling capacitor C_C. In Fig. 11.15(a) the gain drops from its midfrequency value to $0.707A_{\text{mid}}$ (or by 3 dB) around 70 Hz. The upper 3-dB frequency (f_2) is shown to be about 500 kHz in Fig. 11.15(a) and is dependent mainly on stray capacitances and device frequency limitations. The bandwidth (BW) is then defined as the difference between the lower and upper frequencies:

$$\text{BW} = f_2 - f_1 \tag{11.3}$$

(or just f_2 for a direct-coupled amplifier).

General Low-Frequency Response (*RC* Coupling)

For the common type of capacitor coupling of amplifier stages used with FET, or BJT, amplifiers the lower cutoff frequency (f_1 or f_2) is dependent mainly on the value of the coupling capacitor (C) and the input resistance of the amplifier stage (R_i). In the midfrequency range the nonresistive components have slight effect, and so the circuit gain remains essentially constant. A signal V_s is then amplified to V_o, as shown in Fig. 11.16(a), with

$$A_{v_{\text{mid}}} = A_{\text{mid}} = \frac{V_o}{V_s}$$

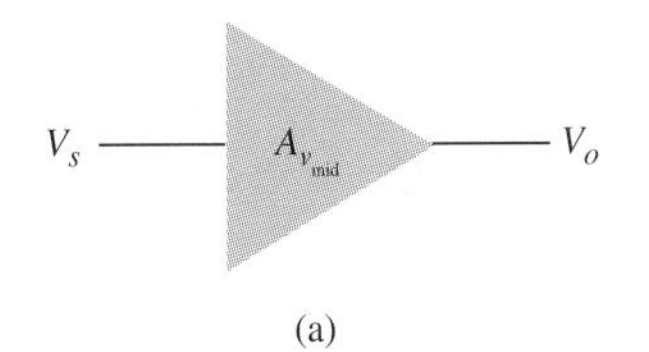

(a)

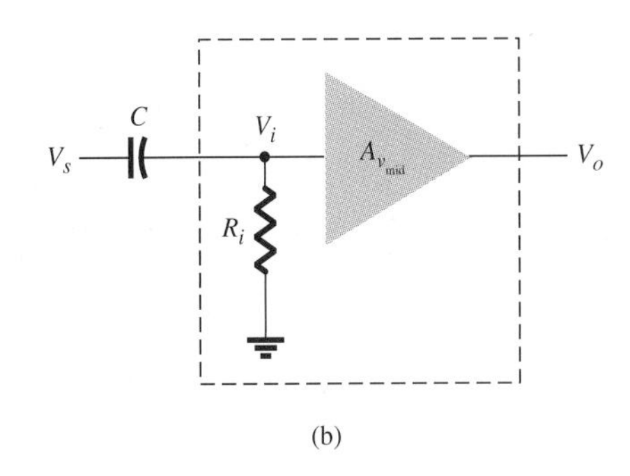

(b)

FIG. 11.16
Amplifier for general low-frequency response: (a) midfrequency; (b) low frequency.

To consider the effect of the capacitor coupling of a signal V_s into an amplifier stage, refer to Fig. 11.16(b), which shows the stage input impedance as an effective resistance R_i. The overall gain is reduced by the voltage division of impedances X_C and R_i, which is a function of the signal frequency. (At midfrequency the impedance X_C is small compared with R_i, and so $V_i \approx V_s$.) To calculate the overall gain we first consider the voltage gain (or attenuation in this case) between V_s and V_i. This attenuation can be obtained using the voltage-divider rule, so that

$$\frac{V_i}{V_s} = \frac{R_i}{R_i - jX_C} = \frac{R_i}{\sqrt{R_i^2 + X_C^2}} \angle -\tan^{-1}\frac{X_C}{R_i}$$

The magnitude of the attenuation factor is thus

$$\text{magnitude of } \frac{V_i}{V_s} = \frac{R_i}{\sqrt{R_i^2 + X_C^2}} = \frac{1}{\sqrt{1 + (X_C/R_i)^2}}$$

In particular, $V_i/V_s = 0.707$* when $X_C/R_i = 1$. This is a notable point to use, since the gain has decreased by 3 dB† from its midfrequency value when it is $0.707A_{\text{mid}}$.

We see then that at the frequency where $X_C = R_i$ the overall gain has dropped by 3 dB from its midfrequency value (or to $0.707A_{\text{mid}}$). Since the 3-dB point is a common one to establish where the gain decrease begins to be considerable, we can identify the lower frequency of the overall amplifier circuit, or the lower 3-dB frequency, as

$$f_1 = f_L = \frac{1}{2\pi R_i C} \tag{11.4}$$

For example, a BJT amplifier having $R_i = 2\ \text{k}\Omega$ and a coupling capacitor $C = 0.5\ \mu\text{F}$ would have a lower cutoff frequency of

$$f_L = \frac{1}{2\pi R_i C} = \frac{1}{6.28(2 \times 10^3)(0.5 \times 10^{-6})} = 159 \cong 160\ \text{Hz}$$

To reduce the lower cutoff frequency to, say, 20 Hz would require using a capacitor of

$$C = \frac{1}{2\pi f R_i} = \frac{1}{6.28(20)(2 \times 10^3)} = 3.98 \times 10^{-6} \cong 4\ \mu\text{F}$$

General High-Frequency Response (*RC* Coupling)

A number of different capacitive effects cause the gain to decrease at frequencies above midfrequency. Parasitic capacitances occur between input or output terminals and ground, and between input and output ter-

---

*$1/\sqrt{1 + (1)^2} = 1/\sqrt{2} = 0.707$.

†$20 \log_{10}(V_i/V_s)$ dB $= 20 \log(1/\sqrt{2})$ dB $= -20 \log \sqrt{2}$ dB $= -20(0.151)$ dB $= -3.01$ dB $\cong -3$ dB.

minals. These capacitances are due to construction and connection of the circuit elements. In addition there is a stray wiring capacitance between amplifier stages due to connecting wires. There is also a decrease in active device gain as frequency increases, which will be considered later.

The effect of parasitic and stray wiring capacitance on the amplifier frequency response is to decrease the overall gain as the frequency increases. Figure 11.17(a) shows the parasitic capacitances of interest at

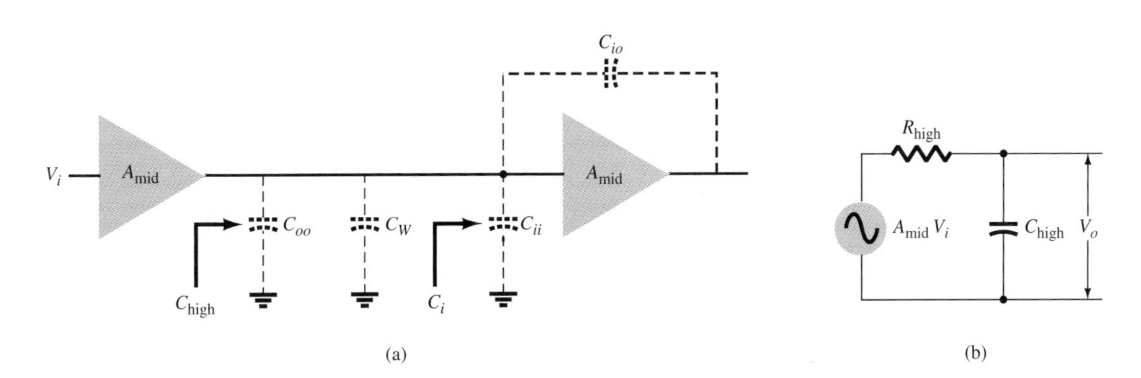

FIG. 11.17
Amplifier circuit for general high-frequency response: (a) distributed parasitic and stray capacitance; (b) effective high-frequency capacitance.

high frequencies—these being a capacitance from input to ground, C_{ii}; a capacitance from output to ground, C_{oo}; and a capacitance between input and output terminals, C_{io}. There is also stray wiring capacitance, C_W. The total capacitive components result in an effective high-frequency capacitance, C_{high}. Figure 11.17(b) shows the output of one amplifier driving a second stage, with the midband output voltage source $A_{mid}V_i$ and output resistance R_{high} driving the effective capacitance C_{high}. The voltage applied to the second amplifier stage is reduced from what it would be at midfrequency by the voltage-divider action of capacitor C_{high} and resistor R_{high} as follows:

$$\frac{V_o}{A_{mid}V_i} = \frac{-jX_{C_{high}}}{R_{high} - jX_{C_{high}}} = \frac{X_{C_{high}}}{\sqrt{X^2_{C_{high}} + R^2_{high}}} \angle -90 - \tan^{-1}\frac{X_{C_{high}}}{R_{high}}$$

In particular, the magnitude of output-to-input voltage $V_o/V_i = 0.707A_{mid}$ when $R_{high} = X_{C_{high}}$, at which frequency the gain is down by 3 dB from the midfrequency gain. That is, the upper 3-dB cutoff frequency due to stray wiring and parasitic capacitance effects is

$$f_2 = f_H = \frac{1}{2\pi R_{high}C_{high}} \qquad \textbf{(11.5)}$$

EXAMPLE 11.10 Calculate the upper cutoff frequency of an amplifier having

$$R_{\text{high}} = 3.3\ \text{k}\Omega \qquad \text{and} \qquad C_{\text{high}} = 220\ \text{pF}$$

Solution: Using Eq. (11.5), we obtain

$$f_2 = f_{\text{high}} = \frac{1}{2\pi(3.3 \times 10^3)(220 \times 10^{-12})} = \mathbf{219.2\ kHz}$$

Determination of C_{high}

To calculate the effective value of capacitance that reduces the gain at higher frequencies, we first shall consider how the parasitic capacitance between input and output, C_{io}, can be reflected to the input terminals. This reflection to the input is due to what is called the *Miller effect*, and the resulting capacitance value is often called the *Miller capacitance*.

Miller Effect

Consider the simplified circuit in Fig. 11.18, which shows the parasitic capacitance from input to ground, C_{ii}, and that between input and output,

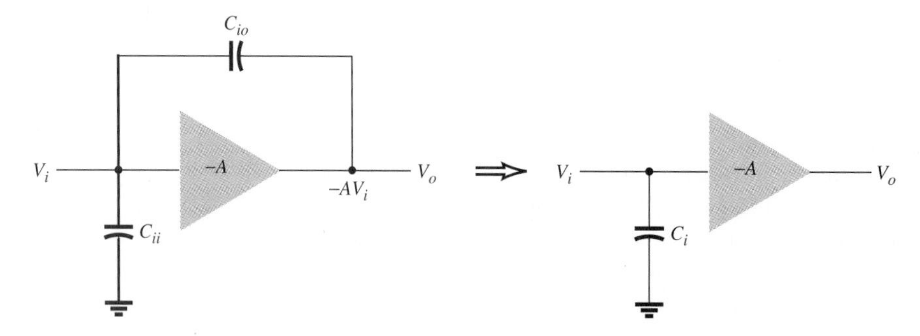

FIG. 11.18
Circuit to calculate effective input capacitance, C_i.

C_{io}. The input voltage is V_i, and the output, AV_i (where $A = A_{\text{mid}}$). These two capacitances can be equated to one equivalent capacitance C_i as follows:

$$I_i = I_{ii} + I_{io}$$
$$j\omega C_i V_i = j\omega C_{ii} V_i + (V_i - AV_i) j\omega C_{io}$$

$$C_i = C_{ii} + \underbrace{(1 - A_v)C_{io}}_{\text{Miller capacitance}} \qquad \textbf{(11.6a)}$$

Since $A_v = -g_m R_L$,

$$C_i = C_{ii} + (1 + g_m R_L)C_{io} \quad \textbf{(11.6b)}$$

Then, including stray wiring between stages and parasitic capacitance of output terminal to ground, we calculate a total high-frequency equivalent capacitance:

$$C_{\text{high}} = C_W + C_{oo} + C_i \quad \textbf{(11.7)}$$

[using C_i from Eq. (11.6a) or (11.6b)]. For example, in an *RC*-coupled amplifier having a midfrequency gain of -55 and an output resistance of 5 kΩ, the parasitic capacitances are determined to be $C_{ii} = 5$ pF, $C_{io} =$ 3 pF, and $C_{oo} = 6$ pF, with 15 pF due to stray wiring. Calculate the upper 3-dB frequency f_H.

Solution: Calculating C_i first, we have

$$C_i = C_{ii} + (1 - A_v)C_{io} = [5 + (1 + 55)3] \text{ pF} = 173 \text{ pF}$$

The equivalent value of C_{high} is then

$$C_{\text{high}} = C_W + C_{oo} + C_i = (15 + 6 + 173) \text{ pF} = 194 \text{ pF}$$

(Note that the Miller capacitance value of 168 pF is most of the total equivalent value.) Since $R_{\text{high}} = 5$ kΩ, the upper cutoff frequency is

$$f_H = \frac{1}{2\pi R_{\text{high}} C_{\text{high}}} = \frac{1}{6.28(5 \times 10^3)(194 \times 10^{-12})} = 164.2 \text{ kHz}$$

EXAMPLE 11.11 Calculate f_H for an amplifier having $R_{\text{high}} = 2.7$ kΩ, $C_{io} = 3.5$ pF, and a gain of $A_v = -85$ (ignore other circuit effects).

Solution:

$$C_{\text{high}} = C_i = (1 - A_v)C_{io} = (1 + 85)(3.5 \text{ pF}) = 301 \text{ pF}$$

$$f_H = \frac{1}{2\pi R_{\text{high}} C_{\text{high}}}$$

$$= \frac{1}{2\pi(2.7 \times 10^3)(301 \times 10^{-12})} = \mathbf{195.8 \text{ kHz}}$$

Asymptotic Gain–Frequency Plot

A simplified approach to plotting a gain versus log of frequency curve uses straight-line asymptotes rather than exact curve values, as we shall see. The magnitude of voltage gain as a function of frequency can be expressed as

$$|A_v| = \frac{K_v}{\sqrt{1 + (f/f_2)^2}}$$

where f_2 is the upper cutoff frequency (upper 3-dB frequency). In decibel units

$$A_{dB}(f) = 20\log|A_v(f)|\text{dB} = 20\log\frac{K_v}{\sqrt{1+(f/f_2)^2}}\text{ dB}$$

$$= 20\log K_v - 20\log\sqrt{1+\left(\frac{f}{f_2}\right)^2}\text{ dB}$$

$$= K_{dB} - 20\log\sqrt{1+\left(\frac{f}{f_2}\right)^2}\text{ dB} \qquad \textbf{(11.8)}$$

To plot this curve of gain versus log of frequency consider the value of A_{dB} for very small and then very large frequency values. For frequencies much less than f_2 ($f << f_2$), Eq. (11.8) reduces to

$$A_{dB}(f) \cong K_{dB} - 20\log(1)\text{ dB} = K_{dB} \qquad \textbf{(11.9a)}$$

Thus, at low frequencies the gain is constant, of value K_{dB}. At high frequencies ($f >> f_2$) Eq. (11.8) can be expressed as

$$A_{dB}(f) = K_{dB} - 20\log\left(\frac{f}{f_2}\right)\text{ dB} \qquad \textbf{(11.9b)}$$

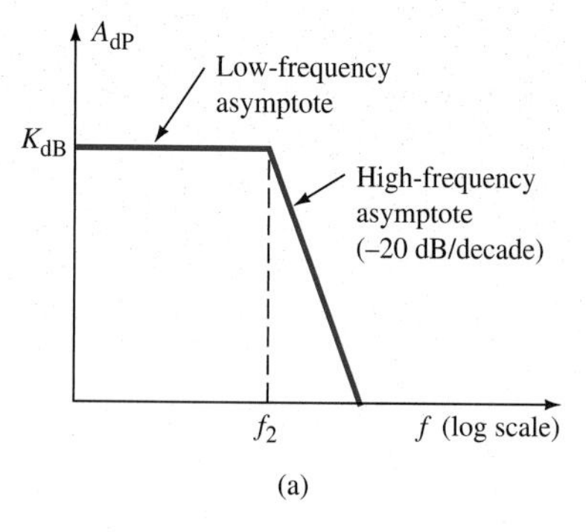

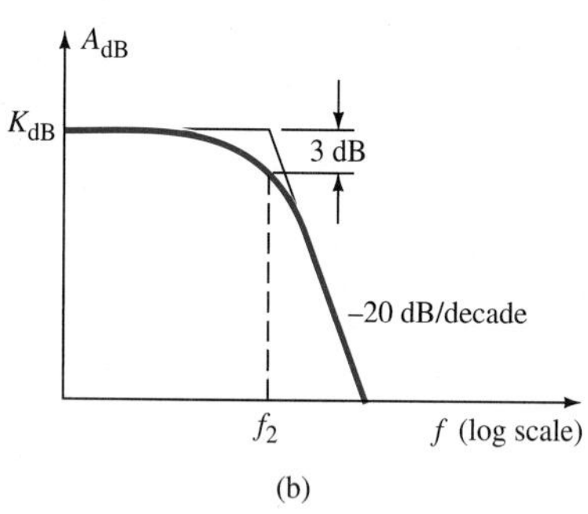

FIG. 11.19
Gain versus log frequency plots: (a) asymptotic; (b) actual.

Consider, for example, the gain decrease from K_{dB} when $f = 10f_2$ (a decade larger frequency than the cutoff value f_2). The gain is then

$$A_{dB}(10f_2) = K_{dB} - 20\log\left(\frac{10f_2}{f_2}\right)\text{ dB} = K_{dB} - 20\text{ dB}$$

or 20 dB down from the low-frequency gain value.* The curve in Eq. (11.8) is plotted in Fig. 11.19(a), showing the asymptotic straight-line gain of K decibels at low frequencies up to a high-frequency break point and then decreasing at a rate of -20 dB per decade from the break point (f_2) frequency.

These asymptotic straight lines are then easily drawn to approximate the actual gain versus log frequency curve in Eq. (11.8). A more exact curve is then easily obtained from the asymptotic plot by noting that the actual gain is 3 dB less than the asymptotic value at frequency f_2[†], as shown in the plot in Fig. 11.19(b).

While the gain of the amplifier decreases at frequencies above f_2, the output signal, originally in phase with the input but of opposite relative

*The slope of the plotted curve for Eq. (11.8) on the log frequency scale is -20 dB per decade or -6 dB per octave, where a decade is a 10:1 frequency change, and an octave is a 2:1 frequency change.

[†]At $f = f_2$ the gain is not K_{dB} as on the asymptotic plot in Fig. 11.19(a) but is reduced from K_{dB} by 3 dB:

$$A_{dB} = K_{dB} - 20\log\sqrt{1+(f/f_2)^2}\text{ dB} = K_{dB} - 20\log\sqrt{2}\text{ dB} \cong K_{dB} - 3\text{ dB}$$

polarity, is shifted in phase from that value as given by the phase angle expression

$$\theta = -\tan^{-1}\left(\frac{f}{f_2}\right)$$

At low frequencies $\theta \cong 0$, and little phase shift (from the 0° midfrequency phase shift) occurs. At frequencies much larger than f_2 the phase shift approaches $-90°$, the phase shift being $-45°$ at f_2. A simple asymptotic plot of phase versus log frequency is shown in Fig. 11.20 as a straight line of slope $-45°$ per decade.

Although not shown here, the *RC* amplifier response is 45° at the low cutoff frequency f_1, with the phase shift increasing to 90° near dc.

The gain–bandwidth relations of an *RC*-coupled amplifier can thus be specified as shown in Fig. 11.21 with the midfrequency gain constant between frequencies f_1 and f_2 and bandwidth equal to the difference, $f_2 - f_1$. Recall that the gain decrease at low frequencies is essentially due to the coupling capacitor, while the gain reduction above f_2 is due to parasitic and stray wiring capacitances. The gain in the frequency band from f_1 to f_2 is the midfrequency gain set by the circuit resistive components and the active device parameters.

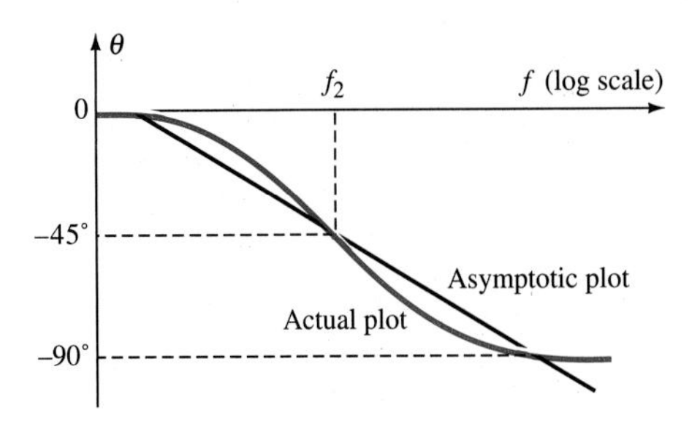

FIG. 11.20
Phase shift versus log frequency.

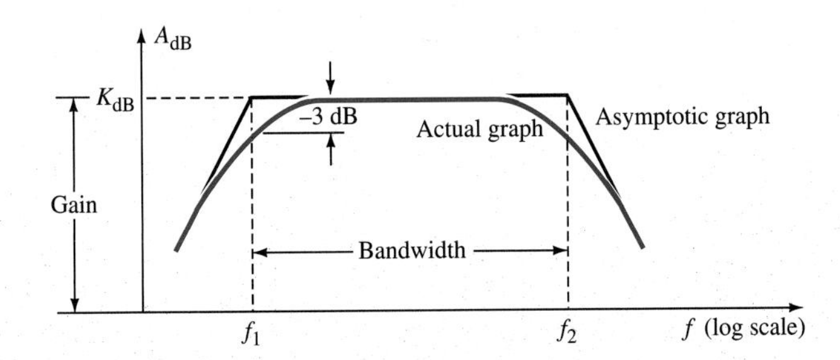

FIG. 11.21
Gain–bandwidth relations for RC-coupled amplifiers.

It is interesting to note that the gain and bandwidth for a particular amplifier circuit cannot both be increased by changing resistive component values and that the gain–bandwidth product is essentially constant for the circuit. Since f_2 is usually much larger than f_1, we can approximately express the gain–bandwidth product as

$$\text{gain–bandwidth} \cong K_v f_2 = (g_m R_{\text{high}}) \frac{1}{2\pi R_{\text{high}} C_{\text{high}}} = \frac{g_m}{2\pi C_{\text{high}}} = f_o$$

where $R_L = R_{\text{high}}$, and the resulting gain–bandwidth product is seen to be determined by the device ideal gain g_m and the circuit high-frequency capacitance C_{high} (which is essentially fixed).

This relationship can be demonstrated graphically by extending the gain–log frequency plot until the gain goes to unity (0 dB), as shown in Fig. 11.22. Since the gain decreases at the rate of 20 dB per decade until the frequency f_o (frequency at 0 dB), the asymptote is completely specified if f_o is known. If the gain is K_{dB} and the upper cutoff frequency is f_2

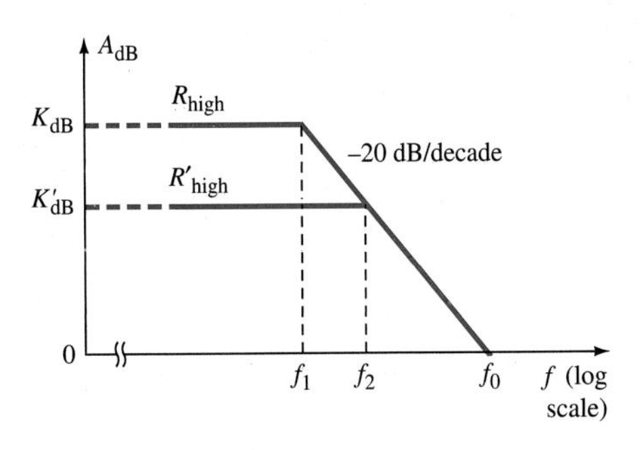

FIG. 11.22
Gain–bandwidth plot showing upper frequency, f_o.

for a circuit high resistance of R_{high}, changing the value of R_{high} to increase f_2, for example, will result in reducing the gain to K_{dB}, as shown in Fig. 11.22. We see that an increase in bandwidth (f_2 greater) was accompanied by a decreased gain—the gain–bandwidth product remained the same value. For all amplifiers having a high-frequency asymptote of -20 dB per decade the gain–bandwidth product can be shown to be the value of f_o, which is then an important figure of merit of the circuit. In practice the designer may trade gain for bandwidth (or vice versa) depending on the particular circuit requirements, with the limit of trade being set by the value f_o.

Practical *RC*-Coupled Amplifier Circuits

Having covered the general action of *RC*-coupled circuits as a function of frequency, it would now be helpful to consider specific *RC*-coupled amplifiers using FETs or BJTs. The general concepts apply to all these circuits, but specific nomenclature will now be used for each type of device.

JFET Amplifier

A practical JFET amplifier is shown in Fig. 11.23. The low-frequency cutoff can easily be obtained using Eq. (11.4), since $R_i = R_g = 1\ \text{M}\Omega$,

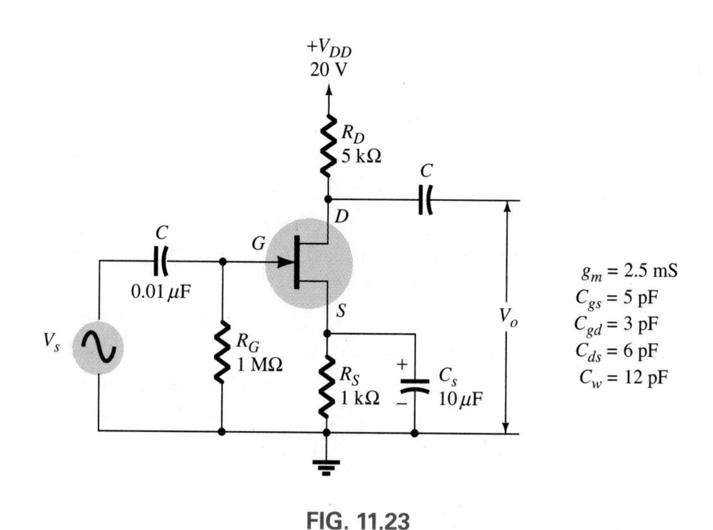

FIG. 11.23
Practical JFET amplifier with capacitor coupling.

$$f_L = \frac{1}{2\pi R_i C} = \frac{1}{6.28(1 \times 10^6)(0.01 \times 10^{-6})} \cong 16\ \text{Hz}$$

The low-frequency gain is also affected by the bypass action of capacitor C_S on bias resistor R_S, this action being omitted here, as it is usually not the primary effect.

At midfrequency the gain is calculated to be

$$A_{\text{mid}} = -g_m R_L = -(2500 \times 10^{-6})(5 \times 10^3) = -12.5$$

where $R_L = R_D = 5\ \text{k}\Omega$.

To determine the upper frequency cutoff point we first calculate C_{high} using Eq. (11.6a) and then Eq. (11.7). Using the JFET device nomenclature gives us

$$C_i = C_{gs} + (1 - A_{\text{mid}})\, C_{gd} = [5 + (1 + 12.5)3]\ \text{pF} = 45.5\ \text{pF}$$

and

$$C_{\text{high}} = C_i + C_{ds} + C_W = (45.5 + 6 + 12)\ \text{pF} = 63.5\ \text{pF}$$

Since the value of $R_{\text{high}} = R_D = 5\ \text{k}\Omega$, we can calculate the high-frequency cutoff using Eq. (11.6):

$$f_H = \frac{1}{2\pi R_{\text{high}} C_{\text{high}}} = \frac{1}{6.28(5 \times 10^3)(6.35 \times 10^{-12})} \cong 5\text{MHz}$$

In decibel units the gain at midfrequency is

$$A_{\text{dB}} = 20 \log 12.5\ \text{dB} = 22\ \text{dB}$$

and is less by 3 dB than that value at the lower 3-dB frequency of 16 Hz and also at the upper 3-dB frequency of 5 MHz. The circuit bandwidth is approximately 5 MHz, since $f_H >> f_2$. The value of R_D can be changed to increase the bandwidth with resulting lower gain, or the gain can be increased with resulting lower bandwidth, as desired.

BJT Amplifier

A practical BJT amplifier circuit is shown in Fig. 11.24. The low-frequency cutoff point can be calculated using Eq. (11.3) after determining the value of R_i:

$$R_i \cong \frac{R_1 R_2}{R_1 + R_2} = \left[\frac{10(40)}{10 + 40}\right] \text{k}\Omega = 8\ \text{k}\Omega$$

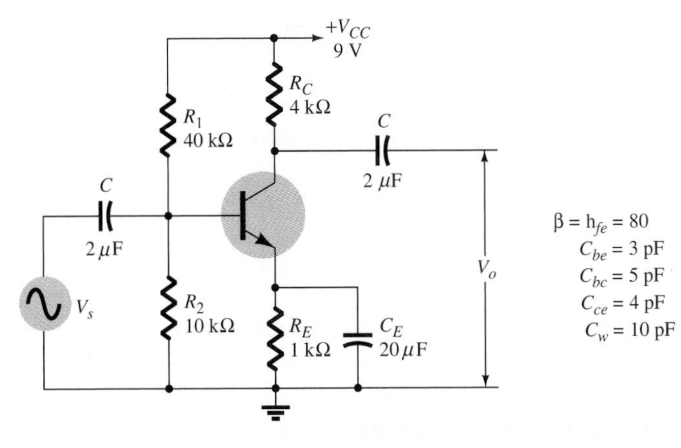

FIG. 11.24

Practical BJT amplifier with capacitor coupling.

[$h_{fe}R_E = 80(1\ \text{k}\Omega) = 80\ \text{k}\Omega >> R_i = 8\ \text{k}\Omega$ and can be ignored]. The lower 3-dB cutoff frequency is then

$$f_L = \frac{1}{2\pi R_i C} = \frac{1}{6.28(8 \times 10^3)(2 \times 10^{-6})} \cong 10\ \text{Hz}$$

To calculate the midfrequency gain we first determine the value r_i for a dc emitter current of 1.3 mA (using dc analysis) to be

$$r_i \cong 1.6\ \text{k}\Omega$$

and the midfrequency gain is then

$$A_{\text{mid}} = -\beta \frac{R_C}{r_i} = -80\left(\frac{4\ \text{k}\Omega}{1.6\ \text{k}\Omega}\right) = -200$$

The upper 3-dB cutoff frequency can now be calculated as

$$C_i = C_{be} + (1 - A_{\text{mid}})C_{bc} = [3 + 201(5)]\ \text{pF} = 1008\ \text{pF}$$
$$C_{\text{high}} = C_i + C_W + C_{ce} = (1008 + 10 + 4)\ \text{pF} = 1022\ \text{pF}$$

so that

$$f_H = \frac{1}{2\pi R_{\text{high}} C_{\text{high}}} = \frac{1}{6.28(4 \times 10^3)(1022 \times 10^{-12})} \cong 39\ \text{kHz}$$

where $R_{\text{high}} \cong R_C = 4\ \text{k}\Omega$.

The high-frequency cutoff is also limited by the variation of h_{fe} with frequency, but in this case (as in many) the circuit limits the cutoff frequency well below the device limitation. The relatively low bandwidth of about 39 kHz can of course be increased at the cost of reduced midfrequency gain from the value 200, which in decibel units is

$$A_{\text{dB}} = 20 \log A_{\text{mid}}\ \text{dB} = 20 \log 200\ \text{dB} = 46\ \text{dB}$$

An increase in the number of cascaded stages can have a pronounced effect on the frequency response. For each additional stage the upper cutoff frequency is determined primarily by the stage having the lowest upper cutoff frequency. The low-frequency cutoff is primarily determined by that stage having the highest lower cutoff frequency. Obviously, therefore, one poorly designed stage can offset an otherwise well-designed cascaded system.

The effect of increasing the number of *identical stages* is demonstrated in Fig. 11.25. In each case the upper and lower cutoff frequencies of each of the cascaded stages is identical. For a single stage the cutoff frequencies are f_1 and f_2, as indicated. For two identical stages in cascade the drop-off occurs at a faster rate, and the bandwidth is considerably decreased.

It should be noted that a decrease in bandwidth is not always associated with an increase in the number of stages of the midband gain and can remain fixed, independent of the number of stages.

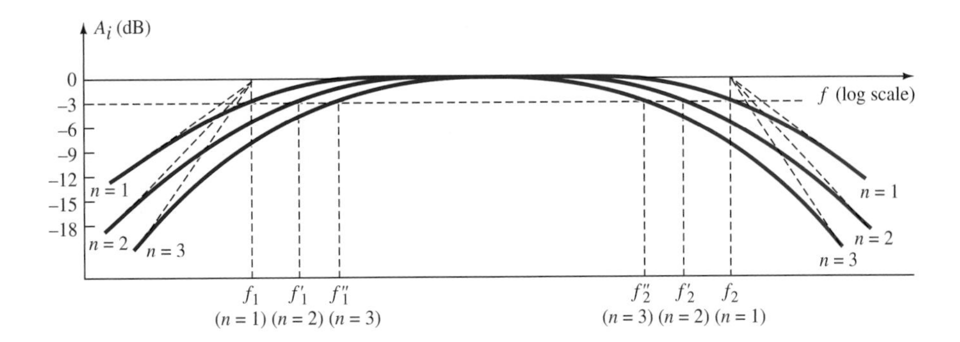

FIG. 11.25
Effect of an increased number of stages on the cutoff frequencies and bandwidth.

11.4 LARGE-SIGNAL (POWER) AMPLIFIERS

After a signal has been sufficiently amplified by a number of small-signal amplifier stages, to a level of a few volts, it can be applied to a large-signal amplifier to drive a power device, such as an audio speaker. A large-signal amplifier, therefore, must be capable of operating efficiently, handling large amounts of power, and delivering that power to the load.

Amplifiers may be classified by their frequency range of operation, by the type of application they are used in, or, in this consideration, by the means by which they are operated to provide power to a load. With regard to efficient coupling of a signal to a load, in terms of power efficiency, amplifiers are specified as either class A or class B.

Power Efficiency

In class A operation the circuit is biased to operate so that the output signal is controlled by the input signal for a full cycle (or for 360° of operation). This is the least efficient operating condition, since the amplifier is biased to operate at about the midpoint of the voltage or current swing of the output. The input signal then causes the circuit to provide an output that varies above and below this operating point as the input signal goes up and down in voltage. Class A amplifiers, which provide transformer coupling to a load, operate at a maximum efficiency of 50%, and usually much less.

In class B operation the circuit is biased off so that no power is dissipated until a signal is applied. The power drawn from the supply and then provided to the load increases as the signal level increases. Maximum theoretical efficiency for a class B circuit is 78.5% and is often close to that value. To achieve a full cycle of output signal swing requires two circuit parts, each operating for half the cycle. This arrangement is called a *push-pull circuit.* It can be accomplished using a transformer to provide each half of the output signal cycle, or as a transformerless circuit, using a circuit in which each half operates for about half of the signal cycle.

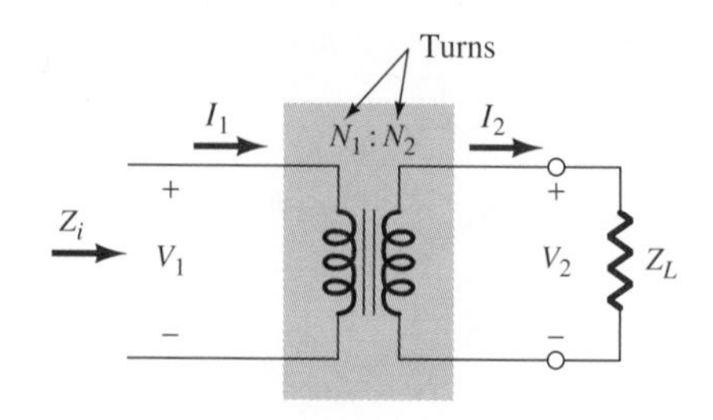

FIG. 11.26
Basic transformer configuration.

Transformer Impedance Transformation

Figure 11.26 shows a basic transformer connection, with applied voltage V_1 and resulting voltage V_2, dependent on the transformer turns ratio.

$$\frac{V_1}{V_2} = \frac{N_1}{N_2} = a \qquad \text{(transformation ratio)} \tag{11.10}$$

In addition, the resulting current I_2 through load R_L causes an input current that is related by the transformer turns ratio

$$\frac{I_1}{I_2} = \frac{N_2}{N_1} = \frac{1}{a} \tag{11.11}$$

The input impedance seen looking into the transformer primary is [Eqs. (11.10d) and (11.11)]

$$R_i = \frac{V_1}{I_1} = \frac{aV_2}{(1/a)I_2} = a^2\frac{V_2}{I_2} = a^2R_L \tag{11.12}$$

The transformer is seen to provide an input impedance that is the factor a^2 times the impedance connected across the secondary winding. Since a is the ratio of primary to secondary turns, a step-down transformer acts to provide an input impedance that is greater by the factor a^2 than the impedance across the secondary.

Distortion

Distortion can occur in an amplifier because the device characteristic is not linear—this being *nonlinear or amplitude distortion.* In addition, the circuit elements and the amplifying devices may respond to the signal differently at different frequencies—this being *frequency distortion.* One technique to provide a means of accounting for a change in the output signal is the method of *Fourier analysis,* which provides a means for describing a periodic signal in terms of its fundamental frequency component and its frequency components at integer multiples. These components are called *harmonic components or harmonic frequencies.* For example, a signal that is originally 1000 Hz could result, after distortion, in a fundamental frequency component at 1000 Hz and harmonic components at 2 kHz (2 × 1000 Hz), at 3 kHz (3 × 1000 Hz), at 4 kHz (4 × 1000 Hz), and so on.

The amount of distortion in a particular waveform may be measured using a spectrum analyzer, which provides measurement of the amplitude of the fundamental and all harmonics present in the measured waveform. The total harmonic distortion of the waveform is then calculated as

$$D_T = \sqrt{D_2^2 + D_3^2 + D_4^2 + \cdots}$$

where D_2 = second harmonic distortion
D_3 = third harmonic distortion
D_4 = fourth harmonic distortion
D_T = total distortion

Class A Power Amplifier

The circuit in Fig. 11.27 is a transformer-coupled class A amplifier. The transformer provides matching of the low value of typical load impedance (4 Ω in the circuit shown) to a much higher impedance as load to the amplifier. The power delivered to the load is

$$P_o\,(\mathrm{ac}) = \frac{V_o^2\,(\mathrm{rms})}{R_L} \qquad \mathbf{(11.13)}$$

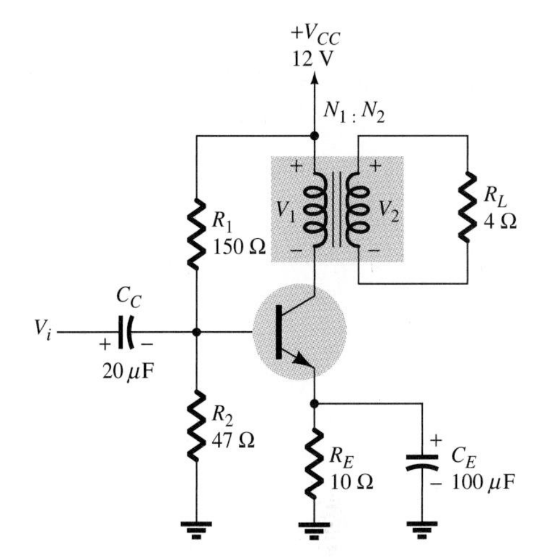

FIG. 11.27
Transformer-coupled class A amplifier.

while the input power to the amplifier is

$$P_i(\mathrm{dc}) = V_{CC}I_{CQ} \qquad \mathbf{(11.14)}$$

The circuit efficiency is then

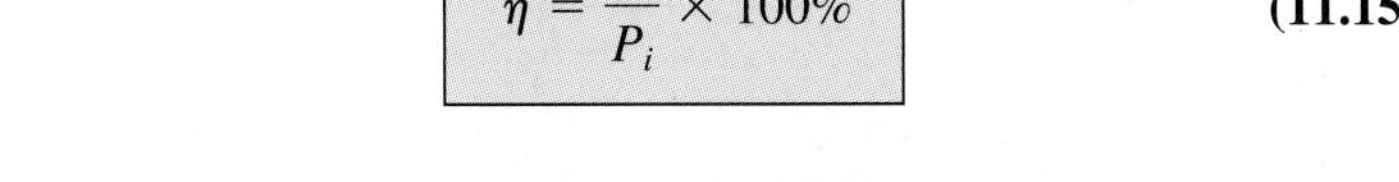

$$\eta = \frac{P_o}{P_i} \times 100\% \qquad \mathbf{(11.15)}$$

EXAMPLE 11.12 Calculate the input power, output power, and circuit efficiency for the class A amplifier in Fig. 11.27 for a turns ratio of 3.46:1 and an output signal of 2 V (rms). The dc bias provides $I_{CQ} = 0.25$ A.

Solution: Power calculations are

$$P_o\,(\mathrm{ac}) = \frac{V_o^2\,(\mathrm{rms})}{R_L} = \frac{(2\text{ V})^2}{4\ \Omega} = \mathbf{1\ W}$$

$$P_i\,(\mathrm{dc}) = V_{CC}I_{CQ} = (12\text{ V})(0.25\text{ A}) = \mathbf{3\ W}$$

$$\eta = \frac{P_o}{P_i} \times 100\% = \frac{1\text{ W}}{3\text{ W}} \times 100\% = \mathbf{33.3\%}$$

Class B Power Amplifier

A class B power amplifier requires two output transistors, with each providing half (or more) of the output signal swing. The simplified circuit in Fig. 11.28 shows a sinusoidal input signal coupled through an amplifier stage to a pair of opposite type (*npn* and *pnp*) or complementary

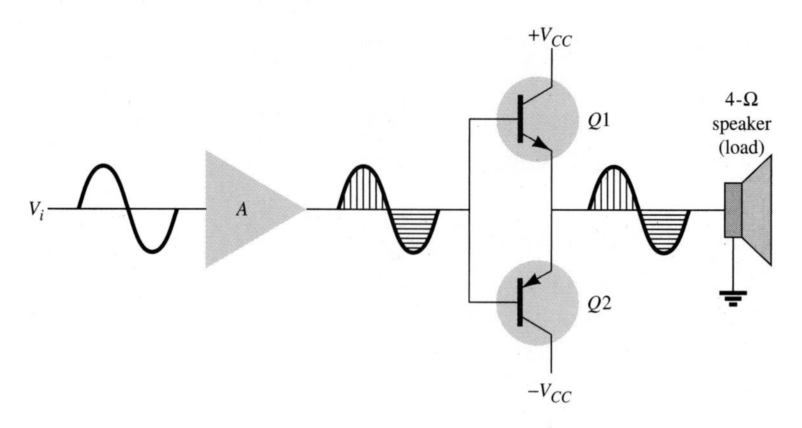

FIG. 11.28
Simple representation of class B push-pull amplifier.

transistors. Transistor Q_1 operates for the positive half-cycle, providing output to the load (a 4-Ω speaker in this example), while transistor Q_2 provides output to the load on the negative half-cycle of the output signal. The power delivered to the load is

$$P_o\,(\text{ac}) = \frac{V_o^2}{R_L} \tag{11.16}$$

The power provided by the supplies is

$$P_i\,(\text{dc}) = V_{CC}I_{\text{dc}} \tag{11.17}$$

where the dc current drawn is

$$I_{\text{dc}} = \frac{2I_{\text{peak}}}{\pi} \tag{11.18}$$

EXAMPLE 11.13 Calculate the efficiency of a class B power amplifier like that shown in Fig. 11.28 for an output signal of 12 V (rms) across a 4-Ω speaker with power supply values of $V_{CC} = \pm 20$ V.

Solution:

$$P_o\,(\text{ac}) = \frac{V_o^2}{R_L} = \frac{(12)^2}{4} = 36\text{ W}$$

$$V_{\text{peak}} = 1.414V_o\,(\text{rms}) = 1.414(12\text{ V}) = 16.97\text{ V}$$

$$I_{\text{peak}} = \frac{V_{\text{peak}}}{R_L} = \frac{16.97\text{ V}}{4\ \Omega} = 4.24\text{ A}$$

$$I_{\text{dc}} = \frac{2I_{\text{peak}}}{\pi} = \frac{2(4.24)}{\pi} = 2.7\text{ A}$$

$$P_i = V_{CC}I_{\text{dc}} = (20\text{ V})(2.7\text{ A}) = 54\text{ W}$$

The circuit efficiency is then

$$\eta = \frac{P_o}{P_i} \times 100\% = \frac{36\text{ W}}{54\text{ W}} \times 100\% = \mathbf{66.7\%}$$

Complementary Symmetry Circuits

Complementary symmetry circuits provide the push-pull operation using an *npn* and a *pnp* to operate on the opposite half-cycles of the input signal. Figure 11.29 shows a simplified version of the circuit with input V_i applied to *both* transistors, with *each* providing an output signal for half the signal cycle. As it is often more difficult to use matched output transistors of different types, the circuit in Fig. 11.30 shows the use of

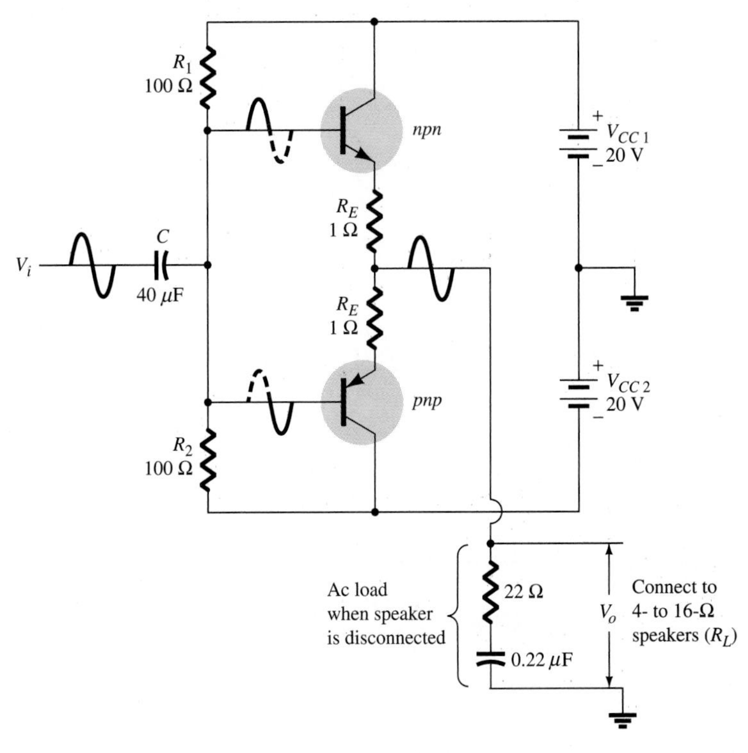

FIG. 11.29
Complementary push-pull power amplifier.

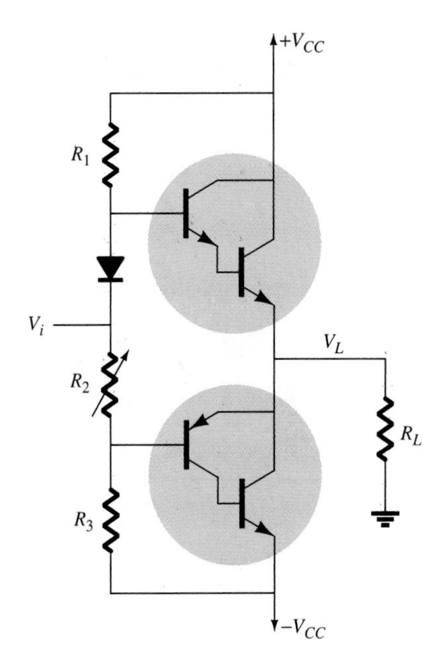

FIG. 11.30
Quasi-complementary push-pull power amplifier (using two supplies).

the same type of transistors in the power output, using complementary transistors to drive the output transistors in push-pull operation. This quasi-complementary connection is the more popular, with the upper transistor pair a *Darlington pair* and the lower transistor pair referred to as a *feedback pair.*

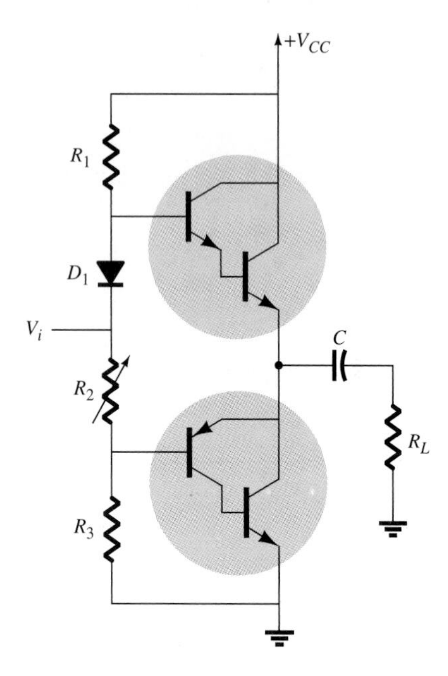

FIG. 11.31
Quasi-complementary push-pull power amplifier.

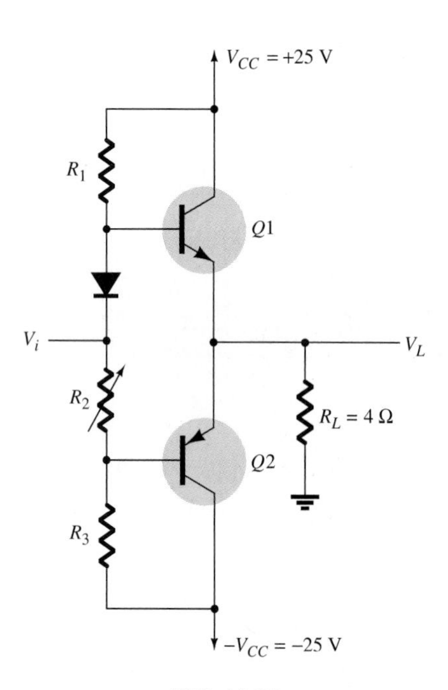

FIG. 11.32
Class B power amplifier for Example 11.14.

If two supply voltages are used, the dc level of the output is near 0 V and may be directly connected to the load. If a single supply voltage is employed, as shown in Fig. 11.31, a capacitor must be used to couple the output signal to the load while blocking the dc level at the output from getting to the load.

EXAMPLE 11.14 For the circuit in Fig. 11.32:

a. Calculate the input and output power handled by the circuit and the power dissipated by *each* output transistor for an input of 12 V (rms).
b. If the input signal is increased to provide the maximum undistorted output, calculate the values of maximum input and output power and the power dissipated by each transistor for this condition.

Solution:

a. The peak input voltage is

$$V_i\,(\text{p}) = \sqrt{2}\,V_i\,(\text{rms}) = \sqrt{2}\,(12\text{ V}) = 16.97\text{ V} = 17\text{ V}$$

Since the voltage across the load is ideally the same as the input signal (the amplifier ideally has a voltage gain of unity),

$$V_L\,(\text{p}) = 17\text{ V}$$

$$P_o\,(\text{ac}) = \frac{V_L^2\,(\text{p})}{2(R_L)} = \frac{(17\text{ V})^2}{2(4\ \Omega)} = \mathbf{36.125\ W}$$

$$I_L\,(\text{p}) = \frac{V_L\,(\text{p})}{R_L} = \frac{17\text{ V}}{4\ \Omega} = 4.25\text{ A}$$

The dc current drawn from the two power supplies is then

$$I_{\text{dc}} = \frac{2I_L\,(\text{p})}{\pi} = \frac{2(4.25\text{ A})}{\pi} = 2.71\text{ A}$$

so that the power supplied to the circuit is

$$P_i\,(\text{dc}) = V_{CC}I_{\text{dc}} = (25\text{ V})(2.71\text{ A}) = \mathbf{67.75\ W}$$

The circuit efficiency [for an input of V_i = 12 V (rms)] is

$$\eta = \frac{P_o}{P_i} \times 100\% = \frac{36.125\text{ W}}{67.75\text{ W}} \times 100\% = 53.3\%$$

and the power dissipated by each output transistor is

$$P_Q = \frac{\text{P}_{2Q}}{2} = \frac{P_i - P_o}{2} = \frac{67.75\text{ W} - 36.125\text{ W}}{2} = \mathbf{15.8\ W}$$

b. If the input signal is increased to V_i = 25 V peak [V_i = 17.68 V (rms)], so that V_L(p) = 25 V, the calculations are

$$\text{maximum } P_o = \frac{V_{CC}^2}{2(R_L)} = \frac{(25\text{ V})^2}{2(4\ \Omega)} = \mathbf{78.125\ W}$$

$$\text{maximum } P_i = \left(\frac{2}{\pi}\right)\left(\frac{V_{CC}^2}{R_L}\right) = \left(\frac{2}{\pi}\right)\frac{(25\text{ V})^2}{4\ \Omega} = \mathbf{99.47\ W}$$

and

$$\eta = \frac{P_o}{P_i} \times 100\% = \frac{78.125\text{ W}}{99.47\text{ W}} \times 100\% = 78.54\%$$

(the maximum circuit efficiency). At this maximum signal condition the power dissipated by *each* output transistor is

$$P_Q = \frac{P_i - P_o}{2} = \frac{99.47\text{ W} - 78.125\text{ W}}{2} = \mathbf{10.67\text{ W}}$$

11.5 COMPUTER ANALYSIS

PSpice Examples

PSpice provides models for many of the popular devices. One can build multistage circuits and active filter circuits using these devices, as the following examples demonstrate.

PSpice EXAMPLE 11.1 Figure 11.33(a) shows a two-stage cascade amplifier. The attributes of the three VPRINT1 components should be set to ac = yes and magnitude = yes. These result in the output files containing the rms values of these voltages. Figure 11.33(b) shows that at a midfrequency of 20 kHz, the input V_i = 2 mV results in a

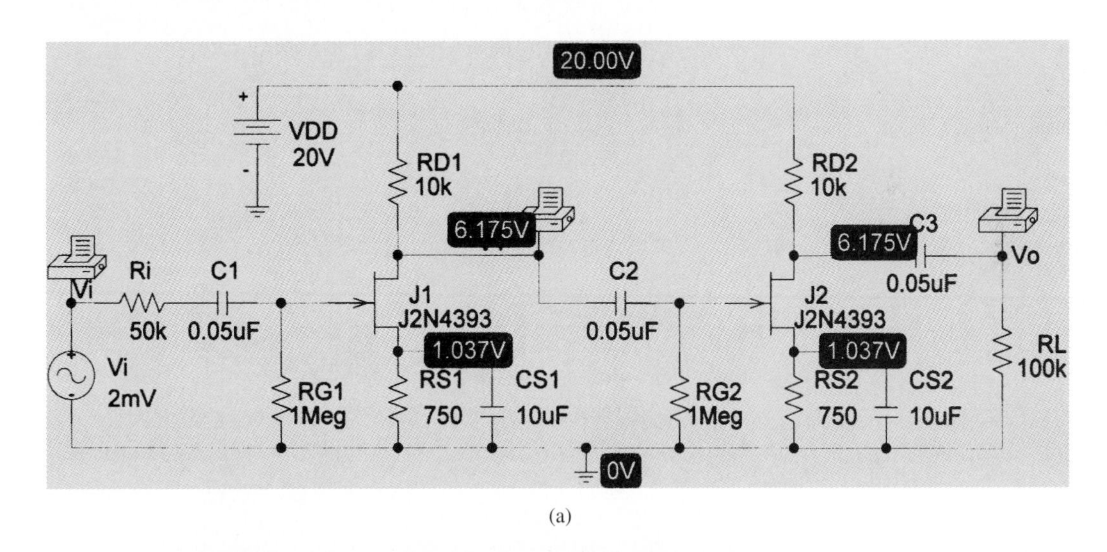

(a)

FREQ	VM(Vi)	VM(V1)	VM(Vo)
2.000E+04	2.000E-03	8.931E-02	5.401E+00

(b)

FIG. 11.33

(a) PSpice circuit for PSpice Example 11.1; (b) selected output for PSpice Example 11.1.

stage 1 drain voltage of $V_1 = 89.31$ mV and an output voltage of $V_o =$ 5.4V. The first stage gain is then

$$A_{v1} = \frac{-V_1}{V_i} = \frac{-89.31 \text{ mV}}{2 \text{ mV}} = -44.66$$

and the second stage gain is

$$A_{v2} = \frac{-V_o}{V_1} = \frac{-5.4 \text{ V}}{89.31 \text{ mV}} = -60.46$$

The overall amplifier gain is then

$$A_v = \frac{-V_o}{V_i} = \frac{5.4 \text{ V}}{2 \text{ mV}} = 2700$$

which is the same as

$$A_v = -(A_{v1})(A_{v2}) = (-44.66)(-60.46) = 2700.1$$

PSpice EXAMPLE 11.2 Figure 11.34 shows a high-pass filter circuit. Using Eq. (11.4) we calculate the cutoff frequency to be

$$f_L = \frac{1}{2\pi R_i C_i} = \frac{1}{2\pi(1 \text{ k}\Omega)(0.5\ \mu F)} = 318.3 \text{ Hz}$$

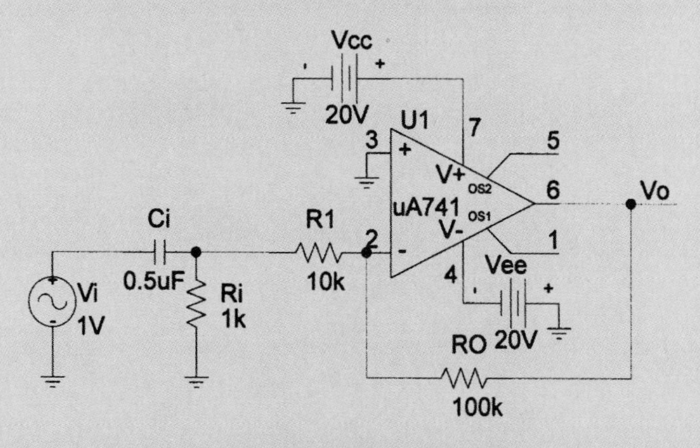

FIG. 11.34
PSpice schematic for PSpice Example 11.2.

The probe output in Fig. 11.35 shows the cutoff frequency at about 349 Hz.

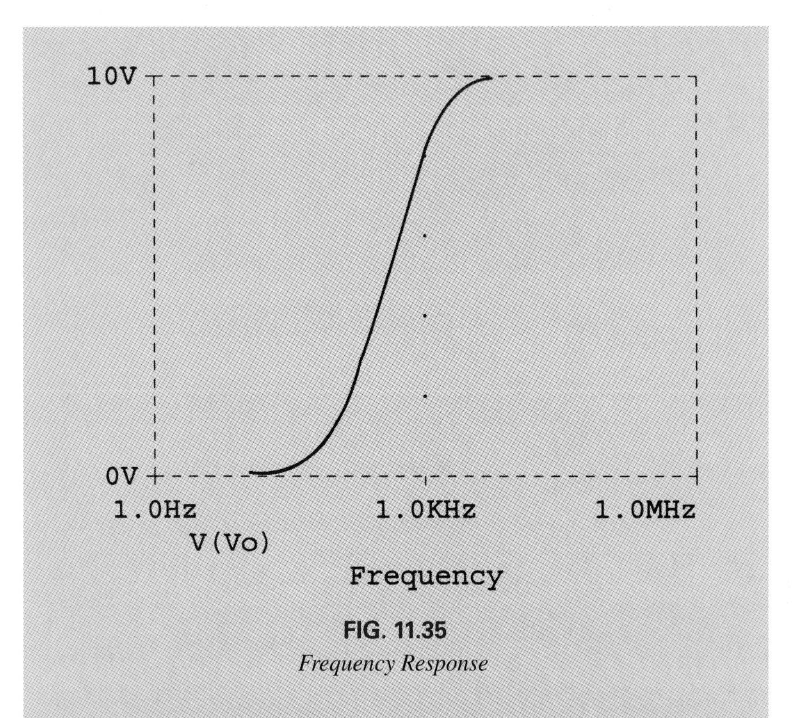

FIG. 11.35
Frequency Response

PSpice EXAMPLE 11.3 A low-pass filter circuit is shown in Fig. 11.36(a). The upper cutoff frequency is calculated using Eq. (11.5):

$$F_H = \frac{1}{2\pi R_{\text{high}} C_{\text{high}}} = \frac{1}{2\pi(10\text{ k}\Omega)(220\text{ pF})} = 72.3\text{ kHz}$$

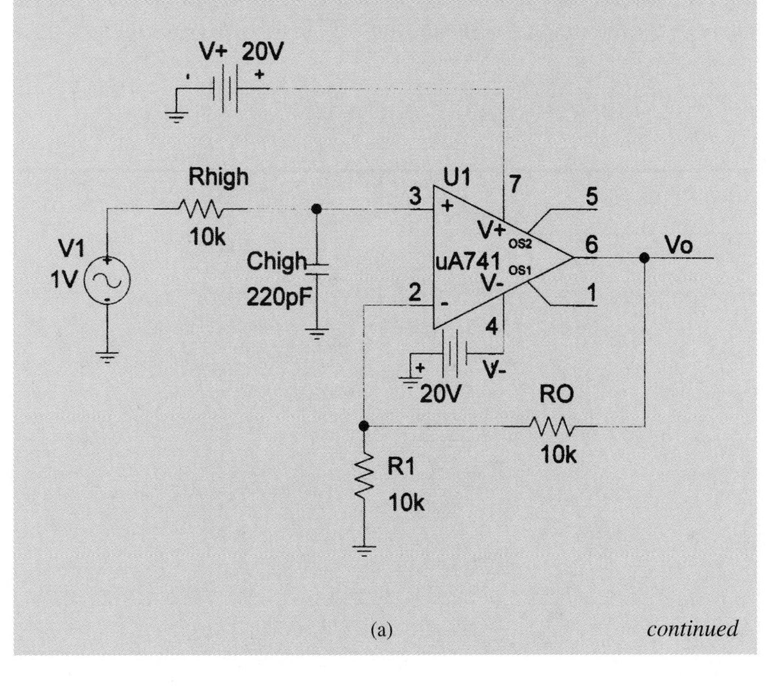

(a)

continued

The probe output in Fig. 11.36(b) shows the cutoff frequency to be about **71.5 kHz.**

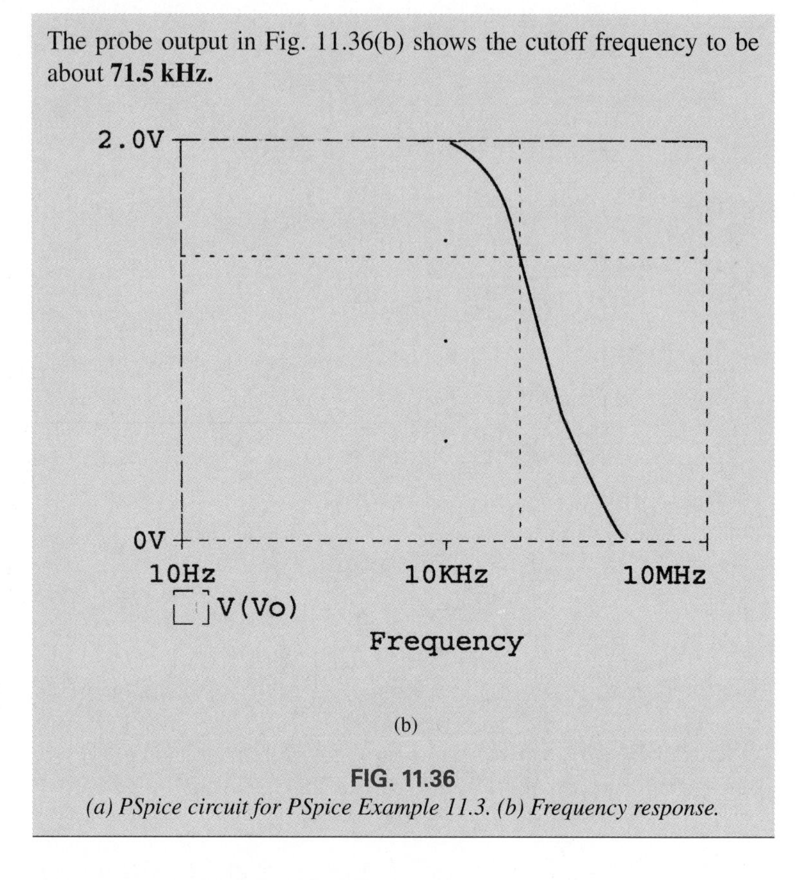

(b)

FIG. 11.36
(a) PSpice circuit for PSpice Example 11.3. (b) Frequency response.

Electronic Workbench Example

An EWB circuit like that of Fig. 11.36(a) is shown in Fig. 11.37(a). Ac analysis for the frequency range from 10 Hz to 1 MHz results in the Ac

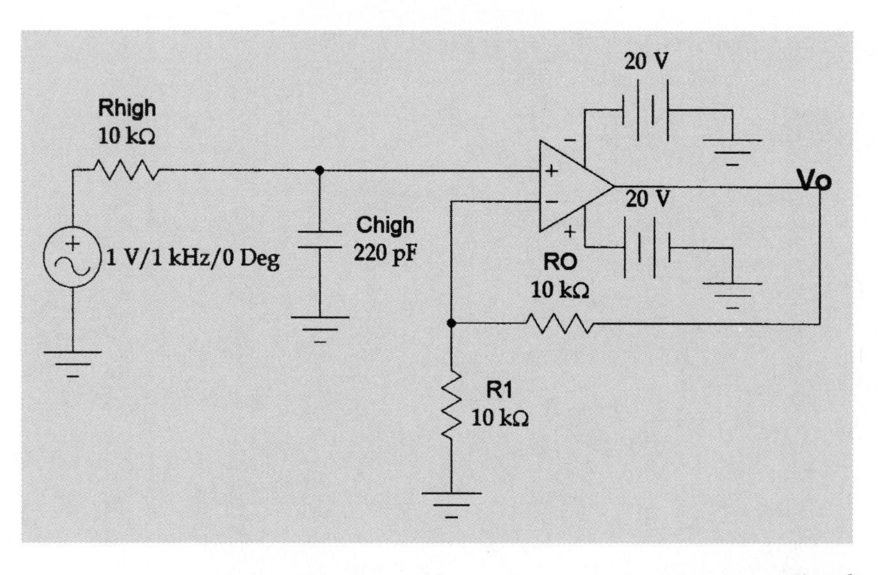

(a)

continued

Analysis output shown in Fig. 11.37(b). The cutoff frequency is seen to be just over 70 kHz.

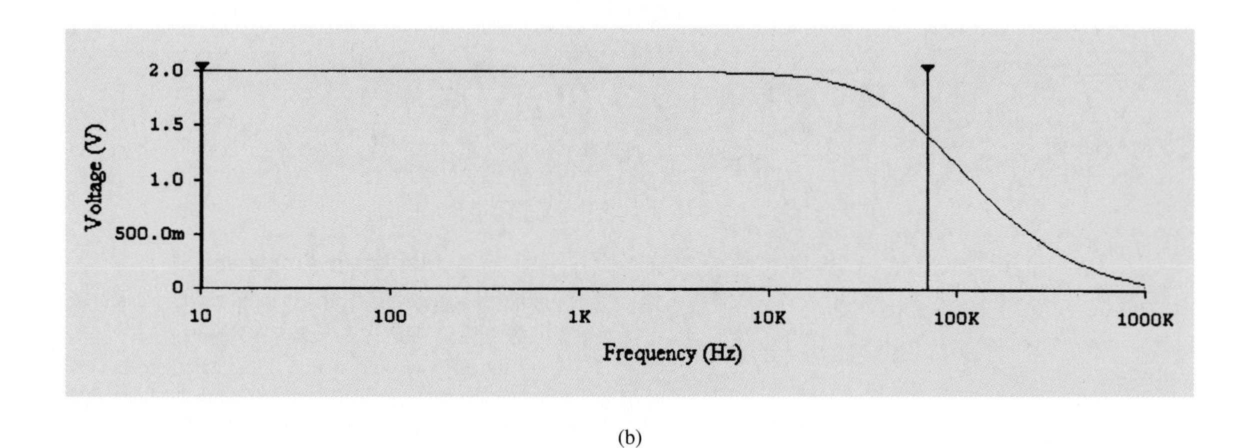

(b)

FIG. 11.37
EWB: (a) circuit. EWB: (b) ac analysis for PSpice.

11.6 SUMMARY

Common-Mode Rejection (CMR):

$$\text{CMR} = 20 \log \frac{A_d}{A_c} \text{ dB}$$

DB Voltage Gain:

$$A_v\,(\text{dB}) = 20 \log \frac{V_o}{V_i} \text{ dB}$$

Cutoff Frequencies:

$$f_L = \frac{1}{2\pi R_i C_i}$$

$$f_H = \frac{1}{2\pi R_{\text{high}} C_{\text{high}}}$$

Power Amplifier Classes:

Class A: $P_i\,(\text{dc}) = V_{CC} I_{CQ}$

$$P_o\,(\text{ac}) = \frac{V_o^2\,(\text{rms})}{R_L}$$

$$\eta = \frac{P_o}{P_i \times 100\%}$$

$$\text{Class B: } P_i\text{ (dc)} = V_{CC}I_{dc}, \text{ where } I_{dc} = \frac{2I_{peak}}{\pi}$$

$$P_o\text{ (ac)} = \frac{V_o^2\text{ (rms)}}{R_L}$$

$$\eta = \frac{P_o}{P_i} \times 100\%$$

PROBLEMS

SECTION 11.1 Multistage Amplifiers

1. For the two-stage FET circuit in Fig. 11.1, calculate the voltage gain of stage 1 if R_D is changed to 7.5 kΩ.

2. Using the FET circuit in Fig. 11.1, calculate the overall voltage gain for $R_{D1} = R_{D2} = 7.5$ kΩ (all other values remaining the same as shown in Fig. 11.1) for $R_L > R_{o2}$.

3. If the two-stage FET amplifier in Fig. 11.1 drives a 50-kΩ load, calculate the output voltage for an input of $V_s =$ 10 mV (rms) using the circuit values shown in Fig. 11.1 except for $R_{D1} = R_{D2} = 7.5$ kΩ.

4. Draw the circuit diagram of a differential amplifier.

5. What is common-mode rejection?

6. If a differential amplifier has a common-mode rejection ratio of 3×10^8 and a differential gain of 55, what is the value of the common-mode gain?

7. What is the value of differential gain for a differential amplifier having CMR = 120 dB and $A_o = 2.5 \times 10^{-3}$?

8. What value of CMR in decibel units results for a differential amplifier with a constant-current source?

9. Draw the circuit diagram of a differential amplifier with a constant-current source.

10. Calculate the gain of the amplifier circuit in Fig. 11.9 with $R_o = 3.3$ MΩ.

11. Calculate the gain of the amplifier in Fig. 11.12 with 10-kΩ resistors changed to 18 kΩ.

12. Draw a diagram showing three multistage amplifiers using a 347 op-amp with gains of 125, 180, and 275, respectively (list all pin connections).

13. Draw a diagram showing a transducer feeding two unity followers to provide two signal sources (list pin connections).

14. Sketch the output waveforms (V_{o1} and V_{o2}) from the circuit in Fig. 11.38.

15. In the circuit in Fig. 11.38, determine the output voltage V_{o1} if V_i is 180 μV.

16. Draw a diagram showing a 347 op-amp used to take a 40-mV input signal and provide 250-mV and 550-mV output signals, both the same polarity as the input.

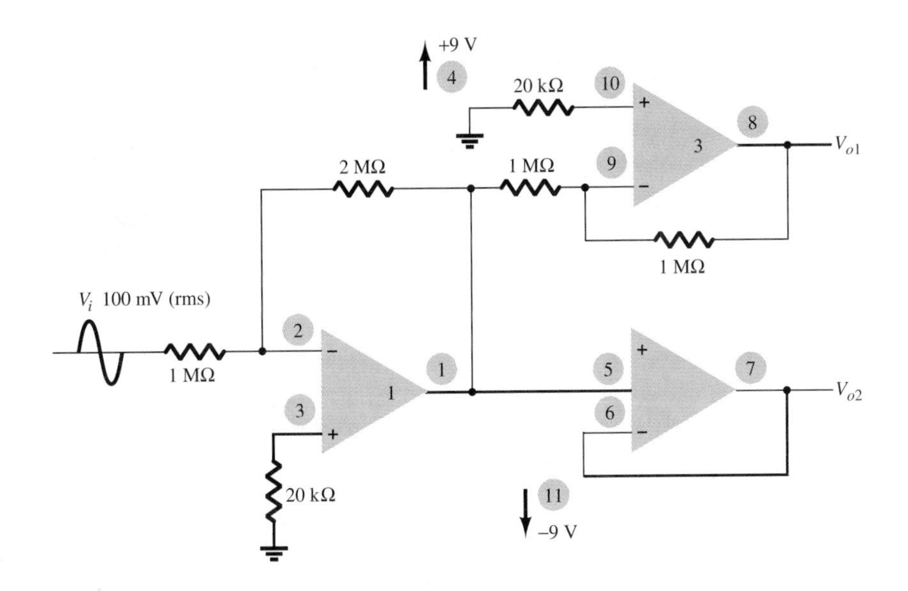

FIG. 11.38
Op-amp circuit for Problems 14 and 15.

SECTION 11.2 Logarithms and Decibel Measurements

17. Calculate the dB gain for an amplifier with input V_i = 3 mV (rms) and an output of:
a. V_o = 140 mV (rms).
b. V_o = 320 mV (rms).
c. V_o = 120 mV (peak).
d. V_o = 550 mV (rms).
e. V_o = 0.65 mV (rms).

18. Calculate the resulting output voltage when an input of 5 mV (rms) is applied to an amplifier with a gain of:
a. 62 dB.
b. 48 dB.
c. 33 dB.
d. 118 dB.

19. Calculate the input voltage needed to provide an output voltage of 2.5 V (rms) for an amplifier having a voltage gain of:
a. 40 dB.
b. 65 dB.
c. 37 dB.
d. 74 dB.

20. Calculate the overall decibel gain for three amplifier stages in series with voltage gains of 125, 78, and 45, respectively.

21. What overall decibel gain results from a series connection of amplifiers with voltage gains of 170, 80, 0.6, and 42?

SECTION 11.3 Frequency Considerations

22. Calculate the low cutoff frequency for an amplifier having an input resistance of R_i = 1.8 kΩ and a coupling capacitor of C = 0.47 μF.

23. What value of coupling capacitor is required to have a cutoff (lower 3-dB) frequency of 25 Hz for an amplifier having an input resistance of R_i = 5 kΩ?

24. Calculate the upper cutoff frequency for an amplifier having values of R_{high} = 2 kΩ and C_{high} = 250 pF.

25. Calculate the value of C_{high} for an amplifier having a midfrequency gain of magnitude $A_{v,\text{mid}} = -120$, stray wiring capacitance of 12 pF, and amplifier terminal capacitance of $C_{ii} = 5$ pF, $C_{oo} = 8$ pF, and $C_{io} = 6$ pF.

26. Calculate the value of the low-frequency cutoff for the circuit in Fig. 11.23 with C changed to 8 μF.

27. For R_D changed to 4.3 kΩ and $g_m = 5.5$ mS in the circuit in Fig. 11.23, calculate the value of A_{mid}.

28. Using the capacitance values in Fig. 11.23 with new values of $R_D = 4.3$ kΩ and $g_m = 5.5$ mS, calculate the upper cutoff frequency.

29. Calculate the midfrequency gain of the BJT amplifier circuit in Fig. 11.24 for new values of $R_C = 5.6$ kΩ and $h_{fe} = 185$.

30. Calculate the lower 3-dB frequency for the circuit in Fig. 11.24 with new values of $C = 0.33$ μF and $h_{fe} = 185$.

31. Calculate the value of the upper 3-dB frequency for the circuit in Fig. 11.24 with new values of $R_C = 5.6$ kΩ and $h_{fe} = 185$ (all other circuit values remaining as listed in Fig. 11.24).

32. Calculate the dB voltage gain of the amplifier in Fig. 11.24 for new values of $R_C = 5.6$ kΩ and $h_{fe} = 185$.

SECTION 11.4 Large-Signal (Power) Amplifiers

33. Calculate the resulting output power delivered to a load if the resulting voltage swing is 15 V (p-p) with a current swing of 120 mA (p-p).

34. Calculate the ac power delivered to an 8-Ω load by a voltage swing of 6-V peak.

35. If an average dc current of 25 mA is drawn from a 9-V supply by an amplifier circuit, calculate the input dc power delivered to the amplifier.

36. Calculate the input power, output power, and circuit efficiency for the amplifier in Fig. 11.27 with a transformer having a turns ratio of 7:1 and an output signal of 1.7 V (rms) (dc bias provides $I_{CQ} = 0.25$ mA).

37. Calculate the efficiency of a class B power amplifier like that shown in Fig. 11.28 for an output signal of 22 V (rms) across a 4-Ω load with power supply values of $V_{CC} = \pm 40$ V.

38. For the circuit in Fig. 11.32 with $R_L = 8$ Ω and $V_{CC} = \pm 30$ V, calculate the input and output power handled by the circuit and the power dissipated by each output transistor for an input of $V_i = 15$ mV (rms).

39. For the circuit in Fig. 11.32 with $R_L = 8$ Ω and $V_{CC} = \pm 30$ V, with the input signal increased to provide the maximum undistorted output, calculate the values of maximum input and output power and the power dissipated by each transistor for this condition.

40. Show that the maximum power dissipated by each transistor occurs when the input signal is $0.636V_{CC}$.

Communications 12

12.1 GENERAL COMMUNICATIONS CONCEPTS

The methods commonly used for information transmission and reception are generally well known. Voice (audio) information is transmitted through space by amplitude modulation (AM) or frequency modulation (FM) techniques, through wire as audio over telephone lines, or by a number of other less common methods of signal transmission. Visual (and audio) information is, of course, sent as TV signals and transmitted over cables or through the air. Digital data are transmitted by a wide variety of methods—over cables, through the air, by pulse coding, by frequency coding, and by amplitude coding, as examples.

Any communication network can be rather simply described by similar basic parts with some generally common characteristics. Figure 12.1 is a block diagram of a general communications system. The *information source* originates the message or information that is to be sent via a transmitter into the *channel.* The transmitter processes the information and converts it to a form suitable for transmission. This signal processing by the transmitter may be as simple as direct correspondence between electrical current over telephone wires and air pressure variations arising from the human voice; it may be an encoding process such as Morse code or more complex encoding such as amplitude modulation, frequency modulation, and phase or pulse modulation, as examples.

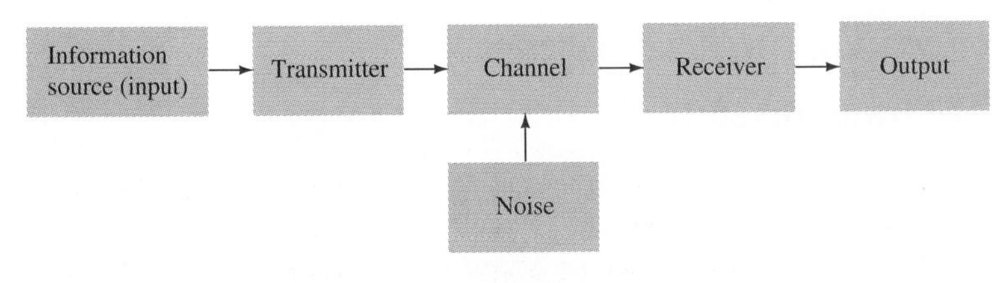

FIG. 12.1
Block diagram of a communication system.

This information (encoded or modulated) is then sent via a channel, which may be simply a pair of wires, the air, or through free space. The channel characteristics are most important in setting the limits of the transmitter or receiver and are of greatest concern in designing the system. Transmission through the channel results in signal attenuation, distortion, and, worst of all, the introduction of *noise* onto the signal. If the channel is a pair of wires, then the wire-distributed impedances attenuate and distort various frequency components of the transmitted signal. Noise can come from signals picked up from other nearby lines or through electromagnetic signals around it. These electromagnetic signals can cause problems in transmission through air or space as well and originate from solar radiation, lightning, motors, or other electrical disturbances. The channel's *signal-to-noise* ratio is quite important in specifying how well the channel operates. The signal and noise picked up by the receiver must then be converted back as close to the original transmitted information as possible with as much of the noise eliminated as possible. The processes of decoding and demodulating are the inverse of those used in transmission.

Some of the major types of circuits used in communications systems can be seen using signal processing in an AM transmitter as an example. The acoustic energy from the person speaking is converted into electrical energy by the microphone. This electrical signal source is generally too weak or low in amplitude for transmission over an air channel or long wire or cable channel. Amplifiers operating in the audio-frequency range signal (about 30 to 18,000 Hz, typically) increase the signal strength so that it can be transmitted. Audio frequencies are, however, not suitable for long-distance transmission, as the air channel tends to attenuate sound in a relatively short distance. Higher-frequency transmission is capable of much longer distances; in fact, it is possible for electromagnetic radiation transmission to go around the globe or out into space.

A radio-frequency (RF) oscillator circuit is used to generate high-frequency signals used to carry audio information. The process of placing the audio signal on the carrier signal is called *modulation,* and in this discussion is AM, or amplitude modulation. The amplitude-modulated RF signal is amplified so that a suitable large signal can be applied to the transmitting antenna for conversion to electromagnetic radiation into the air (or free space).

After an introduction to AM and FM transmission and reception techniques, some of the basic circuits—tuned amplifiers, oscillators, modulation, and demodulating techniques—will be covered and applied to AM and FM transmission and reception.

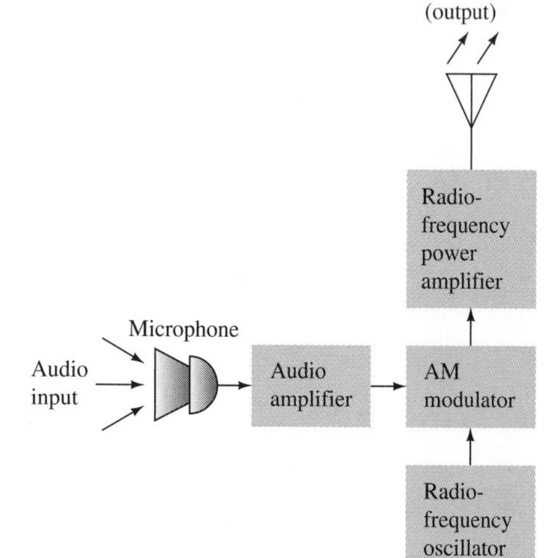

FIG. 12.2
Signal passage through an AM transmitter.

12.2 AM RADIO CONCEPTS

AM Transmission

Figure 12.2 of an AM transmitter indicates the basic parts for obtaining an amplitude-modulated signal. The radio-frequency oscillator generates an RF carrier signal in the range 550 to 1600 kHz for the AM broadcast band on a precise frequency assigned in this broadcast range. The precise oscillator carrier frequency is then a sinusoidal signal as shown in

Fig. 12.3(a). An audio-frequency (AF) signal (30 to 18,000 Hz) is produced by sound waves striking the microphone, and the resulting electrical signal is raised to the level of a few volts by the audio amplifier [Fig. 12.3(b)]. The modulator circuit then uses the audio signal to modulate the RF carrier signal by varying the amplitude of the carrier with the audio signal [Fig. 12.3(c)]. The modulated RF signal is then amplified by amplifier circuits tuned to the RF frequency band and sent out as electromagnetic radiation from the transmitting antenna.

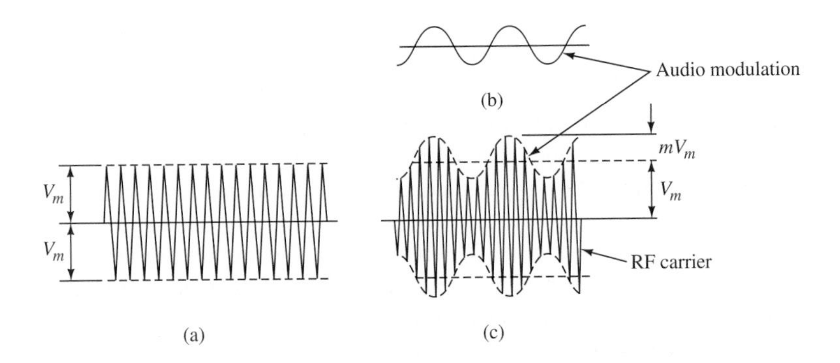

FIG. 12.3
AM signals: (a) carrier (RF) signal; (b) audio-frequency signal; (c) amplitude-modulated carrier signal.

A voltage that varies sinusoidally at the carrier frequency (from the RF oscillator) can be expressed as

$$v = V_c \cos \omega_c t$$

where V_c = rms voltage amplitude
ω_c = carrier radian frequency

This is the signal shown in Fig. 12.3(a). The audio signal [Fig. 12.3(b)] can be expressed as

$$v_m = V_m(m \cos \omega_m t)$$

where m is a modulating factor (with $m = 1$ being 100% modulation) and ω_m is the audio or modulating radian frequency. The modulated carrier signal [Fig. 12.3(c)] can then be expressed as

$$v = V_m(1 + m \cos \omega_m t) \cos \omega_c t \qquad \textbf{(12.1)}$$

which is just $V_m \cos \omega_c t$ for $m = 0$ (no modulation). Using a trigonometric relationship,* it is possible to rework Eq. (12.1) into the form

$$v = V_m \cos \omega_c t + 0.5mV_m \cos (\omega_c + \omega_m)t + 0.5mV_m \cos (\omega_c - \omega_m)t \qquad \textbf{(12.2)}$$

*$\cos A \cos B = \frac{1}{2} [\cos(A + B) + \cos(A - B)]$.

Equation (12.2) more clearly shows that the resulting modulated signal contains three frequency components: the carrier frequency, a frequency ω_m (rad/s) *above* the carrier, and another frequency ω_m (rad/s) *below* the carrier frequency. These upper and lower components around the carrier frequency are referred to as signal *side bands*. For the pure tone assumed here a pair of side bands is generated. When the actual modulating signal contains many frequency components, as in voice or music, then the resulting modulated signal contains side bands extending above and below the carrier frequency by a frequency range set by the highest frequency of the applied modulating signal. Thus, the transmitted signal has a bandwidth of twice the highest modulating frequency. In the AM commercial broadcast band, for example, the station carrier frequencies have to be separated by 40 kHz if the full 20-kHz audio band is desired. This wide separation does not usually occur, especially in an area with many stations, and the station's modulating bandwidth is reduced from this full audio range.

Modulations greater than 100% are not allowed, as they cause objectionable side-band splattering. Figure 12.4 shows the modulated carrier signal for a number of modulating conditions. For the condition of less than 100% modulation [Fig. 12.4(a)], the carrier amplitude peak voltage remains greater than zero. It becomes zero [Fig. 12.4(b)] for 100% modulation at the negative peaks of the modulating signal. For up to 100% modulation the output signal is described by Eq. (12.2). Greater than 100% modulation, however, results in a discontinuous output signal [Fig. 12.4(c)], which is no longer a purely sinusoidal signal. The overmodulated signal results in more than two side bands, and this side-band splattering causes interference in adjacent station channels. Obviously, transmission at greater than 100% modulation is not allowed.

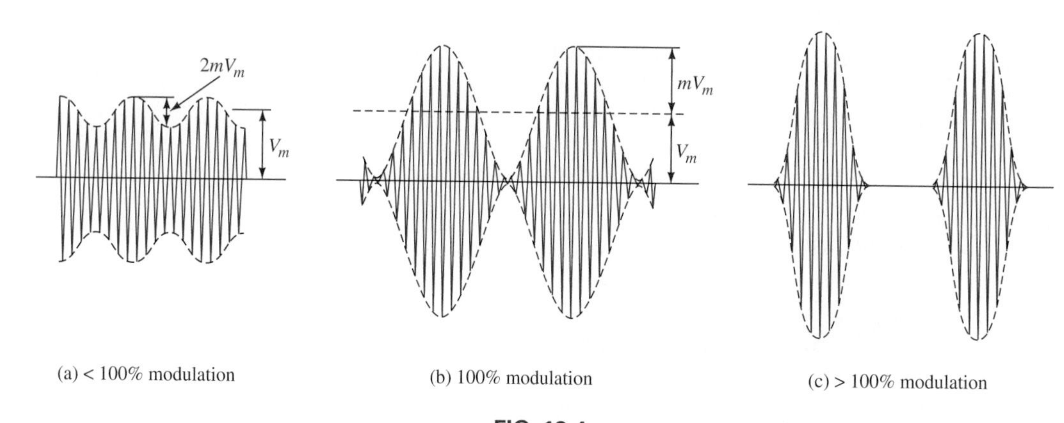

FIG. 12.4
Various modulation conditions: (a) <100% modulation; (b) 100% modulation; (c) >100% modulation.

It is often desirable to measure the amount of modulation of a signal. Using the waveform in Fig. 12.5, we can obtain the amount of modulation from the oscilloscope display by measuring the peak-to-peak voltages at maximum and minimum points of the signal, as shown, and calculating m using*

$$m = \frac{A - B}{A + B} \qquad \textbf{(12.3)}$$

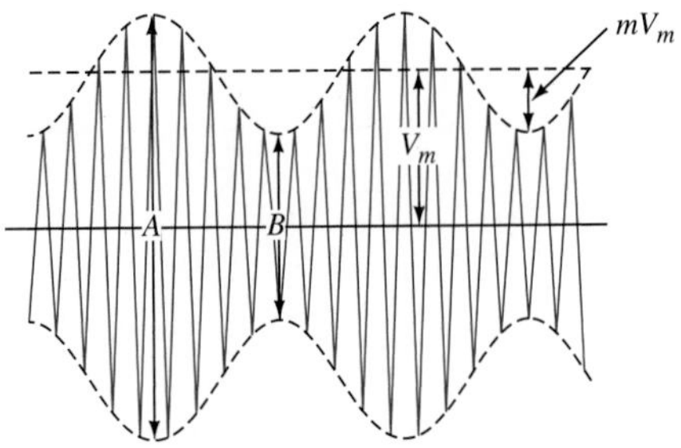

If $B = 0$, $m = 1$ (100% modulation) as expected, while $B = A$ results in $m = 0$ (0% modulation), and no modulation is present.

FIG. 12.5
Waveform for measuring modulation (m).

Amplitude-Modulation Circuit

We have seen that carrier and modulating signals are applied to a circuit called a modulator, resulting in an amplitude-modulated signal that is transmitted at the carrier frequency with a bandwidth of twice the highest modulating signal frequency. To achieve this modulating action on a carrier signal requires a nonlinear circuit. Mixing the carrier and modulating signals with a diode or other nonlinear element or using an amplifying device in its nonlinear operating range will result in the desired modulated carrier signal.

Figure 12.6 shows a circuit that can be used as a small-signal modulator. The RF carrier signal is applied to the base, and the AF signal to the emitter, with a resulting modulated signal in the collector circuit. A tuned circuit at the collector is tuned to the RF carrier frequency and has a bandwidth in the range of twice the highest signal frequency. The modulated signal resulting from this circuit is a small signal that has to be coupled through an RF power amplifier to drive the transmitting antenna.

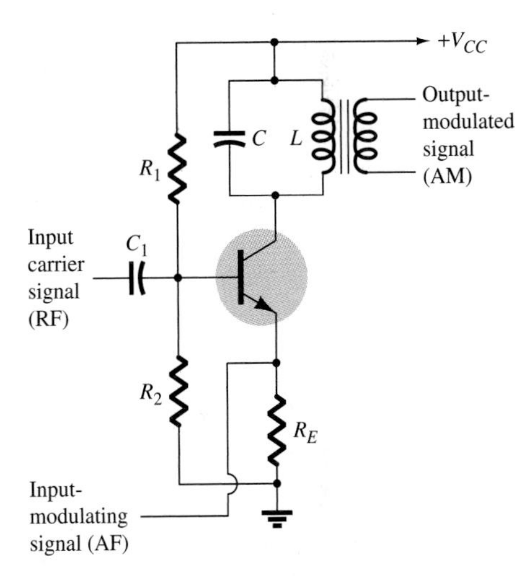

FIG. 12.6
Circuit for small-signal amplitude modulation.

AM Detection (Demodulation)

The received AM signal must be demodulated or detected to extract the AF signal from the RF carrier. A combination detector and filter circuit acts to obtain the audio signal from the modulated carrier and then remove the carrier, leaving the desired signal.

Figure 12.7 shows the modulated RF signal applied first to a diode circuit that clips the signal, resulting in one having an average value. The rectified signal is then applied to a filter circuit that follows the signal peaks or signal *envelope,* reproducing the audio signal. A simple circuit for obtaining demodulation is shown in Fig. 12.8. The AM signal

*From the waveform in Fig. 12.5,

$$mV = \frac{1}{2}\left(\frac{A}{2} - \frac{B}{2}\right) = \frac{1}{4}(A - B)$$

$$V_m = \frac{B}{2} + \frac{1}{2}\left(\frac{A}{2} - \frac{B}{2}\right) = \frac{1}{4}(A + B)$$

$$\frac{mV_m}{V_m} = m = \frac{\frac{1}{4}(A - B)}{\frac{1}{4}(A + B)} = \frac{A - B}{A + B}$$

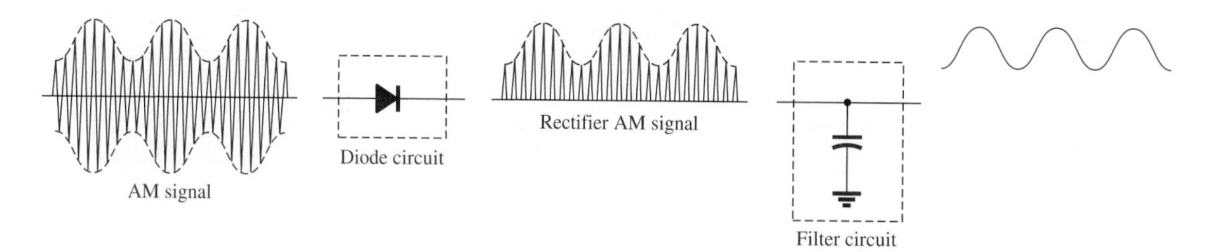

FIG. 12.7
Demodulation process.

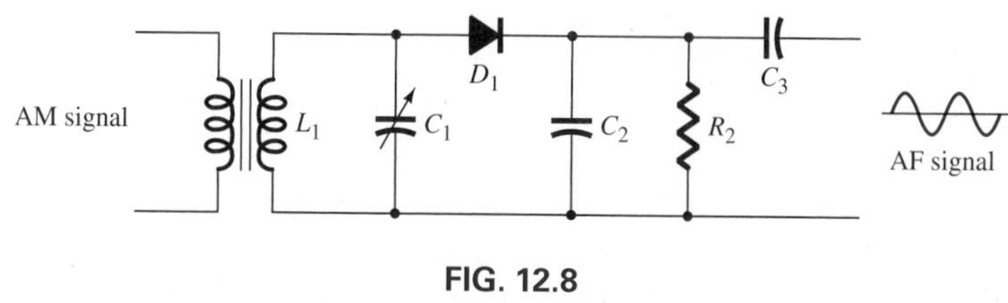

FIG. 12.8
AM demodulation circuit.

received by an antenna is amplified to the level of a few volts and then coupled, through a circuit tuned (L_1, C_1) to the RF carrier frequency, to the detection diode D_1. The rectified signal is then filtered by capacitor C_2 (and resistor R_2) with the resulting audio-frequency signal, coupled through capacitor C_3, the desired audio signal.

AM Receiver

An overall block diagram of the AM receiver is shown in Fig. 12.9. The signal received by the antenna is quite small (microvolts) and is passed through a number of RF amplifier stages, bringing it to the level of a few volts. The AM signal is then demodulated, and the resulting audio signal is again amplified to drive a speaker.

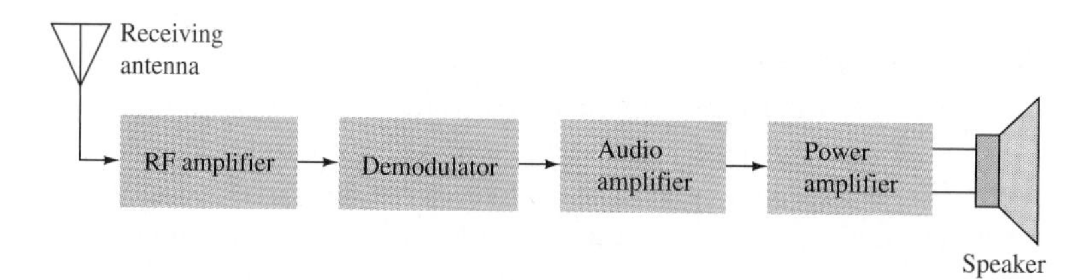

FIG. 12.9
AM receiver, block diagram.

Superheterodyne Operation

Most receiver circuits use *heterodyne* operation to provide better station tuning, selection, and sensitivity. Figure 12.10 is a block diagram of a superheterodyne receiver. The signal received by the antenna is first

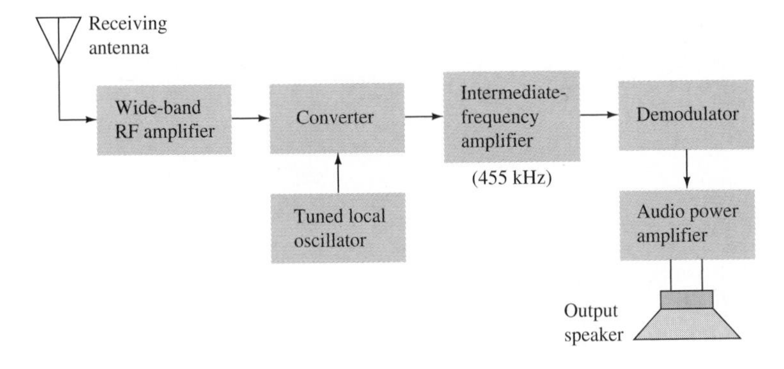

FIG. 12.10
Superheterodyne receiver, block diagram.

amplified by a wide-band RF amplifier, which amplifies any signals received in the full AM frequency band. The RF signal is then heterodyned or beat with an oscillator frequency obtained from an adjustable oscillator controlled by the tuning knob of the receiver. The resulting signal is still a modulated RF signal but at beat frequencies at the sum and difference of the RF and oscillator frequencies. Using the difference frequency from the convertor acts to shift the center frequency of the carrier down to an intermediate frequency, which retains the original modulated information. Using a local oscillator tunable from 1 MHz to 2 MHz results in an intermediate frequency of 455 kHz for RF carrier signals in the range 550 to 1600 kHz. By converting the received signal into a fixed intermediate frequency, it is possible to then use very selective high-gain tuned intermediate-frequency (IF) stages designed to operate at one fixed frequency. Various settings of the local oscillator beat various carrier frequencies down to the IF value, so that tuning the oscillator acts to select the signal to be further amplified and detected. Carrier signals received at frequencies other than that set to be beat by the oscillator to the IF value are also converted but do not pass through the IF stages, which are highly selective to the intermediate frequency.

Automatic Gain Control

The strength of the signal received from different stations or even from the same station at various times can change considerably. Without the automatic gain control (AGC) function, to be described, the output signal volume would be almost constantly changing and would be obviously quite disturbing to listen to. Receivers are therefore provided with automatic volume control (AVC) or automatic gain control circuitry to eliminate this problem. In an AM radio receiver the audio voltage obtained at the detector output is directly related to the strength of the signal received. An AGC control voltage can be obtained from this audio signal by further filtering to develop a voltage that is related to the amplitude of the detected signal. Figure 12.11(a) shows, in block form, the action of the AGC filter. The received signal is converted to IF and amplified, before detection. The larger the amplified IF signal, the larger the

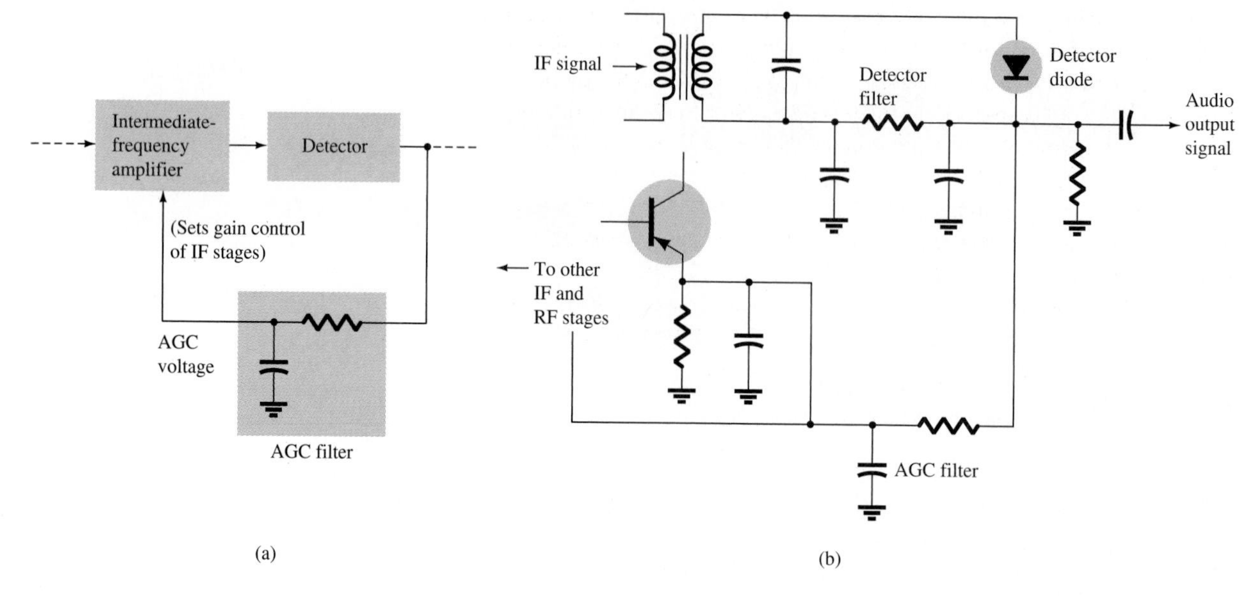

FIG. 12.11
AGC filter: (a) block; (b) circuit.

amplitude of the detected signal and the larger the AGC voltage. This AGC voltage is therefore small for small audio signal amplitudes and larger for larger audio signal amplitudes. AGC voltage is then used to control the bias of the RF and IF amplifier stages, thereby controlling the stage gains. This AGC voltage is used so that larger AGC voltage results in reduced amplifier stage gain. In most actual radio circuits the AGC voltage is developed as delayed AGC, which results only when the audio signal increases above a predetermined signal level to provide a smoother-acting gain control.

12.3 FM RADIO CONCEPTS

Instead of varying the amplitude of a carrier signal with an audio-frequency signal, it is possible to vary the *frequency* of the carrier signal. This concept of *frequency modulation* was conceived about 1933 by Edwin Armstrong, one of the early geniuses of the radio field. Armstrong conceived the most important principles in both AM and FM radio, including the early regenerative receiver, the superheterodyne receiver, and later the concept of FM transmission. The story of Armstrong's life is a most fascinating story of the history of radio electronics up to the early 1950s.* Realizing that the source of most noise in AM transmission and reception is amplitude-sensitive (e.g., lightning, radiation from fluorescent devices, electronic machinery), Armstrong conceived the idea of keeping the carrier amplitude constant and varying the

---

*A most absorbing account of Armstrong's life and work is given in *Man of High Fidelity: Edwin Howard Armstrong,* by Lawrence Lessing, Bantam Books, New York, 1969.

carrier's frequency at the audio signal rate. Not only does this result in the virtual elimination of much unwanted noise but it also provides operation around a high carrier with greater bandwidth for much higher fidelity than is possible in AM operation. Figure 12.12 shows frequency modulation of a carrier signal. Notice that the amplitude of the frequency-modulated signal does not vary, whereas the frequency of the signal is higher for higher audio signal amplitude. If the audio signal amplitude is held constant, then the FM signal remains at its center carrier frequency (f_c). The resulting frequency derivation (f_d) from the carrier follows the amplitude variation of the audio signal (f_s). Not only does the carrier frequency deviate according to the amplitude of the audio signal, but the rate of change of the FM signal also depends on the audio frequency.

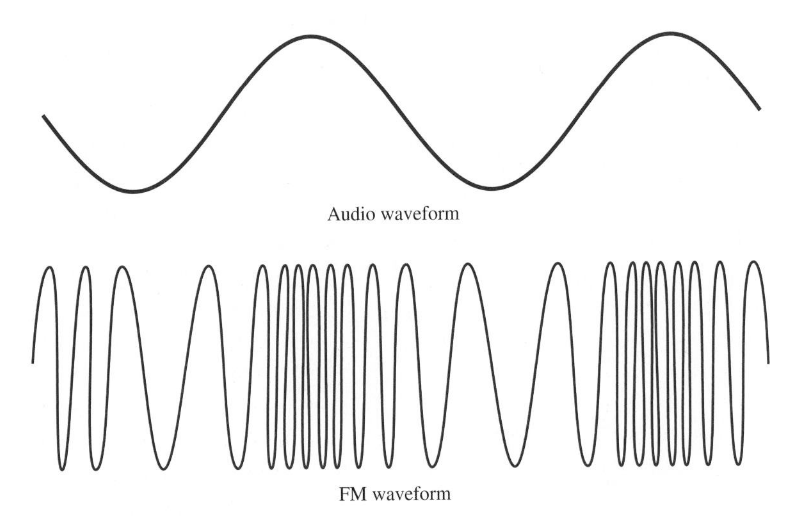

FIG. 12.12
Frequency modulation.

Frequency modulation is usually specified by an index of modulation:

$$m_f \equiv \frac{f_d}{f_s} \tag{12.4}$$

where f_s is the modulating or signal frequency and f_d is the resulting frequency deviation of the carrier. In FM broadcasting the maximum allowable frequency deviation is 75 kHz with a maximum audio frequency of 15 kHz (so that $m_f = \frac{75}{15} = 5$). Indices of modulation greater than unity are considered wide-band frequency modulation.

One important difference between amplitude and frequency modulation is in the amount of transmitted energy. In AM signals 100% modulation occurs when the carrier amplitude varies between zero and twice its (unmodulated) value. The power transmitted has to increase by 50%

from 0 to 100% modulation, making the circuit design more complicated for efficient operation over this power range. FM transmission, on the other hand, is always at the same amplitude or power level, so that the number of side bands produced in FM transmission and the energy at the carrier and side-band frequencies depend on the modulation index m_f. Although the relationship is relatively complex, the number of side bands increases with larger values of modulating index, and the energy transmitted is distributed among the side bands and carrier center frequency so that for some values of modulating index the side-band energy is greater than the carrier energy. Although an FM wave theoretically has an infinite number of side bands even for a single-tone modulating signal, only a limited number of side bands contain sufficient energy to be important. It is usually assumed that the side bands extend around the carrier frequency by an amount equal to the sum of the modulating frequency and carrier frequency deviation. Thus, the bandwidth requirement of an FM system is *greater* than twice the modulating frequency. To allow this larger bandwidth, commercial FM transmission is made on carrier frequencies ranging from 88 to 108 MHz.

FM transmission is provided by a number of circuit arrangements including the reactance transmitter [Fig. 12.13(a)] and the standard Armstrong transmitter [Fig. 12.13(b)]. In the reactance transmitter the audio signal varies reactance, which acts as a trimming capacitor in the oscillator circuit it is connected to. Modern transmitter units use varactor

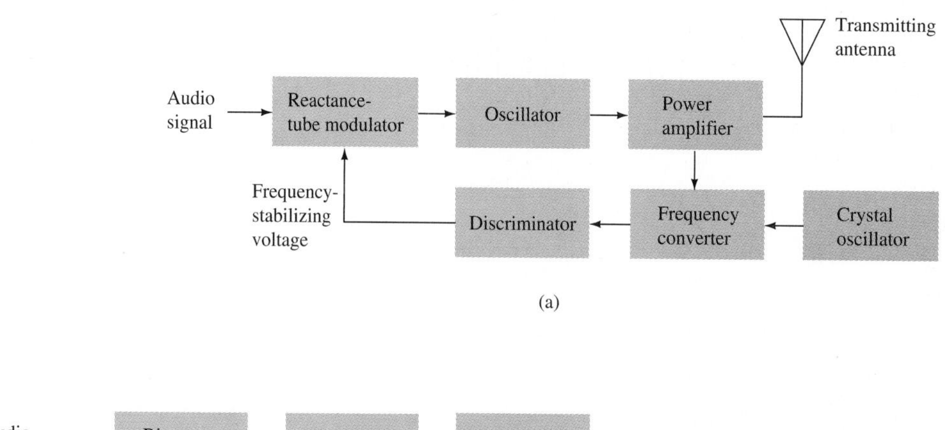

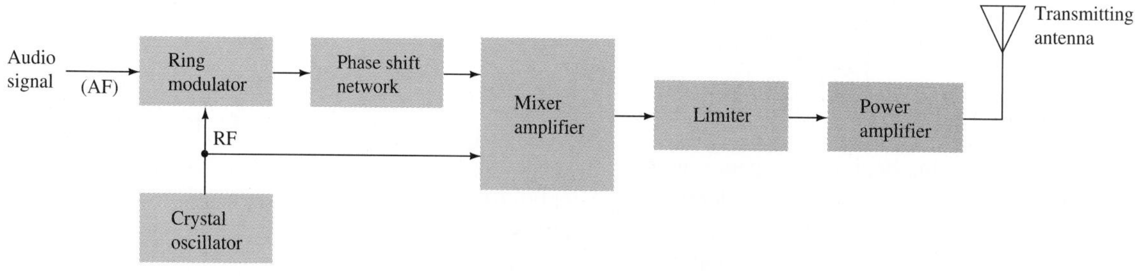

FIG. 12.13
(a) FM transmitter; (b) Armstrong FM transmitter.

diodes as voltage-controlled capacitive reactances. The main effect is the variation of oscillator carrier frequency by the audio signal. Where practical considerations require a lower oscillator frequency than the carrier value, frequency multiplier stages may be used before the power amplifier stage to set the desired carrier frequency. The modulated carrier is transmitted by the antenna. A part of this signal is generally connected back to be used to stabilize the carrier frequency. A crystal oscillator provides the frequency standard which the frequency converter uses to compare with the signal to be transmitted. The frequency difference applied to the discriminator results in a dc voltage to the varactor diode to stabilize the oscillator carrier frequency. This crystal control is quite important in keeping transmission at the allowed carrier frequency within the frequency tolerance permitted. Figure 12.13(b) shows an Armstrong FM transmitter. A crystal-controlled oscillator signal is combined with the input audio signal in a ring modulator whose output signal is phase-shifted and combined (mixed) with the crystal oscillator signal for stabilizing purposes. The resulting FM signal is amplitude-limited to provide a purely frequency-modulated signal to the power amplifier.

FM Reception

The FM radio wave received by an antenna is generally small enough to require some amplification before it can be processed (Fig. 12.14). The FM signal is then combined with a local oscillator signal in a mixer to obtain an IF signal (as in AM reception), with the audio information still carried as frequency variations. A series of IF amplifier stages are used to amplify the selected FM signal to the useful level of a few volts. A limiter circuit is used to ensure that no amplitude variation occurs. This frequency-modulated signal is then applied to discriminator FM detection circuits to convert the FM frequency variations into their original AF variations. The detector essentially provides a voltage output varied by the frequency variation of the FM signal. The audio signal is then amplified and used to drive output speakers.

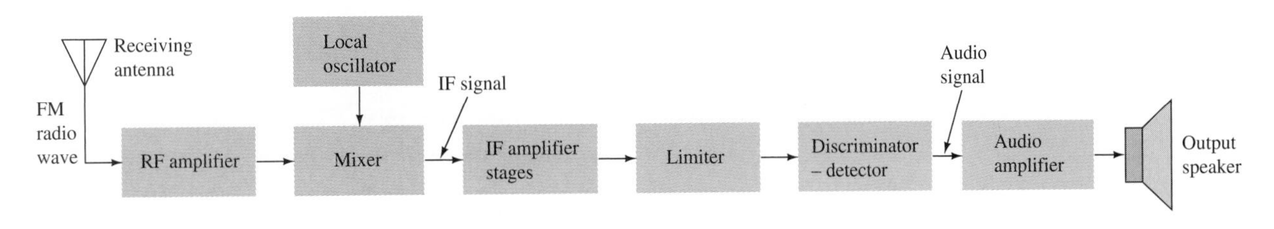

FIG. 12.14
FM receiver, block diagram.

Two basic circuits for converting an FM signal to audio are the discriminator [Fig. 12.15(a)] and the radio detector [Fig. 12.15(b)]. Both provide essentially the same circuit action of developing an output

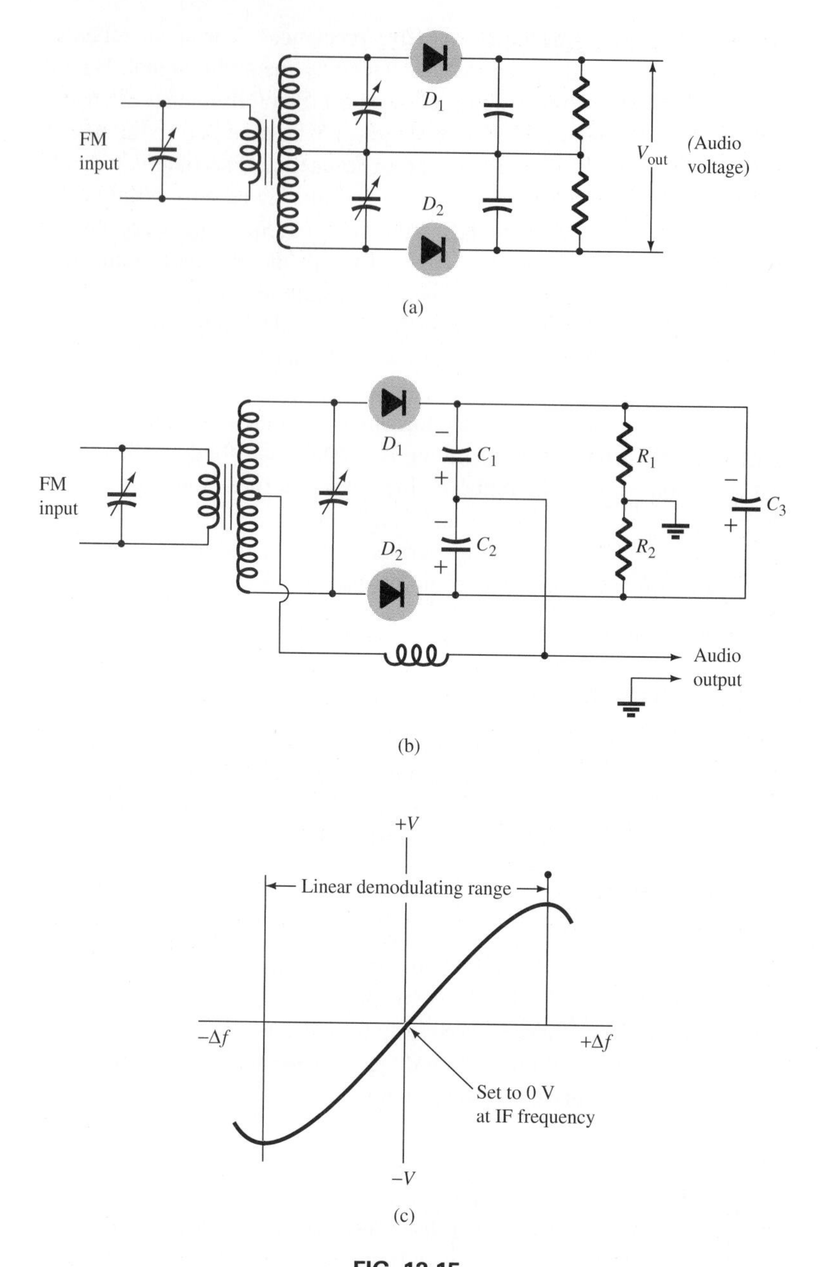

FIG. 12.15
(a) FM discriminator; (b) ratio detector; (c) frequency–response curve.

voltage dependent on the frequency of the input signal. This voltage–frequency relationship is shown in Fig. 12.15(c) as a frequency–response curve. With the demodulator circuit aligned so that no voltage results at the IF frequency, frequency variations above and below the carrier produce directly corresponding voltages. For a range of frequencies around circuit resonance the resulting voltage is linear with change in frequency.

12.4 OSCILLATOR CIRCUITS

An important class of electronic circuits consists of those that provide a sinusoidally varying signal at a frequency that can be adjusted or set to a desired value. An oscillator is a circuit in which part of the output signal is fed back or connected back to the input to add to the input signal (positive feedback) so that the circuit amplification increases until it goes into the desired state of oscillation. Various types of oscillators are built to provide either a very stable fixed frequency, a widely adjustable frequency range, a low-frequency signal, a high-frequency signal, or a nonsinusoidal-type signal.

Feedback Concepts

Part of the output signal from an amplifier circuit can be fed back to the input, with this feedback resulting in modified circuit operation. Feedback can be either voltage or current, and the feedback connection may be made either in series or shunt with the input. Each of these combinations results in somewhat different effects. If the feedback is connected so as to *oppose* the input, then negative feedback results, generally providing lowered amplifier gain and higher input impedance. This type of feedback is often used to provide a more stabilized gain setting. When the output is connected back to the input to aid the input, then positive feedback occurs, and under proper conditions the circuit will go into oscillation, usually varying sinusoidally, at a specific frequency set by particular circuit components.

A basic feedback circuit is shown in Fig. 12.16. The amplifier has a gain A $(=V_o/V_i)$, where V_i is not the direct input signal V_s but that value added to a fraction of the output signal (V_o), and where β is the fraction of the output fed back to the input. Expressing the overall circuit gain (V_o/V_s) as A_f (gain with feedback), we obtain*

$$A_f = \frac{A}{1 - \beta A} \tag{12.5}$$

If the amplifier is a single-stage (or an inverting) amplifier, then the value of A_f is less than A (negative feedback). For example, $A = -100$, $\beta = \frac{1}{10}$:

$$A_f = \frac{-100}{1 - (\frac{1}{10})(-100)} = \frac{-100}{1 + 10} = -9.1$$

The gain, with feedback, is thus reduced from -100 (without feedback) to -9.1 with improvements in gain stability, input, and output impedance.

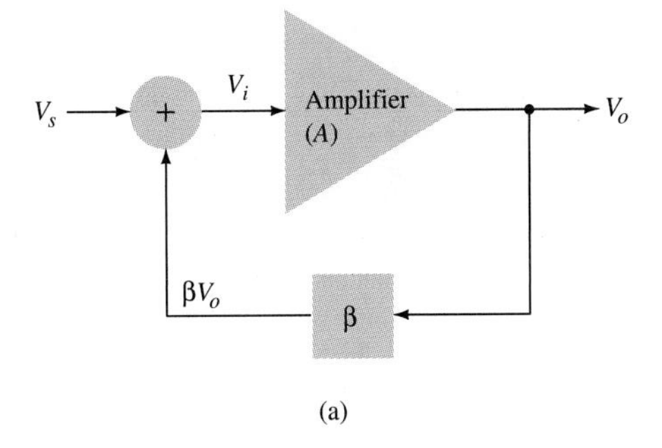

(a)

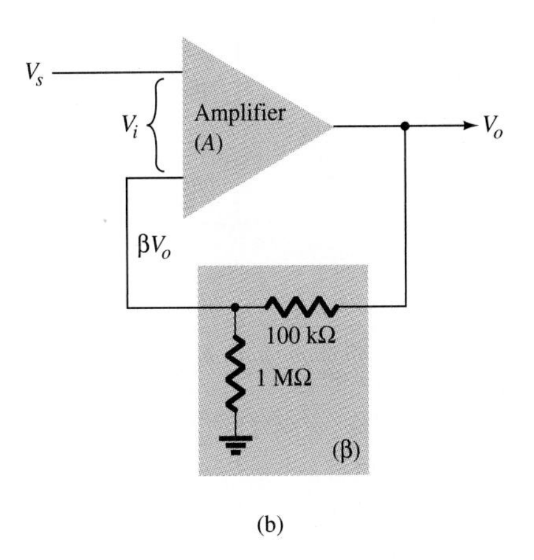

(b)

FIG. 12.16
Feedback circuit: (a) basic block diagram; (b) circuit schematic.

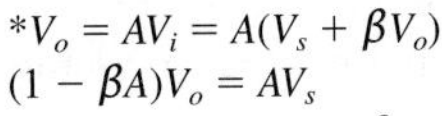

$*V_o = AV_i = A(V_s + \beta V_o)$
$(1 - \beta A)V_o = AV_s$
$A_f \equiv V_o/V_s = A/(1 - \beta A)$

Feedback Conditions

Considering all the possible conditions for the feedback circuit using Eq. (12.5), we can specify

1. If $\beta A < 0$ (negative), then A_f is less than A and the circuit is a negative-feedback amplifier.
2. If $0 < \beta A < 1$, then $A_f > A$ and the circuit acts as a positive-feedback amplifier.
3. If $\beta A \geq 1$, then the circuit goes into oscillation. If $\beta A = 1$, which is the Barkhausen criterion for undamped sinusoidal oscillation, $A_f \rightarrow \infty$.

As the values of β and A are adjusted so that βA approaches 1, the feedback circuit goes into oscillation. Since β and A are usually frequency-dependent, this condition for oscillation may occur over a limited frequency range and can be set to a specific value by specific circuit component values. Of the many types of sinusoidal oscillators some of the more popular ones to be described are the *LC* oscillators (Hartley and Colpitts), the crystal *RC* phase-shift oscillator, and the Wien bridge oscillator.

LC Oscillator Circuits

A basic *LC* oscillator circuit is shown in Fig. 12.17(a). For impedances Z_1 and Z_2 chosen as inductive reactances and impedance Z_3 as a capacitive reactance, the circuit is called a Hartley oscillator [Fig. 12.17(b)], while capacitive reactances used for Z_1 and Z_2 with Z_3 an inductive reactance result in a Colpitts oscillator circuit [Fig. 12.17(c)].

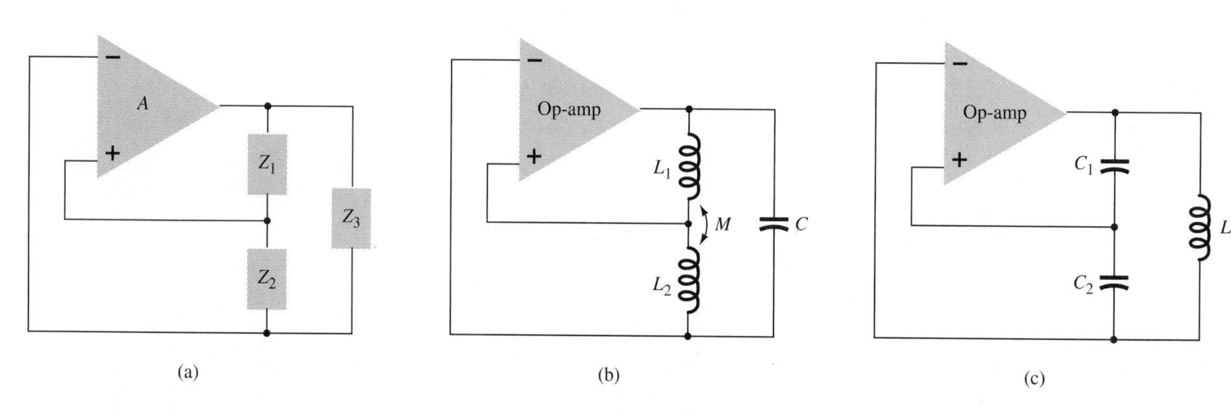

FIG. 12.17
(a) Basic LC oscillator circuit; (b) Hartley oscillator; (c) Colpitts oscillator.

The frequency of oscillation of the Hartley circuit can be determined from

$$f_o \cong \frac{1}{2\pi\sqrt{LC}} \tag{12.6}$$

where $L = L_1 + L_2 + 2M$, the total inductance of the coil (M is the coil mutual inductance). A coil of inductance $L = 800\ \mu\text{H}$ and a 120-pF capacitor used in a Hartley oscillator circuit would provide a sinusoidal signal at a frequency calculated to be

$$f_o = \frac{1}{6.28\sqrt{(800 \times 10^{-6})(120 \times 10^{-12})}} \cong 514\ \text{kHz}$$

For a Colpitts oscillator the resonant (oscillation) frequency is also given by

$$f_o \cong \frac{1}{2\pi\sqrt{LC}}$$

but in this case $C = C_1C_2/(C_1 + C_2)$, the series equivalent of the two capacitors.

Using a 400-μH coil with two 100-pF capacitors would result in an oscillator frequency of

$$f_o = \frac{1}{2\pi\sqrt{(400 \times 10^{-6})(50 \times 10^{-12})}} \cong 1.126\ \text{MHz}$$

where C was calculated to be

$$C = \frac{C_1C_2}{C_1 + C_2} = \frac{(100 \times 10^{-12})(100 \times 10^{-12})}{(100 \times 10^{-12}) + (100 \times 10^{-12})} = 50\ \text{pF}$$

Single-stage circuit versions of Hartley and Colpitts oscillator circuits using FET and bipolar transistor circuits are shown in Fig. 12.18. A

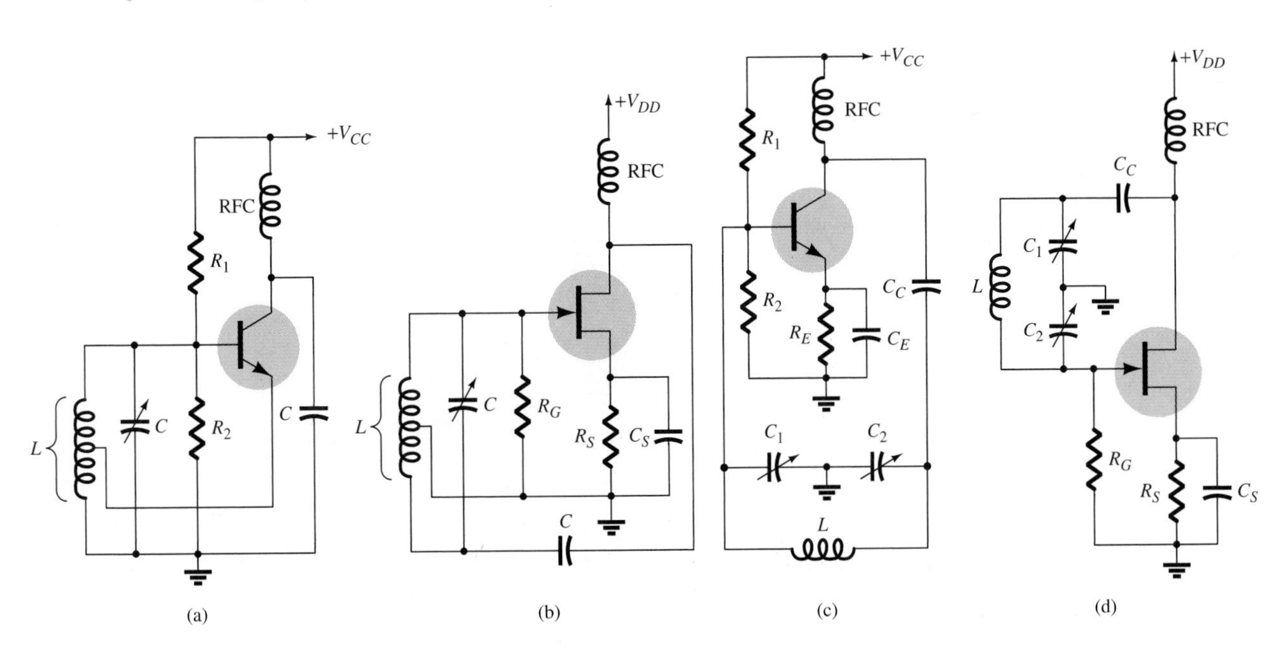

FIG. 12.18
(a) BJT Hartley; (b) JFET Hartley; (c) BJT Colpitts; (d) JFET Colpitts.

radio-frequency coil (RFC) of low dc impedance and very high ac impedance is used as the collector load impedance for the BJT and drain load impedance for the FET circuit. Both Hartley oscillator circuits use a tapped coil for the inductor (L), and an adjustable capacitor (C) is employed to adjust or vary the oscillator frequency. In the Colpitts oscillator circuit form the two capacitors C_1 and C_2 are usually ganged and adjusted as a pair to vary the oscillator frequency. Note the slightly different appearance of the feedback connection in Figs. 12.18(c) and (d), both of which connect output back to input and are electronically the same.

In all these circuits the frequency is essentially set as given in Eq. (12.5). It also is important that the gain A be large enough so that the condition $\beta A = 1$ is met. For a fixed value of β it is necessary that the circuit amplification be large enough so that oscillation occurs.

Crystal Oscillator Circuit

Where very precise frequency is desired, a crystal element is used to set the oscillator circuit frequency. A crystal is cut so that it has electromechanical resonance at a specific frequency. Actually, crystals exhibit series resonance, at which frequency the crystal acts as a very low-impedance element, or parallel resonance, at which frequency the crystal acts as a very high impedance element. There are many ways of connecting the crystal in an oscillator circuit. For example, the Colpitts circuit form can be modified to act as a crystal-controlled oscillator circuit, as shown in Fig. 12.19(a). The crystal operates here in its parallel-impedance mode, providing the highest impedance and greatest feedback at the crystal's parallel-resonance frequency. Figure 12.19(b) shows a Pierce crystal oscillator circuit with the crystal acting in its series-resonance mode, providing the least impedance and greatest feedback at the crystal's series-resonance frequency.

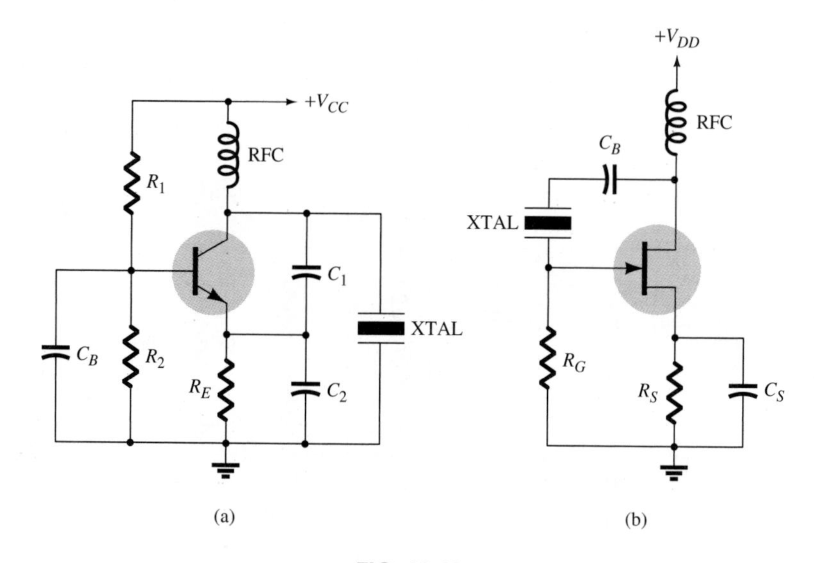

FIG. 12.19
Crystal oscillator circuits: (a) modified Colpitts; (b) Pierce.

Crystal oscillators develop a very stable, precise frequency that can be altered only by changing crystals. Transmission over AM, FM, or other frequency bands must be made at very precise frequencies with great stability maintained over a wide range of supply voltage and temperature, which requires the use of crystal-controlled oscillators. Crystals range in frequency from a few kilohertz to tens of megahertz for operation in the crystal's fundamental mode of oscillation (and in the hundreds of megahertz operating in an overtone mode). Crystals are also manufactured with various-sized and -shaped cuts for operation in the various frequency ranges, different temperature ranges, and filter or oscillator applications.

RC Phase-Shift Oscillation

A popular oscillator circuit using a resistor–capacitor network to set the oscillator frequency is the phase-shift circuit shown in Fig. 12.20. The op-amp circuit provides the gain of $-A$ required for the signal fed back through the *RC* phase-shift network (β) to supply the required $\beta A = 1$ for circuit oscillation. At the conditional $\beta A = 1$ the circuit oscillates at a frequency of

$$f_o = \frac{1}{2\pi\sqrt{6}\,RC} \tag{12.7}$$

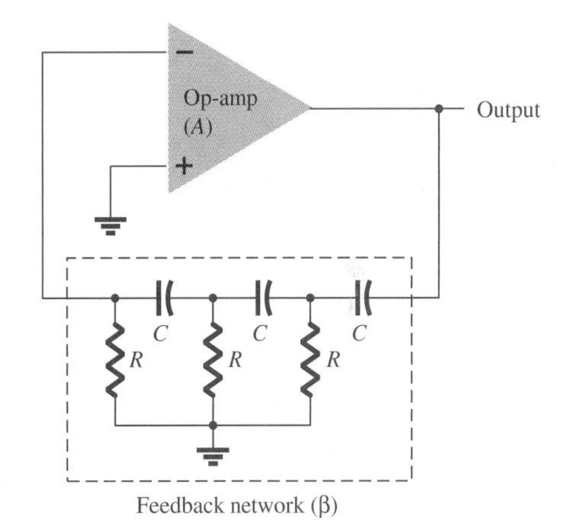

FIG. 12.20
RC phase-shift oscillator using op-amp.

provided that the amplifier gain A is greater than 29 (because $\beta = \frac{1}{29}$ at this frequency). For example, values of $R = 1\ \text{k}\Omega$ and $C = 1000$ pF would set the circuit oscillating at a frequency of

$$f_o = \frac{1}{2\pi(2.45)(1 \times 10^3)(1000 \times 10^{-12})} \cong 65\ \text{kHz}$$

Adjustment of the oscillator frequency can be made by varying either R or C.

The phase-shift network provides a phase shift of 180° to develop the required positive feedback, and the circuit oscillates at this frequency. Other networks may be used to give the required 180° phase shift, although the *RC* circuit in Fig. 12.20 is quite popular.

Practical FET and BJT transistor circuit versions are shown in Fig. 12.21. For the FET circuit [Fig. 12.21(a)] the amplifier will have a gain greater than 29 if the values of g_m and R_D are chosen so that

$$|A_v| = g_m R_D > 29$$

For example, a JFET having $g_m = 2$ mS and a drain resistance of 18 kΩ results in an amplifier gain of -36, enough to provide for oscillation at the frequency set by the feedback network. If the feedback components are $R = 10\ \text{k}\Omega$ and $C = 0.01\ \mu\text{F}$, then the oscillator will operate at

$$f_o = \frac{1}{2\pi\sqrt{6}\,RC} = \frac{1}{6.28(2.45)(10 \times 10^3)(0.01 \times 10^{-6})} \cong 650\ \text{Hz}$$

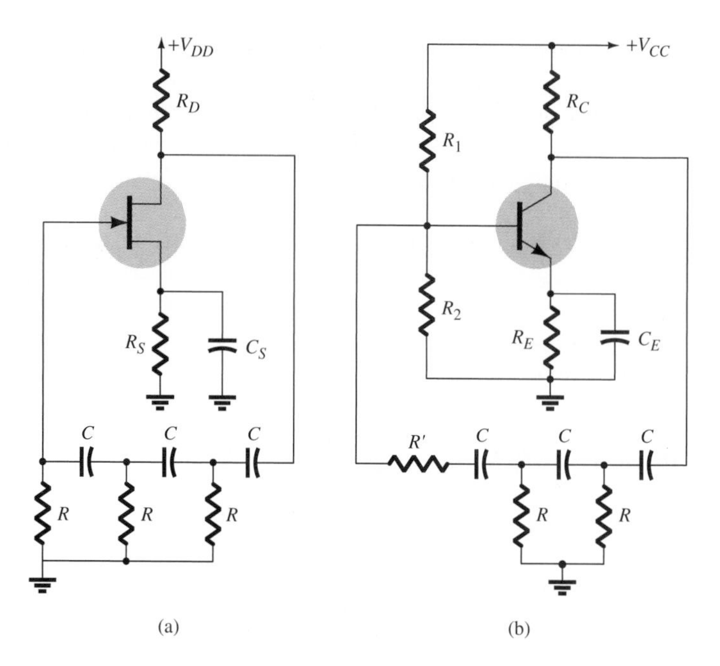

FIG. 12.21
Transistor RC phase-shift oscillator circuits: (a) FET; (b) BJT.

Changing C to 1000 pF would change the operating frequency by a factor of 10, to 6.5 kHz. Changing C to 100 pF and R to 1 kΩ would result in an operating frequency of 65 kHz. It should be noted here that reducing R any lower might result in loading the amplifier, thereby reducing the gain, and no oscillation would occur. The capacitor value could still be lowered to, say, 10 pF, but the stray circuit capacitance would affect the circuit operation. At $C = 10$ pF the frequency of operation would be 650 kHz, showing that upper frequencies of about 1 MHz begin to cause practical circuit effects, limiting the upper range to this low-megahertz limit. At the lower frequency the values of R up to about 10 MΩ and C to, say, 1000 μF would result in cutoff frequencies below 1 Hz, at which value the effect of C_E would result in lowered gain, possibly preventing the circuit from oscillating. Although the calculated frequency can be for any given set of values for R and C, there are practical circuit limits that must always be taken into consideration.

The BJT oscillator circuit [Fig. 12.21(b)] is seen to be slightly modified. The third RC section does not have a direct connection to ground. Since the ac impedance of R_1 in parallel with R_2 provides an impedance to ground, the resistor R' is chosen so that in series with $R_1 \parallel R_2$ it provides an effective impedance of R, as desired. The amplifier gain must be greater than 29, which requires a transistor with gain h_{fe} such that

$$h_{fe} > 23 + 29\frac{R}{R_C} + 4\frac{R_C}{R} \qquad \textbf{(12.8)}$$

For $R = 10\text{ k}\Omega$ and $R_C = 10\text{ k}\Omega$ we would need

$$h_{fe} > 23 + 29\left(\frac{10\text{ k}\Omega}{10\text{ k}\Omega}\right) + 4\left(\frac{10\text{ k}\Omega}{10\text{ k}\Omega}\right) = 23 + 29 + 4 = 56$$

so that a transistor having h_{fe} of, say, 60 would just cause oscillation to occur.

Note here the practical effect of choosing R greater than R_C, requiring very high transistor gain. Typically $R_C \cong R$ for practical values of h_{fe} around 50.

The oscillator frequency for this circuit is also affected by the R/R_C ratio, where

$$f_o = \frac{1}{2\pi RC\sqrt{6 + 4(R_C/R)}} \tag{12.9}$$

which for $C = 1000\text{ pF}$, with $R = R_C = 10\text{ k}\Omega$, would result in

$$f_o = \frac{1}{2\pi(10 \times 10^3)(1000 \times 10^{-12})\sqrt{6 + 4(10\text{ k}\Omega/10\text{ k}\Omega)}} \cong 5\text{ kHz}$$

Wien Bridge Oscillator

Another circuit using resistors and capacitors to set the oscillator frequency is the Wien bridge circuit shown using an op-amp in Fig. 12.22. The circuit form is seen to be a bridge connection. Resistors R_1 and R_2 and capacitors C_1 and C_2 form the frequency-adjustment

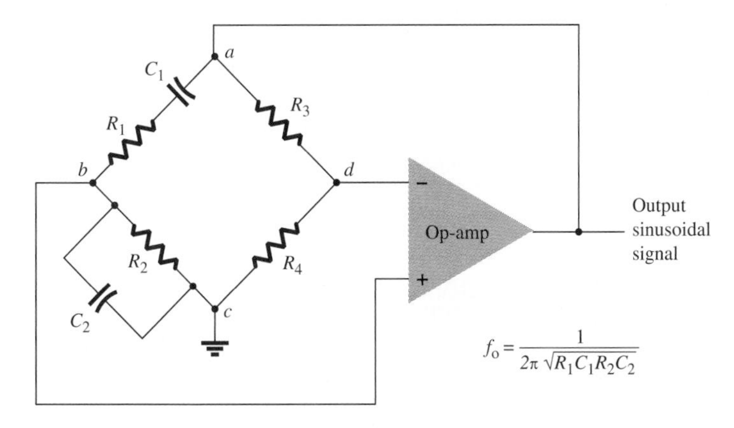

FIG. 12.22
Wien bridge oscillator circuit using op-amp amplifier.

elements, while resistors R_3 and R_4 form part of the feedback path. The op-amp output is connected as the bridge input at points a and c. The bridge circuit output at points b and d is the input to the op-amp.

The frequency of oscillation can be calculated to be

$$f_o = \frac{1}{2\pi\sqrt{R_1C_1R_2C_2}} \tag{12.10a}$$

If the values are chosen identically so that $R_1 = R_2 = R$ and $C_1 = C_2 = C$, then

$$f_o = \frac{1}{2\pi RC} \tag{12.10b}$$

For sufficient loop gain it is necessary that the ratio of R_3 to R_4 be

$$\frac{R_3}{R_4} = \frac{R_1}{R_2} + \frac{C_2}{C_1} \tag{12.11a}$$

which for identical resistors and capacitors is

$$\frac{R_3}{R_4} = 2 \tag{12.11b}$$

For example, $R = 10\text{ k}\Omega$ and $C = 0.001\ \mu\text{F}$ would result in an oscillator frequency of

$$f_o = \frac{1}{6.28(10 \times 10^3)(0.001 \times 10^{-6})} = 15.9\text{ kHz}$$

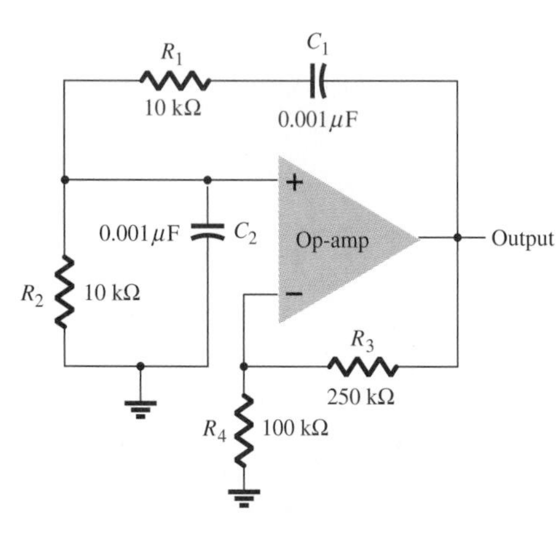

FIG. 12.23
Practical Wien bridge oscillator circuit.

A practical Wien bridge oscillator circuit is shown in Fig. 12.23. This is exactly the same bridge connection as in Fig. 12.22 but redrawn in a more common form.

12.5 VOLTAGE-CONTROLLED OSCILLATOR CIRCUIT

A voltage-controlled oscillator (VCO) is an IC device that provides an output frequency that is a linear function of a controlling voltage. The 566 IC is a popular unit containing circuitry to generate both square-wave and triangular-wave signals whose frequency is set by an external resistor and capacitor and then varied by an applied dc voltage. Figure 12.24 shows that the 566 contains current sources to charge and discharge an external capacitor C_1 at a rate set by external resistor R_1 and the modulating dc input voltage.

A Schmitt trigger circuit is used to switch the current sources between charging and discharging the capacitor, and the triangular volt-

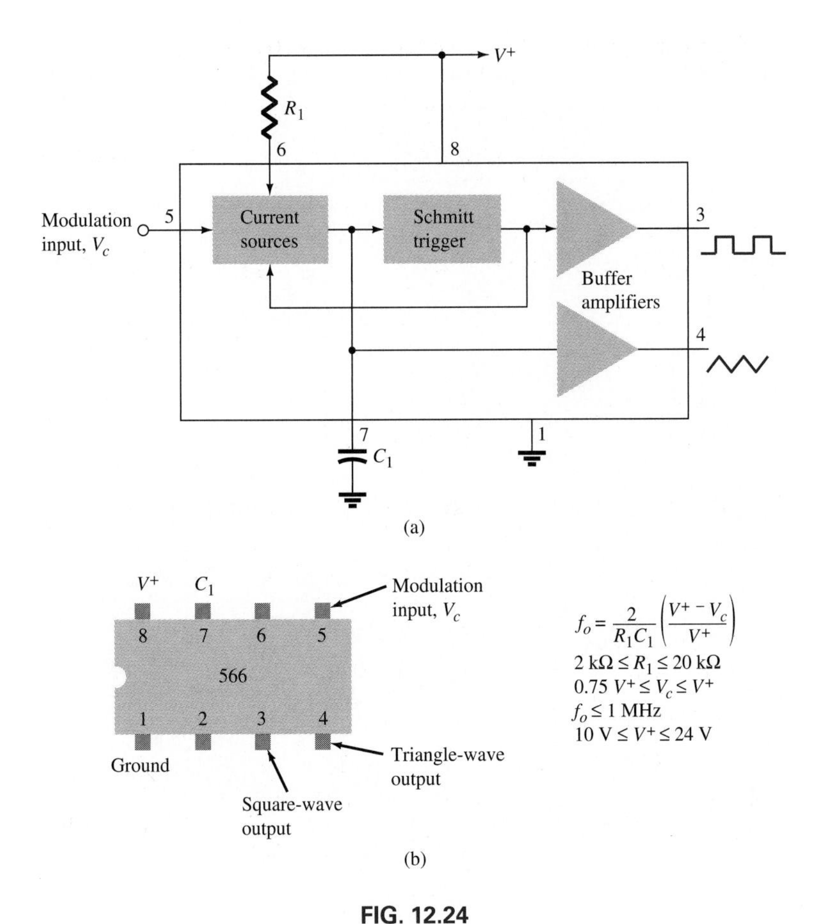

FIG. 12.24

566 function generator: (a) block diagram; (b) pin configuration and summary of operating data.

ages developed across the capacitor and the square wave from the Schmitt trigger are provided as outputs through buffer amplifiers. Figure 12.24 shows the pin connections of the 566 unit and a summary of formula and value limits. The oscillator can be programmed over a 10-to-1 frequency range by proper selection of an external resistor and capacitor and then modulated over an additional 10-to-1 frequency range by a control voltage, V_C.

A free-running or center-operating frequency, f_o, can be calculated from

$$f_o = \frac{2}{R_1C_1}\left(\frac{V^+ - V_C}{V^+}\right) \qquad \textbf{(12.12)}$$

Figure 12.25 shows an example in which the 566 function generator is used to provide both square-wave and triangular-wave signals at a fixed frequency set by R_1, C_1, and V_C.

EXAMPLE 12.1 Calculate the center frequency for the VCO in Fig. 12.25 when V_C is set at 10.4 V.

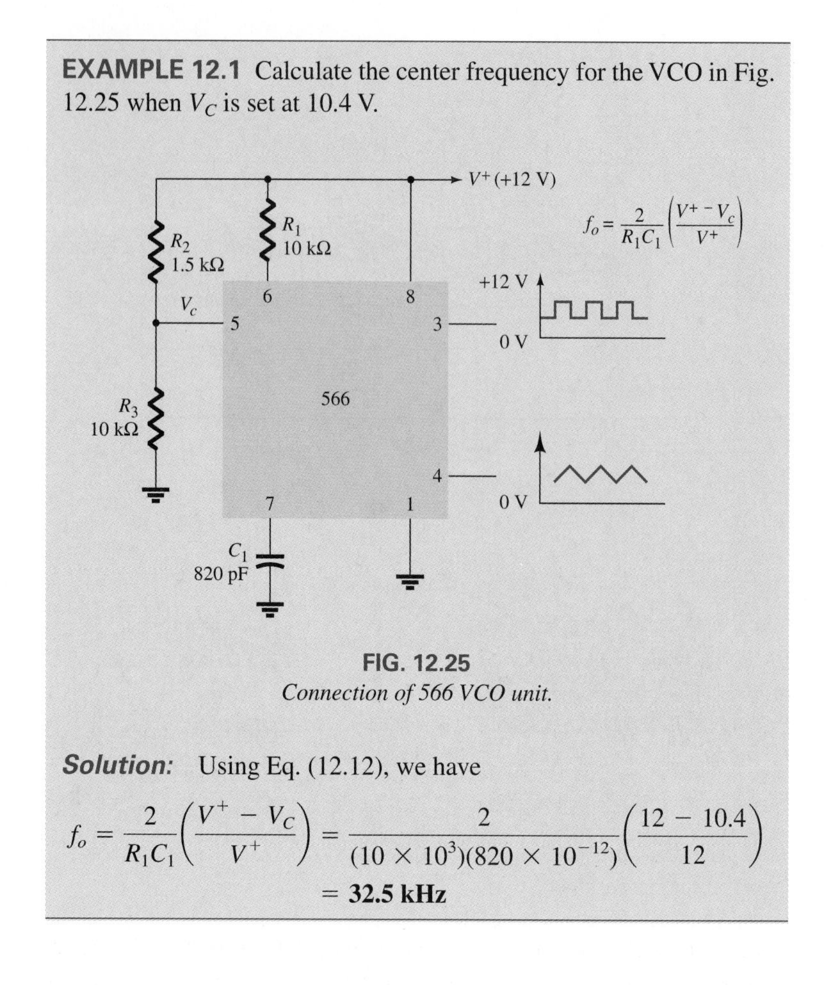

FIG. 12.25
Connection of 566 VCO unit.

Solution: Using Eq. (12.12), we have

$$f_o = \frac{2}{R_1C_1}\left(\frac{V^+ - V_C}{V^+}\right) = \frac{2}{(10 \times 10^3)(820 \times 10^{-12})}\left(\frac{12 - 10.4}{12}\right)$$
$$= \mathbf{32.5\ kHz}$$

12.6 PHASE-LOCKED LOOP

A phase-locked loop (PLL) is an electronic circuit that consists of a phase detector, a low-pass filter, and a voltage-controlled oscillator connected as shown in Fig. 12.26. Common applications of a PLL include (1) frequency synthesizers that provide multiples of a reference signal frequency (for example, the carrier frequency for the multiple channels of a citizens-band (CB) unit or marine-radio-band unit can be generated using one crystal-controlled frequency and its multiples generated using a PLL); (2) FM demodulation networks for FM operation with excellent linearity between the input signal frequency and the PLL output voltage; (3) demodulation of the two data transmission or carrier frequencies in digital data transmission used in frequency-shift keying (FSK) operation; and (4) a wide variety of areas including modems, telemetry receivers and transmitters, tone decoders, AM detectors, and tracking filters.

An input signal, V_i, and that from a VCO, V_o, are compared by a phase comparator (refer to Fig. 12.26), providing an output voltage, V_e, that represents the phase difference between the two signals. This voltage is then fed to a low-pass filter that provides an output voltage (amplified if necessary) that can be taken as the output voltage from the

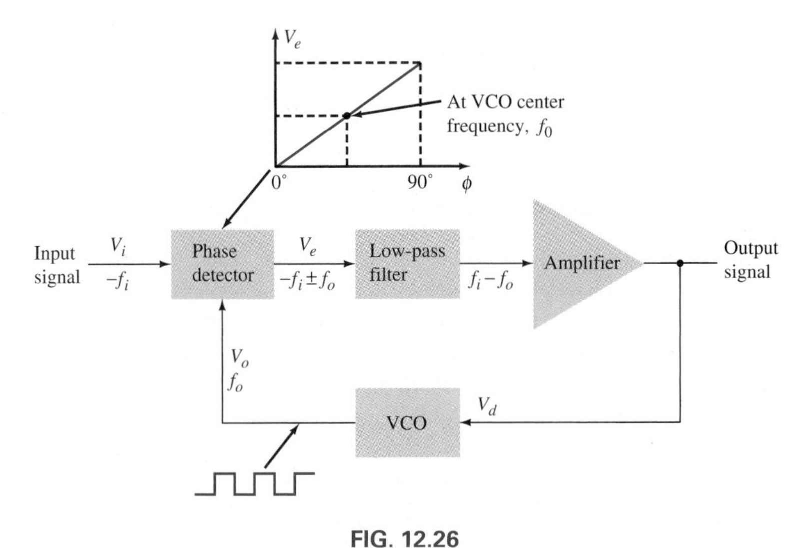

FIG. 12.26
Block diagram of basic phase-locked loop.

PLL and is used internally as the voltage to modulate the VCO's frequency. The closed-loop operation of the circuit is to maintain the VCO frequency *locked* to that of the input signal frequency.

Basic PLL Operation

The basic operation of a PLL circuit can be explained using the circuit in Fig. 12.26 as a reference. We shall first consider the operation of the various circuits in the phase-locked loop when the loop is operating in lock (input signal frequency and VCO frequency are the same). When the input signal frequency is the same as that from the VCO to the comparator, the voltage, V_d, taken as output is the value needed to hold the VCO in lock with the input signal. The VCO then provides output of a fixed-amplitude square-wave signal at the frequency of the input. The best operation is obtained if the VCO center frequency, f_o, is set with the dc bias voltage midway in its linear operating range. The amplifier allows this adjustment in dc voltage from that obtained as output of the filter circuit. When the loop is in lock, the two signals to the comparator are of the same frequency although not necessarily in phase. A fixed phase difference between the two signals to the comparator results in a fixed dc voltage to the VCO. Changes in the input signal frequency then result in a change in the dc voltage to the VCO. Within a capture-and-lock frequency range, the dc voltage drives the VCO frequency to match that of the input.

While the loop is trying to achieve lock, the output of the phase comparator contains frequency components at the sum and difference of the signals compared. A low-pass filter passes only the lower-frequency component of the signal so that the loop can obtain lock between input and VCO signals.

Owing to the limited operating range of the VCO and the feedback connection of the PLL circuit, there are two important frequency bands specified for a PLL. The *capture range* of a PLL is the frequency range centered about the VCO free-running frequency, f_o, over which the loop can *acquire* lock with the input signal. Once the PLL has achieved capture, it can maintain lock with the input signal over a somewhat wider frequency range called the *lock range*.

Frequency Demodulation

FM demodulation or detection can be directly achieved using the PLL circuit. If the PLL center frequency is selected or designed at the FM carrier frequency, the filtered or output voltage in the circuit in Fig. 12.26 is the desired demodulated voltage, varying in value proportionally to the variation of the signal frequency. The PLL circuit thus operates as a complete intermediate-frequency strip, limiter, and demodulator as used in FM receivers.

One popular PLL unit is the 565 shown in Fig. 12.27(a). The 565 contains a phase detector, amplifier, and voltage-controlled oscillator, which are only partially connected internally. An external resistor and capacitor, R_1 and C_1, are used to set the free-running or center frequency of the VCO. Another external capacitor, C_2, is used to set the low-pass filter pass band, and the VCO output must be connected back as input to the phase detector to close the PLL loop. The 565 typically uses two power supplies, V^+ and V^-.

Figure 12.27(b) shows the 565 PLL connected to work as an FM demodulator. Resistor R_1 and capacitor C_1 set the free-running frequency,

$$f_o = \frac{0.3}{R_1 C_1} \tag{12.13}$$

therefore

$$f_o = \frac{0.3}{(10 \times 10^3)(220 \times 10^{-12})} = 136.36 \text{ kHz}$$

with the limitation $2\text{ k}\Omega \leq R_1 \leq 20\text{ k}\Omega$. The lock range is then

$$f_L = \pm \frac{8 f_o}{V} \tag{12.14}$$

so

$$f_L = \pm \frac{8(136.36 \times 10^3)}{6} = \pm 181.8 \text{ kHz}$$

and the capture range is

$$f_C = \pm \frac{1}{2\pi}\sqrt{\frac{2\pi f_L}{(3.6 \times 10^3)C_2}} \tag{12.15}$$

therefore

$$f_C = \pm \frac{1}{2\pi}\sqrt{\frac{2\pi(181 \times 10^3)}{(3.6 \times 10^3)(330 \times 10^{-12})}} = 156.1 \text{ kHz}$$

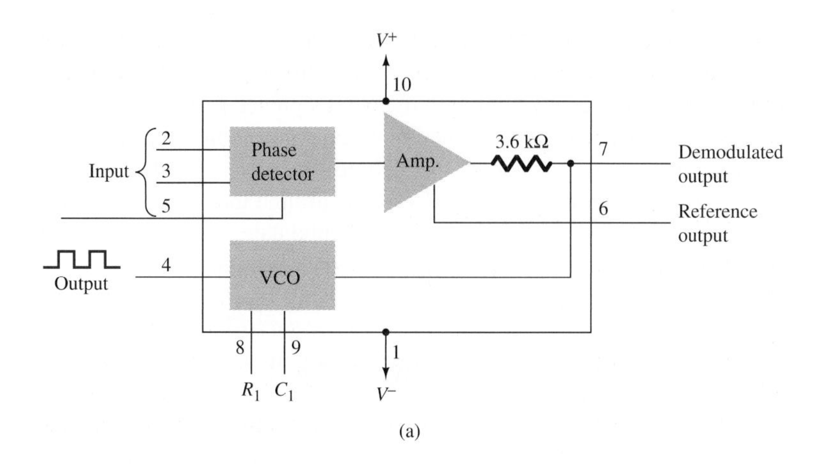

(a)

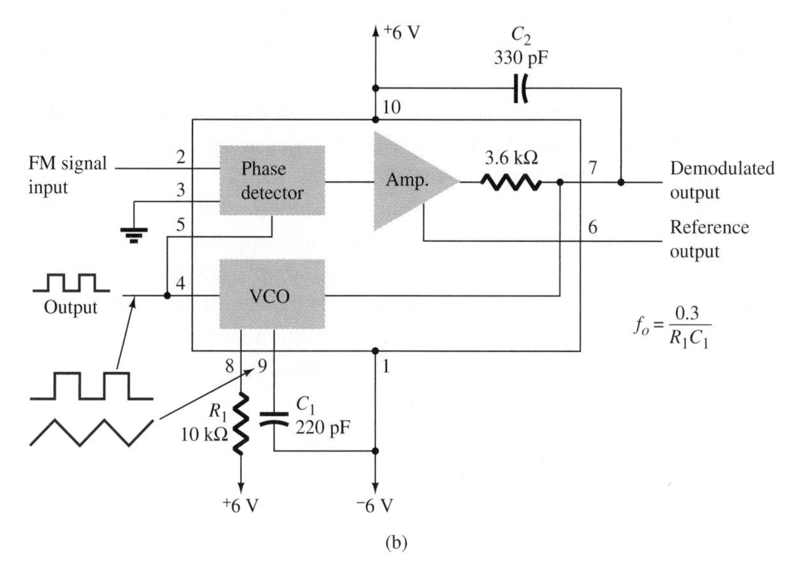

(b)

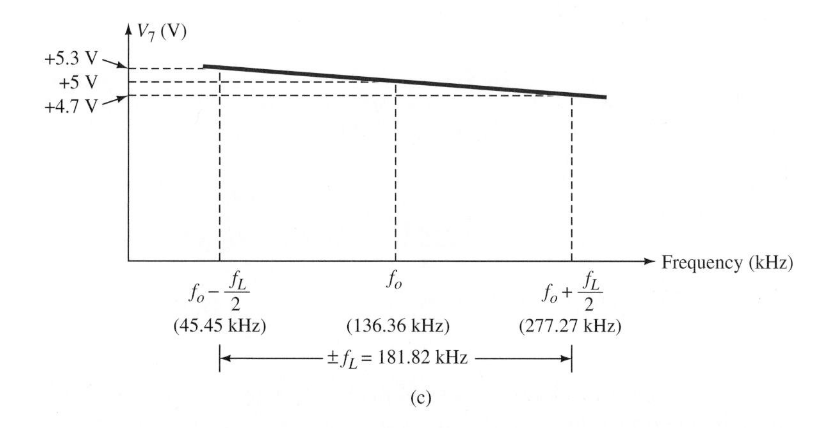

(c)

FIG. 12.27

(a) Block diagram of 565 PLL unit; (b) connection as FM demodulator; (c) output voltage–frequency relation.

The signal at pin 4 is a 136.36-kHz square wave. An input within the lock range of 181.8 kHz will cause the output voltage at pin 7 to vary around its dc voltage level set with the input signal at f_o. Figure 12.27(c) shows the output at pin 7 as a function of the input signal frequency. The dc voltage at pin 7 is linearly related to the input signal frequency within the frequency range $f_L = 181.8$ kHz around the center frequency 136.36 kHz. The output voltage is the demodulated signal that varies with frequency within the operating range specified.

12.7 DIGITAL COMMUNICATIONS CONCEPTS

Digital communication uses a modulator to transmit the digital data (bits) over a channel (typically a phone line or microwave link). At the receiving end a demodulator extracts the digital data to complete the transmission. The rate of data transmission over standard telephone lines is typically at data rates of 56,000 (56K) bits per second. Long-distance transmission is accomplished over microwave links or over satellite links. In the latter cases the data may be encoded using encryption standards that permit secure data communications.

A modem or modulator–demodulator is the unit that converts digital data to the phone frequencies for transmission and then back to digital data at the receiver end. A simple serial modem transmission scheme transmits binary (two-level) data. In that case the baud rate is the same as the number of bits per second. If the modulation scheme provides, for example, two bits transmitted at a time, the baud rate will be half the bit transmission rate. Baud rate may be defined as

$$\text{baud rate} = \frac{\text{total data rate (bits per second)}}{\text{number of simultaneous bits}} \quad \textbf{(12.16)}$$

The bandwidth of the data line then is

$$\text{bandwidth} = 0.5 \times \text{baud rate} \quad \textbf{(12.17)}$$

EXAMPLE 12.2 Calculate the baud rate and bandwidth of a modem providing 2 bits at a time at a total data rate of 19,200 bps.

Solution:

$$\text{band rate} = \frac{19{,}200}{2} = \mathbf{9600\ baud}$$

$$\text{bandwidth} = 0.5 \times (9600\ \text{baud}) = \mathbf{4800\ Hz}$$

When data are transmitted via a modem, they may be sent in *simplex* (only in one direction), in *half-duplex* (one channel used to send data in forward and in reverse directions), or in *full-duplex* (one channel used to transmit data in both directions at the same time).

Low-speed modems transmit data at rates from 4800 bps to about 115,200 bps. Higher-speed modems must carry more than 2 bits simultaneously to maintain the bandwidth with that of the phone line (typically from 300 Hz to about 3000 Hz).

Frequency Synthesis

A frequency synthesizer can be built around a PLL as shown in Fig. 12.28. A frequency divider is inserted between the VCO output and the phase comparator so that the loop signal to the comparator is at frequency f_o, while the VCO output is Nf_o. This output is a multiple of the input frequency as long as the loop is in lock. The input signal can be crystal-stabilized at f_1 with the resulting VCO output at Nf_1 if the loop is set up to lock at the fundamental frequency (when $f_o = f_1$).

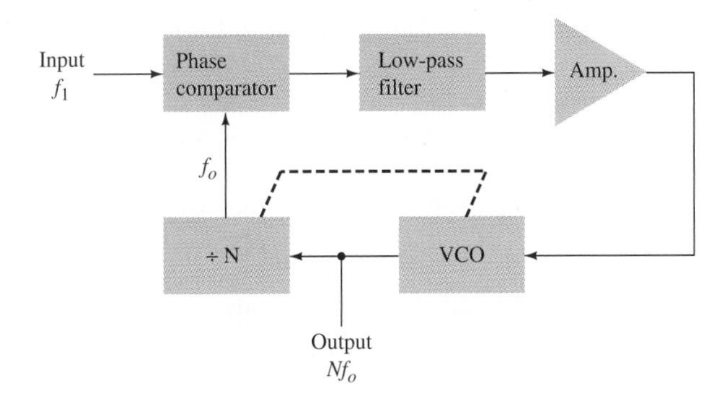

FIG. 12.28
Connection of PLL as frequency synthesizer.

12.8 SUMMARY

am Concepts:

Carrier Voltage:

$$v = V_c \cos \omega_c t$$

Audio Signal:

$$v_m = V_m (m \cos \omega_m t)$$

Modulated Carrier Signal:

$$\begin{aligned} v &= V_m (1 + m \cos \omega_m t) \cos \omega_c t \\ &= V_m \cos \omega_c t + 0.5\, mV_m \cos(\omega_c + \omega_m)t + 0.5\, mV_m \cos(\omega_c - \omega_m)t \end{aligned}$$

Frequency Modulation:

Index of Modulation:

$$m_f = \frac{f_d}{f_s}$$

Feedback:

$$A_f = \frac{A}{1 - \beta A}$$

LC and Colpitts Oscillator:

$$f_o = \frac{1}{2\pi\sqrt{LC}}$$

RC Phase-shift Oscillator:

$$f_o = \frac{1}{2\pi\sqrt{6}\,RC}$$

Wien Bridge Oscillator:

$$f_o = \frac{1}{2\pi\sqrt{R_1C_1R_2C_2}}$$

For $R_1 = R_2$ and $C_1 = C_2$

$$f_o = \frac{1}{2\pi RC}$$

Baud Rate:

$$\text{Baud rate} = \frac{\text{Total data rate (bits per second)}}{\text{Number of simultaneous bits}}$$

PROBLEMS

SECTION 12.2 AM Radio Concepts

1. Show how the trigonometric relationship

$$\cos A \cos B = \tfrac{1}{2}[\cos(A + B) + \cos(A - B)]$$

is used to express Eq. (12.1) in the form of Eq. (12.2).

2. What are the side bands expressed in Eq. (12.2)?

3. If $f_c = 900$ kHz and $f_m = 10$ kHz, what are the side-band frequencies?

4. If an unmodulated carrier is 25 V peak and the modulating carrier signal is 4 V peak, what is the modulation index?

5. What is the difference between the received RF signal and the IF signal?

6. For an AM radio intermediate frequency of 455 kHz, what local oscillator frequency will tune in a 1.6-MHz radio station signal?

7. For an AM-modulated wave described by

$$v(t) = [60 + 30 \cos 2\pi(6 \times 10^3)t] \cos 2\pi(880 \times 10^3)t$$

a. What is the carrier frequency and the modulating frequency of the AM wave?
b. What is the value of the modulation index m?
c. What is the amplitude of the unmodulated carrier?

8. What is a superheterodyne receiver? What are its advantages?

9. Draw the block diagram of a superheterodyne AM receiver and list typical signal frequency values.

SECTION 12.4 Oscillator Circuits

10. What does the value of β represent in a feedback circuit?

11. How do positive and negative feedback differ?

12. How does the gain with feedback differ from the gain without feedback?

13. How do Colpitts and Hartley oscillators differ?

14. What is the resonant frequency of a Hartley oscillator having total inductance $L = 650\ \mu\text{H}$ and capacitance $C = 240\ \text{pF}$?

15. What is the resonant frequency of a Colpitts oscillator having $C_1 = 150\ \text{pF}$, $C_2 = 250\ \text{pF}$, and $L = 300\ \mu\text{H}$?

16. Calculate the resonant frequency of an *RC* Wien bridge oscillator circuit having $R_1 = R_2 = R = 2.4\ \text{k}\Omega$ and $C_1 = C_2 = C = 2500\ \text{pF}$.

17. What values of equal capacitors are needed to provide a Wien bridge oscillator frequency of $f_o = 80\ \text{kHz}$ using $R_1 = R_2 = R = 4.4\ \text{k}\Omega$?

18. Calculate the oscillator frequency of a phase-shift oscillator circuit having $R = 7.2\ \text{k}\Omega$ and $C = 0.002\ \mu\text{F}$.

19. What value capacitor is needed in a phase-shift oscillator to provide a resonant frequency of 12.5 kHz using $R = 5\ \text{k}\Omega$?

20. Calculate the resonant frequency of a Wien bridge oscillator circuit having $R_1 = R_2 = R = 3.6\ \text{k}\Omega$ and $C_1 = C_2 = C = 1200\ \text{pF}$.

21. What value of equal capacitors is needed to provide a resonant frequency of 75 kHz using equal resistors $R = 4.7\ \text{k}\Omega$ in a Wien bridge oscillator circuit?

22. Draw the circuit diagram of a phase-shift oscillator using FETs.

23. Draw the circuit diagram of a bipolar *RC* phase-shift oscillator circuit.

24. Draw the circuit diagram of an op-amp Wien bridge oscillator.

SECTION 12.5 Voltage-Controlled Oscillator Circuit

25. Calculate the center frequency for the VCO in Fig. 12.25 when V_C is set at 6.8 V (with $V_{CC} = 9\ \text{V}$).

26. If a 5-kΩ potentiometer is connected in series between resistors R_2 and R_3 in Fig. 12.25 with the wiper arm providing V_C to pin 5, what range of frequency is provided when the potentiometer is adjusted between 0 Ω and 5 kΩ?

27. What value of capacitor should be used to provide a center frequency of about 100 kHz in the circuit in Fig. 12.25? ($V_C = 10.4\ \text{V}$).

SECTION 12.6 Phase-Locked Loop

28. Calculate the free-running frequency of the circuit in Fig. 12.27(b) using $C_1 = 0.00005\ \mu\text{F}$.

29. Calculate the lock and capture frequencies for the circuit in Fig. 12.27(b) using capacitor $C_1 = 0.00005\ \mu\text{F}$.

30. What value of capacitor C_1 is required to obtain a free-running frequency of 1 MHz in the circuit in Fig. 12.27(b) using $R_1 = 1\ \text{k}\Omega$?

13 Digital Computers

13.1 INTRODUCTION

Digital concepts and digital circuits have become an essential part of any electronic product. While analog circuits deal with voltages that can vary throughout any desired range from negative to positive values, a digital circuit essentially is at one of two possible levels. At present these levels are typically 0 V and +5 V, which represent logical-0 and logical-1, respectively.

Digital circuits deal with binary (base 2) numbers represented as 0 or 1. Grouping of these binary numbers provides decimal or hexadecimal (base 16) values. These various number systems can all be accounted for using only binary digits (*bits*).

Alphabetic and other numeric values can be represented using special groupings or code forms of these 1s and 0s. For example, a group of 7 bits is used in the American Standard Code for Information Interchange (ASCII) to represent all the various keyboard characters. Another code form, the seven-segment code is used to represent the numbers in a light-emitting diode or liquid-crystal diode display. Code forms are also used to provide security.

Logic circuits are AND gates, OR gates, and inverters (or combinations such as NAND gates, NOR gates, and exclusive-OR gates). Using these basic circuits one can build any desired logical operation. In fact, all digital units, microprocessors, and complete computer units are built using these basic circuits or gates.

Connections of logic circuits to form multivibrator circuits provide for bistable multivibrator or flip-flop circuits, monostable multivibrator circuits or single-shot circuits, and astable multivibrator or clock circuits. These circuits are used to build registers, counters, and the various parts of a microprocessor unit.

Memory circuits provide either dynamic or static random access memory (RAM), which, along with a microprocessor, forms the basic part of a computer. Memory is typically organized using 8 bits, or a *byte*. Memory ICs from 1 MB (1 million bytes) to 256 MB are currently quite popular. At present these circuits typically operate at speeds of under 50 ns (50×10^{-9} s).

The addition of input–output units—keyboards, magnetic disks, printers, and so on—completes a digital computer system. Also, circuits that convert signals from digital to analog and analog to digital are used to connect a computer to various equipment, as in cars, CD players, and appliances.

13.2 NUMBER SYSTEMS

Whereas we are first taught to use decimal numbers, computer circuits operate in binary (base 2) numbers. Digital circuits provide voltages that are high or low and devices that are on or off, all by supplying logic signals representing logical-1 or logical-0.

Decimal Number System

When the decimal number 982, for example, is read, it stands for

9 hundreds, 8 tens, and 2 units

This can be expressed more systematically as

$$9 \times 100 + 8 \times 10 + 2 \times 1$$

or also as

$$9 \times 10^2 + 8 \times 10^1 + 2 \times 10^0$$

A general expression for any integer decimal number is

$$N = d_n 10^n + \cdot \cdot \cdot + d_3 10^3 + d_2 10^2 + d_1 10^1 + d_0 10^0$$

The *digits* of the base 10 number system are

0, 1, 2, 3, 4, 5, 6, 7, 8, 9

(Digits number in value to 1 less than the base—10 in the case of decimal numbers.)

Binary Number System

A binary number is made up of the digits 0 and 1 (base 2). Since binary numbers represent values in the base 2 number system, they may be expressed, in general, in the form

$$N = d_n 2^n + \cdot \cdot \cdot + d_3 2^3 + d_2 2^2 + d_1 2^1 + d_0 2^0$$

The binary number 110101 is then

$$
\begin{aligned}
N &= 1 \times 2^5 + 1 \times 2^4 + 0 \times 2^3 + 1 \times 2^2 + 0 \times 2^1 + 1 \times 2^0 \\
&= 1 \times 32 + 1 \times 16 + 0 \times 8 + 1 \times 4 + 0 \times 2 + 1 \times 1 \\
&= 32 + 16 + 4 + 1 = 53_{10}
\end{aligned}
$$

EXAMPLE 13.1 Write the binary numbers from 1 to 20.

Solution:

Decimal Number	Binary Number	Decimal Number	Binary Number
0	00000	11	01011
1	00001	12	01100
2	00010	13	01101
3	00011	14	01110
4	00100	15	01111
5	00101	16	10000
6	00110	17	10001
7	00111	18	10010
8	01000	19	10011
9	01001	20	10100
10	01010		

Converting Binary to Decimal

Converting a binary number to decimal only requires adding the values for each position of the number having a binary digit of 1.

EXAMPLE 13.2 Determine the decimal value of binary 10011011.

Solution:

$$
\begin{aligned}
N = 1 \times 2^7 + 0 \times 2^6 + 0 \times 2^5 + 1 \times 2^4 + 1 \times 2^3 \\
+ 0 \times 2^2 + 1 \times 2^1 + 1 \times 2^0 \\
= 128 + 0 + 0 + 16 + 8 + 0 + 2 + 1 = \mathbf{155_{10}}
\end{aligned}
$$

Converting Decimal to Binary

Converting a decimal number to binary can be carried out by repeatedly dividing by 2 and keeping track of the remainder, as demonstrated in the following example.

EXAMPLE 13.3 Convert decimal 228 to binary.

Solution:

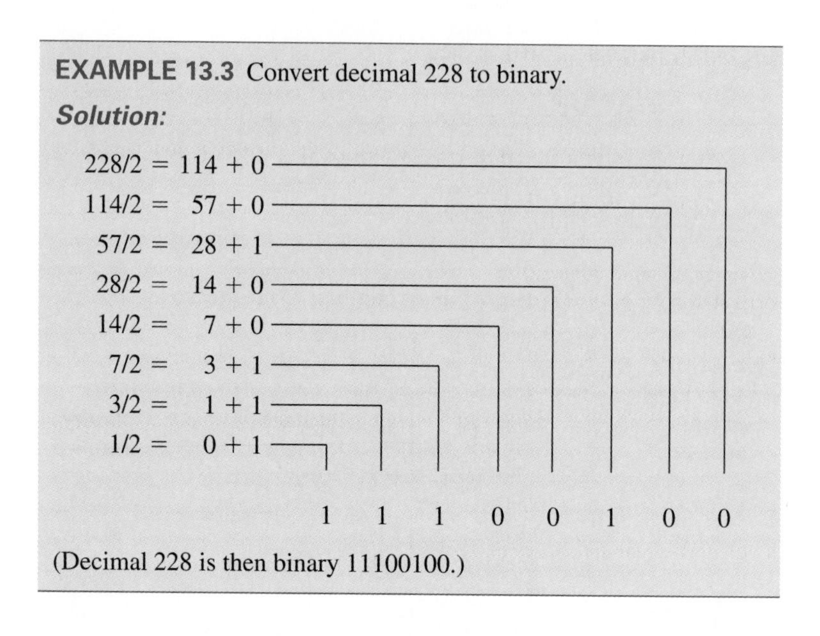

(Decimal 228 is then binary 11100100.)

Hexadecimal Number System

The *hexadecimal* or base 16 number system is used to represent binary digits handled by a computer. The use of hexadecimal comes from representing each group of 4 bits as a hexadecimal digit. Table 13.1 shows the relation among binary, hexadecimal, and decimal.

Table 13.1 shows that all combinations of four binary digits can be represented by a single hexadecimal digit, with digits from 10 to 16 represented as A through F. Thus, hexadecimal digit C is binary 1100, which is decimal 12; hexadecimal digit F is binary 1111, which is decimal 15; and so on.

TABLE 13.1
Hexadecimal, binary, and decimal numbers

Decimal	Binary	Hexadecimal
0	0000	0
1	0001	1
2	0010	2
3	0011	3
4	0100	4
5	0101	5
6	0110	6
7	0111	7
8	1000	8
9	1001	9
10	1010	A
11	1011	B
12	1100	C
13	1101	D
14	1110	E
15	1111	F

Converting between Hexadecimal and Binary

Converting a hexadecimal number to binary is quite direct.

EXAMPLE 13.4 Convert hexadecimal A8C5 to binary.

Solution:

A	8	C	5
1010	1000	1100	0101

Thus

$$A8C5_{16} = 1010100011000101_2.$$

Converting binary to hexadecimal is also direct.

EXAMPLE 13.5 Convert binary 1001110010001111_2 to hexadecimal.

Solution:

1001	1100	1000	1111
9	C	8	F

Thus $1001110010001111_2 = 9C8F_{16}$.

Converting between Hexadecimal and Decimal

While hexadecimal digits number from 0 to F, hexadecimal place values number from 16^0 to 16^1 to 16^2, and so on, as follows:

$$\begin{matrix} 16^n \cdots & 16^4 & 16^3 & 16^2 & 16^1 & 16^0 \\ \cdots & 65536 & 4096 & 256 & 16 & 1 \end{matrix}$$

Read from right to left, hexadecimal place values are 1s, 16s, 256s, and so on. Thus, the hexadecimal number 9A5 is

$$9 \times 16^2 + A \times 16^1 + 5 \times 16^0$$
$$= 9 \times 256 + 10 \times 16 + 5 \times 1 = 2469_{10}$$

EXAMPLE 13.6 Convert hexadecimal 3D9 to decimal.

Solution:

$$3D9_{16} = 3 \times 16^2 + 13 \times 16^1 + 9 \times 16^0$$
$$= 768 + 208 + 9 = 985_{10}$$

13.3 CODES

Computers use binary to represent numbers, alphabetic characters, and any other necessary data. Various codes have evolved, some to represent numeric data and others to represent alphabetic data.

Binary Coded Decimal

Numbers may be represented using binary digits with the 8421 binary coded decimal (BCD) code. Table 13.2 defines this code form.

As an example, the decimal number 483 can be represented using BCD as

$$483 = 0100\ 1000\ 0011$$

As shown, it takes 12 binary digits to represent 3 decimal digits. Each decimal digit is represented by 4 bits in the BCD code form.

TABLE 13.2
8421 BCD code

Decimal Digit	8421 BCD Code
	8421
0	0000
1	0001
2	0010
3	0011
4	0100
5	0101
6	0110
7	0111
8	1000
9	1001

EXAMPLE 13.7 Write decimal 8051 in BCD code form.

Solution:

$$8051_{10} = 1000\ 0000\ 0101\ 0001_{BCD}$$

Hexadecimal Code

Binary data can be represented as hexadecimal values, that being the primary purpose of hexadecimal code. Each group of 4 bits can be expressed in hexadecimal to make it easier to express large binary values. Refer to Table 13.1 for the binary–hexadecimal code.

EXAMPLE 13.8 Write the hexadecimal value of 00111010111011110100.

Solution:

$$00111010111011110100$$

can be expressed in hexadecimal as

$$0011 \quad 1010 \quad 1110 \quad 1111 \quad 0100 = 3AEF4$$

The 20-bit binary value is more easily expressed as an equivalent hexadecimal value.

Alphanumeric Code

The keyboard characters, including alphabetic, numeric, and various special characters, can all be represented using binary digits. The most popular code used in personal computers is that formulated as ASCII

TABLE 13.3
ASCII code

MSB\LSB

Hex	0	1	2	3	4	5	6	7	8	9	A	B	C	D	E	F
Binary	0000	0001	0010	0011	0100	0101	0110	0111	1000	1001	1010	1011	1100	1101	1110	1111
0 000								bell			LF		FF	CR		
1 001																
2 010	space	!	“	#	$	%	&	'	(	)	*	+	,	-	.	/
3 011	0	1	2	3	4	5	6	7	8	9	:	;	<	=	>	?
4 100		A	B	C	D	E	F	G	H	I	J	K	L	M	N	O
5 101	P	Q	R	S	T	U	V	W	X	Y	Z	[	\	]	^	–
6 110		a	b	c	d	e	f	g	h	i	j	k	l	m	n	o
7 111	p	q	r	s	t	u	v	w	x	y	z	{	|	}	~	DEL

Example: A = 41 = 100 0001

code. This code uses 7 bits to represent $2^7 = 128$ possible characters. Table 13.3 summarizes this code.

EXAMPLE 13.9 Write the hexadecimal form of the ASCII coding for the alphanumeric data

ABCD 123 abcd

Solution:

A	B	C	D		1	2	3		a	b	c	d
41	**42**	**43**	**44**	**20**	**31**	**32**	**33**	**20**	**61**	**62**	**63**	**64**

For hexadecimal convenience computers store each character using 8 bits, the eighth bit usually being a parity bit (to be discussed next).

Parity

Parity is the state of being odd or even. This property permits one or more bits to be added to a selected group of bits to detect bit error and in some cases to correct errors.

A parity bit added to a fixed group of bits provides bit parity.
A parity word added to a group of words provides word parity.
If the parity bit added makes the total number of 1s *even,* then *even* parity exists.
If the parity bit added makes the total number of 1s *odd,* then *odd* parity exists.

When a parity bit is added to a 7-bit number, it can be selected to provide either odd or even parity. As an example, each byte or character in RAM comprises 8 bits. A ninth parity bit is internally added to every byte stored in RAM. When the RAM data are read back, each byte is checked for even parity. If a single byte is found to not have even parity, the computer stops and displays a message that a parity error has occurred.

EXAMPLE 13.10 Write the binary form of the ASCII message HELLO, including odd bit parity.

Solution:

$$\begin{aligned}
H &= 48 = 100\ \ 1000 \Rightarrow 1\ \ 100\ \ 1000\\
E &= 45 = 100\ \ 0101 \Rightarrow 0\ \ 100\ \ 0101\\
L &= 4C = 100\ \ 1100 \Rightarrow 0\ \ 100\ \ 1100\\
L &= 4C = 100\ \ 1100 \Rightarrow 0\ \ 100\ \ 1100\\
O &= 4F = 100\ \ 1111 \Rightarrow 0\ \ 100\ \ 1111
\end{aligned}$$

parity bit

Gray Code

When binary position data are placed on a disk on a rotating shaft so that they can be read to provide information on the position of the shaft, they use a code in which only 1 bit changes going from one position to the next. This technique reduces errors when reading that position. One such code is the Gray code. It has the property just described and additionally can easily be converted to values of the binary number system.

Gray Code to Binary Code Conversion

Conversion of a Gray code to binary can be accomplished using an exclusive-OR operation. The exclusive-OR operation on 2 bits is defined as follows:

Two bits opposite result in output of 1.
Two bits the same result in output of 0.

Conversion of a Gray code number to binary is then carried out as follows:

Starting with the most significant bit (MSB) of the Gray code number, exclusive-OR each Gray code bit with the preceding binary bit.

EXAMPLE 13.11 Convert the Gray code number 01011101 to binary.

Solution:

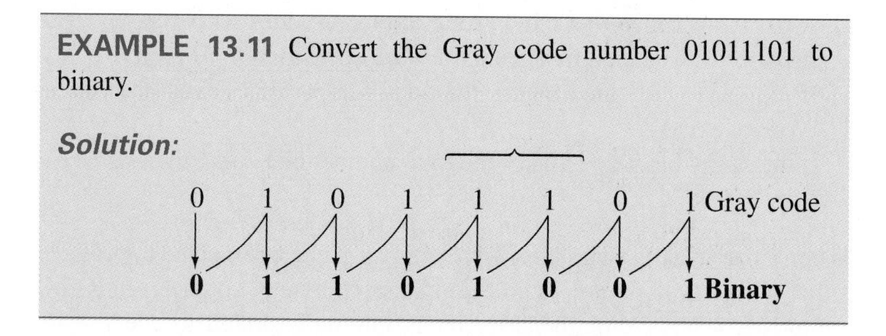

Binary Code to Gray Code Conversion

The procedure for converting a binary number to Gray code is quite similar.

Starting with the MSB of the binary number, exclusive-OR each **succeeding** ***pair to obtain the corresponding Gray code bit.***

EXAMPLE 13.12 Convert the binary number 011010111101 into Gray code.

Solution:

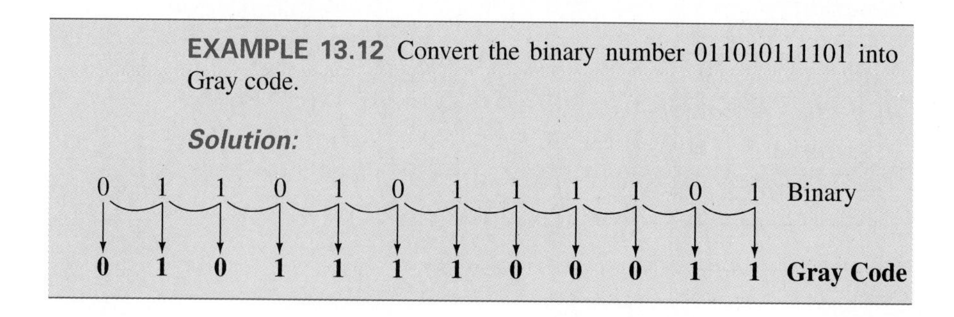

13.4 BOOLEAN ALGEBRA AND LOGIC GATES

The mathematics for dealing with two-valued logic is called *Boolean algebra** or logical algebra. The values in this algebra are either 0 or 1, true or false, and so on. Among the operations performed in this area are

Inverse operation: Changing of a 0 to a 1, or a 1 to a 0.
AND operation: Combining two or more inputs with the output being 1 only if *all* inputs are 1.
OR operation: Combining two or more inputs with the output being 1 if *any* input is 1.

Table 13.4 summarizes these logical operations.

TABLE 13.4
Logical operations

Logical Inverse Operation		**Logical AND Operation**			**Logical OR Operation**		
A	$\overline{A}$	A	B	$A \cdot B$	A	B	$A + B$
0	1	0	0	0	0	0	0
1	0	0	1	0	0	1	1
		1	0	0	1	0	1
		1	1	1	1	1	1

There are a few logic operations derived using the basic ones defined in Table 13.4. These include

Logical NAND operation: AND operation followed by inversion.
Logical NOR operation: OR operation followed by inversion.
Exclusive-OR operation: Output is logical-1 only if one input is logical-1, but not if both are.

Table 13.5 summarizes these logical operations.

TABLE 13.5
Combined logical operations

Logical NAND Operation			**Logical NOR Operation**			**Exclusive-OR Operation**		
A	B	$\overline{A \cdot B}$	A	B	$\overline{A + B}$	A	B	$A \oplus B$
0	0	1	0	0	1	0	0	0
0	1	1	0	1	0	0	1	1
1	0	1	1	0	0	1	0	1
1	1	0	1	1	0	1	1	0

*Named after George Boole, a British mathematician in the mid-1800s.

Logic Block Gates

Figure 13.1 shows the symbols used to indicate the circuits or gates representing the various logic operations.

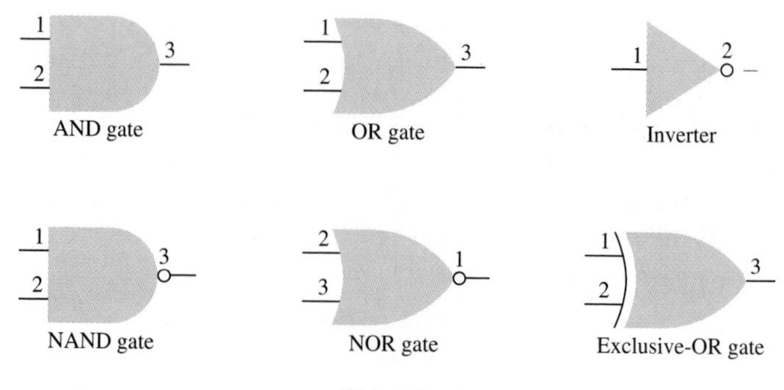

FIG. 13.1
Standard logic symbols.

To see how logic gates are used to build logic operations consider the following examples.

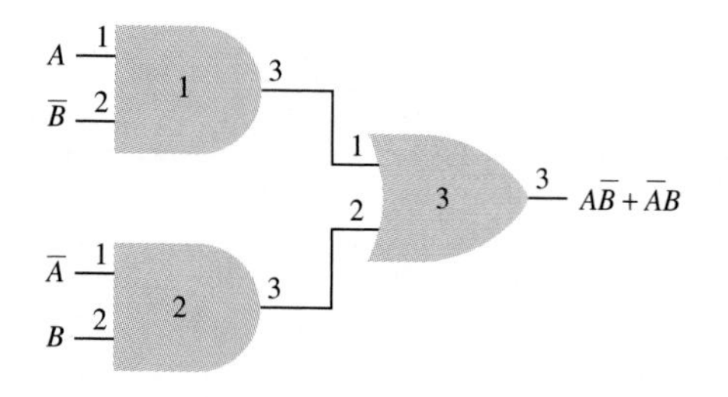

FIG. 13.2
Connection for Example 13.13.

EXAMPLE 13.13 Write the logic expression for the logic connection in Fig. 13.2.

Solution:
The output of AND gate 1 is A ANDed with $\overline{B}$.
The output of AND gate 2 is $\overline{A}$ ANDed with B.
The output of OR gate 3 is then $A\,\overline{B} + \overline{A}\,B$.

EXAMPLE 13.14 Write the logic expression for the connection in Fig. 13.3.

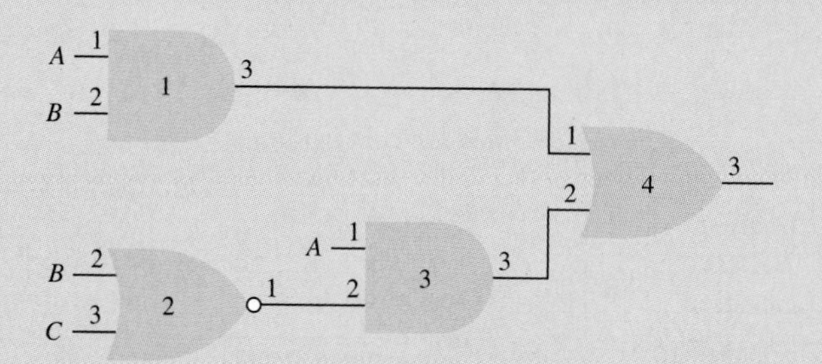

FIG. 13.3
Logic connection for Example 13.14.

Solution: The logic expression for the connection in Fig. 13.3 is

$$A \bullet B + A \bullet \overline{(B + C)}$$

EXAMPLE 13.15 Draw the logic connection for the expression

$$(\overline{AB}) + \overline{AB\overline{C}} + (\overline{ABC})$$

Solution: The resulting logic circuit is drawn in Fig. 13.4.

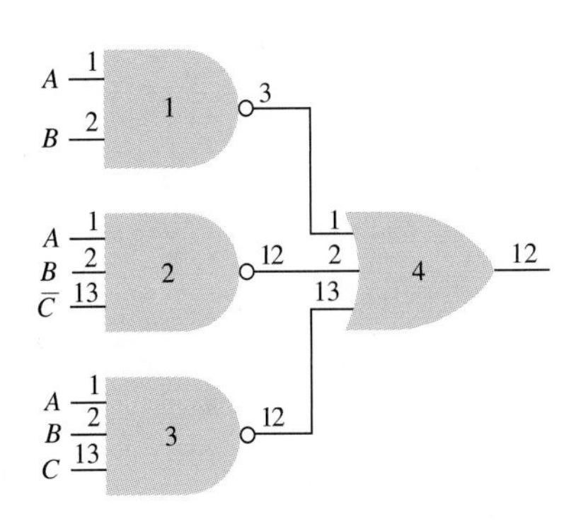

FIG. 13.4
Logic circuit for Example 13.15.

Adder Logic

Logic gates can be used to build a circuit that adds two binary numbers. When the bits of the number are all added at the same time, a parallel adder unit is built containing as many identical 3-bit adders as the number of bits to be added. Each 3-bit adder sums the 2 bits of that column and any carry bit from the lower column.

$$\begin{array}{rl} & 1 \leftarrow \text{carry from lower column} \\ & 1\ 0\ 0\ 1\ 1\ 1\ 0\ 1 \\ + & 0\ 1\ 0\ 0\ 1\ 1\ 0\ 0 \\ \hline & 0 \leftarrow \text{sum of bits in that column} \end{array}$$

As shown in the preceding partial example, 3 bits must be added in each column to get the sum for that column. In addition, a carry must be determined for use by the next higher column. Figure 13.5 shows the block diagram of a 3-bit (full) adder. The logic operation of one full-adder stage is summarized in Table 13.6. Adding 3 bits, A, B, and carry-in (C_i) results in sum and carry-out (C_o).

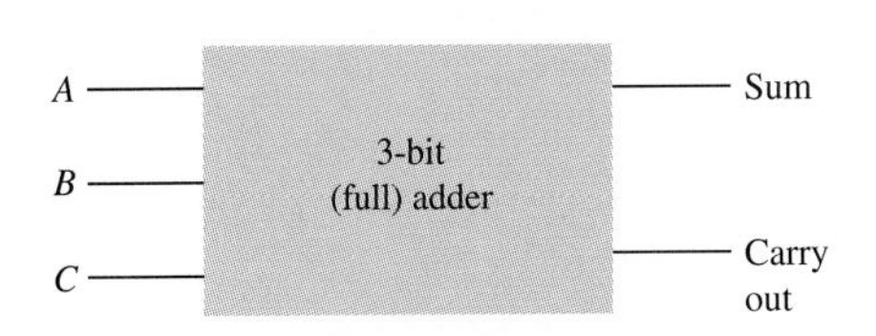

FIG. 13.5
Block diagram of full adder.

TABLE 13.6
Full adder

A	B	C_i	Sum	C_o
0	0	0	0	0
0	0	1	1	0
0	1	0	1	0
0	1	1	0	1
1	0	0	1	0
1	0	1	0	1
1	1	0	0	1
1	1	1	1	1

The logic expression for the sum 1 can be written as

$$\text{sum} = \overline{A}\,\overline{B}C + \overline{A}B\overline{C} + A\overline{B}\,\overline{C} + ABC$$

The logic expression for the carry-out 1 can be written as

$$\text{carry-out} = \overline{A}BC + A\overline{B}C + AB\overline{C} + ABC$$

Figure 13.6 shows one form of a full-adder circuit using only NAND gates.

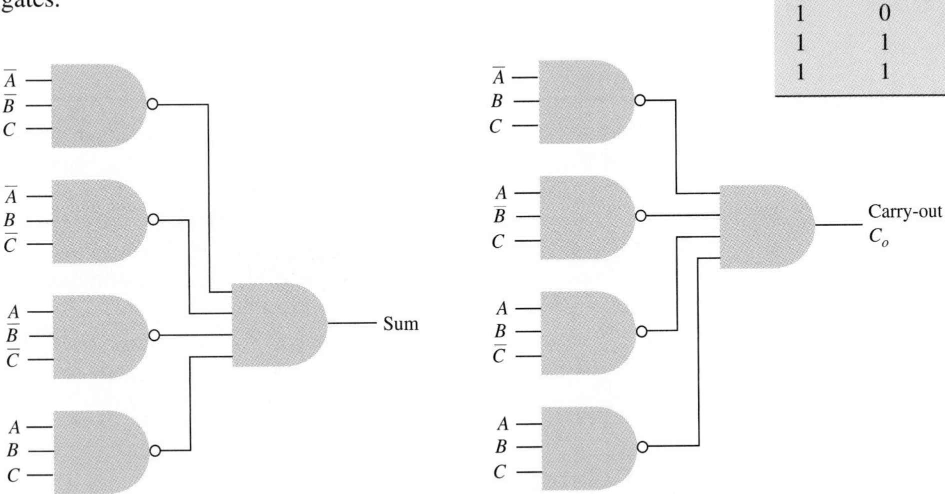

FIG. 13.6
Logic circuit of full adder using NAND gates.

The logic expression for sum and carry-out can be simplified and expressed as

$$\text{sum} = A \oplus B \oplus C$$
$$\text{carry-out} = \overline{A}BC + AB + AC$$

The simplified logic diagram for these equations is shown in Fig. 13.7.

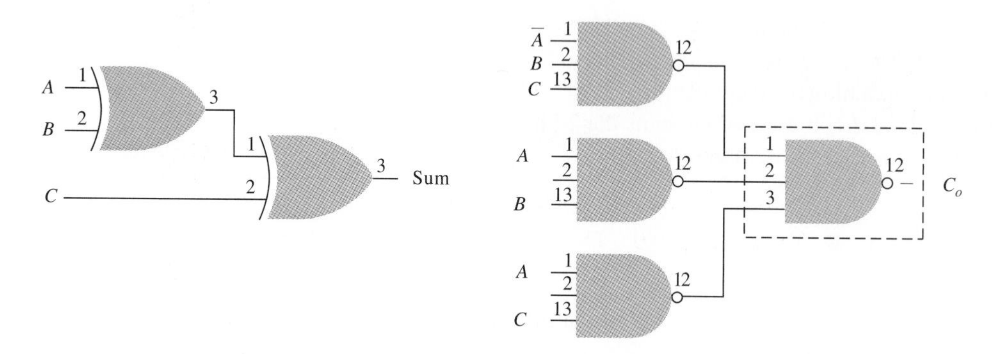

FIG. 13.7
Simplified logic circuit implementation of full adder.

13.5 COMPUTER LOGIC CIRCUITS

Computer logic circuits are provided in integrated circuit packages. They may be built using bipolar transistors in transistor–transistor logic (TTL) form or using MOSFETs in nMOS, pMOS, or CMOS form, among the more popular types. These circuits are provided in small-scale integrated (SSI) packages for basic gates, in medium-scale integrated (MSI) packages for simple functional parts, and in large-scale integrated (LSI) or very-large-scale integrated (VLSI) packages for whole functional units.

TTL Gates

BJT devices can be used to form a basic NAND gate as shown in the connection in Fig. 13.8. As shown, inputs are provided using multiple emitters on the input transistor Q_1.

If any input is *low* (0 V), transistor Q_1 is biased *on,* resulting in transistor Q_2 biased *off* and finally Q_3 *on* and Q_4 *off,* so that the output goes *high.*
Only if all inputs are *high* (or left open) does Q_1 cause Q_2 to be biased *on* so that Q_3 is biased *off* and Q_4 *on,* so that the output goes *low.*

MOSFET Gates

With the use of *n*-channel MOSFET transistors, an nMOS circuit can be built as in Fig. 13.9. Each input requires one *n*-channel MOSFET with a

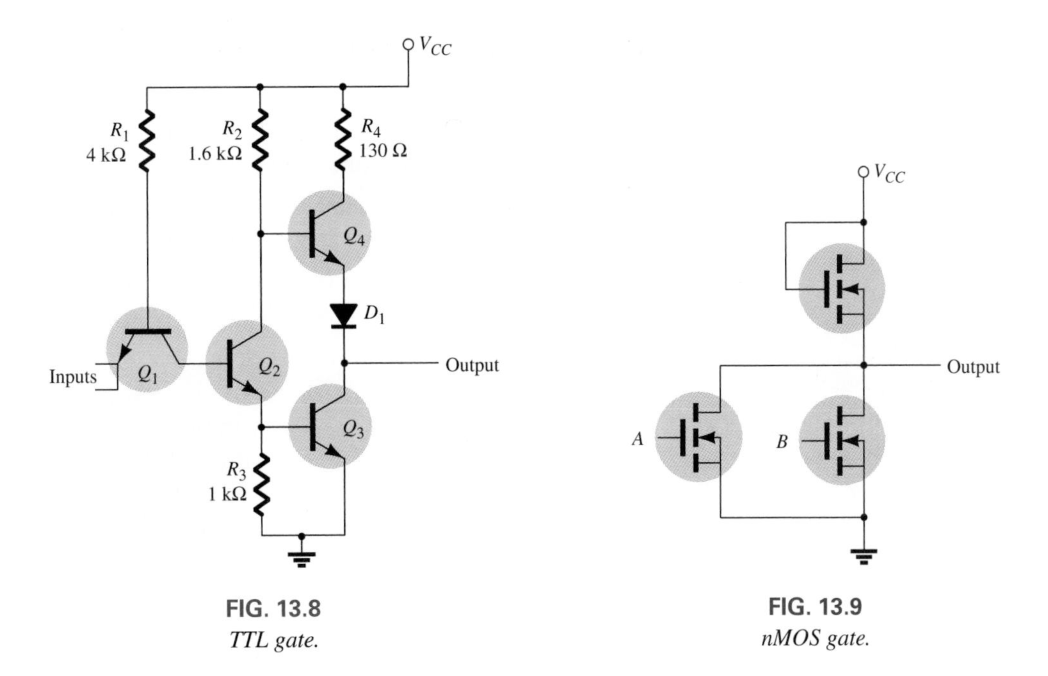

FIG. 13.8
TTL gate.

FIG. 13.9
nMOS gate.

common nMOS transistor (Q_3) acting as a resistive load. The circuit operates as a NOR gate as follows.

If the input to either Q_1 or Q_2 is *high,* that device turns *on,* and the output then goes *low.*
Only if both inputs are *low* and both transistors *off* does the output go *high.*

CMOS Gate

Probably the most popular form of gate using MOSFET devices is the complementary MOS circuit. As shown in Fig. 13.10, each input requires a pair of transistors—one nMOS and one pMOS, thus the complementary nature of the circuit. The connection shown is a NOR gate. The gate operates as follows:

If either input is *high,* the corresponding nMOS transistor is driven *on,* with the output going *low.* Since both pMOS transistors are *off,* no current is drawn from the voltage supply.
Only if both inputs are *low,* resulting in both nMOS transistors *off* and both pMOS transistors *on,* does the output go *high.*

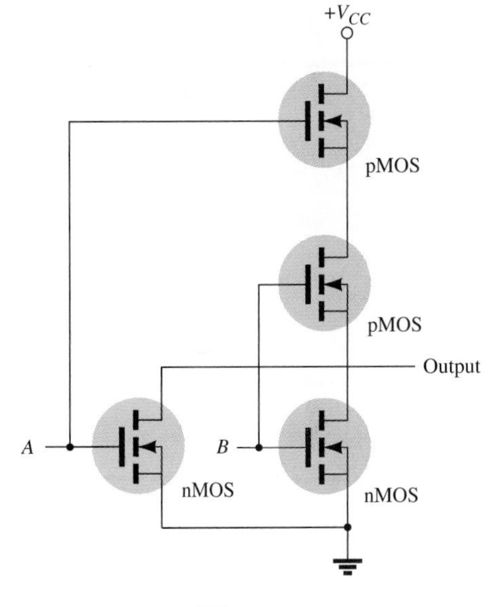

FIG. 13.10
CMOS gate.

13.6 MULTIVIBRATOR CIRCUITS

Multivibrator circuits are used to store bits, to count up or down, to provide timing signals, and generally to provide for storage and transfer of binary data. The most popular block is the bistable multivibrator or flip-flop, a circuit used to store a bit, or used as part of a counter unit. Basically, a multivibrator unit provides opposite outputs; the bistable circuit provides either output state.

Bistable Multivibrator (Flip-Flop)

Figure 13.11 shows the block form of a bistable circuit. The Q output is the reference output of the circuit. The $\overline{Q}$ output is always the opposite of that of Q. The output state of the circuit is defined as

Set state when Q output is high or 1 and $\overline{Q}$ is low or 0.
Reset state when Q output is low or 0 and $\overline{Q}$ is high or 1.

A clocked flip-flop is either set or reset at the occurrence of a clocking pulse. Figure 13.12 shows a clocked *JK* flip-flop. In this unit the action of the flip-flop depends on the inputs at J and K at the occurrence of a negative clock pulse edge. The operation for each possible input combination for a *JK* flip-flop is listed in Table 13.7.

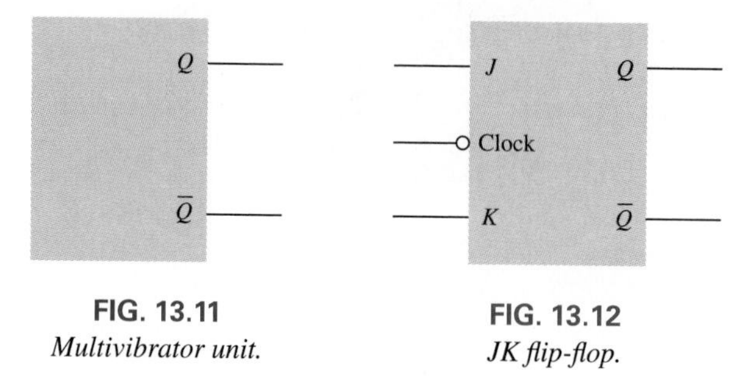

FIG. 13.11
Multivibrator unit.

FIG. 13.12
JK flip-flop.

TABLE 13.7
Operation of JK flip-flop

J	K	Q (after clock)
0	0	No change
0	1	Reset
1	0	Set
1	1	Toggle

What Table 13.7 shows is how the flip-flop changes (or not) after the correct clock pulse edge. For the unit in Fig. 13.12, this clocking takes place when the clock edge goes negative (from *high* to *low*). If the J and K inputs previous to the clock edge are both 0, then no change in the flip-flop will occur. If, however, the inputs are both 1, then the stage will toggle—the Q output going *low* if it is presently high, or going *high* if it is presently low.

EXAMPLE 13.16 Draw the output waveform resulting from the input clock and data waveforms in Fig. 13.13(a).

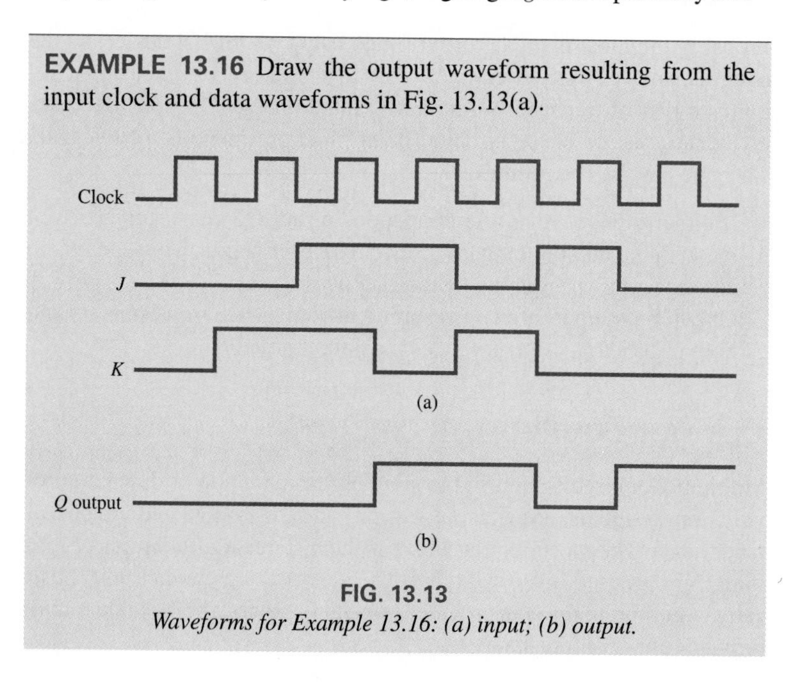

FIG. 13.13
Waveforms for Example 13.16: (a) input; (b) output.

Solution: The resulting Q-output waveform is shown in Fig. 13.13.

Using a Flip-Flop as a Data Stage

A flip-flop can be used to store a bit of data. A number of identical stages can be used to store an 8-, 16-, or 32-bit number, for example. A popular type of flip-flop used to store data is the D type shown in Fig. 13.14.

A bit applied to the D input is stored in the flip-flop stage when the proper clock edge occurs, the negative edge in the present example. This connection can be repeated for as many bits as the unit handles. An 8-bit data register is shown in Fig. 13.15. When the transfer pulse goes low, the input data are transferred into the eight register stages and thereby stored or held in the 8-bit register.

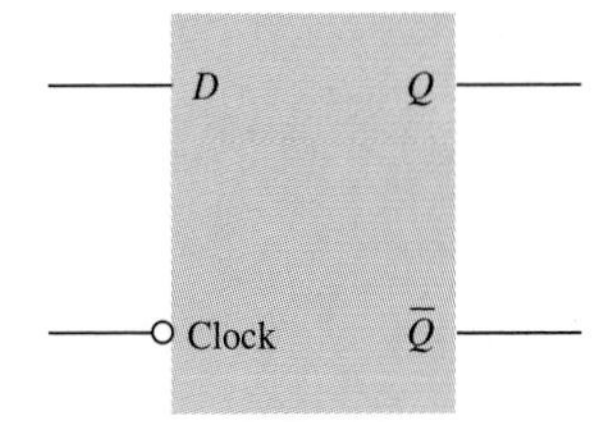

FIG. 13.14
D-type flip-flop.

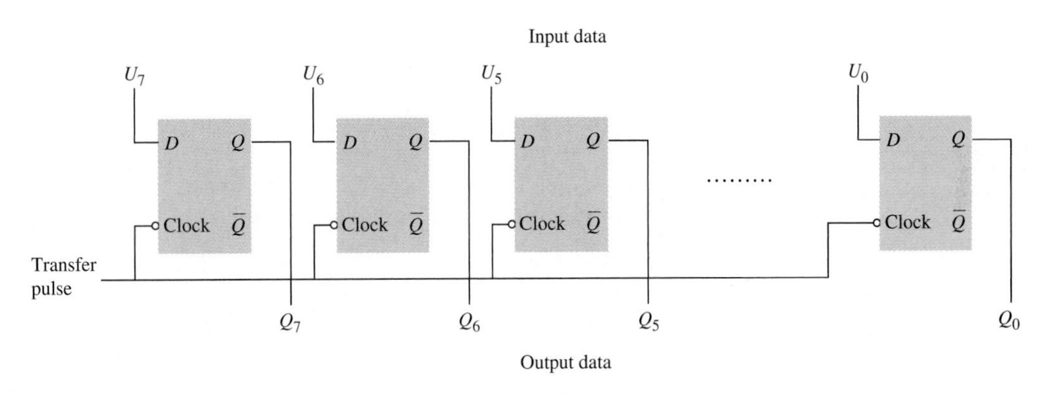

FIG. 13.15
Eight-bit data register using D-type flip-flops.

Using a Flip-Flop as a Toggle Stage

A JK flip-flop can be used as a toggle stage, as is done in counters. A JK- or D-type flip-flop can be used to toggle each clock pulse, providing a counter operation. Figure 13.16(a) shows a JK flip-flop operating as a toggle stage—both J and K inputs are logical-1. The clock input cycling, as shown in Fig. 13.16(b), toggles the flip-flop on each negative edge (for the present flip-flop example). The resulting Q-output signal shows that it cycles at half the clock rate of the input. One use of a toggle stage is therefore to divide the clock frequency. This is used in clocks to divide a reference clock frequency into the hours, minutes, and seconds required for a watch unit.

A D flip-flop can also be used as a toggle stage. This requires connecting the D-type flip-flop as shown in Fig. 13.17(a). Connecting the $\overline{Q}$ output as D input causes the stage to toggle each clock pulse. In the present example, this toggling action takes place on the positive clock edge. The resulting stage waveforms are shown in Fig. 13.17(b).

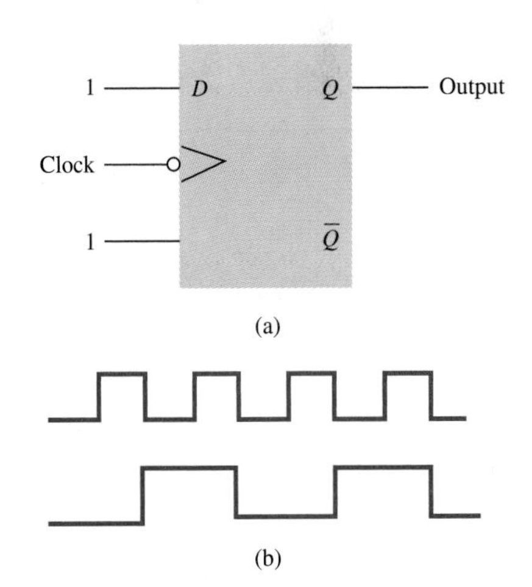

FIG. 13.16
Toggle stage using JK flip-flop: (a) connection; (b) waveforms.

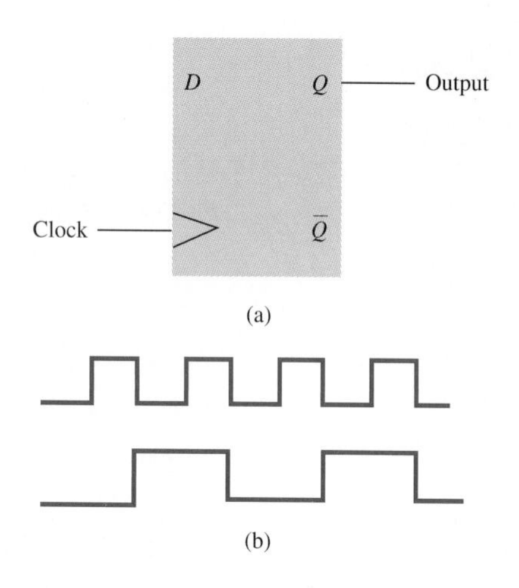

FIG. 13.17
Toggle stage using D flip-flop: (a) connection; (b) waveform.

TABLE 13.8
Counter stages and counts

Number of Stages (n)	Counts (2^n)
3	8
4	16
5	32
8	256
10	1,024
12	4K (4,096)
16	65K (65,536)
20	1M (1,048,576)
24	16M (16,777,216)
32	4G (4,294,967,296)

Using Flip-Flops as Counters

A number of flip-flop stages can be connected to form various base or modulus counters. For example, 3 stages will provide 8 (2^3) counts, 10 stages will provide 1024 (2^{10}) counts, and so on. Table 13.8 provides a summary of count totals and the number of stages needed.

For example, Fig. 13.18 shows a 3-stage (modulus 8) counter using *JK* flip-flops.

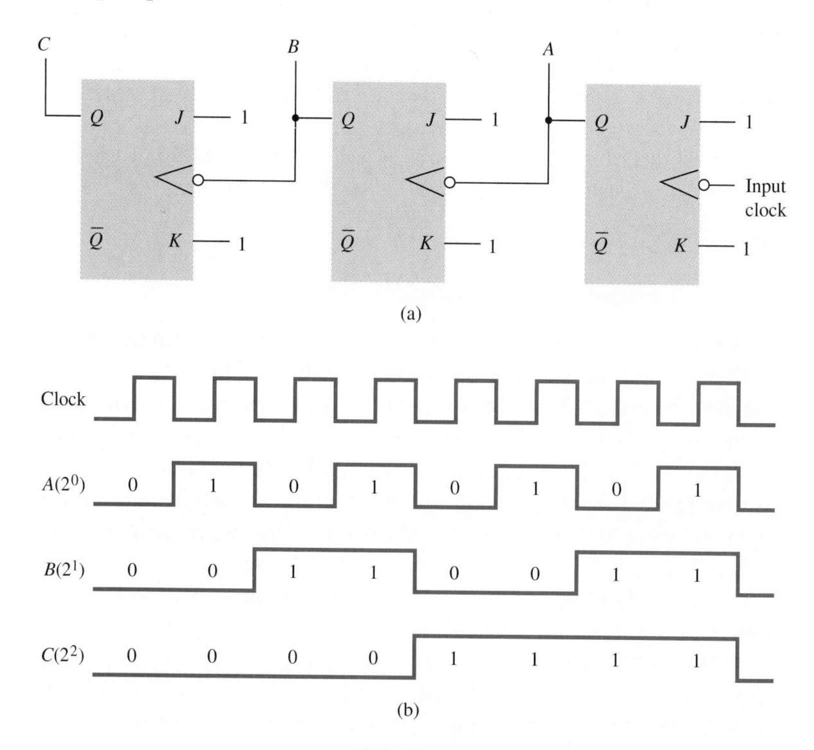

FIG. 13.18
Three-stage counter using JK flip-flops: (a) logic diagram; (b) waveforms.

IC Decade Counter

Counter circuits are typically manufactured as IC units supplying the complete counter operation. Figure 13.19(a) shows one popular counter that provides counts from 0 to 9—a decade or mod-10 counter. Figure 13.19(b) shows the connection of the 74390 wired to operate as a decade counter. Each clock pulse advances the count one step from 0000 to 1001, the counter resetting to 0000 on the next clock pulse.

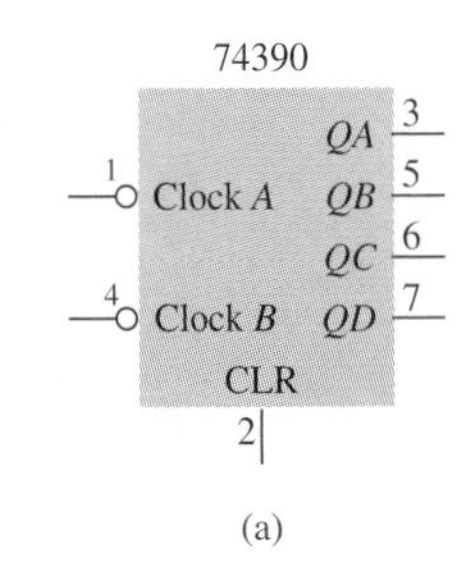

(a)

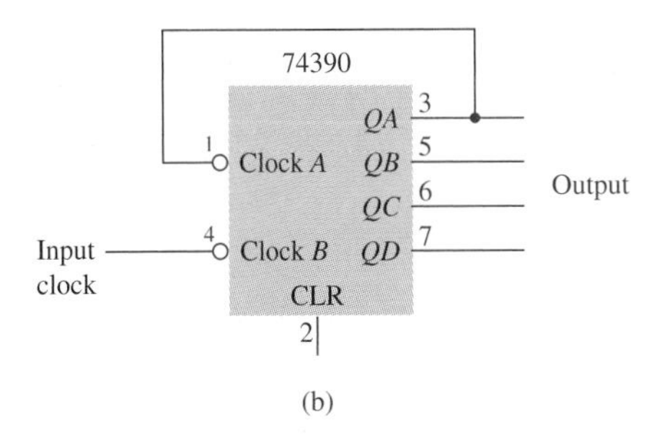

(b)

FIG. 13.19
74390 decade counter.

13.7 MICROPROCESSOR UNITS

A microprocessor is a single IC unit containing the registers, counters, and logic gates required to perform various operations. The microprocessor is essentially the heart of a computer unit. A microprocessor can be surrounded with memory ICs and input and output circuitry to form a computer. Microprocessor ICs (chips) have advanced from handling 4 bits at a time in the late 1970s to currently handling 32 or 64 bits at a time internally. Operating clock speeds have increased from about 1 MHz in the late 1970s to over 1 GHz currently. Two popular microprocessor lines have come from Intel and from Motorola. The Intel line has been used to build IBM and IBM-compatible systems, while the Motorola line has been used by Apple.

Figure 13.20 shows the basic connection of a microprocessor to memory, input, and output units. The microprocessor receives or transmits data along a set of wires called a *data bus*. The microprocessor uses another set of wires called the *address bus* to select which memory cell to read from or write to. The address bus and data bus are typically 16, 32, or 64 bits in the personal computers used today. The various bus sizes, clock speeds, and internal configurations provide a variety of popular microprocessor ICs.

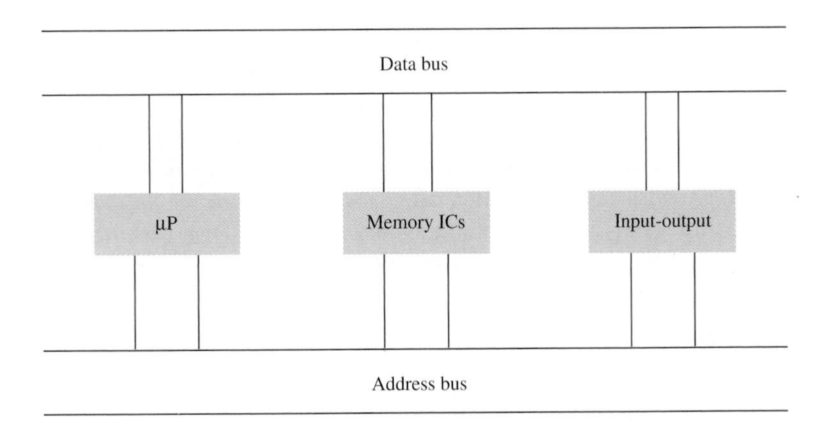

FIG. 13.20
Microprocessor block diagram.

13.8 MEMORY CIRCUITS

A major part of any digital computer is the memory unit. Two basic memory types are read-only memory (ROM), which is fixed or relatively permanent memory, and random-access memory (RAM), which can be read or written to. They serve different purposes, as will be explained. ROM and RAM units are considered main memory in that they are connected directly to the address and data buses. These forms of memory also provide the fastest operation and hold the programs and data used to operate the computer. This type of memory is different from that stored, for example, on a disk; disk memory is read more slowly and is usually used to fill the RAM unit as fast as the disk can operate. Cache memory is any faster memory unit holding a small portion of memory data to speed up the overall operation.

Read-Only Memory

The purpose of ROM is to hold programs and data that is not lost when power is removed. When a computer unit is first turned on, only the program in ROM is available to operate it. The ROM chip can be prepared at the time of manufacture or programmed at a later time by the user (programmable ROM or PROM). While PROM can be programmed only once, ultraviolet-erasable PROM (uv-EPROM) or electrically erasable PROM (EE-PROM) can be reprogrammed numerous times.

Random-Access Memory

Two types of RAM are used in computers—static RAM (SRAM) and dynamic RAM (DRAM). SRAM is built using bipolar transistors, is physically larger, and operates much faster than DRAM, making it popular for cache-type operation. DRAM is built using MOSFET devices, is much smaller and somewhat slower than SRAM, and is used to provide the large RAM needs of a computer memory.

Memory Size

Memory ICs (sometimes called chips) are often packaged in single in-line memory modules (SIMMs). SIMM sizes range from 16 MB to 256 MB. These SIMM packages usually contain DRAM ICs.

Additionally, SRAM ICs are being packaged with a computer motherboard to provide caching for various parts of the system. A cache of ROM provides faster operation after startup. Caching of the DRAM speeds up access to the volatile memory operation. Disk caching provides a speed-up of the disk data access.

Address and Data Bus

Of greatest importance when considering memory ICs are the number of lines used to handle data—the data bus—and the number of lines used to handle the memory address—the address bus. The data bus is typically 32 or 64 bits in current PCs. Address bus size is typically from 32 to 64 bits.

There are, in general, 2^n combinations for n bits. Table 13.9 lists the number of bits versus the number of combinations. For example, a 12-bit address provides 4096 or 4K combinations, while a 24-bit address provides 16M address locations. Present-day PCs with 32 address bits provide for up to 4 terabytes (4×10^9 bytes).

Data bits are generally provided in bytes—8, 16, 32, or 64 bits in size. A 32-bit unit provides 4 bytes per memory location and allows the reading of 4 bytes of data at each read or write operation.

TABLE 13.9
Number of combinations

n	2^n
8	256
9	512
10	1,024 (1K)
11	2,048 (2K)
12	4,096 (4K)
16	65,536 (65K)
20	1,048,576 (1M)
24	16,777,216 (16M)
32	4,294,967,296 (4T)

13.9 INPUT–OUTPUT UNITS

The most used parts of a computer system are the various input and output units. Input units include

keyboard
floppy disk
hard disk
mouse
CD-ROM

while output units include

printer
monitor
hard disk and floppy disk
speakers

Input Units

Data are entered as 8-bit ASCII characters using a keyboard. Each keystroke provides an 8-bit input that is stored in RAM as 1 byte of data. These data are first entered into RAM and then may be stored on a hard disk or a floppy disk to maintain the data after power is removed. Data previously stored on floppy disks or hard drives are then available as input to the computer to be read into RAM and used by various programs. A mouse provides input for selecting an item from the monitor and activating programs, selecting operations, or performing graphical operations. CD-ROMs provide tremendous amounts of data (typically 660 MB), which can be text, graphical data, sound, movies, or animation, for example.

Output Units

Data are provided typically to either a monitor or a printer. Monitors and their associated interface cards provide color pixels from

640×480 (VGA)
800×600, 1280×1024 (SVGA)

with varying numbers of colors.

Printed data are created by various types of printers, including

laser
ink jet

Monitor Operation

The monitor or cathode ray tube (CRT) display is created by an electron beam striking the face of the screen. This beam is electronically scanned across the screen in repeated fashion, dropping one line lower on each sweep until the entire face of the tube has been covered. The refresh or repeated full sweep of the CRT face occurs typically 50 to 72 times per second. Some monitors interlace the sweep by omitting every other line, sweeping first the odd-numbered lines and then repeating for the even-numbered lines. While the speed of these repeated sweeps is fast enough to create a full display, a noninterlaced sweep provides less flicker, as is preferable.

As the beam is swept across the face, the data cause each dot or pixel to be turned on or left off, and additionally to strike any of three color spots (red, green, or blue) to create the desired color. The size of the pixels is typically 0.28-mm dot pitch.

Printer Operation

Two types of printers are popular for small computer systems. The laser printer first creates the image to be printed on an electrostatically charged rotating drum and then transfers it to the paper using toner that adheres to the paper as determined by the drum image. Laser printers provide from 4 pages per second (4 pps) to 17 pps, typically.

Ink jet printers have either black ink or a series of colored ink reservoirs ejecting or spraying very fine drops of ink that are used to create the various characters or images on different types of paper. While the ink dries quickly, the printing speed is much slower than that of laser printers. An ink jet printer typically costs less than a laser printer.

13.10 DIGITAL–ANALOG CONVERSION

A major area of operation involves converting analog signals to digital form for use in digital computer circuitry, and converting digital signals back to analog form for real-world operations. An audio CD-ROM, for example, contains music data in digital form and uses digit-to-analog converter circuitry to obtain the analog form of the music signal to provide to the player's amplifier and speakers. When music or sound information is input to a computer system, it may first be converted into digital form using an analog-to-digital converter.

Analog-to-Digital Converter (ADC)

Various schemes are used to convert an analog signal to digital. Figure 13.21 is the block diagram of an ADC using a ladder network. A clock generator circuit drives the binary counter. As the binary count increases, the ladder network output voltage increases by one step—providing a staircase voltage. An input voltage to be converted to a digital value is compared with the staircase voltage by a dc voltage comparator circuit.

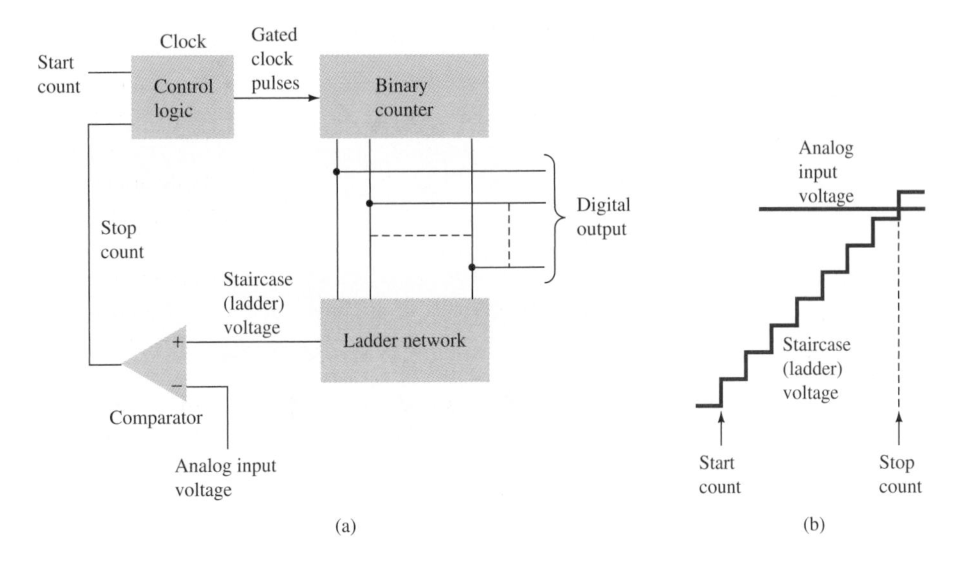

FIG. 13.21
ADC using ladder network: (a) block diagram; (b) comparator voltages.

When the staircase voltage equals or just exceeds the input voltage, the count is stopped and the digital value is then the desired value. The time needed to complete a single conversion depends on the speed of the clock signal and the number of stages of the binary counter. The *resolution* of the circuit depends on the number of count steps and reference voltage used. The *accuracy* of the conversion depends on the dc comparator circuit.

EXAMPLE 13.17 Determine (a) the maximum conversion time and (b) the resolution for a 12-stage counter and ladder network ADC using a 100-kHz clock.

Solution:

a. Since $T = 1/f$,

$$\text{clock period} = T = 1/(100 \times 10^3) = 10 \times 10^{-6} = 10\ \mu\text{s}$$

The number of count steps for 12 stages is

$$2^{12} = 4096$$

Therefore, the maximum conversion time is

$$4096 \text{ steps} \times 10\ \mu\text{s/step} = 40{,}960\ \mu\text{s} \approx 40\text{ ms}$$

b. The resolution provided by using 12 stages is 1 part in 4096 or $(1/4096) \times 100\% = 0.024\%$

Digital-to-Analog Converter (DAC)

The block diagram in Fig. 13.22 shows how a ladder network can be used to convert a digital value to analog form. A digital value is transferred into a binary register. The ladder network sums the voltages

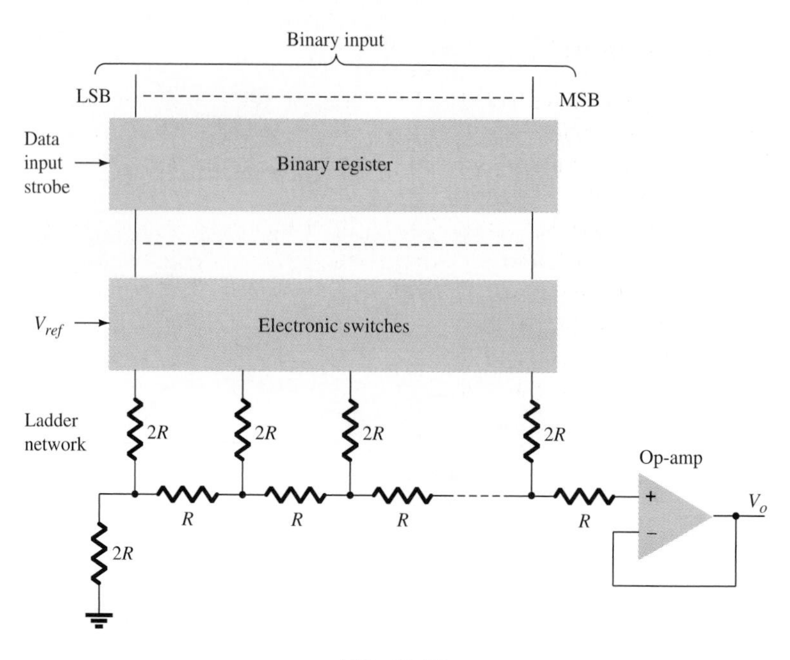

FIG. 13.22
DAC using R-2R ladder network.

depending on the voltage outputs from each bit position, and the unity-gain op-amp then provides the dc voltage to any output device. One type of converter uses resistor values that are of value R or $2R$, as shown in Fig. 13.23. Another uses weighted resistors, each a binary multiple of the other, to provide the voltage parts to be summed to achieve the desired output voltage.

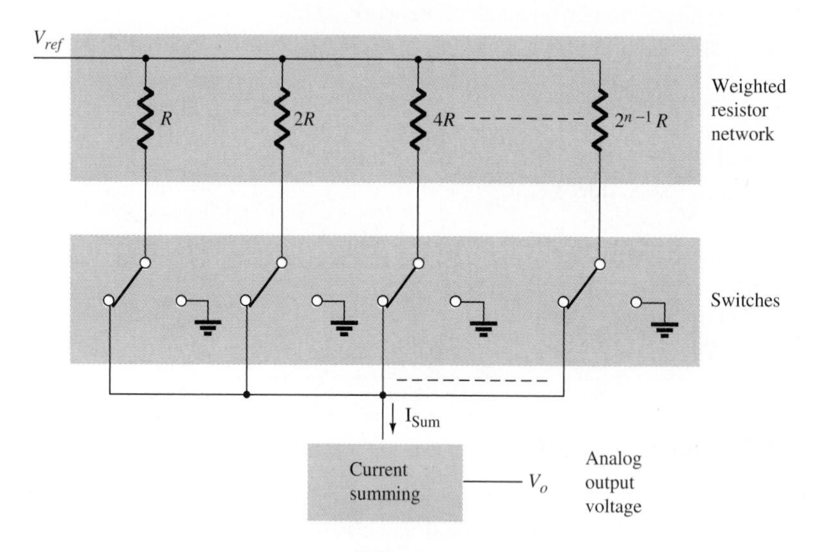

FIG. 13.23
DAC using weighted resistor network.

Dual-Slope ADC

Another type of conversion uses a dual-slope scheme as implemented in the block diagram in Fig. 13.24(a). The analog voltage to be converted is applied through an electronic switch to an integrator or ramp circuit. The digital output is obtained from a counter operated during both positive and negative slope intervals of the integrator. Figure 13.24(b) shows typical waveforms for a few input voltages. The larger the input voltage, the larger the resulting digital count. The integrator charges a capacitor to larger values for a larger input signal during the fixed charge-time interval. During the discharge interval the count continues until the capacitor is discharged to 0 V, and a larger count results.

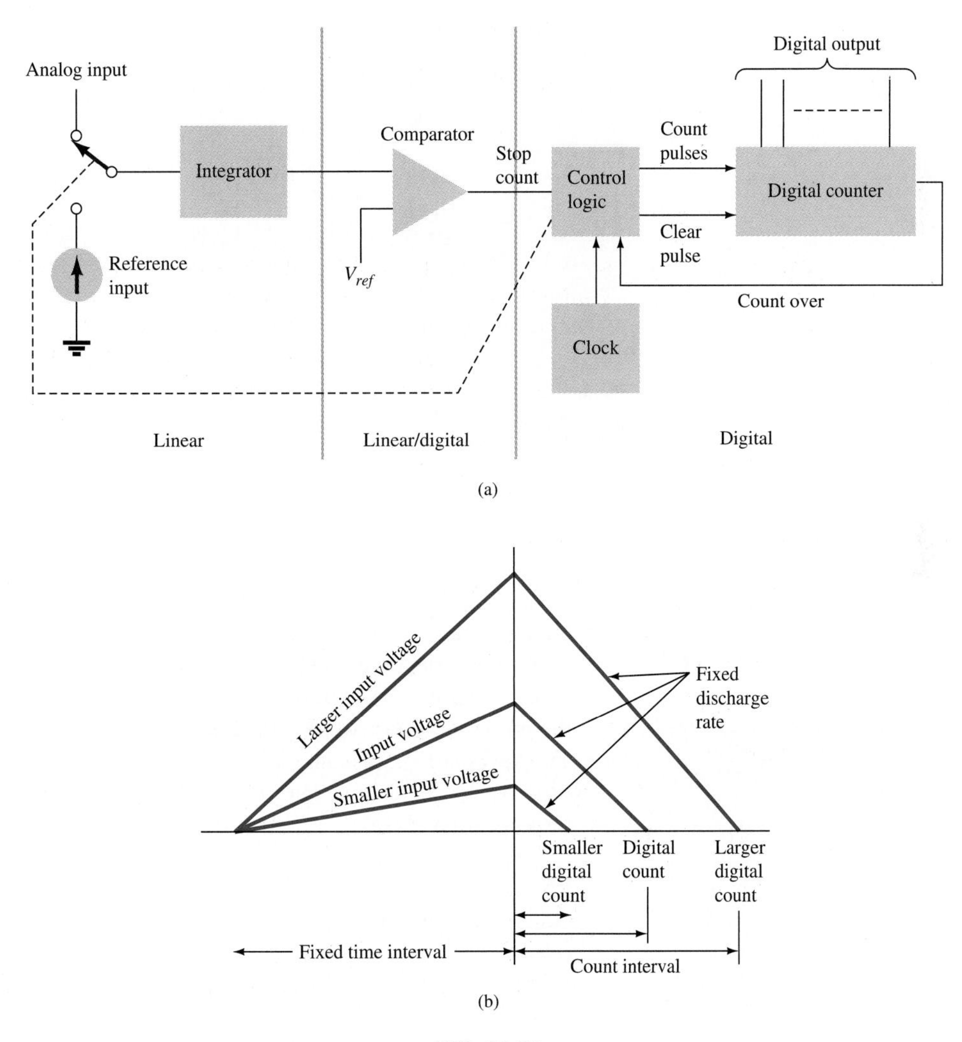

FIG. 13.24
Dual-slope ADC: (a) block diagram; (b) waveforms.

13.11 SUMMARY

Number Systems:

Decimal (base 10)

$$N = d_n 10^n + \cdots + d_2 10^2 + d_1 10^1 + d_0 10^0$$

Binary (base 2)

$$N = d_n 2^n + \cdots + d_2 2^2 + d_1 2^1 + d_0 2^0$$

Hexadecimal (base 16)

$$N = d_n 16^n + \cdots + d_2 16^2 + d_1 16^1 + d_0 16^0$$

Decimal	Binary	Hexadecimal
0	0000	0
1	0001	1
2	0010	2
3	0011	3
4	0100	4
5	0101	5
6	0110	6
7	0111	7
8	1000	8
9	1001	9
10	1010	A
11	1011	B
12	1100	C
13	1101	D
14	1110	E
15	1111	F

Codes:

Decimal	BCD Code
0	0000
1	0001
2	0010
3	0011
4	0100
5	0101
6	0110
7	0111
8	1000
9	1001

- Numbers may be represented using binary digits with the 8421 **binary coded decimal** (BCD) code.
- Binary data can be represented as hexadecimal values, that being the primary purpose of **hexadecimal code.**
- The keyboard characters, including alphabetic, numeric, and various special characters, can all be represented using binary digits **(alphanumeric code).** ASCII code is the most popular code used in personal computers.
- **Parity** is the state of being odd or even.
  Odd parity—has a parity bit added to make the total number of 1s *odd.*
  Even parity—has a parity bit added to make the total number of 1s *even.*
- **Gray code** is one of the number of codes having only one bit change from one position to the next.

Boolean Algebra and Logic Gates:

AND $A \cdot B$
OR $A + B$
XOR $A \oplus B$
Adder: Sum $= A \oplus B \oplus C$ Carry $= A\,B\,C + A\,B + A\,C$

Multivibrator Circuits

Astable–clock
Monostable–timer
Bistable–storage and counter stages

Memory Circuits

Some considerations are size, speed, and address and data busses.

PROBLEMS

SECTION 13.2 Number Systems

1. Determine the decimal number for the following binary numbers:
  a. 10101100
  b. 01101101
  c. 1101101101100111

2. Convert the following decimal numbers to binary:
  a. 987
  b. 478
  c. 628

3. Determine the decimal number for the following hexadecimal numbers:
  a. 5AD
  b. FE4
  c. 3CB

4. Convert the following decimal numbers to hexadecimal:
  a. 987
  b. 478
  c. 628

SECTION 13.3 Codes

5. Write the BCD form for:
  a. 9275
  b. 8042
  c. 6183

6. Write the hexadecimal value for:
  a. 1010111100110101
  b. 1010010111011011
  c. 1010111101001001

7. Write the hexadecimal form of the ASCII coding for the alphabetic data: DIGITAL COMPUTERS

8. Write the binary form of the ASCII message HAPPY DAYS including even bit parity.

9. Convert the following Gray code numbers to binary:
   a. 11000101
   b. 10001101
   c. 10110010

10. Convert the following binary numbers to Gray code:
    a. 01101010
    b. 11000011
    c. 01111011

SECTION 13.4 Boolean Algebra and Logic Gates

11. Write the logic expression for the logic connection in Fig. 13.25.

12. Write the logic expression for the logic connection in Fig. 13.26.

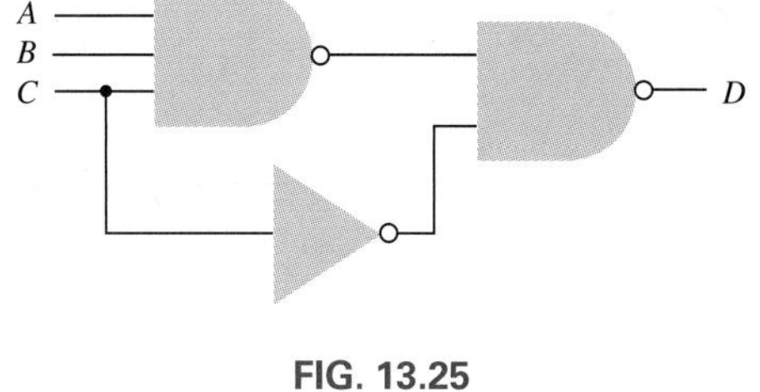
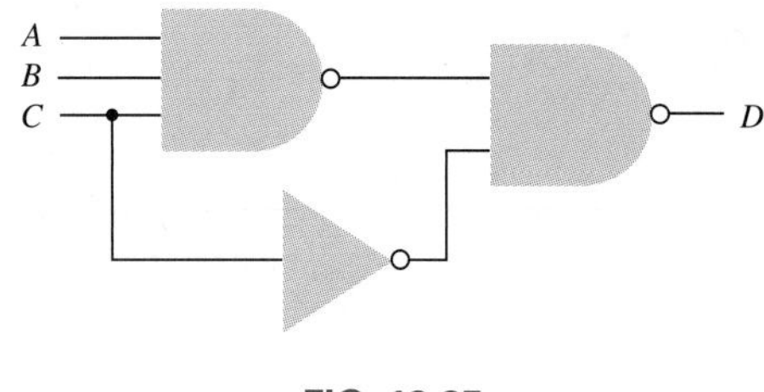

FIG. 13.25

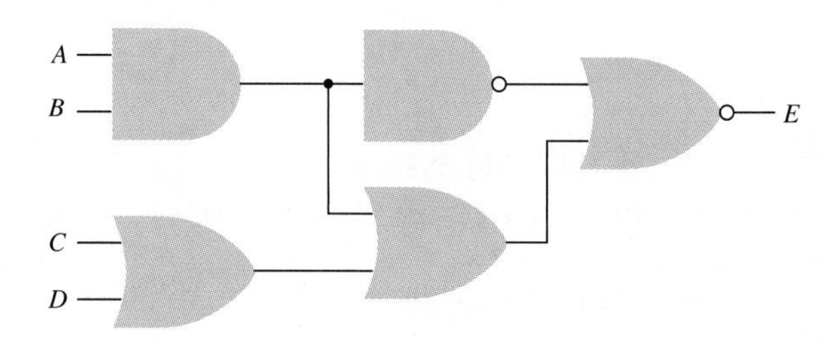

FIG. 13.26

13. Draw the logic connection for the expression:

$$ABC + \overline{ABC} + (\overline{A} + C)$$

14. Draw the logic connection for the expression:

$$A\overline{B}C + \overline{(A + B)}C + \overline{A}B\overline{C}$$

15. Draw the logic connection for the expression:

$$\overline{W}X\overline{Y}Z + \overline{(W + Z)} + \overline{X}Z$$

SECTION 13.6 Multivibrator Circuits

16. Draw the output waveform resulting from the input clock and data waveforms in Fig. 13.27.

17. Draw the connection of a four-stage data register using *D*-type flip-flops.

18. Draw the connection of a four-stage data register using *JK* flip-flops.

19. Draw the connection of a four-stage count-down counter using *JK* flip-flops.

20. Draw the connection of a four-stage count-up counter using *D*-type flip-flops.

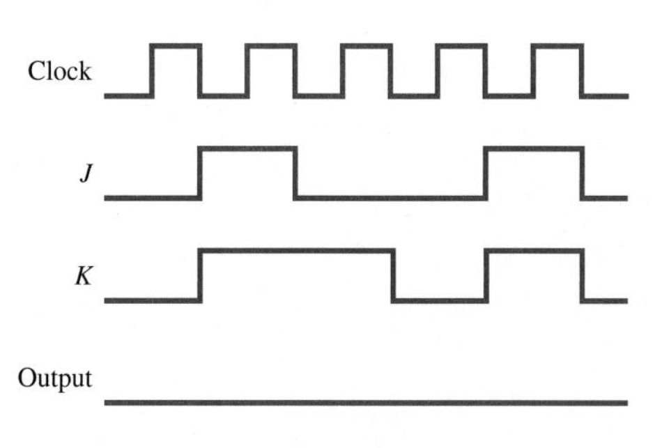

FIG. 13.27

14 Control Systems

14.1 INTRODUCTION

Electric circuits and mechanical devices must both be studied to understand the functioning of a large variety of control systems. A group of components that operate a mechanical device constitute a *control system.* A system that drives a telescope to a desired position, for example, is a control system. This control system can operate either closed loop or open loop. In an *open-loop* system the telescope is driven to the desired position, but no facility is included to check if the instrument actually reaches this position. A *closed-loop* system additionally provides signal *feedback* from the telescope drive so that some comparison can be made between the desired position and the present position and an *error signal* developed to correct for this difference. An open-loop system can be symbolically represented by a block diagram such as that in Fig. 14.1(a), while a closed-loop system can be represented by the block connection in Fig. 14.1(b).

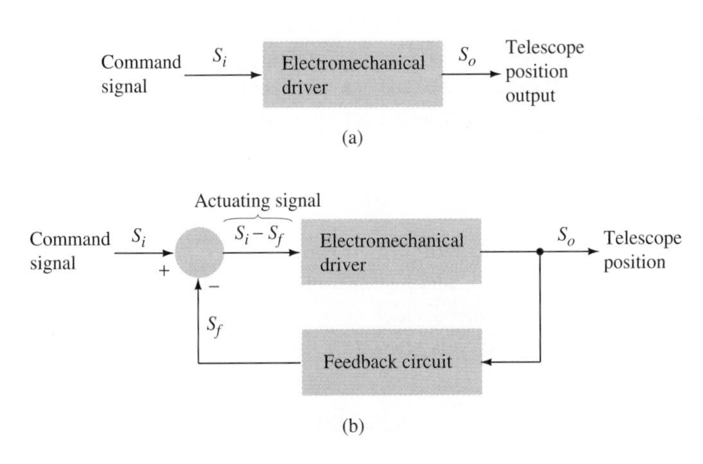

FIG. 14.1
Open- and closed-loop control systems: (a) open loop; (b) closed loop.

In Fig. 14.1(a) the actuating signal or input signal S_i is provided to operate the electromechanical driving components. This command signal may be produced simply by moving a dial to a desired setting, thereby generating a command signal, or the input signal can be generated by some other system (such as a computer). The driver unit then positions the telescope to an output position, indicated by S_o. There is no way for the system to check that S_o is the desired value of S_i after the driver moves the telescope. The more common arrangement [Fig. 14.1(b)] derives a feedback signal S_f from the output signal. A feedback circuit develops a signal from the output that is then compared with the input signal. In the usual arrangement a comparator circuit subtracts the feedback from the input to provide a signal to actuate the telescope position driver. When the feedback signal is quite different from the input signal, a large actuating signal is present. As the output signal approaches the desired position, the actuating signal decreases.

14.2 BLOCK DIAGRAMS

One popular way of describing the interaction of various parts of a control system is representation by a block diagram. Each part of the control system may be represented by a transfer function (Fig. 14.2). The simple

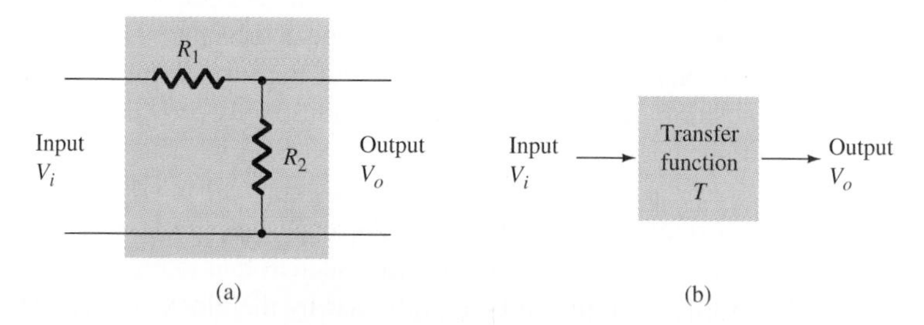

FIG. 14.2
Block transfer function: (a) circuit; (b) block.

resistor divider in Fig. 14.2(a) may be represented by the block diagram in Fig. 14.2(b), where the transfer function of the block is

$$T = \frac{R_2}{R_1 + R_2}$$

A circuit may be broken up into more than one block as shown in Fig. 14.3(a). If we assume an ideal amplifier of voltage gain A and infinite input impedance, then the divider network and amplifier can be represented by separate blocks as shown in Fig. 14.3(b). The overall gain is such that

$$V_o = TAV_i$$

and the overall network can be simplified to a single block as shown in Fig. 14.3(c) if this simplified form is useful.

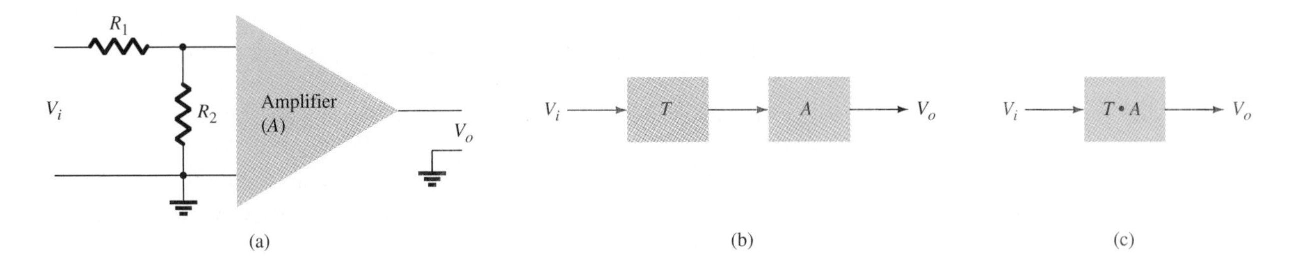

FIG. 14.3
Two block stages in series: (a) circuit; (b) blocks; (c) single block.

Feedback

When part or all of the output signal is fed back and subtracted from the input, the usual form of feedback circuit results. In most control systems various parts or blocks are electrical or mechanical elements or components. A more generalized diagram for representing a feedback control circuit is shown in Fig. 14.4.

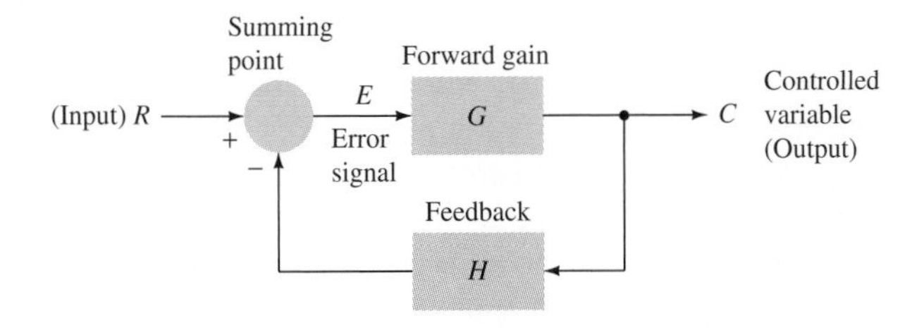

FIG. 14.4
Typical feedback control loop.

The overall transfer function of the feedback control loop can be obtained by solving for the ratio of output to input, C/R. This ratio is calculated as

$$C = GE \tag{14.1}$$

$$E = R - HC \tag{14.2}$$

Multiplying both sides of Eq. (14.2) by G and then inserting Eq. (14.1), we have

$$C = G(R - HC)$$

Solving for C, we obtain

$$C = \frac{GR}{1 + GH} \tag{14.3}$$

so that

$$\frac{C}{R} = \frac{G}{1 + GH} \qquad (14.4)$$

Equation (14.4) shows that the effect of the feedback is a transfer function that reduces the gain from the forward gain G by the factor $1 + GH$ or 1 *plus* the forward gain (G) times the feedback (H). Consider a network having a forward gain $G = 1000$ and feedback of $H = 0.1$. We find that the overall gain is then

$$\frac{C}{R} = \frac{G}{1 + GH} = \frac{1000}{1 + 1000(0.1)} = 9.9$$

If the product of forward gain times feedback is much larger than 1,

$$\frac{G}{1 + GH} \rightarrow \frac{G}{GH} = \frac{1}{H} \quad (\text{for GH} \gg 1) \qquad (14.5)$$

The gain for such a circuit is approximately $1/H$ ($1/0.1 = 10$ in the preceding example). This type of circuit is employed to obtain precise amplifier gain using precision resistors to set the feedback factor H. In many control systems where it may not be possible to set $GH \gg 1$, the factor 1 is still important. If the loop gain GH approaches unity and if the phase shift around the loop approaches 180°, then the denominator 1 + GH approaches zero and the ratio C/R goes to infinity, which in practical terms means that the system becomes unstable and goes into oscillation. Since this is undesirable in a control system, it is necessary to investigate under what conditions this can occur for the particular system and adjust various parts so that it is avoided. System stability is covered more fully later in this chapter.

Breaking down a complex system into its various components and representing them by a feedback block diagram allows for analysis and possible modification or adjustments in a well-defined manner. The techniques of block diagram manipulation and simplification have been well developed. While a number of block diagram simplification rules can be defined and used for relatively simple systems, the transfer function of larger systems can easily be obtained using signal flow graph analysis.

Block Diagram Simplification

A block diagram representing the interconnections of various parts of a system including any existing or added feedback is a means of describing systems in a standardized manner. Table 14.1 shows a number of block diagram simplifications that may be applied to help simplify more complex systems so that the overall system transfer function expression can be obtained. Figure 14.5 shows the steps in reducing a given system

TABLE 14.1

Transformation	Block diagram	Equivalent block diagram	Equation
Cascaded blocks	$R \rightarrow G_1 \rightarrow G_2 \rightarrow C$	$R \rightarrow G_1G_2 \rightarrow C$	$\frac{C}{R} = G_1G_2$
Eliminating a forward loop	R, G_1, G_2, $+$, $\pm$, C	$R \rightarrow G_1 \pm G_2 \rightarrow C$	$\frac{C}{R} = G_1 \pm G_2$
Eliminating a feedback loop	R, $+$, $\pm$, G_1, H_1	$R \rightarrow \frac{G_1}{1 \mp G_1H_1} \rightarrow C$	$\frac{C}{R} = \frac{G_1}{1 \pm G_1H_1}$
Moving a pickoff point beyond a block	R, G, C	R, G, C; R, $\frac{1}{G}$	$\frac{C}{R} = G$
Moving a pickoff point a block ahead	R, G, C; C	R, G, C; C, G	$\frac{C}{R} = G$
Moving a summing point beyond a block	R_1, $+$, $\pm$, R_2, G, C	R_1, G, $+$; R_2, G, $\pm$; C	$\frac{C}{R_1 \pm R_2} = G$
Moving a summing point ahead of a block	R_1, G, $+$, $\pm$, R_2, C	R_1, $+$, $\pm$, G, C; $\frac{1}{G}$, R_2	$C = (R_1G \pm R_2)$

block diagram so that the transfer function expression for the overall system can be obtained. Starting with the loop G_1H_1 in Fig. 14.5(a), we recall that such a feedback loop can be expressed by

$$\frac{G_1}{1 + G_1H_1} \tag{14.6}$$

Figure 14.5(b) shows the G_1H_1 loop replaced by a single block whose transfer function is given by Eq. (14.6). Combining the two blocks in series results in a single block [Fig. 14.5(c)] with the transfer function, we have

$$\frac{G_1G_2}{1 + G_1H_1} \tag{14.7}$$

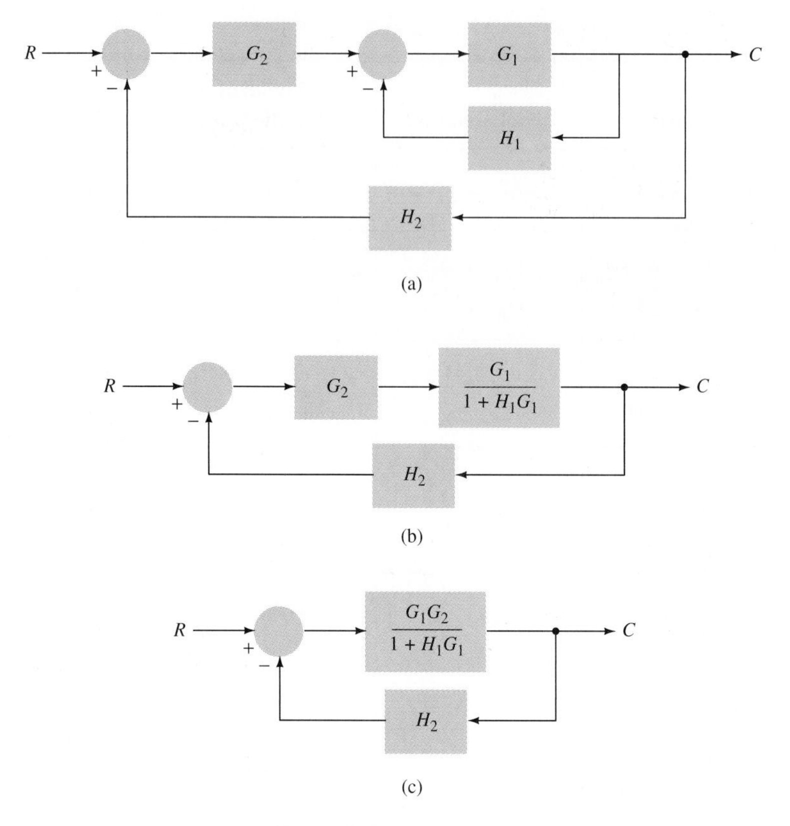

FIG. 14.5
Block diagram simplification.

Figure 14.5(c) again forms a simple feedback loop, and the overall system transfer function can be obtained:

$$\frac{C}{R} = \frac{G_1G_2/(1 + G_1H_1)}{1 + H_2[G_1G_2/(1 + G_1H_1)]} = \frac{G_1G_2}{1 + G_1H_1 + G_1G_2H_2} \quad \textbf{(14.8)}$$

Using the equivalent block diagrams shown in Table 14.1 is also sometimes helpful in rearranging the block diagram so that a modified circuit arrangement still results in the same overall transfer function. For example, the circuit in Fig. 14.6(a) can be modified to that in Fig. 14.6(c), the latter system having the same transfer function as the original one. Obviously, considerable manipulation is possible, with practical systems and component factors dictating choice. Where more complex systems are involved, the signal flow graph technique is more systematic in simplifying and manipulating the system block diagram.

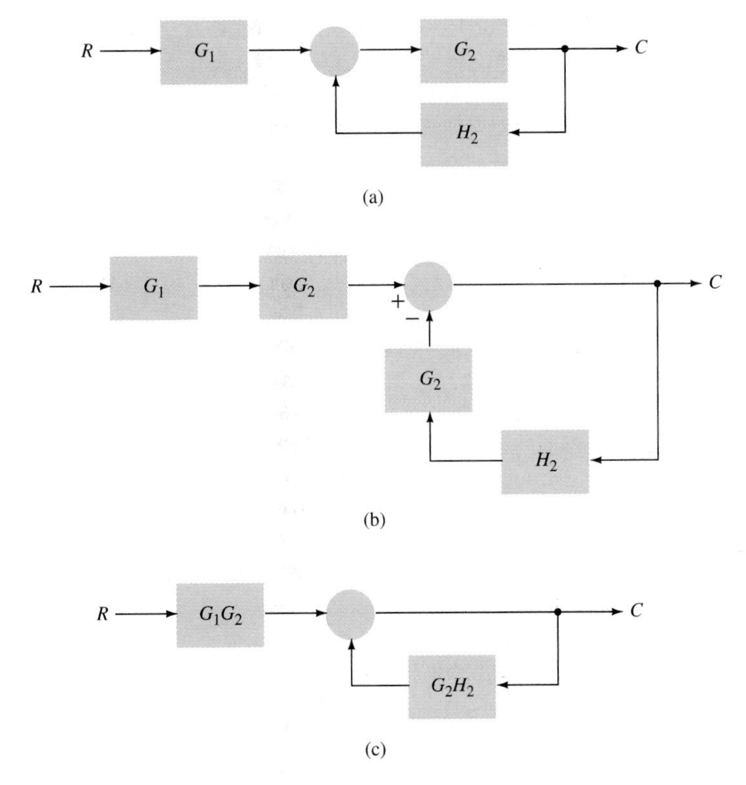

FIG. 14.6
Manipulating a system block diagram.

14.3 BODE PLOTS

Having obtained a transfer function of a system, it is possible to investigate the frequency response of the system. In particular, it is important to determine whether the loop gain approaches unity with a 180° phase shift to examine the system's stability. One way of graphically describing the system transfer function is to use Bode plots—two graphs, one of gain (or magnitude) versus frequency and the other of phase angle versus frequency. To allow for a large frequency scale the $\log_{10}$ of frequency is used as the abscissa axis. For example, consider the simple RC circuit in Fig. 14.7. The transfer function can be expressed using Laplace notation as

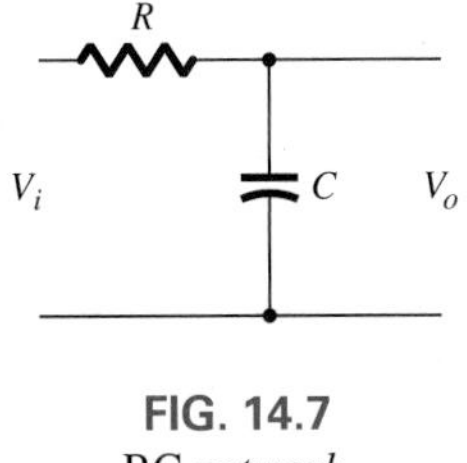

FIG. 14.7
RC *network.*

$$G(s) = \frac{1}{1 + sRC} = \frac{1}{1 + s\tau} \tag{14.9}$$

where $\tau = RC$. Setting $s = j\omega$ (ω is the radian frequency) results in

$$G(j\omega) = \frac{1}{1 + j\omega\tau} \tag{14.10}$$

a complex transfer function having magnitude and phase angle as a function of frequency:

$$|G(j\omega)| = \frac{1}{\sqrt{1 + (\omega\tau)^2}}$$
$$\measuredangle G(j\omega) = \theta = -\tan^{-1}\omega\tau \qquad \textbf{(14.11)}$$

Thus, the transfer function in Eq. (14.10) can be expressed by the two factors in Eq. (14.11). A Bode plot is a plot of each of the expressions in Eq. (14.11) on a $\log_{10}$ frequency scale. Choosing values of R and C in Fig. 14.7 of $R = 10\ \text{k}\Omega$ and $C = 10\ \mu\text{F}$ results in a Bode plot as shown in Fig. 14.8. An exact curve can be obtained by calculating and plotting the corresponding values for a number of frequency points. Simplified techniques are available for drawing the Bode plots using straight-line and asymptotic approximations to the actual curves.

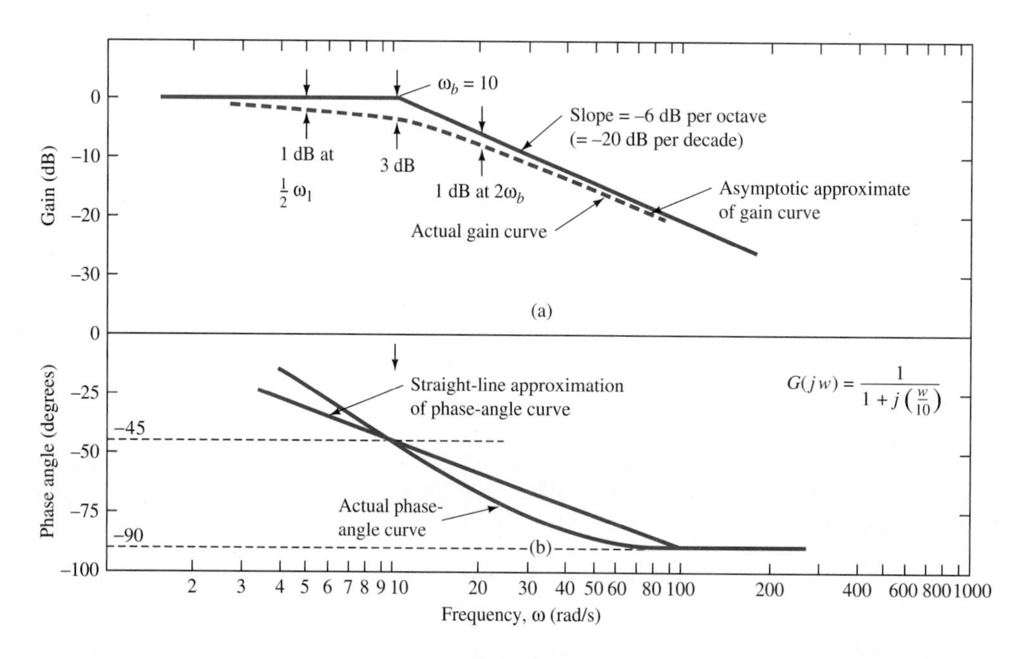

FIG. 14.8
Graphs of Bode magnitude and phase-angle plots: (a) gain; (b) phase.

Gain Asymptotic Approximations

A critical point for the magnitude plot in Fig. 14.8(a), called a *breakpoint* (or corner frequency), is determined for Eq. (14.11) as the point where $\omega\tau = 1$. For frequencies much less than $\omega\tau = 1$ the expression $[1 + (\omega r)^2]^{1/2}$ approaches 1, and $\log_{10} 1$ is 0 dB. For frequencies for which $\omega\tau$ is much greater than 1 ($\omega\tau \gg 1$) the magnitude expression reduces to $1/\omega\tau$, and $-\log_{10}\omega\tau$ is plotted as a sloped line on the log gain–log frequency Bode plot. The critical point to determine is where this sloped line starts, this being the breakpoint $\omega_b\tau = 1$ or $\omega_b = 1/\tau$. For the present example,

$$\tau = RC = (10 \times 10^3)(10 \times 10^{-6}) = 0.1$$

and so $\omega_b = 10$ rad/s. The Bode plot in Fig. 14.8(a) shows this breakpoint, with log gain 0 dB before the breakpoint and dropping at a constant slope of either 6 dB per octave or 20 dB per decade.* In the present example the curve drops off by 6 dB at 20 rad/s and 20 dB at 100 rad/s.

The transfer function in Eq. (14.10) is then easily plotted on a Bode plot by determining ω_b and drawing a fixed slope line (slope of 6 dB per octave or 20 dB per decade) from this frequency point.

This straight-line approximation plot is a quick, easy way of drawing the Bode magnitude plot. The exact magnitude plot can also be easily sketched using the approximate lines plotted as asymptotes. In particular, Fig. 14.8(a) shows that at $\frac{1}{2}\omega_b$ the actual gain plot should be 1 dB less than the asymptotic line, at ω_b the actual gain is 3 dB less, and at $2\omega_b$ it is again 1 dB less than the asymptotic curve.

Phase-Shift Asymptotic Curve

The asymptotic approximation Bode plot of the phase angle or phase shift for Eq. (14.11) is shown in Fig. 14.8(b). For frequencies $\omega\tau$ much less than 1 the value $\tan^{-1}\omega\tau$ is very small and is asymptotically shown as a 0° phase-shift line. The first critical point to note is at $0.1\omega_b$ at which a straight line at slope 45° per decade is started. Notice, then, that at ω_b the phase shift is 45° (when $\omega_b\tau = 1$, $\tan^{-1}\omega_b\tau = \tan^{-1} 1 = 45°$) and that at $10\omega_b$ the phase shift has reached 90°, at which point the asymptotic curve becomes a constant phase-angle line. The actual phase angle curve can then be plotted by noting the phase-angle corrections in Fig. 14.8(b) and listed in Table 14.2.

The preceding rules apply to the Bode plot only for a transfer function such as that in Eq. (14.10), which happens to be very common. Other basic equation forms can also be examined and a set of Bode diagram building blocks described for use in plotting a very wide variety of transfer function equations.

TABLE 14.2
Corrections to asymptotic phase-angle curve

ω	Phase-Angle correction
$0.05\omega_b$	−3°
$0.1\omega_b$	−6°
$0.3\omega_b$	+5°
$0.5\omega_b$	+5°
$1.0\omega_b$	0°
$2.0\omega_b$	−5°
$3.0\omega_b$	−5°
$10\omega_b$	+6°
$20\omega_b$	+3°

Stability Information from Bode Diagrams

The Bode diagram provides a good graphical description of the control system transfer functions. In particular, information on system stability can easily be read off the curves. Two stability factors are commonly obtained from Bode diagrams. An example of these stability factors is given in Fig. 14.9, which combines both magnitude and phase-angle curves on a single graph for the transfer function

$$\frac{4}{(j\omega)(1 + j0.125\omega)(1 + j0.5\omega)} = \frac{4}{(j\omega)(1 + j\omega/8)(1 + j\omega/2)}$$

*An octave is a doubling (halving) of the frequency, while a decade is a 10-fold increase (decrease) in frequency.

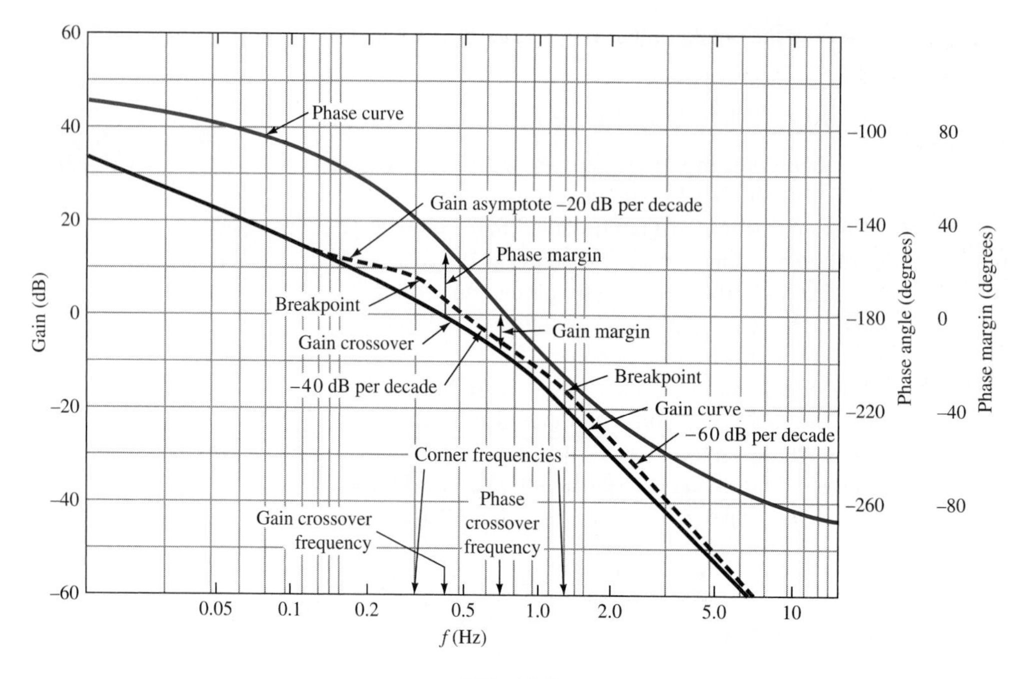

FIG. 14.9
Graph of stability factors.

A *gain margin,* defined as the amount the gain differs from 0 dB occurring at a frequency point when the phase angle is $-180°$, is shown in Fig. 14.9 to be about 8 dB. The frequency at which this occurs is called the *phase crossover* frequency. Since a gain of 1 at a phase angle of 180° is indicative of system stability, the gain margin reflects how much the gain differs from a gain of 1 (0 dB) at a phase shift of 180°.

A *phase margin,* defined as the amount of phase shift from 180° occurring when the gain curve crosses 0 dB, is shown in Fig. 14.9 to be about 30°. The frequency at which this occurs is called the *gain crossover* frequency. The phase margin indicates how much the phase shift may vary before reaching the critical value of 180° at unity gain. The phase margin gives designers a good description of system stability with acceptable phase margins usually around 40° to 60°, much less being too close to instability or too jittery and much more having too much stability or being too sluggish.

14.4 CONTROL SYSTEM COMPONENTS

Control systems are built using many different parts, some electrical, some mechanical, some electromechanical. A number of components are used more frequently, and they are covered next. They include potentiometers, gear trains, synchros, and servos. Our concern here will be with those components more common in instrument servomechanisms—systems that control position, speed, or acceleration. In addition to the basic control system components mentioned already, the complete system usually contains amplifiers, feedback elements, power activators, and modulator–demodulator networks.

Potentiometers

Potentiometers used in control systems are usually precision wire-wound circular components, often with gearing. As a slider-arm pickoff is rotated from a starting point, the amount of resistance between the slider arm and either extreme of the resistor varies. Using dc voltage excitation, it is possible to pick off a dc voltage proportional to the angular shaft rotation. Figure 14.10(a) shows a potentiometer assembly, Fig. 14.10(b) the potentiometer symbol with dc excitation, and Fig. 14.10(c) a block diagram

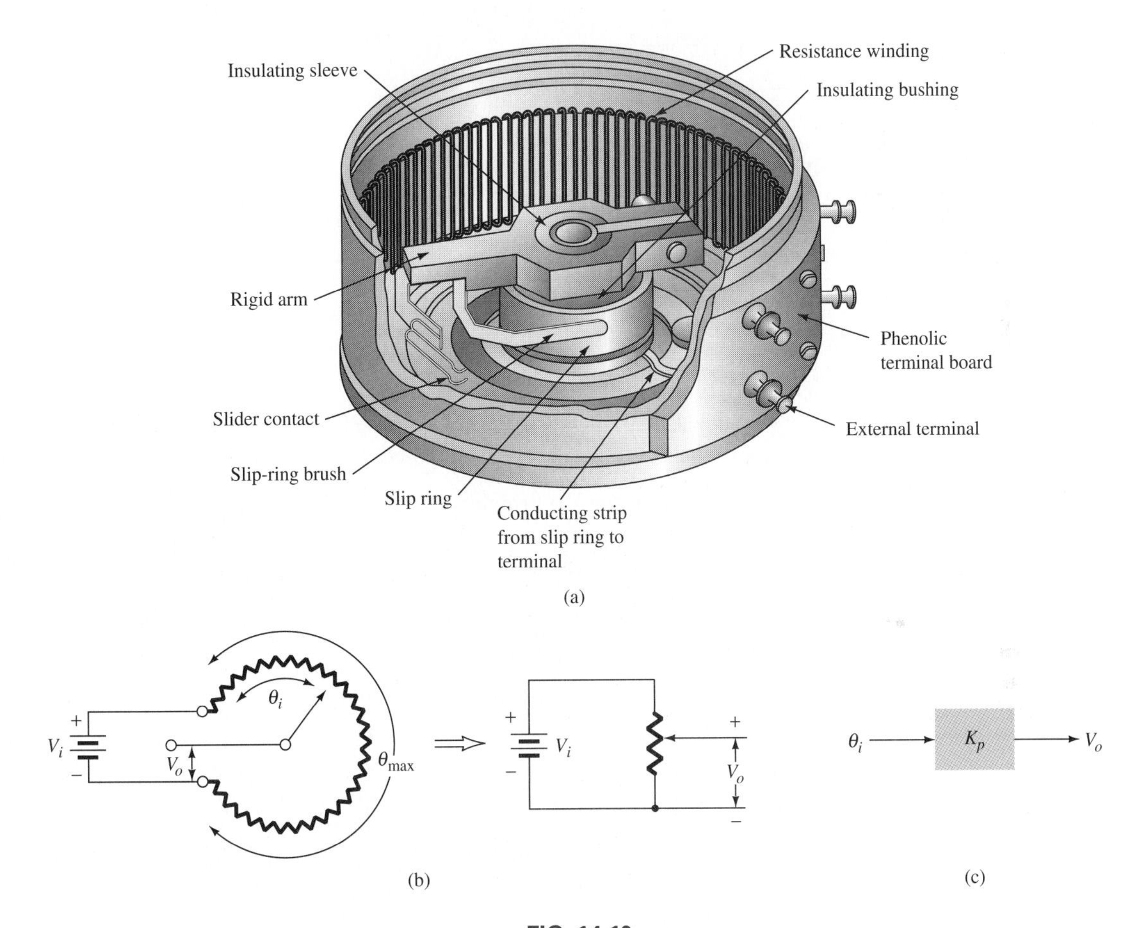

FIG. 14.10
Potentiometer: (a) assembly; (b) schematic; (c) block symbol.

representation. If we assume linear operation, the output voltage is proportional to the amount of rotation and excitation voltage, so that

$$V_o = V_i \frac{\theta_i}{\theta_{max}} = K_P \theta_i \tag{14.12}$$

where K_p, as shown in Fig. 14.10(c), is the potentiometer transfer function in volts per radian. If the potentiometer shaft is connected to some device, for example, a valve that rotates open, the resulting potentiometer voltage will be an indication of the amount of valve opening that can be transmitted as an electrical dc voltage—the potentiometer acting as a transducer device to convert angular rotation into electrical voltage.

A pair of potentiometers may be used as an error detector to sense when the positions of two different shafts are not the same and provide an electrical signal proportional to this shaft displacement. Figure 14.11(a) shows such a circuit setup with the representative block symbol shown in Fig. 14.11(b). The circuit acts as a bridge circuit providing 0-V

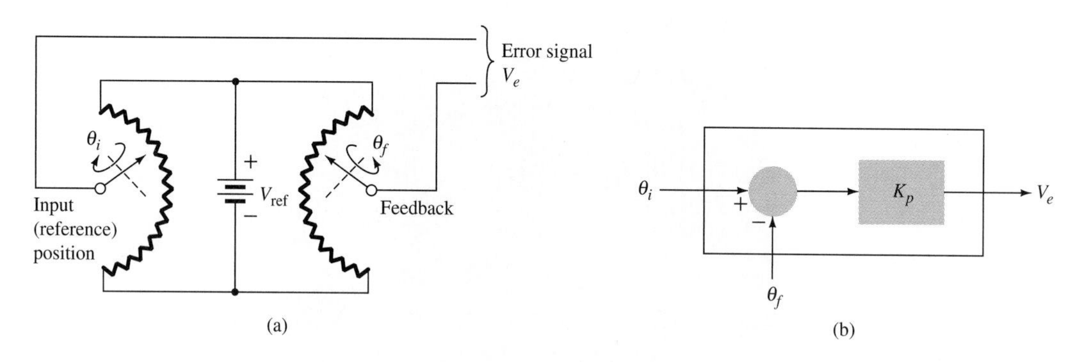

FIG. 14.11
(a) Two-potentiometer error detector; (b) block diagram.

output when the bridge is balanced. The voltage due to a difference between the input and feedback potentiometer settings is used as an error signal to drive the mechanism connected to the feedback shaft. In a properly adjusted system any change in reference position results in an error voltage, which then drives some mechanical device until the feedback shaft position again aligns with the reference shaft and no error voltage occurs. A diagram of a complete servomechanism system using potentiometers as error detectors is shown in Fig. 14.12.

An input shaft position setting, θ_i, resulting in a positional difference between the wiper-arm settings of input and feedback potentiometers, produces a dc voltage to a dc amplifier circuit. The magnitude and polarity of this error signal drive the dc motor to realign the feedback potentiometer wiper arm. The dc motor also repositions the device being controlled to produce the desired overall servomechanism action.

The bridge reference or excitation voltage can be either a dc voltage as previously shown or ac voltage excitation. With ac excitation the error signal magnitude is still an indicator of the amount of shaft position difference between input and feedback signals. The direction of the error signal is indicated, however, by the phase reversal of the ac signal instead of a voltage polarity reversal, as in dc operation. The amplifier can then be an ac circuit and the motor an ac motor unit. When it is desired to operate partly with dc devices, demodulator–modulator circuits can be used to go from ac to dc and back to ac again.

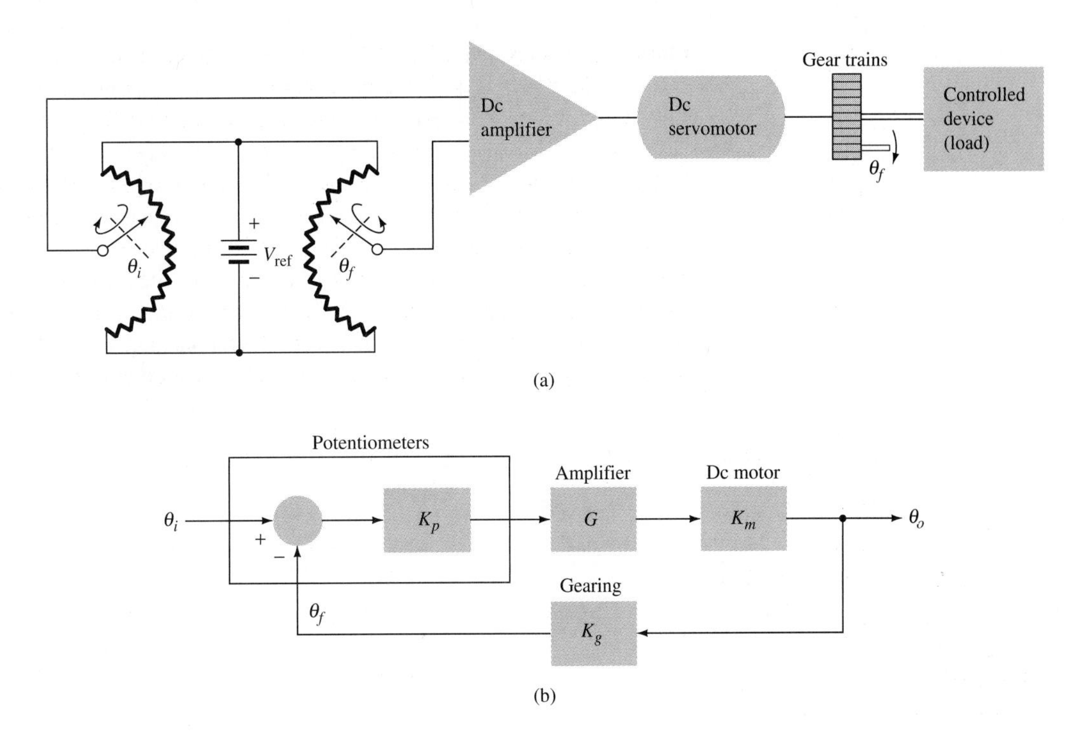

FIG. 14.12
Position servomechanism: (a) system diagram; (b) block diagram.

Synchros

Another very popular unit used as an error detector in control systems is the synchro (Fig. 14.13). Control or instrument synchros (as opposed to torque or driving synchros) provide for transmission of shaft position

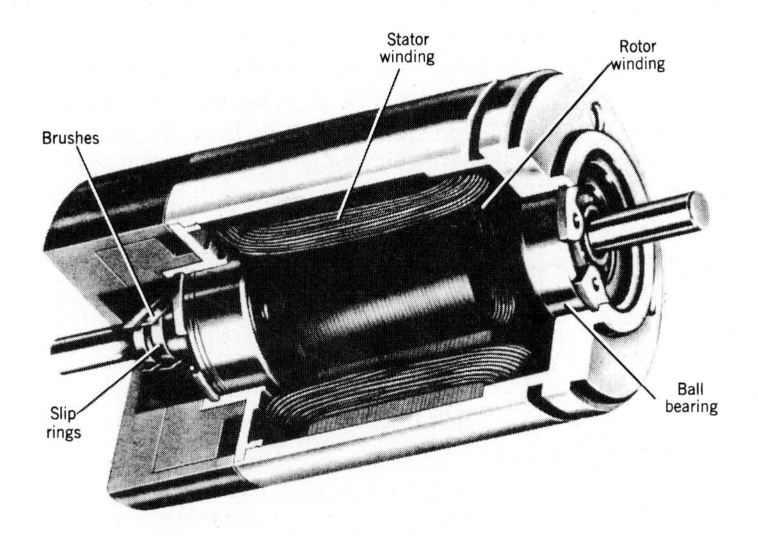

FIG. 14.13
Cutaway view of a typical synchro. (Courtesy of Weston Instruments, Inc., Transicoil Division.)

information for input and output with the ability to provide an error signal if the two shafts are not aligned. The usual connection of a transmitter–receiver (CX-CR) synchro pair, shown in Fig. 14.14, can be

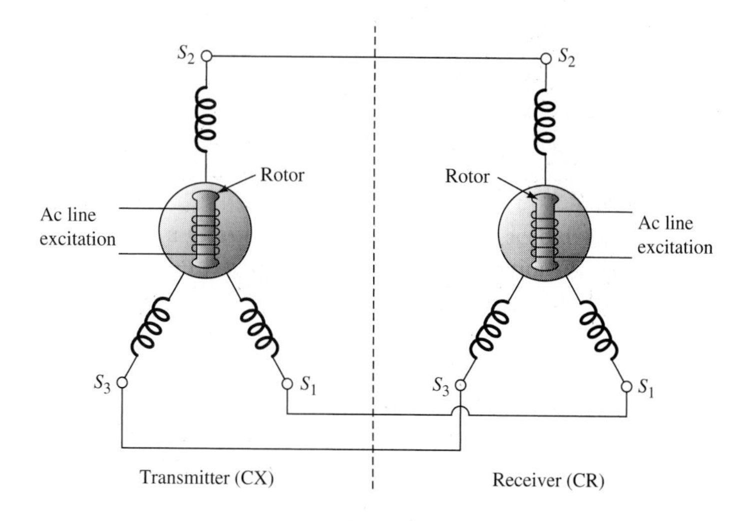

FIG. 14.14
Synchro transmitter–receiver pair.

used to remotely transmit shaft position information. The synchro consists of a rotor, shown in Fig. 14.14, to be excited by the ac line (typically 400 Hz in aircraft systems). The ac voltage is connected to the single salient-pole rotor windings by means of slip rings and brushes. The magnetic field set up by the rotor and ac excitation is coupled to three separate stator windings positioned 120° apart. One terminal of each winding is connected to a common (central) point, while each of the other winding terminals forms a three-phase voltage for connection to the receiver synchro. The transmitter rotor is held fixed at the desired (or command) setting. With stator winding S_2 as reference, the zero rotor position is defined as the rotor aligned with winding S_2. The three stator-induced voltages can be mathematically expressed with reference to the central point as*

$$\begin{aligned} V_{S2} &= KV\cos\theta \\ V_{S3} &= KV\cos(\theta + 120^\circ) \\ V_{S1} &= KV\cos(\theta - 120^\circ) \end{aligned} \qquad \textbf{(14.13)}$$

where θ is the rotor angle with respect to winding S_2.

*The stator voltages can also be expressed across pairs of terminals S_1, S_2, S_3 as

$$\begin{aligned} V_{S1S2} &= \sqrt{3}\,KV\cos(30^\circ - \theta) \\ V_{S2S3} &= -\sqrt{3}\,KV\cos(30^\circ + \theta) \\ V_{S1S3} &= KV\cos\theta \end{aligned}$$

The excited rotor winding of the receiver synchros similarly induces voltages in the three stator windings of the receiver. If the receiver and transmitter rotors are aligned, these induced voltages are equal and there is no net current in the stator windings. If, however, the receiver rotor is at an angle different from the transmitter rotor, the transmitter-induced voltages and net currents circulate in the stator windings. Since the transmitter rotor is held fixed and the receiver rotor is, at most, lightly loaded, the resulting receiver stator and rotor magnetic fields interact to pull the receiver rotor into line with the transmitter rotor position. If the transmitter rotor is later moved again, the receiver rotor will be driven to line up with the transmitter rotor. Since the only connection between the transmitter and the rotor is the electric wires, the transmitter can be mounted to the shaft under observation, and the receiver synchro can be used to drive a light load (indicator dial) at a distant display console.

If the load to be driven is not light, then a synchro transformer can be used in place of the synchro receiver, as shown in Fig. 14.15. The

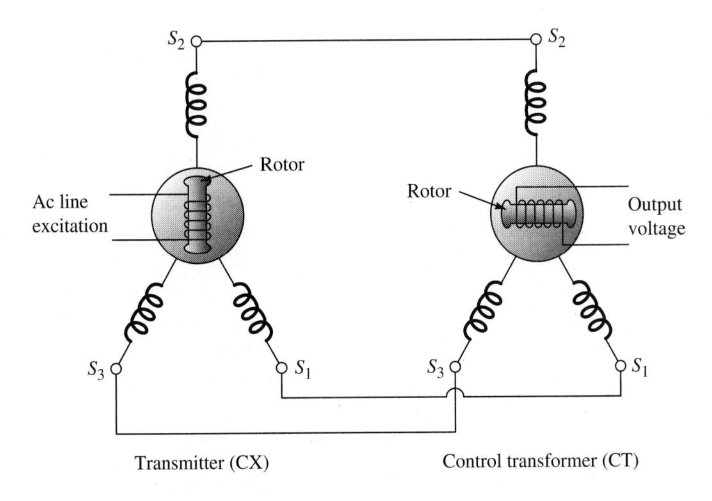

FIG. 14.15
Synchro transmitter-control transformer pair.

synchro transmitter rotor may be positioned as desired with a set of voltages developed in the three windings as expressed in Eq. (14.13). These voltages are connected through a set of information lines to the stator windings of a control transformer. The control transformer rotor is, however, not the same as that of a receiver (the CT rotor is cylindrical) nor is any ac excitation applied to it. The control transformer rotor is set to a fixed zero-angle position, typically perpendicular to the S_2 winding. A net voltage is induced in the CT rotor windings, which is a function of the CX stator voltages and thereby a function of the CX rotor position. The output voltage of the control transformer can be expressed as

$$V_o = K_s \sin \mathrm{v}_i \qquad \textbf{(14.14)}$$

where K_s is the control transformer sensitivity measured in volts per radian or volts per degree and θ_i is the angle or position of the transmitter shaft position rather than a corresponding shaft position as with a CX-CR pair. The resulting control transformer output voltage [Eq. (14.14)] can be used to drive some power device to position some other shaft or display.

Servomotors

A servomotor may be either dc or ac, with the latter type, most common in low-power applications, to be covered here. An ac servomotor is essentially a two-phase induction motor, of typically small size, for use in an instrument servomechanism system. Figure 14.16 shows the schematic diagram of an ac servomotor. A reference stator receives excitation from the fixed ac line voltage. A control stator winding is excited by the ac drive signal from an amplifier. An error signal obtained by, say, a synchro transmitter-control transformer pair is applied to an ac amplifier. The amplifier signal driving the servomotor control winding varies in both amplitude and signal voltage phase. The servomotor speed is proportional to the control winding signal amplitude, while the direction of rotation depends on the phase of the drive signal. A typical control system using a servomotor is shown schematically in Fig. 14.17(a) and in block diagram form in Fig. 14.17(b). A command or input shaft position that determines the load shaft position is connected to the rotor of the synchro transmitter. The CX-CT pair then provides an output voltage

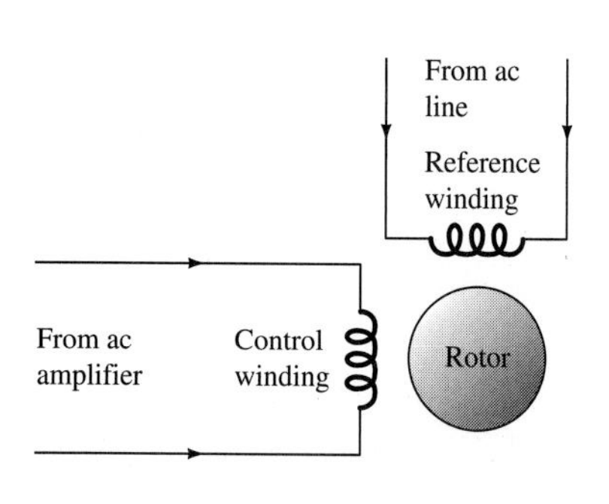

FIG. 14.16
Schematic diagram of ac servomotor.

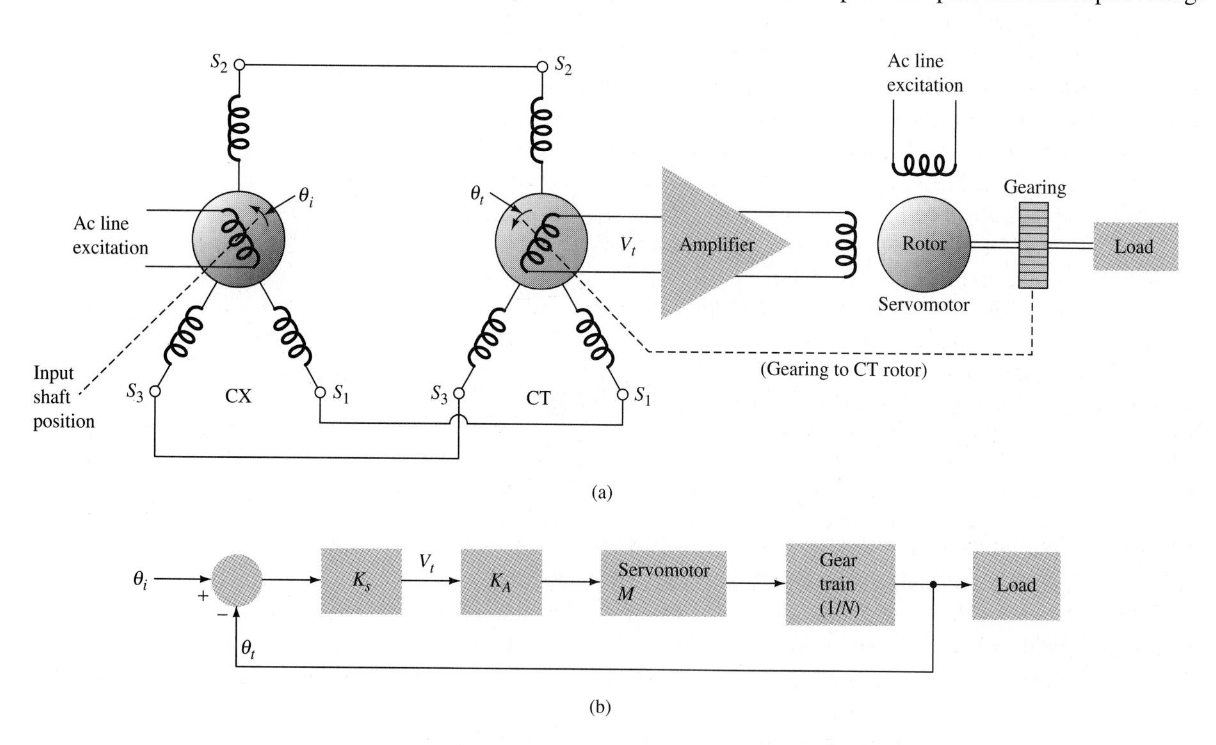

FIG. 14.17
Position control system: (a) circuit; (b) block.

from the control whose amplitude and phase depend on the input shaft position (command) and the CT rotor shaft position (feedback). An error signal in the form of ac voltage V_t from the control transformer is applied to an ac amplifier whose output signal drives the servomotor control winding. The servomotor then drives the load (through gearing, in this example). As the load moves to the desired position the CT rotor is repositioned through the gearing, and the error voltage driving the ac amplifier is reduced until a null position is reached. This load position is maintained unless the input (command) position is changed.

Gear Trains

Connection of various shafts is often made through gear trains rather than directly. If the number of teeth on each gear is different, a gearing ratio results. This gearing can be used in a servomotor system to increase the torque driving the load, since the servomotor itself is a high-speed, low-torque device. The gear-train transfer function [see Fig. 14.17(b)] is given by the teeth ratio as

$$\text{transfer function} = \frac{\theta_o}{\theta_i} = \frac{N_o}{N_i} = \frac{1}{N} \tag{14.15}$$

where $1/N$ is the ratio of teeth in output gears to teeth in input gears.

The control system in Fig. 14.17 compares the command shaft position with the load shaft position, develops an error voltage, and drives the servomotor so that the load is moved to the command position. If the error is great, a large error signal will result, and the servomotor will be driven very fast, so fast that it may overshoot the desired position and then oscillate back and forth around the desired position until it finally settles down. One means of compensating for the drive at great position error without making the small error signal drive too sluggish is to use a rate generator or tachometer in a second feedback path. An ac tachometer, as shown in Fig. 14.18, is a two-phase induction device. In the tachometer an ac reference voltage is applied to one stator winding, and the rotor is driven by the servomotor, resulting in an output voltage proportional to the servomotor speed from a second stator winding. The resulting output ac voltage can be expressed as

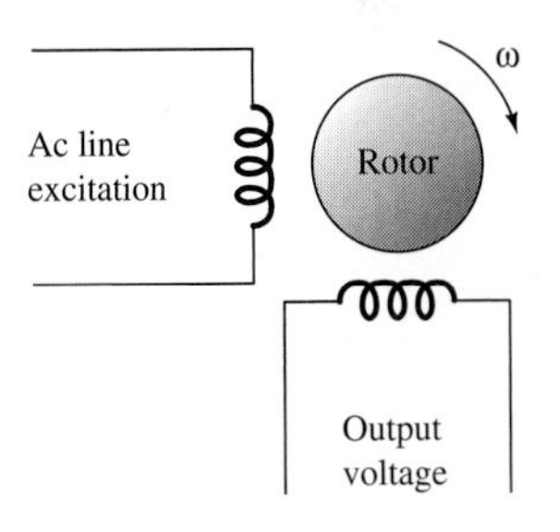

FIG. 14.18
AC tachometer schematic diagram.

$$V_o = K_g\omega \tag{14.16}$$

where K_g is the rate generator sensitivity in volts per radian per minute (usually rated at 1000 rad/min), and ω is the rotational speed of the rate generator rotor shaft in radians per minute. The rate generator transfer function can then be expressed as

$$\frac{V_o}{\theta_i} = K_g s \tag{14.17}$$

where θ_i = input angular position (rotor position)
$\omega = s\theta$ in Laplace notation

Figure 14.19(a) shows a positional servo system schematic diagram. A command position input θ_i positions the synchro transmitter rotor, resulting in an error voltage from the synchro transformer dependent on the position of the CT rotor, θ_t, which is mechanically connected to the load. The error voltage is summed with the rate tachometer output voltage to give a resulting error signal that combines both position feedback via CT and rate feedback via the tachometer. The amplified error voltage then drives the load through the servomotor to the desired position. Figure 14.19(b) shows the positional servo system in block diagram form, with two feedback loops, one positioned and one rate-controlled.

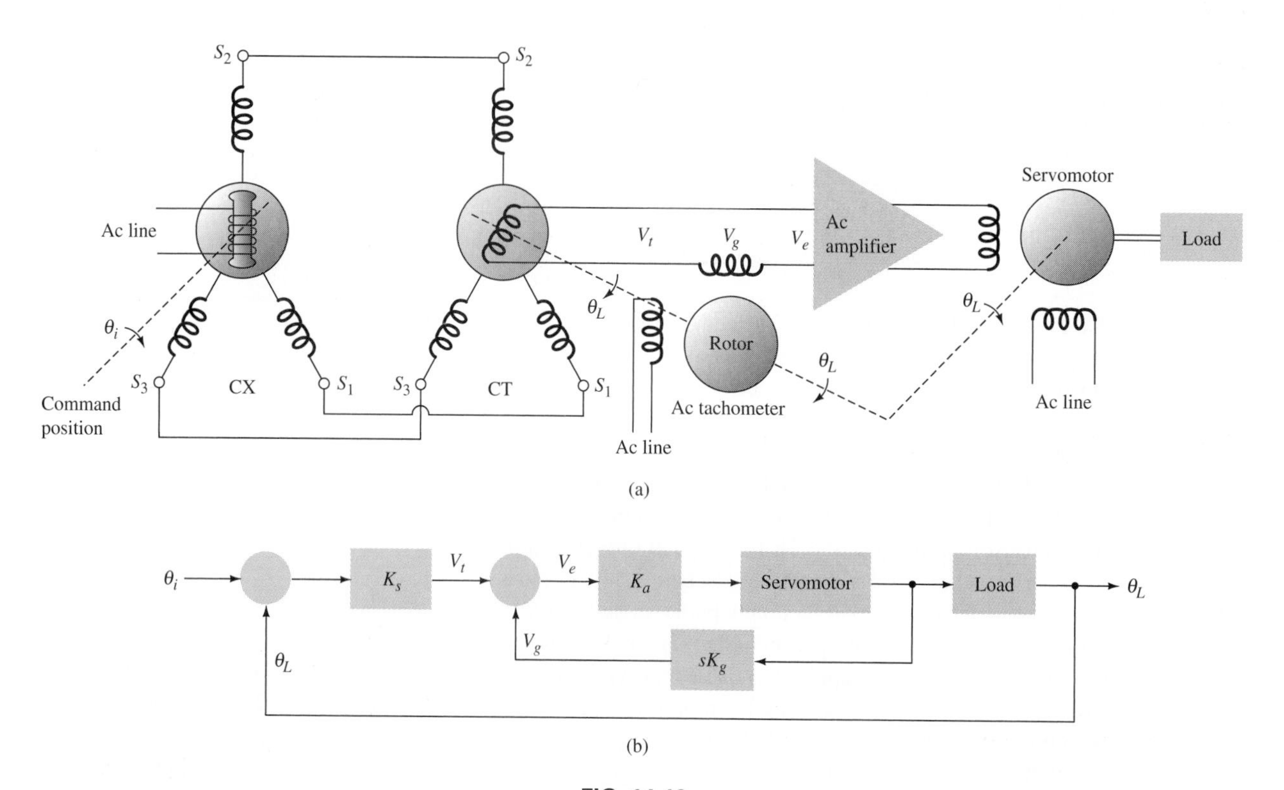

FIG. 14.19
Positional servo system with rate feedback: (a) circuit; (b) block.

14.5 SUMMARY

Block Diagrams:

$$\frac{C}{R} = \frac{G}{1 + GH}$$

Bode Plots:

Gain vs Frequency
Phase Angle vs Frequency
Gain Margin: gain difference from 0 dB occurring at frequency when the phase angle is 180°
Phase Margin: amount of phase shift from **180°** occurring when the gain curve crosses 0 dB
Synchro: typically a rotor excited by the 400 Hz frequency in aircraft
Servo: essentially a two-phase induction motor controlled by an ac voltage

PROBLEMS

SECTION 14.2 Block Diagrams

1. Determine the transfer function of the network shown in Fig. 14.20 and calculate the value of the transfer function for $R_1 = 10\ \text{k}\Omega$, $R_2 = 20\ \text{k}\Omega$.

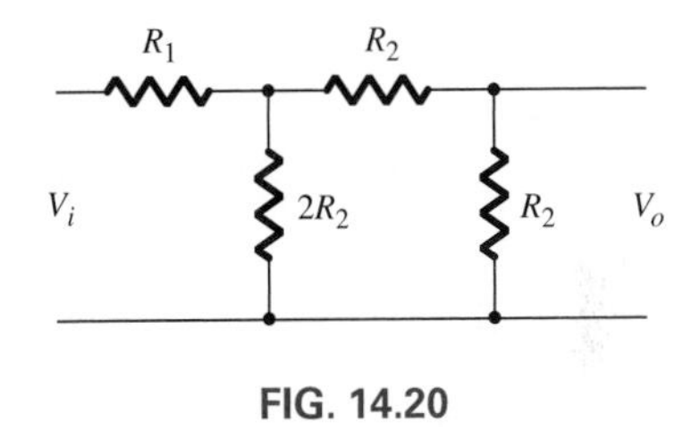

FIG. 14.20

2. Determine the transfer function for each of the networks shown in Fig. 14.21.

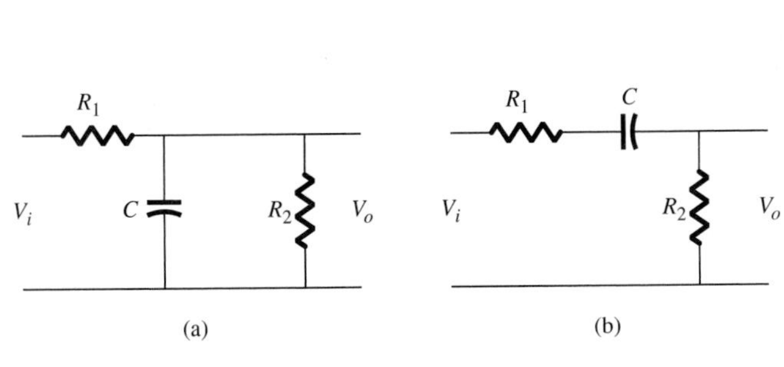

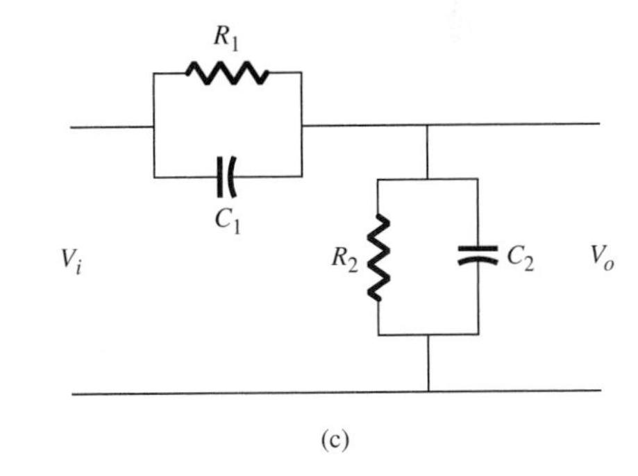

FIG. 14.21

3. What is the difference between open-loop gain and closed-loop gain?

4. Calculate the transfer function of the feedback network in Fig. 14.22.

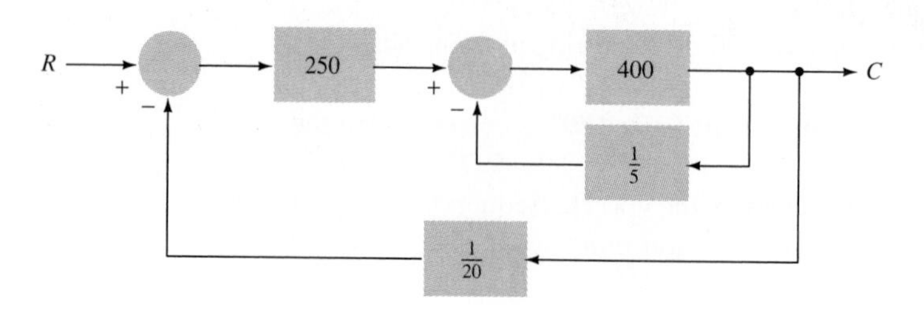

FIG. 14.22

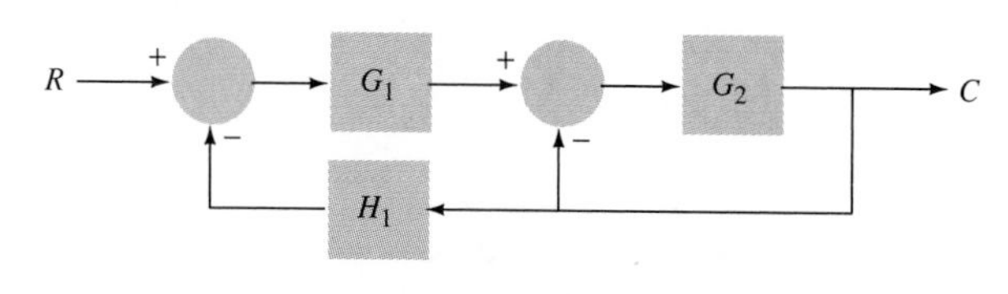

FIG. 14.23

5. Derive the ratio C/R for the network in Fig. 14.23.
6. Draw the block diagram of the network having the transfer function

$$\frac{C}{R} = \frac{G_1 G_2}{1 + G_1 G_2 H}$$

SECTION 14.3 Bode Plots

7. What does a Bode plot represent?
8. Sketch the magnitude Bode plot for $30/[j10(1 + j20)]$.
9. Sketch the phase Bode plot for $2(j20)/(1 + j40)$.
10. Define gain margin.
11. Define phase margin.
12. Sketch the magnitude Bode plot for $15/[(j\omega/\omega_a)(1 + j\omega/\omega_b)]$, where $\omega_a = 12$ rad/s and $\omega_b = 18$ rad/s.
13. Sketch the magnitude and phase Bode plots for $10/[(0.3j)(1 + j1.5)]$.
14. Sketch the Bode phase plot for the transfer function $1/[(j\omega/\omega_a)(1 + j\omega/\omega_b)]$, where $\omega_a = 20$ rad/s and $\omega_b = 30$ rad/s.

15 Power Supplies: Linear ICs and Regulators

15.1 INTRODUCTION

A block diagram containing the parts of a typical power supply and the voltages at various points in the unit is shown in Fig. 15.1. The ac voltage, typically 120 V (rms), is connected to a transformer that steps that voltage up or, more typically, down to the level for the desired dc output. A diode rectifier then provides a half-wave—or, more typically, full-wave—rectified voltage that is applied to a filter to smooth the varying signal. A simple capacitor filter is often sufficient to provide this smoothing action. The resulting dc voltage with some ripple or ac voltage variation is then provided as input to an IC regulator that provides as output a well-defined dc voltage level with extremely low ripple voltage over a range of load.

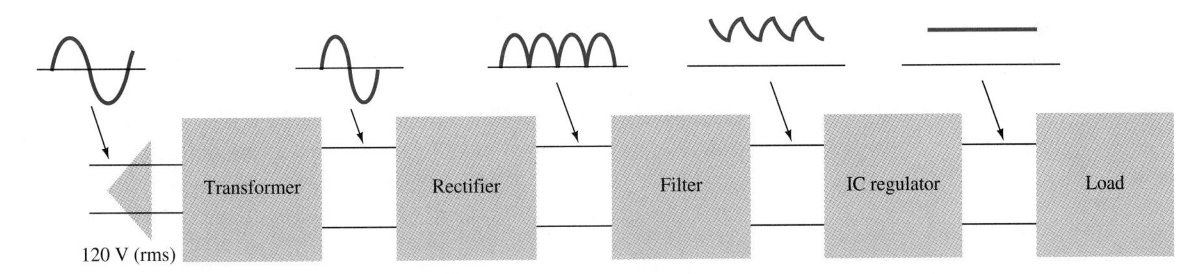

FIG. 15.1
Block diagram showing parts of a power supply.

15.2 GENERAL FILTER CONSIDERATIONS

A rectifier circuit is necessary to convert a signal having zero average value to one that has a nonzero average. However, the resulting pulsating dc signal is not pure direct current or even a good representation of it. Of course, for a circuit such as a battery charger the pulsating nature of the signal is no great detriment as long as the dc level provided will result in charging of the battery. On the other hand, for voltage supply circuits for a tape recorder or radio the pulsating dc will result in a 60-(or 120-)Hz signal appearing in the output, thereby making the operation of the overall circuit poor. For these applications, as well as for many more, the output dc developed will have to be much "smoother" than that of the pulsating dc obtained directly from half-wave or full-wave rectifier circuits.

Filter Voltage Regulation and Ripple Voltage

Before going into the details of the filter circuit it is appropriate to consider the usual method of rating the circuits so that we are able to compare a circuit's effectiveness as a filter. Figure 15.2 shows a typical filter output voltage, which will be used to define some of the signal factors. The filtered output voltage in Fig. 15.2 has a dc value and some ac variation *(ripple)*. Although a battery has essentially a constant or dc output voltage, the dc voltage derived from an ac source signal by rectifying and filtering has some variation (ripple). The smaller the ac variation *with respect to* the dc level, the better the filter circuit operation.

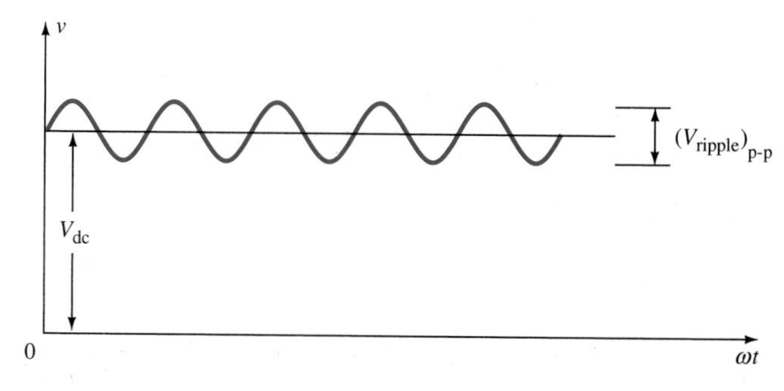

FIG. 15.2
Filter voltage waveform showing dc and ripple voltages.

Consider measuring the output voltage of the filter circuit using a dc voltmeter and an ac (rms) voltmeter. The dc voltmeter reads only the average or dc level of the output voltage. The ac (rms) meter reads only the rms value of the ac component of the output voltage (assuming the signal is coupled to the meter through a capacitor to block out the dc level).

$$r = \text{ripple} = \frac{\text{ripple voltage (rms)}}{\text{dc voltage}} = \frac{V_r\,(\text{rms})}{V_{\text{dc}}} \times 100\% \qquad \textbf{(15.1)}$$

EXAMPLE 15.1 Using a dc and an ac voltmeter to measure the output signal from a filter circuit, a dc voltage of 25 V and an ac ripple voltage of 1.5 V (rms) are obtained. Calculate the ripple of the filter output.

Solution:

$$r = \frac{V_r\,(\text{rms})}{V_{dc}} \times 100\% = \frac{1.5\text{ V}}{25\text{ V}}\,100 = \mathbf{6\%}$$

Voltage Regulation

Another factor of importance in a voltage supply is the amount of change in the output dc voltage over the range of the circuit operation. The voltage provided at the output at no load (no current drawn from the supply) is reduced when load current is drawn from the supply. How much this voltage changes with respect to either the loaded or unloaded voltage value is of considerable interest to anyone using the supply. This voltage change is described by a factor called *voltage regulation:*

$$\text{voltage regulation} = \frac{\text{voltage at no load} - \text{voltage at full load}}{\text{voltage at full load}}$$

$$\text{VR} = \frac{V_{NL} - V_{FL}}{V_{FL}} \times 100\% \qquad \textbf{(15.2)}$$

EXAMPLE 15.2 A dc voltage supply provides 60 V when the output is unloaded. When a full-load current is drawn from the supply, the output voltage drops to 56 V. Calculate the value of the voltage regulation.

Solution:

$$\text{VR} = \frac{V_{NL} - V_{FL}}{V_{FL}} \times 100\% = \frac{60\text{ V} - 56\text{ V}}{56\text{ V}} \times 100\% = \mathbf{7.14\%}$$

If the value of full-load voltage is the same as the no-load voltage, the VR calculated is 0%, which is the best that can be expected. This value means that the supply is a true voltage source for which the output voltage is independent of current drawn from the supply. The output voltage from most supplies decreases as the amount of current drawn from the voltage supply is increased. The smaller the voltage decreases, the smaller the percent of VR and the better the operation of the voltage supply circuit.

Ripple Factor of Rectified Signal

Although the rectified voltage is not a filtered voltage, it nevertheless contains a dc component and a ripple component. We can calculate these values of dc voltage and ripple voltage (rms) and from them obtain the ripple factor for the half-wave- and full-wave-rectified voltages. The calculations will show that the full-wave-rectified signal has less percent of ripple and is therefore a better rectified signal than the half-wave-rectified signal if the lowest percent ripple is desired. The percent ripple is not always the most important concern. If circuit complexity or cost considerations are important (and the percent ripple is secondary), then a half-wave rectifier may be satisfactory. Also, if the filtered output supplies only a small amount of current to the load and the filtering circuit is not critical, then a half-wave-rectified signal may be acceptable. On the other hand, when the supply must have as low a ripple as possible, it is best to start with a full-wave-rectified signal, since it has a smaller ripple factor, as will now be shown.

For a half-wave-rectified signal the output dc voltage is $V_{dc} = 0.318V_m$ (Chapter 4). The rms value of the ac component of an output signal is $V_r\,(\text{rms}) = 0.385V_m$, as determined using calculus techniques. The percent ripple is

$$r = \frac{V_r\,(\text{rms})}{V_{dc}}(100\%) = \frac{0.385V_m}{0.318V_m}(100\%) = 1.21(100\%) = 121\% \quad \text{(half-wave)} \quad \textbf{(15.3)}$$

For the full-wave rectifier the value of V_{dc} is $0.636V_m$. For a full-wave-rectified signal $V_r\,(\text{rms}) = 0.308V_m$ with a percent ripple of

$$r = \frac{V_r\,(\text{rms})}{V_{dc}}(100\%) = \frac{0.308V_m}{0.636V_m}(100\%) = 48\% \quad \text{(full-wave)} \quad \textbf{(15.4)}$$

The ripple factor of the full-wave-rectified signal is about 2.5 times smaller than that of the half-wave-rectified signal and provides a better filtered signal. Note that these values of ripple factor are absolute values and do not depend at all on the peak voltage. If the peak voltage is made larger, the dc value of the output increases but then so does the ripple voltage. The two increase in the same proportion, so that the ripple factor stays the same.

15.3 SIMPLE-CAPACITOR FILTER

A popular filter circuit is the simple-capacitor filter circuit shown in Fig. 15.3. The capacitor is connected across the rectifier output, and the dc

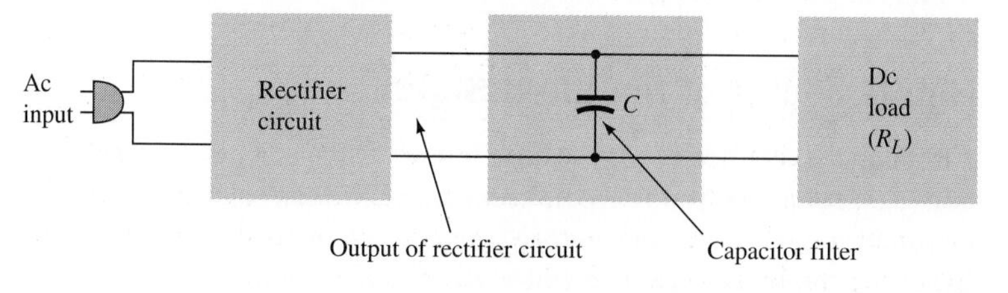

FIG. 15.3
Simple-capacitor filter.

output voltage is available across the capacitor. Figure 15.4(a) shows the rectifier output voltage of a full-wave-rectifier circuit before the signal is filtered. Figure 15.4(b) shows the resulting waveform after the capacitor has been connected across the rectifier output. As shown, this filtered voltage has a dc level with some ripple voltage riding on it.

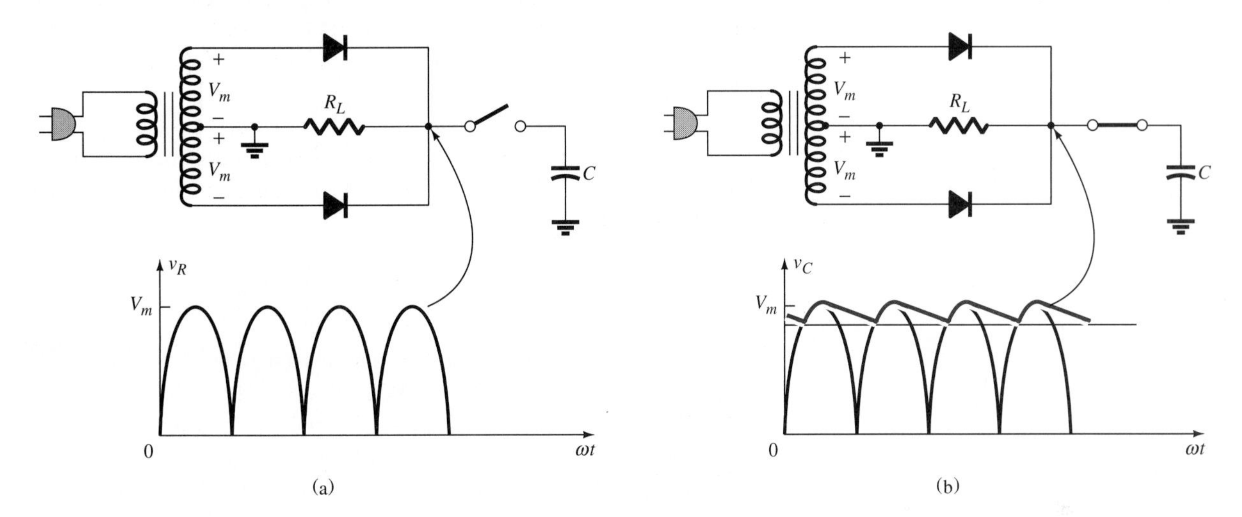

FIG. 15.4

Capacitor filter operation: (a) full-wave-rectifier voltage; (b) filtered output voltage.

Figure 15.5(a) shows a full-wave rectifier and the output waveform obtained from the circuit when connected to an output load. If no load was connected to the filter, the output waveform would ideally be a constant dc level equal in value to the peak voltage (V_m) from the rectifier circuit. However, the purpose of obtaining a dc voltage is to provide this voltage for use by other electronic circuits, which then constitute a load on the voltage supply. Since there will always be some load on the filter, we must consider this practical case in our discussion. For the full-wave-rectified signal indicated in Fig. 15.5(b) there are two intervals of time indicated. T_1 is the time during which a diode of the full-wave rectifier conducts and charges the capacitor up to the peak rectifier output voltage (V_m). T_2 is the time during which the rectifier voltage drops below the peak voltage and the capacitor discharges through the load.

If the capacitor was to discharge only slightly (because of a light load), the average voltage would be very close to the optimum value of V_m. The amount of ripple voltage would also be small for a light load. This shows that the capacitor filter circuit provides a large dc voltage with little ripple for light loads (and a smaller dc voltage with larger ripple for heavy loads). To appreciate these quantities better we must further examine the output waveform and determine some relations among the input signal to be rectified, the capacitor value, the resistor (load) value, the ripple factor, and the regulation of the circuit.

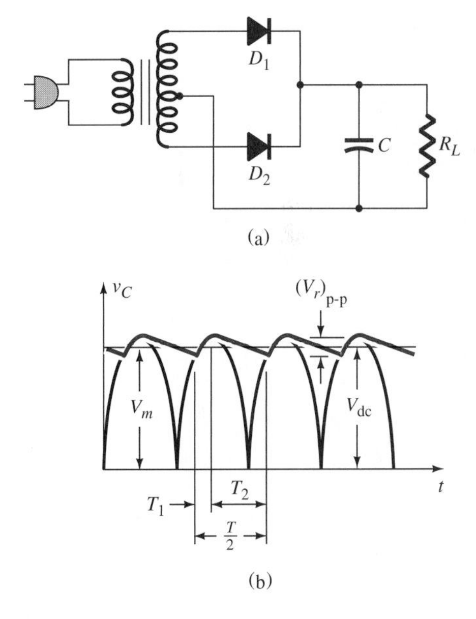

FIG. 15.5

Capacitor filter: (a) capacitor filter circuit; (b) output voltage waveform.

Figure 15.6 shows the output waveform approximated by straight-line charge and discharge. This is reasonable, since the analysis with the

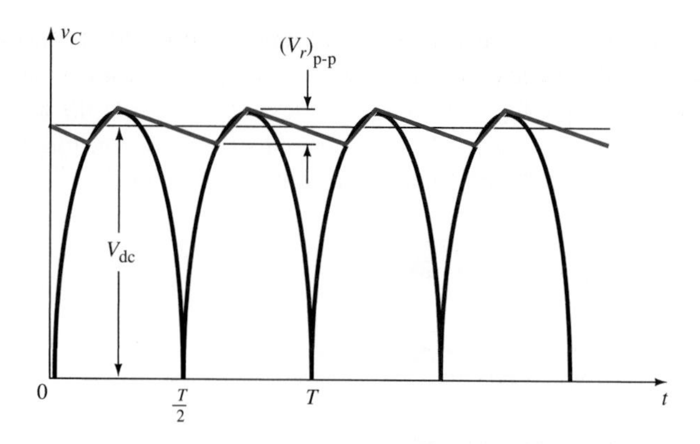

FIG. 15.6
Approximate output voltage of capacitor filter circuit.

nonlinear charge and discharge that actually takes place is complex to analyze and because the results obtained will yield values that agree well with actual measurements made on circuits. The waveform in Fig. 15.6 shows the approximate output voltage waveform for a full-wave-rectified signal. From an analysis of this voltage waveform the following relations can be obtained:

$$V_{dc} = V_m - \frac{V_r(\text{p-p})}{2} \quad \text{(half-wave} \qquad \textbf{(15.5)}$$

and

$$V_r\,(\text{rms}) = \frac{V_r(\text{p-p})}{2\sqrt{3}} \quad \text{full-wave)} \qquad \textbf{(15.6)}$$

These relations, however, are only in terms of the waveform voltages, and we must further relate them to the different components in the circuit. Since the form of the ripple waveform for half-wave is the same as for full-wave, Eqs. (15.5) and (15.6) apply to both rectifier-filter circuits.

Ripple Voltage, V_r (rms)

In terms of the other circuit parameters the result obtained for V_r (rms) is

$$V_r\,(\text{rms}) \cong \frac{I_{dc}}{4\sqrt{3}fC} \times \frac{V_{dc}}{V_m} \quad \text{(full-wave)} \qquad \textbf{(15.7a)}$$

where f = frequency of the sinusoidal ac power supply voltage (usually 60 Hz)
I_{dc} = average current drawn from the filter by the load
C = filter capacitor value

Another simplifying approximation that can be made is to assume that when used typically for light loads, the value of V_{dc} is only slightly less than V_m, so that $V_{dc} \cong V_m$, and the equation can be written as

$$V_r\,(\text{rms}) \cong \frac{I_{dc}}{4\sqrt{3}fC} \qquad \text{(full-wave, light load)} \qquad \textbf{(15.7b)}$$

Finally, we can include the typical value of line frequency (f = 60 Hz) and the other constants in the simpler equation

$$V_r\,(\text{rms}) = \frac{2.4I_{dc}}{C} = \frac{2.4V_{dc}}{R_L C} \qquad \text{(full-wave, light load)} \qquad \textbf{(15.7c)}$$

where I_{dc} is in milliamperes, C is in microfarads, and R_L is in kilohms.

EXAMPLE 15.3 Calculate the ripple voltage of a full-wave rectifier with a 100-μF filter capacitor connected to a load of 50 mA.

Solution: Using Eq. (15.7c), we obtain

$$V_r\,(\text{rms}) = \frac{2.4(50)}{100} = \mathbf{1.2\ V}$$

DC Voltage, V_{dc}

Using Eqs. (15.5), (15.6), and (15.7a), we see that the dc voltage of the filter is

$$V_{dc} = V_m - \frac{V_r(\text{p-p})}{2} = V_m - \frac{I_{dc}}{4fC} \times \frac{V_{dc}}{V_m} \qquad \text{(full-wave)} \qquad \textbf{(15.8a)}$$

Again, using the simplifying assumption that V_{dc} is about the same as V_m for light loads, we get an approximate value of V_{dc} (which is less than V_m) of

$$V_{dc} = V_m - \frac{I_{dc}}{4fC} \qquad \text{(full-wave, light load)} \qquad \textbf{(15.8b)}$$

which can be written (using f = 60 Hz) as

$$V_{dc} = V_m - \frac{4.17I_{dc}}{C} \qquad \text{(full-wave, light load)} \qquad \textbf{(15.8c)}$$

where V_m = peak rectified voltage in volts
I_{dc} = load current in milliamperes
C = filter capacitor in microfarads

EXAMPLE 15.4 If the peak rectified voltage for the filter circuit in Example 15.3 is 30 V, calculate the filter dc voltage.

Solution: Using Eq. (15.8c), we obtain

$$V_{dc} = V_m - \frac{4.17 I_{dc}}{C} = 30 - \frac{4.17(50)}{100} = \mathbf{27.9\ V}$$

The value of dc voltage is less than the peak rectified voltage. Note also from Eq. (15.8c) that the larger the value of average current drawn from the filter, the less the value of output dc voltage, and the larger the value of the filter capacitor, the closer the output dc voltage approaches the peak value of V_m.

Filter-Capacitor Ripple

Using the definition of ripple [Eq. (15.1)] and the equation for ripple voltage [Eq. (15.7c)], we obtain the expression for the ripple factor of a full-wave capacitor filter as

$$r = \frac{V_r\,(\text{rms})}{V_{dc}} \times 100\% \cong \frac{2.4 I_{dc}}{C V_{dc}} \times 100\% \qquad \text{(full-wave, light load)} \qquad \textbf{(15.9a)}$$

Since V_{dc} and I_{dc} relate to the filter load R_L, we can also express the ripple as

$$r = \frac{2.4}{R_L C} \times 100\% \qquad \text{(full-wave, light load)} \qquad \textbf{(15.9b)}$$

where I_{dc} is in milliamperes, C is in microfarads, V_{dc} is in volts, and R_L is in kilohms.

This ripple factor is seen to vary directly with the load current (larger load current, larger ripple factor) and inversely with the capacitor size. This agrees with the previous discussion of the filter circuit operation.

EXAMPLE 15.5 A load current of 50 mA is drawn from a capacitor filter circuit (C = 100 μF). If the peak rectified voltage is 30 V, calculate r.

Solution: Using the results in Examples 15.3 and 15.4 in Eq. (15.9a), we obtain

$$r = \frac{2.4I_{dc}}{CV_{dc}} \times 100\% = \frac{2.4(50)}{(100)(27.9)} \times 100\% = \mathbf{4.3\%}$$

From the basic definition of r we could also calculate

$$r = \frac{V_r\,(\text{rms})}{V_{dc}} \times 100\% = \frac{1.2}{27.9} \times 100\% = \mathbf{4.3\%}$$

Diode Conduction Period and Peak Diode Current

From the previous discussion it should be clear that larger values of capacitance provide less ripple and higher average voltages, thereby providing better filter action. From this one may conclude that to improve the performance of a capacitor filter it is necessary only to increase the size of the filter capacitor. However, the capacitor also affects the peak current through the rectifying diode and, as will now be shown, the larger the value of capacitance used, the larger the peak current through the rectifying diode.

Referring back to the operation of the rectifier and capacitor filter circuit, we see that there are two periods of operation to consider. After the capacitor is charged to the peak rectified voltage [see Fig. 15.5(b)], a period of diode nonconduction elapses (time T_2) while the output voltage discharges through the load. After T_2 the input rectified voltage becomes greater than the capacitor voltage, and for a time T_1 the capacitor charges back up to the peak rectified voltage. The average current supplied to the capacitor and load during this charge period must equal the average current drawn from the capacitor during the discharge period. Figure 15.7 shows the diode current waveform for half-wave-rectifier operation. Notice that the diode conducts for only a short period of the cycle. In fact, it should be seen that the larger the capacitor, the less the amount of voltage decay and the shorter the interval during which charging takes place. In this shorter charging interval the diode has to pass the same amount of *average current* and can do so only by passing larger peak current. Figure 15.8 shows the output current and voltage waveforms for small and large capacitor values. The important factor to note is the increase in peak current through the diode for the larger values of capacitance. Since the average current drawn from the supply must equal the average of the current through the diode during the charging period, the following relation can be derived from Fig. 15.8 using rectangular pulses as an approximation for the actual waveform.

$$I_{dc} = \frac{T_1}{T} I_{peak} \qquad \textbf{(15.10a)}$$

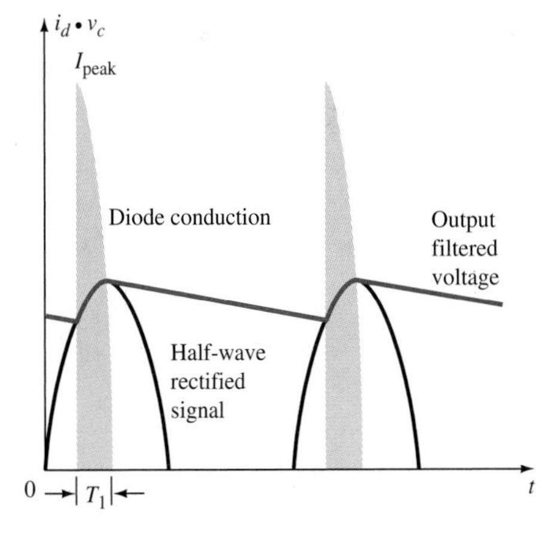

FIG. 15.7
Diode conduction during charging part of cycle.

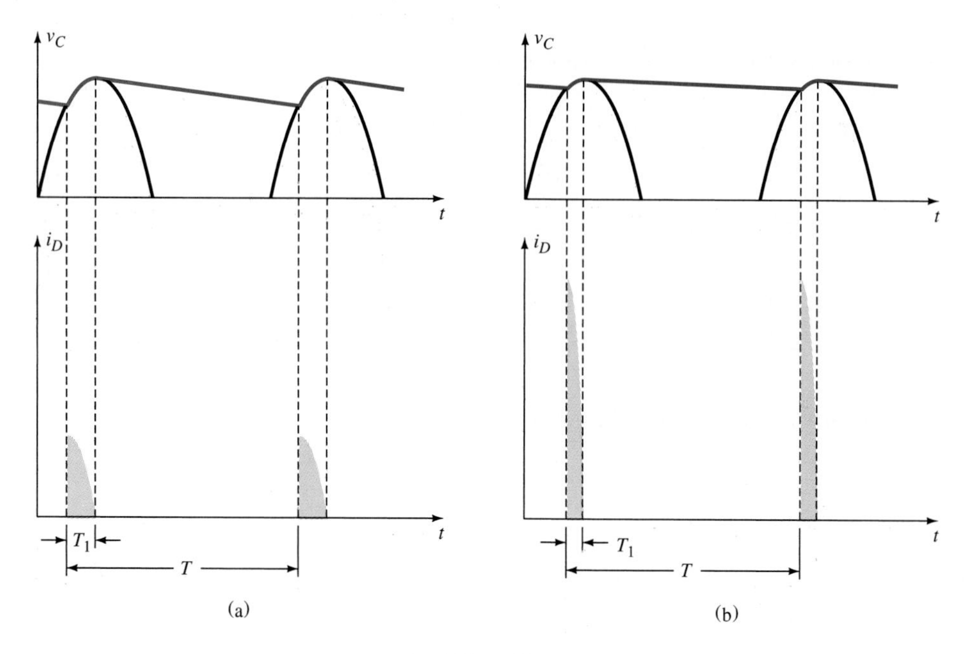

FIG. 15.8
Output voltage and diode current waveforms: (a) small C; (b) large C.

from which we obtain

$$I_{\text{peak}} = \frac{T}{T_1} I_{\text{dc}} \tag{15.10b}$$

where T_1 = diode conduction time
$T = 1/f = \frac{1}{60}$ for usual line 60-Hz voltage
I_{dc} = average current drawn from the filter circuit
I_{peak} = peak current through the conducting diode

15.4 *RC* FILTERS

It is possible to further reduce the amount of ripple across a filter capacitor while reducing the dc voltage by using an additional *RC* filter section, as shown in Fig. 15.9. The purpose of the added network is to pass as much of the dc component of the voltage developed across the first filter capacitor C_1 and to attenuate as much of the ac component of the ripple voltage developed across C_1 as possible. This action reduces the amount of ripple in relation to the dc level, providing better filter operation than with the simple-capacitor filter. There is a price to pay for this improvement, as will be shown; this includes a lower dc output voltage due to the dc voltage drop across the resistor and the cost of the two additional components in the circuit.

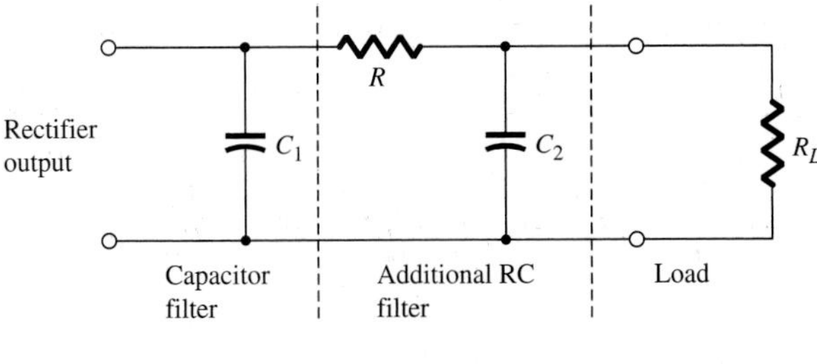

FIG. 15.9
RC filter stage.

Figure 15.10 shows the rectifier filter circuit for a full-wave operation. Since the rectifier feeds directly into a capacitor, the peak currents through the diodes are many times the average current drawn from the supply. The voltage developed across capacitor C_1 is then further filtered by the resistor–capacitor section (RC_2), providing an output voltage having less percent of ripple than that across C_1. The load, represented by resistor R_L, draws dc current through resistor R with a somewhat smaller output dc voltage across the load than that across C_1 because of the voltage drop across R. This filter circuit, like the simple-capacitor filter circuit, provides best operation at light loads, with considerably poorer voltage regulation and a higher percent ripple at heavy loads.

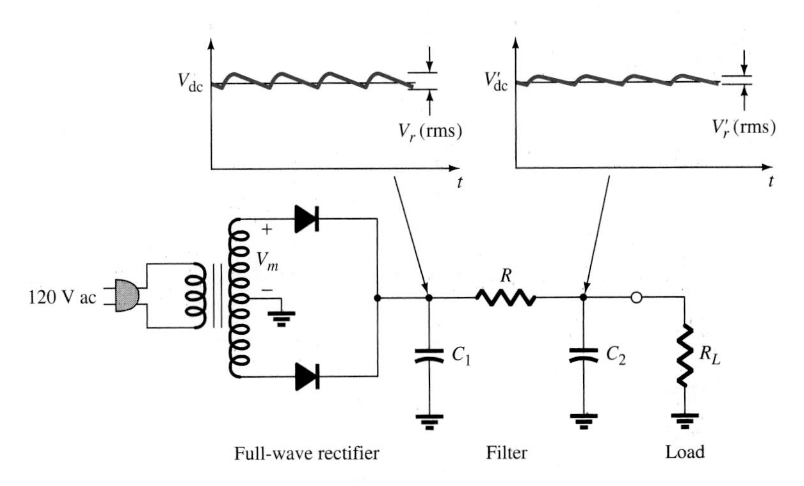

FIG. 15.10
Full-wave rectifier and RC filter circuit.

The analysis of the resulting ac and dc voltages at the output of the filter from that obtained across capacitor C_1 can be carried out by using superposition. We can separately consider the *RC* circuit acting on the dc level of the voltage across C_1 and then the *RC* circuit action on the ac (ripple) portion of the signal developed across C_1. The resulting values can then be used to calculate the overall circuit voltage regulation and ripple.

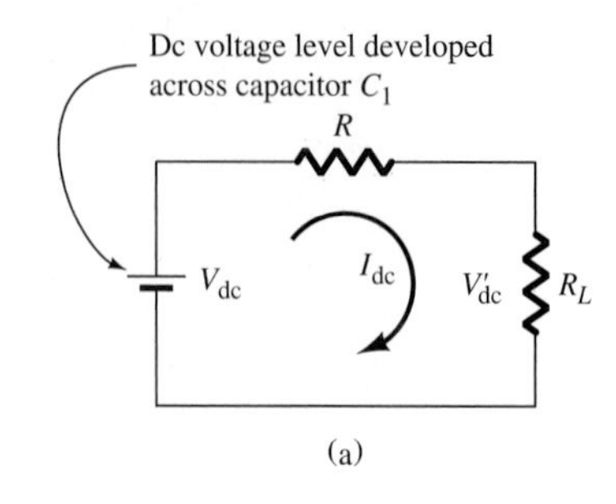

(a)

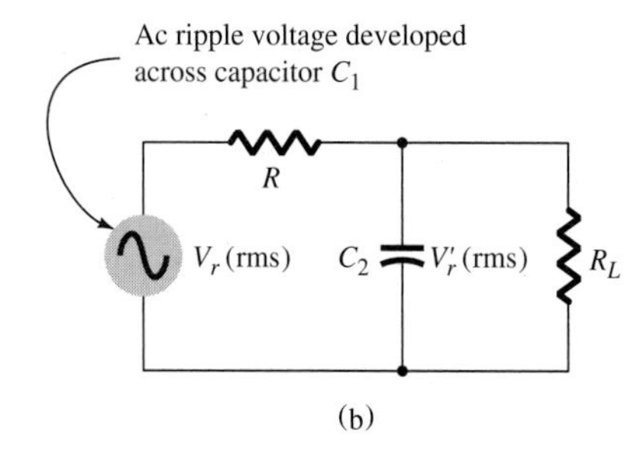

(b)

FIG. 15.11
Dc and ac equivalent circuits of RC filter: (a) dc equivalent circuit; (b) ac equivalent circuit.

DC Operation of *RC* Filter Section

Figure 15.11(a) shows the equivalent circuit to use when considering the dc voltage and current in the filter and load. The two filter capacitors are open circuit for dc and are thus removed from consideration at this time. Calculation of the dc voltage across filter capacitor C_1 was essentially covered in Section 15.3, and the treatment of the additional *RC* filter stage will proceed from there. Knowing the dc voltage across the first filter capacitor (C_1), we can calculate the dc voltage at the output of the additional *RC* filter section. From Fig. 15.11(a) we see that the voltage V_{dc} across capacitor C_1 is attenuated by a resistor-divider network of R and R_L (the equivalent load resistance), the resulting dc voltage across the load being V'_{dc}:

$$V'_{dc} = \frac{R_L}{R + R_L} V_{dc} \tag{15.11}$$

EXAMPLE 15.6 The addition of an *RC* filter section with $R = 120\ \Omega$ reduces the dc voltage across the initial filter capacitor from 60 V (V_{dc}). If the load resistance is 1 kΩ, calculate the value of the output dc voltage (V'_{dc}) from the filter circuit.

Solution: Using Eq. (15.1), we have

$$V'_{dc} = \frac{R_L}{R + R_L} V_{dc} = \frac{1000}{120 + 1000} \times 60 = \mathbf{53.6\ V}$$

In addition, we may calculate the drop across the filter resistor and the load current drawn:

$$V_R = V_{dc} - V'_{dc} = 60 - 53.6 = 6.4\ \text{V}$$

$$I_{dc} = \frac{V'_{dc}}{R_L} = \frac{53.6}{1 \times 10^3} = 53.6\ \text{mA}$$

AC Operation of *RC* Filter Section

Figure 15.11(b) shows the equivalent circuit for analyzing the ac operation of the filter circuit. The input to the filter stage from the first filter capacitor (C_1) is the ripple or ac signal part of the voltage across C_1, V_r (rms), which is approximated now as a sinusoidal signal. Both the *RC* filter stage components and the load resistance affect the ac signal at the output of the filter.

For a filter capacitor (C_2) value of 10 μF at a ripple voltage frequency (f) of 60 Hz, the ac impedance of the capacitor is*

$$X_C = \frac{1}{\omega C} = \frac{1}{2\pi f C} = \frac{1}{6.28(60)(10 \times 10^{-6})} = 0.265\ \text{k}\Omega$$

*X_C is understood here to represent only the *magnitude* of the capacitor's ac impedance.

Referring to Fig. 15.11(b), we see that this capacitive impedance is in parallel with the load resistance. For a load resistance of 2 kΩ, for example, the parallel combination of the two components would yield an impedance of magnitude:

$$Z = \frac{R_L X_C}{\sqrt{R_L^2 + X_C^2}} = \frac{2(0.65)}{\sqrt{2^2 + (0.265)^2}} = \frac{2}{2.02}(0.265) = 0.263 \text{ k}\Omega$$

This is close to the value of the capacitive impedance alone, as expected, since the capacitive impedance is much less than the load resistance, and the parallel combination of the two would be smaller than the value of either. As a rule of thumb we can consider neglecting the loading by the load resistor on the capacitive impedance as long as the load resistance is at least five times as large as the capacitive impedance. Because of the limitation of light loads on the filter circuit the effective value of load resistance is usually large compared with the impedance of capacitors in the range of microfarads.

In the preceding discussion it was stated that the frequency of the ripple voltage was 60 Hz. If we assume that the line frequency is 60 Hz, the ripple frequency will also be 60 Hz for the ripple voltage from a half-wave rectifier. The ripple voltage from a full-wave rectifier, however, will be double, since there are twice the number of half-cycles, and the ripple frequency will then be 120 Hz. Referring to the relation for capacitive impedance $X_C = 1/\omega C$, we have a value of $\omega = 377$ for 60 Hz and of $\omega = 754$ for 120 Hz. Using values of capacitance in microfarads, we can express the relation for capacitive impedance as

$$X_C = \frac{2653}{C} \quad \text{(half-wave)} \tag{15.12a}$$

$$X_C = \frac{1326}{C} \quad \text{(full-wave)} \tag{15.12b}$$

where C is in microfarads and X_C is in ohms.

EXAMPLE 15.7 Calculate the impedance of a 15-μF capacitor used in the filter section of a circuit using full-wave rectification.

Solution:

$$X_C = \frac{1326}{C} = \frac{1326}{15} = \mathbf{88.4\ \Omega}$$

Using the simplified relation that the parallel combination of the load resistor and the capacitive impedance approximately equals the capacitive impedance, we can calculate the ac attenuation in the filter stage:

$$V_r'\text{(rms)} \cong \frac{X_C}{\sqrt{R^2 + X_C^2}} V_r\text{(rms)} \tag{15.13a}$$

The use of the square root of the sum of the squares in the denominator is necessary, since the resistance and capacitive impedance must be added vectorially, not algebraically. If the value of the resistance is larger (by a factor of 5) than that of the capacitive impedance, a simplification of the denominator may be made, yielding the following result:

$$V'_r\,(\text{rms}) \cong \frac{X_C}{R} V_r\,(\text{rms}) \tag{15.13b}$$

EXAMPLE 15.8 The output of a full-wave rectifier and capacitor filter is further filtered by an *RC* filter section (see Fig. 15.12). The component values of the *RC* section are $R = 500\ \Omega$ and $C = 10\ \mu\text{F}$. If the initial capacitor filter develops 150 V dc with a 15-V ac ripple voltage, calculate the resulting dc and ripple voltage across a 5-kΩ load.

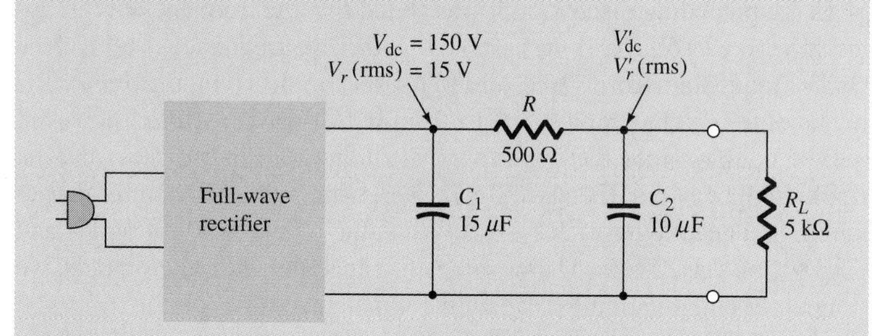

FIG. 15.12
RC filter circuit for Example 15.8.

Solution: *Dc calculations:* Calculating the value of V'_{dc} from Eq. (15.11), we have

$$V'_{dc} = \frac{R_L}{R + R_L} V_{dc} = \frac{5000}{500 + 5000}(150) = \frac{5000}{5500}(150) = \mathbf{136.4\ V}$$

Ac calculations: Calculating the value of the capacitive impedance first (for full-wave operation), we have

$$X_C = \frac{1326}{C} = \frac{1326}{10} = 133\ \Omega$$

Since this impedance is not quite five times smaller than that of the filter resistor ($R = 500\ \Omega$), we shall use Eq. (15.13a) for the calculation and then repeat the calculation to show what the difference would have been using Eq. (15.13b) (since the components are almost five times different in size). Using Eq. (15.13a), we have

$$V'_r\,(\text{rms}) = \frac{X_C}{\sqrt{R^2 + X_C^2}} V_r\,(\text{rms}) = \frac{0.133}{\sqrt{(0.5)^2 + (0.133)^2}}(15)$$

$$= \frac{0.133}{0.517}(15) = \mathbf{3.86\ V}$$

Now using Eq. (15.13b), we obtain

$$V'_r\,(\text{rms}) = \frac{X_C}{R} V_r\,(\text{rms}) = \frac{0.133}{0.500}(15) = \mathbf{3.99\ V}$$

Comparing the results of 3.86 V and 3.99 V, we see that the use of Eq. (15.13b) would have yielded an answer within 3.5% of the more exact solution.

15.5 VOLTAGE-MULTIPLIER CIRCUITS

Voltage Doubler

A modification of the capacitor filter circuit allows building up a larger voltage than the peak rectified voltage (V_m). The use of this type of circuit allows the transformer peak voltage rating to be kept low while the peak output voltage stepped up to two, three, four, or more times the peak rectified voltage.

Figure 15.13 shows a half-wave voltage doubler. During the positive-voltage half-cycle across the transformer, secondary diode D_1 conducts (and diode D_2 is cut off), charging capacitor C_1 up to the peak rectified voltage (V_m). Diode D_1 is ideally a short during this half-cycle, and the input voltage charges capacitor C_1 to V_m with the polarity shown in Fig. 15.14(a). During the negative half-cycle of the secondary voltage, diode

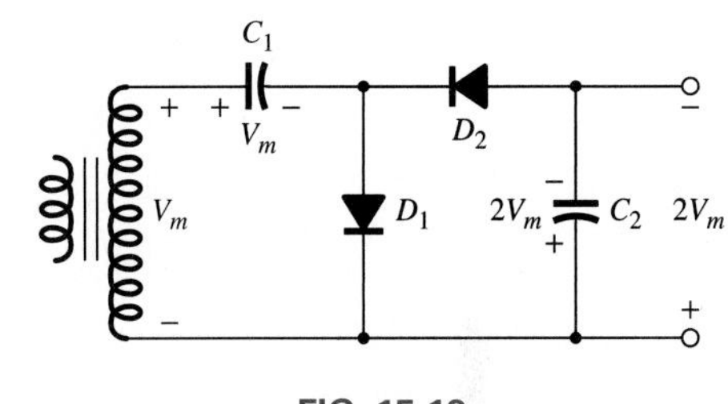

FIG. 15.13
Half-wave voltage doubler.

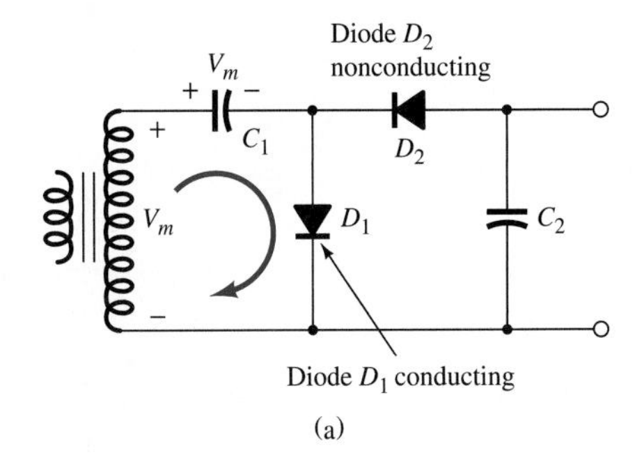

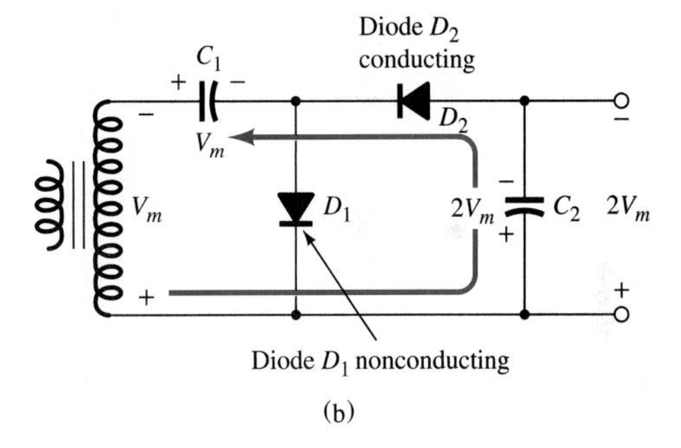

FIG. 15.14
Double operation, showing each half-cycle of operation: (a) positive half-cycle; (b) negative half-cycle.

D_1 is cut off, and diode D_2 conducts, charging capacitor C_2. Since diode D_2 acts as a short during the negative half-cycle (and diode D_1 is open), we can sum the voltages around the outside loop [see Fig. 15.14(b)]:

$$-V_{C_2} + V_{C_1} + V_m = 0$$
$$-V_{C_2} + V_m + V_m = 0$$

from which

$$V_{C_2} = 2V_m$$

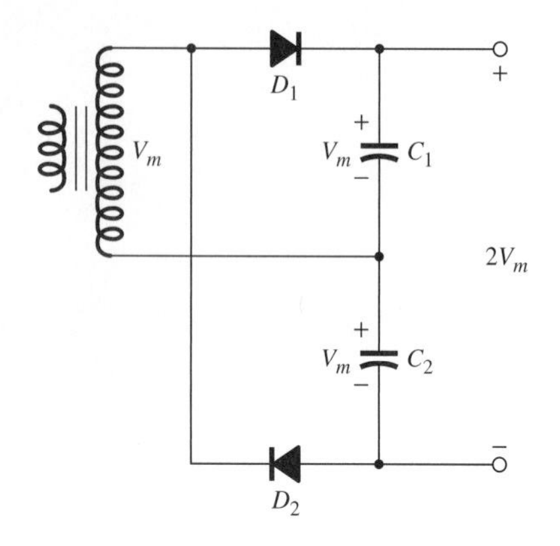

FIG. 15.15
Full-wave voltage doubler.

On the next positive half-cycle, diode D_2 is nonconducting, and capacitor C_2 discharges through the load. If no load is connected across capacitor C_2, both capacitors stay charged—C_1 to V_m and C_2 to $2V_m$. If, as would be expected, there is a load connected to the output of the voltage doubler, the voltage across capacitor C_2 drops during the positive half-cycle (at the input) and the capacitor is recharged up to $2V_m$ during the negative half-cycle. The output waveform across capacitor C_2 is that of a half-wave signal filtered by a capacitor filter. The peak inverse voltage across each diode is $2V_m$.

Another doubler circuit is the full-wave doubler in Fig. 15.15. During the positive half-cycle of transformer secondary voltage [see Fig. 15.16(a)], diode D_1 conducts, charging capacitor C_1 to a peak voltage V_m. Diode D_2 is nonconducting at this time.

During the negative half-cycle [see Fig. 15.16(b)], diode D_2 conducts, charging capacitor C_2, while diode D_1 is nonconducting. If no load current is drawn from the circuit, the voltage across capacitors C_1 and C_2 is $2V_m$. If load current is drawn from the circuit, the voltage across capacitors C_1 and C_2 is the same as that across a capacitor fed by a full-wave-rectifier circuit. One difference is that the effective capacitance is that of C_1 and C_2 in series, which is less than the capacitance of either C_1 or C_2 alone. The lower capacitor value provides poorer filtering action than the simple-capacitor filter circuit.

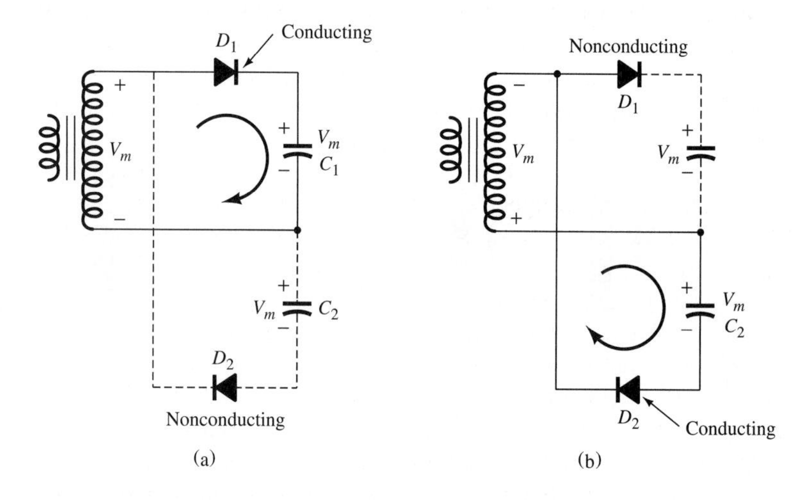

FIG. 15.16
Alternate half-cycles of operation for full-wave voltage doubler: (a) positive half-cycle; (b) negative half-cycle.

The peak inverse voltage across each diode is $2V_m$, as it is for the filter capacitor circuit. In summary, half-wave and full-wave voltage doubler circuits provide twice the peak voltage of the transformer secondary while requiring only a $2V_m$ PIV rating for the diodes and not requiring a no center-tapped transformer.

Voltage Tripler and Quadrupler

Figure 15.17 shows an extension of the half-wave voltage doubler, which develops three and four times the peak input voltage. It should be obvious from the pattern of the circuit connection how additional diodes and capacitors can be connected so that the output voltage can also be five, six, seven, and so on, times the basic peak voltage (V_m).

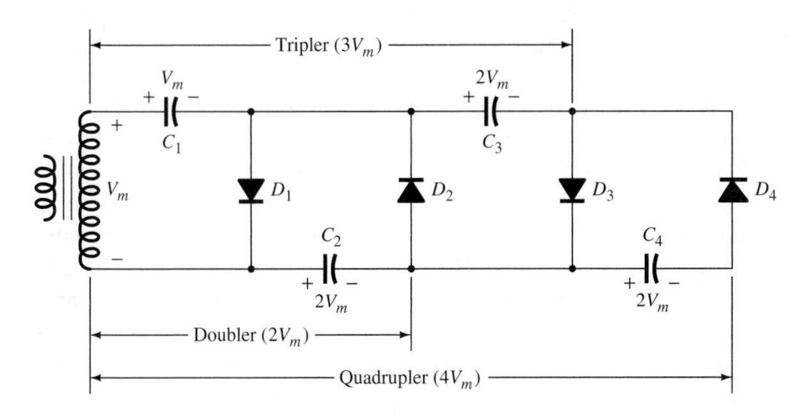

FIG. 15.17
Voltage tripler and quadrupler.

In operation capacitor C_1 charges through diode D_1 to a peak voltage V_m during the positive half-cycle of the transformer secondary voltage. Capacitor C_2 charges to twice the peak voltage $2V_m$ developed by the sum of the voltages across capacitor C_1 and the transformer during the negative half-cycle of the transformer secondary voltage.

During the positive half-cycle, diode D_3 conducts, and the voltage across capacitor C_2 charges capacitor C_3 to the same $2V_m$ peak voltage. On the negative half-cycle, diodes D_2 and D_4 conduct, with capacitor C_3, charging C_4 to $2V_m$.

The voltage across capacitor C_2 is $2V_m$, across C_1 and C_3 it is $3V_m$, and across C_2 and C_4 it is $4V_m$. If additional sections of diode and capacitor are used, each capacitor will be charged to $2V_m$. Measuring from the top of the transformer winding (Fig. 15.17) provides odd multiples of V_m at the output, whereas measuring from the bottom of the transformer the output voltage provides even multiples of the peak voltage V_m.

The transformer rating is only V_m, maximum, and each diode in the circuit must be rated at $2V_m$ PIV. If the load is small and the capacitors have little leakage, extremely high dc voltages may be developed by this type of circuit, using many sections to step up the dc voltage.

15.6 IC VOLTAGE REGULATORS

Voltage regulators are a class of widely used ICs. These units contain the circuitry for reference source, error amplifier, control device, and overload protection all in a single IC chip. Although the internal construction is somewhat different from that described for discrete voltage regulator circuits, the external operation is much the same. We shall examine operation using some of the popular three-terminal fixed-voltage regulators (for both positive and negative voltages) and those allowing an adjustable output voltage.

A power supply can be built very simply using a transformer connected to the ac supply to step the voltage to a desired level, then rectifying with a half- or full-wave circuit, filtering the voltage using a simple-capacitor filter, and finally regulating the dc voltage using an IC voltage regulator.

A basic category of voltage regulators includes those used with only positive voltages, those used with only negative voltages, and those further classified as having fixed or adjustable output voltages. These regulators can be selected for operation with load currents from hundreds of milliamperes to tens of amperes corresponding to power ratings from milliwatts to tens of watts. A representation of various types of voltage-regulator ICs is presented next. Various terms common to this area of electronics also are introduced and defined.

Three-Terminal Voltage Regulators

Voltage regulators that provide a positive fixed regulated voltage over a range of load currents are schematically represented in Fig. 15.18. The fixed voltage regulator has an unregulated voltage V_{in} applied to one terminal, delivers a regulated output voltage, V_o, from a second terminal, with the third terminal connected to ground. For a particular IC unit, device specifications list a voltage range over which the input voltage can vary to maintain the regulated output voltage, V_o, over a range of load current, I_o. An output–input voltage differential must be maintained for the IC to operate, which means that the varying input voltage must always be kept large enough to maintain a voltage drop across the IC to permit proper operation of the internal circuit. The device specifications also list the amount of output voltage change, V_o, resulting from changes

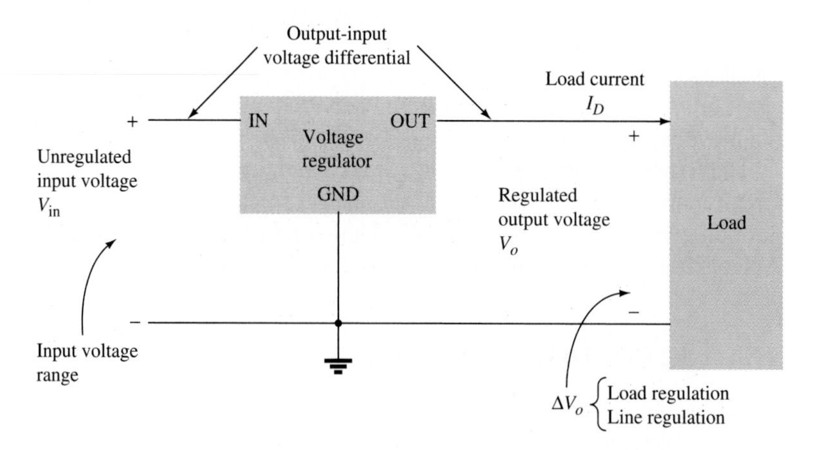

FIG. 15.18
Block representation of three-terminal voltage regulator.

in load current (load regulation) and also from changes in input voltage (line regulation).

A group of fixed positive voltage regulators is the series 78, which provide fixed voltages from 5 V to 24 V. Figure 15.19(a) shows how many of these regulators are connected. A rectified and filtered unregulated dc voltage is the input, V_{in}, to pin 1 of the regulator IC. Capacitors connected from input or output to ground help to maintain the dc voltage and additionally to filter any high-frequency voltage variation. The output voltage from pin 2 is then available to connect to the load. Pin 3 is the IC circuit reference or ground. When the desired fixed regulated output voltage is selected, the two digits after the 78 prefix indicate the regulator output voltage. Table 15.1 lists some typical data.

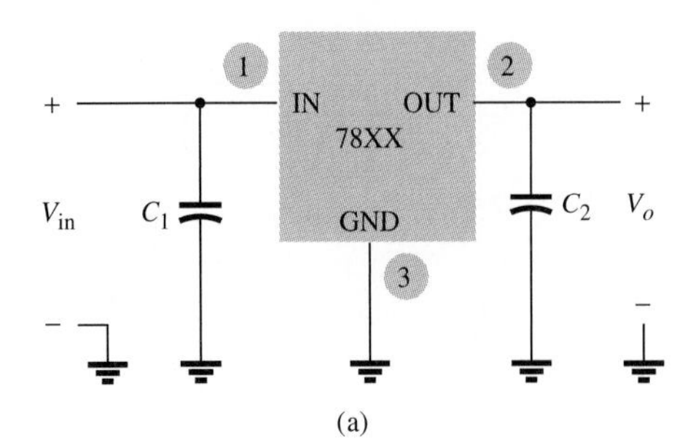

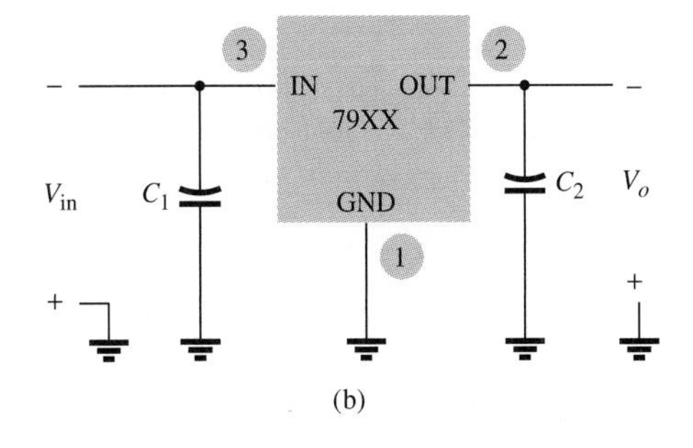

FIG. 15.19
(a) Series 78XX positive voltage regulator; (b) series 79XX negative voltage regulator.

TABLE 15.1
Positive series 78XX voltage regulator ICs

IC part number	Regulated positive voltage (V)	Minimum V_{in} (V)
7805	+5	7.3
7806	+6	8.35
7808	+8	10.5
7810	+10	12.5
7812	+12	14.6
7815	+15	17.7
7818	+18	21
7824	+24	27.1

Negative voltage regulator ICs are available in the 79 series [see Fig. 15.19(b)], which provide a series of ICs similar to the 78 series but operating on negative voltages, providing a regulated negative output voltage. Table 15.2 lists the 79XX series of fixed negative-voltage regulators and their corresponding regulated voltages.

TABLE 15.2
Fixed negative-voltage regulators in the 79XX series

IC part number	Regulated output voltage (V)	Minimum V_{in} (V)
7905	−5	−7.3
7906	−6	−8.4
7908	−8	−10.5
7909	−9	−11.5
7912	−12	−14.6
7915	−15	−17.7
7918	−18	−20.8
7924	−24	−27.1

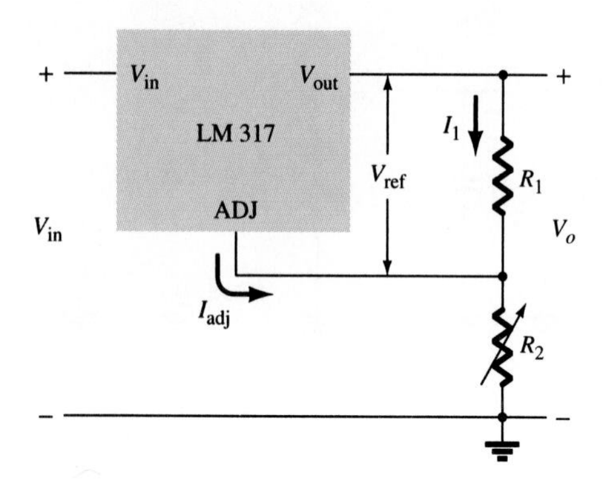

FIG. 15.20
Connection of LM317 adjustable voltage regulator.

Voltage regulators are also available in circuit configurations that allow the user to set the output voltage to a desired regulated value. The LM317, for example, can be operated with output voltage regulated at any setting over the range of voltage from 1.2 V to 37 V. Figure 15.20 shows a typical connection using the LM317 IC.

Selection of resistors R_1 and R_2 allows the setting of the output to any desired voltage over the adjustment range (1.2 to 37 V). The output voltage desired can be calculated using

$$V_o = V_{ref}\left(1 + \frac{R_2}{R_1}\right) + I_{adj}R_2 \qquad \textbf{(15.14)}$$

with typical values of

$$V_{ref} = 1.25\text{ V} \qquad \text{and} \qquad I_{adj} = 100\ \mu\text{A}$$

EXAMPLE 15.9
Determine the regulated output voltage using an LM317, as in Fig. 15.20, with $R_1 = 240\ \Omega$ and $R_2 = 2.4\text{ k}\Omega$.

Solution: Using Eq. (15.14), we obtain

$$V_o = 1.25\text{ V}\left(1 + \frac{2.4\text{ k}\Omega}{240\ \Omega}\right) + 100\ \mu\text{A}\ (2.4\text{ k}\Omega)$$
$$= 13.75\text{ V} + 0.24\text{ V} = \mathbf{13.99\ V}$$

15.7 PRACTICAL POWER SUPPLIES

A practical power supply can be built to convert the 120-V supply voltage into a desired regulated dc voltage. The standard circuit includes a transformer to step the voltage to a desired ac level, a diode rectifier to half-wave or full-wave rectify the ac signal, and a capacitor filter to develop an unregulated dc voltage. The unregulated dc voltage is then connected as input to an IC voltage regulator, which provides the desired regulated output dc voltage. A few examples will show how a dc voltage supply can be built and how it operates.

EXAMPLE 15.10
Analyze the operation of the +12-V voltage supply shown in Fig. 15.21 connected to a load drawing 400 mA.

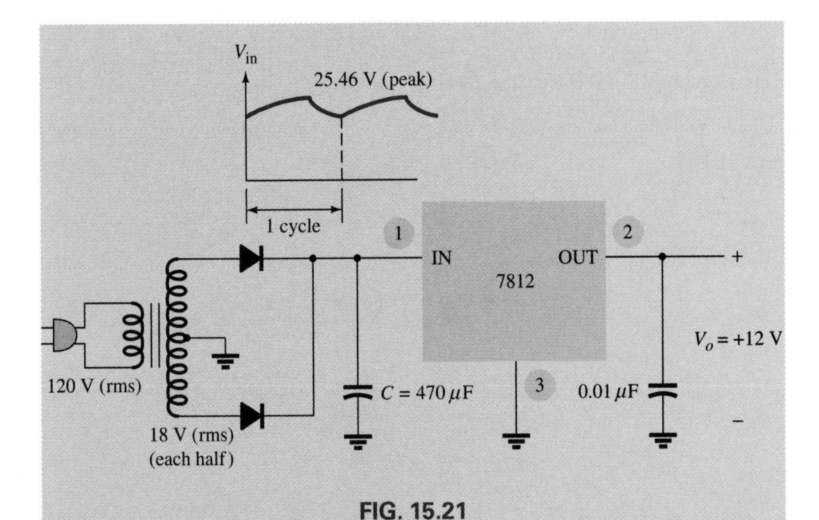

FIG. 15.21

Solution: The transformer steps down the line voltage from 120 V (rms) to a secondary voltage of 18 V (rms) across each transformer half. This results in a peak voltage across the transformer of

$$V_m = \sqrt{2}\,V_{\text{rms}} = \sqrt{2} \times 18\text{ V} = \mathbf{25.456\ V}$$

The ripple voltage [using Eq. (15.7c)] is then

$$V_r\,(\text{rms}) = \frac{2.4I_{\text{dc}}}{C} = \frac{2.4(400)}{470} = \mathbf{2.043\ V}$$

and the peak ripple voltage is [using Eq. (15.6)]

$$V_r\,(\text{peak}) = \sqrt{3}\,V_r\,(\text{rms}) = \sqrt{3}\,(2.043\text{ V}) = \mathbf{3.539\ V}$$

The dc level of the voltage across the 470-μF capacitor C is

$$V_{\text{dc}} = V_m - V_r\,(\text{peak}) = 25.456\text{ V} - 3.539\text{ V} = \mathbf{21.917\ V}$$

The ripple factor of the filter capacitor when operating into a 400-mA load is then [Eq. (15.9a)]

$$r = \frac{2.4I_{\text{dc}}}{CV_{\text{dc}}} \times 100\% = \frac{2.4(400)}{(470)(21.917)} \times 100\% \cong \mathbf{9.3\%}$$

The voltage across filter capacitor C has a ripple of about 9.3% and drops to a minimum voltage of

$$V_{\text{in}_{\text{min}}} = V_m - 2V_r\,(\text{peak}) = 25.456\text{ V} - 2(3.539\text{ V}) = \mathbf{18.378\ V}$$

Device specifications list V_{in} as required to maintain line regulation at 14.6 V. The lowest voltage being maintained across the capacitor is somewhat greater at 18.378 V.

Lowering the value of the filter capacitor or increasing the load current results in greater ripple voltage and lower minimum voltage across the capacitor. As long as this minimum voltage remains above 14.6 V, the 7812 will maintain the output voltage regulated at +12 V.

Device specifications for the 7812 list the maximum voltage change as 60 mV. This means that the output voltage regulation will be less than

$$\text{VR} = \frac{60\text{ mV}}{12\text{ V}} \times 100\% = \mathbf{0.5\%}$$

EXAMPLE 15.11 Analyze the operation of the 5-V supply in Fig. 15.22 operating at a load current of (a) 200 mA and (b) 400 mA.

Solution: The specifications for the 7805 list an input of 7.3 V as the minimum allowable to maintain line regulation.

a. At a load of I_{dc} = 200 mA, the ripple voltage is

$$V_r\,(\text{peak}) = \sqrt{3}V_r\,(\text{rms}) = \sqrt{3} \times \frac{(2.4)I_{dc}}{C} = \sqrt{3} \times \frac{2.4(200)}{(250)} = \mathbf{3.326\ V}$$

and the dc voltage across the 250-μF filter capacitor is

$$V_{dc} = V_m - V_r\,(\text{peak}) = 15\text{ V} - 3.326\text{ V} = \mathbf{11.674\ V}$$

The voltage across the filter capacitor drops to a minimum value of

$$V_{\text{in}_{\text{min}}} = V_m - 2V_r\,(\text{peak}) = 15\text{ V} - 2(3.326\text{ V}) = \mathbf{8.348\ V}$$

Since this is above the rated value of 7.3 V, the output will be maintained at the regulated +5 V.

b. At a load of I_{dc} = 400 mA, the ripple voltage is

$$V_r\,(\text{peak}) = \sqrt{3} \cdot \frac{(2.4)(400)}{250} = \mathbf{6.65\ V}$$

around a dc voltage of

$$V_{dc} = 15\text{ V} - 6.65\text{ V} = \mathbf{8.35\ V}$$

which is above the rated 7.3-V lower level. However, the input swings around this dc level by 6.65 V (peak), dropping during part of the cycle to

$$V_{\text{in}_{\text{min}}} = 15\text{ V} - 2(6.65\text{ V}) = \mathbf{1.7\ V}$$

which is well below the minimum allowed input voltage of 7.3 V. Therefore, the output is not maintained at the regulated +5-V level over the entire input cycle. Regulation is maintained for load currents below 200 mA but not at or above 400 mA.

EXAMPLE 15.12 Determine the maximum value of load current at which regulation is maintained for the circuit in Fig. 15.22.

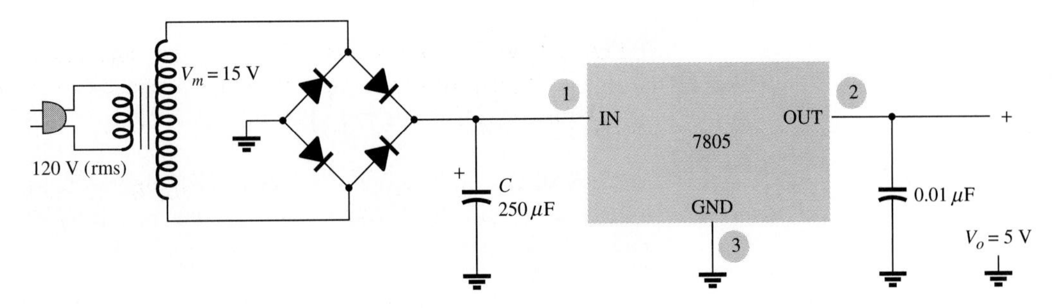

FIG. 15.22
Positive 5-V power supply.

Solution: To maintain $V_{\text{in}} \geq 7.3$ V

$$V_r\,(\text{p-p}) \leq V_m - V_{\text{in}_{\text{min}}} = 15\text{ V} - 7.3\text{ V} = 7.7\text{ V}$$

and so

$$V_r\,(\text{rms}) = \frac{V_r\,(\text{p-p})/2}{\sqrt{3}} = \frac{7.7\text{ V}/2}{\sqrt{3}} = 2.2\text{ V}$$

We can then determine the value of I_{dc} (in milliamperes):

$$I_{\text{dc}} = \frac{V_r\,(\text{rms})\,C}{2.4} = \frac{(2.2)(250)}{2.4} = \mathbf{229.2\text{ mA}}$$

Any current above this value is too large for the circuit to maintain the regulator output at +5 V.

Using a positive adjustable voltage regulator IC, it is possible to set the regulated output voltage to any desired voltage (within the device operating range).

EXAMPLE 15.13 Determine the regulated output voltage of the circuit in Fig. 15.23.

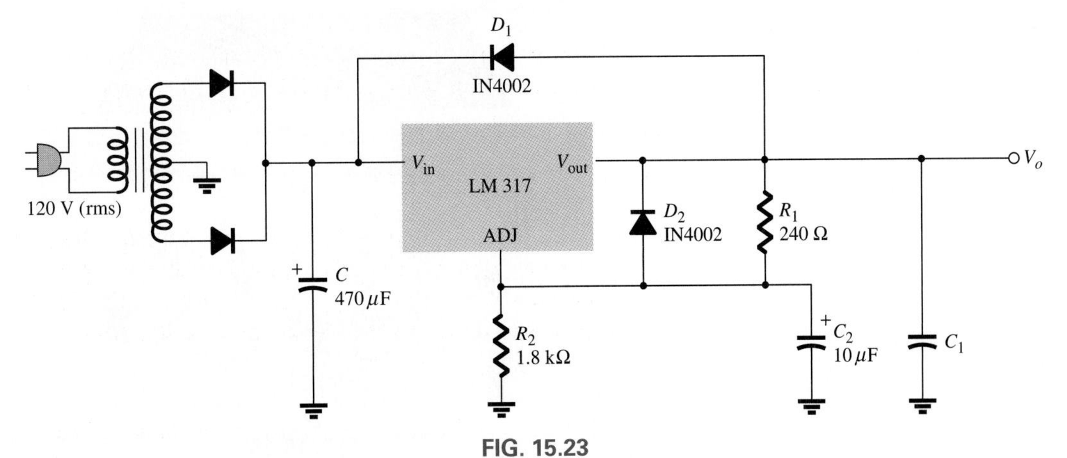

FIG. 15.23
Positive adjustable voltage regulator for Example 15.13.

Solution: The output voltage is

$$V_o = 1.25\text{ V}\left(1 + \frac{1.8\text{ k}\Omega}{240\ \Omega}\right) + 100\ \mu\text{A}(1.8\text{ k}\Omega) \cong \mathbf{10.8\ V}$$

A check of the filter-capacitor voltage shows that an input–output voltage differential of 2 V can be maintained up to at least 200 mA of load current.

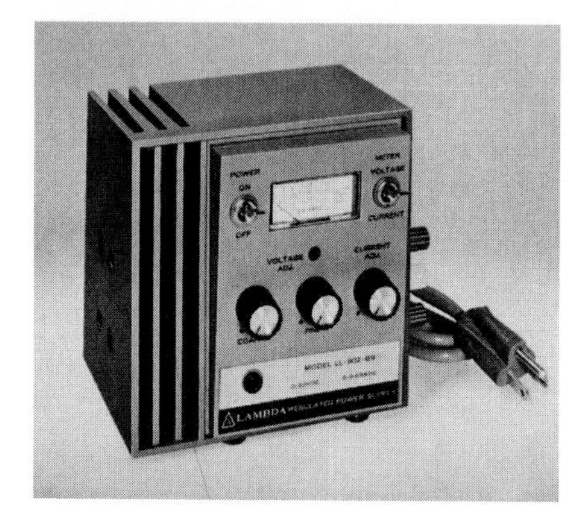

FIG. 15.24
IC-regulated dc supply.

An IC (dual-in-line package) regulated dc power supply appears in Fig. 15.24. The unit shown is a 0 → 20-V dc, 0 → 0.65-A meter. Its ripple content is 1 mV peak to peak or 250 μV (rms). For a 20-V output this would result in a percent ripple of

$$r = \frac{V_r\text{ (rms)}}{V_{\text{dc}}} \times 100\% = \frac{250 \times 10^{-6}}{20} \times 100\%$$
$$= 12.5 \times 10^{-4}\% = \mathbf{0.00125\%}$$

The load regulation is given as 4 mV, indicating that from no-load to full-load conditions the terminal voltage varies only 4 mV. For the 20-V level this provides a voltage regulation of

$$\text{VR} = \frac{V_{NL} - V_{FL}}{V_{FL}} \times 100\% = \frac{20 - (20 - 4\text{ mV})}{(20 - 4\text{ mV})} \times 100\%$$
$$= \frac{20 - 19.996}{19.996} \times 100\% = \mathbf{0.02\%}$$

A dc supply with a range of 0 → 250 V dc appears in Fig. 15.25. The load regulation for this supply is given as 0.0005% + 100 μV for load variations from 0 to full load. For an output of 200 V this represents a variation in terminal voltage of (0.0005/100)200 + 100 μV = 1000 μV +

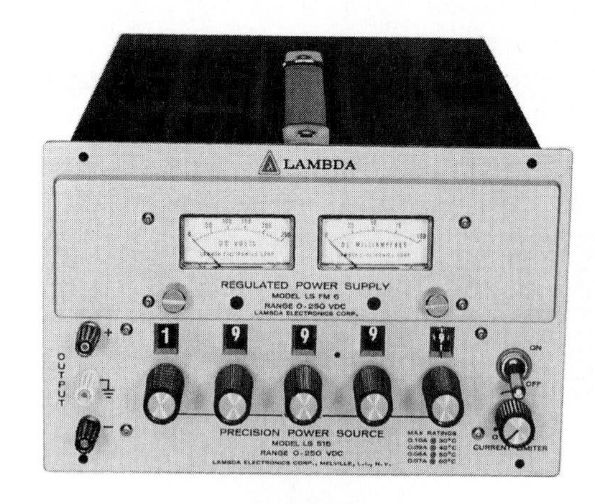

FIG. 15.25
High-voltage dc laboratory supply.

100 μV = 1100 μV (0.0011 V) from no-load to full-load conditions. Its output level is set by the five-digit system shown in the figure. The maximum current is determined by temperature levels. At 30°C it is 0.1 A, while at 60°C it is 0.07 A. It can be set for constant current and result in less than 2-mA regulation for input variations of $105V_{ac}$ to $132V_{ac}$ and from 0 to rated V_{dc} load voltage change.

The internal construction of a dc supply appears in Fig. 15.26. Note the use of a printed circuit board for packaging and increased availability of components. This particular unit is designed specifically for rack mounting.

FIG. 15.26
Internal construction of a dc supply.

Power kits are available for constructing your own power supply such as shown in Fig. 15.27. After the desired output dc voltage and current levels are chosen, the components are provided for construction of the supply. The series regulator units in Fig. 15.28, which are also available in design packages such as the preceding, employ a dual-in-line package IC for regulation control on the output (also provided).

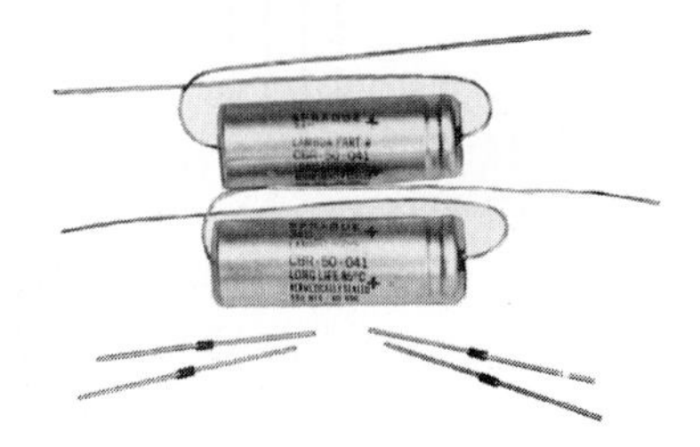

FIG. 15.27
Power kit.

FIG. 15.28
Series regulator unit.

15.8 DC/AC INVERTERS AND DC/DC CONVERTERS

Our interest thus far has been to convert an ac signal into one with an average or dc level. The block diagram in Fig. 15.29 indicates the sequence of operations required to convert a dc input to a higher dc level or an ac sinusoidal voltage. The latter process is called an *inversion* operation, while the former is a *conversion* process. Note that for either process the input dc must first be "chopped" as shown in the figure using any one of the techniques listed. It can then be transformed to a higher level through transformer action, followed by one of the two paths indicated for the type of output desired. Recall that transformers can operate only on a changing signal, thereby requiring the chopping action indicated. In the inversion process, the filter *(LC)* provides the smoothing action to convert the input waveform to one that more closely resembles the sinusoidal pattern.

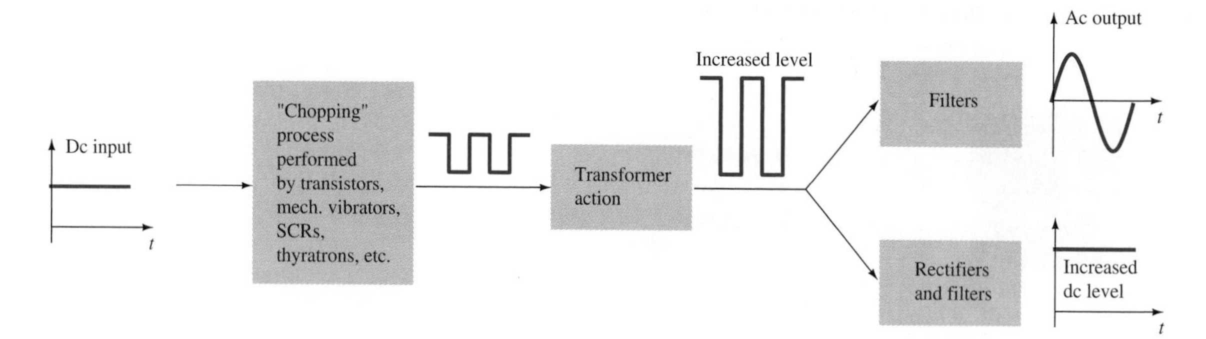

FIG. 15.29
Inversion and conversion process.

One network configuration capable of performing the chopping function on the dc input appears in Fig. 15.30. Although the two transistors chosen have the same stock number and manufacturer, they can never be *exactly* alike in characteristics. When the dc input is applied, the basic biasing for each transistor by the resistors R_1 and R_2 and dc voltage V_{CC} turn that transistor on with the least base-to-emitter resistance. The other, as we shall see, enters the nonconduction state. Let us assume for sake of discussion that Q_1 is turning on. The resulting collector current for the transistor passes through the primary winding of the transformer and, since it is a changing level, induces a voltage across the other "feedback" primary windings through the flux induced in the core. For the on transistor this induced voltage at the base of the device turns the device heavily on, causing a sharp increase in I_{B_1} and I_{C_1}. The induced voltage at the base of the other transistor has a polarity to further turn that device off. For the on transistor, the collector current reaches a sufficiently high level to saturate the core. That is, the flux in the core is an absolute maximum, and any further increase in primary current will have almost no effect on the flux in the core. When this occurs, the induced voltage in the primary windings is zero, since Faraday's law [$e = N(d\phi/dt)$] depends on a changing flux. The on transistor finds itself starting to turn off, causing a decrease in collector current. A decreasing collector current in the primary induces a voltage of reverse polarity in the base windings, further turning Q_1 off and starting to turn Q_2 on. Eventually, Q_2 enters the full on state with Q_1 off. However, when Q_2 saturates the primary as Q_1 did, the process reverses itself. The reversing collector currents result in reversing induced voltages in the secondary, as shown in Fig. 15.30. The nonlinearity of the magnetic core of the transformer

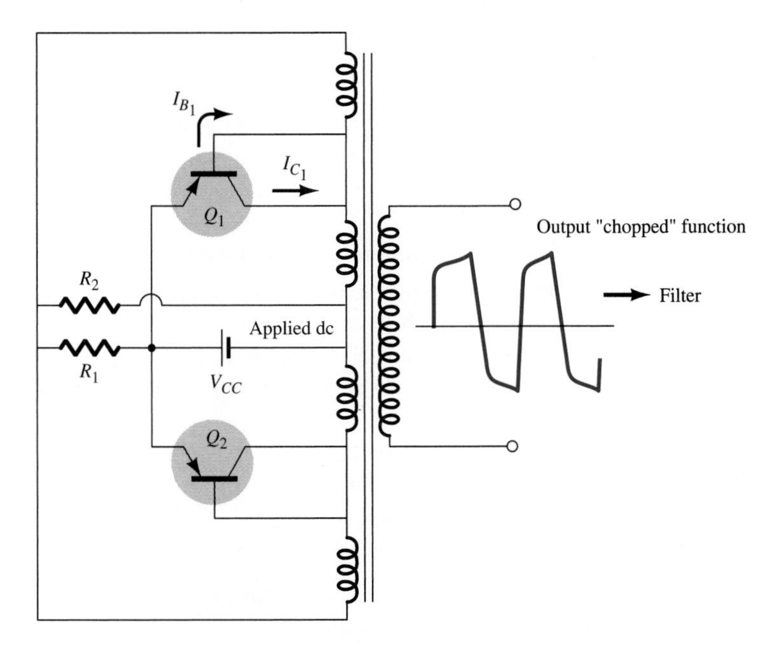

FIG. 15.30
Inverter network.

FIG. 15.31
Dc/dc converter.

causes the major portion of the distortion to appear in the almost square wave signal. For an ac output a choke-input filter will smooth the edges for an acceptable output sinusoidal signal. For the dc output the signal is rectified and filtered as described in this chapter. The turns ratio of the transformer determines the output swing of the "square" wave as compared with the primary voltage.

A commercially available dc/dc converter appears in Fig. 15.31 with its terminal data.

15.9 SUMMARY

$$\% \text{ ripple} = \%r = \frac{\textit{ripple voltage (rms)}}{\textit{dc voltage}} \times 100\%$$

$$\% \text{ voltage regulation} = \frac{V_{NL} - V_{FL}}{V_{FL}} \times 100\%$$

Filter-Capacitor Equations:

Ripple Voltage, V_r(rms):

$$V_r\,(\text{rms}) = \frac{4.17 I_{\text{dc}}}{4\sqrt{3}fC} \times \frac{V_{\text{dc}}}{V_m} \quad \text{(full-wave)}$$

$$\approx \frac{2.4 V_{\text{dc}}}{R_L C} \quad \text{(full-wave, light load)}$$

DC Voltage, V_{dc}

$$V_{\text{dc}} = V_m - \frac{4.17 I_{\text{dc}}}{C} \quad \text{(full-wave, light load)}$$

Ripple, r

$$r = \frac{V_r\,(\text{rms})}{V_{\text{dc}}} \times 100\% \cong \frac{2.4 I_{\text{dc}}}{C V_{\text{dc}}} \times 100\%$$

$$= \frac{2.4}{R_L C} \times 100\% \quad \text{(full-wave, light load)}$$

PROBLEMS

SECTION 15.2 General Filter Considerations

1. What is the ripple factor of a sinusoidal signal having a peak ripple of 2 V on an average of 50 V?
2. A filter circuit provides an output of 28 V unloaded and 25 V under full-load operation. Calculate the percent voltage regulation.
3. A half-wave rectifier develops 20 V dc. What is the rms value of the ripple voltage?
4. What is the rms ripple voltage of a full-wave rectifier whose output voltage is 8 V dc?

SECTION 15.3 Simple-Capacitor Filter

5. A simple-capacitor filter fed by a full-wave rectifier develops 14.5 V dc at an 8.5% ripple factor. What is the output ripple voltage (rms)?

6. A full-wave-rectified signal of 18 V peak is fed into a capacitor filter. What is the voltage regulation of the filter circuit if the output dc is 17 V dc at full load?

7. A full-wave-rectified voltage of 18 V peak is connected to a 400-μF filter capacitor. What is the dc voltage across the capacitor at a load of 100 mA?

8. A full-wave rectifier operating from a 60-Hz ac supply produces a 20-V peak rectified voltage. If a 200-μF filter capacitor is used, calculate the ripple at a load of 120 mA.

9. A capacitor filter circuit ($C = 100\ \mu$F) develops 12 V dc when connected to a load of 2.5 kΩ. Using a full-wave rectifier operating from the 60-Hz supply, calculate the ripple of the output voltage.

10. Calculate the size of the filter capacitor needed to obtain a filtered voltage with 15% ripple at a load of 150 mA. The full-wave-rectified voltage is 24 V dc, and supply is 60 Hz.

11. A 500-μF filter capacitor provides a load current of 200 mA at 8% ripple. Calculate the peak rectified voltage obtained from the 60-Hz supply and the dc voltage across the filter capacitor.

12. Calculate the size of the filter capacitor needed to obtain a filtered voltage with 7% ripple at a load of 200 mA. The full-wave-rectified voltage is 30 V dc, and supply is 60 Hz.

13. Calculate the percent ripple for the voltage developed across a 120-μF filter capacitor when providing a load current of 80 mA. The full-wave rectifier operating from the 60-Hz supply develops a peak rectified voltage of 25 V.

14. Calculate the amount of peak diode current through the rectifier diode of a full-wave rectifier feeding a capacitor filter when the average current drawn from the filter is 100 mA when diode conduction is for $\frac{1}{10}$ cycle.

SECTION 15.4 *RC* Filters

15. An *RC* filter stage is added after a capacitor filter to reduce the percent of ripple to 2%. Calculate the ripple voltage at the output of the *RC* filter stage providing 80 V dc.

16. An *RC* filter stage ($R = 33\ \Omega$, $C = 120\ \mu$F) is used to filter a signal of 24 V dc with 2 V rms operating from a full-wave rectifier. Calculate the percent ripple at the output of the *RC* section for a load of 100 mA. Calculate also the percent ripple of the filtered signal applied to the *RC* stage.

17. A simple-capacitor filter has an input of 40 V dc. If this voltage is fed through an *RC* filter section ($R = 50\ \Omega$, $C = 40\ \mu$F), what is the load current for a load resistance of 500 Ω?

18. Calculate the ripple voltage (rms) at the output of an *RC* filter section that feeds a 1-kΩ load when the filter input is 50 V dc with 2.5 V (rms) ripple from a full-wave rectifier and capacitor filter. The *RC* filter section components are $R = 100\ \Omega$ and $C = 100\ \mu$F.

19. If the output no-load voltage for the circuit in Problem 18 is 60 V, calculate the percent of voltage regulation with the 1-kΩ load.

SECTION 15.5 Voltage-Multiplier Circuits

20. Draw the circuit diagram of a voltage doubler. Indicate the value of the diode PIV rating in terms of the transformer peak voltage V_m.

21. Draw a voltage tripler circuit. Indicate diode PIV ratings and voltage across each circuit capacitor. Include polarity.

22. Repeat Problem 21 for a voltage quadrupler.

SECTION 15.6 IC Voltage Regulators

23. Determine the regulated output voltage using an LM317, as in Fig. 15.20, with $R_1 = 240\ \Omega$ and $R_2 = 1.8\ \text{k}\Omega$. (Assume $V_{\text{ref}} = 1.25$ V and $I_{\text{adj}} = 100\ \mu\text{A}$.)

24. What output voltage results in a circuit as in Fig. 15.20 with $R_1 = 240\ \Omega$ and $R_2 = 3.3\ \text{k}\Omega$? (Assume $V_{\text{ref}} = 1.25$ V and $I_{\text{adj}} = 100\ \mu\text{A}$).

SECTION 15.7 Practical Power Supplies

25. Determine the percent ripple across the filter capacitor of a voltage supply as in Fig. 15.21 operating into a load that draws 250 mA. Assume that $V_m = 25.5$ V.

26. If a voltage supply, as in Fig. 15.21, has a ripple of 12% at 20 V dc across capacitor C, to what lowest value does V_{in} drop during the cycle? Assume that $V_m = 25.5$ V.

27. For a 5-V supply, as in Fig. 15.22, with $C = 500\ \mu\text{F}$ and load $I_{\text{dc}} = 150$ mA, determine:

a. V_{dc} across capacitor C.

b. V_r (peak) across capacitor C.

28. A +5-V supply as in Fig. 15.22 with $C = 330\ \mu\text{F}$ and a load of 300 mA has what value of $V_{\text{in}_{\text{min}}}$? Will output be maintained at a regulated +5-V level? (Assume that $V_m = 15$ V.)

Appendix A
Determinants

Determinants are employed to find the mathematical solutions for the variables in two or more simultaneous equations. Once the procedure is properly understood, solutions can be obtained with a minimum of time and effort and usually with fewer errors than when using other methods.

Consider the following equations, where x and y are the unknown variables and a_1, a_2, b_1, b_2, c_1, and c_2 are constants:

Col. 1		Col. 2		Col. 3	
a_1x	$+$	b_1y	$=$	c_1	**(A.1a)**
a_2x	$+$	b_2y	$=$	c_2	**(A.1b)**

It is certainly possible to solve for one variable in Eq. (A.1a) and substitute into Eq. (A.1b). That is, solving for x in Eq. (A.1a), we have

$$x = \frac{c_1 - b_1y}{a_1}$$

and substituting the result in Eq. (A.1b), we obtain

$$a_2\left(\frac{c_1 - b_1y}{a_1}\right) + b_2y = c_2$$

It is now possible to solve for y, since it is the only variable remaining, and then substitute into either equation for x. This is acceptable for two equations, but it becomes a very tedious and lengthy process for three or more simultaneous equations.

Using determinants to solve for x and y requires that the following formats be established for each variable:

$$\begin{array}{cc} \text{Col. 1} \quad \text{Col. 2} & \text{Col. 1} \quad \text{Col. 2} \\ \hline x = \dfrac{\begin{vmatrix} c_1 & b_1 \\ c_2 & b_2 \end{vmatrix}}{\begin{vmatrix} a_1 & b_1 \\ a_2 & b_2 \end{vmatrix}} & y = \dfrac{\begin{vmatrix} a_1 & c_1 \\ a_2 & c_2 \end{vmatrix}}{\begin{vmatrix} a_1 & b_1 \\ a_2 & b_2 \end{vmatrix}} \end{array} \tag{A.2}$$

First note that only constants appear within the vertical brackets and that the denominator of each is the same. In fact, the denominator is simply the coefficients of x and y in the same arrangement as in Eqs. (A.1a) and (A.1b). When solving for x, the coefficients of x in the numerator are replaced by the constants to the right of the equal sign in Eqs. (A.1a) and (A.1b), and the coefficients of the y variable are simply repeated. When solving for y, the y coefficients in the numerator are replaced by the constants to the right of the equal sign, and the coefficients of x are repeated.

Each configuration in the numerator and denominator of Eqs. (A.2) is referred to as a *determinant* (D), which can be evaluated numerically in the following manner:

$$\text{Determinant} = D = \overset{\text{Col. 1} \quad \text{Col. 2}}{\begin{vmatrix} a_1 & b_1 \\ a_2 & b_2 \end{vmatrix}} = a_1b_2 - a_2b_1 \tag{A.3}$$

The expanded value is obtained by first multiplying the top left element by the bottom right and then subtracting the product of the lower left and upper right elements. This particular determinant is referred to as a *second-order* determinant, since it contains two rows and two columns.

It is important to remember when using determinants that the columns of the equations, as indicated in Eqs. (A.1a) and (A.1b), must be placed in the same order within the determinant configuration. That is, since a_1 and a_2 are in column 1 of Eqs. (A.1a) and (A.1b), they must be in column 1 of the determinant. (The same is true for b_1 and b_2.)

Expanding the entire expression for x and y, we have the following:

$$x = \frac{\begin{vmatrix} c_1 & b_1 \\ c_2 & b_2 \end{vmatrix}}{\begin{vmatrix} a_1 & b_1 \\ a_2 & b_2 \end{vmatrix}} = \frac{c_1b_2 - c_2b_1}{a_1b_2 - a_2b_1} \tag{A.4a}$$

$$y = \frac{\begin{vmatrix} a_1 & c_1 \\ a_2 & c_2 \end{vmatrix}}{\begin{vmatrix} a_1 & b_1 \\ a_2 & b_2 \end{vmatrix}} = \frac{a_1c_2 - a_2c_1}{a_1b_2 - a_2b_1} \qquad \textbf{(A.4b)}$$

EXAMPLE A.1 Evaluate the following determinants:

a. $\begin{vmatrix} 2 & 2 \\ 3 & 4 \end{vmatrix} = (2)(4) - (3)(2) = 8 - 6 = \mathbf{2}$

b. $\begin{vmatrix} 4 & -1 \\ 6 & 2 \end{vmatrix} = (4)(2) - (6)(-1) = 8 + 6 = \mathbf{14}$

c. $\begin{vmatrix} 0 & -2 \\ -2 & 4 \end{vmatrix} = (0)(4) - (-2)(-2) = 0 - 4 = \mathbf{-4}$

d. $\begin{vmatrix} 0 & 0 \\ 3 & 10 \end{vmatrix} = (0)(10) - (3)(0) = \mathbf{0}$

EXAMPLE A.2 Solve for x and y:

$$\begin{aligned} 2x + y &= 3 \\ 3x + 4y &= 2 \end{aligned}$$

Solution:

$$x = \frac{\begin{vmatrix} 3 & 1 \\ 2 & 4 \end{vmatrix}}{\begin{vmatrix} 2 & 1 \\ 3 & 4 \end{vmatrix}} = \frac{(3)(4) - (2)(1)}{(2)(4) - (3)(1)} = \frac{12 - 2}{8 - 3} = \frac{10}{5} = \mathbf{2}$$

$$y = \frac{\begin{vmatrix} 2 & 3 \\ 3 & 2 \end{vmatrix}}{5} = \frac{(2)(2) - (3)(3)}{5} = \frac{4 - 9}{5} = \frac{-5}{5} = \mathbf{-1}$$

Check:

$$\begin{aligned} 2x + y &= (2)(2) + (-1) \\ &= 4 - 1 = 3 \qquad \text{(checks)} \\ 3x + 4y &= (3)(2) + (4)(-1) \\ &= 6 - 4 = 2 \qquad \text{(checks)} \end{aligned}$$

EXAMPLE A.3 Solve for x and y:

$$\begin{aligned} -x + 2y &= 3 \\ 3x - 2y &= -2 \end{aligned}$$

Solution: In this example, note the effect of the minus sign and the use of parentheses to ensure that the proper sign is obtained for each product:

$$x = \frac{\begin{vmatrix} 3 & 2 \\ -2 & -2 \end{vmatrix}}{\begin{vmatrix} -1 & 2 \\ 3 & -2 \end{vmatrix}} = \frac{(3)(-2) - (-2)(2)}{(-1)(-2) - (3)(2)}$$

$$= \frac{-6 + 4}{2 - 6} = \frac{-2}{-4} = \mathbf{\frac{1}{2}}$$

$$y = \frac{\begin{vmatrix} -1 & 3 \\ 3 & -2 \end{vmatrix}}{-4} = \frac{(-1)(-2) - (3)(3)}{-4}$$

$$= \frac{2 - 9}{-4} = \frac{-7}{-4} = \mathbf{\frac{7}{4}}$$

EXAMPLE A.4 Solve for x and y:

$$x = 3 - 4y$$
$$20y = -1 + 3x$$

Solution: In this case, the equations must first be placed in the format of Eqs. (A.1a) and (A.1b):

$$x + 4y = 3$$
$$-3x + 20y = -1$$

$$x = \frac{\begin{vmatrix} 3 & 4 \\ -1 & 20 \end{vmatrix}}{\begin{vmatrix} 1 & 4 \\ -3 & 20 \end{vmatrix}} = \frac{(3)(20) - (-1)(4)}{(1)(20) - (-3)(4)}$$

$$= \frac{60 + 4}{20 + 12} = \frac{64}{32} = \mathbf{2}$$

$$y = \frac{\begin{vmatrix} 1 & 3 \\ -3 & -1 \end{vmatrix}}{32} = \frac{(1)(-1) - (-3)(3)}{32}$$

$$= \frac{-1 + 9}{32} = \frac{8}{32} = \mathbf{\frac{1}{4}}$$

The use of determinants is not limited to the solution of two simultaneous equations; determinants can be applied to any number of simultaneous linear equations. First we shall examine a shorthand method that is applicable to third-order determinants only, since most of the problems in the text are limited to this level of difficulty. We shall then investigate the general procedure for solving any number of simultaneous equations.

Consider the three following simultaneous equations:

Col. 1		Col. 2		Col. 3		Col. 4
a_1x	$+$	b_1y	$+$	c_1z	$=$	d_1
a_2x	$+$	b_2y	$+$	c_2z	$=$	d_2
a_3x	$+$	b_3y	$+$	c_3z	$=$	d_3

in which x, y, and z are the variables, and $a_{1,2,3}$, $b_{1,2,3}$, and $d_{1,2,3}$ are constants.

The determinant configuration for x, y, and z can be found in a manner similar to that for two simultaneous equations. That is, to solve for x, find the determinant in the numerator by replacing column 1 with the elements to the right of the equal sign. The denominator is the determinant of the coefficients of the variables (the same applies to y and z). Again, the denominator is the same for each variable.

$$x = \frac{\begin{vmatrix} d_1 & b_1 & c_1 \\ d_2 & b_2 & c_2 \\ d_3 & b_3 & c_3 \end{vmatrix}}{D}, \quad y = \frac{\begin{vmatrix} a_1 & d_1 & c_1 \\ a_2 & d_2 & c_2 \\ a_3 & d_3 & c_3 \end{vmatrix}}{D}, \quad z = \frac{\begin{vmatrix} a_1 & b_1 & d_1 \\ a_2 & b_2 & d_2 \\ a_3 & b_3 & d_3 \end{vmatrix}}{D}$$

where

$$D = \begin{vmatrix} a_1 & b_1 & c_1 \\ a_2 & b_2 & c_2 \\ a_3 & b_3 & c_3 \end{vmatrix}$$

A shorthand method for evaluating the third-order determinant consists simply of repeating the first two columns of the determinant to the right of the determinant and then summing the products along specific diagonals as shown here:

$$\begin{matrix} & & & & 4(-) & 5(-) & 6(-) \\ D = & \begin{vmatrix} a_1 & b_1 & c_1 \\ a_2 & b_2 & c_2 \\ a_3 & b_3 & c_3 \end{vmatrix} & \begin{matrix} a_1 & b_1 \\ a_2 & b_2 \\ a_3 & b_3 \end{matrix} \\ & & & & 1(+) & 2(+) & 3(+) \end{matrix}$$

The products of diagonals 1, 2, and 3 are positive and have the following magnitudes:

$$+a_1b_2c_3 + b_1c_2a_3 + c_1a_2b_3$$

The products of diagonals 4, 5, and 6 are negative and have the following magnitudes:

$$-a_3b_2c_1 - b_3c_2a_1 - c_3a_2b_1$$

The total solution is the sum of diagonals 1, 2, and 3 minus the sum of diagonals 4, 5, and 6:

$$+(a_1b_2c_3 + b_1c_2a_3 + c_1a_2b_3) - (a_3b_2c_1 + b_3c_2a_1 + c_3a_2b_1) \quad \textbf{(A.5)}$$

Warning: **This method of expansion is good only for third-order determinants!** It cannot be applied to fourth- and higher-order systems.

EXAMPLE A.5 Evaluate the following determinant:

$$\begin{vmatrix} 1 & 2 & 3 \\ -2 & 1 & 0 \\ 0 & 4 & 2 \end{vmatrix} \Rightarrow \begin{vmatrix} 1 & 2 & 3 \\ -2 & 1 & 0 \\ 0 & 4 & 2 \end{vmatrix} \begin{matrix} 1 & 2 \\ -2 & 1 \\ 0 & 4 \end{matrix}$$

(−)(−)(−) above; (+)(+)(+) below

Solution:

$$[(1)(1)(2) + (2)(0)(0) + (3)(-2)(4)] - [(0)(1)(3) + (4)(0)(1) + (2)(-2)(2)]$$
$$= (2 + 0 - 24) - (0 + 0 - 8) = (-22) - (-8)$$
$$= -22 + 8 = \mathbf{-14}$$

EXAMPLE A.6 Solve for x, y, and z:

$$1x + 0y - 2z = -1$$
$$0x + 3y + 1z = +2$$
$$1x + 2y + 3z = 0$$

Solution:

$$x = \frac{\begin{vmatrix} -1 & 0 & -2 \\ 2 & 3 & 1 \\ 0 & 2 & 3 \end{vmatrix} \begin{matrix} -1 & 0 \\ 2 & 3 \\ 0 & 2 \end{matrix}}{\begin{vmatrix} 1 & 0 & -2 \\ 0 & 3 & 1 \\ 1 & 2 & 3 \end{vmatrix} \begin{matrix} 1 & 0 \\ 0 & 3 \\ 1 & 2 \end{matrix}}$$

$$= \frac{[(-1)(3)(3) + (0)(1)(0) + (-2)(2)(2)] - [(0)(3)(-2) + (2)(1)(-1) + (3)(2)(0)]}{[(1)(3)(3) + (0)(1)(1) + (-2)(0)(2)] - [(1)(3)(-2) + (2)(1)(1) + (3)(0)(0)]}$$

$$= \frac{(-9 + 0 - 8) - (0 - 2 + 0)}{(9 + 0 + 0) - (-6 + 2 + 0)}$$

$$= \frac{-17 + 2}{9 + 4} = \mathbf{-\frac{15}{13}}$$

$$y = \frac{\begin{vmatrix} 1 & -1 & -2 \\ 0 & 2 & 1 \\ 1 & 0 & 3 \end{vmatrix} \begin{matrix} 1 & -1 \\ 0 & 2 \\ 1 & 0 \end{matrix}}{13}$$

$$= \frac{[(1)(2)(3) + (-1)(1)(1) + (-2)(0)(0)] - [(1)(2)(-2) + (0)(1)(1) + (3)(0)(-1)]}{13}$$

$$= \frac{(6 - 1 + 0) - (-4 + 0 + 0)}{13}$$

$$= \frac{5 + 4}{13} = \mathbf{\frac{9}{13}}$$

$$z = \frac{\begin{vmatrix} 1 & 0 & -1 \\ 0 & 3 & 2 \\ 1 & 2 & 0 \end{vmatrix} \begin{matrix} 1 & 0 \\ 0 & 3 \\ 1 & 2 \end{matrix}}{13}$$

$$= \frac{[(1)(3)(0) + (0)(2)(1) + (-1)(0)(2)] - [(1)(3)(-1) + (2)(2)(1) + (0)(0)(0)]}{13}$$

$$= \frac{(0 + 0 + 0) - (-3 + 4 + 0)}{13}$$

$$= \frac{0 - 1}{13} = -\frac{\mathbf{1}}{\mathbf{13}}$$

or from $0x + 3y + 1z = +2$,

$$z = 2 - 3y = 2 - 3\left(\frac{9}{13}\right) = \frac{26}{13} - \frac{27}{13} = -\frac{\mathbf{1}}{\mathbf{13}}$$

Check:

$$\left.\begin{array}{r} 1x + 0y - 2z = -1 \\ 0x + 3y + 1z = +2 \\ 1x + 2y + 3z = 0 \end{array}\right\} \begin{array}{l} -\frac{15}{13} + 0 + \frac{2}{13} = -1 \\ 0 + \frac{27}{13} + \frac{-1}{13} = +2 \\ -\frac{15}{13} + \frac{18}{13} + \frac{-3}{13} = 0 \end{array} \left\} \begin{array}{l} -\frac{13}{13} = -1 \checkmark \\ \frac{26}{13} = +2 \checkmark \\ -\frac{18}{13} + \frac{18}{13} = 0 \checkmark \end{array}\right.$$

The general approach to third- or higher-order determinants requires that the determinant be expanded in the following form. There is more than one expansion that will generate the correct result, but this form is typically employed when the material is first introduced.

$$D = \begin{vmatrix} a_1 & b_1 & c_1 \\ a_2 & b_2 & c_2 \\ a_3 & b_3 & c_3 \end{vmatrix} = a_1\left(+\begin{vmatrix} b_2 & c_2 \\ b_3 & c_3 \end{vmatrix}\right) + b_1\left(+\begin{vmatrix} a_2 & c_2 \\ a_3 & c_3 \end{vmatrix}\right) + c_1\left(+\begin{vmatrix} a_2 & b_2 \\ a_3 & b_3 \end{vmatrix}\right)$$

Labels: each 2×2 determinant — Minor; each parenthesized term — Cofactor; a_1, b_1, c_1 — Multiplying factor.

This expansion was obtained by multiplying the elements of the first row of D by their corresponding cofactors. It is not a requirement that the first row be used as the multiplying factors. In fact, any *row* or *column* (not diagonals) may be used to expand a third-order determinant.

The sign of each cofactor is dictated by the position of the multiplying factors (a_1, b_1, and c_1 in this case) as in the following standard format:

$$\begin{vmatrix} + \rightarrow & - & + \\ \downarrow & & \\ - & + & - \\ & & \\ + & - & + \end{vmatrix}$$

Note that the proper sign for each element can be obtained by simply assigning the upper left element a positive sign and then changing sign as you move horizontally or vertically to the neighboring position.

For the determinant D, the elements would have the following signs:

$$\begin{vmatrix} a_1^{(+)} & b_1^{(-)} & c_1^{(+)} \\ a_2^{(-)} & b_2^{(+)} & c_2^{(-)} \\ a_3^{(+)} & b_3^{(-)} & c_3^{(+)} \end{vmatrix}$$

The minors associated with each multiplying factor are obtained by covering up the row and column in which the multiplying factor is located and writing a second-order determinant to include the remaining elements in the same relative positions that they have in the third-order determinant.

Consider the cofactors associated with a_1 and b_1 in the expansion of D. The sign is positive for a_1 and negative for b_1, as determined by the standard format. Following the procedure outlined above, we can find the minors of a_1 and b_1 as follows:

$$a_{1\,(\text{minor})} = \begin{vmatrix} \not{a_1} & \not{b_1} & \not{c_1} \\ \not{a_2} & b_2 & c_2 \\ \not{a_3} & b_3 & c_3 \end{vmatrix} = \begin{vmatrix} b_2 & c_2 \\ b_3 & c_3 \end{vmatrix}$$

$$b_{1\,(\text{minor})} = \begin{vmatrix} \not{a_1} & \not{b_1} & \not{c_1} \\ a_2 & \not{b_2} & c_2 \\ a_3 & \not{b_3} & c_3 \end{vmatrix} = \begin{vmatrix} a_2 & c_2 \\ a_3 & c_3 \end{vmatrix}$$

It was pointed out that any row or column may be used to expand the third-order determinant, and the same result will still be obtained. Using the first column of D, we obtain the expansion

$$D = \begin{vmatrix} a_1 & b_1 & c_1 \\ a_2 & b_2 & c_2 \\ a_3 & b_3 & c_3 \end{vmatrix} = a_1\left(+\begin{vmatrix} b_2 & c_2 \\ b_3 & c_3 \end{vmatrix}\right) + a_2\left(-\begin{vmatrix} b_1 & c_1 \\ b_3 & c_3 \end{vmatrix}\right) + a_3\left(+\begin{vmatrix} b_1 & c_1 \\ b_2 & c_2 \end{vmatrix}\right)$$

The proper choice of row or column can often effectively reduce the amount of work required to expand the third-order determinant. For example, in the following determinants, the first column and third row, respectively, would reduce the number of cofactors in the expansion:

$$D = \begin{vmatrix} 2 & 3 & -2 \\ 0 & 4 & 5 \\ 0 & 6 & 7 \end{vmatrix} = 2\left(+\begin{vmatrix} 4 & 5 \\ 6 & 7 \end{vmatrix}\right) + 0 + 0 = 2(28 - 30)$$

$$= \mathbf{-4}$$

$$D = \begin{vmatrix} 1 & 4 & 7 \\ 2 & 6 & 8 \\ 2 & 0 & 3 \end{vmatrix} = 2\left(+\begin{vmatrix} 4 & 7 \\ 6 & 8 \end{vmatrix}\right) + 0 + 3\left(+\begin{vmatrix} 1 & 4 \\ 2 & 6 \end{vmatrix}\right)$$

$$= 2(32 - 42) + 3(6 - 8) = 2(-10) + 3(-2)$$

$$= \mathbf{-26}$$

EXAMPLE A.7 Expand the following third-order determinants:

a. $$D = \begin{vmatrix} 1 & 2 & 3 \\ 3 & 2 & 1 \\ 2 & 1 & 3 \end{vmatrix} = 1\left(+\begin{vmatrix} 2 & 1 \\ 1 & 3 \end{vmatrix}\right) + 3\left(-\begin{vmatrix} 2 & 3 \\ 1 & 3 \end{vmatrix}\right) + 2\left(+\begin{vmatrix} 2 & 3 \\ 2 & 1 \end{vmatrix}\right)$$

$$= 1[6 - 1] + 3[-(6 - 3)] + 2[2 - 6]$$

$$= 5 + 3(-3) + 2(-4)$$

$$= 5 - 9 - 8$$

$$= \mathbf{-12}$$

b. $$D = \begin{vmatrix} 0 & 4 & 6 \\ 2 & 0 & 5 \\ 8 & 4 & 0 \end{vmatrix} = 0 + 2\left(-\begin{vmatrix} 4 & 6 \\ 4 & 0 \end{vmatrix}\right) + 8\left(+\begin{vmatrix} 4 & 6 \\ 0 & 5 \end{vmatrix}\right)$$

$$= 0 + 2[-(0 - 24)] + 8[(20 - 0)]$$

$$= 48 + 160$$

$$= \mathbf{208}$$

Appendix B
Computer Analysis: PSpice and Electronic Workbench

Computer analysis is available from a number of software packages. Among the more popular are PSpice and Electronic Workbench. All the analysis used by these programs is based on the programs developed at the University of California at Berkeley during the early 1970s. Various versions have been developed for use on a variety of computer platforms—Windows, Macintosh, SUN systems, among the larger users. SPICE is an acronym for *S*imulation *P*rogram with *I*ntegrated *C*ircuit *E*mphasis. While PSpice was introduced by the Microsim Corp., that product is now part of the ORCAD Corp. family of products. The Electronic Workbench (and MultiSIM) family of software is produced by Interactive Image Technologies, Ltd. and continues to be provided in new (and higher numbered) versions. Both PSpice and Electronic Workbench provide a graphical schematic input front end, with the user selecting components and then placing them on a drawing page or sheet to construct the desired circuit. While PSpice has some means of showing calculated results, its PROBE section provides the output of various analyses. Electronic Workbench, as the name implies, provides a number of input and output components (multimeters, oscilloscopes, and the like), as one would use in the lab, for analyzing the results of circuit operations.

Both software packages come with ideal component values—resistors, diodes, transistors, and a selection of commerical components. A good number of digital integrated circuits and a handful of analog integrated circuits are also provided for student use. The more expensive commercial packages include a very large array of integrated circuits and components that are used in industry. The examples in the text are limited to the versions used by students.

Basic System Requirements

The PSpice and EWB packages can be run on suitably configured IBM systems, Macintosh computers, and SUN workstations, among the more popular equipment. Both of these programs are supplied on CD, so a typical computer system must operate in Windows, have at least 32 MB of RAM, and have close to 100 MB of hard drive space. A number of data files are produced when these programs are run, including the schematic circuit and analysis output(s).

Operating With PSpice

The PSpice program starts with the schematic capture. A blank sheet (screen) is provided along with a number of toolbars. Figure B.1 shows the usual screen.

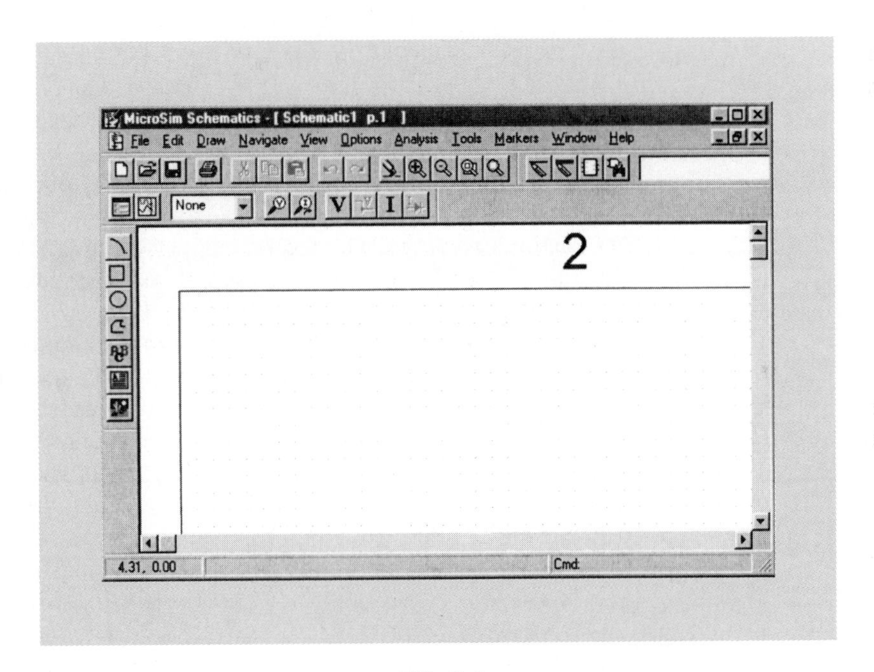

FIG. B.1

Figure B.2 shows the basic parts browser. A component is selected directly and then placed on the schematic sheet, as shown for a few components in Fig. B.3.

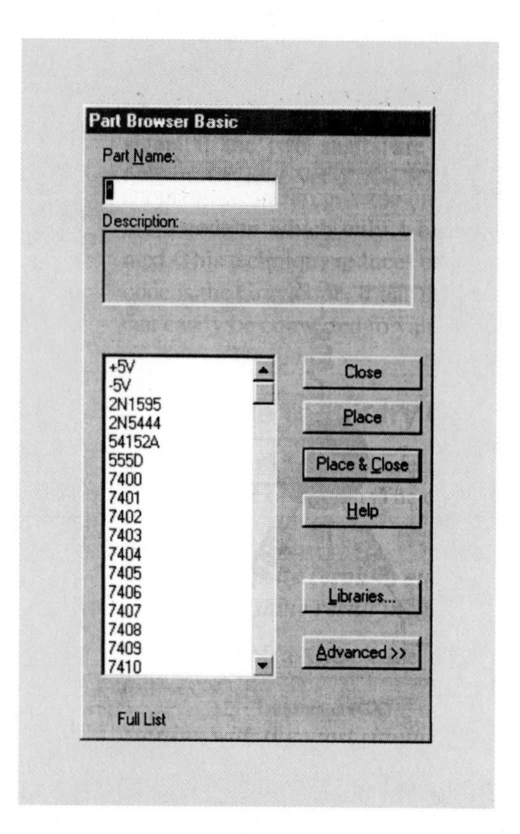
Part Browser Basic
Part Name:
Description:
+5V
-5V
2N1595
2N5444
54152A
555D
7400
7401
7402
7403
7404
7405
7406
7407
7408
7409
7410
Close
Place
Place & Close
Help
Libraries...
Advanced >>
Full List

FIG. B.2

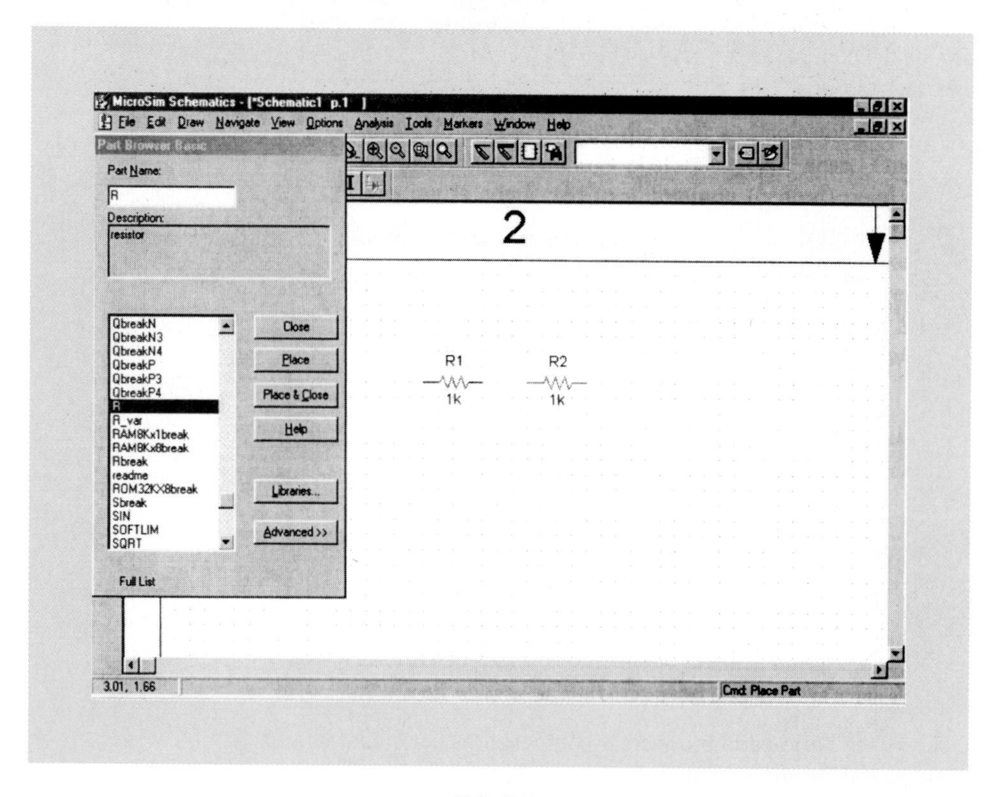
MicroSim Schematics - [*Schematic1 p.1]
File Edit Draw Navigate View Options Analysis Tools Markers Window Help
Part Browser Basic
Part Name:
R
Description:
resistor
QbreakN
QbreakN3
QbreakN4
QbreakP
QbreakP3
QbreakP4
R
R_var
RAM8Kx1break
RAM8Kx8break
Rbreak
readme
ROM32KX8break
Sbreak
SIN
SOFTLIM
SQRT
Close
Place
Place & Close
Help
Libraries...
Advanced >>
Full List
2
R1
1k
R2
1k
3.01, 1.66
Cmd: Place Part

FIG. B.3

The components are then connected using the Draw Wire tool (Fig. B.4).

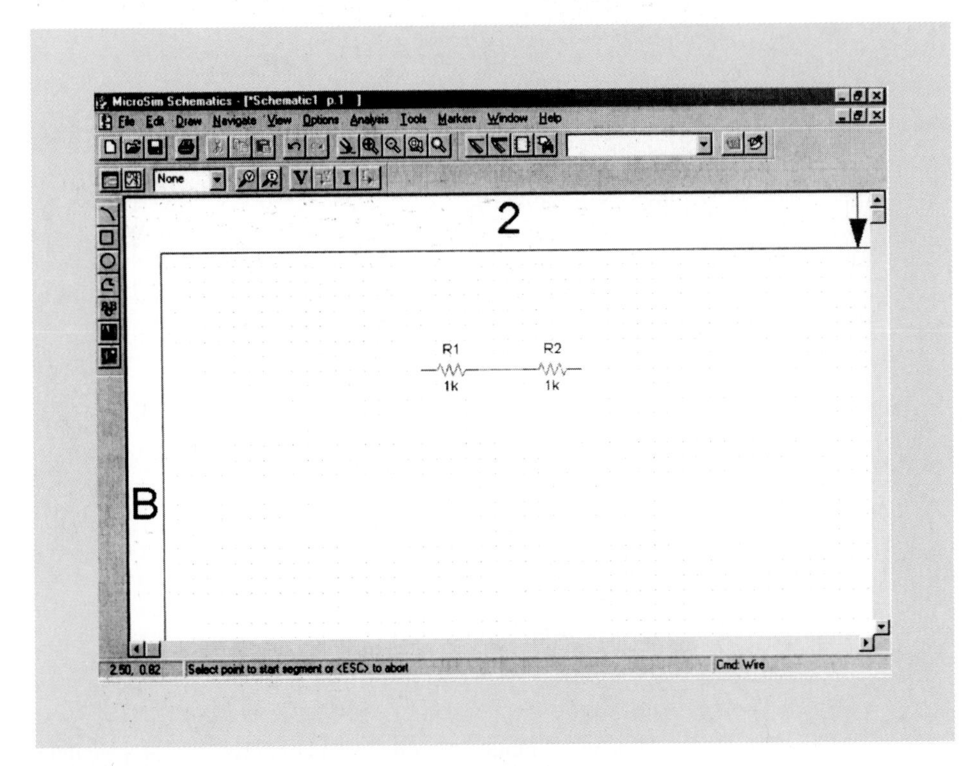

FIG. B.4

Double-clicking on the component brings up its parameters. For resistor R2, for example, the starting value of 1k, provided by PSpice is changed to 3k, as shown in Fig. B.5.

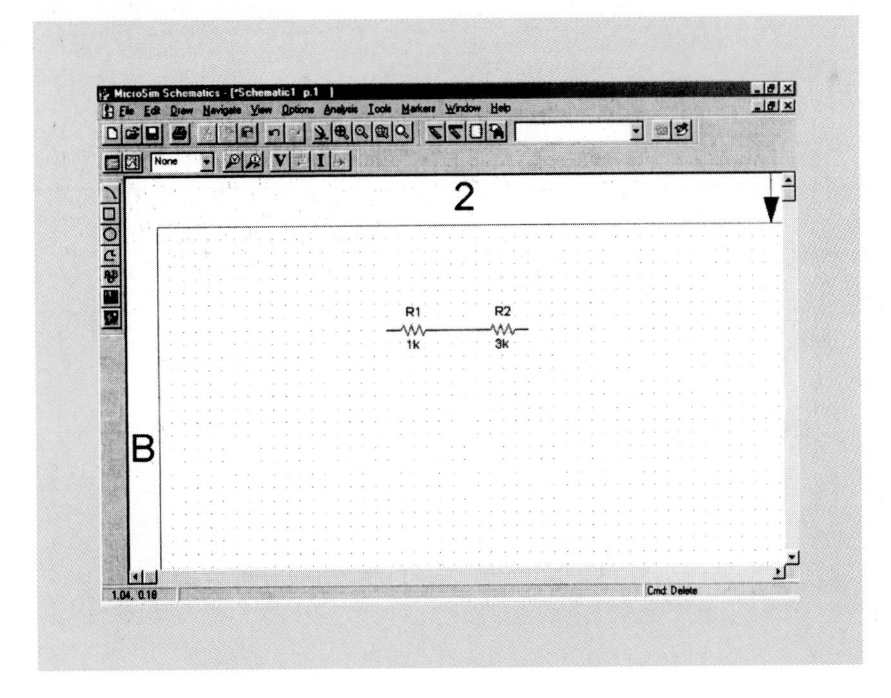

FIG. B.5

After the voltage is set to 12 V, the circuit to be analyzed is that shown in Fig. B.6.

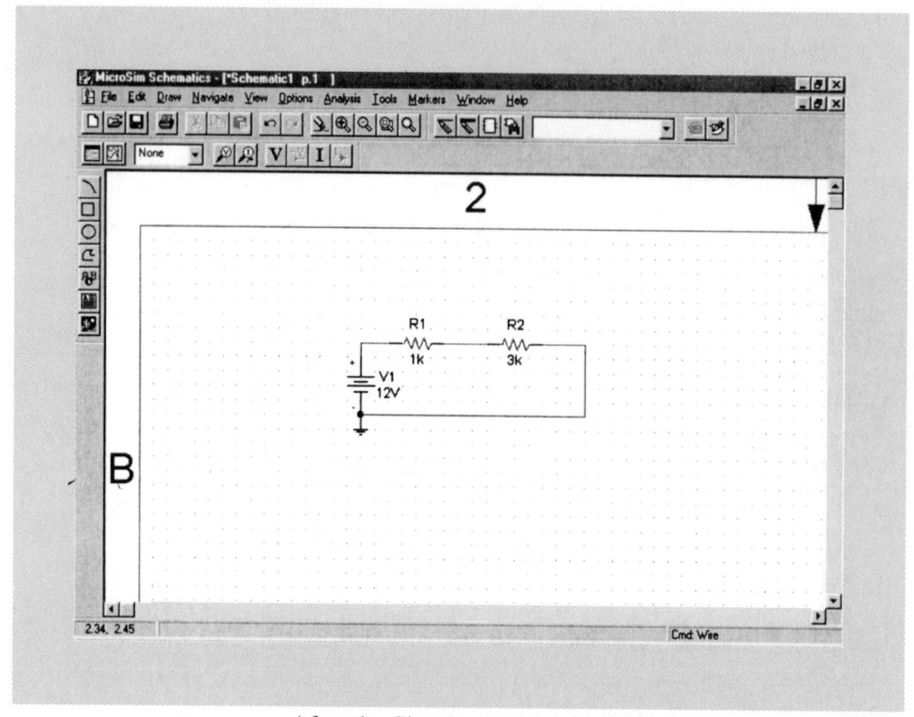

FIG. B.6

After the Simulate button (or F11) is pressed, the calculated circuit currents are displayed as in Fig. B.7. Notice that the toolbar button for current in depressed (or selected).

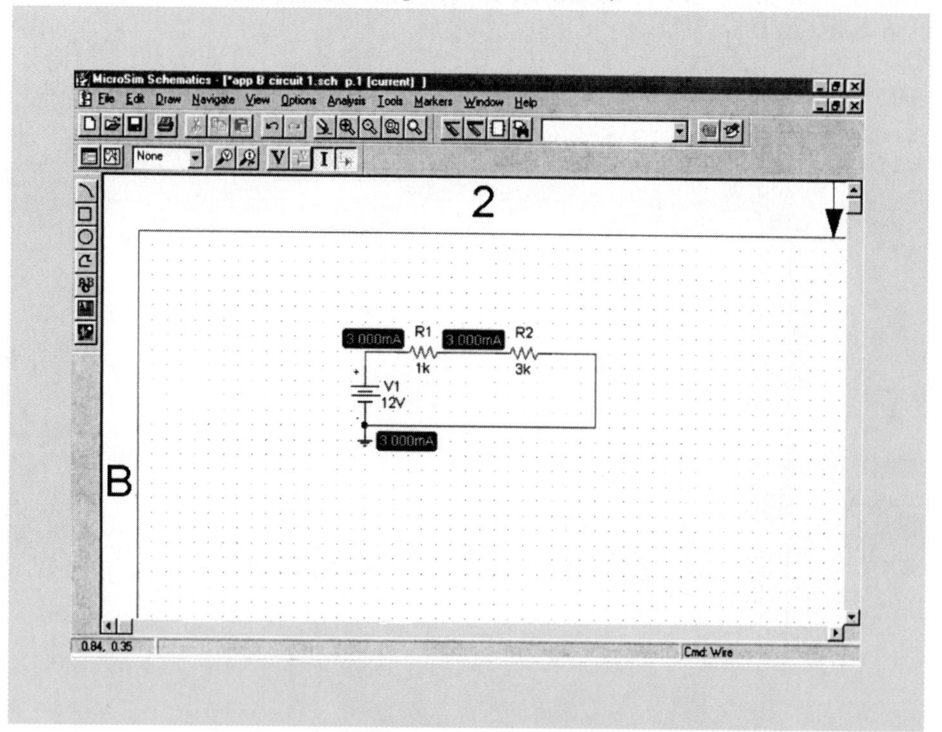

FIG. B.7

Because there are numerous books and manuals for PSpice this text provides examples in appropriate chapters, with comments on the circuit results, leaving the details of drawing and analyzing the circuit to material provided with the software package or suitable reference texts.

File Names

Files used with PSpice have a number of standard three-letter extensions:

.sch Schematic capture file—file of circuit drawing
.out Output results file—text file with output results listed
.net Net list output file—shows connection information
.als Schematic aliases—alias names of components or output nodes
.cir Circuit setup information
.lib Library information on some components

Other file types exist; see the manuals for additional information.

Electronic Workbench

The starting screen for EWB is shown in Fig. B.8.

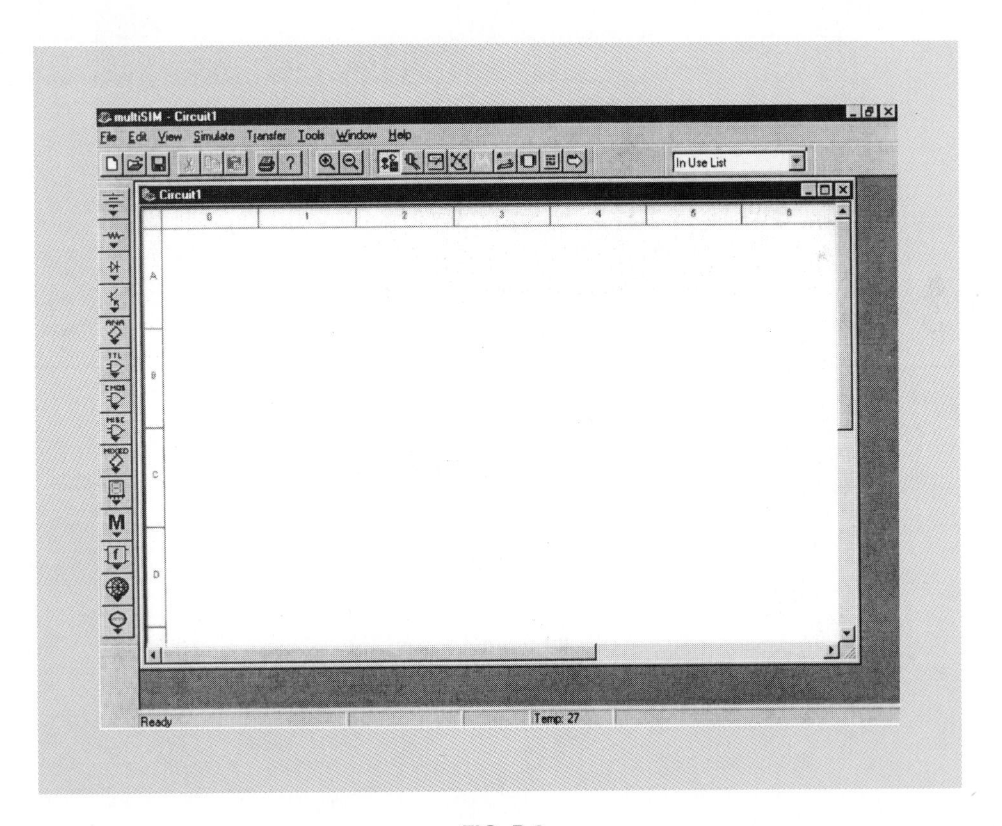

FIG. B.8

As in PSpice, the user selects components, sets their values, and then connects the components as shown in the simple circuit of Fig. B.9. The ammeter shows the circuit current to be 3 mA.

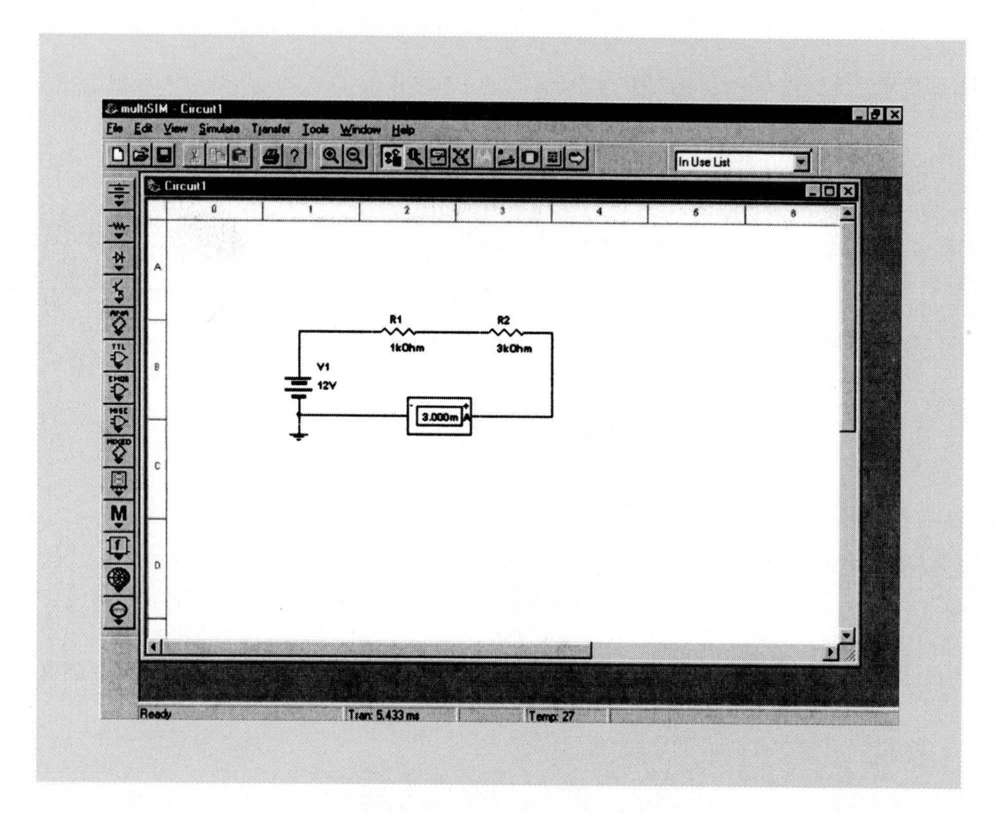

FIG. B.9

A multimeter, set for current, can be used to measure the current, as shown in Fig. B.10.

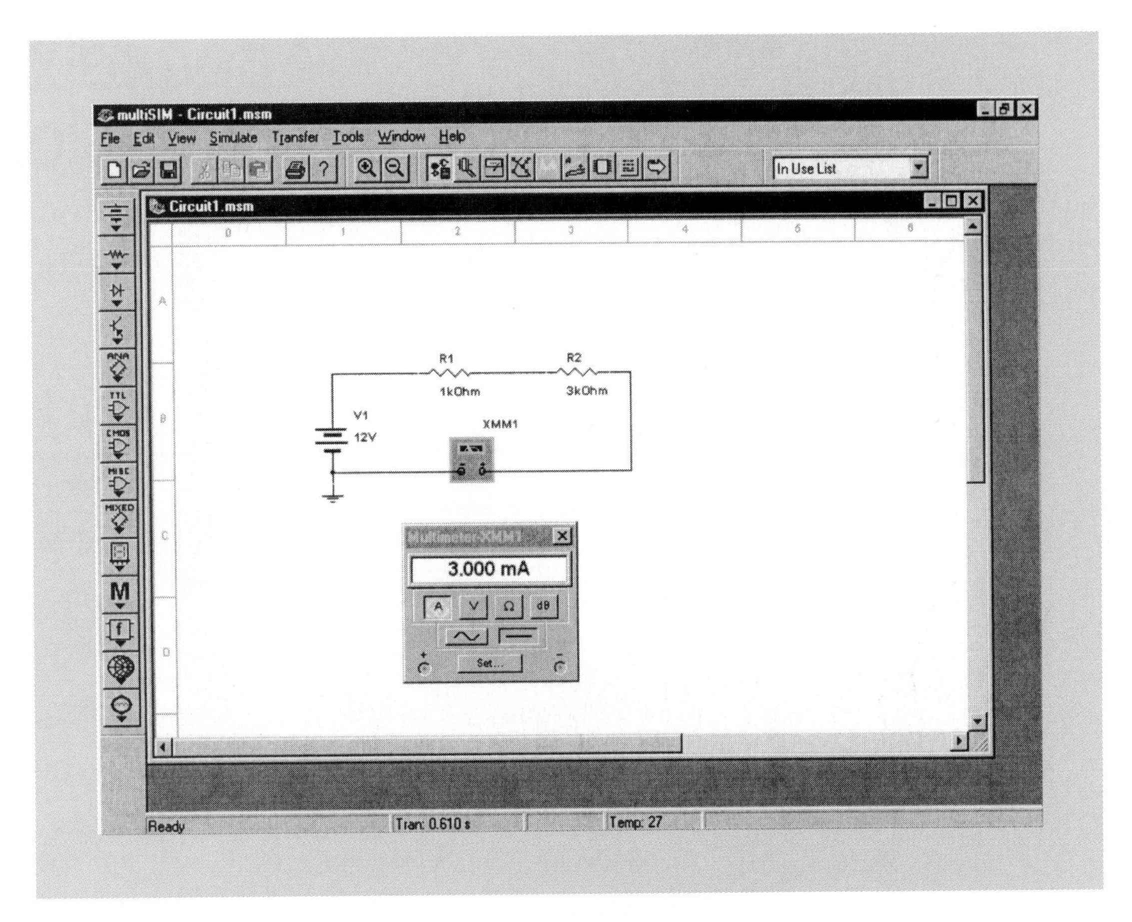

FIG. B.10

EWB provides a large number of instruments, as found in many labs, and a very large number of components to build and simulate most linear and digital circuits.

Appendix C
Magnetic Parameter Conversions

	SI (MKS)	CGS	English
Φ	webers (Wb)	maxwells	lines
	1 Wb	$= 10^8$ maxwells	$= 10^8$ lines
B	Wb/m^2	gauss (maxwells/cm^2)	lines/in.^{2}
	1 Wb/m^2	$= 10^4$ gauss	$= 6.452 \times 10^4$ lines/in.^{2}
A	1 m^2	$= 10^4$ cm^2	$= 1550$ in.^{2}
μ_o	$4\pi \times 10^{-7}$ Wb/Am	$= 1$ gauss/oersted	$= 3.20$ lines/Am
$\mathscr{F}$	NI (ampere-turns, At)	$0.4\pi NI$ (gilberts)	NI (At)
	1 At	$= 1.257$ gilberts	1 gilbert $= 0.7958$ At
H	NI/l (At/m)	$0.4\pi NI/l$ (oersteds)	NI/l (At/in.)
	1 At/m	$= 1.26 \times 10^{-2}$ oersted	$= 2.54 \times 10^{-2}$ At/in.
H_g	$7.97 \times 10^5 B_g$ (At/m)	B_g (oersteds)	$0.313B_g$ (At/in.)

Solutions to Odd-Numbered Exercises

Chapter 2

1. **(a)** 38.45 mC; **(b)** 1.282 mA
3. 160 μC
5. **(a)** 50 mA; **(b)** 0.4 mV; **(c)** 30 kV; **(d)** 1.2 kV; **(e)** 0.7 μA; **(f)** 32 MV
7. **(a)** 976.56 CM; **(b)** 14,400 CM; **(c)** 1550 CM
9. 4.8 J
11. 34.6 J
13. 833.33 Ω
15. 66 V, 42.67 Ω
17. 1.515 Ω
19. 8.38 Ω
21. 274.5° C
23. **(a)** 220 Ω ± 5%; **(b)** 209 Ω→ 231 Ω
25. orange, white, gold, silver
27. 35.2 mW
29. 294 kJ
31. $6.31
33. 372.09 W
35. **(a)** 72.9% **(b)** 16.2%
37. **(a)** 7 kΩ; **(b)** 6 kΩ
39. **(a)** $I = 2$ A, $R_1 = 4\ \Omega$, $R_T = 11\ \Omega$, $E = 22$ V; **(b)** $R_1 = 4\ \Omega$, $R_2 = 5\ \Omega$, $E_2 = 71$ V
41. $V_3 = 3.6$ V, $V_4 = 9.6$ V
43. $V_1 = 6$ V, $V_3 = 1$ V, $E = 9$ V
45. **(a)** $R_1 = 6$ kΩ; **(b)** $R_1 = 1$ kΩ
47. **(a)** $R_2 = 6$ kΩ, $E = 24$ V, $I = 8$ mA; **(b)** $R_1 = 9\ \Omega$, $R_3 = 8.471\ \Omega$, $V_3 = 24$ V, $I_3 = 2.833$ A, $I_2 = 0.5$ A, $I_1 = 3.333$ A
49. $I_1 = 8.333$ mA, $I_2 = 1.667$ mA, the inverse
51. $I_1 = 8$ mA, $I_3 = 80$ mA, $I = 90$ mA

Chapter 3

1. **(a)** 11 Ω; **(b)** 6 A; **(c)** $I_1 = 6$ A, $I_2 = 3$ A, $I_3 = 2$ A; **(d)** 54 W
3. 1 A
5. 16 mA
7. 0.82%
9. 202.4 mA
11. 0.9 A
13. $I_1(R_1) = 4.929$ A (down); $I_2(R_2) = 1.214$ A (down); $I_3(R_3) = 6.143$ A (ccw)
15. 5.161 mA (cw)
17. **(a)** $V_1 = -5.143$ V (left), $V_2 = 12.571$ V; **(b)** $I_{R_1} = 5.143$ A, $I_{R_2} = 8.857$ mA, $I_{R_3} = 3.143$ mA
19. $V_1 = -2.972$ V (left), $V_2 = 13.255$ V (middle), $V_3 = -19.07$ V
21. **(a)** $\frac{4}{3}\ \Omega$; **(b)** 2.297 W; **(c)** 1.721 W
23. 9 Ω
25. balanced, 1.25 A
27. 30 μF
29. 177 pF
31. **(a)** $v_C = 13.33$ V $(1 - e^{-t/0.16\,s})$; **(b)** —
33. **(a)** 20 μs; **(b)** $i_L = 10$ mA $(1 - e^{-t/20\,\mu s})$, $v_L = 10\,\text{V}e^{-t/20\,\mu s}$, $v_R = 10$ V $(1 - e^{-t/20\,\mu s})$; **(c)** —; **(d)** 6.32 V; **(e)** 4.46 μs; **(f)** 1 μJ
35. **(a)** 100 mH; **(b)** 9 mH
37. 0→3 ms: 0 V; 3→7 ms: 1.25 mV; 7→12 ms: 0.5 mV; 12→14 ms: 0 V

Chapter 4

5. $t = 0$ s
7. **(a)** i leads v by 56°; **(b)** i leads v by 140°
9. $v = 48 \times 10^{-3} \sin(\omega t - 90°)$
11. **(a)** $i = 50.9 \times 10^{-3} \sin(6283.2t + 60°)$; **(b)** $v = 11.31 \sin(377t - 10°)$
13. 0.25 V
15. 18 A^2
17. **(a)** $i = 57.2 \times 10^{-3} \sin 6.283 \times 10^3 t$; **(b)** $v = 125.84 \sin 6.283 \times 10^3 t$
19. $i_C = 0.4 \times 10^{-3} \sin(2000t + 120°)$
21. **(a)** 1:320 W, 2:8.66 W; **(b)** 1:0.5 leading, 2:0.866 lagging
23. **(a)** $46.910\angle 31.581°$; **(b)** $6.376\angle -33.291°$; **(c)** $126.49\angle 131.565°$; **(d)** $5\angle 0°$; **(e)** $10.211\angle 30.11°$; **(f)** $339.41\angle 105°$
25. $i_L = 0.2 \sin(1000t - 80°)$
27. $v_1 = 91.64 \sin(377t - 10.893°)$
29. **(a)** 10 kΩ; **(b)** $10.198\ \text{k}\Omega\angle -78.69°$; **(c)** $11.767\ \text{mA}\angle 78.69°$; **(d)** $\mathbf{V}_R = 23.534\ \text{V}\angle 78.69°$, $\mathbf{V}_C = 117.67\ \text{V}\angle -11.31°$; **(e)** —; **(f)** 276.925 mW;

(g) 276.925 mW; **(h)** —;
(i) 0.196 leading;
(j) $i = 16.639 \times 10^{-3} \sin(1000t + 78.69^\circ)$, $v_R = 33.277 \sin(1000t + 78.69^\circ)$, $v_C = 166.385 \sin(1000t - 11.31^\circ)$

31. $v_L = 109.89 \sin(377t + 49.635^\circ)$

33. $\mathbf{V}_C = 41.18 \text{ mV} \angle -107.1^\circ$

35. **(a)** —; **(b)** $\mathbf{Z}_T = 70.711\ \Omega \angle 45^\circ$, $\mathbf{Y}_T = 14.142 \text{ mS} \angle -45^\circ$; **(c)** $707.1 \text{ mA} \angle -45^\circ$; **(d)** $\mathbf{I}_R = 500 \text{ mA} \angle 0^\circ$, $\mathbf{I}_L = 500 \text{ mA} \angle -90^\circ$; **(e)** 25 W; **(f)** 0.7071 lagging; **(g)** —;

37. **(a)** —; **(b)** $\mathbf{Y}_T = 1.118 \text{ mS} \angle -26.565^\circ$, $\mathbf{Z}_T = 0.894 \text{ k}\Omega \angle 26.565^\circ$; **(c)** $\mathbf{I} = 55.93 \text{ mA} \angle 3.435^\circ$, $\mathbf{I}_R = 50 \text{ mA} \angle 30^\circ$, $\mathbf{I}_L = 50 \text{ mA} \angle -60^\circ$, $\mathbf{I}_C = 25 \text{ mA} \angle 120^\circ$; **(d)** 2.5 W; **(e)** 0.894 lagging; **(f)** $i = 79.09 \times 10^{-3} \sin(\omega t + 3.435^\circ)$, $i_R = 70.7 \times 10^{-3} \sin(\omega t + 30^\circ)$, $i_L = 70.7 \times 10^{-3} \sin(\omega t - 60^\circ)$, $i_C = 35.35 \times 10^{-3} \sin(\omega t + 120^\circ)$; **(g)** e leads i_L by 90°

39. 10 A, 100 V

Chapter 5

1. **(a)** $\mathbf{I} = 15.29 \text{ mA} \angle 8.44^\circ$, $\mathbf{I}_L = 21.18 \text{ mA} \angle -47.87^\circ$; **(b)** 1.346 W

3. $6.37 \text{ mA} \angle -45.57^\circ$

5. $6.37 \text{ mA} \angle -45.63^\circ$

7. $6.37 \text{ mA} \angle -45.63^\circ$

9. **(a)** $6.37 \text{ mA} \angle -45.63^\circ$; **(b)** $25.48 \text{ V} \angle 44.37^\circ$

11. $13.2 \text{ mA} \angle 4.05^\circ$

13. $\mathbf{V}_1 = 13.133 \text{ V} \angle 72.58^\circ$, $\mathbf{V}_2 = 36.72 \text{ V} \angle -172.47^\circ$; $\mathbf{V}_{R_2} = 43.91 \text{ V} \angle 23.27^\circ$

15. **(a)** $10.61 \text{ k}\Omega + j4.8 \text{ k}\Omega$; **(b)** 544 mW

17. **(a)** $\mathbf{Z}_{\text{Th}} = 10.07\ \Omega \angle -65.56^\circ$, $\mathbf{E}_{\text{Th}} = 17.7 \text{ V} \angle -135^\circ$; **(b)** $4.17\ \Omega + j9.17\ \Omega$; **(c)** 18.78 W

19. $15.25 \text{ mA} \angle -32.16^\circ$

21. —

23. **(a)** $\mathbf{I}_R = 30 \text{ mA} \angle 0^\circ$, $\mathbf{I}_L = 12 \text{ mA} \angle -90^\circ$, $\mathbf{I}_C = 15 \text{ mA} \angle 90^\circ$; **(b)** 3.6 W; **(c)** 0.36 VAR(C); **(d)** 3.62 VA; **(e)** 0.994 leading

25. **(a)** $\mathbf{I}_{R_1} = 0.894 \text{ A} \angle 26.565^\circ$, $\mathbf{I}_{L_1} = 0.632 \text{ A} \angle -18.435^\circ$, $\mathbf{I}_{L\text{-}C} = 0.632 \text{ A} \angle 71.565^\circ$; **(b)** 15.98 W; **(c)** 7.988 VAR(C); **(d)** 17.865 VA; **(e)** 0.894 leading

27. **(a)** $V_g = V_{\phi L} = 115.47$ V; **(b)** 16.01 mA; **(c)** 16.01 mA; **(d)** 5.546 W

29. **(a)** $V_g = V_{\phi L} = 115.47$ V; **(b)** 9.33 A; **(c)** 9.33 A; **(d)** 3.23 kW

31. **(a)** 4 kΩ; **(b)** 40; **(c)** 125 Hz; **(d)** $P_{\text{max}} = 4\ \mu\text{W}$, $P_{\text{HPF}} = 2\ \mu\text{W}$; **(e)** $L = 127.32$ mH, $C = 7.96$ nF

33. **(a)** 2 kΩ; **(b)** 40 kΩ; **(c)** 7957.75 Hz; **(d)** $f_1 = 7558.8$ Hz, $f_2 = 8156.69$ Hz; **(e)** —

Chapter 6

3. 1.51×10^{-3} Wb/Am

5. **(a)** 110 mA; **(b)** no

7. **(a)** 80 At; **(b)** 400 At/m; **(c)** $\cong 1$ T; **(d)** 4×10^{-4} Wb

9. 0.375 A

11. 3.97 A

13. 4.56 A

15. **(a)** $R_{\text{shunt}} = 0.1\ \Omega$; **(b)** $R_s = 180 \text{ k}\Omega$

17. **(a)** 0.4 A; **(b)** $\frac{1}{50}$; **(c)** step-up; **(d)** 2t

19. **(a)** 3 A; **(b)** 0 W; **(c)** yes, $I_L > I_s$ (rated)

21. 560 VA > 480 VA

Chapter 7

1. **(a)** 12 V; **(b)** 24 V

3. **(a)** 1.5 A; **(b)** 1.5 A

5. 2100 rpm

7. 192 V

9. **(a)** $I_A = 6.26$ A, $I_F = 1.04$ A, $I_L = 5.22$ A; **(b)** 15.65 V; **(c)** 104.4 V; **(d)** 545 W

11. **(a)** $I_A = I_S = I_L = 4$A; **(b)** 72.0 V

13. 0.159 lb

15. 133.33 V

17. **(a)** 1790.4 W; **(b)** 2400 W; **(c)** 74.6%

19. **(a)** 65.65 ft-lb; **(b)** 7.29 ft-lb

21. **(a)** series; **(b)** differential

23. **(a)** 241.92 W; **(b)** 0.946 ft-lb; **(c)** 3.784 ft-lb

25. $I_L = I_\phi = 14.43$ A

27. 153.61 V

29. 139.3 V

31. 1755 rpm

33. drops 50%

35. **(b)** 1600 kVAR

37. $\mathbf{I}'_p = 1.98 \text{ A} \angle 0^\circ$, $\mathbf{I}'_F = 1.18 \text{ A} \angle -90^\circ$

39. —

41. no

Chapter 8

5. **(a)** 10 mA : 8 Ω, 30 mA : 2Ω; **(b)** 10 mA : 2.6 Ω, 30 mA : 0.867 Ω

7. **(a)** 0.3 V : 0 mA, −0.5 V : 0mA; **(b)** 1 mA : 0.7 V, 3.6 mA : 0.7 V

9. 7.22 mA

11. —

13. **(a)** $V_o = 9.515$ V, $I_D = 4.325$ mA; **(b)** $V_o = 5.7$ V, $I_D = 3.04$ mA

15. **(a)** $V_o = 11.3$ V, $I_D = 1.2$ mA; **(b)** $V_o = 14.3$ V, $I_D = 8.773$ mA

17. $V_{o_1} = 0.7$ V, $V_{o_2} = 0.3$ V, $I = 1.86$ mA

19. 0 V

21. 0.7 V

23. −9.3 V

25. —

27. $V_{dc} = -49.46$ V

29. **(a)** 20 mA; **(b)** 36.24 mA; **(c)** 18.12 mA; **(d)** yes; **(e)** 36.24 mA > 20 mA

31. PIV = 19.3 V

33. —

35. —

37. **(a)** 34 ms; **(b)** $T/2 = 0.5 \text{ ms} \ll 34$ ms; **(c)** —

39. max = +8 V, min = −8 V

41. **(a)** $R_s = 20\ \Omega$, $V_Z = 12$ V; **(b)** 2.4 W

43. $R_s = 0.5 \text{ k}\Omega$, $I_{ZM} = 60$ mA

45. 28.57 mA

47. **(a)** 41.84 pF; **(b)** 0.014×10^{12}

49. 18 mW

51. **(a)** 80 mW/cm^2; **(b)** —; **(c)** 133.33 mA

53. $0^\circ = 10^4\ \Omega$-cm, $100^\circ = 200\ \Omega$-cm

55. 20 kΩ

57. 0.5 lm = 26.67 kΩ, 0.8 lm = 11.43 kΩ
59. —
61. —

Chapter 9

1. 80
3. $I_E = 4.04$ mA, $I_B = 40\ \mu$A
5. —
7. $V_{CB_Q} = 4.575$ V, $R_E = 1.767$ kΩ
9. **(a)** 0.779 mA; **(b)** 33.376 Ω; **(c)** 33.213 Ω; **(d)** 42.242; **(e)** 10.518; **(f)** 4.7 kΩ
11. $V_{CE_Q} = 4.5$ V, $R_B = 226$ kΩ
13. **(a)** $I_B = 28.974\ \mu$A, $I_C =$ 5.215 mA, $I_E = 5.24$ mA; **(b)** 4.96 Ω; **(c)** $Z_i = 890.76$ Ω, $Z_o = 3.3$ kΩ; **(d)** -266.13; **(e)** -81.97; **(f)** 107.75
15. **(a)** $Z_i = 1.815$ kΩ, $Z_o = 3.9$ kΩ; **(b)** -121.88; **(c)** 56.72
17. **(a)** 21.41 μA; **(b)** $I_{E_Q} =$ 2.162 mA, $r_e = 12.03$ Ω; **(c)** —; **(d)** -99.75
19. 4.5 mA
21. —
23. **(a)** $V_{GS_Q} = -2.7$ V, $I_{D_Q} = 2.4$ mA; **(b)** $Z_i = 10$MΩ, $Z_o = 3.3$ kΩ; **(c)** -2.845; **(d)** -2.845
25. **(a)** $V_{GS_Q} = 9.5$ V, $I_{D_Q} = 6.2$ mA; **(b)** $Z_i = 2.36$ MΩ, $Z_o = 2.4$ kΩ; **(c)** -3.24
27. **(a)** 12.4 V; **(b)** 13.1 V; **(c)** 6.2 kΩ
29. **(a)** false; **(b)** true
31. —
33. 1 kΩ = 8.6 μs, 100 Ω = 2 μs, no
35. $R_{B_2} = 6.43$ kΩ, $V_{BB} = 16$ V

Chapter 10

1. $V_o/V_i = -10.91$
3. $R_i = 8$ kΩ
5. $V_o/V_i = 7$
9. $V_o = -1.8$ V
11. $R_1 = 21.5$ kΩ

Chapter 11

1. $A_{v_1} = -15$
3. $V_{OL} = 1.86$ V
7. $A_d = 2.5 \times 10^3$
11. $A_v = 6.17 \times 10^3$
15. $V_{01} = 360\ \mu$V
17. **(a)** 33.38 dB; **(b)** 40.56 dB; **(c)** 29 dB; **(d)** 45.26 dB; **(e)** -13.28 dB
19. **(a)** $V_i = 0.25$ mV, rms; **(b)** $V_i =$ 1.4 mV, rms; **(c)** $V_i = 35.3$ mV, rms; **(d)** $V_i = 498.8\ \mu$V, rms
21. $A = 110.7$ dB
23. $C = 1.27\ \mu$F
25. $C_{high} = 751$ pF
27. $A_{mid} = -23.65$
29. $A_{mid} = -671.15$
31. $f_{high} = 8.42$ kHz
33. $P_o = 1.8$ W
35. $P_i(dc) = 225$ mW
37. $\eta = 38.9\%$
39. max $P_o = 56.25$ W, max $P_i =$ 71.62 W, $P_Q = 7.685$ W

Chapter 12

3. 890 kHz, 910 kHz
7. **(a)** $f_c = 880$ kHz, $f_m = 6$ kHz; **(b)** m = 0.5; **(c)** $V = 60$ V
15. $f_o = 949$ kHz
17. $C = 452$ pF
19. $C = 1040$ pF
21. $C = 451.5$ pF
25. $f_o = 59.6$ kHz
27. $C_1 = 266.7$ pF
29. $f_L = 800$ kHz; $f_C = 327.38$ kHz

Chapter 13

1. **(a)** 172; **(b)** 109; **(c)** 56,167
3. **(a)** 1453; **(b)** 4068; **(c)** 971
5. **(a)** 1001 0010 0111 0101 **(b)** 1000 0000 0100 0010 **(c)** 0110 0001 1000 0011
7. 44 49 47 49 54 41 4C 43 4F 40 50 55 54 45 52 53
9. **(a)** 10000110; **(b)** 11110110; **(c)** 11011100
11. $(\overline{\overline{A \cdot B \cdot C \cdot \overline{C}}})$

Chapter 14

1. Transfer function = 0.33

Chapter 15

1. 2.8 %
3. 22.4 V
5. 1.233 V
7. 16.958 V
9. 0.960%
11. $V_{dc} = 12$ V, $V_m = 1.663$ V
13. 7.20%
15. 1.6 V
17. 73 mA
19. 32.01%
23. 10.805 V
25. 5.485%
27. **(a)** 13.749 V **(b)** 1.247 V

Index

P

Q

R

S

T

U

V

W

Y

Z